AF360708

Elemental Concentration and Pollution in Soil, Water, and Sediment

Elemental Concentration and Pollution in Soil, Water, and Sediment

Editors

Ana Romero-Freire

Hao Qiu

MDPI • Basel • Beijing • Wuhan • Barcelona • Belgrade • Manchester • Tokyo • Cluj • Tianjin

Editors
Ana Romero-Freire
Department of Soil Science and
Agricultural Chemistry
University of Granada
Granada
Spain

Hao Qiu
Material Building F523
Shanghai Jiao Tong University
Shanghai
China

Editorial Office
MDPI
St. Alban-Anlage 66
4052 Basel, Switzerland

This is a reprint of articles from the Special Issue published online in the open access journal *Minerals* (ISSN 2075-163X) (available at: www.mdpi.com/journal/minerals/special_issues/ECPSWS).

For citation purposes, cite each article independently as indicated on the article page online and as indicated below:

LastName, A.A.; LastName, B.B.; LastName, C.C. Article Title. *Journal Name* **Year**, *Volume Number*, Page Range.

ISBN 978-3-0365-3738-2 (Hbk)
ISBN 978-3-0365-3737-5 (PDF)

Contents

About the Editors

Ana Romero-Freire

Dr Romero Freire holds a Double PhD in Earth and Life Sciences (University of Granada and VU Amsterdam, 2015). She gained postdoctoral research experience at the University of Lorraine (France), The Spanish National Research Council (CSIC, Spain), Univ. of Naples (Italy) and Univ. Aveiro (Portugal). Her scientific research studies have been mainly focused on metal pollution, ecotoxicology and geochemistry, with her recent work being dedicated to investigating technology-critical elements (TCEs), a group of emergent elements mainly used for the production of high-tech devices. In her research, she has studied metals and TCEs in different environmental fractions such as water bodies, sediments and soils. Currently, she works at the Department of Soil Science and Agricultural Chemistry of the University of Granada (UGR) in Spain.

Hao Qiu

Dr. Hao Qiu is currently an associate professor (tenured) at Shanghai Jiao Tong University and a research professor at Shanghai Institute of Pollution Control and Ecological Safety. He earned his Ph.D. from Leiden University, the Netherlands, and later worked as a postdoctoral researcher at KU Leuven, Belgium. His research interest centers around soil–biota interactions and creating a sustainable environment. He has successfully secured over 10 projects from the National Natural Science Foundation of China, Ministry of Science and Technology of China, Ministry of Agriculture and Rural Affairs of China, and Shanghai Science and Technology Commission. He has published more than 80 SCI papers in high-quality journals including {Environmental Science and Technology}, {Water Research, Environmental Science: Nano, Applied Catalysis B: Environmental, Journal of Hazardous Materials}, etc.

Editorial

Editorial for Special Issue "Elemental Concentration and Pollution in Soil, Water, and Sediment"

Ana Romero-Freire [1,*] and Hao Qiu [2,*]

[1] Department Soil Science and Agricultural Chemistry, University of Granada, 18071 Granada, Spain
[2] School of Environmental Science and Engineering, Shanghai Jiao Tong University, Shanghai 200240, China
* Correspondence: anaromerof@ugr.es (A.R.-F.); haoqiu@sjtu.edu.cn (H.Q.)

Citation: Romero-Freire, A.; Qiu, H. Editorial for Special Issue "Elemental Concentration and Pollution in Soil, Water, and Sediment". *Minerals* **2022**, *12*, 338. https://doi.org/10.3390/min12030338

Received: 16 February 2022
Accepted: 5 March 2022
Published: 9 March 2022

Publisher's Note: MDPI stays neutral with regard to jurisdictional claims in published maps and institutional affiliations.

Certain elements are essential to the growth and health of living organisms with specific biochemical functions in their metabolic processes. However, these elements can become toxic when their amounts exceed certain thresholds. Therefore, the environmental levels of various trace elements in soil, water, or sediment are of major concern due to their dose-dependent effects on living organisms. The aim of this Special Collection is to bring researchers from different fields together, involving biogeochemistry and ecotoxicology in various environmental media, in order to provide a more comprehensive understanding of the environmental fate of trace elements in their biogeochemical cycles for different ecosystems.

In this context, several contributions have been made to study the presence and effect of trace elements in different environmental sinks. In the soil system, Horváth and co-authors [1] monitored the contamination levels and executed a comparative assessment of soil properties and dwelling mesofauna in a mid-sized Hungarian city in two different years, and they found a correlation between specific chemical parameters and soil microarthropods. Pan and co-authors [2] studied the concentrations of potentially toxic elements in 27 surface soil samples from areas where coal-mining activities ceased nine years ago (Guizhou Province, China). Cadmium was the only element that showed a mean concentration higher than the national soil quality standard, with levels that could be harmful to live organisms. Korzeniowska and Krąż [3] studied heavy metal pollution caused by anthropogenic activities in the natural Tatra National Park (Poland). They observed the presence of Cd and Pb from human activities and noted that with the increase in the altitude of the terrain, the concentration of metals in soils decreased. Sutkowska and co-authors [4] defined pollution indices as an efficient tool for distinguishing anthropogenic soil pollution and geogenic contamination in two polish areas differing in geological setting and type of land use. In the attempts to establish remediation techniques for polluted soils, Baek and co-authors [5] evaluated the feasibility of using a practical chemical washing method for the removal of fluorine from an enriched soil, and they stated that chemical washing might not be effective for remediating soils containing chemically stable forms of fluorine. Aguilar-Garrido and co-authors [6] evaluated the potential remediation of peat in different As-polluted soils by assessing the decrease of As solubility and its toxicity through bioassays. Peat addition decreased As mobility, but less effective in buffering As pollution was observed in calcareous soils. Moreover, Zaragüeta and co-authors [7] observed that the use of sewage sludge, which can be used as an organic amendment in agricultural soils, increased the concentration of some trace metals in the soil, in bioavailable forms, and in the crop. Interesting results also presented in the work done by Yang and co-authors [8], who studied the adsorption behavior of p-arsanilic acid (ASA), a kind of organic arsenic feed additive that contains the arsenic group in a chemical structure, on three kinds of pure iron oxides and nine soils to elucidate its implication for contaminated soils. Their findings help to understand the environmental transportation behavior of organoarsenicals by evaluating the potential hazards associated with the usage of organic arsenic feed additives. Last but not least, in studies done with soil fractions, Suazo-Hernández and co-authors [9] analyzed the effect of metal (Cu, Ag) engineered nanoparticles on phosphorous availability in an agricultural Andisol and observed that the incorporation of the studied nanoparticles into the selected soil generated an increase in P retention, which may affect agricultural crop production.

Several studies were focused on the anthropopressure on sediments; Warta River studied by Jaskuła and co-authors [10] and Wigry Lake surveyed by Kostka and Leśniak [11], both located in Poland, showed signs of heavy metal pollution. In Warta River, the third-longest river in Poland, heavy metal contaminated sediments can act as point sources in urbanized areas and fluvial processes. The results obtained from Wigry Lake showed signs of metal pollution, mainly by Pb, although it is located in a pristine region. In the same line, the work performed by Ramírez-Pérez and co-authors [12] investigated the geochemistry, enrichment, and pollution of trace metals in anoxic sediments, pointing to a possible ecotoxicological risk to organisms for Pb, Cu and Zn in superficial sediment layers in the San Simon Bay (Spain), contaminated in the surface mainly due to anthropogenic inputs, especially in the case of Pb, reflecting the enormous human pressure on these ecosystems. Huang and co-authors [13] studied the causes of copper and other common heavy metals input in sediments of irrigation canals in Taiwan province (China). They observed that sediments were polluted mainly due to the highest masses of pollutants released into drainage wastewater of the county and the return flow from irrigation, and the illegal discharge of wastewater. Dinis and co-authors [14] assessed the ecological risk of cadmium in karst lake sediment at Yelang reservoir in Guizhou province (China), also incorporating an ecotoxicology approach, and found strong to extremely strong ecological risks of Cd in sediments, but low ecotoxicology for the organism investigated, due, mainly, to water properties (pH, and Ca and Mg content). Matabane and co-authors [15] did a sequential extraction and risk assessment of potentially toxic elements in river sediments of Blood River (South Africa), to assess a possible trend of mobilization of these elements from sediment to water, they found a high toxicity-risk level, which could cause a threat to organisms dwelling in sediments and local residents via consumption of crops irrigated with the polluted river water. Finally, the sediment research is completed by a review article of Nawrot and co-authors [16], which provides different methods in assessing the status of the trace metal contamination in sediments affected by anthropogenic interference by applying geochemical and ecotoxicological assessment and classification indices.

In water systems, two contributions determined pollutants in Russia: Menshikova and co-authors [17] analyzed the water balance at the Kachkanar Mining and Processing Plant tailings dump (Russia, Ural Region), and they highlighted that increasing the volume of seepage discharge will lead to an increase in natural water pollution within the area. Novikov and co-authors [18] identified the chemical composition of water from natural springs near large cities in the Arctic region of Russia, which is used instead of tap water. They found hazardous pollutants exceeding Russian hygienic limits in half of the tested waters. Interesting findings were also shown by Moskovchenko and co-authors [19], who studied the chemical composition of snow from Tyumen (Russia), which allowed for a quantitative determination of pollutants deposited from the atmosphere. With this determination, they identified the most polluted parts of the city, which were in the center and along the roads with the most intensive traffic. Kozina and co-authors [20] studied the conditions of sedimentation in the Caspian Sea, taking into account lithological, mineralogical and geochemical data and observed that hydrogen sulfide contamination, recorded in the bottom layer of the water column of the deep-water basins, affects the formation of authigenic sulfides, sulfates and carbonates associated with the activity of sulfate-reducing bacteria.

The research in water systems was completed by Fedoročková and co-authors [21], who designed and verified a laboratory method for the testing of alkaline, magnesite-based reactive materials for permeable reactive barriers to remove heavy metals from contaminated groundwaters.

Above mentioned studies especially focused on the effect and distribution of heavy metals investigated in soil, sediment and water systems and demonstrated that several regions in the World present potential risks from trace element pollution. Furthermore, different ecological approaches (indices, ecotoxicological studies, geochemistry), different forms of pollutants (nanoparticles, bulk metals, organic metals) and potential remediation or techniques for trace element pollution controls are also illustrated in this volume and provide a thorough comprehension of the status, control, and remediation of elemental pollution in the ecosystems under anthropogenic disturbances.

Conflicts of Interest: The authors declare no conflict of interest.

References

1. Horváth, A.; Csáki, P.; Szita, R.; Kalicz, P.; Gribovszki, Z.; Bidló, A.; Bolodár-Varga, B.; Balázs, P.; Winkler, D. A Complex Soil Ecological Approach in a Sustainable Urban Environment: Soil Properties and Soil Biological Quality. *Minerals* **2021**, *11*, 704. [CrossRef]
2. Pan, L.; Guan, X.; Liu, B.; Chen, Y.; Pei, Y.; Pan, J.; Zhang, Y.; Hao, Z. Pollution Characteristics, Distribution and Ecological Risk of Potentially Toxic Elements in Soils from an Abandoned Coal Mine Area in Southwestern China. *Minerals* **2021**, *11*, 330. [CrossRef]
3. Korzeniowska, J.; Krąż, P. Heavy Metals Content in the Soils of the Tatra National Park Near Lake Morskie Oko and Kasprowy Wierch—A Case Study (Tatra Mts, Central Europe). *Minerals* **2020**, *10*, 1120. [CrossRef]
4. Sutkowska, K.; Teper, L.; Czech, T.; Hulok, T.; Olszak, M.; Zogala, J. Quality of Peri-Urban Soil Developed from Ore-Bearing Carbonates: Heavy Metal Levels and Source Apportionment Assessed Using Pollution Indices. *Minerals* **2020**, *10*, 1140. [CrossRef]
5. Baek, D.-J.; Kim, Y.-E.; Jung, M.-Y.; Yoon, H.-O.; An, J. Feasibility of a Chemical Washing Method for Treating Soil Enriched with Fluorine Derived from Mica. *Minerals* **2021**, *11*, 134. [CrossRef]
6. Aguilar-Garrido, A.; Romero-Freire, A.; García-Carmona, M.; Martín Peinado, F.J.; Sierra Aragón, M.; Martínez Garzón, F.J. Arsenic Fixation in Polluted Soils by Peat Applications. *Minerals* **2020**, *10*, 968. [CrossRef]
7. Zaragüeta, A.; Enrique, A.; Virto, I.; Antón, R.; Urmeneta, H.; Orcaray, L. Effect of the Long-Term Application of Sewage Sludge to A Calcareous Soil on Its Total and Bioavailable Content in Trace Elements, and Their Transfer to the Crop. *Minerals* **2021**, *11*, 356. [CrossRef]
8. Yang, Y.; Tao, S.; Dong, Z.; Xu, J.; Zhang, X.; Pan, G. Adsorption of *p*-Arsanilic Acid on Iron (Hydr)oxides and Its Implications for Contamination in Soils. *Minerals* **2021**, *11*, 105. [CrossRef]
9. Suazo-Hernández, J.; Klumpp, E.; Arancibia-Miranda, N.; Poblete-Grant, P.; Jara, A.; Bol, R.; de La Luz Mora, M. Describing Phosphorus Sorption Processes on Volcanic Soil in the Presence of Copper or Silver Engineered Nanoparticles. *Minerals* **2021**, *11*, 373. [CrossRef]
10. Jaskuła, J.; Sojka, M.; Fiedler, M.; Wróżyński, R. Analysis of Spatial Variability of River Bottom Sediment Pollution with Heavy Metals and Assessment of Potential Ecological Hazard for the Warta River, Poland. *Minerals* **2021**, *11*, 327. [CrossRef]
11. Kostka, A.; Leśniak, A. Natural and Anthropogenic Origin of Metals in Lacustrine Sediments; Assessment and Consequences—A Case Study of Wigry Lake (Poland). *Minerals* **2021**, *11*, 158. [CrossRef]
12. Ramírez-Pérez, A.M.; Álvarez-Vázquez, M.A.; De Uña-Álvarez, E.; de Blas, E. Environmental Assessment of Trace Metals in San Simon Bay Sediments (NW Iberian Peninsula). *Minerals* **2020**, *10*, 826. [CrossRef]
13. Huang, S.-H.; Chang, T.-C.; Chien, H.-C.; Wang, Z.-S.; Chang, Y.-C.; Wang, Y.-L.; Hsi, H.-C. Comprehending the Causes of Presence of Copper and Common Heavy Metals in Sediments of Irrigation Canals in Taiwan. *Minerals* **2021**, *11*, 416. [CrossRef]
14. Dinis, F.; Liu, H.; Liu, Q.; Wang, X.; Xu, M. Ecological Risk Assessment of Cadmium in Karst Lake Sediments Based on *Daphnia pulex* Ecotoxicology. *Minerals* **2021**, *11*, 650. [CrossRef]
15. Matabane, D.L.; Godeto, T.W.; Mampa, R.M.; Ambushe, A.A. Sequential Extraction and Risk Assessment of Potentially Toxic Elements in River Sediments. *Minerals* **2021**, *11*, 874. [CrossRef]
16. Nawrot, N.; Wojciechowska, E.; Mohsin, M.; Kuittinen, S.; Pappinen, A.; Rezania, S. Trace Metal Contamination of Bottom Sediments: A Review of Assessment Measures and Geochemical Background Determination Methods. *Minerals* **2021**, *11*, 872. [CrossRef]
17. Menshikova, E.; Fetisov, V.; Karavaeva, T.; Blinov, S.; Belkin, P.; Vaganov, S. Reducing the Negative Technogenic Impact of the Mining Enterprise on the Environment through Management of the Water Balance. *Minerals* **2020**, *10*, 1145. [CrossRef]
18. Novikov, A.I.; Shirokaya, A.A.; Slukovskaya, M.V. Elemental Concentrations of Major and Trace Elements in the Spring Waters of the Arctic Region of Russia. *Minerals* **2022**, *12*, 8. [CrossRef]
19. Moskovchenko, D.; Pozhitkov, R.; Zakharchenko, A.; Tigeev, A. Concentrations of Major and Trace Elements within the Snowpack of Tyumen, Russia. *Minerals* **2021**, *11*, 709. [CrossRef]
20. Kozina, N.; Reykhard, L.; Dara, O. Authigenic Minerals of the Derbent and South Caspian Basins (Caspian Sea): Features of Forms, Distribution and Genesis under Conditions of Hydrogen Sulfide Contamination. *Minerals* **2022**, *12*, 87. [CrossRef]
21. Fedoročková, A.; Raschman, P.; Sučik, G.; Švandová, M.; Doráková, A. Reactive, Sparingly Soluble Calcined Magnesia, Tailor-Made as the Reactive Material for Heavy Metal Removal from Contaminated Groundwater Using Permeable Reactive Barrier. *Minerals* **2021**, *11*, 1153. [CrossRef]

Article

A Complex Soil Ecological Approach in a Sustainable Urban Environment: Soil Properties and Soil Biological Quality

Adrienn Horváth [1,*], Péter Csáki [2], Renáta Szita [2,3], Péter Kalicz [2], Zoltán Gribovszki [2], András Bidló [1], Bernadett Bolodár-Varga [1], Pál Balázs [1] and Dániel Winkler [4]

[1] Institute of Environmental and Earth Sciences, University of Sopron, 9400 Sopron, Hungary; bidlo.andras@uni-sopron.hu (A.B.); varga.bernadett@uni-sopron.hu (B.B.-V.); balazs.pal@uni-sopron.hu (P.B.)
[2] Institute of Geomatics and Civil Engineering, University of Sopron, 9400 Sopron, Hungary; csaki.peter@uni-sopron.hu (P.C.); szita.reni@gmail.com (R.S.); kalicz.peter@uni-sopron.hu (P.K.); gribovszki.zoltan@uni-sopron.hu (Z.G.)
[3] Directory of Fertő-Hanság National Park, 9435 Sarród, Hungary
[4] Institute of Wildlife Management and Wildlife Biology, University of Sopron, 9400 Sopron, Hungary; winkler.daniel@uni-sopron.hu
[*] Correspondence: horvath.adrienn@uni-sopron.hu

Citation: Horváth, A.; Csáki, P.; Szita, R.; Kalicz, P.; Gribovszki, Z.; Bidló, A.; Bolodár-Varga, B.; Balázs, P.; Winkler, D. A Complex Soil Ecological Approach in a Sustainable Urban Environment: Soil Properties and Soil Biological Quality. *Minerals* **2021**, *11*, 704. https://doi.org/10.3390/min11070704

Academic Editors: Ana Romero-Freire and Hao Qiu

Received: 31 May 2021
Accepted: 27 June 2021
Published: 29 June 2021

Publisher's Note: MDPI stays neutral with regard to jurisdictional claims in published maps and institutional affiliations.

Abstract: The main purpose of the present study was to monitor actual contamination levels and execute a comparative assessment of results in a mid-sized Hungarian city for two different years. The first citywide soil investigations were completed in 2011. In 2018, the most prominent properties (pH, $CaCO_3$, texture, and trace metals Cd, Co, Cu, Ni, Pb, and Zn) were reanalyzed and were supplemented with mesofauna on selected sites. The available trace metal elements of urban soils showed the following tendency in 2011: Zn > Cu > Pb > Cd > Cr = Ni = Co. In 2018, the previous order changed to Zn > Pb > Cu > Cr > Cd = Ni = Co. Cd and Pb enrichments were found, especially near the M7 motorway. The comparison between 2011 and 2018 revealed soil contamination was, on average, higher in 2011. Soil microarthropod communities were sampled and assessed using abundance data and diversity measurements. Soil biological quality was evaluated with the help of the Soil Biological Quality (QBS-ar) index. Acari and Collembola appeared to be the most abundant, ubiquitous taxa in the samples. Simultaneously, important groups like Symphyla, Protura, and Chilopoda were completely absent from the most polluted sites. For the most part, lower taxa richness, diversity, and QBS-ar index were observed with higher available Cu Zn, and Pb concentrations.

Keywords: soil properties; urban ecology; trace metal pollution; soil organisms; diversity

1. Introduction

The impact of anthropogenic effects and environmental pollution creates an urgent need to investigate complex urban ecosystems. Soil, water, and sediment analysis complemented with biological indices (hydrobiological and mesofauna) can detect all prominent anthropogenic influences in the urban environment. Deficiencies within the national regulations of Hungary hinder the mitigation of adverse human effects. Moreover, complex investigations are often not well interpreted. Appropriate reference sources at the global level are also difficult to find.

Extensive evaluation of smaller aspects, like air quality monitoring with mosses [1] or detection of heavy metal content in dust [2], is available. Still, a knowledge gap exists regarding the long-term effects of these factors. The complexity of the current research requires a structured literature background for the parts of the study performed. The present study is part of a more complex analysis measuring urbanization effects on soil, water, sediment, mesofauna, and aquatic invertebrates.

Evaluations of the means, mobility, and interactions of heavy metal pollution in urban soils [3–7] have proven effective globally, both for soil development and ecosystem services [8]. Utilizing better knowledge of urban land use [9–11] or urban parks, schoolyards [12], or urban industry effects [13,14] not only improves the quality of life for urban residents but also aids in evaluating the impacts of different reclaimed land uses on topsoil properties. Soil development under different land use/cover is critical for restoring eco-environmental integrity and providing vital functions and services for city dwellers [15–17].

Urbanization is usually linked with changes in the soil environment, including habitat alteration, fragmentation and loss, and pollution, which also have strong negative effects on soil biota [18,19]. The diversified role of microarthropod communities in soil quality and health has been widely reported (e.g., [20,21]). Owing to their stability, relatively sedentary lifestyle, and sensitivity to soil property changes, specific taxa are often used as bioindicators of stressed environments [22,23]. The most frequently studied soil microarthropod groups in urban environments are Collembola and Acari (e.g., [24–27]); however, a more comprehensive evaluation of soil biological quality involving other microarthropod groups is also worth considering [19,28]. The novelty of our research is related to the fact that no complex ecological soil assessment (physicochemical and biological) has yet been carried out in urban environments in Hungary.

Despite the numerous urban soil studies in every continent that investigated the actual and increasing level of potential toxic elements (e.g., China [29], Poland [10], New York [30], Brazil [31], Angola [32]), we still do not have soil monitoring networks even at national levels. Particularly, many studies dealt with polycyclic aromatic hydrocarbon (PAH) and polychlorinated biphenyl (PCB) [33,34], or total element accumulation of urban parks [35], but less information has been published about available element contents [36]. Noticeable garden (fruit and vegetable) utilization is rather widespread in small and medium-sized cities. Gardens and urban parks or playgrounds could also bind airborne pollutants. Therefore, people may regularly come into contact with toxic elements accumulated by plants or soils of public parks or gardens. Thus, the determination of available trace metal contents is of particular importance. Due to the shortcomings mentioned above, our research has several goals, some of which are long term.

- The top priority of our research is to create a comprehensive database to complete the national system. Our national soil monitoring network does not examine the health conditions of human settlements.
- The long-term aim is to propose a modification of the existing soil limit system based on our results. Hungarian law only sets limits for total element content, e.g., for toxic trace metals. In our opinion, the limit values should also be completed with available toxic element limits to protect human health in urban areas.
- In addition, it is worth noting that in Hungary, the preparation of 4–6-year-long environmental programs for every settlement has been mandatory since 2006. Most of these programs are prepared with the involvement of a team of experts coordinated by the local government, but the prepared documents rely on unmeasured data. The lack of specific databases or municipal monitoring networks for cities is the reason for this. On the other hand, experts often rely on literature in their attempts to identify local problems affecting cities and offer suggestions from these literature-based findings. Most of these suggestions can only be called "symptom management" and do not attempt to uncover the "root problem". Therefore, experts can only make modest suggestions and lack the information needed to take definite steps.

Székesfehérvár, the city selected for the current paper, is the third in a closely linked citywide ecological status survey. It is worth noting that no complex citywide investigation had been completed in urbanized areas in Hungary before. The citywide urban soil investigation by Horváth et al. [37], recently conducted in Sopron and Szombathely, is a related research study. That study hypothesized that the Szombathely soils are more polluted than Sopron soils, but the investigation revealed Sopron soils to be more contaminated on average. The method employed in the current study differs from the one used for the two abovementioned cities in so far as the comparison of our results to the base year 2011 that was re-examined in 2018 also incorporates factors influenced by human activity. The city government has taken several municipal interventions aimed at sustainability. Therefore, the results of this current study reflect these positive effects of the interventions. We have already investigated these effects in water quality [38] and soil aspects as well. In 2018, the method was complemented by biological supporting measurements; we did not utilize these here.

Based on former experiences and conclusions, this study hypothesizes that pollution in Székesfehérvár has increased over the last seven years, with expected contamination increases in green areas. High pollution levels are also anticipated in the downtown area where heavy traffic is typical, especially near the M7 motorway. During the comparison of soil microarthropod groups, significant differences were expected in abundance between the less polluted suburban sites and the urbanized area sites. Based on our fact-finding, the population of Székesfehérvár started to increase from 1990. The area of the city also increased due to construction and occupation. General migration processes (from the east to the west) have occurred in the country since 2010. The continuous migration of the capital's population to the agglomeration is also an observable phenomenon. Both processes affected the city due to its favorable location. The city has operated nine industrial parks in the suburbs since the 2000s, where significant investments and developments (e.g., road network development) have been made to increase job opportunities. In addition, since the handover of the M7 motorway in 2008, commuter traffic and through traffic increased by 25% between 2010 and 2020. Based on previous expectations, the aims of the study were the following:

- to analyze the basic soil properties and available concentration of trace metals (Cd, Co, Cr, Cu, Ni, Pb, Zn) of the separate study years (2011 and 2018) and compare the results with suggested or legal limits;
- to compare and evaluate the results of 2011 to 2018 and estimate the changes in urban soils;
- to determine the degree of accumulation with enrichment factor (EF) calculations;
- to evaluate the quality and health of urban soil using the QBS approach;
- to clarify the directions of city development: Is Székesfehérvár still a livable city? Or do anthropogenic impacts have increasingly negative effects on soil and edaphon?

2. Materials and Methods

2.1. Study Area

Székesfehérvár, the city selected for this study, is located in Fejér County and has a population of 100,500 inhabitants. The city covers an area of 170.89 km^2 and is situated where Sárrét and South-Mezőföld merge. The elevations are between 103 and 222 m above sea level. On the eastern border of the city are the Velencei Mountains that consist of Carboniferous granite. During the rock formation process, the rocks of the mountain gradually sank deep into the east. The city is located in the northern part of the most extensive loess cover of the country; therefore, most of its territory is covered by thick loess. In the eastern part of the city, watercourses from the mountains to the north formed a significant alluvial cone on the Pannonian clayey sediments during the Pleistocene period. Due to the continuous subsidence, peat formation also appeared in some parts of this area in the Holocene [39]. The city's climate is warm and dry. The annual mean temperature is 10.2–10.4 °C, and annual precipitation is less than 540 mm. The most common winds are from the northwest, with an average speed of 2.5–3 m/s. The most important watercourse is the Gaja Brook, which runs through the city with many channels connecting to it. A natural saline lake (called "Sóstó") and some artificial lakes (a boating lake, fish ponds, and quarry lakes) and two large fish ponds can also be found in the city. The township is surrounded by degraded forests and grasslands or mostly utilized for agriculture. There are nine soil types in the investigated area, of which the productivity of Vertisols and Chernozems are the most favorable in the suburbs [40,41]. The natural conditions favor field crop cultivation [39,42]. Based on geological circumstances, loess-bedded forest soils (Luvisols) transformed into Technosols (Figure 1). These soils, mainly the topsoil layer formed by centuries-old human activities, are typical (IUSS Working Group WRB, 2014). The city is located on a busy travel route between the capital city Budapest and Lake Balaton, the biggest lake in Central Europe. The M7 motorway passes through the township from the south side; therefore, the city and its surroundings receive high traffic levels. During the analysis, an extremely high trace metal level was expected in the southeastern part of the city. In addition, the city's government has participated in many green programs and in recent years has received funding for urban greening, urban planning, soil remediation, road and park renovation, and dredging projects.

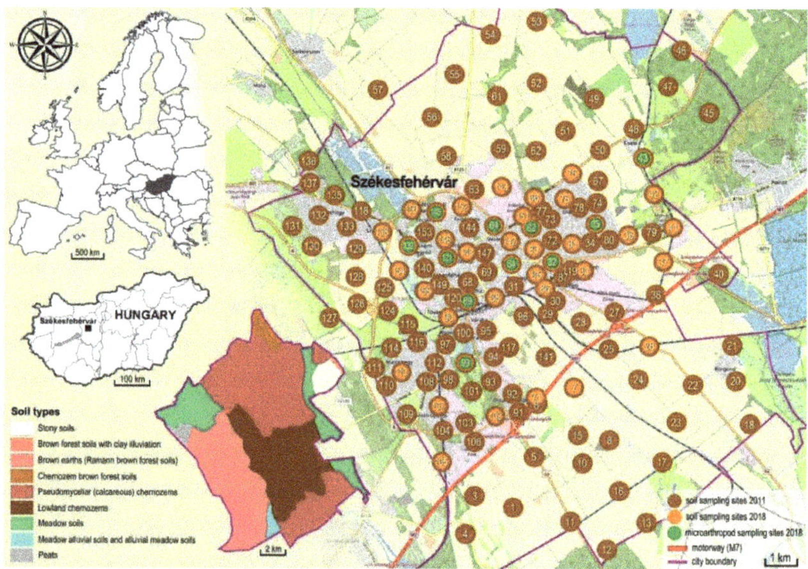

Figure 1. Distribution of sampling sites with soil type information of the study area (© OpenStreetMap contributors, soil map source: MTA TAKI 1991, https://www.openstreetmap.org, accessed on 29 June 2021).

2.2. Methods and Data Analysis

The effects of urbanization on the natural environment were the focus in the three western Hungarian cities (Sopron, Szombathely, and Székesfehérvár) selected as case study regions in 2010. The results of Sopron and Szombathely have already been published [37,43,44]. In addition, water quality analyses were carried out for the watercourses of Székesfehérvár [38]. The current paper will present the soil quality of the city. The chemical and physical characteristics of 144 soil surface samples were analyzed in Székesfehérvár and its surroundings in 2011. In 2018, 42 monitoring sites were revisited (Figure 1). During the research of pedological media, the following methods and measurements were implemented.

2.2.1. Soil Analysis

In 2011, 154 topsoil samples were collected from 0–10 and 10–20 cm depths; the sample number was eventually reduced to 144. In 2018, 42 topsoil samples were recollected from sites where some soil property or condition measured in 2011 warranted a re-examination. Altogether, 372 soil samples were analyzed during the research. Each test site was represented by three replications of randomly collected and then thoroughly mixed soil samples. The air-dried samples were sieved through a 2 mm mesh. The guidelines of the Hungarian Standards were used for sample preparation and methods of soil analysis (Table 1). These standards are in accordance with the methods of Van Reeuwijk [45]. Soil

pH_{H2O} (soil:solution ratio 1:2.5) was examined potentiometrically after 12 h of mixing to determine the mobility readiness of trace metals [46]. $CaCO_3$ content was measured with a Scheibler-type calcimeter [46]. Based on the standard [47], a sedimentation method (via the pipette method) was used for the determination of soil texture. A soil/water suspension was mixed in a sedimentation cylinder, and then sampled with a pipette to collect particles of a given size and calculate the ratio of clay (<0.002 mm), silt (0.002–0.02 mm), fine sand (0.02–0.2 mm), and coarse sand (0.2–2.0 mm). The skeletal percent (>2.0 mm) was separated from fine fractions with a sieve. Soil organic matter (SOM) content was determined following the FAO method [48,49]. As the 2011 results of the following measurement were negligible, they were not monitored in 2018. Kjeldahl total nitrogen, K_2O, P_2O_5, KCl extractable Ca-Mg, and EDTA/DTPA-extractable Fe, Mn, Cu, Zn analyses were not repeated, and their results will not be discussed in this study (Table 1).

Table 1. The summary of the pedological examinations.

Type of Measurements (Units)	Standard	2011/2018
Skeletal percent (%)	MSZ-08-0205/2:1978 [46]	+/+
pH (H_2O, KCl)—potentiometrically	MSZ-08-0205/2:1978 [46]	+/+
Total salinity (%)	MSZ-08-0205/2:1978 [46]	+/+
$CaCO_3$ (%)—Schleibler method	MSZ-08-0205/2:1978 [46]	+/+
Humus (%)—($K_2Cr_2O_7$ + cc. H_2SO_4)	MSZ-08-0452:1980 [49]	+/+
Texture (2–0.002 mm—%)	MSZ-08-0206:1978 [47]	+/+
Total nitrogen (%)	MSZ-EN-16169:2013 [50]	+/−
Potassium (K_2O, g/kg)—photometrically	MSZ-20135:1999 [51]	+/−
Phosphorus (P_2O_5, g/kg)—photometrically	MSZ-20135:1999 [51]	+/−
KCl-extractable calcium, magnesium (g/kg)—AAS	MSZ-20135:1999 [51]	+/−
EDTA/DTPA Fe, Mn, Cu, Zn (mg/kg)—AAS	MSZ-20135:1999 [51]	+/−
Available toxic element content (NH_4-acetate + EDTA)—ICP	MSZ 21470-50:2006 [52]	+/+
Mesofauna analysis (Soil Biological Quality—QBS)	Menta et al. [53]	−/+

+ analysis was carried out; − no analysis was carried out.

The available soil element fraction was measured in a 0.5 M NH4-acetate + 0.02 M EDTA extract [52,54]. The element concentrations were measured using the ICP-OES method (ICAP 6000 Series, Thermo Fisher Scientific, Waltham, MA, USA). Merck calibration standards were used for the element analyses, and the measurements were taken according to the manufacturer's instructions. Concentrations of Cd, Co, Cu, Ni, Pb, and Zn were determined with detection limits of 0.05, 0.14, 0.26, 0.28, 1.21, and 2.10 mg/kg, respectively. Each measurement session included the extract of a standard soil sample as a control. The calibration curves were determined after every twelfth sample. In order to ensure the required measuring accuracy, calibration was carried out in the international survey of the Forest Soil Co-ordination Center (FSCC)-INBO. The unit of measurement was mg/kg for Cd, Co, Cu, Ni, Pb, and Zn, with detection limits of 0.04, 0.12, 0.21, 0.28, 1.23, and 2.11 mg/kg, respectively. The results were evaluated based on the limit values of suggested thresholds by Kádár [55]. These limits are in accordance with contamination levels set by Hungarian law [56,57]. The field and laboratory results were processed using geospatial methods and were made compatible with the previous database (Microsoft Office vers. 2016).

2.2.2. Enrichment Factor Calculation

For pollution designation, total element fractions were also measured where mesofauna abundance was tested. The total trace metal amounts were analyzed (cc. 5 cm^3 HNO_3 + 2 cm^3 H_2O_2) in microwave Teflon bombs [54] using ICP-OES. Altogether, nine elements were measured (Al, Cd, Co, Cr, Cu, Pb, and Zn), but we focused on the most common urban pollutants that are prominent for mesofauna: Co, Cu, Ni, Pb, and Zn. The total fraction results were needed for enrichment factor (EF) calculations according to the concentration rate between the measured contamination of mesofauna samples and naturally occurring reference metal source concentrations (e.g., Al, Li, Ti, Zr [58]) in the crust and were modified by the following formula and categories [43].

$$EF\ sediment = \frac{\frac{X}{Al}\ sediment}{\frac{X}{Al}\ crust}$$

The applied reference element was Al, which is a common rock-forming element [59]. The crust averages for the investigated metals were studied by Taylor and McLennan [60]. There is no kind of pollution enrichment where EF is near to 1. Pollution enrichment (increasing anthropogenic impact) is expected in the case of EF rate >2 [61].

2.2.3. Soil Microarthropod Sampling and Identification

For the soil microarthropod surveys, three undisturbed soil samples of 100 cm^3 were taken from a depth of 10 cm using a cylindrical soil corer from each selected survey site. Soil microarthropods were extracted from the total of 33 soil samples into 70% ethanol within a two-week period using Berlese–Tullgren funnels. Microarthropod specimens were counted and identified at the major taxonomic group level using a stereomicroscope. Taxa diversity was determined for each plot using the Shannon formula [62], while evenness was calculated by Pielou's index [63]. Soil biological quality was assessed using the QBS-ar index [53,64], based on a classification of microarthropod groups present in the soil sample. Each biological form was assigned a value (ecological-morphological index (EMI)) ranging from 1 to 20 according to its adaptation level to the soil environment. The QBS-ar index summarizes the abovementioned EMI values derived from the actual sample [64]. In order to analyze the connection between soil parameters (pH, SOM, available trace metal content) and soil microarthropod communities, a canonical correspondence analysis (CCA) was performed using Canoco ver. 4.5 [65]. Taxa appeared in only one sample, and those that occurred at a low level (<5 individuals) were excluded due to possible uncertain relationships. A Monte Carlo permutation test with 1000 randomizations was performed to evaluate the significance of the canonical axes.

3. Results

3.1. Soil Data

The 144 sampling sites of 2011 were classified into land use categories by the most characteristic anthropogenic or environmental conditions according to the land registry. After the categorization, 40 residential, 37 agricultural, 19 traffic zone, 17 miscellaneous, 9 park, 7 forest, 7 creek bank and lake shore, 6 industrial zone, and 2 viticulture areas were separated (Table 2). In 2018, the selected 42 monitoring sites were reduced to 15 traffic zone sites, 9 park sites, 6 creek bank and lake shore sites, and 5 residential zone sites; the sampling network was completed with three forest sites, one industrial zone site, one agricultural site, and one grassland for reference. The distribution of land use categories is worth noting because only one sampling site was included in some categories, which greatly influenced the statistical analysis. However, the monitoring of these sampling sites was necessary as a control for confirming subsequent trace metal contents. Before trace metal evaluation, the general soil properties of the base year had to be introduced. Table 2 shows the distribution of watery soil pH, CaCO$_3$, texture, and SOM in 2011 in the study area by land use categories.

The pH of soil mostly determines soil characterization [66]. In Székesfehérvár, slightly alkaline soils were found in most cases. The average values were 7.3–8.1, and the lower results appeared in forested areas. In Table 2, the averages of CaCO$_3$ content clearly show the presence of calcareous sediments that appeared in categories with high anthropogenic activities. The maximum value (33% CaCO$_3$) was found in a traffic zone site in the city center. According to the particle size distribution, the clay (<0.002 mm) and silt fraction (0.002–0.02 mm) were less than sand (0.02–2.0 mm) in the urbanized areas. Nevertheless, the clay fraction did not exceed 22% and the samples were predominantly loamy. In a citywide context, there was a slight radial decrease in texture as the distance from the city center increased. In the case of SOM, the highest values were found in semi-natural categories. In summary, there were no notable differences in basic soil characteristics.

Table 2. Properties of the investigated soils of 2011: average values of particle size distribution, soil pH H$_2$O, CaCO$_3$ content, soil organic matter (SOM) in soils categorized by land use types and soil layer depth, n: number of samples.

Land Use Category	Sampling Depth	n qty	Clay%	Silt%	Fine Sand%	Coarse sand%	pH (H$_2$O)	CaCO$_3$ %	SOM %
Forested area	0–10 cm	7	13	9	37	41	7.3	12	5.88
	10–20 cm	7	11	11	35	43	7.4	12	4.33
Viticulture area	0–10 cm	2	14	16	34	36	7.9	8	2.97
	10–20 cm	2	13	12	36	39	7.9	8	3.19
Agricultural area	0–10 cm	37	22	21	47	10	8.0	13	4.26
	10–20 cm	37	22	20	47	11	8.1	15	3.70
Residential area	0–10 cm	40	13	13	44	30	7.9	25	3.99
	10–20 cm	40	13	14	44	29	7.9	17	3.94
Traffic zone	0–10 cm	19	15	13	46	26	7.9	15	3.67
	10–20 cm	19	14	13	46	27	8.0	19	3.08
Industrial area	0–10 cm	6	22	19	43	16	7.9	21	4.32
	10–20 cm	6	20	21	41	18	8.1	16	4.74
Creek and lake bank	0–10 cm	7	18	19	40	23	7.9	12	5.65
	10–20 cm	7	16	22	40	22	8.0	15	4.58
Park	0–10 cm	9	13	16	43	28	7.9	14	6.52
	10–20 cm	9	14	15	41	30	8.0	15	4.54
Miscellaneous	0–10 cm	17	16	20	46	18	7.8	21	3.36
	10–20 cm	17	18	18	44	20	7.9	19	2.92

Nevertheless, when the 2011 results were compared to the values of 2018 on the monitored sites, changes in soil properties became apparent. Over this period, the built environment of the city increased by 10%. The average pH$_{H2O}$ of monitoring sites decreased to 7.7 from 7.9 in the 0–10 cm depth during the seven years. At the 10–20 cm depth, the average pH$_{H2O}$ of monitoring sites decreased to 7.9 from 8.0. Figure 2a clearly shows that the number and value of outliers increased in 2018 in both layers that appeared in forested areas in the suburbs. The average CaCO$_3$ results also decreased by ~3% in 2018 in both layers (Figure 2b).

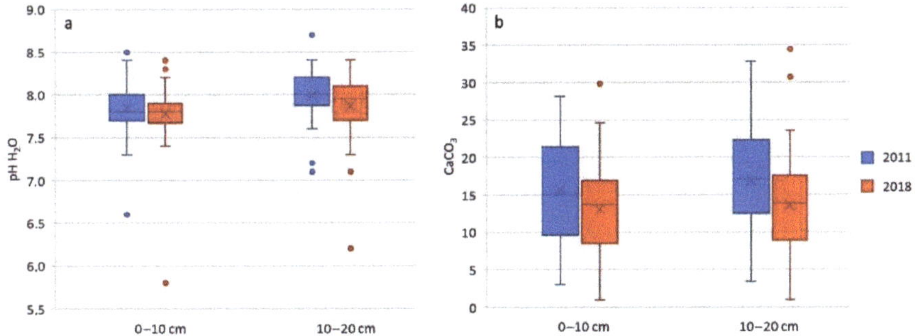

Figure 2. The distribution of pH$_{H2O}$ (**a**) and CaCO$_3$ (**b**) content on the monitoring sites, 2011 vs. 2018.

In the case of texture, there were no significant changes between the averages. Soil texture is still loamy, but the averages hide the growth of sand fractions at each site (Figure 3). Changes in texture were observed in traffic zone sites (S88, S138, S151), in park sites (S121, S139), and creek/lake banks (S35, S154).

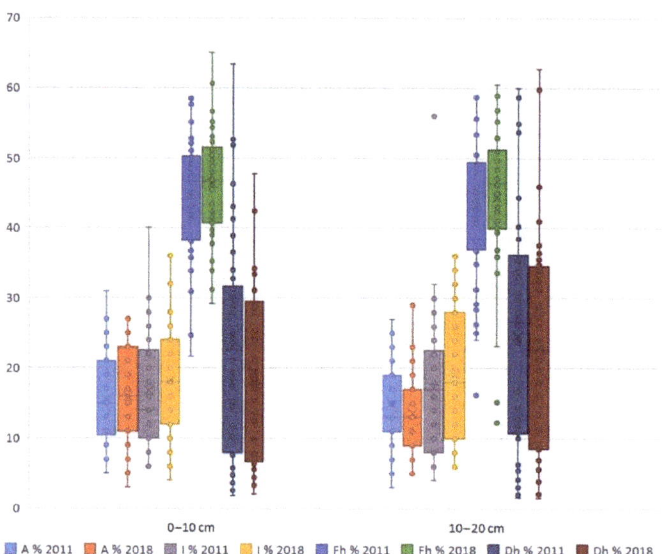

Figure 3. The distribution of soil fractions at 0–10 cm soil depth.

Table 3 summarizes the most important results of available trace metal analysis in both years where significant changes were detected. The available Co, Cr, and Ni values were lower than or close to the suggested natural background limits in both years. Extremes above the interventional pollution limits were observed at several sampling sites in 2011. At sampling site S86 (industrial area), the highest result was found in the case of Cu (408.4 mg Cu/kg) at a 0–10 cm depth. In addition, the highest Pb (97.5 mg Pb/kg) and Zn (247.5 mg Zn/kg) contents were found at this site as well. These extreme results were decreased in the 10–20 cm depth, but they still stand out. Sampling site S152 (miscellaneous) also showed high contents of Cd (2.5 mg Cd/kg), Cu (163.1 mg Cu/kg), Pb (38.2 mg Pb/kg), and Zn (63.3 mg Zn/kg) in the lower layer. In these mentioned sites, the pollution disappeared from samples in 2018. Considering the element occurrence, Zn exceeded the B, C1, and C2 limits most often in the base year. The available toxic elements of urban soils showed the following tendency in 2011: Zn > Cu > Pb > Cd > Cr = Ni = Co.

In 2018, samples taken alongside busy roads—especially near the M7 motorway—were contaminated with Zn and exceeded the "C1" limit (>40 mg Zn/kg) and "C2" limit (>80 mg Zn/kg). In the suburb, the amount of Pb decreased, especially in the southern part of the city. The measured elements showed the following quantitative order in 2018: Zn > Pb > Cu > Cr > Cd = Ni = Co (Figure 4).

The most significant negative changes were detected in samples of traffic zones (S102, S107, S138), which are near the M7 motorway. Furthermore, the trace metal content of a park (S90) and a creek/lake bank (S99) sample increased. In summary, the results showed that soils were more polluted in the base year.

3.2. Enrichment Factor Calculation

To complete the previous determinations, the collected samples on mesofauna sampling sites were evaluated and normalized with Al content. Enrichment factors were calculated and assessed according to the terms in the table below (Table 4).

Minerals **2021**, *11*, 704

Table 3. Available trace metal concentration of soil samples on selected sites classified by pollution limits.

2011 Cd	2011 Co	2011 Cr	2011 Cu	2011 Ni	2011 Pb	2011 Zn	Site Nr.	2018 Cd	2018 Co	2018 Cr	2018 Cu	2018 Ni	2018 Pb	2018 Zn	Δ Cd	Δ Co	Δ Cr	Δ Cu	Δ Ni	Δ Pb	Δ Zn	Land Use Category
0.23	0.77	0.01	3.92	1.50	13.79	29.41	S7	0.24	2.30	0.20	6.31	3.94	18.62	12.36	0.01	1.53	0.20	2.39	2.43	4.83	−17.05	traffic zone
0.53	1.15	0.47	10.99	2.71	21.95	44.05		0.20	2.18	0.17	6.08	3.82	13.64	11.49	−0.32	1.02	−0.30	−4.90	1.11	−8.31	−33.01	
0.28	1.30	0.03	14.92	1.51	66.34	53.81	S32	0.18	0.47	0.32	17.85	0.90	20.90	54.66	−0.09	−0.83	0.32	2.93	−0.60	−45.44	0.85	traffic zone
0.23	1.31	0.06	13.40	1.42	70.51	23.65		0.10	0.29	0.41	8.30	0.84	14.64	22.31	−0.13	−1.02	0.41	−5.09	−0.57	−55.87	−1.34	
0.19	1.08	0.21	8.09	1.72	9.10	42.13	S81	0.24	1.86	0.11	6.71	3.53	10.97	12.22	0.043	0.78	−0.09	−1.38	1.81	1.86	−29.91	residential area
0.17	1.06	0.20	7.21	1.78	7.68	31.82		0.17	1.50	0.10	5.02	2.72	9.32	8.95	0.01	0.44	−0.09	−2.19	0.93	1.63	−22.86	
0.18	0.63	0.21	7.14	0.86	17.06	24.95	S85	0.14	0.77	0.14	7.42	0.99	11.12	22.71	−0.04	0.13	−0.06	0.25	0.13	−5.94	−2.24	traffic zone
0.20	0.69	0.24	5.90	0.87	16.29	23.55		0.12	0.71	0.14	4.88	1.24	13.52	13.34	−0.07	0.01	−0.09	−1.01	0.36	−2.77	−10.21	
2.36	1.20	1.25	408.4	2.51	97.45	247.5	S86	0.31	1.10	0.30	9.54	1.55	47.95	25.96	−2.05	−0.09	−0.94	−398.85	−0.95	−49.5	−221.54	industrial area
1.45	1.03	0.95	180.9	1.36	73.45	93.86		0.25	0.89	0.33	6.98	1.34	29.49	14.81	−1.20	−0.13	−0.61	−173.91	−0.02	−43.96	−79.05	
0.36	0.43	0.56	13.14	1.08	28.71	58.09	S90	0.58	0.66	0.63	27.55	1.79	23.80	65.54	0.22	0.22	0.07	14.41	0.71	25.09	7.45	park
0.41	0.39	0.58	11.78	0.94	30.74	59.69		0.33	0.42	0.39	22.59	1.16	35.24	42.17	−0.07	0.03	−0.18	10.81	0.22	4.51	−17.52	
0.17	0.33	0.20	1.88	0.49	2.96	8.96	S99	0.62	0.43	0.43	11.52	0.99	25.92	35.93	0.44	0.09	0.22	9.63	0.50	22.95	26.96	creek and lake bank
0.11	0.29	0.26	1.93	0.50	2.76	4.91		0.56	0.32	0.41	11.62	0.97	25.02	34.66	0.45	0.02	0.15	9.68	0.46	22.26	29.74	
0.11	0.94	0.10	3.01	1.71	3.63	4.98	S102	0.15	0.97	0.21	17.44	1.90	5.09	52.98	0.04	0.03	0.10	14.42	0.18	1.46	47.99	traffic zone
0.10	0.87	0.10	2.61	1.68	3.18	2.60		0.16	1.11	0.14	10.47	2.18	4.83	28.15	0.05	0.24	0.04	7.85	0.49	1.65	25.54	
0.35	0.49	0.46	8.99	0.95	16.88	36.47	S107	0.52	0.40	0.93	15.54	0.74	43.02	52.43	0.17	−0.08	0.47	6.54	−0.21	26.14	15.96	traffic zone
0.46	0.36	0.80	13.11	0.88	27.37	44.28		0.56	0.45	1.27	21.52	0.89	39.89	69.18	0.10	0.09	0.46	8.41	0.01	12.52	24.90	
0.15	0.34	0.28	2.93	0.44	5.27	29.24	S138	0.28	0.75	0.30	5.18	0.63	10.06	80.40	0.13	0.41	0.01	2.25	0.18	4.78	51.16	traffic zone
0.24	0.29	0.47	2.91	0.43	5.47	24.40		0.31	0.54	0.32	4.33	0.55	9.87	98.67	0.07	0.25	−0.15	1.41	0.12	4.40	73.77	
0.27	0.42	0.18	6.28	1.14	13.65	31.52	S139	0.36	0.59	0.23	6.40	1.46	22.75	26.38	0.09	0.16	0.04	0.11	0.31	9.10	−5.14	park
0.44	0.38	0.22	7.94	1.15	16.24	28.33		0.44	0.47	0.28	7.72	1.39	27.34	26.22	−0.01	0.09	0.08	−0.22	0.23	11.10	−2.11	
2.51	0.62	0.53	13.03	1.14	9.85	22.32	S150	0.23	0.18	0.33	6.39	0.43	12.77	10.93	−2.27	−0.44	−0.20	−6.63	−0.70	2.92	−11.39	forested area
0.33	0.40	0.28	80.36	0.93	23.35	30.93		0.25	0.16	0.33	6.69	0.41	13.38	11.41	−0.07	−0.24	0.05	−73.66	−0.52	−9.97	−19.52	
2.19	1.46	1.81	120.6	2.59	43708	53.40	S152	0.17	0.54	1.02	3.83	1.02	8.51	8.88	−2.02	−0.92	−0.79	−116.76	−1.57	−22.56	−44.51	miscella neous
2.52	0.90	1.19	163.10	2.12	38.15	63.32		0.25	0.48	0.36	5.75	1.19	11.73	17.02	−2.27	−0.42	−0.82	−157.34	−0.93	−26.42	−46.3	
0.11	0.32	0.25	2.77	0.55	3.10	4.01	S155	0.50	0.34	0.47	5.26	0.49	8.05	16.39	0.39	0.02	0.22	2.48	−0.05	4.94	12.37	residential area
0.11	0.37	0.30	3.27	0.73	2.56	2.32		0.58	0.36	0.52	5.65	0.50	10.43	16.54	0.46	−0.01	0.22	2.38	−0.23	7.86	14.21	
0–0.5	0–5	0–0.5	0–10	0–10	0–10	0–5	<A	0–0.5	0–5	0–0.5	0–10	0–10	0–10	0–5								
0.5–1	5–10	0.5–3	10–40	10–20	10–25	5–20	A < X < B	0.5–1	5–10	0.5–3	10–40	10–20	10–25	5–20								
1–2	10–20	3–6	40–90	20–60	25–70	20–40	B < X < C1	1–2	10–20	3–6	40–90	20–60	25–70	20–40			increase					
	20–30	6–18	90–140	60–90	70–150	40–80	C1 < X < C2		20–30	6–18	90–140	60–90	70–150	40–80								
	30–40	18–36	140–190	90–120	150–300	80–160	C2 < X < C3		30–40	18–36	140–190	90–120	150–300	80–160			decrease					
	40<	36<	190<		300<	160<	C3 < X		40<		190<	120<	300<	160<								

Notes: "A": background concentration, "B": pollution limit, "C1": first interventional limit, "C2": second interventional limit, "C3": third interventional limit based on the suggested limits for the method of Lakanen-Erviö [54] by Kádár [55] for Hungarian soils. All concentrations are indicated in mg/kg.

13

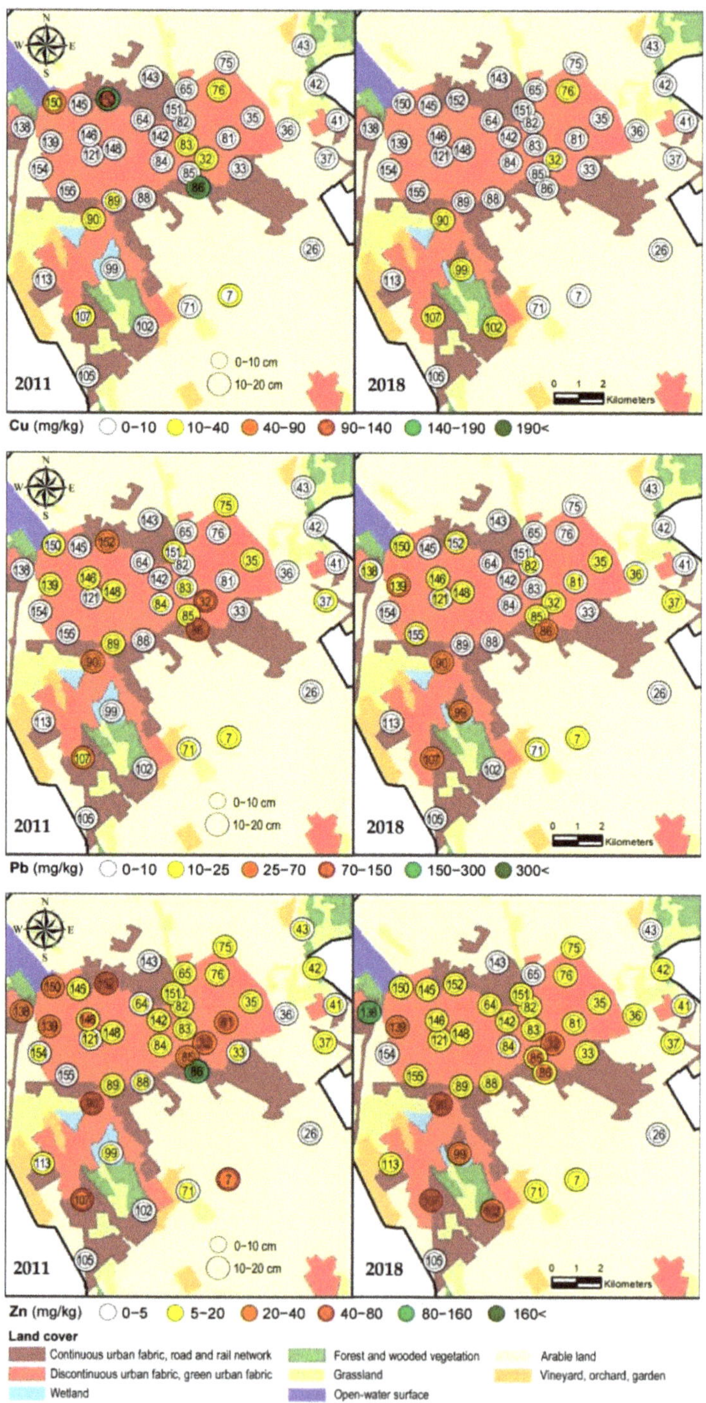

Figure 4. Comparison of the spatial distribution of available Cu, Pb, and Zn results in selected sites in 2011 vs. 2018 (EEA 2012, EEA 2018).

Table 4. Toxic element enrichment of soil samples on selected sites classified by pollution limits.

Site Nr.	Available							Total							Enrichment Factor						
	Cd	Co	Cr	Cu	Ni	Pb	Zn	Cd	Co	Cr	Cu	Ni	Pb	Zn	Cd	Co	Cr	Cu	Ni	Pb	Zn
S32	0.18	0.47	0.32	17.85	0.90	20.90	54.66	0.25	4.25	25.10	42.22	12.95	33.65	119.81	2	0	0	1	0	4	2
S35	0.44	0.91	0.20	3.46	1.01	17.56	11.79	0.49	7.66	28.65	15.58	14.05	44.25	102.24	2	0	0	0	0	2	1
S43	0.13	1.45	0.36	4.55	1.85	4.84	2.73	0.15	6.22	28.94	15.65	17.33	9.87	41.30	1	0	0	0	0	1	0
S64	0.22	1.12	0.31	6.42	2.05	6.56	11.64	0.28	7.92	38.18	24.70	21.36	15.44	62.82	1	0	0	0	0	1	1
S82	0.27	1.11	0.26	8.17	2.16	11.31	9.70	0.33	8.16	35.79	27.99	22.13	21.32	68.03	1	0	0	0	0	1	1
S84	0.16	0.91	0.53	3.48	0.93	8.02	6.17	0.19	5.70	25.15	15.64	14.35	15.12	45.25	1	0	0	0	0	1	1
S89	0.18	0.34	0.32	6.42	0.36	9.15	12.67	0.18	3.86	18.13	23.79	9.94	17.47	47.55	1	0	0	1	0	2	1
S99	0.62	0.43	0.43	11.52	0.99	25.92	35.93	0.67	3.99	19.06	27.97	12.35	33.49	94.44	4	0	0	1	0	3	2
S121	0.13	0.56	0.23	5.63	1.06	12.37	7.97	0.16	5.55	25.19	19.45	15.27	20.89	47.28	1	0	0	0	0	2	1
S139	0.36	0.59	0.23	6.40	1.46	22.75	26.38	0.57	4.67	22.93	19.57	11.97	33.21	81.87	3	0	0	0	0	3	1
S145	0.25	1.08	0.42	4.13	1.49	9.91	9.3	0.29	7.21	34.54	22.10	22.84	16.23	64.41	1	0	0	0	0	1	1
A>	0–0.5	0–5	0–0.5	0–10	0–10	0–10	0–5	0–0.5	0–15	0–30	0–30	0–25	0–25	0–100	≤1	no enrichment					
A< X < B	0.5–1	5–10	0.5–3	10–40	10–20	10–25	5–20	0.5–1	15–30	30–75	30–75	25–40	25–100	100–200	≤3	minor enrichment					
B < X < C1	1–2	10–20	3–6	40–90	20–60	25–70	20–40	1–2	30–100	75–150	75–200	40–150	100–150	200–500	3–5	moderate enrichment					
C1 < X < C2		20–30	6–18	90–140	60–90	70–150	40–80	2–5	100–200	150–400	200–300	150–200	150–500	500–1000	5–10	moderate enrichment					
C2 < X < C3		30–40	18–36	140–190	90–120	150–300	80–160	5–10	200–300	400–800	300–400	200–250	500–600	1000–2000	10–25	severe enrichment					
>C3		40<	36<	190<	120<	300<	160<	10<	300<	800<	400<	250<	600<	2000<	25–50	very severe enrichment					

Note: According to the risk substance concentration levels for pseudo-total fraction set forth in legislation [56,57], the term "A" (background concentration) indicates the typical particular substance reflected under natural conditions in soil. The term "B" (background concentration) represents a risk substance, with due regard, in the case of groundwater, to the requirements of drinking quality and the aquatic ecosystem and, in the case of the geological medium, to the full range of soil functions and the sensitivity of groundwater to pollution. The term "C" refers to the interventional pollution limit level.

Table 5. EMI scores and soil microarthropod abundance (ind./m^2), number of taxa, Shannon diversity, equitability, and QBS-ar index values (mean $\pm$ SE) of the soil samples collected from selected sites.

Microarthropod Taxa (EMI Scores)	Site Nr.										
	S32	S35	S43	S64	S82	S84	S89	S99	S121	S139	S145
Acari (20)	4415 (1103)	9122 (2450)	9392 (3212)	7130 (1460)	5141 (1419)	8500 (2507)	5519 (1219)	6644 (1600)	5563 (1844)	6781 (2260)	10,133 (2970)
Araneae (1–5)	33 (19)	67 (51)	133 (69)	0 (0)	0 (0)	189 (89)	56 (40)	0 (0)	0 (0)	0 (0)	78 (48)
Chilopoda (10)	0 (0)	89 (73)	56 (40)	0 (0)	22 (11)	0 (0)	33 (33)	44 (29)	0 (0)	0 (0)	89 (73)
Coleoptera (1–20)	33 (19)	56 (40)	78 (29)	56 (40)	0 (0)	0 (0)	0 (0)	56 (40)	0 (0)	0 (0)	67 (40)
Collembola (1–20)	822 (323)	4085 (946)	4481 (1084)	3200 (869)	1626 (454)	2270 (679)	3019 (940)	1578 (551)	2578 (1039)	3544 (1126)	4526 (930)
Diplopoda (10–20)	0 (0)	0 (0)	0 (0)	0 (0)	0 (0)	0 (0)	0 (0)	67 (33)	0 (0)	0 (0)	78 (48)
Diplura (20)	0 (0)	222 (142)	256 (78)	78 (48)	56 (40)	67 (38)	0 (0)	156 (91)	0 (0)	89 (59)	44 (29)
Hemiptera (1–10)	122 (68)	0 (0)	0 (0)	89 (72)	0 (0)	133 (58)	0 (0)	89 (40)	0 (0)	0 (0)	111 (48)
Hymenoptera (1–5)	644 (367)	0 (0)	111 (68)	522 (185)	944 (244)	722 (193)	211 (87)	0 (0)	467 (168)	800 (150)	1089 (330)
Isopoda (10)	0 (0)	56 (40)	44 (29)	0 (0)	11 (11)	44 (29)	0 (0)	0 (0)	0 (0)	0 (0)	56 (40)
Pauropoda (20)	0 (0)	0 (0)	89 (29)	67 (19)	0 (0)	89 (56)	78 (56)	0 (0)	0 (0)	0 (0)	56 (22)
Protura (20)	0 (0)	0 (0)	56 (11)	0 (0)	0 (0)	0 (0)	0 (0)	0 (0)	0 (0)	0 (0)	122 (59)
Pseudoscorpionida (20)	0 (0)	56 (29)	0 (0)	0 (0)	0 (0)	0 (0)	0 (0)	0 (0)	0 (0)	0 (0)	144 (87)
Psocoptera (1)	0 (0)	0 (0)	0 (0)	33 (33)	0 (0)	111 (62)	0 (0)	0 (0)	67 (38)	0 (0)	0 (0)
Symphyla (10)	0 (0)	56 (40)	189 (172)	0 (0)	78 (29)	0 (0)	78 (48)	0 (0)	56 (29)	0 (0)	89 (48)
Thysanoptera (1)	0 (0)	22 (11)	0 (0)	0 (0)	0 (0)	0 (0)	0 (0)	0 (0)	0 (0)	89 (89)	0 (0)
Coleoptera larvae (10)	33 (33)	178 (40)	111 (44)	0 (0)	0 (0)	67 (67)	0 (0)	122 (59)	0 (0)	111 (44)	133 (51)
Diptera larvae (10)	167 (107)	56 (22)	356 (59)	244 (91)	89 (29)	0 (0)	56 (40)	0 (0)	111 (95)	144 (80)	22 (22)
Hymenoptera larvae (10)	0 (0)	56 (56)	0 (0)	0 (0)	0 (0)	0 (0)	0 (0)	0 (0)	0 (0)	0 (0)	0 (0)
Taxa richness	6.00 (1.15)	10.0 (0.58)	11.67 (0.88)	7.33 (0.67)	6.67 (0.67)	8.67 (0.33)	6.00 (0.58)	7.00 (0.58)	5.33 (0.88)	6.00 (0.58)	13.33 (0.33)
Shannon index	0.64 (0.21)	0.74 (0.11)	0.80 (0.05)	0.76 (0.09)	0.78 (0.04)	0.76 (0.13)	0.73 (0.07)	0.59 (0.15)	0.68 (0.08)	0.82 (0.05)	0.86 (0.10)
Pielou's index	0.51 (0.12)	0.40 (0.05)	0.43 (0.02)	0.49 (0.04)	0.53 (0.03)	0.47 (0.08)	0.49 (0.05)	0.38 (0.08)	0.52 (0.06)	0.57 (0.04)	0.45 (0.05)
QBS-ar index	61.0 (6.4)	125.7 (7.5)	153.3 (7.3)	92.7 (8.3)	88.3 (8.8)	95.3 (6.7)	77.0 (4.4)	94.3 (9.3)	59.0 (3.0)	78.7 (6.8)	162.7 (10.9)

During the trace metal investigations, site S32 (traffic zone) showed moderate enrichment of Pb and minor enrichment of Cd, Cu, and Zn. Moderate Cd and Pb enrichment were also typical for site S99 (creek and lake bank) and S139 (park). At these sites, extremely high results were detected during the soil monitoring as well. In detail, Cd, Pb, and Zn showed minor enrichment at almost every site. Cr, Co, and Ni enrichments were negligible. The rank order of mobility for these elements was Pb > Cd > Zn > Cu = Cr = Ni = Co.

3.3. Soil Microarthropods

A total of 11,014 microarthropod specimens belonging to 19 different groups were extracted from soil samples (Table 5). Cumulative taxa richness per site ranged from a minimum of six (site S121) to a maximum of 16 at site S145, whereas total microarthropod abundance ranged from a minimum of 5222 ind. m^{-2} at site S82 to a maximum of 21,789 ind. m^{-2} at site S145. Acari and Collembola appeared to be the two most abundant taxa, representing 68.6% and 27.8% of the extracted microarthropods, respectively. Percentages of the other groups were markedly lower, and with the exception of Hymenoptera (Formicidae), none of them reached 1% of the total arthropod number. Among the microarthropod taxa, Acari and Collembola were ubiquitous; moreover, the frequency of Diplura and Hymenoptera was considerably high.

The Shannon index indicated the highest diversity at site S145 and the lowest at site S99. Pielou's index suggested the highest evenness at site S139 and the lowest at site S99. The values of the QBS-ar index varied within a wide range, from 51 to 176; the lowest and highest were associated with sites S32 and S145, respectively. Examining the relationship between the indices calculated above and the soil physicochemical properties, only Shannon diversity showed a significant correlation with Cu and Zn content ($p < 0.05$; r = -0.66 and r = -0.62, respectively).

The CCA analysis explored a more detailed relationship between soil fauna and soil environmental factors (Figure 5). The first two axes together explained 70.7% (46.3% and 24.4%, respectively) of the total variance in microarthropod taxa distribution. As the Monte Carlo permutation test proved, eigenvalues of the first two canonical axes were significant ($p < 0.01$ and $p < 0.05$, respectively). The first axis of this data set mainly represents the trace metals Pb, Cu, and Zn, and, furthermore, soil pH and organic matter content (SOM), whereas the second axis is mainly governed by metals Cd, Co, and Ni. Along axis 1, the dispersion of microarthropod taxa distinguished the more polluted plots and those less affected by trace metal accumulation. Chilopoda, Diplura, Pauropoda, and Symphyla appeared to be the most sensitive groups, projected on the negative side of axis 1. The groups Pseudoscorpionida, Protura, Isopoda, and Collembola were more prevalent in less polluted or moderately polluted soils. At the same time, Acari, Psocoptera, Hemiptera, and Hymenoptera (Formicidae) were also present in considerable numbers in soils markedly polluted by Pb, Zn, or Cu.

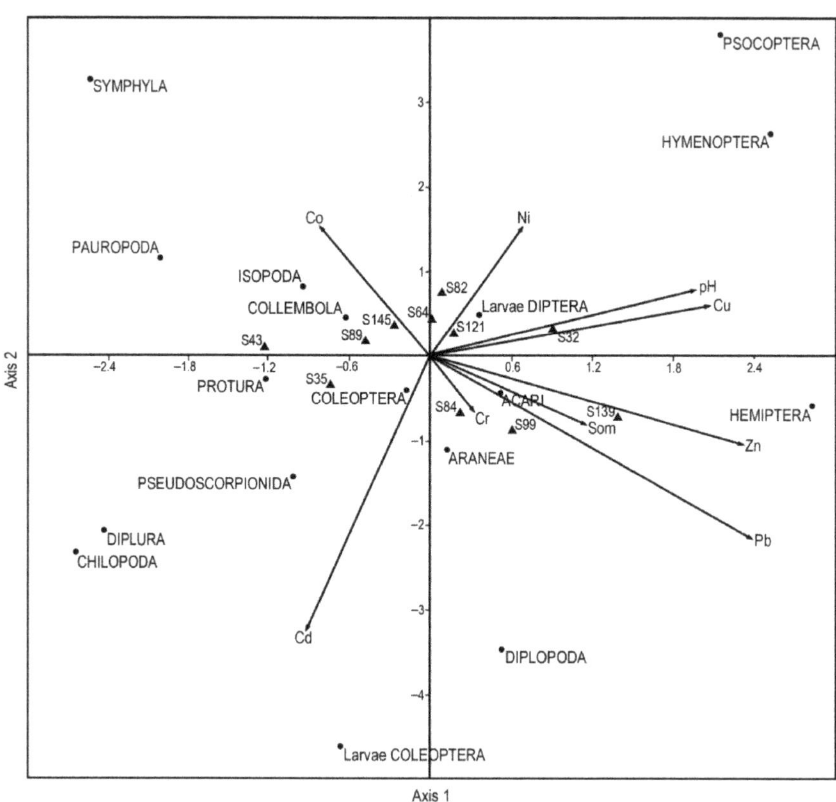

Figure 5. Ordination biplot of canonical correspondence analysis (CCA) of soil physical and chemical properties and soil microarthropod taxa (dots). Triangles represent the sampling sites (for site numbers, see Figure 1).

4. Discussion

4.1. Evaluation of Soil Analysis

Despite the heterogeneity of the city, the comparison of soil properties was difficult. However, similarities and frequencies can be detected by performing measurements. The impact of geological processes can be demonstrated with bedrock, which is related to the spatiality of urbanized areas. In the case of Székesfehérvár, the acidic parent material—granite—of the Velence Mountains affected the peripheral areas of the northeastern part. However, soils of the urbanized areas were influenced not only by the mountains' acidic forest soils but also by sediments and deposits as a result of geological soil formation and transformed by anthropogenic activities. Therefore, soil pH was generally slightly alkaline in both years; hence, high trace metal concentrations were not characteristic. Stream deposits cover the suburban and urban sites located in the Mezőföld, the sediment of which has a watery pH that is slightly alkaline (pH_{H2O} 7.9–8.1). Thus, the results of the settlement were similar to values found in the city of Sopron at about pH 8.0 [37]. The arable lands of Székesfehérvár are still under cultivation throughout the suburbs of the city. The use of fertilizers or pesticides is typical [67,68]. Based on pH results of 2011 versus 2018, a small decreasing tendency was detected. The decrease in average pH_{H2O} values means that a soil acidification process was already apparent in the area. However, the earlier average of 0.42 increased to 0.48. A value of 0.8 was barely measured over a span of seven years but, currently, a value of 0.9 shows an increase in susceptibility to soil acidification. Alkalinity of soil impedes the mobility of anthropogenic contaminants in urban soils because pollutants tend to bind within this range; therefore, they cannot leach into lower soil layers or water bases. However, if this acidification tendency continues, pollutants will become more mobile.

Consequently, $CaCO_3$ content was high in samples due to geological conditions. Despite the significant increase in the skeleton percentage—debris, artefacts, and construction waste—lime content was lower in 2018 than it was in 2011. The city was a coronation and burial place for Hungarian kings in the Middle Ages. The building materials originate from the nearby mountains (e.g., limestone, dolomite, granite). However, the highest concentrations of $CaCO_3$ were found in creek and lake banks. The reason for this is Gaja Brook, which flows through the city area. The catchment area of the Gaja is located in the karstic, limestone–dolomite–marlic Bakony Mountains. The brook drains off the excess water until reaching Székesfehérvár. Soil texture in Székesfehérvár is quite varied, which may be due to the different types of deposits and the characteristics of the Mezőföld. Between the soil samples, loam, clay loam, and sandy loam were typical in both layers in both years. The soils around the city are suitable for agricultural cultivation because of favorable climatic conditions in the area. The presence of sand is attributed to to the origin of the soils, while the rivers and streams that were active in the lower and middle Pleistocene deposited Pannonian clay sediments in the area [42], and loess was deposited as well. The other reason is that sand is a popular filler and is often mixed into the soil in the course of construction and renovation projects. These characteristics were typical in the city of Szombathely as well [37].

The increasing environmental risk of a particular toxin is demonstrated by amounts exceeding pollution limits. The available Co, Cr, and Ni contents were negligible in the investigated topsoil layer. Based on Li et al. [69], plants can accumulate small amounts of Co, Co, or Ni that do not bind to humus [70]; therefore, these two mobile elements appear in deeper layers near the parent material. The city is located in Hungary's most significant geochemical region where the covered sediment of streams made excellent arable lands, and most of the elements can enter into the soil with fertilizers and pesticides, Cd^{2+}, Cu^{2+}, Pb^{2+}) [71]. According to previous studies [72], Cu and Cd deposition on the surface of the soil was mainly due to vehicle corrosion, tire wear, or air pollution. Cd accumulation was found near the M7 motorway, and extremes also appeared in the southern part of the city, where busy road traffic is typical. According to Meuser [73], heavy metal concentration decreases with distance from roads and with soil depth. It is estimated that about 10% of the lead released is deposited within 100 m [74]. In Székesfehérvár, Cu values did not exceed the pollution limit (40 mg Cu/kg), but extreme limit excesses (e.g., 408 mg Cu/kg) were more characteristic in 2011. The extremely high results were detected in an industrial zone and in a grassland site in the suburb. The mentioned areas are located next to viticulture areas and arable lands where pesticide use is already characteristic.

Pb is only slightly mobile in alkaline soils, and it binds to clay mineral or humic substances. In the acidic soils of the suburban areas, the mobility of Pb increased similarly to Cd, where low contamination already carried risks for the environment. Previously, the greatest source of Pb was fuel, but the use of leaded fuel was banned in the 1990s. In addition, lead-rich paints were used previously to treat buildings externally, with the aging and dusting of soils in the nearby soils, which resulted in several studies in the 1980s (e.g., [75]). Many houses contain lead-rich paint in Hungary, even today. The city experienced an expansion period in 2011, but the expected population and economic growth experienced then has stagnated today. In addition to migration into cities, labor migration has emerged as a new factor in recent years. Samples taken alongside busy roads, especially in the city center, were contaminated with Pb due to the continuous traffic. Only a small part of the Zn compounds originated from traffic and transport. Zn and its compounds are a constant accompaniment to anthropogenic effects because the element exists in many household, industrial, and agricultural materials [73].

The comparison of results is difficult because other authors have analyzed total element content in general. Therefore, the results of Székesfehérvár were compared to trace metal contents of Sopron and Szombathely, which were measured by the authors. Székesfehérvár is clearly more contaminated with Cd (2011: 0.33 mg Cd/kg on average, 2018: 0.26 mg Cd/kg on average), Co (2011: 0.76 mg Co/kg on average, 2018: 0.87 mg Co/kg on average), Cr (2011: 0.31 mg Cr/kg on average, 2018: 0.34 mg Cr/kg on average), Cu 2011: 17.00 mg Cu/kg on average, 2018: 6.38 mg Cu/kg on average), Ni (2011: 1.31 mg Ni/kg on average, 2018: 1.42 mg Ni/kg on average), Pb (2011: 13.01 mg Pb/kg on average, 2018: 11.98 mg Pb/kg on average), and Zn (2011: 19.07 mg Zn/kg on average, 2018: 15.70 mg Zn/kg on average) by mean values. Both cities (Sopron and Szombathely) are characterized by more green areas and fewer industrial zones (only 2–2 industrial parks due to city locations). The diverse industrial structure of Székesfehérvár is provided by nearly 12,000 companies operating in the city. Among the largest companies, the most significant processors of the Hungarian aluminum industry must also be mentioned. Sectors represented in the settlement include: materials technology, mechatronics, machine/component manufacturing, electronics, informatics, and the food industry. In contrast, Sopron and Szombathely can be characterized by light industrial activities and tourism. In Szombathely, the Cu values did not exceed the pollution limit (40 mg Cu/kg), but extreme limit excesses (e.g., 130 Cu mg/kg) were characteristic in viticulture and garden areas of Sopron. Pb pollution was higher in Sopron soils than in Szombathely soils, but Pb accumulation was characteristic in the eastern part of Szombathely, where intense industrial and transportation activities occurred. Based on the results of the suburban sites, the Zn results were higher in the urban area of Szombathely, where intensive agricultural activity is still an ongoing process [37]. In other Hungarian contexts, trace metals in green fields were measured in Szeged, which proved to be less contaminated with Cd (0.21 mg Cd/kg on average) and Zn (14.49 mg Zn/kg) [76].

4.2. Soil Microarthropods and Soil Biological Quality

Urban soils can be considered a sink for pollutants, including trace metals [77,78]. Pollution can affect soil organisms directly and indirectly. Soil microarthropods come into direct contact with contaminants through their consumed food [79]. In urban soils polluted with trace metals, toxicity for soil-inhabiting animals is mainly related to pollutant concentration in the pore water [80–83]. However, the total concentrations of heavy metals linked to the soil solid phase might also contribute to contaminant uptake by specific groups of soil microarthropods [84,85]. The studied microarthropod groups have specialized morpho-physiological systems for absorbing water, like the water-conducting system unique to Isopoda; the rectal uptake systems in Diplopoda; the ventral tube in Collembola; or the eversible bladders in Diplura, Pauropoda, Protura, Symphyla, and Thysanura [79,86]. Indirect toxic effects of urban soil contamination may also be related to changes in food resources and food webs or to the physical changes in habitats [22,79,84]. A decrease in the number of microbial populations or the quantity of fungal hyphae as a response to soil contamination can change Collembola communities by affecting groups feeding on soil fungi and bacteria [87]. The hyphae of fungi may accumulate extremely high metal concentrations in urban environments [88–90]; consequently, fungi are an important route for food chain transfer of metals through their consumers, such as Collembola, all the way through to their predators, e.g., specific groups of Pseudoscorpionida, Araneae, Acari, and Coleoptera [91].

In our study, the soil of the sites sampled differed widely both in microarthropod abundance and taxa richness. Although we found the highest total abundance in a less polluted site (S145), no correlation was observed with soil trace metal content, suggesting that total abundance is an unsuitable characteristic for assaying metal pollution [92,93]. Certain species or groups that are more tolerant to trace metal pollution can benefit from the disappearance or abundance decrease in less tolerant sensitive taxa, compensating or even over-compensating the number of individuals [94]. The two ubiquitous taxa, Acari and Collembola, occurred in relatively high numbers in all samples and are, therefore, unsuitable indicators of soil condition differences at the group level [19,95]. Acari represent a major arthropod group, the one found to be the most abundant in every site. Though metal pollution can negatively affect the abundance and composition of mite communities (e.g., [96–98]), the discovered mite communities proved quite tolerant to trace metal pollution. From the mentioned ubiquitous taxa, Collembola showed higher variation in abundance among the studied sites and, as the CCA ordination revealed, showed slight sensitivity to the content of certain available trace metals (Cu, Zn, and Pb). However, for a more accurate evaluation, the necessity of species-level analyses cannot be neglected [22,23]. On the other hand, the behavior of rare microarthropod groups can provide useful information on soil condition and health even at a higher taxonomic level [19,93]. According to our results, Symphyla, Pauropoda, Chilopoda, and Diplura appeared to be the most sensitive groups. However, Protura, Isopoda, and Pseudoscorpionida also showed avoidance of soils with higher concentrations of Cu, Zn, and Pb. Nevertheless, available information concerning the metal tolerance of the mentioned groups is limited. In agreement with our results, some authors [93,99] observed a decrease in Symphyla abundance in trace metal-contaminated soils, while other studies [19,100] found this group rather tolerant to metals. In sound agreement with our results, Protura appeared to be particularly sensitive to metals in a lead-polluted area [101]. With regard to Isopoda, several authors (e.g., [102,103]) noted that the absence or decreased abundance of this group is mainly related to the increased content of Zn. Our findings also support this. On the other hand, it is worth noting that isopods are capable of accumulating heavy metals [104,105] and can, therefore, also occur with considerable abundance in urban environments [19]. Bogyó et al. [106] reported on the significant impact of urbanization on Diplopoda communities. Diplopoda were sampled in only two sites in Székesfehérvár, although one of them (site S99) presented relatively high Pb, Cu, and Zn concentrations. Hymenoptera, mainly represented by Formicidae, which show relatively high tolerance to metal pollution, occurred in all sampling sites except two. Similar results were obtained by McIntyre et al. [95] and Eeva et al. [107]. By contrast, Santorufo et al. [19] found Formicidae to be sensitive to metal contamination in their study, which was executed in a Mediterranean city.

The diversity measurement (Shannon index) used showed low taxa diversity in two sites with higher available Cu, Zn, and Pb concentrations, but the opposite (relatively high diversity in a polluted site) was also observed. As several authors emphasized, no evident relationship between diversity and trace metal contamination exists; therefore, diversity measurement indices should be applied with caution when evaluating pollution [19,105,108].

The advantage of the QBS approach lies in taking into account the functional role and adaptation level of organisms of different trophic levels, which makes it a useful method to evaluate soil degradation [64]. The QBS-ar indexes obtained showed high variation among the studied sites, with the highest values associated with less-polluted soils. Usually, a more stable soil environment is characterized by QBS-ar values between 100 and 200 [109]. The higher values in the central area of the city (sites S35 and S145) might indicate isolated habitat remnants of favorable conditions where soil organisms are able to tolerate a greater degree of urban development and pollution [18].

5. Conclusions

Székesfehérvár is surrounded by agricultural lands; thus, toxic elements also possibly exist in the suburbs due to fertilization. The city has several busy roads. Consequently, large amounts of Pb, Zn, Cu, and Cd are expected in the traffic zones and green spaces. In addition, the M7 motorway passes through the township from the south side. Therefore, an extremely high trace metal occurrence was expected. The soils were mostly alkaline; hence, high $CaCO_3$ was typical. Available toxic elements (Cd, Co, Cu, Ni, Pb, and Zn) showed extremes in the case of Cu (value: 408 mg Cu/kg, interventional pollution limit C1 $\geq$ 90 mg Cu/kg) and Zn (value: 47.54 mg/kg, interventional pollution limit C1 $\geq$ 40 mg Cu/kg). Based on these previous findings, sharp increases were expected between the samples of 2011 and 2018. In 2018, samples taken alongside busy roads—especially near the M7 motorway or other busy roads—were contaminated with Zn and exceeded the "C1" limit (>40 mg Zn/kg) and "C2" limit (>80 mg Zn/kg). The Pb concentration decreased slightly towards the peripheral areas in the southern part of the city. Near the M7 motorway, the most polluted sites were detected, and moderate enrichment was calculated. In addition, the trace metal content of green areas definitely increased. Evaluation of soil biological quality showed that soil biota has clear responses to the degree of trace metal pollution in urban environments. Sites characterized by high metal concentrations are usually associated with lower diversity and QBS-ar values. Important groups such as Symphyla, Protura, Pauropoda, and Chilopoda were completely absent from the most polluted sites. Nevertheless, higher QBS-ar values were also obtained in a few city center sites, indicating stable soil ecosystems.

The novelty of the present study is the fact that not only chemical but also biological surveys and thus a more complex evaluation were performed. There was a demonstrable correlation between specific chemical parameters and soil microarthropod groups. Analyses complemented with biological indices related to, e.g., aquatic macroinvertebrate and soil mesofauna communities can detect all prominent anthropogenic influences in the urban environment. The close relationship between biological and chemical properties suggests the necessity of similar soil biological monitoring in the future in urban environments.

In summary, our results have shown that soils were generally more polluted in 2011. Currently, Székesfehérvár is a livable urban environment, one which is still being formed, but is developing in a suitable way.

Author Contributions: Conceptualization, A.H., A.B., and D.W.; methodology, A.H. and D.W.; software, P.B.; validation, Z.G., P.K., and R.S.; formal analysis, D.W.; investigation, B.B.-V., A.H., and D.W.; resources, P.C.; data curation, P.K.; writing—original draft preparation, A.H. and D.W.; writing—review and editing, A.H. and D.W.; visualization, P.B.; supervision, A.B.; project administration, A.H.; funding acquisition, A.B. All authors have read and agreed to the published version of the manuscript.

Funding: This article was written in the framework of the "EFOP-3.6.1-16-2016-00018—Improving the role of research+development+innovation in the higher education through institutional developments assisting intelligent specialization in Sopron and Szombathely".

Acknowledgments: Special thanks to Tamás Hofmann, Frank Berger, and all of the project contributors.

Conflicts of Interest: The authors declare no conflict of interest. The funders had no role in the design of the study; in the collection, analyses, or interpretation of data; in the writing of the manuscript, or in the decision to publish the results.

References

1. Zinicovscaia, I.; Aničić Urošević, M.; Vergel, K.; Vieru, E.; Frontasyeva, M.V.; Povar, I.; Duca, G. Active moss biomonitoring of trace elements air pollution in Chisinau, Republic of Moldova. *Ecol. Chem. Eng. S* **2018**, *25*, 361–372. [CrossRef]
2. Cai, K.; Li, C. Street dust heavy metal pollution source apportionment and sustainable management in a typical city—Shijiazhuang, China. *Int. J. Environ. Res. Public Health* **2019**, *16*, 2625. [CrossRef]
3. Liu, D.; Li, Y.; Ma, J.; Li, C.; Chen, X. Heavy metal pollution in urban soil from 1994 to 2012 in Kaifeng city, China. *Water Air Soil Pollut.* **2016**, *227*, 154. [CrossRef]
4. Bezuglova, O.S.; Tagiverdiev, S.S.; Gorbov, S.N. Physical properties of urban soils in Rostov agglomeration. *Eurasian Soil Sci.* **2018**, *51*, 1105–1110. [CrossRef]
5. Cipullo, S.; Snapir, B.; Tardif, S.; Campo, P.; Prpich, G.; Coulon, F. Insights into mixed contaminants interactions and its implication for heavy metals and metalloids mobility, bioavailability and risk assessment. *Sci. Total Environ.* **2018**, *645*, 662–673. [CrossRef]
6. Eghbal, N.; Nasrabadi, T.; Karbassi, A.R.; Taghavi, L. Investigating the pattern of soil metallic pollution in urban areas (case study: A district in Tehran city). *Int. J. Environ. Sci. Technol.* **2019**, *16*, 6717–6726. [CrossRef]
7. Tume, P.; González, E.; King, R.W.; Monsalve, V.; Roca, N.; Bech, J. Spatial distribution of potentially harmful elements in urban soils, city of Talcahuano, Chile. *J. Geochem. Explor.* **2018**, *184*, 333–344. [CrossRef]
8. O'Riordan, R.; Davies, J.; Stevens, C.; Quinton, J.N.; Boyko, C. The ecosystem services of urban soils: A review. *Geoderma* **2021**, *395*, 115076. [CrossRef]
9. Adimalla, N.; Qian, H.; Wang, H. Assessment of heavy metal (HM) contamination in agricultural soil lands in northern Telangana, India: An approach of spatial distribution and multivariate statistical analysis. *Environ. Monit. Assess.* **2019**, *191*, 246. [CrossRef]

10. Ciupa, T.; Suligowski, R.; Kozłowski, R. Trace metals in surface soils under different land uses in Kielce city, south-central Poland. *Environ. Earth Sci.* **2020**, *79*, 14. [CrossRef]
11. Novák, T.J.; Balla, D.; Kamp, J. Changes in anthropogenic influence on soils across Europe 1990–2018. *Appl. Geogr.* **2020**, *124*, 102294. [CrossRef]
12. Tume, P.; Acevedo, V.; Roca, N.; Ferraro, F.X.; Bech, J. Potentially toxic elements concentrations in schoolyard soils in the city of Coronel, Chile. *Environ. Geochem. Health* **2021**. [CrossRef]
13. Konstantinova, E.; Minkina, T.; Sushkova, S.; Konstantinov, A.; Rajput, V.D.; Sherstnev, A. Urban soil geochemistry of an intensively developing Siberian city: A case study of Tyumen, Russia. *J. Environ. Manag.* **2019**, *239*, 366–375. [CrossRef]
14. Wieczorek, K.; Turek, A.; Szczesio, M.; Wolf, W.M. Comprehensive Evaluation of Metal Pollution in Urban Soils of a Post-Industrial City—A Case of Łódź, Poland. *Molecules* **2020**, *25*, 4350. [CrossRef] [PubMed]
15. Zhou, W.; Yang, K.; Bai, Z.; Cheng, H.; Liu, F. The development of topsoil properties under different reclaimed land uses in the Pingshuo opencast coalmine of Loess Plateau of China. *Ecol. Eng.* **2017**, *100*, 237–245. [CrossRef]
16. Romzaykina, O.N.; Vasenev, V.I.; Khakimova, R.R.; Hajiaghayeva, R.; Stoorvogel, J.J.; Dovletyarova, E.A. Spatial variability of soil properties in the urban park before and after reconstruction. *Soil Environ.* **2017**, *36*, 155–165. [CrossRef]
17. Gabarrón, M.; Zornoza, R.; Martínez-Martínez, S.; Muñoz, V.A.; Faz, Á.; Acosta, J.A. Effect of land use and soil properties in the feasibility of two sequential extraction procedures for metals fractionation. *Chemosphere* **2019**, *216*, 266–272. [CrossRef] [PubMed]
18. McIntyre, N.E. Ecology of urban arthropods: A review and a call to action. *Ann. Entomol. Soc. Am.* **2000**, *93*, 825–835. [CrossRef]
19. Santorufo, L.; Van Gestel, C.A.M.; Rocco, A.; Maisto, G. Soil invertebrates as bioindicators of urban soil quality. *Environ. Pollut.* **2012**, *161*, 57–63. [CrossRef] [PubMed]
20. Jouquet, P.; Dauber, J.; Lagerlof, J.; Lavelle, J.P.; Lepage, M. Soil invertebrates as ecosystem engineers: Intended and accidental effects on soil and feedback loops. *Appl. Soil Ecol.* **2006**, *32*, 153–164. [CrossRef]
21. Lavelle, P.; Decaens, T.; Aubert, M.; Barot, S.; Blouin, M.; Bureau, F.; Margerie, P.; Mora, P.; Rossi, J.P. Soil invertebrates and ecosystem services. *Eur. J. Soil Biol.* **2006**, *42*, 3–15. [CrossRef]
22. Van Straalen, N.M. The Use of Soil Invertebrates in Ecological Surveys of Contaminated Soils. In *Developments in Soil Science*; Doelman, P., Eijsackers, H.J.P., Eds.; Elsevier: Amsterdam, The Netherlands, 2004; pp. 159–195. [CrossRef]
23. Sterzyńska, M.; Nicia, P.; Zadrożny, P.; Fiera, C.; Shrubovych, J.; Ulrich, W. Urban springtail species richness decreases with increasing air pollution. *Ecol. Indic.* **2018**, *94*, 328–335. [CrossRef]
24. Stamou, G.P.; Argyropoulou, M.D. A preliminary study on the effect of Cu, Pb and Zn contamination of soils on community structure and certain life-history traits of oribatids from urban areas. *Exp. Appl. Acarol.* **1995**, *19*, 381–390. [CrossRef]
25. Eitminaviciute, I. Microarthropod communities in anthropogenic urban soils. 1. Structure of microarthropod complexes in soils of roadside lawns. *Entomol. Rev.* **2006**, *86* (Suppl. S2), S128–S135. [CrossRef]
26. Fiera, C. Preliminary data on the species diversity of Collembola (Hexapoda, Collembola) along an urban gradient in Bucharest. *Trav. Mus. Natl. Hist. Nat. Grigore Antipa* **2008**, *51*, 363–367.
27. Santorufo, L.; Cortet, J.; Arena, C.; Goudon, R.; Rakoto, A.; Morel, J.-L.; Maisto, G. An assessment of the influence of the urban environment on collembolan communities in soils using taxonomy- and trait-based approaches. *Appl. Soil Ecol.* **2014**, *78*, 48–56. [CrossRef]
28. Menta, C.; Remelli, S. Soil Health and Arthropods: From Complex System to Worthwhile Investigation. *Insects* **2020**, *11*, 54. [CrossRef]
29. Zhang, W.; Han, J.; Molla, A.; Zuo, S.; Ren, Y. The Optimization Strategy of the Existing Urban Green Space Soil Monitoring System in Shanghai, China. *Int. J. Environ. Res. Public Health* **2021**, *18*, 4820. [CrossRef]
30. Huot, H.; Joyner, J.; Córdoba, A.; Shaw, R.K.; Wilson, M.A.; Walker, R.; Muth, T.R.; Cheng, Z. Characterizing urban soils in New York City: Profile properties and bacterial communities. *J. Soils Sediments* **2017**, *17*, 393–407. [CrossRef]
31. Da Silva Ferreira, M.; Fontes, M.P.F.; Pacheco, A.A.; Ker, J.C.; Lima, H.N. Health risks of potentially toxic trace elements in urban soils of Manaus city, Amazon, Brazil. Environ. *Geochem. Health* **2021**. [CrossRef]
32. Silva, M.M.V.G.; Carvalho, P.C.S.; António, A.; Luís, A.C.M. Geochemistry of leptosols and fluvisols in the fast growing city of Benguela (Angola) and assessment of potential risks. *Geoderma Reg.* **2020**, *20*, e00257. [CrossRef]
33. Hindersmann, B.; Förster, A.; Achten, C. Novel and specific source identification of PAH in urban soils: Alk-PAH-BPCA index and "V"-shape distribution pattern. *Environ. Pollut.* **2020**, *257*. [CrossRef] [PubMed]
34. Fazeli, G.; Karbassi, A.; Khoramnejadian, S.; Nasrabadi, T. Evaluation of Urban Soil Pollution: A Combined Approach of Toxic Metals and Polycyclic Aromatic Hydrocarbons (PAHs). *Int. J. Environ. Res.* **2019**, *13*, 801–811. [CrossRef]
35. Zhang, Y.; Chen, Q. Contents of heavy metals in urban parks and university campuses. *IOP Conf. Ser. Earth Environ. Sci.* **2018**, *108*, 42060. [CrossRef]
36. Cheng, H.; Li, L.; Zhao, C.; Li, K.; Peng, M.; Qin, A.; Cheng, X. Overview of trace metals in the urban soil of 31 metropolises in China. *J. Geochem. Explor.* **2014**, *139*, 31–52. [CrossRef]
37. Horváth, A.; Kalicz, P.; Farsang, A.; Balázs, P.; Berki, I.; Bidló, A. Influence of human impacts on trace metal accumulation in soils of two Hungarian cities. *Sci. Total Environ.* **2018**, *637–638*, 1197–1208. [CrossRef]
38. Szita, R.; Horváth, A.; Winkler, D.; Kalicz, P.; Gribovszki, Z.; Csáki, P. A complex urban ecological investigation in a mid-sized Hungarian city—SITE assessment and monitoring of a liveable urban area, PART 1: Water quality measurement. *J. Environ. Manag.* **2019**, *247*, 78–87. [CrossRef] [PubMed]
39. Dövényi, Z. (Ed.) *Microregions of Hungary*; Hungarian Academy of Science: Budapest, Hungary, 2010. (In Hungarian)

40. IUSS Working Group WRB. *World Reference Base for Soil Resources 2006*; First update 2007; World Soil Resources Reports No. 103; FAO: Rome, Italy, 2014.
41. Novák, T.J. *Soil Practicum*; Meridián Alapítvány: Debrecen, Hungary, 2013. (In Hungarian)
42. Farkas, G. (Ed.) *Handbook of Fejér Country. Country Handbooks of Hungary*; Ceba Publisher: Budapest, Hungary, 1997. (In Hungarian)
43. Horváth, A.; Szűcs, P.; Bidló, A. Soil condition and pollution in urban soils: Evaluation of the soil quality in a Hungarian town. *J. Soils Sediments* **2015**, *15*, 1825–1835. [CrossRef]
44. Horváth, A.; Szita, R.; Bidló, A.; Gribovszki, Z. Changes in soil and sediment properties due the impact of the urban environment. *Environ. Earth Sci.* **2016**, *75*, 1211. [CrossRef]
45. Van Reeuwijk, L.P. (Ed.) *Procedures for Soil Analysis*, 6th ed.; Technical Paper 9; ISRIC: Wageningen, The Netherlands, 2002.
46. Hungarian Standard MSZ-08-0205. *Determination of soil pH, Total Salinity and CaCO3 Content*; Hungarian Standard Association: Budapest, Hungary, 1978. (In Hungarian)
47. Hungarian Standard MSZ-08-0206. *Determination of Particle Size Distribution of Soils*; Hungarian Standard Association: Budapest, Hungary, 1978. (In Hungarian)
48. FAO. *Guidelines for Soil Description*, 3rd ed.; Soil Resources, Management and Conservation Service, Land and Water Development Division; FAO: Rome, Italy, 1990.
49. Hungarian Standard MSZ-08-0452. *Determination of Humus Content in Soils*; Hungarian Standard Association: Budapest, Hungary, 1980. (In Hungarian)
50. Hungarian Standard MSZ-EN-16169. *Determination of Total Nitrogen Content in Soils*; Hungarian Standard Association: Budapest, Hungary, 2013. (In Hungarian)
51. Hungarian Standard MSZ-20135. *Determination of the Soluble Nutrient Element Content of the Soil*; Hungarian Standard Association: Budapest, Hungary, 1999. (In Hungarian)
52. Hungarian Standard MSZ 21470-50. *Environmental Testing of Soils. Determination of Total and Soluble Toxic Element, Heavy Metal and Chromium (VI) Content*; Hungarian Standard Association: Budapest, Hungary, 2006. (In Hungarian)
53. Menta, C.; Conti, F.D.; Pinto, S. Microarthropods biodiversity in natural, seminatural and cultivated soils—QBS-ar approach. *Appl. Soil Ecol.* **2018**, *123*, 740–743. [CrossRef]
54. Lakanen, E.; Erviö, R. A comparison of eight extractants for the determination of plant available micronutrients in soil. *Acta Agr. Fenn.* **1971**, *123*, 223–232.
55. Kádár, I. *Remediation Handbook 2. Investigation of Contaminated Soils*; Ministry of Environment: Budapest, Hungary, 1998. (In Hungarian)
56. Joint Decree No. 10/2000. *(VI. 2) KöM-EüM-FVM-KHVM of the Ministers of Environmental Protection, Public Health, Agriculture and Regional Development, and of Traffic, Communication and Water Management on the Limit Values Necessary to Protect the Quality of Groundwater and the Geological Medium*; Hungarian Government: Budapest, Hungary, 2000. (In Hungarian)
57. Joint Decree No. 6/2009. *(IV. 14) KvVM-EüM-FVM of the Ministers of Environmental Protection and Water Management, Public Health, Agriculture and Regional Development on the Limit Values Necessary to Protect the Quality of Geological Medium and the Groundwater and on Measurement of Pollution*; Hungarian Government: Budapest, Hungary, 2009. (In Hungarian)
58. Sterckeman, T.; Douay, F.; Baize, D.; Fourrier, H.; Proix, N.; Schvartz, C. Trace elements in soils developed in sedimentary materials from Northern France. *Geoderma* **2006**, *136*, 912–929. [CrossRef]
59. Kabata-Pendias, A.; Pendias, H. *Trace Elements in Soils and Plants*, 3rd ed.; CRC Press: Boca Raton, FL, USA, 2001.
60. Taylor, S.R.; McLennan, S.M. The geochemical evolution of the continental crust. *Rev. Geophys.* **1995**, *33*, 241–265. [CrossRef]
61. Facchinelli, A.; Sacchi, E.; Mallen, L. Multivariate statistical and GIS-based approach to identify heavy metal sources in soils. *Environ. Pollut.* **2001**, *114*, 313–324. [CrossRef]
62. Shannon, C.E.; Weaver, W. *The Mathematical Theory of Communication*; University of Illionis Press: Urbana, IL, USA, 1949.
63. Pielou, E.C. The measurement of diversity in different types of biological collections. *J. Theor. Biol.* **1966**, *13*, 131–144. [CrossRef]
64. Parisi, V.; Menta, C.; Gardi, C.; Jacomini, C.; Mozzanica, E. Microarthropod communities as a tool to assess soil quality and biodiversity: A new approach in Italy. *Agric. Ecosyst. Environ.* **2005**, *105*, 323–333. [CrossRef]
65. Ter Braak, C.J.F.; Šmilauer, P. *CANOCO Reference Manual and Canodraw for Windows User's Guide: Software for Canonical Community Ordination*; Version 4.5; Microcomputer Power: Ithaca, NY, USA, 2002.
66. Tume, P.; González, E.; King, R.W.; Cuitiño, L.; Roca, N.; Bech, J. Distinguishing between natural and anthropogenic sources for potentially toxic elements in urban soils of Talcahuano, Chile. *J. Soils Sediments* **2018**, *18*, 2335–2349. [CrossRef]
67. Rékási, M.; Filep, T. Factors determining Cd, Co, Cr, Cu, Ni, Mn, Pb and Zn mobility in uncontaminated arable and forest surface soils in Hungary. *Environ. Earth Sci.* **2015**, *74*, 6805–6817. [CrossRef]
68. Wang, H.; Marshall, C.W.; Cheng, M.; Xu, H.; Li, H.; Yang, X.; Zheng, T. Changes in land use driven by urbanization impact nitrogen cycling and the microbial community composition in soils. *Sci. Rep.* **2017**, *7*, 44049. [CrossRef] [PubMed]
69. Li, H.F.; Gray, C.; Mico, C.; Zhao, F.J.; McGrath, S.P. Phytotoxicity and bioavailability of cobalt to plants in a range of soils. *Chemosphere* **2009**, *75*, 979–986. [CrossRef]
70. González-Costa, J.J.; Reigosa, M.J.; Matías, J.M.; Fernández-Covelo, E. Analysis of the Importance of Oxides and Clays in Cd, Cr, Cu, Ni, Pb and Zn Adsorption and Retention with Regression Trees. *PLoS ONE* **2017**, *12*, e0168523. [CrossRef]

71. Ódor, L.; Horváth, I. Element distribution and environmental conditions in geochemical mapping. In *Material Flows and Their Impact on Nature*; Pantó, G., Ed.; Hungarian Academy of Sciences: Budapest, Hungary, 2003; pp. 151–188. (In Hungarian)

72. Napier, F.; D'Arcy, B.; Jefferies, C. A review of vehicle related metals and polycyclic aromatic hydrocarbons in the UK environment. *Desalination* **2008**, *226*, 143–150. [CrossRef]

73. Meuser, H. *Contaminated Urban Soils*; Springer: Dordrecht, The Netherlands, 2010.

74. Thorton, I. Metal contamination of soils in urban areas. In *Soils in the Urban Environment*; Bullock, P., Gregory, P.J., Eds.; Blackwell: Oxford, UK, 1991.

75. Chaney, R.L.; Sterret, S.B.; Mielke, H.W. The potential for heavy metal exposure from urban gardens and soils. In *Proceedings of the Symposium on Heavy Metal in Urban Gardens*; Preer, J.R., Ed.; University of the District of Columbia Extension Service: Washington, DC, USA, 1984; pp. 37–84.

76. Farsang, A.; Puskás, I. Characteristics of soils in urban ecosystems—A Complex analysis of soils in Szeged. *Földr. Köz.* **2009**, *133*, 397–409. (In Hungarian)

77. Biasioli, M.; Barberis, R.; Ajmone-Marsan, F. The influence of a large city on some soil properties and metals content. *Sci. Total Environ.* **2006**, *356*, 154–164. [CrossRef]

78. Guagliardi, I.; Cicchella, D.; De Rosa, R.; Buttafuoco, G. Assessment of lead pollution in topsoils of a southern Italy area: Analysis of urban and peri-urban environment. *J. Environ. Sci.* **2015**, *33*, 179–187. [CrossRef]

79. Rusek, J.; Marshall, V.G. Impacts of airborne pollutants on soil fauna. *Ann. Rev. Ecol. Syst.* **2000**, *31*, 395–423. [CrossRef]

80. Smit, C.E.; van Gestel, C.A.M. Effects of soil type, prepercolation, and ageing on bioaccumulation and toxicity of zinc for the springtail *Folsomia candida*. *Environ. Toxicol. Chem.* **1998**, *17*, 1132–1141. [CrossRef]

81. Martikainen, E.A.T.; Krogh, P.H. Effects of soil organic matter content and temperature on toxicity of dimethoate to *Folsomia fimetaria* (Collembola: Isotomiidae). *Environ. Toxicol. Chem.* **1999**, *18*, 865–872. [CrossRef]

82. Lock, K.; Janssens, F.; Janssen, C.R. Effects of metal contamination on the activity and diversity of springtails in an ancient Pb-Zn mining area at Plombières, Belgium. *Eur. J. Soil Biol.* **2003**, *39*, 25–29. [CrossRef]

83. Domene, X.; Colón, J.; Uras, M.V.; Izquierdo, R.; Àvila, A.; Alcañiz, J.M. Role of soil properties in sewage sludge toxicity to soil collembolans. *Soil Biol. Biochem.* **2010**, *42*, 1982–1990. [CrossRef]

84. Hopkin, S.P. *Biology of the Springtails (Insecta: Collembola)*; Oxford University Press: Oxford, UK, 1997.

85. Vijver, M.; Jager, T.; Posthuma, L.; Peijnenburg, W. Impact of metal pools and soil properties on metal accumulation in *Folsomia candida* (Collembola). *Environ. Toxicol. Chem.* **2001**, *20*, 712–720. [CrossRef]

86. Eisenbeis, G. Physiological absorption of liquid water by Collembola: Absorption by the ventral tube at different salinities. *J. Insect Physiol.* **1982**, *28*, 11–20. [CrossRef]

87. Filser, J. The effect of the systemic fungicide Aktuan on Collembola under field conditions. *Acta Zool. Fenn.* **1994**, *195*, 32–34.

88. Cromack, K.; Sollins, P.; Todd, R.L.; Crossley, D.A.; Fender, W.M.; Fogel, R.; Todd, A.W. Soil Microorganism-Arthropod Interactions: Fungi as Major Calcium and Sodium Sources. In *The Role of Arthropods in Forest Ecosystems*; Proceedings in Life, Sciences; Mattson, W.J., Ed.; Springer: Berlin/Heidelberg, Germany, 1977; pp. 78–84. [CrossRef]

89. Bååth, E. Tolerance of copper by entomogenous fungi and the use of copper-amended media for isolation of entomogenous fungi from soil. *Mycol. Res.* **1991**, *95*, 1140–1142. [CrossRef]

90. Urík, M.; Bujdoš, M.; Milová-Žiaková, B.; Mikušová, P.; Slovák, M.; Matúš, P. Aluminium leaching from red mud by filamentous fungi. *J. Inorg. Biochem.* **2015**, *152*, 154–159. [CrossRef] [PubMed]

91. Van Straalen, N.M.; Verhoef, H.A.; Joosse, E.N.G. Functionele classificatie van bodemdieren en de ecologische functie van de bodem. *Vakbl. Biol.* **1985**, *65*, 131–135.

92. Bargagli, R. *Trace Elements in Terrestrial Plants*; Springer and Landes Company: Berlin, Germany, 1998.

93. Migliorini, M.; Piginoa, G.; Carusob, T.; Fanciullia, P.P.; Leonzio, C.; Berninia, F. Soil communities (Acari Oribatida; Hexapoda Collembola) in a clay pigeon shooting range. *Pedobiologia* **2004**, *49*, 1–13. [CrossRef]

94. Gillet, S.; Ponge, J.F. Changes in species assemblages and diets of Collembola along a gradient of metal pollution. *Appl. Soil Ecol.* **2003**, *22*, 127–138. [CrossRef]

95. McIntyre, N.E.; Rango, J.; Fagan, W.F.; Faeth, S.H. Ground arthropod community structure in a heterogeneous urban environment. *Landsc. Urban Plan.* **2001**, *52*, 257–274. [CrossRef]

96. Tarman, K. Oribatid fauna in polluted soil. *Biol. Vestn. Ljubl.* **1973**, *21*, 153–158.

97. Bielska, I. Communities of moss mites (Acari, Oribatei) of grasslands under the impact of industrial pollution. III. Communities of moss mites of wastelands. *Pol. Ecol. Stud.* **1989**, *15*, 101–110.

98. Mangová, B.; Krumpál, M. Oribatid mites (Acari) in urban environments—Bratislava city, West Slovakia. *Entomofauna Carpathica* **2017**, *29*, 27–50.

99. Strojan, C.L. The impact of zinc smelter emissions on forest litter arthropods. *Oikos* **1978**, *31*, 41–46. [CrossRef]

100. Visioli, G.; Menta, C.; Gardi, C.; Conti, F.D. Metal toxicity and biodiversity in serpentine soils: Application of bioassay tests and microarthropod index. *Chemosphere* **2013**, *90*, 1267–1273. [CrossRef]

101. Hågvar, S.; Abrahamsen, G. Microarthropods and Enchytraeidae (Oligochaeta) in naturally lead-contaminated soil: A gradient study. *Environ. Entomol.* **1990**, *19*, 1263–1277. [CrossRef]

102. Joosse, E.N.G.; Wulffraat, K.J.; Glas, H.P. Tolerance and acclimation to zinc of the isopod *Porcellio scaber* Latr. In Proceedings of the International Conference: Heavy Metals in the Environment, Amsterdam, The Netherlands, 15–18 September 1981; CEP Consultants Ltd.: Edinburgh, UK, 1981; pp. 425–428.
103. Grelle, C.; Fabre, M.-C.; Leprêtre, A.; Descamps, M. Myriapod and isopod communities in soil contaminated by heavy metals in northern France. *Eur. J. Soil Sci.* **2000**, *51*, 425–433. [CrossRef]
104. Hopkin, S.P.; Martin, M.H. The distribution of zinc, cadmium, lead and copper within the woodlouse Oniscus asellus (Crustacea, Isopoda). *Oecologia* **1982**, *54*, 227–232. [CrossRef]
105. Cortet, J.; De Vaufleury, A.; Poinsotbalaguer, N.; Gomot, L.; Texier, C.; Cluzeau, D. The use of invertebrate soil fauna in monitoring pollutant effects. *Eur. J. Soil Biol.* **1999**, *35*, 115–134. [CrossRef]
106. Bogyó, D.; Magura, T.; Simon, E.; Tóthmérész, B. Millipede (Diplopoda) assemblages alter drastically by urbanisation. *Landsc. Urban Plan.* **2015**, *133*, 118–126. [CrossRef]
107. Eeva, T.; Sorvari, J.; Koivunen, V. Effects of heavy metal pollution on red wood ant (Formica s. str.) populations. *Environ. Pollut.* **2004**, *132*, 533–539. [CrossRef]
108. Van Straalen, N. Community structure of soil arthropods as a bioindicator of soil health. In *Biological Indicators of Soil Health*; Pankhurst, C.E., Doube, B.M., Gupta, V.V.S.R., Eds.; CAB International: Wallingford, UK, 1997; pp. 235–264.
109. Wahsha, M.; Bini, C.; Nadimi-Goki, M. The impact of olive mill wastewater on the physicochemical and biological properties of soils in Northwest Jordan. *Int. J. Environ. Qual.* **2014**, *15*, 25–31. [CrossRef]

minerals

Article

Pollution Characteristics, Distribution and Ecological Risk of Potentially Toxic Elements in Soils from an Abandoned Coal Mine Area in Southwestern China

Libo Pan [1], Xiao Guan [1,*], Bo Liu [1], Yanjun Chen [1], Ying Pei [1], Jun Pan [2], Yi Zhang [2] and Zhenzhen Hao [1]

[1] Ecological Department, Chinese Research Academy of Environmental Sciences, Beijing 100012, China; panlb@craes.org.cn (L.P.); liubo@craes.org.cn (B.L.); chenyj@craes.org.cn (Y.C.); peiying@craes.org.cn (Y.P.); haozhenzhen2021@126.com (Z.H.)

[2] Water Department, Guiyang Bureau of Ecology and Environment, Guiyang 550000, China; panjun2021@126.com (J.P.); sthjjshjc@guiyang.gov.cn (Y.Z.)

* Correspondence: guanxiao@craes.org.cn

Abstract: Acid mine drainage (AMD) from abandoned coal mines can lead to serious environmental problems due to its low pH and high concentrations of potentially toxic elements. In this study, soil pH, sulfur (S) content, and arsenic (As), cadmium (Cd), chromium (Cr), copper (Cu), lead (Pb), nickel (Ni), zinc (Zn), iron (Fe), manganese (Mn), and mercury (Hg) concentrations were measured in 27 surface soil samples from areas in which coal-mining activities ceased nine years previously in Youyu Catchment, Guizhou Province, China. The soil was acidic, with a mean pH of 5.28. Cadmium was the only element with a mean concentration higher than the national soil quality standard. As, Cd, Cu, Ni, Zn, Mn, Cr, and Fe concentrations were all higher than the background values in Guizhou Province. This was especially true for the Cd, Cu, and Fe concentrations, which were 1.69, 1.95, and 12.18 times their respective background values. The geoaccumulation index of Cd and Fe was present at unpolluted to moderately polluted and heavily polluted levels, respectively, indicating higher pollution levels than for the other elements in the study area. Spatially, significantly high Fe and S concentrations, as well as extremely low pH values, were found in the soils of the AMD sites; however, sites where tributaries merged with the Youyu River (TM) had the highest Cd pollution level. Iron originated mainly from non-point sources (e.g., AMD and coal gangues), while AMD and agricultural activity were the predominant sources of Cd. The results of an eco-risk assessment indicated that Cd levels presented a moderate potential ecological risk, while the other elements all posed a low risk. For the TM sites, the highest eco-risk was for Cd, with levels that could be harmful for aquatic organisms in the wet season, and may endanger human health via the food chain.

Keywords: acid mine drainage; abandoned coal mine; potentially toxic elements; pollution level; potential ecological risk

Citation: Pan, L.; Guan, X.; Liu, B.; Chen, Y.; Pei, Y.; Pan, J.; Zhang, Y.; Hao, Z. Pollution Characteristics, Distribution and Ecological Risk of Potentially Toxic Elements in Soils from an Abandoned Coal Mine Area in Southwestern China. *Minerals* **2021**, *11*, 330. https://doi.org/10.3390/min11030330

Academic Editors: Ana Romero-Freire and Hao Qiu

Received: 29 January 2021
Accepted: 16 March 2021
Published: 22 March 2021

Publisher's Note: MDPI stays neutral with regard to jurisdictional claims in published maps and institutional affiliations.

1. Introduction

Mining activity has important economic and social benefits; however, inadequately treated acid mine drainage (AMD) is globally recognized as one of the environmental issues, due to its acidic nature (extremely low pH of <3), and the presence of different potentially toxic elements, such as cadmium (Cd), copper (Cu), iron (Fe), manganese (Mn), lead (Pb), and zinc (Zn), toxic metalloids (e.g., arsenic (As) and selenium (Se), and sulfates) [1,2], posed high threat to soil, water, and the local inhabitants in the mine-affected areas. Acid mine drainage (AMD) occurs in sulfide-bearing mine waste due to the oxidation of iron sulfides, especially pyrite (FeS_2), pyrrhotite (FeS) and marcasite (FeS_2) leading to the creation of several soluble hydrous iron sulfates, the generation of acidity and subsequent leaching of metals [3–6]. Acid mine drainage can lead to soil acidification, which reduces soil enzyme activity, and inhibits plant respiration. Simultaneously, the low pH promotes the growth of acidophilic microorganisms, further accelerating the generation of AMD [7]. However, the potentially toxic elements in AMD can be ingested by plants, animals, and humans through the food chain, where their enrichment can harm human health [8]. Another prominent feature of AMD is its high Fe concentration, which is of great concern because it can influence the biogeochemistry of the ambient environment [9]. A diverse range of microorganisms capable of metabolizing Fe compounds can be leached from AMD into soils, and may play an important role in Fe cycling in soils. Therefore, the Fe cycle can potentially lead to unique microbial community assemblages in AMD-impacted soils compared to pristine soils [10].

Guizhou is located in southwestern China and has some of the country's largest coal resources. The coal reserves of Guizhou, which total 100 million tons, are distributed over an area of more than 70,000 km^2, accounting for 40% of the total land area. It was reported that as of December 31, 2019, there were 249 coal mines in production in Guizhou Province, with a total production capacity is 133.52 million tons per year. High sulfur (S) content in the main feature of the coal in Guizhou province, which can be in excess of 3% [11]. Among the sulfide minerals, FeS_2 is the most widespread in nature and is regarded as the main cause of AMD [12,13]. Youyu Catchment in Huaxi District, locating in the upstream of Aha reservoir, which is an important water resource for the Guiyang City, has a 60-year history of coal mining. The coal resources in Huaxi District are shallow, with most of the coal seams being exposed to the ground. It has therefore been proven difficult to conduct large-scale open-pit mining, and only underground mining has been established. During mining activities undertaken in the region's small coal mines, coal bunkers, liver dumps, and sewage treatment tanks are usually established close to the main cave. Coal gangue waste piles are typically disordered, and most have not been treated or protected after mine closure; moreover, the management of coal gangue piles is even less effective. In rainy season, waste piles have been washed away by rainwater, potentially toxic elements may continue to migrate via runoff water. In early 2010, the large-scale mining activities with regard to the environmental and economic aspects had been stopped by the local authority and the government. However, Guizhou is one of the most typical areas of karst landform in China, distributing 1710 springs on the ground with an average flow of 121.23 billion m^3/a. Even after abandonment, the continuously produced AMD with the presence the O_2, microorganism, and water, the AMD could still pollute the surrounding environment via springs in rainy days. Therefore, the environmental pollution and potentially ecological risks left by the decades of mining activity may last for very long time. There have been several studies focused on Fe-cycling, metals behavior, and remediation technology in the river, sediment, and soils contaminated by AMD [2,14,15]. To date, relatively few studies have reported comprehensively for soil properties and pollution level in Youyu Catchment after a long period time of closure. Even after nine years of self-restoration, Youyu Catchment's pollution is still not removed and the situation is unclear, so the aims of this study were to do the following: (1) estimate the soil acidity, soil sulfur, and levels of potentially toxic elements in the soil nine years after mine closure; (2) identify the sources of potentially toxic elements in the soil; and (3) evaluate the eco-risk posed by the distribution of potentially toxic elements in the soil.

2. Materials and Methods

2.1. Soil Sampling

Youyu Catchment is located in Huaxi District, Guiyang City, which is in the central area of Guizhou Province, China. The catchment has a total population of approximately 14,227 and covers an area of 53.33 km^2. The terrain is mostly mountains and hills. The area has a humid plateau monsoon climate, characterized by a lack of severe cold in winter and of extreme heat in summer, and by abundant rainfall and high humidity. The average annual temperature and rainfall are 14.9 °C and 1178.3 mm, respectively. The altitude ranges between 1090 and 1450 m. Silico-aluminum yellow soil is the main soil type in the study area. Youyu Catchment is characterized by well-developed agricultural activities, with the main products cultivated being tea, vegetables, and fruit. The major industries include coking, cement production, and metal foundries [16]. The soil sample collection method used herein was described by Pan et al. [17]. In June 2019, we collected soil samples surrounding eight coal gangues and six AMD effluents. Briefly, the soil samples were divided into four types: (1) soil samples surrounding (within 100 m) coal gangues (SCG); (2) samples within 100 m, where AMD emerges from underground (AMD); (3) soil samples from where tributaries merge into the Youyu River (TM); and (4) soil samples from control areas (CK). A 2 m × 2 m grid spacing was used for soil sampling, and a total of 27 surface (0–20 cm) soil samples were collected, using the five-point sampling method. At each sampling site, five subsamples were taken from the same area (approximately 4 m^2) and pooled to form one composite sample. The sampling sites are illustrated in the Figure 1.

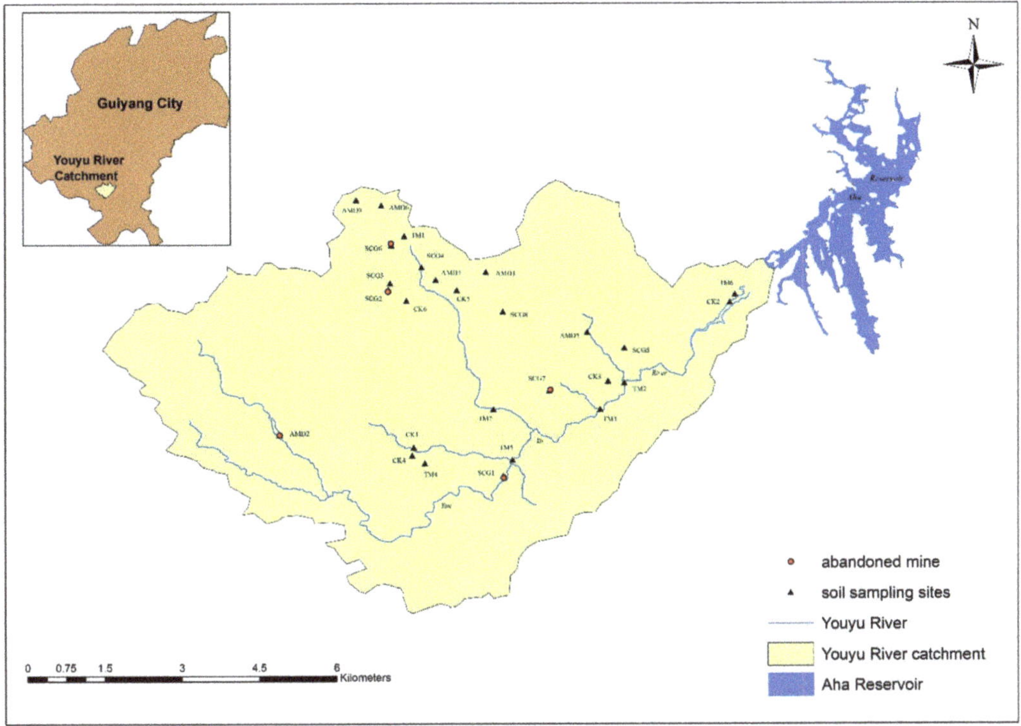

Figure 1. Locations of sampling sites in the studied area.

2.2. Sample Preparation and Analysis

All samples were air-dried for 1 week, at room temperature, in a storage room, passed through a 60-mesh sieve after removing stones, residual roots, and other unwanted materials, and then sealed in brown glass bottles and conserved in a refrigerator, at −4 °C, until analysis. Samples were analyzed by following the 3050B and 6010Cmethods established by the United States Environmental Protection Agency (USEPA) [18,19]. In brief, approximately 10 g of sample was air-dried, gently ground with an agate mortar, and passed through a 100-mesh nylon sieve. Then, 1 g of the sample was placed into an Anton PVC (Polyvinyl chloride) digestion vessel together with 9 mL concentrated nitric acid (HNO$_3$) and 3 mL hydrogen peroxide (H$_2$O$_2$). The vessel was sealed and heated at 180 °C for 15 min. The concentrations of eight potentially toxic elements (As, Cd, Cu, Pb, Ni, Zn, Mn, Cr, and Hg) in the digestion solution were determined by using inductively coupled plasma–mass spectrometry (ICP–MS; POEMS3; Thermo Fisher Scientific, Waltham, MA, USA).

Iron levels in soil were determined, using the national standard method (HJ 804-2016), with 10.0 g samples (accurate to 0.01 g) placed in a 100 mL conical flask and 20.0 mL extract liquid (c5(TEA, Triethanolamine) = 0.1 mol /L, c (CaCl$_2$) = 0.01 mol /L, c (DTPA, Diethylenetriaminepentaacetic acid) = 0.005 mol /L); pH = 7.3) then added. The solution was shaken at 20 $\pm$ 2 °C at a speed of 160~200 r/min for 2 h. The extraction solution was then slowly poured into a centrifuge tube and centrifuged for 10 min. The supernatant was isolated within 48 h of gravity filtration, using a medium-speed quantitative filter paper. Inductively coupled plasma–atomic emission spectrometry was used to detective iron levels.

For measuring pH, 10 g of soil was mixed with 25 mL of deionized water (1:2.5 (m/v) soil-to-water ratio), and the supernatant was measured by using a pH meter (HQ30d; Hach, Loveland, CO, USA), according to the methods of the People's Republic of China National Environmental Protection Standards (HJ962–2018) [20]. The S content in soil was determined by using an elemental analyzer (PE 2400-II; Perkin Elmer, Waltham, MA, USA).

2.3. Quality Assurance and Quality Control

Soil standard reference materials (GBW07401 and GSS-1), obtained from the Center of National Standard Reference Material of China, was used for quality assurance and quality control (QA/QC). The accepted recoveries ranged from 81.0% to 109%. Analysis methods were evaluated by using blank (n = 7) and duplicate samples (n = 13) for each set of samples. The relative deviation of the duplicate samples was <7% for all batch treatments.

2.4. Data Analysis

The data analysis was performed, using SPSS software (version 18.0; SPSS Inc., Chicago, IL, USA). Descriptive statistics generated for analysis of the potentially toxic elements in the soil samples included the maximum, minimum, median, and mean values, and the standard deviation (SD). Relationships between concentrations of potentially toxic elements were investigated, using Spearman correlation coefficients. The spatial distribution of the concentrations and eco-risks of potentially toxic elements in Huaxi District was visualized, using ArcGIS software (version 10.3; ESRI, Redlands, CA, USA).

2.5. The Geoaccumulation Index (Igeo)

The Igeo is a geochemical criterion introduced by Müller [21]. It can be used to evaluate soil contamination by comparing the differences between current and preindustrial concentrations of potential contaminants. Unlike other methods of pollution assessment, the Igeo takes the natural diagenesis process into account, making the assessments more practical. The Igeo was calculated by using the following equation:

$$\text{Igeo} = \log_2 \left(C_n / 1.5 \times B_n \right) \tag{1}$$

where C_n is the measured concentration of the elements in soil (mg kg^{-1}), and B_n is the geochemical background value (mg kg^{-1}). A coefficient of 1.5 was applied due to potential variations in the baseline data [22]. The relationships between the Igeo and the pollution levels are given in Supplementary Materials Table S1.

2.6. Potential Ecological Risk Index (RI)

The RI, proposed by Hakanson [23], is a relatively rapid, simple, and standard method for assessing the ecological risk posed by potentially toxic elements in soils or sediments [24]. The RI can reflect the potential ecological risk according to the overall pollution level. The formula is as follows:

$$\text{RI} = \sum E_r^i = \sum T_r^i (C_s^i / C_B^i) \tag{2}$$

The RI is the sum of ecological risk indices of various potentially toxic elements; E_r^i is the ecological risk factor for individual elements; C_s^i (mg/kg) is the concentration of a target element in a sample soil; C_B^i (mg/kg) is the background value of the element in Guizhou Province (Table 1); and T_r^i is the toxicity response factor of each metal. Liu et al. [25] calculated the toxicity response coefficient of Tl in a potential ecological risk assessment, with the value set to 10. The toxic response coefficients of As, Cd, Cr, Cu, Pb, and Zn were 10, 30, 2, 5, 5, and 1, respectively. The classification standard for the RI of the potentially toxic elements is also shown in Supplementary Materials Table S1.

3. Results and Discussion

3.1. pH

Soil pH is usually classified into the following five categories: strongly acidic (pH < 4.5), acidic (4.5 ≤ pH < 5.5), slightly acidic (5.5 ≤ pH < 6.5), neutral (6.5 ≤ pH < 7.5), alkaline (7.0 ≤ pH < 8.5), and strongly alkaline (pH ≥ 8.5) [26]. As shown in Table 1, the range of soil pH values in the study area was 0.94–7.77 (mean, 5.44 ± 1.46, i.e., acidic). In contrast to the soil background pH value (6.2) in Guizhou Province, 60% of the samples were classified as strongly acidic. The sample site with the most acidic soil conditions was AMD4 (0.94), which was located in an area affected by the largest flows of AMD effluents. Most of the other sites with strongly acidic soil conditions (pH range of 3.27–5.89) were located on or surrounding coal gangues, e.g., AMD1, SCG3, and SCG5. In contrast, the soil pH in sites relatively remote from the coal gangues (and thus from the impact of AMD effluents) was neutral to alkaline. Similar trends were also observed in other studies conducted in Youyu Catchment. For example, the pH was found to be 4.48 ± 0.93 in farmland soil from Youyu Catchment, and irrigation by AMD was considered to be the main reason for soil acidification [27]. The farmland and paddy soil pH values were reported to be 4.52–6.40 and 4.71–5.55, respectively, in a study conducted in the same area ago, and the soil pH within 100 m of coal gangues was lower than that at sites 300 m away [28]. Studies in other coal-mining areas in Guizhou Province have also revealed a strongly acidic soil pH, e.g., Zhijin mine in Bijie City (4.78–5.38 for paddy soil and 4.995–5.77 for farmland soil), with AMD resulting from high-S ores and gangues identified as the dominant source. The study area was located in the karst landform region of Guizhou Province, which is dominated by carbonate rocks, and the background soil pH is neutral to slightly alkaline. Since the early 1980s, coal resources have been exploited on a large scale in Youyu Catchment, with numerous small coal mines and several larger private mines forming large-scale goafs. Communication between coal bearing strata and the overlying karst strata has led to karst water flowing into pits and interacting with coal seams, which has generated acidic wastewater containing rusty, yellow-colored granular suspended matter. Because most coal mines are located along the Youyu River, the acid waste water from mining has typically been discharged into the river along the wellhead.

3.2. Concentrations of Potentially Toxic Elements in Soils

Descriptive statistics for As, Cd, Cu, Pb, Ni, Zn, Mn, Cr, Hg, Fe, and S in the 27 soil samples from Youyu Catchment are presented in Table 1 and Figure 2. The background pH and concentrations of potentially toxic elements in Guizhou Province and the Chinese Environmental Quality Standards for Soils [29] are listed together for reference. Potentially toxic elements were found to be widespread in the study area, with concentration ranges of 4.85–12.94, 0.10–2.85, 4.04–167.92, 5.15–43.11, 3.39–77.73, 10.60–264.58, 6.54–3524.02, 56.37–118.25, 0.006–0.015, 3764.24–185,519.41 mg/kg for As, Cd, Cu, Pb, Ni, Zn, Mn, Cr, Hg, and Fe, and 0.003–2.91% for S; the respective mean values were 8.74, 1.12, 62.68, 18.40, 25.95, 72.52, 593.60, 86.06, 0.010, and 35,575.95 mg/kg, and 0.44%. In general, except for Cd, the concentrations of the potentially toxic elements (As, Cu, Pb, Ni, Zn, Cr, and Hg) were all below the Chinese environmental quality standards for soils. For Cd, the mean value was 1.12 mg/kg, which was 2.73 times higher than the Chinese standard (0.30 mg/kg). The As, Cd, Cu, Ni, Zn, Mn, Cr, and Fe concentrations were higher than the soil background values in Guizhou Province [30], with the Cd, Cu, and Fe concentrations being particularly high at 1.12, 62.68, and 35,575.95 mg/kg, which were 1.69, 1.95, and 12.18 times higher than their respective background values. There are no published records of soil background values for S in Guizhou Province, but when compared to a mountain soil in the adjacent area (0.11%) [31], the mean S content in the study area was 3.4 times higher (0.44%).

There have been several other studies reporting the levels of potentially toxic elements in the soils of the studied area. According to a study conducted in Youyu Catchment in 2014, the Cr, Cu, Zn, As, Cd, and Pb concentrations in the soil surrounding the coal mine were 79.09 ± 16.48, 87.07 ± 13.28, 127.5 ± 36.23, 30.53 ± 10.82, 0.36 ± 0.15, and 29.05 ± 9.04 mg/kg, respectively [27]. In a study in Huaxi District conducted in 2010, the Hg, As, Cr, Pb, Cd, Fe, and Mn concentrations were 0.15, 29.60, 62.42, 26.56, 0.61, 31,300, and 437.52 mg/kg, respectively [32]. Gao et al. [15] determined the total Fe concentration in an abandoned coal mine in Youyu Catchment and reported an AMD volume of >7000 m^3, and a mean Fe concentration of 20,929.65 mg/kg, which was lower than the value reported in this study. Since 2010, the As, Hg, and Pb concentrations in soils in Youyu Catchment have displayed a downward trend, while the Cd, Fe, Mn, and Cr concentrations have displayed upward trends.

There have been several studies of other high-S coal mines in Guizhou Province [33] determined the As, Cd, Cr, Cu, Pb, and Zn concentrations in the soils of Xingren County, which is located in southwest Guizhou where numerous high-S coal mines are located. The As, Cd, Cr, Cu, Pb, and Zn concentrations were 28.82, 133.75, 1.14, 52.11, 79.63, 44.39, and 118.23 mg/kg, respectively. According to Li [34], at the Zhijin coal mine in western Guizhou Province, the Hg, As, Cu, Zn, Cd, Pb, Cr, Fe, and Mn concentrations in the soil surrounding coal gangues were 0.128, 6.94, 126.49, 85.92, 0.56, 17.62, 157.69, 61,689, and 435.1 mg/kg, respectively. In comparison to these coal-mining areas in Guizhou Province, the concentrations of potentially toxic elements in this study were relatively low. This disparity reflects the variation in potentially toxic element concentrations among different regions in coal-mining areas, as well as the enrichment of potentially toxic elements in soils by anthropogenic activities.

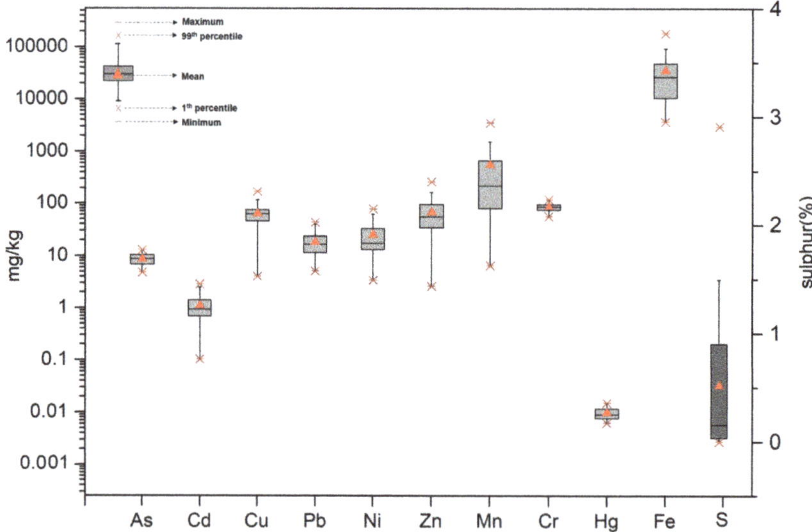

Figure 2. The statistical characteristics (maximum, 99th percentile, median, 1st percentile, and minimum values) of potentially toxic elements in the soil from Youyu Catchment. The concentrations of As, Cd, Cu, Pb, Ni, Zn, Mn, Cr, Hg, Fe refer to the left axis; the percentage of sulfur refers to right axis.

Table 1. The concentrations, coefficient of variation, and geoaccumulation index (Igeo) values of potentially toxic elements in soils of Youyu Catchment.

-	pH	As (mg/kg)	Cd (mg/kg)	Cu (mg/kg)	Pb (mg/kg)	Ni (mg/kg)	Zn (mg/kg)	Mn (mg/kg)	Cr (mg/kg)	Hg (mg/kg)	Fe (mg/kg)	S (%)
Mean	5.44	8.74	1.12	62.68	18.40	25.95	72.52	593.60	86.06	0.010	35,575.95	0.44
SD	2.06	0.70	27.52	9.05	18.88	49.12	775.69	14.30	0.00	33,710.01	0.59	SD
CV	23.59	62.17	43.90	49.19	72.76	72.65	130.68	16.62	24.76	94.76	132.53	CV
Igeo	-	−0.13	0.22	−1.26	−0.79	−1.05	−1.51	−0.11	−2.65	2.47	-	Igeo
CEQS	-	25.00	0.30	200.00	350.00	50.00	300.00	-	250.00	1.00	-	-
Background values in Guizhou province	6.2	8.6	0.21	32.0	26.2	23.4	67.0	482	61.0	0.040	2920	-

CEQS: Chinese Environmental Quality Standards for Soils (CEPA 1995); Background Values in Guizhou Province: CNEMC, 1990.

3.3. Spatial Distribution of Potentially Toxic Elements

In general, the sample types were divided into four classes, to represent the abandoned coal area: (1) soil samples surrounding (within 100 m) the abandoned coal mines (SCG), (2) samples from where AMD emerges from underground (AMD), (3) soil samples from where tributaries merge into the Youyu River (TM), and (4) soil samples from control areas (CK). As shown in Figures 3 and 4, different soil sample sites displayed varying degrees of pollution. Generally, the soil sampled at TM sites had the highest Cd, Cu, Ni, Zn, and Mn concentrations, while the As, Pb, Cr, and Fe concentrations were highest in the soils at AMD sites; the Hg concentrations were similar across the five sampling sites. The soil at SGC sites had the highest S content, and conversely had the lowest levels of all potentially toxic elements, except for Fe. Therefore, Cd and Fe were considered to be the elements of most concern; however, these two elements had different pollution characteristics. The Cd concentrations followed the order of TM > AMD > SCG > CK, while the Fe concentrations followed the order of AMD > SCG > CK > TM.

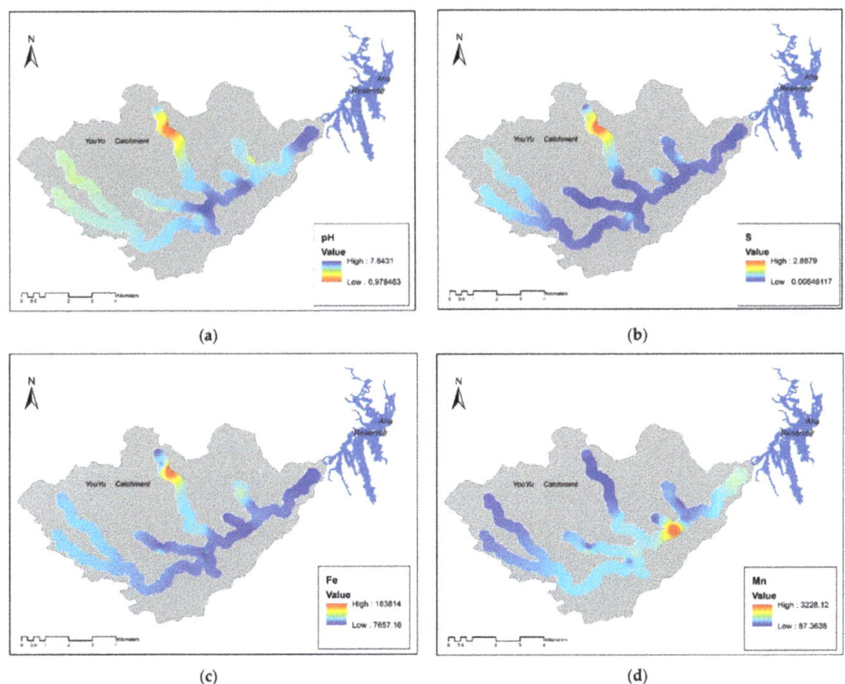

(a)

(b)

(c)

(d)

Figure 3. *Cont.*

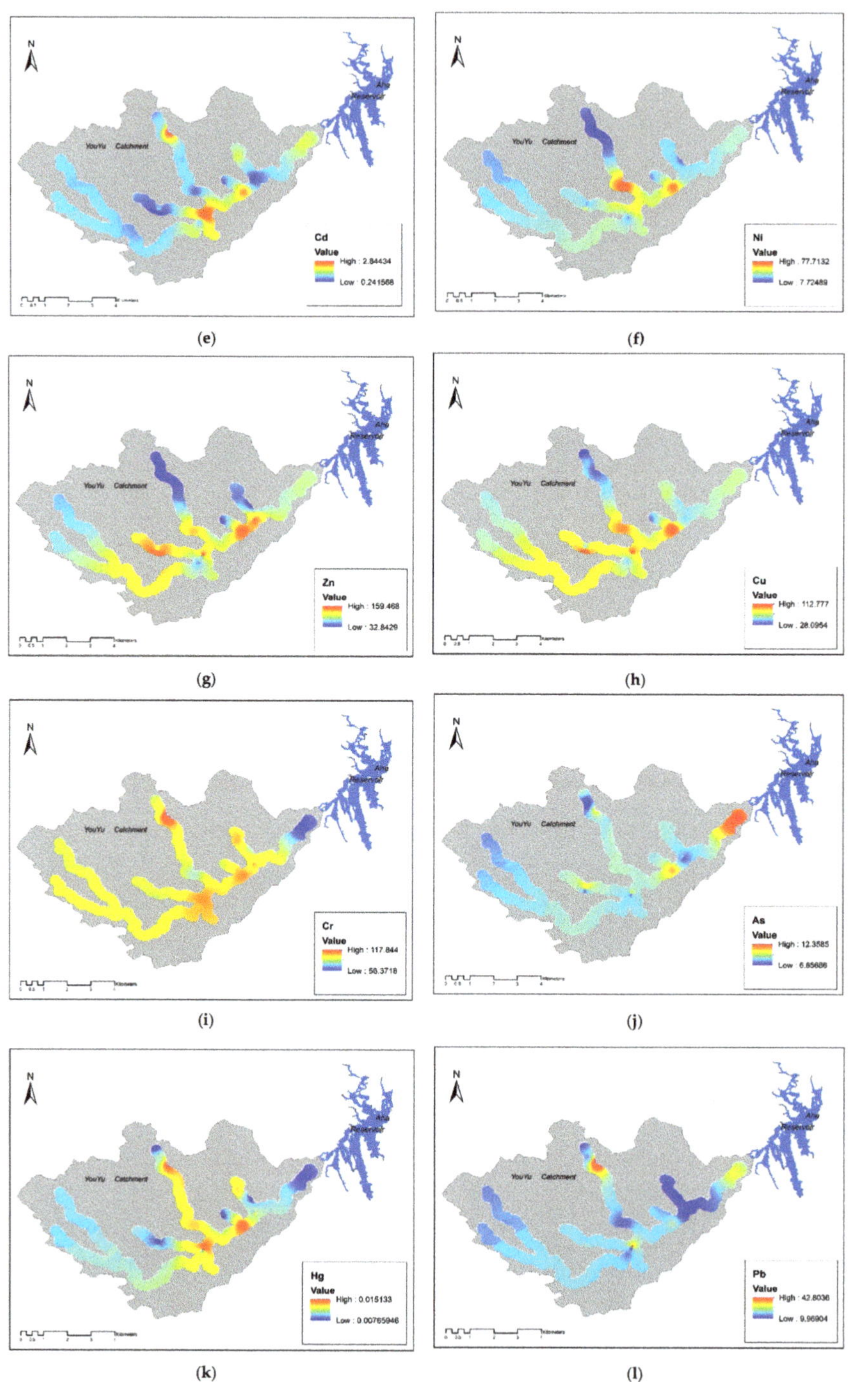

Figure 3. Spatial distribution of the concentrations of potential toxic elements in Youyu catchment. (**a**) pH value; (**b**) S value; (**c**) Fe value; (**d**) Mn value; (**e**) Cd value; (**f**) Ni value; (**g**) Zn value; (**h**) Cu value; (**i**) Cr value; (**j**) As value; (**k**) Hg value; (**l**) Pb value.

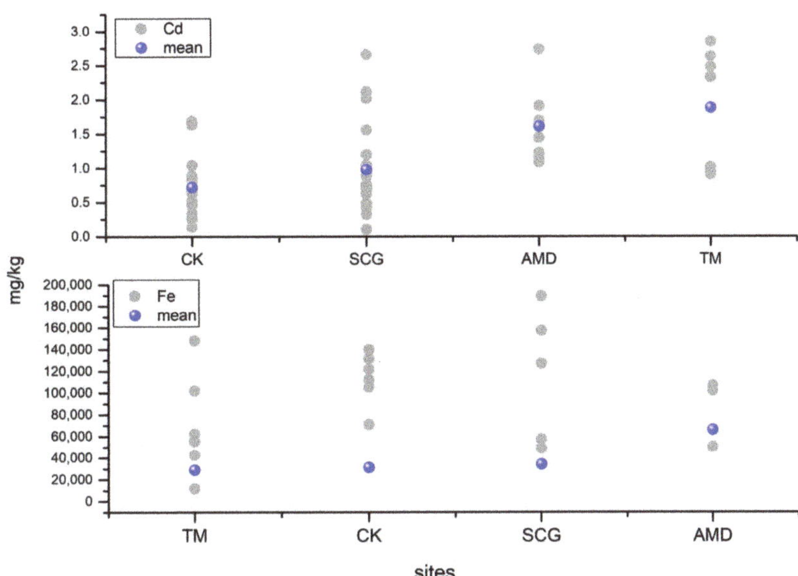

Figure 4. The concentrations of Cd and Fe in the soil from different sites in Youyu Catchment.

3.4. Ecological Risk of Potentially Toxic Elements

The method of Hakanson [23] was adopted to assess the ecological risk posed by heavy metal pollution in the topsoil of the study area. Hakanson developed a quantitative approach that has been used in many other studies, to assess the ecological risk posed by heavy metals in soil [35,36]. Supplementary Materials Table S2 shows the potential ecological risk values, E(i), of eight potentially toxic elements in the topsoil in Youyu Catchment. The mean E(i) values of As, Cr, Cu, Ni, Pb, Zn, and Hg were all less than 40, indicating that these metals posed a low ecological risk. Cadmium had a higher mean E(i) value (51.16) than the other seven heavy metals, thus presenting a moderate potential ecological risk. Overall, 61.8% of all samples had an E(Cd) value in the range of 40 to 80, while 14.5% were in the range of 80 to 160, representing moderate and considerable potential ecological risk, respectively. The RIs were calculated to assess the risk of multiple potential toxic elements in soils. The RI values for 91.0% of all samples were <160, indicating a low potential ecological risk, while the other 9.0% of samples posed a moderate ecological risk. As shown in Figures 5 and 6, the spatial distribution of E(Cd) is consistent with that of RI, indicating that the potentially ecological risks in the studied area was dominated by Cd.

The oxidation of sulfides releases H^+, Fe^{2+}, and SO_4^{2+}, which is the first step in the environmental accumulation of potentially toxic elements in the environment. However, the subsequent migration of elements is also controlled by a series of complex precipitation, hydrolysis, adsorption and desorption, co-precipitation, and ion exchange reactions. In general, the highest concentrations of Fe, S and the lowest pH values were all observed in the AMD sites. At the AMD sites, soil Fe and S concentrations had significant negative correlations with soil pH (P > 0.01, Fe:R = −0.83, S:R = −0.93), suggesting that the Fe and S concentrations were higher at lower pH values. This was the karst landscape facilitated AMD outflows, resulting in massive amounts of Fe^{3+}, H^+, and SO_4^{2-} being leached into the soil, thus causing significantly high Fe and S concentrations and extremely low pH values in the soil at AMD sites. It was also apparent that the Fe concentration displayed a decreasing trend with distance from AMD and SCG sites, while pH displayed the opposite trend.

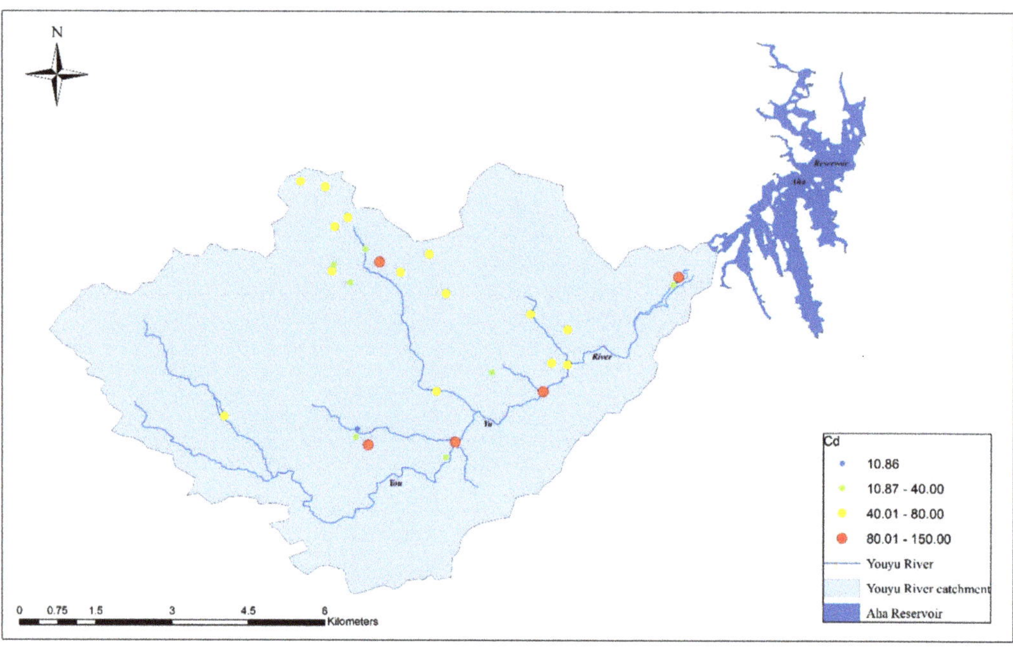

Figure 5. Spatial distribution of the potential ecological risks (E(i)) for Cd in the soil of the studied area.

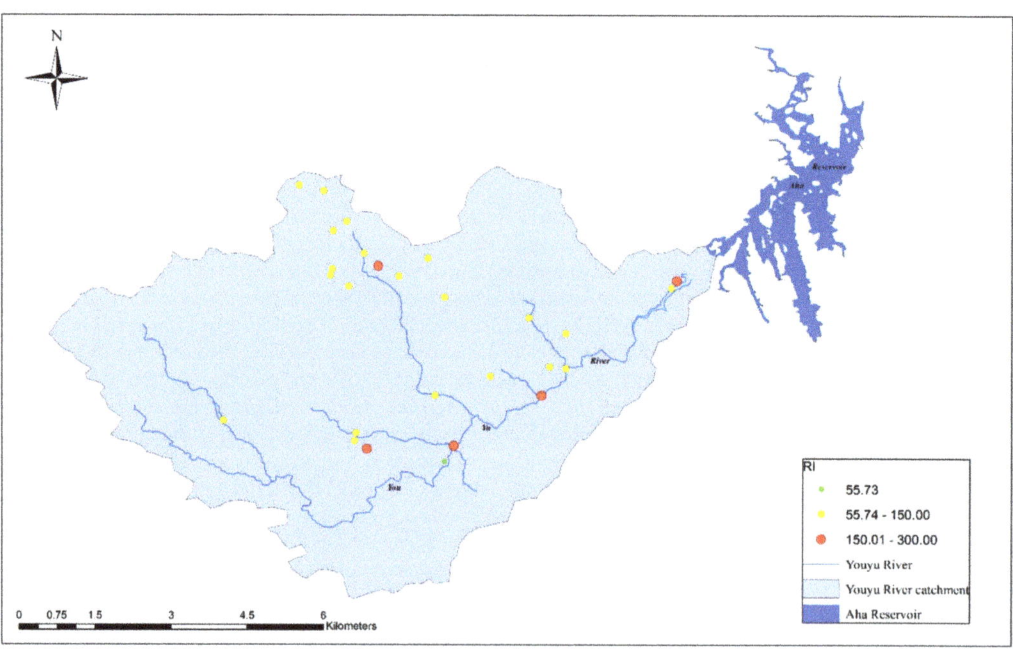

Figure 6. Spatial distribution of the potential ecological risk indices (RIs) in the soil of the studied area.

However, compared to Fe, the Cd concentrations displayed the opposite trend, being highest in the TM sites and thus implying other pollution sources. Based on the field investigation, coal mining at the AMD sites and abandoned coal mines was responsible for the direct introduction of Cd into the surface environment. However, there were also numerous exposed coal seams upstream of TM sites. The local karst area was dominated by a lithologic soil that has been substantially affected by the carbonate rocks. The soil was shallow and neutral to slightly alkaline due to the lack of a transition layer, and the affinity and adhesion between the underlying rock and soil was poor. For a long period of time, the local area has had a tropical and subtropical climate, and has therefore experienced humid and hot conditions with high precipitation, resulting in strong chemical leaching activities. Leaching or weathering of these residues may transfer Cd and other elements downhill to the TM soils, which can then migrate onto farmland and into surface waters. It has been reported that soil Cd is more likely to migrate under acidic conditions (pH < 6) [14,37], but is more likely to deposit as $Cd(OH)_2$ or other precipitates in an alkaline environment (pH > 6). In contrast, Fe is more inclined to precipitate under acidic conditions [38], which was the case in this study. In the TM sites, the Cd concentration showed a significantly negative correlation ($p < 0.01$, R = 0.97) with pH (mean value, 6.76), while in the SCG and AMD sites, which had the highest soil Fe concentrations, there was a significant positive correlation ($p < 0.01$, R = 0.96) between the Fe concentration and pH (mean value, 4.35). The above results indicated that with distance from AMD and SCG sites, the pH tended to be higher and soil Cd tended to precipitate, while Fe tended to precipitate under a low soil pH. However, it should be recognized that the migration of various elements in the mining area varied with the chemical properties of solid minerals and solutions. There are likely to be multiple migration mechanisms, which should be studied further in combination with an analysis of the various chemical forms of soil elements. The TM sites showed higher Cd concentrations compared to other sites in the upper reach; as mentioned above, Cd is more likely migrate under acid condition and absorbed into sediment under opposite condition. Yuan et al. [39] reported that, under flood conditions, sediments are easily mixed with water, and potentially toxic elements-rich sediments are enriched in the soil around the river and in the lower-lying terrain. It can be seen that river sediments are not only the enrichment area of potentially toxic elements but also the secondary pollution sources with potential hazards to water quality and the surrounding soil. In addition, coal wastes are often used by local residents to increase the height of riverbanks, and can also be enriched in river sediment. Under flood conditions, the sediments will mix with river water, and the soil around the river and surrounding low-lying terrain can become enriched with sediment [39]. Runoff from tributaries of the Youyu River has decreased annually, especially in the dry season, and the bare riverbed has formed a soil over a long period. The levels of Cd and other potentially toxic elements in soil may also be influenced by agricultural activities, such as irrigation with contaminated water and fertilization. In addition, due to the prohibition of mining activities, tea plantations have been established by local famers in recent years. Phosphate fertilizer is used by local farmers, to improve soil fertility, and has had an important influence on soil quality.

After 2010, the mines in this area were successively closed, and the pits were all sealed. However, AMD formation became more extensive after mine closure, because the pumps used to keep the water table near open-pit and/or underground mines artificially low, to facilitate mining activities, were turned off. This has resulted in a rise in the groundwater level, which then comes into contact with the exposed sulfide minerals on wall rocks, causing AMD to be continuously generated. In addition, even if all the pits were sealed, groundwater would flow in the form of karst gaps, pores, karst caves, pipes, and underground rivers, due to the karst landforms. Rock cutting and fragmentation create favorable conditions for groundwater storage, migration, and dissolution. Surface water and groundwater are exchanged frequently, and karstification is strong. Water usually emerges from the ground in the form of springs, seriously contaminating the surrounding aquatic and soil environment. At the same time, coal gangues from the coal mines have been randomly deposited in the area, and harmful components are leached out by rainfall leaching, which then pollutes surface water, groundwater, and soil. According to the spatial distribution of potentially ecological risk posed by Cd (Figure 5 and Table 2), the sites where tributaries merged with the main steam of the river had the highest pollution levels. Soil pollutants can enter rivers through leaching, where they may harm aquatic organisms in the wet season, and further endanger human health via the food chain.

Table 2. The potential ecological risks of potentially toxic elements in soils from different sites of Youyu Catchment.

Sampling Sites	As	Cd	Cu	Pb	Ni	Zn	Cr	Hg	RI
TM	10.39	85.60	11.66	3.87	10.71	1.66	3.08	14.50	141.47
AMD	11.27	73.17	8.93	3.61	5.11	0.91	3.26	12.30	118.56
SCG	9.63	44.78	8.95	3.11	3.69	0.69	2.75	9.45	83.06
CK	10.78	32.84	11.29	3.62	5.96	1.35	2.66	9.00	77.50
Mean	10.17	51.16	9.79	3.51	5.55	1.01	2.82	9.87	93.87
Median	10.46	43.84	9.82	3.14	3.91	0.82	2.82	9.12	82.15
Maximum	15.04	129.70	26.24	8.23	16.61	3.95	3.88	15.24	199.51
Minimum	5.64	4.69	0.63	0.98	0.72	0.04	1.85	6.32	38.96
SD	2.40	31.81	4.30	1.73	4.04	0.73	0.47	2.44	36.25

TM: Youyu River; AMD: acid mine drainage; SCG: surrounding (within 100 m) coal gangues; CK: control areas.

Cadmium is considered to be one of the most toxic trace elements in the environment, and can cause serious health problems in animals and humans [40]. Cadmium accumulation in grains and vegetables has received much attention [41]. Cadmium pollution of the edible parts of grains and vegetables is particularly serious in China, due to rapid industrialization and the wide application of agrochemicals in agricultural activities. It has been reported that Cd is the main element responsible for soil heavy-metal contamination in China, and poses the highest potential ecological risk in both rural and urban areas. The highest soil Cd concentrations have been reported in south China, e.g., the Yunnan–Guizhou Plateau. However, there are many other hotspots in China, due to mining and smelting activities, with the soil in mining and smelting areas having much higher Cd concentrations than in areas where other land uses predominate, such as irrigation, urban, and remote areas [42–44].

4. Conclusions

Obvious acidification and enrichment of potentially toxic elements was observed in the soils of Youyu Catchment, especially Cd, S, and Fe. Among all elements in the study area, the highest concentrations were observed for Cd and Fe, with Cd and Fe being present at unpolluted to moderately polluted and heavily polluted levels, respectively. Significantly high Fe and S concentrations and extremely low pH values were found in the soils of AMD sites, and the sites where tributaries merged with the TM had the highest Cd concentrations. Iron contamination originated mainly from the non-point sources of AMD and coal gangues, while the adsorption and desorption of AMD and agricultural activity resulted in high Cd levels in the estuary. Cadmium levels posed a moderate ecological risk, while the other elements, such as As, Cr, Pb, Zn, Cu, Ni, and Hg, all posed a low eco-risk. Spatially, the TM sites had the highest Cd eco-risk; moreover, they tended to pose a significant risk of harm to aquatic ecosystems, particularly in the wet season, and endangered human health via the food chain.

Supplementary Materials: The following are available online at https://www.mdpi.com/2075-163X/11/3/330/s1, Table S1: Contamination categories on the basis of Igeo values and ecological risk indices; Table S2: Potentially ecological risks values of the elements of As, Cd, Cu, Pb, Ni, Zn, Cr, and Hg.

Author Contributions: Conceptualization, L.P. and X.G.; methodology, L.P.; software, L.P. and B.L.; validation, L.P. and Y.C.; formal analysis, L.P.; investigation, X.G., B.L., Y.C., Z.H., J.P., Y.P., Y.Z.; resources, L.P.; data curation, L.P.; writing—original draft preparation, L.P. and B.L.; writing—review and editing, L.P.; visualization, L.P.; supervision, X.G. and L.P.; project administration, Chinese Research Academy of Environmental Sciences; funding acquisition, X.G. All authors have read and agreed to the published version of the manuscript.

Funding: This study was supported by the National Key R&D Plan (2018YFC1801302) and Chinese Research Academy of Environmental Sciences, State Key Laboratory of Environmental Criteria and Risk Assessment (201520 and 201542).

Institutional Review Board Statement: Not applicable.

Informed Consent Statement: Not applicable.

Data Availability Statement: Not applicable.

Acknowledgments: We would like to thank all those who helped with the sampling.

Conflicts of Interest: The authors declare no conflict of interest.

References

1. Simate, G.S.; Ndlovu, S. Acid mine drainage: Challenges and opportunities. *J. Environ. Chem. Eng.* **2014**, *2*, 1785–1803. [CrossRef]
2. Park, I.; Tabelin, C.B.; Jeon, S.; Li, C.L.; Seno, K.; Ito, M.; Hiroyoshi, N. A review of recent strategies for acid mine drainage prevention and mine tailings recycling. *Chemosphere* **2019**, *219*, 588–606. [CrossRef]
3. Gray, N. Environmental impact and remediation of acid mine drainage: A management problem. *Environ. Geol.* **1997**, *30*, 62–71. [CrossRef]
4. Manzano, M.; Ayora, C.; Domenech, C.; Navarrete, P.; Antonio, G.; Maria-Jesús, T. The impact of the Aznalcóllar mine tailing spill on groundwater. *Sci. Total Environ.* **1999**, *242*, 189–209. [CrossRef]
5. Wates, J.A.; Rykaart, E.M. *The Performance of Natural Soil Covers in Rehabilitating Opencast Mines and Waste Dumps in South Africa*; Research Report 575/1/99; Water Research Commission: Pretoria, South Africa, 1999.
6. Hua, C.H.; Zhou, G.Z.; Yin, X.; Wang, C.Z.; Chi, B.R.; Cao, Y.Y.; Wang, Y.; Zheng, Y.; Cheng, Z.R.; Li, R.Y. Assessment of heavy metal in coal gangue: Distribution, leaching characteristic and potential ecological risk. *Environ. Sci. Pollut. Res.* **2018**, *25*, 32321–32331. [CrossRef] [PubMed]
7. Druschel, G.K.; Baker, B.J.; Gihring, T.M.; Banfield, J.F. Acid mine drainage biogeochemistry at iron mountain, California. *Geochem. Trans.* **2004**, *5*, 13–32. [CrossRef]
8. Azhari, A.E.; Rhoujjati, A.; Hachimi, M.L.E.; Ambrosi, J. Pollution and ecological risk assessment of heavy metals in the soil-plant system and the sediment-water column around a former Pb/Zn-mining area in NE Morocco. *Ecotoxicol. Environ. Saf.* **2017**, *144*, 464–474. [CrossRef]
9. Hallberg, K.B. New perspectives in acid mine drainage microbiology. *Hydrometallurgy* **2010**, *104*, 448–453. [CrossRef]
10. Johnson, D.B.; Hallberg, K.B. Acid mine drainage remediation options: A review. *Sci. Total Environ.* **2005**, *338*, 3–14. [CrossRef]
11. Department of Natural Resource of Guizhou Province. 2019. Available online: http://zrzy.guizhou.gov.cn/zfxxgk/zfxxgkml/tjsj_81192/gtzyddjcjb/201907/t20190702_25808845.html (accessed on 2 June 2019).
12. Evangelou, V.P. *Pyrite Oxidation and its Control*; CRC Press: Boca Raton, FL, USA, 1995.
13. Tabelin, C.B.; Igarashi, T.; Villacorte-Tabelin, M.; Park, I.; Opiso, E.M.; Ito, M.; Hiroyoshi, N. Arsenic, selenium, boron, lead, cadmium, copper, and zinc in naturally contaminated rocks: A review of their sources, modes of enrichment, mechanisms of release, and mitigation strategies. *Sci. Total Environ.* **2018**, *645*, 1522–1553. [CrossRef] [PubMed]
14. Equeenuddin, S.M.; Tripathy, S.; Sahoo, P.K.; Panigrahi, M.K. Metal behavior in sediment associated with acid mine drainage stream: Role of pH. *J. Geochem. Explor.* **2013**, *124*, 230–237. [CrossRef]
15. Gao, P.; Sun, X.X.; Xiao, E.Z.; Xu, Z.X.; Li, B.Q.; Sun, W.M. Characterization of iron-metabolizing communities in soils contaminated by acid mine drainage from an abandonded coal mine in Southwest China. *Environ. Sci. Pollut. Res.* **2019**, *26*, 9585–9598. [CrossRef] [PubMed]
16. Huaxi Annals Compilation Committee. *Huaxi Annals*; Guizhou People's Publishing House: Bejing, China, 2007. (In Chinese)
17. Pan, L.B.; Ma, J.; Wang, X.L.; Hou, H. Heavy metals in soils from a typical county in Shanxi Province, China: Levels, sources and spatial distribution. *Chemosphere* **2016**, *148*, 248–254. [CrossRef]
18. USEPA. *Method 3050B: Acid Digestion of Sediments, Sludges and Soils*; United States Environmental Protection Agency: Washington, WA, USA, 1996.
19. USEPA. *Method 6010C: Inductively Coupled Plasma Atomic Emission Spectrometry*; United States Environmental Protection Agency: Washington, WA, USA, 1996.
20. CEPA (Chinese Environmental Protection Administration). Soil-Determination of pH-Potentiometry (HJ962-2018). In *National Environmental Protection Standards of the People's Republic of China*; China Environmental Science Press: Beijing, China, 2018. (In Chinese)
21. Müller, G. Index of geoaccumulation in sediments of the Rhine River. *Geojournal* **1969**, *2*, 108–118.
22. Solgi, E.; Esmailisari, A.; Riyahibakhtiari, A.; Hadipour, M. Soil contamination of metals in the three industrial estates, Arak, Iran. *Bull. Environ. Contam. Toxicol.* **2012**, *88*, 634–638. [CrossRef]

23. Hakanson, L. An ecological risk index for aquatic pollution control, a sedimentological approach. *Water Res.* **1980**, *14*, 975–1001. [CrossRef]

24. Wu, Q.; Qi, J.; Xia, X. Long-term variations in sediment heavy metals of a reservoir with changing trophic states: Implications for the impact of climate change. *Sci. Total Environ.* **2017**, *609*, 242–250. [CrossRef]

25. Liu, J.; Luo, X.W.; Wang, J.; Xiao, T.F.; Chen, D.Y.; Sheng, G.D.; Yin, M.L.; Lppold, H.; Wang, C.L.; Chen, Y.H. Thallium contamination in arable soils and vegetables around a steel plant—A newly-found significant source of tl pollution in South China. *Environ. Pollut.* **2017**, *224*, 445–453. [CrossRef] [PubMed]

26. Institute of Soil Science; Chinese Academy of Sciences. *Chinese Soil*; Scientific Publisher: Beijing, China, 1978.

27. Zhu, H.L.; Liu, H.Y.; Long, J.H.; Yan, Z.Y. Pollution characteristics of heavy metals in soils in typical polluted areas of Guizhou Province. *Earth Environ.* **2014**, *4*, 505–512. (In Chinese)

28. Yang, L.J.; Long, J.; Li, J.; Wu, D.; Liu, F.; Liao, H.K.; Han, S. Effect of Acidic Waste Water from Coal Mine on Cu Form Change in Coal Gangue and Surrounding Farmland Soils. *Guizhou Agri. Sci.* **2011**, *9*, 88–91.

29. CEPA (Chinese Environmental Protection Administration). *Environmental Quality Standard for Soils (GB 15618-1995)*; China Environmental Science Press: Beijing, China, 1995. (In Chinese)

30. CNEMC (China National Environmental Monitoring Center). *Soil Elements Background Values in China*; China Environmental Science Press: Beijing, China, 1990. (In Chinese)

31. Yang, C.; Liu, C.Q.; Song, Z.L.; Liu, Z.M.; Zheng, H.Y. Distribution characteristics of carbon, nitrogen and sulphur of plants and soils in Guizhou karst mountain area, southwestern China. *J. Beijing For. Univ.* **2008**, *30*, 45–51. (In Chinese)

32. Li, J. *Investigation and Study on Heavy Metal Pollution in Abandoned Coal Mine and Surrounding Soil in Huaxi District of Guiyang*; Guizhou University: Guizhou, China, 2010. (In Chinese)

33. Jiang, F.; Ren, B.Z.; Andrew, H.; Deng, R.J.; Wang, Z.H. Distribution, source indentification, and ecological-health risks of potentially toxic elements (PTEs) in soil of Thallium mine area (southwest Guizhou, China). *Environ. Sci. Pollut. Res* **2019**, *26*, 16556–16567. [CrossRef]

34. Li, J. *Study on Pollution Characters and Migration Rules of Heavy Metals on High-Sulfur Coal Gangue Abandoned Land*; China University of Mining and Technology: Beijing, China, 2012. (In Chinese)

35. Sun, Y.; Zhou, Q.; Xie, X.; Liu, R. Spatial, sources and risk assessment of heavy metal contamination of urban soils in typical regions of Shenyang, China. *J Hazard Mater* **2010**, *174*, 455–462. [CrossRef] [PubMed]

36. Wang, H.; Lu, S. Spatial distribution, source identification and affecting factors of heavy metals contamination in urban-suburban soils of Lishui City, China. *Environ. Earth Sci.* **2011**, *64*, 1921–1929. [CrossRef]

37. Yang, W.J.; Ding, K.B.; Zhang, P.; Qiu, H.; Cloquet, C.; Wen, H.J.; Morel, J.L.; Qiu, R.L.; Tang, Y.T. Cadmum stable isotope variation in a mountain are impacted by acid mine drainage. *Sci. Total Environ.* **2019**, *646*, 696–703. [CrossRef]

38. Cathes, L.M. Attempts to model the industrial-scale leaching of copper-bearing mine waste in environmental geochemistry of sufide oxidation. *Ame Chem. Soc.* **1994**, *550*, 123–131.

39. Yuan, H.; Song, S.; An, S.; Liu, E. Ecological risk assessment of potentially toxic elements (PTEs) in the soil-plant system after reclamation of dredged sediment. *Environ. Sci. Pollut. Res.* **2018**, *25*, 29181–29191. [CrossRef]

40. Wang, M.; Chen, W.P.; Peng, C. Risk assessment of Cd polluted paddy soils in the industrial and township areas in Hunan, Southern China. *Chemosphere* **2016**, *144*, 346–351. [CrossRef]

41. Kosolsaksakul, P.; Farmer, J.G.; Oliver, I.W.; Graham, M.C. Geochemical associations and availability of cadmium (Cd) in a paddy field system, northwestern Thailand. *Environ. Pollut.* **2014**, *187*, 153–161. [CrossRef] [PubMed]

42. Chen, H.Y.; Teng, Y.G.; Lu, S.J.; Wang, Y.Y.; Wang, J.S. Contamination features and health risk of soil heavy metals in China. *Sci. Total Environ.* **2015**, *512*, 143–153. [CrossRef]

43. Zhang, X.Y.; Chen, D.M.; Zhong, T.Y.; Zhang, X.M.; Cheng, M.; Li, X.H. Assessment of cadmium(Cd) concentration in arable soil in China. *Environ. Sci. Pollut. Res.* **2015**, *22*, 4932–4941. [CrossRef] [PubMed]

44. Qi, G.X.; Jia, Y.F.; Liu, W.J.; Wei, Y.H.; Du, B.; Fang, W.; Guo, Y.M.; Guo, F.; Wu, Y.H.; Zou, Q.; et al. Leaching behavior and potential ecological risk of heavy metals in Southwestern China soils applied with sewage sludge compost under acid precipitation based on lysimeter trials. *Chemosphere* **2020**, *249*, 126212. [CrossRef] [PubMed]

Article

Heavy Metals Content in the Soils of the Tatra National Park Near Lake Morskie Oko and Kasprowy Wierch—A Case Study (Tatra Mts, Central Europe)

Joanna Korzeniowska * and Paweł Krąż *

Institute of Geography, Pedagogical University of Krakow, 30-084 Krakow, Poland
* Correspondence: joanna.korzeniowska@up.krakow.pl (J.K.); pawel.kraz@up.krakow.pl (P.K.); Tel.: +48-12-662-62-61 (J.K. & P.K.)

Received: 16 November 2020; Accepted: 11 December 2020; Published: 13 December 2020

Abstract: This paper presents the content of selected heavy metals (Cd, Cr, Cu, Ni, Pb and Zn) in the soils of the Tatra National Park (TNP). In order to determine the anthropogenic impact on the environment, the following coefficients were calculated: enrichment factors (EF), geoaccumulation index (Igeo), contamination factor (Cf), degree of contamination (C_d), and modified degree of contamination (mC_d). It turned out that in the Kasprowy Wierch and Lake Morskie Oko test areas, the content of metals in the soil decreases with the increasing altitude above sea level. In both regions, the highest concentrations of cadmium and lead were found, for which the coefficients indicated significant environmental pollution. These metals, since they persist in the atmosphere for a long time and have a small particle diameter, can be moved over long distances. Long-range emission contributes to environmental contamination on a global scale. Under the influence of such emissions, even protected areas such as the Tatra National Park, considered to be of natural value, are exposed to the effects of human activities (industry in general, automotive industry in particular).

Keywords: heavy metals; soil; enrichment factor; geoaccumulation index; contamination factor

1. Introduction

Heavy metals pose a great threat to the natural environment [1,2]. The toxicity of metals results not only from the degree of environmental contamination, but above all from the biochemical role they play in metabolic processes. Heavy metals, such as: Cd, Cr, Cu, Ni, Pb, and Zn, pose a particularly high risk of chemical imbalance in ecosystems, if they are introduced in significant amounts [3].

Dust in the atmosphere, containing heavy metals, gets into the soil and falls on the above-ground parts of plants. As a result, the concentration of heavy metals in these elements of the environment increases. Natural contents of heavy metals in soils are closely related to the type of soil [3–5].

The sources of metals in the soil are bedrock material and atmospheric pollution caused by anthropogenic activities (emissions from the industrial and automotive sector). Heavy metals can travel long distances from emitters [6,7]. The spread of heavy metals over long distances is related to the long duration of dust pollution in the atmosphere. The length of time of particulate matter remaining in the atmosphere depends on the size of these particles, the terrain configuration, and the weather conditions. Low pressure, strong wind, considerable cloud cover and high precipitation contribute to the spread of pollutants over long distances [8]. The length of time of heavy metals remaining in the environment differs for individual metals. The metals that remain in the atmosphere for a long time and have a very small diameter of particles are lead and cadmium. Lead and cadmium are easily transported over long distances and, therefore, cause environmental contamination on a global scale [8,9].

Research on the content of heavy metals in the Tatra National Park (TNP) area was conducted by numerous teams. The Tatra National Park is a good place to conduct research aimed at assessing the scale of pollution introduced into this area, as for many years it has not been directly affected by human activity. The research area is diversified in terms of natural conditions determined by large differences in altitude. In Poland, the authors who investigated the relationship between height above sea level and the content of heavy metals in soils included Kubica et al. [10,11], Miechówka et al. [12,13]. Kubica et al. [10,11] investigated the concentrations of selected radionuclides and heavy metals (Zn, Pb, Cd, Ni, and Fe) in soil samples in the Kościeliska, Rybi Potok, and Chochołowska Valleys. The concentration of heavy metals in forestless soils was investigated by, e.g., Miechówka et al. [12]. The same team [13] analyzed the concentrations of heavy metals in rocky initial rendzinas (Lithic Leptosols). Wieczorek and Zadrożny [14] conducted an analysis of the content of Cd, Pb, and Zn only in podzolic soils in selected areas of the Tatra National Park. In the Slovak part of the Tatra National Park, related research was carried out by, among others, Barančoková et al. [15]. Similar studies have been conducted also in other protected areas in the Polish Carpathians, including in the Bieszczady National Park [16,17] and other parts of the main ridge of the Flysch Carpathians [18].

The aim of this study is to determine the effect of altitude on the content of heavy metals (Cd, Cr, Cu, Ni, Pb and Zn) in soils in mountain areas.

2. Materials and Methods

2.1. Study Area

The Tatra National Park is one of 23 national parks in Poland, where high-mountain relief is provided, and valuable species of plants and animals (including endemics and relics) are protected. The research area is located in the Polish part of the Central Western Carpathians, in the northern part of the Tatra Range macroregion [19] and it is the highest part of the entire Carpathians. The specificity of this area is the complex geological structure [20–22], land relief heterogeneity (fluvial-denudation, karst, and glacial) [23–25], climatic conditions changing with the increase in altitude above sea level (air temperature, total precipitation, etc.). The specificity of the climate of the Tatra Range is determined by the incidence of different air masses. Arctic maritime air masses (PPm) have the largest share in the formation of weather, i.e., 65% of days a year, while continental polar air masses (PPk) approximately 20% of days a year [26,27]. The above elements determine the specificity of water circulation (spatially diversified possibility of water retention, the volume of runoff, water chemistry, etc.). The soil cover of the Tatra Mountains is strongly related to, among others, their geological substrate, morphogenetic processes, and climatic conditions, and its characteristic feature is openwork, as well as poorly developed soils (i.e., initial soils) [28]. All the physico-geographical zones, characteristic of high mountain areas, have developed in the Tatra Mountains. Two test areas (Figure 1) in the Tatra National Park in Poland, on the northern slope of the Tatra Mountains, were selected for the study. These areas were selected owing to the diversity of the natural environment, including the physico-geographical location, landscape zone, and geological structure. The test areas were given working names—Kasprowy Wierch (KW) and Morskie Oko (MO).

2.1.1. Kasprowy Wierch (KW) Tested Area

The area is located within two physico-geographical mesoregions—the Reglowe Tatras (collection points 1–4) and the Western Tatras (collection points 5–8) [29], and it ranges from the forest zone to the alpine zone. It is characterized by high lithological and tectonic diversity. This affects, among other matters, the lack of overlapping of the topographic watershed with the underground watershed. The area belongs to the Bystra catchment (with the Potok Jaworzynka sub-catchment) and the Sucha Woda Gąsienicowa catchment, which is part of the Dunajec basin. Depending on the altitude, the mean annual air temperature ranges from 0 to 6 °C [30], the annual total precipitation ranges from 800 to 1800 mm, and the length of the snow cover deposition ranges from 100 to 200 days a year [31]. The soil cover is varied and dominated by the following soils: Fluvisols, Rendzic Leptosols, Folic Rendzic Leptosols, Cambic Rendzic Leptosols, Haplic Cambisols (Eutric), Haplic Podzols (Skeletic), Entic Podzols, Leptic Podzols, and Folic Leptosols [32]. The height difference within the sampling points is 750 m (1100–1850 m above sea level).

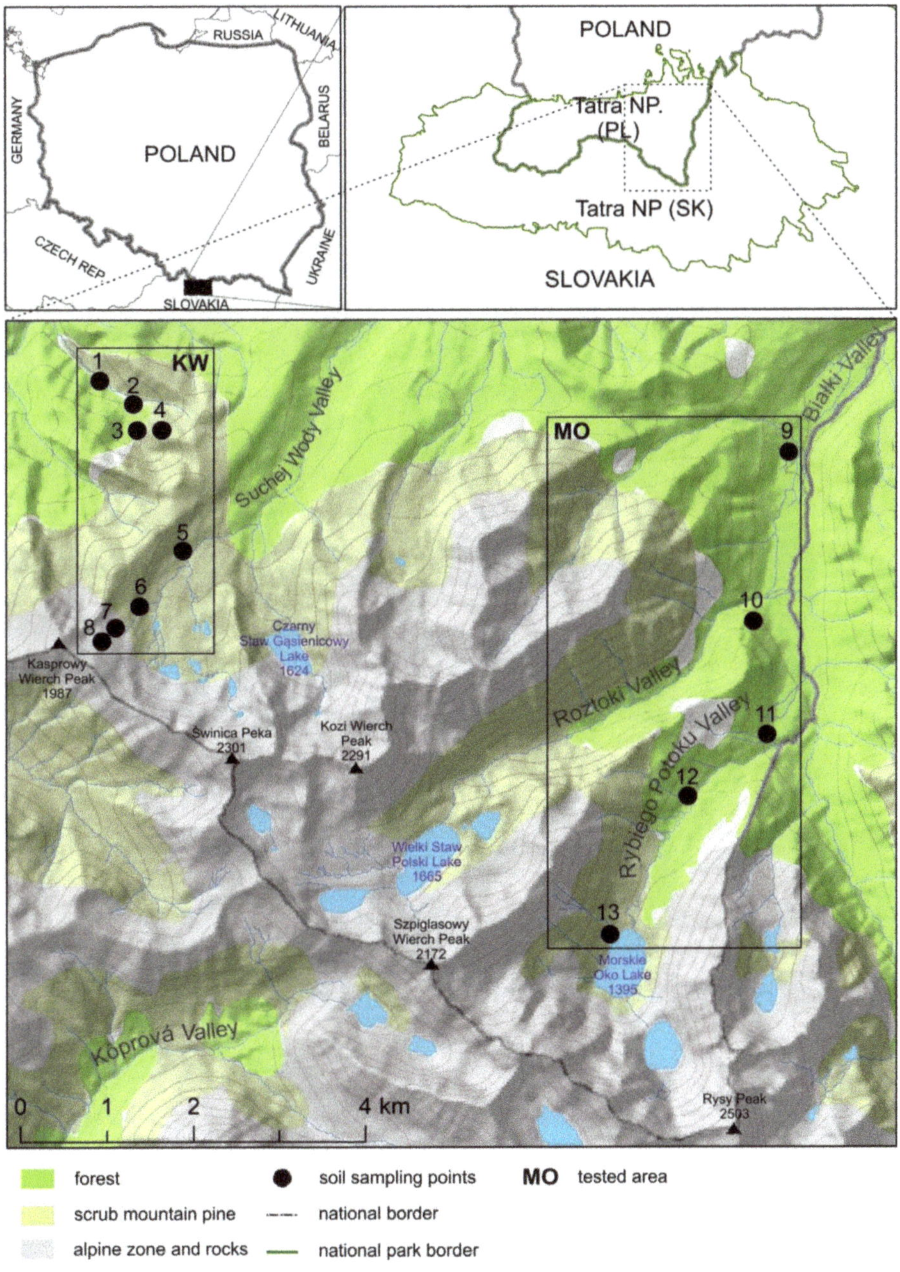

Figure 1. Location of the tested area on the background of the map of Poland and Tatra National Parks.

2.1.2. Morskie Oko (MO) Tested Area

The area is located within the High Tatras in the Białka catchment (the Dunajec river basin) drained by the Rybi Potok, Roztoka, and Białka. With regard to the zonality of the environment, it is entirely located within the forest zone. It is part of one of the largest U-shaped valleys in the Tatras. Depending on the altitude, the mean annual air temperature ranges from 2 to 4 °C [30], the annual total precipitation ranges from 1000 to 1400 mm, and the length of snow cover deposition ranges from 120 to 160 days a year [31]. The dominant soils in this part are, among others: Haplic Podzols (Skeletic), Haplic Cambisols (Dystric, Skeletic), Lithic Leptosols, and Regosols (Hyperskeletic) [32]. The height difference within the sampling points is 400 m (1000–1400 m above sea level).

2.2. Sampling and Analysis

2.2.1. Sampling

Top soil samples (0–10 cm) were taken from the Kasprowy Wierch and Morskie Oko tested areas in the Tatra National Park. The samples were taken every 100 m of altitude, starting from 1100 m above sea level for KW and from 1000 m above sea level for MO. During the field tests, a total of 130 soil samples were collected, 10 from each point (80 for KW, 50 for MO). The field tests were carried out on 4 September 2019 (KW), and 5 September 2019 (MO). The average temperature during the sampling was 10 °C and no precipitation was recorded. The characteristics of the sampling sites and the adopted designations are presented in Table 1.

Table 1. Selected characteristics of soil sampling points in the Kasprowy Wierch (KW) and Morskie Oko (MO) tested areas.

Test Area	Sample No.	Altitude (m asl)	Geographical Coordinates	Geological Structure [21,22]	Physico-Geographical Mesoregion [19]
KW	1	1100	49°15.572′ N 19°59.322′ E	Boulders, gravel, sand, and silts of stones and river terraces 0.5–3.0 m high, e.g., rivers (Holocene)	Reglowe Tatras
	2	1200	49°15.424′ N 19°59.645′ E	Dolomites, limestones, siltstones, and breccia (Lower Triassic)	Reglowe Tatras
	3	1300	49°15.254′ N 19°59.681′ E	Dolomites, limestones, siltstones, and breccia (Lower Triassic)	Reglowe Tatras
	4	1400	49°15.252′ N 19°59.908′ E	Dolomites and limestones, undivided (Middle Triassic)	Reglowe Tatras
	5	1550	49°14.497′ N 20°00.097′ E	Boulders, moraine rock debris, clayey (Pleistocene)	Western Tatras
	6	1650	49°14.133′ N 19°59.671′ E	Porphyry granites (Carbon)	Western Tatras
	7	1750	49°14.013′ N 19°59.446′ E	Boulders and rock debris of rubble cones (screes) (Quaternary)	Western Tatras
	8	1850	49°13.927′ N 19°59.301′ E	Boulders and rock debris of rubble cones (screes) (Quaternary)	Western Tatras
MO	9	1000	49°15.065′ N 20°05.898′ E	Boulders, gravel, sand, clayey sands and silts of cones, of fluvioglacial levels and terraces 12.0–15.0 m high, e.g., rivers (Pleistocene)	High Tatras
	10	1100	49°13.984′ N 20°05.524′ E	Granodiorites and tonalities, equal grained, grey (Carbon)	High Tatras
	11	1200	49°13.270′ N 20°05.647′ E	Boulders, moraine rock debris, clayey (Pleistocene)	High Tatras
	12	1300	49°12.893′ N 20°04.867′ E	Boulders, rock debris and silts of dump and alluvial cones (Pleistocene–Holocene)	High Tatras
	13	1400	49°12.021′ N 20°04.115′ E	Boulders, rock debris and silts of dump and alluvial cones (Pleistocene–Holocene)	High Tatras

2.2.2. Chemical Analysis

In order to determine the content of heavy metals (Cd, Cr, Cu, Ni, Pb, and Zn) in the sampled soil material (topsoil, up to 10 cm), the following laboratory work was carried out, in accordance with the methodology used for collecting and preparing samples for chemical analyses [33,34]:

- Cleaning the collected samples by removing foreign material (dry leaves, twigs, grass, etc.);
- Drying the samples at 70 °C;
- Grinding soil samples in a ceramic mortar and sieving through a sieve with a mesh diameter of 2 mm;
- Mineralization, which is performed to completely break down soil samples into simple, solid compounds—1 g of the dried sample material was treated with a mixture of acids and aqua regia in the proportion (1:1:1 HNO_3:HCl:H_2O). The resulting solution was filtered and stored in sealed polyethylene containers until sent for spectrometric analysis;
- Determination of the total content of heavy metals using the inductively coupled plasma mass spectrometry (ICP-MS) method in the Bureau Veritas laboratory. The use of the Bureau Veritas methodology made it possible to accurately determine the metal content in the soil material, with the following detection limits (μg/g dm) for Cd: 0.01, Cr: 0.5, Cu: 0.01, Ni: 0.1, Pb: 0.01 and Zn: 0.1.

2.2.3. Soil Pollution Indicators

There are many indices for determining the degree of soil contamination with heavy metals [35]. This work focused on the enrichment factor, geoaccumulation index, and contamination factor. As reference values (background) for the obtained results, the concentrations of metals in the soils of Europe given in the Geochemical Atlas of Europe [36,37] were adopted. On the other hand, to calculate the enrichment factor, the average Earth's crust values for granite rocks were used, given by Turekian and Wedepohl [38] and Wedepohl [39]. The average concentrations of metals in the soil (background) for the selected research area and Earth's crust for granitic rocks are presented in Table 2.

Table 2. Mean metal concentrations (background values) in soils according to Salminen et al. [37] and average Earth's crust according to Turekian and Wedepohl [38].

Element	Background Values ($mg \cdot kg^{-1}$)	Average Earth's Crust for Granitic Rocks ($mg \cdot kg^{-1}$)
Cd	0.14	0.13
Cr	21.5	13
Cu	15.8	14
Ni	21.9	15
Pb	22.8	17.5
Zn	54.6	50
Fe (%)	2.50	2.3

Enrichment Factor (EF)

The enrichment factor (EF) is used to determine the anthropogenic effect on soil contamination with heavy metals [40]. The recommended reference element—iron (Fe) [41], is used to calculate the index. Fe is recommended because it does not take an active part in biogeochemical cycles.

The enrichment factor is calculated according to the following equation:

$$EF = \frac{Mx \cdot Feb}{Mb \cdot Fex},\tag{1}$$

where Mx and Fex are the soil sample concentrations of the heavy metal and Fe (or another normalizing element), while Mb and Feb are their concentrations in a suitable background (average Earth's crust) [40].

The enrichment factor is classified as follows:

EF < 2: deficient to minimal enrichment;
2 ≤ EF < 5: moderate enrichment;
5 ≤ EF < 20: significant enrichment;
20 ≤ EF < 40: very high enrichment;
EF ≥ 40: extremely high enrichment.

Geoaccumulation Index (Igeo)

In order to assess the degree of contamination of environmental components, Müller [42] suggested using the geoaccumulation index. He provided seven categories of contamination depending on the index value.

Geoaccumulation index is calculated as follows:

$$Igeo = \log_2\left(\frac{CmSample}{1.5 \times CmBackground}\right) \tag{2}$$

where CmSample is the concentration of the element in the enriched samples, and CmBackground is the background value of the element (Table 2). The contamination categories given by Müller [42] are as follows:

Igeo > 5: extremely contaminated;
4 < Igeo < 5: strongly to extremely contaminated;
3 < Igeo < 4: strongly contaminated;
2 < Igeo < 3: moderately to strongly contaminated;
1 < Igeo < 2: moderately contaminated;
0 < Igeo < 1: uncontaminated to moderately contaminated;
Igeo = 0: uncontaminated.

Contamination Factor (Cf)

The third factor determining soil contamination with heavy metals is the contamination factor (Cf). It is calculated according to the following formula:

$$Cf = \frac{CmSample}{CmBackground}, \tag{3}$$

where the contamination factor Cf < 1 refers to low contamination; $1 \le Cf < 3$ means moderate contamination; $3 \le Cf \le 6$ indicates considerable contamination, and Cf > 6 indicates very high contamination.

Degree of Contamination (C_d)

Håkanson [43] proposed an overall indicator of contamination based on integrating data for a series of seven specific heavy metals (As, Cd, Cu, Cr, Hg, Pb, Zn) and the organic pollutant PCB. This method is based on the calculation for each pollutant of a contamination factor (Cf). The degree of contamination is calculated using the following formula:

$$C_d = \sum_{i=1}^{i=n} C_f^i, \tag{4}$$

where n—number of analyzed elements, i—i^{th} element (or pollutant), and Cf—contamination factor.

To determine the level of soil contamination, the following ranges were adopted according to Håkanson [43]:

$C_d < 8$: low degree of contamination;

$8 \leq C_d < 16$: moderate degree of contamination;

$16 \leq C_d < 32$: considerable degree of contamination;

$C_d \geq 32$: very high degree of contamination indicating serious anthropogenic pollution.

Modified Degree of Ontamination (m C_d)

Abrahim [44] extended Håkanson's formula [43] by introducing a modified degree of contamination and an appropriate scale of contamination. The modification of Håkanson's formula, carried out by Abrahim [44], made it possible to determine the influence of all metals on soils, without quantitative restrictions. The modified degree of contamination is the sum of all contamination factors (Cf) for soil pollutants divided by the number of analyzed pollutants. The equation for calculating the degree of contamination is provided below:

$$mC_d = \frac{\sum_{i=1}^{i=n} C_f^i}{n} \tag{5}$$

Due to the obtained values of the mC_d coefficient, the following scale of the degree of soil contamination was adopted:

$mC_d < 1.5$: very low degree of contamination;

$1.5 \leq mC_d < 2$: low degree of contamination;

$2 \leq mC_d < 4$: moderate degree of contamination;

$4 \leq mC_d < 8$: high degree of contamination;

$8 \leq mC_d < 16$: very high degree of contamination;

$16 \leq mC_d < 32$: extremely high degree of contamination;

$mC_d \geq 32$: ultra-high degree of contamination.

3. Results

The average content of heavy metals in the soils of the test areas (KW and MO) and the pH value are presented in Table 3.

Table 3. Heavy metal content (average values) and soil reaction in the Kasprowy Wierch and Morskie Oko test areas.

Tested Area	Sample No.	Mean Metal Concentrations (mg·kg^{-1})							pH H$_2$O
		Cd	Cr	Cu	Ni	Pb	Zn	Fe (%)	
KW	1	1.3	49.2	12.9	15.6	117.8	122.5	4.2	7.5
	2	1.2	41.8	12.4	14.2	115.1	118.0	4.1	7.6
	3	1.1	40.5	10.9	12.5	113.7	96.8	3.7	7.7
	4	0.9	31.6	6.4	7.7	99.1	86.0	3.6	7.8
	5	0.7	31.9	5.6	6.2	85.4	79.8	2.9	5.2
	6	0.7	31.1	4.2	6.5	65.8	76.6	2.6	4.3
	7	0.5	26.0	4.0	5.5	54.2	74.1	2.3	4.9
	8	0.3	25.3	3.1	4.6	53.5	61.1	2.0	4.9
MO	9	1.3	32.6	14.8	5.8	161.1	125.4	4.9	4.0
	10	1.0	31.1	12.2	5.0	149.8	105.5	4.4	3.5
	11	0.9	31.3	11.6	5.1	129.3	84.1	3.9	3.6
	12	0.6	27.9	10.8	3.1	118.3	67.5	3.7	3.4
	13	0.5	27.5	9.2	2.7	114.6	52.2	3.3	3.8

On the basis of the conducted research, it was found that the highest average heavy metal content in soils for both test areas was recorded for Zn and Pb, and the lowest for Cd, Cu and Ni. Metal contents in the soils for the KW and MO test areas decrease with the increase in the height in meters above sea level. By comparing the metal contents between the test areas at the same altitudes, higher lead contents for MO can be seen compared to the Pb content in KW. On the other hand, the content of Cr and Ni in the soil of the test areas is higher in the case of KW in relation to the content of these metals in the MO test area.

Soil reaction (pH) is the highest (from 7.5 to 7.8) for the soils sampled at an altitude of 1100–1400 m in the vicinity of KW. The lowest pH values were found for the soils of the MO test area (from 3.4 to 4.0) and for the soils of the KW area sampled at an altitude of 1550–1850 m above sea level. The reaction of the soils for the MO test area was similar to that of the soils sampled from the KW test area for altitudes higher or equal to 1550 m above sea level (for the highest absolute heights of the KW). Different values of soil reaction were observed for the KW test area, from very acidic to alkaline, which is related to the land cover.

Based on the data from Table 3 concerning the average content of metals in the soils in the study areas, as well as the background values and the average Earth's crust, the indices of soil contamination with metals were calculated. Table 4 and Figure 2 present the values of the given coefficients for soil samples. Sample numbers from 1 to 8 refer to the samples taken from KW, while numbers 9–13 refer to the samples taken from MO. On the basis of the conducted research, it can be concluded that in MO the concentrations of metals in the soil are higher than in KW. This relationship is illustrated by the coefficients presented below. Another regularity is the decrease in the content of metals in soil samples with the increase in altitude. This regularity occurs for all collected soil samples, both in KW and MO.

Table 4. Enrichment factors (EFs) for soil samples from the Kasprowy Wierch and Morskie Oko test areas.

Tested Area	Sample No.	Cd	Cr	Cu	Ni	Pb	Zn
KW	1	5.5	2.1	0.5	0.6	3.7	1.3
	2	5.4	1.8	0.5	0.5	3.7	1.3
	3	5.1	1.9	0.5	0.5	4.0	1.2
	4	4.3	1.6	0.3	0.3	3.6	1.1
	5	4.1	1.9	0.3	0.3	3.9	1.2
	6	5.0	2.1	0.3	0.4	3.3	1.4
	7	3.8	2.0	0.3	0.4	3.1	1.5
	8	2.7	2.2	0.3	0.4	3.5	1.4
MO	9	4.5	1.2	0.5	0.2	4.3	1.2
	10	4.0	1.3	0.5	0.2	4.5	1.1
	11	3.9	1.4	0.5	0.2	4.4	1.0
	12	2.7	1.3	0.5	0.1	4.2	0.8
	13	2.7	1.5	0.5	0.1	4.6	0.7
Average		4.1	1.7	0.4	0.3	3.9	1.2

In Table 4, the values for the enrichment factor (EF) are presented as the first factor. The highest EF values are for cadmium (about 5) and for lead (about 4) in KW and MO. EF values for other metals are much lower and do not exceed the value of 2.0, except for Cr, which means that they do not pose a threat to the environment. The EF for chromium for samples 1, 6, 7 and 8 was greater than or equal to 2.0 only for KW. Even for Cd and Pb, EF values fall within the range of up to 5, which classifies them as moderate enrichment. Only in the case of cadmium is there significant enrichment (EF $\geq$ 5) for samples 1, 2, 3, and 6.

The other calculated index, presented in Table 5, is the geoaccumulation index (Igeo). As in the case of enrichment factors, this coefficient is highest for cadmium and lead. For cadmium for samples 1, 2, 3, 4, 9, 10, and 11, and for lead for samples 9 and 10, the Igeo values indicate moderate to strong contamination. For the other samples, in the case of cadmium and lead, the Igeo values are on the level of moderate contamination. For the remaining measured metals, the Igeo index values are very low (<1), and thus do not indicate strong environmental pollution.

Table 5. Geoaccumulation index (Igeo) for soil samples from the Tatra National Park area (KW and MO).

Tested Area	Sample No.	Geoaccumulation Index (I_{geo})					
		Cd	Cr	Cu	Ni	Pb	Zn
KW	1	2.63	0.61	−0.88	−1.07	1.78	0.58
	2	2.58	0.37	−0.94	−1.22	1.75	0.53
	3	2.35	0.33	−1.13	−1.40	1.73	0.24
	4	2.07	−0.03	−1.88	−2.10	1.53	0.07
	5	1.68	−0.02	−2.08	−2.42	1.32	−0.04
	6	1.80	−0.05	−2.49	−2.35	0.94	−0.10
	7	1.25	−0.31	−2.58	−2.58	0.66	−0.15
	8	0.56	−0.35	−2.95	−2.84	0.64	−0.42
MO	9	2.58	0.02	−0.68	−2.51	2.24	0.61
	10	2.26	−0.05	−0.96	−2.71	2.13	0.36
	11	2.02	−0.05	−1.04	−2.70	1.92	0.04
	12	1.42	−0.21	−1.13	−3.42	1.79	−0.28
	13	1.25	−0.23	−1.37	−3.63	1.74	−0.65

The next calculated coefficients indicating the degree of environmental pollution with heavy metals are the contamination factors (Cf), the degree of contamination (C_d), and modified degree of contamination (m C_d). The values of these coefficients are presented in Table 6. The highest values of the Cf > 6 coefficient, indicating very high contamination, were obtained for cadmium (samples No. 1, 2, 3, 4, 9, 10, and 11) and for lead (samples No. 9 and 10). The values of the Cf coefficient in the range of 3–6 indicating considerable contamination were obtained for cadmium (samples No. 5, 6, 7, 12, and 13) and lead (4, 5, 11, 12 and 13). For chromium and zinc, in the case of all the samples, the values of Cf indicate moderate contamination, and for samples No. 1, 2, 3, 9, 10, and 11 the values of the factor are the highest. For copper and nickel, the contamination factor (Cf < 1) refers to low contamination.

Table 6. Degree of contamination (C_d) and contamination factors (Cf) for soil samples from KW and MO.

Tested Area	Sample No.	Contamination Factors (C_f)						C_d
		Cd	Cr	Cu	Ni	Pb	Zn	
KW	1	9.3	2.3	0.8	0.7	5.2	2.2	20.5
	2	8.9	1.9	0.8	0.6	5.0	2.2	19.5
	3	7.6	1.9	0.7	0.6	5.0	1.8	17.5
	4	6.3	1.5	0.4	0.4	4.3	1.6	14.4
	5	5.0	1.5	0.4	0.3	3.7	1.5	12.3
	6	5.0	1.4	0.3	0.3	2.9	1.4	11.3
	7	3.6	1.2	0.3	0.3	2.4	1.4	9.0
	8	2.2	1.2	0.2	0.2	2.3	1.1	7.3
MO	9	9.0	1.5	0.9	0.3	7.1	2.3	21.1
	10	7.2	1.4	0.8	0.2	6.6	1.9	18.1
	11	6.1	1.5	0.7	0.2	5.7	1.5	15.7
	12	4.0	1.3	0.7	0.1	5.2	1.2	12.6
	13	3.6	1.3	0.6	0.1	5.0	1.0	11.5
Average		6.0	1.5	0.6	0.3	4.6	1.6	14.7

The C_d coefficient is the highest in the case of soil samples No. 9, 1, 2, 10, and 3, which indicate a considerable degree of contamination. Cadmium for samples No. 4, 5, 6, 7, 11, 12, and 13 constitutes a moderate degree of contamination. Only one sample (No. 8) with $C_d < 8$ indicates a low degree of contamination.

The values of the modified degree of contamination (mC_d) coefficient are presented in Figure 2. The figure shows that the mC_d values are the highest for samples No. 1, 2, 3, 4, and 9, 10 and 11. These samples are the soil material collected at altitudes of 1100, 1200, 1300, and 1400 m above sea level in the KW and at altitudes of 1000, 1100, and 1200 m above sea level in the MO. For the above-mentioned samples, the mC_d coefficient is in the range of 2–4, denoting a moderate degree of contamination. The remaining samples, i.e., samples No. 5–7 (1550, 1650, 1750 m asl), as well as 12 and 13 (1300 and 1400 m asl), for mC_d in the range of 1.5–2, indicate a low degree of contamination. Only sample No. 8 (1850 m asl), for $mC_d < 1.5$, indicates a very low degree of contamination.

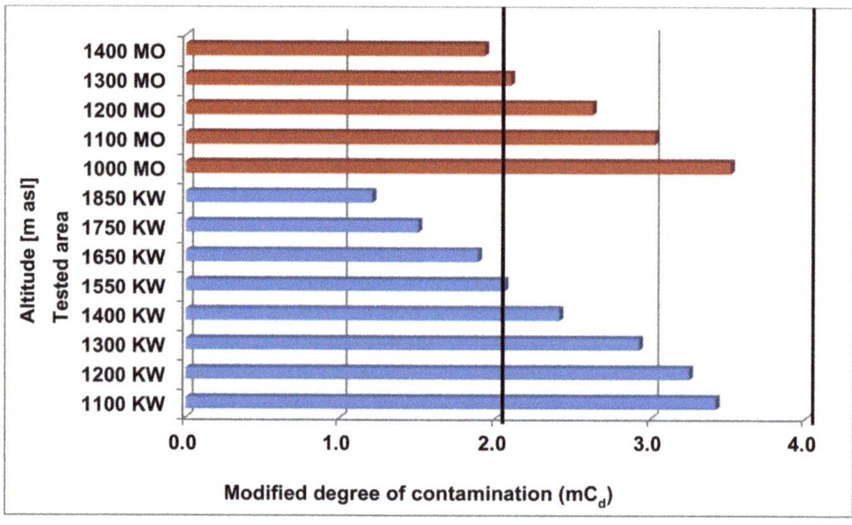

Figure 2. Histogram of the modified degree of contamination (mC_d) using Cd, Cr, Cu, Ni, Pb, and Zn in soil samples from the Tatra National Park. The two vertical lines represent the boundaries between low ($mC_d < 2$) and moderate ($2 \leq mC_d < 4$) degrees of contamination. Morskie Oko tested area is marked in red, Kasprowy Wierch tested area is marked in blue.

All the calculated coefficients in this study, except for EF, show that there is a trend for the content of metals in soils to decrease together with an increase in altitude.

4. Discussion

The conducted research concerning the determination of the content of selected heavy metals in the soils of the Tatra National Park (TNP) showed that, with the increase in the altitude of the terrain, the concentration of metals in soils decreases. This tendency was observed for all tested metals in the Kasprowy Wierch and Morskie Oko test areas. A decrease in the metal content in soils in mountainous regions with an increase in altitude was found by Kubica et al. [11]. The Pb content in the soil at an altitude of 1720 m asl was lower than the Pb content at altitudes of 1180 m asl and 1355 m asl. Similar relationships were found for Cr, the content of which in the soil at an altitude of 1720 m asl was lower than at altitudes of 1000 m asl and 1180 m asl. Miechówka et al. [13] found a decrease in the Cd content together with an increase in altitude above the upper forest limit. They observed lower Cd contents for the altitude range of 1850–2200 m asl than for the altitude range of 1100–1550 m asl. Similar to the authors, Stobiński et al. [45] found a decrease in Zn and Cr in the TNP soils together with an increase in altitude (the Zn content at an altitude of 1852 m asl was almost two times lower than at an altitude of 1200 m asl; while the content of Cr at an altitude of 1850 m asl was lower than at an altitude of 1780 m asl). In the area of the Gonga Mountains (Eastern Tibetan Plateau), Bing et al. [46] found the lowest heavy metal content in soils for the altitude range of 3500–3700 m asl (above the upper limit of timberline forest). In the case of the conducted research, a similar relationship was observed for the KW area (in the Tatras, the upper forest limit runs at an average altitude of 1500 m asl).

The determined metal contents in the TNP soils (KW and MO test areas) are similar to those found by Stobiński et al. [45] for Cr and Zn in the soil samples collected in the High Tatras, and by Kubica et al. [10] for Cd, Pb, Zn in the soil samples collected in the Kościeliska and Rybi Potok Valleys. Niemyska-Łukaszuk et al. [47] obtained comparable Pb contents for the rankers of the non-forest areas of the TNP. In the Morskie Oko and Kasprowy Wierch test areas, the contents of Pb and Zn were found to be almost twice as high as those determined by Wieczorek and Zadrożny [14] in podsol soils (Podzols) in the TNP area, and by Kubica et al. [11] for Zn in the soils in the Chochołowska Valley. Similar lower concentrations of the metals Cu, Pb, and Zn were obtained by Skwaryło-Bednarz [48] in the forest soils of the Roztocze National Park.

The indices calculated in both test areas were not compared with each other due to the local differences in the geological structure affecting the soil pH. In the case of the MO, the bedrock consists of acidic crystalline rocks (e.g., granitoids and moraine formations) and the acid reaction of the soil was found (pH in the range of 3.4–4.0). On the other hand, in the KW area (predominant carbonate rocks, mainly limestones and dolomites) the acid reaction of soils was found (pH in the range of 4.3–5.2 for an altitude of 1550–1850 m above sea level) and a slightly alkaline reaction (pH in the range of 7.5–7.8 for an altitude of 1100–1400 m above sea level). The acid reaction of the soil is most likely the result of the falling of mountain pine needles and has favored the accumulation of metals. We observe high acidification of soils in the areas located higher, at an altitude of 1550 m above sea level, which may also result from the impact of long-range emissions on the condition of the natural environment (industrial and communication pollution, e.g., SO_2, CO_2, and NO_x). It is difficult to compare the results from the area of the Tatra Mountains with other high-mountain areas, because they are characterized by a different specificity determined by their geographical location, affecting the climatic and geological conditions, and also by an often unique type of plant communities (in the mountains that change zonally). The vicinity of high-mountain areas is also diversified in terms of land use, e.g., the range and intensity of impact of industrial, communication and highly urbanized areas. Therefore, the impact of the anthropogenic factor on mountain areas will be incomparable. Hence, the authors limited the comparison of their research results to the results of works from the same area.

The tests carried out in the MO and KW test areas showed that, among the six metals determined, the highest concentrations in the soil were found for cadmium and lead. This is owing to the small diameter of the particles of both elements and, consequently, the long duration of their lingering in the atmosphere. Such particles are easily transported, in the direction of prevailing winds, over long distances, up to 100–200 km (long-range emission). In the case of the studied areas (the Tatra National Park), taking into account the dominant south-western direction of winds, the content of metals in soils may be influenced by emissions from nearby countries, i.e., the Czech Republic and Slovakia, as well as from the Upper Silesian agglomeration. During their research in the Morskie Oko region, they found a high impact of long-range emissions on the content of heavy metals in soils and plants [49]. At the same time, they pointed out the slight influence of local pollution sources (including hotels, transport). The sampling sites were more exposed to the effects of long-range emissions as they are located on the northern slope of the Tatra Mountains. Long-range emission contributes to metal contamination of important natural areas and is currently a problem not only on a local (national), but also global (world) scale. Therefore, when examining environmental pollution with heavy metals, one should also take into account the emissions from neighboring areas, according to the prevailing wind directions.

5. Conclusions

The studied coefficients helped to determine the degree of anthropogenic impact on different natural environments of Poland, according to their altitude. The highest values of the calculated coefficients, i.e., the greatest impact of human activity on the environment, were found in the case of cadmium and lead. For the other elements (Cr, Cu, Ni, and Zn), the calculated coefficients were low, which proves that the environment was not contaminated with these metals at all, or only to a small degree. Long-range emissions influence the increased content of heavy metals in the TPN soils. The height above sea level, as well as the proximity of the main mountain ridge, affects the spread of pollution. Specific terrain conditions affect the increase in wind velocity, therefore, soils at higher altitudes and closer to the ridge contained fewer metals than those located lower and further from the ridge.

Author Contributions: Conceptualization, J.K.; methodology, J.K., P.K.; software, J.K., P.K.; validation, J.K., P.K.; formal analysis, J.K.; investigation, J.K.; resources, J.K.; data curation, J.K., P.K.; writing—original draft preparation, J.K.; writing-review and editing, J.K., P.K.; visualization, J.K., P.K.; supervision, J.K.; project administration, J.K., P.K.; funding acquisition, J.K., P.K. All authors have read and agreed to the published version of the manuscript.

Funding: The work was financed by Pedagogical University of Krakow research project No. BN.610-192/PBU/2020

Acknowledgments: We would like to thank the Director of the Tatra National Park for making it possible to take soil samples and for their positive opinion on the research. This was greatly appreciated.

Conflicts of Interest: The authors declare no conflict of interest.

References

1. Nguyen, H.T.H.; Nguyen, B.Q.; Duong, T.T.; Bui, A.T.K.; Nguyen, H.T.A.; Cao, H.T.; Mai, N.T.; Nguyen, K.M.; Pham, T.T.; Kim, K.-W. Pilot-Scale Removal of Arsenic and Heavy Metals from Mining Wastewater Using Adsorption Combined with Constructed Wetland. *Minerals* **2019**, *9*, 379. [CrossRef]
2. García-Lorenzo, M.L.; Crespo-Feo, E.; Esbrí, J.M.; Higueras, P.; Grau, P.; Crespo, I.; Sánchez-Donoso, R. Assessment of Potentially Toxic Elements in Technosols by Tailings Derived from Pb–Zn–Ag Mining Activities at San Quintín (Ciudad Real, Spain): Some Insights into the Importance of Integral Studies to Evaluate Metal Contamination Pollution Hazards. *Minerals* **2019**, *9*, 346. [CrossRef]
3. Kabata-Pendias, A.; Pendias, H. *Trace Elements in Soils and Plants*; CRC Press: Boca Raton, FL, USA; London, UK, 2001; pp. 15–204.
4. Ciepał, R. *Kumulacja Metali Ciężkich i Siarki w Roślinach Wybranych Gatunków Oraz Glebie Jako Wskaźnik Stanu Skażenia Środowiska Terenów Chronionych Województw Śląskiego i Małopolskiego [Accumulation of Heavy Metals and Sulphur in Plants of Selected Species and Soil as an Indicator of the State of Environmental Contamination of Protected Areas in Silesia and Małopolska Provinces]*; Wydawnictwo Uniwersytetu Śląskiego: Katowice, Poland, 1999; pp. 9–35.
5. Migaszewski, Z.; Gałuszka, A. *Geochemia Środowiska [Geochemistry of the Environment]*; Wydawnictwo Naukowe PWN: Warszawa, Poland, 2016; pp. 84–289.
6. Bergbäck, B.; Johansson, K.; Mohlander, U. Urban metal flows—A case study of Stockholm. *Water Air Soil Pollut.* **2001**, *1*, 3–24. [CrossRef]
7. Sörme, L.; Bergbäck, B.; Lohm, U. Goods in the antroposphere as a metal emission source. *Water Air Soil Pollut.* **2001**, *1*, 213–227. [CrossRef]
8. Lehndorf, E.; Schwark, L. Accumulation histories of major and trace elements on pine needles in the Cologne Conurbation as function of air Quality. *Atmos. Environ.* **2008**, *42*, 833–845. [CrossRef]
9. Steinnes, E.; Friedland, A.J. Metal contamination of natural surface soils from long-range atmospheric transport: Existing and missing knowledge. *Environ. Rev.* **2006**, *14*, 169–186. [CrossRef]
10. Kubica, B.; Mietelski, J.W.; Gołaś, J.; Skiba, S.; Tomankiewicz, E.; Gaca, P.; Jasińska, M.; Tuteja-Krysa, M. Concentration of 137Cs, 40K, 238Pu and 239 + 240Pu radionuclides and some heavy metals in soil samples from the two main valleys from Tatra National Park. *Pol. J. Environ. Stud.* **2002**, *11*, 537–545.

11. Kubica, B.; Kwiatek, W.M.; Stobiński, M.; Skiba, S.; Skiba, M.; Gołaś, J.; Kubica, M.; Tuleja-Krysa, M.; Wrona, A.; Misiak, R. Concentration of 137Cs, 40K radionuclides and some heavy metals in soil samples from Chochołowska Valley from Tatra National Park. *Pol. J. Environ. Stud.* **2007**, *16*, 735–741.

12. Miechówka, A.; Niemyska-Łukaszuk, J.; Ciarkowska, K. Heavy metals in selected non-forest soils from the Tatra National Park. *Chem. Inżynieria Ekol.* **2002**, *9*, 1433–1438.

13. Miechówka, A.; Niemyska-Łukaszuk, J. Content diversity of Zb, Pb and Cd in Lithic Leptosols of the Tatra National Park (Poland). *Oecologia Mont.* **2004**, *13*, 1–5.

14. Wieczorek, J.; Zadrożny, P. Content of Cd, pPb and Zn in Podzols Tatra National Park. *Proc. ECOpole* **2013**, *7*, 421–426. [CrossRef]

15. Barančoková, M.; Barančok, P.; Mišovičová, D. Heavy metal loading of the Belianske Tatry Mts. *Ekológia Bratisl.* **2009**, *28*, 255–268. [CrossRef]

16. Skiba, S.; Drewnik, M.; Prędki, R.; Szmuc, R. Zawartość metali ciężkich w glebach Bieszczadzkiego Parku Narodowego [Heavy metals content in the soils of the Bieszczady National Park]. *Rocz. Bieszcz.* **1995**, *4*, 111–116.

17. Skiba, S.; Michalik, M. Heavy metals in soils and sulphate minerals on rock surfaces as indicator of pollution of the environment (on the example of the Bieszczady Mts., Eastern Carpathians). *Pol. J. Soil Sci.* **2000**, *33*, 57–66.

18. Kubica, B.; Szarlowicz, K.; Stobinski, M.; Skiba, S.; Reczynski, W.; Gołaś, J. Concentrations of 137Cs and 40K radionuclides and some heavy metals in soil samples from the eastern part of the Main Ridge of the Flysch Carpathians. *J. Radioanal. Nucl. Chem.* **2014**, *299*, 1313–1320. [CrossRef] [PubMed]

19. Solon, J.; Borzyszkowski, J.; Bidłasik, M.; Richling, A.; Badora, K.; Balon, J.; Brzezińska-Wójcik, T.; Chabudziński, Ł.; Dobrowolski, R.; Grzegorczyk, I.; et al. Physico-geographical mesoregions of Poland: Verifi cation and adjustment of boundaries on the basis of contemporary spatial data. *Geogr. Pol.* **2018**, *91*, 143–170. [CrossRef]

20. Bac-Moszaszwili, M.; Burchart, J.; Głazek, J.; Iwanow, A.; Jaroszewski, W.; Kotański, Z.; Lefeld, J.; Mastella, L.; Ziomkowski, W.; Roniewicz, P.; et al. *Mapa Geologiczna Tatr Polskich [Geological Map of the Polish Tatras]*; 1:30 000; Wydawnictwo Geologiczne: Warszawa, Poland, 1979.

21. Piotrowska, K.; Kotański, Z.; Gawęda, A.; Piotrowski, J.; Rączkowski, W. *Szczegółowa Mapa Geologiczna Polski. Arkusz Tatry Zachodnie [Detailed Geological Map of Poland. The Western Tatras sheet]*; NAG, PGI-NRI: Warsaw, Poland, 2008.

22. Piotrowska, K.; Michalik, M.; Rączkowski, W.; Iwanow, A.; Wójcik, A.; Derkacz, M.; Wasiluk, R. *Szczegółowa Mapa Geologiczna Poslki. Arkusz Tatry Wysokie [Detailed Geological Map of Poland. The High Tatras sheet]*; NAG, PIG-PIB: Warszawa, Poland, 2013.

23. Kotarba, A.; Kaszowski, L.; Krzemień, K. *High-Mountain Denudational System in the Polish Tatra Mountains: Geographical Studies*; Ossolineum: Wrocław, Poland, 1987; pp. 1–106.

24. Klimaszewski, M. *Rzeźba Tatr Polskich 1988 [Geomorphologic Relief in the Polish Part of Tatras]*; Wydawnictwo PWN: Warszawa, Poland, 1988; pp. 1–707.

25. Hreško, J.; Boltizar, M.; Bugar, G. The present-day development of landforms and landcover in alpine environment—Tatra Mts (Slovakia). *Studia Geomorphol. Carpatho-Balc.* **2005**, *39*, 23–48.

26. Niedźwiedź, T. *Sytuacje Synoptyczne i ich Wpływ na Zróżnicowanie Przestrzenne Wybranych Elementów Klimatu w Dorzeczu Górnej WISŁY [Synoptic Situations and Their Impact on the Spatial Variability of Selected Climate Elements in the Upper Vistula Basin, a Habilitation Dissertation at the Jagiellonian University]*; Jagiellonian University: Kazimierz, Poland, 1981; Volume 58, 165p.

27. Niedźwiedź, T. Climate of the Tatra Mountains. *Mt. Res. Dev.* **1992**, *12*, 131–146. [CrossRef]

28. Drewnik, M. *Geomorfologiczne Uwarunkowania Rozwoju Pokrywy Glebowej w Obszarach Górskich na Przykładzie Tatr [Geomorphological Conditions of the Development of the Soil Cover in Mountain Areas on the Example of the Tatra Mountains]*; Wydawnictwo UJ: Kraków, Poland, 2008.

29. Balon, J.; Jodłowski, M.; Krąż, P. The Tatra mountains: Physico-geographical regions. In *Atlas of The Tatr Mountains—Abiotic Nature*; Dąbrowska, K., Guzik, M., Eds.; plate I.4; Tatra National Park: Tatra County, Poland, 2015.

30. Żmudzka, E.; Nejedlik, P.; Mikulova, K. Temperature, thermal indices. In *Atlas of The Tatr Mountains—Abiotic Nature*; Dąbrowska, K., Guzik, M., Eds.; plate III.2; Tatra National Park: Tatra County, Poland, 2015.

31. Ustrnul, Z.; Walawender, E.; Czekierda, D.; Štasny, P.; Lapin, M.; Mikulova, K. Precipitation and snow cover. In *Atlas of the Tatr Mountains—Abiotic Nature*; Dąbrowska, K., Guzik, M., Eds.; plate III.3; Tatra National Park: Tatra County, Poland, 2015.

32. Skiba, S.; Koreň, M.; Drewnik, M.; Kukla, J. Soils. In *Atlas of The Tatr Mountains—Abiotic Nature*; Dąbrowska, K., Guzik, M., Eds.; plate V; Tatra National Park: Tatra County, Poland, 2015.

33. Gajec, M.; Król, A.; Kukulska-Zając, E.; Mostowska-Stąsiek, J. Pobieranie próbek gleby w kontekście prowadzenia oceny zanieczyszczenia powierzchni ziemi [Soil sampling in the context of soil pollution assessment]. *Nafta-Gaz* **2018**, *3*, 215–225. [CrossRef]

34. Namieśnik, J.; Łukasiak, J.; Jamrógiewicz, Z. *Pobieranie Próbek Środowiskowych do Analizy [Collecting of Environmental Samples for Analysis]*; Wydawnictwo Naukowe PWN: Warsaw, Poland, 1995; pp. 67–96.

35. Ridgway, J.; Shimmield, G. Estuaries as Repositories of Historical Contamination and Their Impact on Shelf Seas. *Estuar. Coast. Shelf Sci.* **2002**, *55*, 903–928. [CrossRef]

36. Salminen, R.; Batista, M.J.; Bidovec, M.; Demetriades, A.; De Vivo, B.; De Vos, W.; Duris, M.; Gilucis, A.; Gregorauskiene, V.; Halamic, J.; et al. *FOREGS Geochemical Atlas of Europe, Part 1: Background Information, Methodology and Maps*; Geological Survey of Finland: Espoo, Finland, 2005; pp. 140–347.

37. De Vos, W.; Tarvainen, T. (Eds.) Geochemical Atlas of Europe, Part 2. Interpretation of Geochemical Maps, Additional Tables, Figures, Maps and Related Publications. electronic version. Available online: http://weppi.gtk.fi/publ/foregsatlas/part2.php (accessed on 8 August 2020).

38. Turekian, K.K.; Wedepohl, D.H. Distribution of the elements in some major units of the earth_s crust. *Bull. Geol. Soc. Am.* **1961**, *72*, 175–192. [CrossRef]

39. Wedepohl, K.H. The composition of the continental crust. *Geochim. Cosmocheimica Acta* **1995**, *59*, 1217–1232. [CrossRef]

40. Salomons, W.; Förstner, U. *Metals in the Hydrocycle*; Springer: Berlin/Heidelberg, Germany, 1984; pp. 55–225.

41. Migaszewski, Z.; Gałuszka, A. *Podstawy Geochemii środowiska [Fundamentals of Environmental Geochemistry]*; Wydawnictwo Naukowe PWN: Warsaw, Poland, 2007; pp. 11–158.

42. Muller, G. Index of geoaccumulation in sediments of the Rhine river. *Geol. J.* **1969**, *2*, 108–118.

43. Håkanson, L. An Ecological Risk Index for Aquatic Pollution Control—A Sedimentological Approach. *Water Res.* **1980**, *14*, 975–1001. [CrossRef]

44. Abrahim, G.; Parker, R. Assessment of heavy metal enrichment factors and the degree of contamination in marine sediments from Tamaki Estuary, Aukland, New Zealand. *Environ. Monit. Assess.* **2008**, *136*, 227–238. [CrossRef]

45. Stobiński, M.; Kubica, B.; Misiak, R.; Kwiatek, W.M.; Gołaś, J.; Reczyński, W.; Hajduk, R.; Dutkiewicz, E.; Bartyzel, M.; Fleischer, P.; et al. *Przestrzenny Rozkład Zawartości Zn Oraz Cr w Próbkach Gleby Pobranej w Tatrach Wysokich [The Spatial Distribution of Zn and Cr Content in the Soil Samples Collected in the High Tatras]*; REPORT 2012/C 2008; Instytut Fizyki Jądrowej: Kraków, Poland, 2008; pp. 1–12.

46. Bing, H.; Wu, Y.; Zhou, J. Vegetation and cold trapping modulating elevation-dependent distribution of trace metals in soils of a High Mountain in Eastern Tibetan Plateau. *Sci. Rep.* **2016**, *6*, 1–14. [CrossRef]

47. Niemyska-Łukaszuk, J.; Miechówka, A.; Zadrożny, P. Całkowita zawartość ołowiu w profilach rankerów Tatrzańskiego Parku Narodowego [Total lead content in the profiles of rankers of the Tatra National Park]. *Zesz. Probl. Postępów Nauk Rol.* **1999**, *467*, 429–437.

48. Skwaryło-Bednarz, B. Ogólna zawartość wybranych metali ciężkich w glebach leśnych Roztoczańskiego Parku Narodowego (RPN) [General content of selected heavy metals in the forest soils of Roztocze National Park (RNP)]. *Acta Agrophysica* **2006**, *8*, 727–733.

49. Paukszto, A.; Mirosławski, J. Using stinging nettle (*Urtica dioica* L.) to assess the influence of long term emission upon pollution with metals of the Tatra National Park area (Poland). *Atmos. Pollut. Res.* **2019**, *10*, 73–79. [CrossRef]

Publisher's Note: MDPI stays neutral with regard to jurisdictional claims in published maps and institutional affiliations.

 *minerals*

Article

Quality of Peri-Urban Soil Developed from Ore-Bearing Carbonates: Heavy Metal Levels and Source Apportionment Assessed Using Pollution Indices

Katarzyna Sutkowska [1,*], Leslaw Teper [1], Tomasz Czech [2], Tomasz Hulok [1], Michał Olszak [1] and Jan Zogala [1]

[1] Institute of Earth Sciences, University of Silesia in Katowice, Bedzinska 60, 41-200 Sosnowiec, Poland; leslaw.teper@us.edu.pl (L.T.); Thulok@op.pl (T.H.); michal.olszak@outlook.com (M.O.); jan_zogala@o2.pl (J.Z.)

[2] Department of Agricultural and Environmental Chemistry, University of Agriculture in Krakow, Aleja Adama Mickiewicza 21, 31-120 Kraków, Poland; Tomasz.Czech@ur.krakow.pl

* Correspondence: katarzyna.sutkowska@us.edu.pl

Received: 20 November 2020; Accepted: 16 December 2020; Published: 19 December 2020

Abstract: Pollution indices are used to assess the influence of the bedrock as a natural source of heavy-metal (*HM*), and anthropogenic pollution from ore mining in soils developed from ore-bearing carbonates. The research was conducted in two areas differing in geological setting and type of land use in the Upper Silesia Industrial Region, Southern Poland. Physical properties such as pH, total sulfur, total carbon and total organic carbon values, as well as total Zn, Pb, and Cd contents (ICP-OES) for 39 topsoil samples were measured. Contamination factor (C_f), degree of contamination (C_{deg}), pollution load index (*PLI*) and geoaccumulation index (I_{geo}), were used to determine the deterioration of topsoil due to *HM* pollution. The *HM* content exceeded geochemical background levels by 2.5–18.1 times. Very high to moderate topsoil contamination was determined. In a shallow historical mining zone, the relative influence of particular *HM* was found to be in the order of Pb > Cd > Zn and, in a deep mining zone, Zn > Cd > Pb. In the topsoil developed over shallow ore bodies, the *HM* content was mainly (60%) due to naturally occurring *HM*. In the area of deeply buried ore bodies, 90% of the *HM* load was related to anthropogenic sources. Zn, Pb and Cd vertical distributions and the patterns of topsoil pollution differ in terms of types of mined ores, mining methods and times elapsed since mining ceased. Pollution indices are an efficient tool for distinguishing soil anthropogenic pollution and geogenic contamination.

Keywords: pollution indices; heavy metals; soil contamination; geogenic and anthropogenic origin

1. Introduction

Heavy metals (*HM*) in the environment originate from geological, industrial, agricultural, atmospheric, and waste sources. Soil is one of the most important environmental components at risk of *HM* contamination as a result of anthropogenic activities. A high concentration of *HM*s and metalloids (Zn, Pb, Cd, As, Tl, etc.) can be found in and around active and abandoned mines or smelting plants [1–3] due to the emission and dispersion of pollutants into air [4,5], water [6,7], soil [8–17], plants [4,18–22], and fauna [23,24]. The global average concentration of *HM*s in soils varies for Zn (10–300 mg·kg^{-1}), Pb (10–150 mg·kg^{-1}) and Cd (0.06 mg·kg^{-1}) [25]. In general, the *HM* concentrations in soils are increasing over time, and the highest concentrations are observed in industrial cities, due to traffic, power plants, and other industrial activities [25]. Soil contains baseline or background concentrations of *HM*. The *HM* content is determined by the composition of the parent rock material from which the soil is derived.

The effective assessment of soil *HM* contamination is an important global issue [10,25–29]. Indicators for the geochemical assessment of the soil environment include contamination factor (C_f), degree of contamination (C_{deg}), pollution load index (*PLI*), and geoaccumulation index (I_{geo}). These indicators enable the estimation of environmental risk and soil degradation due to accumulation of *HM*s [28–30]. Moreover, they facilitate differentiation between the accumulation of *HM*s produced by natural processes and anthropogenic activities (e.g., [31]). Essential to any assessment of the degree of soil contamination is the selection of an appropriate reference value. Although this issue has already been widely discussed [26,32–36], there is still a lack of unambiguous methodological findings.

The aim of the current study is to assess the quality of soil developed from Zn-Pb ore-bearing carbonates, in terms of (1) the influence of the bedrock, as a natural/geogenic source of *HM*s, and (2) anthropogenic pollution from ore mining and processing. We attempted a source apportionment of *HM*s (Zn, Pb, Cd) based on the pollution indices and using the local/on-site geochemical background. The research was carried out on the topsoil layers that are usually expected to accumulate trace elements of anthropogenic origin.

The main novelty of our work consists in taking into account data linking the indices to a broader geological context and mining history which has scarcely been published. We compare two areas located in the same Zn-Pb MVT-type deposit, which differ in terms of types of mined ores, ore-body depth, mining method and the time of mining cessation. The presented approach can be applied to the analyses of environmental risk and *HM* source apportionment in the abandoned mining sites worldwide.

2. Study Area

Two areas located in the north-east part of the Upper Silesia Basin (Figure 2) in the Dlugoszyn and Wilkoszyn Synclines in Jaworzno City were chosen for the study. Both synclines belong to the superior synclinorial structure with an NW-NE trending axis resulting from the Alpine orogenic movements. The inclination of the rock layers varies from 4° to 15° NW in the Dlugoszyn Syncline, and from 4° to 20° SE in the North-West limb of the Wilkoszyn Syncline. These tectonic structures comprise the Middle Triassic carbonate formation, which is composed of dolomites, limestones, and marls. The carbonate rock profile is partially altered due to the epigenetic fluid flow resulting in dolomitisation [37]. The Zn-Pb mineralisation followed the dolomitisation episode, which developed the ore-bearing dolomites with Zn-Pb ore deposits [38]. The Zn-Pb ore deposits in the area belong to the stratabound type. The ore minerals (galena, sphalerite, pyrite, marcasite, and secondary nonsulfides) form bodies of varying horizontal- and vertical extension ranging from several tens of centimetres up to several tens of meters. They consist of metasomatic-, dispersed- and cavity-filling-ores, the latter including crusted-, veined-, drusy- and breccia varieties [39]. Most of the ore bodies have a typical tabular form concordant with the rock bedding. The boundaries between the ore-bearing dolomites and the host carbonate rocks are rarely defined, but more commonly form a broad transition zone into weakly mineralized dolostones.

The areas selected for the study vary in terms of the geological setting. In the Dlugoszyn area (see point 1 in Figures 1 and 2), the ore-bearing dolomites occur near the surface (0–1 m), while in the Wilkoszyn syncline (see point 2 in Figures 1 and 2) the ore bodies occur at a depth of 20 to 90 m. The relatively shallow occurrence of ore mineralisation favours the development of mining. There is evidence of outcrop and underground Zn-Pb ore mining and smelting from the 12th [40] to the 20th century. The shallow ore exploitation in the Dlugoszyn area was completed at the end of the 19th century, while the deep mines and processing plants in the Wilkoszyn area were abandoned in the middle of the 20th century [41,42]. There is evidence of the historical mining and industrial activity in the landscape of the town [43]. Local ore mining left its mark on natural habitats impacting air [44], water [45,46] and soil [2,3,47–49] quality and imposing imprints onto flora [50,51] and fauna [52,53].

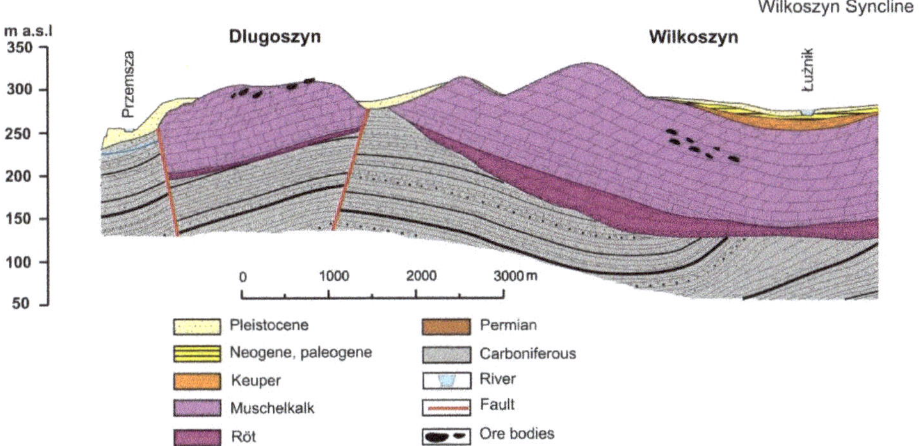

Figure 1. Schematic geological cross-section of the study area showing location of ore bodies, without Quaternary sediments (after [54]).

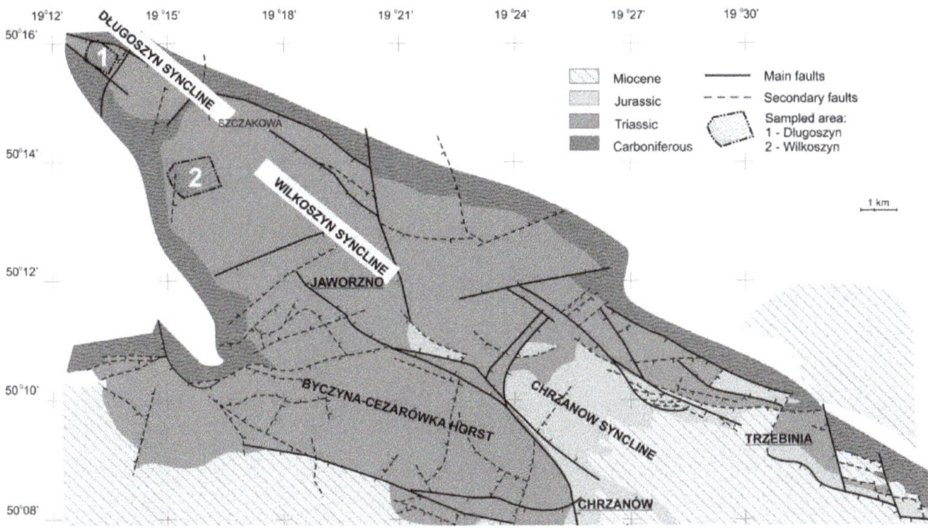

Figure 2. Geological context of the study area, without Quaternary sediments (after [54]).

3. Material and Analytical Methods

To investigate the relationship between the ore-bearing bedrock and content of *HM*s (Zn, Pb, and Cd) in the soils (mostly skeletal Rendzic Leptosol), 15 soil profiles were studied (Figure 3). We tried to select homogenous soil profiles. Each soil profile was divided into separate horizons, providing 54 soil samples. Only the topsoil layers (up to 30 cm depth), characterised by the highest concentrations of *HM*, were considered further; these are represented by 39 soil samples in Table 1. Moreover, bedrock specimens (from each study area) were sampled and analysed to evaluate the local/natural chemical pedogenic enrichment.

Table 1. Geochemical properties of the topsoil samples. D—Dlugoszyn area, W—Wilkoszyn area.

Sample Point	Coordinates	Sample Number	Depth [cm]	pH	TC [%]	TOC [%]	TS [%]	HM Content [mg·kg^{-1}]		
								Zn	Pb	Cd
D1	50°14′19.2″ N 19°14′30″ E	a	0–2	6.40	10.08	9.83	0.11	3329 ± 84	1674 ± 30	37.58 ± 3.27
		b	2–25	7.04	3.29	2.78	0.03	3319 ± 79	1877 ± 39	32.40 ± 1.63
D2	50°14′58.5″ N 19°14′18.7″ E	a	0–2	4.26	20.92	20.92	0.14	701 ± 17	341 ± 6	11.75 ± 0.56
		b	2–9	5.26	9.88	9.88	0.08	907 ± 8	353 ± 3	14.72 ± 0.36
		c	9–19	6.23	1.35	1.35	0.01	482 ± 48	97 ± 10	4.54 ± 0.46
		d	19–26	7.24	0.21	0.20	0.00	277 ± 30	55 ± 8	2.14 ± 0.27
D3	50°15′7.2″ N 19°14′25″ E	a	0–3	6.04	22.01	21.22	0.13	1438 ± 66	234 ± 12	7.46 ± 0.51
		b	3–11	6.65	9.89	7.38	0.04	1291 ± 14	329 ± 0.2	11.29 ± 0.05
		c	11–22	7.03	2.06	1.72	0.01	1170 ± 21	248 ± 0.3	6.99 ± 0.07
D4	50°15′7.1″ N 19°14′33″ E	a	0–2	6.41	8.55	8.50	0.07	1113 ± 12	275 ± 3	9.59 ± 0.45
		b	2–14	6.71	7.14	6.28	0.07	1037 ± 16	269 ± 8	9.50 ± 0.13
		c	14–34	6.91	1.41	1.39	0.01	937 ± 23	227 ± 6	8.11 ± 0.3
D5	50°15′18.2″ N 19°14′25.6″ E	a	0–2	5.55	23.12	23.12	0.17	926 ± 32	283 ± 11	10.46 ± 0.47
		b	2–14	6.06	7.16	7.16	0.07	797 ± 18	267 ± 2	10.45 ± 0.04
		c	14–19	5.96	0.20	0.20	0.00	68 ± 8	18 ± 2	0.31 ± 0.01
W1	50°13′36.1″ N 19°16′48.3″ E	a	0–2	6.74	5.76	5.76	0.06	2975 ± 45	440 ± 2	26.83 ± 0.23
		b	2–18	6.97	3.40	3.40	0.04	3077 ± 54	456 ± 1	27.66 ± 0.13
W2	50°13′36.2″ N 19°16′58.1″ E	a	0–2	6.93	11.57	11.08	0.10	1861 ± 4	673 ± 8	14.45 ± 0.25
		b	2–17	7.36	6.24	5.07	0.05	1831 ± 9	711 ± 6	11.86 ± 0.19
W3	50°13′28.6″ N 19°16′58.7″ E	a	0–2	6.34	9.22	9.22	0.07	2455 ± 34	885 ± 13	17.93 ± 0.14
		b	2–12	6.67	7.47	7.47	0.07	2761 ± 34	1035 ± 18	21.44 ± 0.34
		c	12–24	6.69	0.68	0.68	0.01	1388 ± 2	658 ± 14	3.30 ± 0.01
W4	50°13′27.7″ N 19°16′44.5″ E	a	0–2	6.74	13.89	13.84	0.12	1956 ± 3	329 ± 19	13.70 ± 1.06
		b	2–20	7.08	7.11	7.10	0.07	2547 ± 9	502 ± 6	20.57 ± 0.30
		c	20–30	7.10	2.03	2.03	0.02	2771 ± 27	328 ± 3	12.59 ± 0.31
W5	50°13′27.8″ N 19°16′35.4″ E	a	0–2	7.05	3.49	3.49	0.04	1063 ± 9	403 ± 2	7.42 ± 0.05
		b	2–18	7.24	3.65	3.65	0.04	1052 ± 6	400 ± 14	7.32 ± 0.30
		c	18–28	7.60	0.83	0.82	0.01	1124 ± 48	207 ± 8	4.39 ± 0.21
W6	50°13′9″ N 19°16′11″ E	a	0–3	6.21	2.04	2.04	0.02	630 ± 15	290 ± 4	6.28 ± 0.29
		b	3–11	6.40	1.65	1.65	0.02	583 ± 17	275 ± 5	4.34 ± 0.01
W7	50°13′8″ N 19°15′56″ E	a	0–3	6.99	6.33	6.11	0.07	1063 ± 11	289 ± 1	5.81 ± 0.14
		b	3–19	7.00	3.79	3.70	0.05	1169 ± 1	307 ± 2	7.09 ± 0.02
		c	19–26	7.20	1.45	1.17	0.01	639 ± 1	118 ± 1	2.34 ± 0.01
W8	50°13′15″ N 19°15′56″ E	a	0–4	6.47	3.44	3.44	0.04	1003 ± 3	242 ± 1	4.38 ± 0.03
		b	4–20	6.50	2.78	2.78	0.03	1080 ± 7	298 ± 1	6.05 ± 0.02
W9	50°13′15″ N 19°16′10″ E	a	0–3	6.01	1.59	1.59	0.02	480 ± 9	146 ± 5	3.90 ± 0.04
		b	3–13	6.12	1.60	1.60	0.02	497 ± 2	169 ± 1	3.95 ± 0.01
W10	50°13′25″ N 19°15′59″ E	a	0–3	6.53	2.34	2.34	0.02	652 ± 7	207 ± 2	3.71 ± 0.01
		b	3–17	6.56	2.31	2.31	0.03	727 ± 3	219 ± 1	4.20 ± 0.02

Prior to the analysis, the soil samples were oven-dried at 105 °C to constant weight, sieved to 2 mm through a stainless-steel sieve, and milled into a fine powder, while the bedrock samples were crushed and ground. The soil pH was determined using a 1:5 (g:mL) ratio of soil and 1 M KCl solution, with pH meter ELMETRON CP-315 m. The concentration of total carbon (TC), total organic carbon (TOC), and total sulfur (TS) were determined using an Eltra CS-530 IR-analyser with a TIC module. In order to determine the total content of Zn, Pb, and Cd, the soil and rock samples were wet digested in a closed system in a mixture of 6 cm^3 of concentrated nitric acid (65% Suprapur), and 2 cm^3 of hydrochloric acid (30% Suprapur). Digestion proceeded with the use of a Perkin Elmer Multiwave 3000 Microwave Digestion, in two steps according to the program of mineralisation recommended by the manufacturer—power: 1400 W, recovery time: 5 min, hold: 25 min in the first step, 10 min in the second, fan speed: 1 in the first step, 3 in the second. After mineralisation, the samples were transferred to measuring flasks (10 cm^3) with a 1% solution of Suprapur nitric acid. The *HM* content in the prepared solution was determined using an atomic emission spectrometer ICP-OES Optima 7300 Dual View Perkin Elmer. Each sample was analysed twice. The quality control procedure was performed using internal laboratory standard for soil material. The percentage recovery for observed elements ranged from 86.5% to 100% (Supplementary Materials Table S1). The ICP-OES analyses were performed at the Department of Agricultural and Environmental Chemistry, University of Agriculture in Krakow.

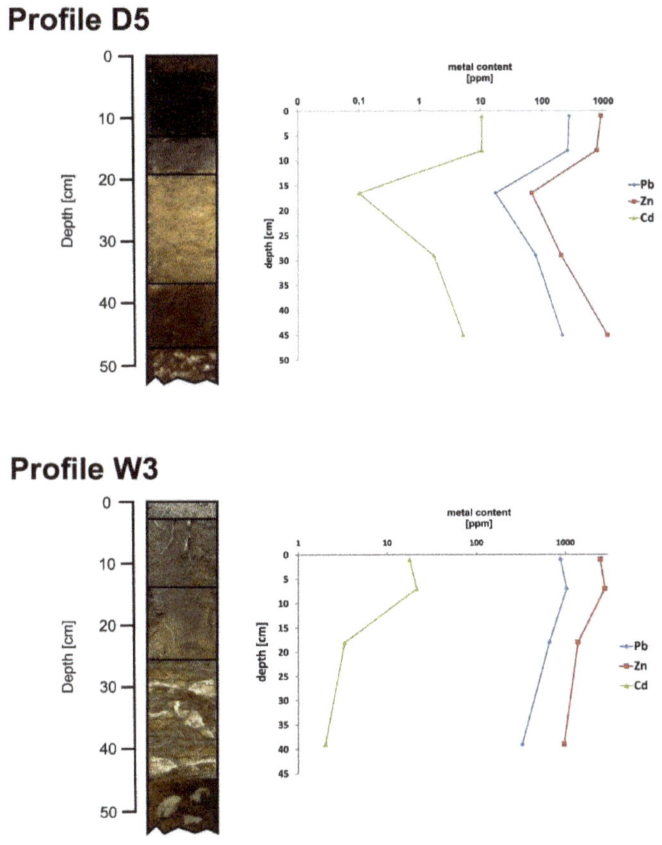

Figure 3. Characteristic examples of studied soil profiles in elevated- (profile D5) and buried (profile W3) ore-bearing dolomitic areas (photographs: T. Hulok).

The phase composition was provided by X-ray diffraction (XRD). The analyses were performed on powdered samples using a PANalytical X'Pert Pro MPD (multipurpose diffractometer) powered by a Philips PW3040/60 X-ray generator and fitted with a 1D silicon strip detector (X'Celerator). The measurements were performed using Co Kα-radiation with a wavelength of 0.1789010 nm, an acceleration voltage of 40 kV, a current of 40 mA, and with 0.02° 2θ step sizes between the angles of 5° and 70° 2θ and a 300 s measurement time per step. The data obtained were processed using HighScore+ software and the ICSD database and PDF4+ ICDD database. All XRD analyses were performed at the Institute of Earth Sciences, Faculty of Natural Sciences, University of Silesia, Sosnowiec.

4. Soil Pollution Assessment

The concentration of *HMs* in the soil is related either to the natural abundance if they come from parent rocks or the anthropogenic load if they come from mining- and processing activities. The distinction between the background of the site level and the pollution load is essential to determine the nature of any anomaly, especially in industrial areas such as the selected study area.

The *HM* contamination in the soil samples is evaluated by applying the C_f, C_{deg}, *PLI*, and I_{geo}.

The C_f is employed to determine the degree of soil pollution with a particular potentially toxic element. The calculation of this factor follows Håkonson's [55] definition (Equation (1)):

$$C_f = \frac{C_{HM}}{C_B},\tag{1}$$

where C_{HM} is the *HM* concentration in soil samples (mg·kg^{-1}) and C_B is reference concentration of *HMs* in soil in the study area (mg·kg^{-1}). According to Håkonson [55], four contamination categories can be distinguished: $C_f < 1$, low; $1 < C_f < 3$, moderate; $3 < C_f < 6$, considerable; and $C_f > 6$, very high contamination.

The estimation of the degree of total contamination of surface layers in a particular core or sampling site was proposed by Håkanson [55] as a sum of the C_f for each sample, or C_{deg} (Equation (2)):

$$C_{deg} = \sum_{i=1}^{i=n} C_f,\tag{2}$$

The classification of the C_{deg} in sediments is as follows [55]: $C_{deg} < 6$, low; $6 < C_{deg} < 12$, moderate; $12 < C_{deg} < 24$, considerable; and $C_{deg} > 24$, high.

The status of the *HM* pollution in the studied soils is assessed through the *PLI* [56]. The *PLI* is calculated by obtaining the *n*-root from the product of *n*- C_f s for all the *HMs*, where n is the number of *HMs* studied (Equation (3)):

$$PLI = \sqrt[n]{(CF_{HM1} \times CF_{HM2} \ldots \times CF_{HMn})},\tag{3}$$

A *PLI* value < 1 indicates minimal or no metal pollution, a *PLI* = 1 indicates that the level of pollutants is equal to the baseline content of *HM*, whereas *PLI* > 1 means that there is pollution and the value indicates the severity of the pollution. For the latter case, Zhang et al. [57] classify soils into the following five categories: $1 < PLI < 2$, moderately; $2 < PLI < 3$, moderately to highly; $3 < PLI < 4$, highly $4 < PLI < 5$, very highly; and $PLI > 5$, extremely polluted.

The I_{geo} proposed by Müller [58] is a common criterion to evaluate the *HM* pollution in sediments/soil. It is defined as follows (Equation (4)):

$$I_{geo} = log_2\left(\frac{C_{HM}}{1.5\,C_B}\right),\tag{4}$$

where C_{HM} is the *HM* concentration in soil samples (mg·kg^{-1}); C_B is reference concentration of *HMs* in the soil in the study area (mg·kg^{-1}), and factor 1.5 is incorporated in the relationship to account for possible variation in background data due to the lithogenic effect. Müller [58] distinguished seven classes of soil quality based on *HM* enrichment: Class 0–$I_{geo} < 0$, uncontaminated; Class 1–$0 < I_{geo} < 1$, uncontaminated to moderately; Class 2–$1 < I_{geo} < 2$, moderately; Class 3–$2 < I_{geo} < 3$, moderately to highly; Class 4–$3 < I_{geo} < 4$, highly; Class 5–$4 < I_{geo} < 5$, highly to extremely; Class 6–$I_{geo} > 5$, extremely contaminated. I_{geo} Class 0 suggests the lack of contamination, while the highest Class 6 shows the extreme enrichment of the *HMs* relative to background values.

The above reference concentration, symbolised as C_B, depends on the local natural conditions, and differs widely from one geologic unit to another. The metal background depends on the composition of the parent rock from which the soil was derived [13]. Thus, use of Clark values or off-site reference methods are obviously not appropriate, especially in assessing weakly contaminated sites [34]. The on-site reference value is more specific and sensitive in the case of trace-element soil contamination [33]. Because of the bedrock of ore-bearing dolomites in the research area, we decided to use the *HM* content in the carbonate bedrock (Table 2) as a local reference value. In addition, three more off-site reference values (Table 2) were applied to further define the role played by different sources in the topsoil contamination detected.

Table 2. HM concentrations in topsoil samples and baseline CB values (mg·kg^{-1} dry weight).

Sample Point	Weighted Arithmetic Mean (mg·kg^{-1} Dry Weight)			Statistical Parameters				
	Zn	Pb	Cd			Zn	Pb	Cd
D1	3320	1861	32.82	content	min.	68	18	0.31
D2	558	173	7.19		max.	3329	1877	37.58
D3	1251	275	8.62	weighted arithmetic mean		1351	550	13.01
D4	982	244	8.69	median		937	269	9.59
D5	619	203	7.73	standard deviation		910	535	9.83
				coefficient of variation [%]		77	123	83.29
W1	3066	454	27.57					
W2	1159	492	7.04					
W3	1032	361	4.14					
W4	2582	432	17.45			Zn	Pb	Cd
W5	1054	340	7.34	content	min.	480	118	2.34
W6	595	279	4.87		max.	3077	1035	27.66
W7	1014	254	5.66	weighted arithmetic mean		1549	410	10.43
W8	1149	331	4.43	median		1102	317	6.68
W9	287	87	2.29	standard deviation		849	232	7.53
W10	714	217	4.11	coefficient of variation [%]		1102	317	6.68
	baseline C_B values							
	Zn	Pb	Cd					
1	552	61	3.41	carbonate bedrock for Dlugoszyn area				
2	86	81	1.22	carbonate bedrock for Wilkoszyn area				
3	104	44	1.30	median for topsoil for S Poland [59]				
4	200	84	2	median for topsoil in Cracow-Silesia region [60]				
5	48	15	0.15	median for topsoil in Europe [61]				

5. Results and Discussion

The pH values of the soil in the Dlugoszyn area (Table 1) fluctuate between 4.26 and 7.24 (median 6.57), while in the Wilkoszyn area, the pH varies from 6.01 to 7.6 (median 6.77), indicating that the examined soil samples are extremely acidic to slightly alkaline, but mostly neutral. Generally, the pH value decreases with depth. The TOC values obtained for the soil range from 0.2% to 23.12% (median 1.72%) in the Dlugoszyn area and from 0.68% to 13.84% (median 2.04%) in the second study area, while TS varies from 0 to 0.14% (median 0.01%) and 0.01 to 0.12% (median 0.02%), in the two study areas, respectively (Table 1). Both parameters diminish with depth. The TOC levels observed are characteristic for rendzina skeletal soils.

The concentrations of Zn, Pb, and Cd in the soil differ significantly between the elevated (Dlugoszyn) and buried (Wilkoszyn) ore-bearing dolomite areas (Table 2). The Dlugoszyn soil samples contain 68–3329 mg·kg⁻¹ d.m. of Zn, 18–1877 mg·kg⁻¹ d.m. of Pb, and 0.31–37.58 mg·kg⁻¹ d.m. of Cd, with weighted arithmetic means 1351 mg·kg⁻¹ d.m., 550 mg kg⁻¹ d.m., and 13.01 mg·kg⁻¹ d.m., respectively. The pollution load of the same elements in the Wilkoszyn area ranges as follows: 480–3077 mg·kg⁻¹ d.m. for Zn, 118–1035 mg·kg⁻¹ d.m. for Pb, and 2.34–27.66 mg·kg⁻¹ d.m. for Cd, with weighted arithmetic means 1549 mg·kg⁻¹ d.m., 410 mg·kg⁻¹ d.m., and 10.43 mg·kg⁻¹ d.m., respectively. For the soil derived from ore-bearing dolomites occurring close to the surface in the Dlugoszyn area, the mean concentrations of the studied metals exceed 2.5-, 9- and 3.8-fold of the Zn, Pb, and Cd content in bedrock, respectively. For the soil developed in the Wilkoszyn area from deeper ore-bearing dolomites, the mean values calculated are 18.1- (for Zn), 5.1- (for Pb), and 8.6-fold (for Cd) higher than the baseline level (compare Table 2), indicating intensive soil contamination in both study areas.

The main crystalline phases in the topsoil are in the 79.5–84 wt.% (Table 3) represented by quartz (Figure 4). Goethite is abundant, especially in the Wilkoszyn area (11.5 wt.%). The large concentration of dolomite (3–5%), and some calcite (to 0.5 wt.%) were also detected. The K-feldspars (microcline and orthoclase), albite, plagioclases, and kaolinite occur there as the accessory minerals. The low-crystalline goethite (7.16Å in Figure 4), is accompanied by amorphous iron oxides or sulfides (marcasite, hematite—2.695 Å) and/or iron oxide/hydroxide. Additionally, the pyroxene (augite-diopside) or spinel Zn-Al-Fe family phases were documented by diffraction peak 4.675 Å.

Table 3. Semiquantitative mineralogical analysis of XRD results.

Mineral Name	D Area	W Area
	[%]	
Quartz	84.0	79.5
Dolomite	4.5	3.0
Goethite	0.5	11.5
Microcline	3.0	2.5
Orthoclase	3.0	1.0
Albite	2.5	2.5
Kaolinite	2.0	-
Calcite	0.5	-

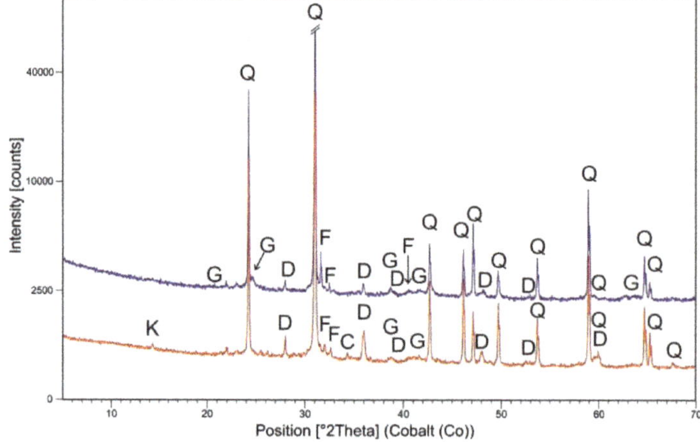

Figure 4. XRD patterns of topsoil from D-Dlugoszyn (red line) and W-Wilkoszyn (blue line) area. Symbol explanations: C—calcite, D—dolomite, F—feldspars, G—goethite, Q—quartz.

On the basis of the trace-element contents of the sampled soil, the natural concentration and the contamination from anthropogenic sources cannot be distinguished. To assess the degree of anthropogenic influence on the soils, the selected pollution indices (C_f, C_{deg}, *PLI*, and I_{geo}) were calculated.

The C_f was calculated for each soil sample relative to the bedrock value for both study areas (Table 2). The C_f of Zn, Pb, and Cd in the Dlugoszyn area is in the range of 1.0–6.0, 2.8–30.6, and 2.1–9.6, with a mean value of 2.4, 9.1, and 3.8, respectively (Table 4). This indicates considerable to very high (80% of results for Pb) and moderate to considerable (100% for Zn and 80% for Cd) topsoil contamination level (Table 5). In the Wilkoszyn area, C_f has values of 6.9–35.8, 1.1–6.1 and 1.9–22.6 for Zn, Pb, and Cd, respectively, with a mean value of 14.8, 4.1, and 7.0, respectively (Table 4). Such C_f levels point to considerable to very high contamination degree with Zn (100% of results) and Cd (90%), and moderate to considerable pollution with Pb (80%) (Table 5).

Table 4. Assessment of *HM* pollution indices in studied topsoil in relation to the local carbonate bedrock in the D-Dlugoszyn and W-Wilkoszyn areas as a background value.

Sample Point	C_f			C_{deg}	*PLI*	I_{geo}		
	Zn	**Pb**	**Cd**			**Zn**	**Pb**	**Cd**
D1	**6.0**	**30.6**	**9.6**	46.2	**12.1**	2.0	**4.3**	2.7
D2	1.0	2.8	2.1	6.0	1.8	−0.6	0.9	0.5
D3	2.3	**4.5**	2.5	9.3	3.0	0.6	1.6	0.8
D4	1.8	**4.0**	2.5	8.3	2.6	0.2	1.4	0.8
D5	1.1	**3.3**	2.3	6.7	2.0	−0.4	1.2	0.6
W1	35.8	**5.6**	22.6	64.0	16.5	4.6	1.9	3.9
W2	13.5	**6.1**	5.8	25.4	7.8	3.2	2.0	1.9
W3	12.0	**4.5**	3.4	19.9	5.7	3.0	1.6	1.2
W4	30.1	**5.3**	14.3	49.8	13.2	4.3	1.8	3.3
W5	12.3	**4.9**	6.0	23.2	7.1	3.0	1.7	2.0
W6	6.9	**3.4**	4.0	14.4	4.6	2.2	1.2	1.4
W7	11.8	**3.1**	4.6	19.6	5.6	3.0	1.1	1.6
W8	13.4	**4.1**	3.6	21.1	5.8	3.2	1.4	1.3
W9	3.3	1.1	1.9	6.3	1.9	1.2	−0.5	0.3
W10	8.3	2.7	3.4	14.4	4.2	2.5	0.8	1.2

In bold—strongly contaminated.

The C_{deg} index determines the degree of overall contamination of a particular sample in the study areas. Based on the C_{deg} (see Tables 4 and 5), 80% of the calculated values for the Dlugoszyn samples are recognised as moderately contaminated, while 20% are very highly contaminated. In the Wilkoszyn samples, 60% indicate very high (30%) and moderate (10%) pollution of the topsoil (Table 5).

The PLI indicates deterioration of the soil quality due to metal pollution. In the current study, the PLI obtained for the Dlugoszyn samples indicates moderate (20% of results), moderate to high (40%), high (20%) and extreme (20%) topsoil pollution (Table 5), with values ranging from 1.8 to 12.1, and mean equal to 4.3 (Table 4). In the Wilkoszyn samples, the PLI values range from 1.9 to 16.5, with a mean of 7.2 (Table 4) pointing to extremely (70% of results), very high (20%), and moderately (10%) polluted topsoil (Table 5).

The pollution indices indicate the moderately to extremely high level of topsoil contamination and reveal a difference in the level of topsoil contamination between the study areas (Table 5).

Table 5. Percentage of class distribution of topsoil pollution for indices considered. D—Dlugoszyn area, W—Wilkoszyn area.

Pollution Index	Classes of Soil Quality	Studied Area					
		D			W		
		Zn	Pb	Cd	Zn	Pb	Cd
C_f [%]	very high contamination	0	20	20	90	10	20
	considerable contamination	20	60	0	10	70	70
	moderate contamination	80	20	80	0	20	10
	low contamination	0	0	0	0	0	0
C_{deg} [%]	very high contamination		20			30	
	considerable contamination		0			60	
	moderate contamination		80			10	
	low contamination		0			0	
PLI [%]	extremely high pollution		20			70	
	very high pollution		0			20	
	high pollution		20			0	
	moderate to high pollution		40			0	
	moderate pollution		20			10	
	unpolluted		0			0	
I_{geo} [%]	Class 6 extremely polluted	0	0	0	0	0	0
	Class 5 highly to extremely polluted	0	20	0	20	0	0
	Class 4 heavy polluted	0	0	0	50	0	0
	Class 3 moderately to heavily polluted	20	0	20	20	10	30
	Class 2 moderately polluted	0	60	0	10	70	60
	Class 1 unpolluted to moderately polluted	40	20	80	0	10	10
	Class 0 unpolluted	40	0	0	0	10	0

The I_{geo} ranges from −0.6 to 2.0 for Zn, with a mean value of 0.4; 0.9 to 4.3 for Pb, with a mean value of 1.9, and 0.5 to 2.7 for Cd, with a mean value of 1.1 in the Dlugoszyn area (Table 4). The values of I_{geo} for *HM*s decrease in the sequence Pb > Cd > Zn. The results classify the area as highly polluted with Pb and moderately polluted with Zn and Cd (Table 5). In contrast, in the Wilkoszyn area, the I_{geo} varies widely from 1.2–4.6 for Zn, with mean value 3.0; −0.5–2.0 for Pb, with mean value 1.3 and 0.3–3.9 for Cd, with mean value 1.8 (Table 4). The *HM* pollution follows the order Zn > Cd > Pb. Using I_{geo}, we can determine the potential source of pollution. The I_{geo} values obtained for the elevated ore-bearing dolomites in the Dlugoszyn area indicate mainly baseline levels of pollution for Zn and Cd, as 80% Zn and 80% Cd belong to Classes 0 and 1. The only exception is pollution with Pb (see Table 5). Most of the Zn and Cd are derived from geogenic sources, and most of the Pb (80%) is derived from anthropogenic sources. As such, in the Dlugoszyn area, the deterioration in soil quality is due to Pb. In the Wilkoszyn area, with buried ore-bearing dolomites, moderate to heavy pollution by Pb and Cd is observed, as 80% Pb and 90% Cd belong to Class 2 and 3 (Table 5). Heavy to extreme pollution by Zn is identified, as 70% Zn belongs to Class 4 and 5 (Table 5). Based on these results and the fact that concentrations of *HM*s decrease with depth, the *HM*s in the topsoil of the Wilkoszyn location is of anthropogenic origin. The graphical model of *HM* immissions into the topsoil of the Dlugoszyn and Wilkoszyn areas is presented in the Figure 5, while the complete set of soil profiles studied is put into the Supplementary Materials (Figure S1).

When calculating pollution indices, the selection of an appropriate reference value is an important and complicated issue. Opinions about the best value to use are divided. Some researchers favour the trace-element contents in the deeper soil horizon [32,36,62] or the average values in soil from the study area [30,63] as correct. Others consider the trace-element contents in the bedrock [64,65] as best. The accuracy and sensitivity of the on-site reference value in identifying trace-element soil contamination was evaluated by Desaules [33]. The specific geological setting of both study areas forced us to calculate the selected pollution indices with reference to the bedrock.

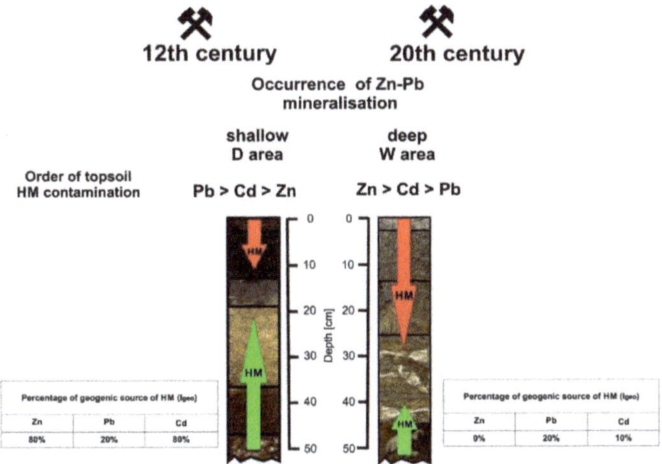

Figure 5. Graphical model of anthropogenic (red arrows) and geogenic (green arrows) *HM* immissions into the topsoil of the D-Dlugoszyn and W-Wilkoszyn areas.

We also take off-site reference values defined as the median for topsoil in (1) the Cracow-Silesia region [60], (2) S Poland [59] and (3) Europe [61] into consideration. Outcomes of these examination are listed in Tables 6–8. In applying the off-site reference values to topsoil in the area where shallow ore-bearing dolomites occur (the Dlugoszyn area), a significantly higher degree of contamination is observed. Additionally, anthropogenic sources are mainly responsible for the *HM* presence in the soil. The percentage of class distribution shows (Tables 6–8) that 80–100% of I_{geo} values belong to classes 2–6. The only exception refers to Pb when the S Poland median value is used. In the case of deeper situated Zn-Pb mineralization (the Wilkoszyn area), comparable results were obtained using the off-site and on-site reference values. Significant differences were noted when the reference value for European topsoil was considered. Using the latter reference value, the extremely high soil contamination (C_f, C_{deg}) here is attributed mainly to human activity (I_{geo}; see Table 8). In our opinion, and especially in the case where the ore-bearing dolomite is near surface, use of the local reference value to assess topsoil contamination provides more reliable results than use of off-site reference values.

Table 6. Assessment of *HM* pollution indices in studied topsoil in relation to the median for topsoil in the Cracow-Silesia region as a background value [60].

Sample Point	C_f			C_{deg}	PLI	I_{geo}		
	Zn	Pb	Cd			Zn	Pb	Cd
D1	31.9	42.3	25.2	99.5	32.4	4.4	4.8	4.1
D2	5.4	3.9	5.5	14.8	4.9	1.8	1.4	1.9
D3	12.0	6.3	6.6	24.9	7.9	3.0	2.1	2.1
D4	9.4	5.6	6.7	21.7	7.1	2.7	1.9	2.2
D5	5.9	4.6	5.9	16.5	5.5	2.0	1.6	2.0
W1	29.5	10.3	21.2	61.0	18.6	4.3	2.8	3.8
W2	11.1	11.2	5.4	27.7	8.8	2.9	2.9	1.9
W3	9.9	8.2	3.2	21.3	6.4	2.7	2.5	1.1
W4	24.8	9.8	13.4	48.1	14.8	4.0	2.7	3.2
W5	10.1	9.1	5.6	24.9	8.0	2.8	2.6	1.9
W6	5.7	6.3	3.7	15.8	5.1	1.9	2.1	1.3
W7	9.7	5.8	4.4	19.9	6.3	2.7	1.9	1.5
W8	11.0	7.5	3.4	22.0	6.6	2.9	2.3	1.2
W9	2.8	2.0	1.8	6.5	2.1	0.9	0.4	0.2
W10	6.9	4.9	3.2	15.0	4.7	2.2	1.7	1.1

In bold—strongly contaminated.

Table 7. Assessment of *HM* pollution indices in studied topsoil in relation to the median for topsoil in S Poland as a background value [59].

Sample Point	C_f			C_{deg}	PLI	I_{geo}		
	Zn	**Pb**	**Cd**			**Zn**	**Pb**	**Cd**
D1	**16.6**	**22.1**	**16.4**	55.2	**18.2**	3.5	3.9	3.5
D2	2.8	2.1	**3.6**	8.4	2.7	0.9	0.5	1.3
D3	6.3	3.3	4.3	13.8	4.5	2.1	1.1	1.5
D4	4.9	2.9	4.3	12.2	4.0	1.7	1.0	1.5
D5	3.1	2.4	3.9	9.4	3.1	1.0	0.7	1.4
W1	5.4	13.8	15.3	34.5	10.5	3.4	1.9	3.2
W2	5.9	3.5	5.8	15.2	4.9	2.0	2.0	1.2
W3	4.3	2.1	5.2	11.5	3.6	1.8	1.5	0.5
W4	5.1	8.7	12.9	26.8	8.3	3.1	1.8	2.5
W5	4.8	3.7	5.3	13.7	4.5	1.8	1.7	1.3
W6	3.0	3.3	2.4	8.7	2.9	1.0	1.1	0.7
W7	5.1	3.0	2.8	10.9	3.5	1.8	1.0	0.9
W8	5.7	3.9	2.2	11.9	3.7	1.9	1.4	0.6
W9	1.4	1.0	1.1	3.6	1.2	−0.1	−0.5	−0.4
W10	3.6	2.6	2.1	8.2	2.7	1.3	0.8	0.5

In bold—strongly contaminated.

Table 8. Assessment of *HM* pollution indices in studied topsoil in relation to the median for European topsoil as a background value [61].

Sample Point	C_f			C_{deg}	PLI	I_{geo}		
	Zn	**Pb**	**Cd**			**Zn**	**Pb**	**Cd**
D1	69.2	124.0	218.8	412.0	123.4	5.5	6.4	7.2
D2	11.6	11.5	47.9	71.1	18.6	3.0	2.9	5.0
D3	26.1	18.4	57.4	101.9	30.2	4.1	3.6	5.3
D4	20.5	16.3	57.9	94.7	26.8	3.8	3.4	5.3
D5	12.9	13.5	51.5	78.0	20.8	3.1	3.2	5.1
W1	63.9	30.3	183.8	278.0	70.9	5.4	4.3	6.9
W2	24.2	32.8	46.9	103.9	33.4	4.0	4.5	5.0
W3	21.5	24.1	27.6	73.2	24.3	3.8	4.0	4.2
W4	53.8	28.8	116.4	199.0	56.5	5.2	4.3	6.3
W5	21.9	26.7	48.9	97.5	30.6	3.9	4.2	5.0
W6	12.4	18.6	32.5	63.5	19.6	3.0	3.6	4.4
W7	21.1	16.9	37.8	75.8	23.8	3.8	3.5	4.7
W8	23.9	22.0	29.6	75.5	25.0	4.0	3.9	4.3
W9	6.0	5.8	15.3	27.0	8.1	2.0	1.9	3.3
W10	14.9	14.5	27.4	56.7	18.1	3.3	3.3	4.2

In bold—strongly contaminated.

Diatta et al. [9] observed similar results, with significantly high *HM* levels in soils impacted by the Miasteczko Slaskie Zn smelter, in part of the Upper Silesia Industrial Region. The mean Zn, Pb, and Cd contents in mg·kg^{-1} reported are 1062.97, 781.91, and 12.32, respectively. Using the off-site reference value to I_{geo}, C_f, and C_{deg} calculations, extremely high contamination was observed for Zn, Cd, and Pb, in ascending order [9]. Our findings differ slightly in terms of the influence of particular *HM*s. In our opinion, the difference between the orders of pollutant intensity acquired for individual sites resulted primarily from the local industrial land use and the kind of reference value.

The historical mining of Zn-Pb ores in the Dlugoszyn area was based on the dominant mineral, galena (Pb sulfide), which occurs in shallow ore bodies which are easily recognised. Moreover, Pb has a relatively low melting point, allowing for ease of processing [41]. As such, ore mining in the area started in the early medieval period and resulted in the most severe Pb pollution in the Dlugoszyn topsoil. The levels of Zn and Cd are derived from the metal-rich bedrock.

The very high Zn and Pb contamination observed in the Wilkoszyn area probably results from the 20th century underground Zn-Pb ore mining and processing, when galena, Zn sulfides, and nonsulfides were mined and processed in the area. As such, it seems probable that recent industrial operations are a principal cause of the high levels of Zn and Pb accumulated in the Wilkoszyn topsoil.

Generally, soil samples from arable fields and home gardens across the Upper Silesia Industrial Region are characterised by high levels of toxic *HM*s. This is reflected in the *HM* concentrations in locally cultivated vegetables, which are well above the permissible levels [11]. Both study areas are used as peri-urban agricultural lands. The very high topsoil *HM* levels reported in this paper are not as harmful due to the pH value (Table 1) observed in studied soil. The *HM*s are regarded as highly soluble and more toxic in an acidic soil environment [66,67].

6. Conclusions

The application of various pollution indices (C_f, C_{deg}, *PLI*, and I_{geo}) and the local background value C_B enable the detection of Zn, Pb, and Cd pollution in soils. Depending on the index used, we determined very high to considerable (according to the classification proposed by [55]) topsoil contamination. The influence intensity of particular metals differs between studied areas. In the case of shallow historical mining (the Dlugoszyn area), it follows the order Pb > Cd > Zn, while in the place of deep mining and processing (the Wilkoszyn area) it decreases in the sequence Zn > Cd > Pb. With the geo-accumulation index, it is possible to discriminate between natural and anthropogenic *HM*s in the soils developed over the ore-bearing formation in the two areas. Based on I_{geo} values, we consider that the *HM* content in the topsoil developed over the shallowly occurring ore bodies in dolomites (the Dlugoszyn area) is mainly (60%) connected with the natural presence of metals. In the Wilkoszyn area where ore bodies are more deeply buried, 90% of the *HM* load is related to anthropogenic sources. An accurate assessment of soil quality, based on *HM* content, is possible only with the combined use of various pollution indices.

Supplementary Materials: The following are available online at http://www.mdpi.com/2075-163X/10/12/1140/s1, Table S1. Zn, Pb, Cd content mg. kg^{-1} d. m. determined in the internal laboratory standard for soil material (n = 20), Figure S1. The complete set of soil profiles for D-Dlugoszyn and W-Wilkoszyn areas.

Author Contributions: Conceptualization: K.S., L.T.; Methodology: K.S.; Investigation: K.S., T.C., T.H., M.O., J.Z.; Formal analysis: K.S.; Resources: T.C.; Writing—original draft: K.S.; Writing—review & editing: L.T.; Funding acquisition: L.T.; Supervision: L.T.; Data curation and Visualization: K.S. All authors have read and agreed to the published version of the manuscript.

Funding: This study was funded by the University of Silesia, Institute of Earth Sciences (WNP/INOZ/2020_ZB32).

Acknowledgments: This study was undertaken in the framework of the activities of the University of Silesia in Katowice and was funded by the University of Silesia, Institute of Earth Sciences (WNP/INOZ/2020_ZB32). We wish to thank Tomasz Krzykawski from the University of Silesia in Katowice for his assistance with XRD analysis. We are grateful to Pádhraig S. Kennan from University College Dublin for valuable remarks and suggestions. Two anonymous reviewers are thanked for improving the quality of the paper.

Conflicts of Interest: The authors declare no conflict of interest.

References

1. Norgate, T.E.; Jahanshahi, S.; Rankin, W.J. Assessing the environmental impact of metal production processes of metal production processes. *J. Clean. Prod.* **2007**, *15*, 838–848. [CrossRef]

2. Chrastný, V.; Vaněk, A.; Teper, L.; Cabala, J.; Procházka, J.; Pechar, L.; Drahota, P.; Penížek, V.; Komárek, M.; Novák, M. Geochemical position of Pb, Zn and Cd in soils near the Olkusz mine/smelter, South Poland: Effects of land use, type of contamination and distance from pollution source. *Environ. Monit. Assess.* **2012**, *184*, 2517–2536. [CrossRef]

3. Vaněk, A.; Chrastný, V.; Komárek, M.; Penížek, V.; Teper, L.; Cabala, J.; Drábek, O. Geochemical position of thallium in soils from a smelter-impacted area. *J. Geochem. Explor.* **2013**, *124*, 176–182. [CrossRef]

4. Kalinovic, T.S.; Serbula, S.M.; Radojevic, A.A.; Kalinovic, J.V.; Steharnik, M.M.; Petrovic, J.V. Elder, linden and pine biomonitoring ability of pollution emitted from the copper smelter and the tailings ponds. *Geoderma* **2016**, *262*, 266–275. [CrossRef]

5. Ghayoraneh, M.; Qishlaqi, A. Concentration, distribution and speciation of toxic metals in soils along a transect around a Zn/Pb smelter in the northwest of Iran. *J. Geochem. Explor.* **2017**, *180*, 1–14. [CrossRef]

6. Varol, M. Assessment of heavy metal contamination in sediments of the Tigris River (Turkey) using pollution indices and multivariate statistical techniques. *J. Hazard. Mater.* **2011**, *195*, 355–364. [CrossRef] [PubMed]

7. Ogbeibu, A.E.; Omoigberale, M.O.; Ezenwa, I.M.; Eziza, J.O.; Igwe, J.O. Using pollution load index and geoaccumulation index for the assessment of heavy metal pollution and sediment quality of the Benin River, Nigeria. *Nat. Environ.* **2014**, *2*, 1–9. [CrossRef]

8. Loska, K.; Wiechuła, D.; Korus, I. Metal contamination of farming soils affected by industry. *Environ. Int.* **2004**, *30*, 159–165. [CrossRef]

9. Diatta, J.B.; Chudzinska, E.; Wirth, S. Assessment of heavy metal contamination of soils impacted by a zinc smelter activity. *J. Elementol.* **2008**, *13*, 5–16.

10. Li, Z.; Ma, Z.; van der Kuijp, T.J.; Yuan, Z.; Huang, L. A review of soil heavy metal pollution from mines in China: Pollution and health risk assessment. *Sci. Total Environ.* **2014**, *468–469*, 843–853. [CrossRef]

11. Dziubanek, G.; Piekut, A.; Rusin, M.; Baranowska, R.; Hajok, I. Contamination of food crops grown on soils with elevated heavy metals content. *Ecotoxicol. Environ. Saf.* **2015**, *118*, 183–189. [CrossRef] [PubMed]

12. Ettler, V. Soil contamination near non-ferrous metal smelters: A review. *Appl. Geochem.* **2016**, *64*, 56–74. [CrossRef]

13. Nouri, M.; Haddioui, A. Assessment of metals contamination and ecological risk in ait Ammar abandoned iron mine soil, Morocco. *Ekológia* **2016**, *35*, 32–49. [CrossRef]

14. Obiora, S.C.; Chukwu, A.; Davies, T.C. Heavy metals and health risk assessment of arable soils and food crops around Pb-Zn mining localities in Enyigba, southeastern Nigeria. *J. Afr. Earth Sci.* **2016**, *116*, 182–189. [CrossRef]

15. Xue, S.; Shi, L.; Wu, C.; Wu, H.; Qin, Y.; Pan, W.; Hartley, W.; Cui, M. Cadmium, lead, and arsenic contamination in paddy soils of a mining area and their exposure effects on human HEPG2 and keratinocyte cell-lines. *Environ. Res.* **2017**, *156*, 23–30. [CrossRef] [PubMed]

16. Liu, J.; Wang, J.; Xiao, T.; Bao, Z.; Lippold, H.; Luo, X.; Yin, M.; Ren, J.; Chen, Y.; Linghu, W. Geochemical dispersal of thallium and accompanying metals in sediment profiles from a smelter-impacted area in South China. *Appl. Geochem.* **2018**, *88*, 239–246. [CrossRef]

17. Wu, J.; Long, J.; Liu, L.; Li, J.; Liao, H.; Zhang, M.; Zhao, C.; Wu, Q. Risk assessment and source identification of toxic metals in the agricultural soil around a Pb/Zn mining and smelting area in Southwest China. *Int. J. Environ. Res. Public Health* **2018**, *15*, 1838. [CrossRef]

18. Wiggenhauser, M.; Bigalke, M.; Imseng, M.; Müller, M.; Keller, A.; Murphy, K.; Kreissig, K.; Rehkämper, M.; Wilcke, W.; Frossard, E. Cadmium isotope fractionation in soil-wheat systems. *Environ. Sci. Technol.* **2016**, *50*, 9223–9231. [CrossRef]

19. Dinis, L.; Savard, M.M.; Gammon, P.; Bégin, C.; Vaive, J. Influence of climatic conditions and industrial emissions on spruce tree-ring Pb isotopes analyzed at ppb concentrations in the Athabasca oil sands region. *Dendrochronologia* **2016**, *37*, 96–106. [CrossRef]

20. Wei, X.; Lyu, S.; Yu, Y.; Wang, Z.; Liu, H.; Pan, D.; Chen, J. Phylloremediation of air pollutants: Exploiting the potential of plant leaves and leaf-associated microbes. *Front. Plant Sci.* **2017**, *28*, 1318. [CrossRef]

21. Liu, J.; Wei, X.; Zhou, Y.; Tsang, D.C.W.; Bao, Z.; Yin, M.; Lippold, H.; Yuan, W.; Wang, J.; Feng, Y.; et al. Thallium contamination, health risk assessment and source apportionment in common vegetables. *Sci. Total Environ.* **2020**, *703*, 135547. [CrossRef] [PubMed]

22. Wang, J.; Jiang, Y.; Sun, J.; She, J.; Yin, M.; Fang, F.; Xiao, T.; Song, G.; Liu, J. Geochemical transfer of cadmium in river sediments near a lead-zinc smelter. *Ecotoxicol. Environ. Saf.* **2020**, *196*, 110529. [CrossRef] [PubMed]

23. Durkalec, M.; Szkoda, J.; Kolacz, R.; Opalinski, S.; Nawrocka, A.; Zmudzki, J. Bioaccumulation of Lead, Cadmium and Mercury in Roe Deer and Wild Boars from areas with different levels of toxic metal pollution. *Int. J. Environ. Res.* **2015**, *9*, 205–212. [CrossRef]

24. Kapusta, P.; Sobczyk, Ł. Effects of heavy metal pollution from mining and smelting on enchytraeid communities under different land management and soil conditions. *Sci. Total Environ.* **2015**, *536*, 517–526. [CrossRef]

25. Weissmannová, H.D.; Pavlovský, J. Indices of soil contamination by heavy metals—Methodology of calculation for pollution assessment (minireview). *Environ. Monit. Assess.* **2017**, *189*, 616. [CrossRef]

26. Barbieri, M. The importance of enrichment factor (EF) and geoaccumulation index (I_{geo}) to evaluate the soil contamination. *J. Geol. Geophys.* **2016**, *5*, 237. [CrossRef]

27. Baran, A.; Wieczorek, J. Application of geochemical and ecotoxicity indices for assessment of heavy metals content in soils. *Arch. Environ. Prot.* **2015**, *41*, 54–63. [CrossRef]

28. Baran, A.; Wieczorek, J.; Mazurek, R.; Urbański, K.; Klimkowicz-Pawlas, A. Potential ecological risk assessment and predicting zinc accumulation in soils. *Environ. Geochem. Health* **2018**, *40*, 435–450. [CrossRef]

29. Wieczorek, J.; Baran, A.; Urbański, K.; Mazurek, R.; Klimowicz-Pawlas, A. Assessment of the pollution and ecological risk of lead and cadmium in soils. *Environ. Geochem. Health* **2018**, *40*, 2325–2342. [CrossRef]

30. Zang, Z.; Li, Y.; Li, H.; Guo, Z.; Zhang, R. Spatiotemporal variation and pollution assessment of Pb/Zn from smelting activities in China. *Int. J. Environ. Res. Public Health* **2020**, *17*, 1968. [CrossRef]

31. Mazurek, R.; Kowalska, J.B.; Gąsiorek, M.; Zadrożny, P.; Wieczorek, J. Pollution indices as comprehensive tools for evaluation of the accumulation and provenance of potentially toxic elements in soils in Ojców National Park. *J. Geochem. Explor.* **2019**, *201*, 13–30. [CrossRef]

32. Blaser, P.; Zimmermann, S.; Luster, J.; Shotyk, W. Critical examination of trace element enrichments and depletions in soils: As, Cr, Cu, Ni, Pb, and Zn in Swiss forest soils. *Sci. Total Environ.* **2000**, *249*, 257–280. [CrossRef]

33. Desaules, A. Critical evaluation of soil contamination assessment methods for trace metals. *Sci. Total Environ.* **2012**, *426*, 120–131. [CrossRef]

34. Wu, J.; Teng, Y.; Lu, S.; Wang, Y.; Jiao, X. Evaluation of soil contamination indices in a mining area of Jiangxi, China. *PLoS ONE* **2014**, *9*, e112917. [CrossRef]

35. Maanan, M.; Saddik, M.; Maanan, M.; Chaibi, M.; Assobhei, O.; Zourarah, B. Environmental and ecological risk assessment of heavy metals in sediments of Nador lagoon, Morocco. *Ecol. Indic.* **2015**, *48*, 616–626. [CrossRef]

36. Rachwał, M.; Kardel, K.; Magiera, T.; Bens, O. Application of magnetic susceptibility in assessment of heavy metal contamination of Saxonian soil (Germany) caused by industrial dust deposition. *Geoderma* **2017**, *295*, 10–21. [CrossRef]

37. Leach, D.L.; Viets, J.G.; Powell, J.W. Textures of ores from the Silesian-Cracow zinc-lead deposits, Poland: Clues to the ore-forming environment. *Pol. Geol. Inst. Pap.* **1996**, *154*, 37–50.

38. Szuwarzynski, M. Ore bodies in the Silesian-Cracow Zn-Pb ore district, Poland. *Pol. Geol. Inst. Pap.* **1996**, *154*, 9–24.

39. Górecka, E. Mineral sequence development in the Zn-Pb deposits of the Silesian-Cracow area, Poland. *Pol. Geol. Inst. Pap.* **1996**, *154*, 37–50.

40. Molenda, D. *Mining of Lead and Silver in the Silesia-Cracow Region Up to the Middle of the 16th Century*; Ossolineum: Wrocław, Poland, 1963.

41. Panek, S. The History of Lead and Zinc Ore Mining in Jaworzno-Trzebinia-Chrzanow. Unpublished work, Jaworzno. 1995.

42. Cabala, J.; Sutkowska, K. Past exploitation and processing of Zn-Pb ore influence on the industrial soil minerals composition. Olkusz and Jaworzno district. *Pr. Nauk. Inst Górnictwa Polit Wrocł Studia Mater.* **2006**, *117*, 13–22.

43. Sutkowska, K.; Teper, L. Anthropogenic forms of urban post-mining landscape by example of Jaworzno. *Prz. Górn.* **2012**, *68*, 70–77.

44. Teper, E. Dust-particle migration around flotation tailings ponds: Pine needles as passive samplers. *Environ. Monit. Assess.* **2009**, *154*, 383–391. [CrossRef] [PubMed]

45. Aleksander-Kwaterczak, U.; Helios-Rybicka, E. Contaminated sediments as a potential source of Zn, Pb, and Cd for a river system in the historical metalliferous ore mining and smelting industry area of South Poland. *J. Soils Sediments* **2009**, *9*, 13. [CrossRef]

46. Aleksander-Kwaterczak, U.; Ciszewski, D.; Szarek-Gwiazda, E.; Kwandrans, J.; Wilk-Woźniak, E.; Waloszek, A. The influence of historical activity of the Zn-Pb ore mine in Chrzanów on the aquatic environment quality of the Matylda valley. *Górnictwo Geol.* **2010**, *5*, 21–30.

47. Chlopecka, A.; Bacon, J.R.; Wilson, M.J.; Kay, J. Forms of Cadmium, Lead and Zinc in contaminated soils from Southwest Poland. *J. Environ. Qual.* **1996**, *25*, 69–79. [CrossRef]

48. Cabala, J.; Teper, L. Metalliferous constituents of rhizosphere soils contaminated by Zn-Pb mining in southern Poland. *Water Air Soil Pollut.* **2007**, *178*, 351–362. [CrossRef]

49. Sutkowska, K.; Teper, L.; Vaněk, A.; Czech, T.; Baran, A. Effect of historical zinc processing on soil: A case study in Southern Poland. In Proceedings of the 3rd World Congress on New Technologies (NewTech'17), Rome, Italy, 6–8 June 2017; p. 110. [CrossRef]

50. Vaněk, A.; Chrastný, V.; Teper, L.; Cabala, J.; Penížek, V.; Komárek, M. Distribution of thallium and accompanying metals in tree rings of Scots pine (*Pinus sylvestris* L.) from a smelter-affected area. *J. Geochem. Explor.* **2011**, *108*, 73–80. [CrossRef]

51. Pająk, M.; Halecki, W.; Gąsiorek, M. Accumulative response of Scots pine (*Pinus sylvestris* L.) and silver birch (Betula pendula Roth) to heavy metals enhanced by Pb-Zn ore mining and processing plants: Explicitly spatial considerations of ordinary kriging based on a GIS approach. *Chemosphere* **2017**, *168*, 851–859. [CrossRef]

52. Augustyniak, M.; Orzechowska, H.; Kędziorski, A.; Sawczyn, T.; Doleżych, B. DNA damage in grasshoppers' larvae—Comet assay in environmental approach. *Chemosphere* **2014**, *96*, 180–187. [CrossRef]

53. Dmowski, K.; Rossa, M.; Kowalska, J.; Krasnodębska-Ostręga, B. Thallium in spawn, juveniles, and adult common toads (Bufo bufo) living in the vicinity of a zinc-mining complex, Poland. *Environ. Monit. Assess.* **2015**, *187*, 4141. [CrossRef]

54. Kurek, S.; Sobczyński, P.; Szuwarzyński, M.; Wojnar, E. The characteristics of the Chrzanow region deposits. *Pol. Geol. Inst. Pap.* **1977**, *45*, 45–71.

55. Håkanson, L. An ecological risk index for aquatic pollution controls sediment logical approach. *Water Res.* **1980**, *14*, 975–1001. [CrossRef]

56. Tomlinson, D.L.; Wilson, J.G.; Harris, C.R.; Jeffrey, D.W. Problems in the assessment of heavy-metal levels in estuaries and the formation of a pollution index. *Helgol. Wiss. Meeresunters.* **1980**, *33*, 566–575. [CrossRef]

57. Zhang, C.; Qiao, Q.; Piper, D.; Huang, B. Assessment of heavy metal pollution from a Fe-smelting plant in urban river sediments using environmental magnetic and geochemical methods. *Environ. Pollut.* **2011**, *159*, 3057–3070. [CrossRef] [PubMed]

58. Müller, G. Index of geo-accumulation in sediments of the Rhine River. *Geo. J.* **1969**, *2*, 108–118.

59. Detailed Geochemical Map of Upper Silesia. *Sheet: Sławków, Olkusz, Nowa Góra, Myślachowice, Chrzanów, Dabrowa Górnicza, Strzemieszyce, Jaworzno and Libiąż*; Pasieczna, A., Ed.; Polish Geological Institute: Warszawa, Poland, 2008.

60. Lis, J.; Pasieczna, A. *Geochemical Atlas of Upper Silesia 1:200,000*; Publications of the Polish Geological Institute: Warszawa, Poland, 1995.

61. Salminen, R. (Ed.) *Geochemical Atlas of Europe 2005 Part 1 and 2*; Geol. Survey of Finland: Espoo, Finland, 2005.

62. Sutherland, R.A.; Tolosa, C.A.; Tack, F.M.G.; Verloo, M.G. Characterization of Selected Element Concentrations and Enrichment Ratios in Background and Anthropogenically Impacted Roadside Areas. *Arch. Environ. Contam. Toxicol.* **2000**, *38*, 428–438. [CrossRef] [PubMed]

63. Gupta, S.; Jena, V.; Matic, N.; Kapralova, V.; Solanki, J.S. Assessment of geo-accumulation index of heavy metal and source of contamination by multivariate factor analysis. *Int. J. Hazard. Mater.* **2014**, *2*, 18–22.

64. Szuszkiewicz, M.; Łukasik, A.; Magiera, T.; Mendakiewicz, M. Combination of geo- pedo- and technogenic magnetic and geochemical signals in soil profiles Diversification and its interpretation: A new approach. *Environ. Pollut.* **2016**, *214*, 464–477. [CrossRef]

65. Szuszkiewicz, M.; Petrovsky, E.; Łukasik, A.; Gruba, P.; Grison, H.; Szuszkiewicz, M.M. Technogenic contamination or geogenic enrichment in Regosols and Leptosols? Magnetic and geochemical imprints on topsoil horizons. *Geoderma* **2021**, *381*, 114685. [CrossRef]

66. Kabata-Pendias, A.; Pendias, H. *Biogeochemistry of Trace Elements*; PWN: Warszawa, Poland, 1999.
67. Romero-Baena, A.J.; González, I.; Galán, E. Soil pollution by mining activities in Andalusia (South Spain)—The role of mineralogy and geochemistry in three case studies. *J. Soils Sediments* **2018**, *18*, 2231–2247. [CrossRef]

Publisher's Note: MDPI stays neutral with regard to jurisdictional claims in published maps and institutional affiliations.

 minerals

Article

Feasibility of a Chemical Washing Method for Treating Soil Enriched with Fluorine Derived from Mica

Dong-Jun Baek [1], Ye-Eun Kim [2], Moon-Young Jung [2], Hye-On Yoon [3] and Jinsung An [1,2,*]

[1] Department of Environment Safety System Engineering, Semyung University, 65 Semyung-ro, Jecheon-si 27136, Korea; hwwdong123@naver.com

[2] Department of Biological & Environmental Engineering, Semyung University, 65 Semyung-ro, Jecheon-si 27136, Korea; kimyeeun000123@naver.com (Y.-E.K.); myjung@semyung.ac.kr (M.-Y.J.)

[3] Korea Basic Science Institute, 145 Anam-ro, Seongbuk-gu, Seoul 02841, Korea; dunee@kbsi.re.kr

* Correspondence: jsan@semyung.ac.kr; Tel.: +82-43-649-1335; Fax: +82-43-649-1779

Abstract: High levels of fluorine in soil may pose health risks and require remediation. In this study, the feasibility of using a practical chemical washing method for the removal of fluorine from an enriched soil was evaluated. The chemical washing procedures were optimized through experimental analyses of various washing solutions and washing conditions (i.e., washing solution concentration, solid–liquid ratio, agitation speed, and reaction time). Additionally, the effects of techniques for improving the washing efficiency, such as ultrasonic washing, aeration, and multi-stage washing, were evaluated. Herein, among all applied methodologies, the maximum washing efficiency achieved for the total fluorine present in soil was only 6.2%, which indicated that chemical washing was inefficient in remediating this particular soil. Further sequential extraction analysis showed that the fluorine in this soil was present in a chemically stable form (residual fraction), possibly because of the presence of mica minerals. It was demonstrated that chemical washing may not be effective for remediating soils containing such chemically stable forms of fluorine. In these cases, other physical-based remediation technologies or risk management approaches may be more suitable.

Keywords: natural fluorine-enriched soil; natural sources; soil remediation; chemical extraction resistance; low washing efficiency

Citation: Baek, D.-J.; Kim, Y.-E.; Jung, M.-Y.; Yoon, H.-O.; An, J. Feasibility of a Chemical Washing Method for Treating Soil Enriched with Fluorine Derived from Mica. *Minerals* **2021**, *11*, 134. https://doi.org/10.3390/min11020134

Academic Editor: Ana Romero-Freire

Received: 1 December 2020

Accepted: 26 January 2021

Published: 29 January 2021

Publisher's Note: MDPI stays neutral with regard to jurisdictional claims in published maps and institutional affiliations.

1. Introduction

Fluorine, an element with atomic number 9, has high electronegativity and thus can readily react with other elements, such as Fe, Al, and Ca. Therefore, fluorine mostly exists as compounds in the natural state, and it is difficult to find fluorine in natural settings as a single element [1,2]. Human activities (e.g., fertilizer use, coal use, and aluminum and steel industrial activities) have resulted in the accumulation of fluorine in soil. Coal combustion releases hydrogen fluoride (HF), silicon tetrafluoride (SiF_4), and carbon tetrafluoride (CF_4), which can be accumulated in the soil environment [3,4]. Moreover, fluorine levels in soil can be elevated by fluorine-containing rocks (natural origin) [5,6]. Studies have shown that fluorine-containing mica, fluorite (CaF_2), and apatite can contribute to the accumulation of fluorine in soil and groundwater through weathering [5–12].

High intake of fluorine is known to cause dental fluorosis, skeletal fluorosis, and osteoporosis [13–15]. The World Health Organization (WHO) [16] has reported that the intake and inhalation of fluorine may cause cancer, such as osteosarcomas and bone tumors. Therefore, the concentration of fluorine in soil must be properly managed.

Chemical washing has been widely used as a treatment method for soil contaminated with metals and metalloids [17–24]. The chemical washing process can be implemented within a short period of time compared to solidification and stabilization technologies, and this means that it may be possible to reuse the site quickly. For soil contaminated with

As, Cd, Cu, Cr, Pb, and Zn, removal efficiencies between 56% and 100% were achieved with various washing solutions such as HCl, ethylenediaminetetraacetic acid (EDTA), phosphoric acid, sulfuric acid, and $FeCl_3$ (Table 1). Chemical washing with 3 M HCl was also performed on fluorine-contaminated soil near a chemical plant, and a removal efficiency of 97% was observed [17]. The application of the chemical washing technique to treat natural fluorine-enriched soil has not been previously studied.

Table 1. Chemical washing conditions and efficiencies for the metals, metalloids, and fluorine-contaminated soils reported in previous studies.

Soil	Contaminant	Concentration (mg/kg)	Washing Conditions		Removal Efficiency (%)	Reference
			Washing Agents	Washing Conditions		
Erie County, N.Y. (pH 5.5)	Pb	500–600	0.1 N HCl, 0.01 M EDTA, 1 M $CaCl_2$	900 g soil, acrylic column (upflow 10 mL/min, 24 h)	HCl—85 EDTA—100 $CaCl_2$—78	[19]
Lavrion Technology and Cultural Park (LTCP)—mine, refinery, industrial park (pH 7.0)	Fe Pb Zn As Mn Cu	223,600 64,195 55,900 7540 6500 4100	1 M HCl, 0.1 M Na_2EDTA	Soil/solution = 30 g/L, 150 rpm, 4 h for 1 M HCl, 1 h for 0.1 M Na_2EDTA	HCl Na_2EDTA 45 14 44 44 82 38 77 13 80 42 61 41	[20]
Ibaraki of Kuroboku, Japan—forest area soil (pH 5.94)	As	2830	9.4% H_3PO_4, 11% H_2SO_4	1 g soil: 25 mL solution, 20 °C, 2 h	H_3PO_4—97.9 H_2SO_4—87.7	[21]
Construction site in University Park, TX (pH 7.9)	Pb Cd Zn	742 603 624	0.01 M Na_2EDTA + 0.1 M $Na_2S_2O_5$	1 g soil: 12.5 mL solution, Shaker table operated at 175 rpm for 2 h	56.1 92.3 71.0	[22]
Burnley campus garden at Melbourne University, Australia (pH 6.14)	Pb Cd Cr	200 400 600	0.5 M $FeCl_3$	Shaker table operated at 180 rpm for 1 h	93.8 97.4 81.8	[23]
Fluoride contaminated soil from a chemical company in Changwon, Gyeongsangnam-do, Korea (pH 3.7)	F	740	3 M HCl 2 M NaOH 3 M HNO_3 3 M H_2SO_4 3 M $C_4H_6O_6$	5 g soil: 50 mL solution, Shaking incubator at 200 rpm, 20 °C for 1 h	97 71 91 88 64	[17]
Abandoned metallurgic plant located in Wubu, an old city district of Anhui Province, China (pH 6.7)	PAH Pb Cd Cr Ni F	352.8 839.7 23.7 622.4 432.8 2376.5	carboxymethyl-β-cyclodextrin (CMCD) carboxymethyl chitosan (CMC)	50 g/L CMCD + 5 g/L CMC solution Shaking at 100 rpm, 25 °C for 60 min and centrifugation at 1000 rpm for 30 min, multi-stage washing (3 cycle)	94.3 93.2 85.8 93.4 83.2 97.3	[24]
Fluorine-contaminated soil from Incheon City, South Korea (pH 6.4), No source information provided	F	488	1 M HNO_3 1 M H_2SO_4 1 M NaOH 2 M H_2SO_4	30 g soil: 270 mL solution, Shaking at 25 °C for 30 min	19.5 26.7 10.2 40.1	[25]

In this study, the feasibility of using a chemical washing method to remediate natural fluorine-enriched soil was evaluated. The soil at the target site requires specialized treatment because the fluorine concentration exceeds the soil contamination criterion of South Korea (Area 2: 400 mg/kg, "Area 2" refers to locations containing forests, warehouse sites, etc.) [18], and treatment needs to be completed within a relatively short period of time because the site is located in an urban area. The applicability of various washing solutions and washing conditions (i.e., washing solution concentration, solid–liquid ratio, agitation speed, and reaction time), which have been used previously for chemical washing of heavy metal-contaminated soil, was examined. In addition, the effects of techniques for improving the washing efficiency, such as ultrasonic washing, aeration, and multi-stage washing, were evaluated. X-ray diffraction (XRD) analysis and sequential extraction were performed on the target sample to obtain a better understanding of the nature of fluorine within the soil.

2. Materials and Methods

2.1. Soil Preparation and the Determination of Its Characteristics

In this study, natural fluorine-enriched soil was collected from Seoul (latitude: 37°29'23″, longitude: 127°00'03″), a mega city. Topsoil with a depth of 30 cm or less was collected, air-dried at room temperature, and sifted through a 2-mm sieve; then, the pH, organic matter content (Walkley–Black method), cation exchange capacity (ammonium acetate method), and iron/aluminum/manganese oxide content (dithionite–citrate system buffered with sodium bicarbonate (DCB) method) of the soil were measured [26–29]. Additionally, pellets with a diameter of 34 mm (prepared by compressing soil samples to enable X-ray analysis on a flat surface) were fabricated and subjected to X-ray fluorescence analysis (S8 Tiger, Bruker, Billerica, MA, USA) under vacuum conditions at an output of 40 mA and 40 V. Based on these data, the main components (Si, Al, Fe, Ca, Na, K, Mg, Ti, P, and S) of the soil samples were determined. The types of crystalline minerals in the soil samples were identified through XRD analysis (D8 ADVANCE, Bruker, Billerica, MA, USA). This analysis was conducted with a 2theta range of 3–90°, step of 0.02, scan speed of 0.5 s/step, and wavelength of Cu kα1 = 1.5418 Å at a generator output of 40 kV and 40 mA.

Meanwhile, wet sieving was also performed for soil samples of 2 mm or less to analyze the fluorine concentration by particle size. The sizes of the sieves used were 0.5, 0.15, and 0.075 mm, and the composition ratio was calculated by measuring the dry weight after sieving.

Finally, mica, which was estimated to be the main fluorine-containing mineral in the target soil, was manually collected from the gravel (>2 mm). Then, it was pulverized and the fluorine content was measured using the alkali fusion method described in Section 2.3.

2.2. Chemical Washing Procedures for Natural Fluorine-Enriched Soil

To determine optimal conditions for the chemical washing method used on the natural fluorine-enriched soil, experiments were performed by varying the washing solution type, washing solution concentration, ratio between the soil sample and the washing solution (g:mL), agitation speed, and reaction time. Furthermore, changes in washing efficiency due to aeration, ultrasonic irradiation, and multi-stage washing were also evaluated. First, to determine the optimal washing solution, 1 M solutions of sulfuric acid (H_2SO_4), phosphoric acid (H_3PO_4), sodium hydroxide (NaOH), potassium hydroxide (KOH), oxalic acid ($H_2C_2O_4$), nitric acid (HNO_3), perchloric acid ($HClO_4$), and hydrochloric acid (HCl) were mixed with the soil samples at a solid–liquid ratio of 1:5 (g:mL), and the amount of fluorine eluted was evaluated after a reaction time of 60 min at an agitation speed of 200 rpm. Washing experiments were performed while the washing solution concentration was varied from 1 to 2 and 2.5 M, solid–liquid ratio (g:mL) from 1:2 to 1:3 and 1:5, agitation speed from 100 to 150 and 200 rpm, and reaction time from 10 to 30, 60, 120, and 240 min. In addition, the amount of fluorine eluted was evaluated after aeration and ultrasonic irradiation for 10, 30, and 60 min and two to four cycles of multi-stage (repeated) washing. After chemical washing experiments were carried out under each condition, solid–liquid separation was performed using a 0.45-μm Gelman hydrophilic polypropylene (GHP) syringe filter (Pall, Port Washington, NY, USA). The fluorine concentration in the filtrate was then measured using a fluoride ion electrode (F001502, ISTEK, Seoul, Korea) after mixing the filtrate with total ionic strength adjustment buffer (TISAB) at a 1:1 ratio. The washing efficiency was calculated as the ratio of the fluorine concentration in the washing solution (i.e., the filtrate) (unit: mg/kg) to the total fluorine concentration in soil (unit: mg/kg).

2.3. Determination of the Total Fluorine Concentration in Soil

The alkali fusion method was used to measure the total fluorine concentration in soil [30]. For this procedure, 0.5 g of the dried soil sample was placed in a nickel crucible, and 6 mL of 16 M NaOH was injected. The crucible was placed in a dryer at 150 °C for 1 h and then in a furnace at 300 °C. The temperature was increased to 600 °C, and the reaction was allowed to proceed for 30 min. Approximately 10 mL of distilled water was then added to the residue in the nickel crucible, and the pH was adjusted between 8 and 9 with concentrated HCl. After transferring the sample to a volumetric flask and adding distilled water to reach 100 mL, the solution was filtered through Whatman No. 40 filter paper. After mixing the filtrate and TISAB at a ratio of 1:1, the concentration of fluoride ions in the filtrate was measured using the fluoride ion electrode, and these data were used to determine the fluorine concentration in soil.

Furthermore, the accuracy of the method for analyzing the fluorine concentration in soil was assessed using GSP-2 and NIM-G, which are certified reference materials (CRMs). The fluorine content was 3000 mg-F/kg-soil for GSP-2 and 4200 mg-F/kg-soil for NIM-G.

2.4. Sequential Extraction Procedures for Fluorine in Soil

Sequential extraction was performed to understand the binding pattern between soil components and fluorine. Fluorine was divided into a water-soluble fraction (F1), exchangeable fraction (F2), Mn and Fe oxide bound fraction (F3), organic matter bound fraction (F4), and residual fraction (F5) (Table 2). In step 1, 2.5 g of the dried soil sample and 25 mL of the extractant were mixed [31,32]. After centrifugation, the precipitates and 25 mL of each extractant were mixed in steps 2 to 4. For the analysis in step 5, the alkali fusion method described in Section 2.3 was used. For the separation of the extractant and the soil sample, centrifugation (1580R, Labogene, Seoul, Korea) was performed for 15 min at 2357 g [33].

Table 2. Sequential extraction procedures for fluorine-enriched soil (following Yi et al. [32] with some modifications).

Fraction of Fluorine in Soil	Extractant	Experimental Conditions		
		Temperature (°C)	Incubation time (h)	Agitation speed (rpm)
Water soluble (F1)	Deionized water	70	0.5	
Exchangeable (F2)	1 mol/L MgCl$_2$ (pH 7)	25	1	
Bound to Mn and Fe oxides (F3)	0.04 mol/L NH$_2$OH·HCl dissolved in 20% acetic acid	60	1	30–40 [a]
Bound to organic matter (F4)	0.02 mol/L HNO$_3$ + 30% H$_2$O$_2$ + 3.2 mol/L and ammonium acetate [b]	25	0.5	
Residual (F5)	Alkali fusion with NaOH	600	1.5	None

[a] Cited from Davison et al. [34]. [b] 0.02 M HNO$_3$ 3 mL + 30% H$_2$O$_2$ 10 mL and ammonium acetate 12 mL [35].

3. Results

3.1. Characteristics of Fluorine-Natural Enriched Soil

Table 3 shows the fluorine concentration for various soil particle sizes. The properties of the natural fluorine-enriched soil used in this study are presented in Tables A1 and A2. The total fluorine concentration in the soil samples (<2 mm) was found to be 1078 ± 178 mg/kg, which exceeds the soil contamination criterion of Korea (Area 2: 400 mg/kg) [25]. Thus, these data confirmed that an appropriate remediation technique was required for this natural fluorine-enriched soil. The soil sample with a particle size of 0.15 mm or less exhibited 1.5 times higher fluorine concentration than the total soil sample (i.e., particle size of 2 mm or less). Because pollutants introduced to soil from external sources are easily adsorbed on the surface of silt or clay particles with a large specific surface area, the concentration of heavy metals increases as the particle size of soil decreases [36,37]. Despite the above finding, the distribution of fluorine concentrations was relatively homogeneous in this study (i.e., no significant differences were observed for the different soil particle sizes). This indicated that the pollutants adsorbed on soil particles were not introduced from the outside but were possibly of natural origin (from minerals).

Table 3. Particle size distribution of the soil sample and the corresponding total fluorine concentrations.

Particle Size (mm)	Weight Composition (%)	Total Fluorine Concentration (mg/kg)
<2	100	1078 ± 178
0.5–2	26.3	1126 ± 272
0.15–0.5	26.8	1036 ± 34
0.075–0.15	10.2	1564 ± 159
<0.075	36.7	1594 ± 42

3.2. Chemical Washing Efficiency

The soil samples were mixed with 1 M solutions of H_2SO_4, H_3PO_4, NaOH, KOH, $H_2C_2O_4$, HNO_3, $HClO_4$, and HCl at a solid–liquid ratio of 1:5 (g:mL). For each mixture, the reaction was allowed to proceed for 1 h at an agitation speed of 200 rpm, but the washing efficiency remained as low as 0.6–3.0% (Table 4).

Table 4. Chemical washing efficiencies when using various washing solutions.

Washing Reagent (1 M)	Extracted Fluorine Concentration (mg/kg)	Washing Efficiency (%)
H_2SO_4	16.2 ± 1.5	1.5
H_3PO_4	6.8 ± 0.6	0.6
NaOH	20.3 ± 1.9	1.9
KOH	20.2 ± 1.9	1.9
$H_2C_2O_4$	6.2 ± 0.6	0.6
HNO_3	30.3 ± 2.8	2.8
$HClO_4$	27.2 ± 2.5	2.5
HCl	32.7 ± 3.0	3.0

To improve the washing efficiency, various conditions (i.e., washing solution fixed to HCl, washing solution concentration (1–2.5 M), solid–liquid ratio (1:2–1:5), agitation speed (100–200 rpm), reaction time (10–240 min), aeration (10–60 min), ultrasonic washing (10–60 min), and multi-stage washing (1–4 times)) were tested. The washing efficiency, however, did not exceed 6.2% (Table 5).

According to a previous study [17], a 97% removal efficiency was achieved when chemical washing was performed for 1 h under conditions of 200 rpm and 20 °C at a solid–liquid ratio of 1:5 (g:mL) using 3 M HCl (Table 1); these results are markedly different from the results of this study.

3.3. Origin of Fluorine in Soil

To analyze the causes of the significantly low washing efficiency obtained in this research, even though chemical washing conditions similar to those applied in previous research on soil contaminated with heavy metals and fluorine of an artificial origin were maintained (Table 1), XRD analysis and sequential extraction were performed on the target sample.

Table 5. Chemical washing efficiencies when applying various washing conditions.

Washing Conditions		Extracted Fluorine Concentration (mg/kg)	Washing Efficiency (%)
Solid–liquid ratio (g:mL)	1:2	18.2 ± 1.1	1.7
	1:3	24.4 ± 1.8	2.3
	1:5	32.8 ± 1.1	3.0
Reaction time (min)	10	58.7 ± 2.5	5.4
	30	36.8 ± 0.6	3.4
	60	32.8 ± 1.1	3.0
	120	30.3 ± 1.6	2.8
	240	24.9 ± 0.0	2.3
Agitation speed (rpm)	100	34.5 ± 0.3	3.2
	150	36.6 ± 0.9	3.4
	200	58.7 ± 2.5	5.4
Concentration of HCl (mol/L)	0	1.5 ± 0.0	0.1
	1	32.8 ± 1.1	3.0
	2	26.8 ± 0.2	2.5
	2.5	24.8 ± 1.8	2.3
Aeration time (min)	10	49.6 ± 4.1	4.6
	30	62.3 ± 0.3	5.8
	60	66.7 ± 5.3	6.2
Ultrasonicating time (min)	10	37.7 ± 0.3	3.5
	30	43.4 ± 0.7	4.0
	60	41.9 ± 0.7	3.9
Multi-stage washing (cycle)	1	35.3 ± 2.1	3.3
	2	14.3 ± 1.1	1.3
	3	6.1 ± 0.3	0.6
	4	2.8 ± 0.4	0.3

3.3.1. X-Ray Diffraction Analysis

Figure 1 shows the XRD analysis results. Peaks of biotite, phlogopite, muscovite, and lepidolite, which belong to mica, were detected in the soil sample data. It is known that mica can contain fluorine as a result of the substitution of hydroxide ions (OH^-) and fluoride ions (F^-) [4–11]. Based on the XRD analysis results, gravel-sized fragments rich in mica were manually collected from the > 2 mm sifted soil sample in which a large amount of mica was detected. Then, the mineral samples obtained were pulverized and subjected to fluorine concentration analysis and XRD analysis.

The XRD analysis results for the minerals that were estimated to be mica exhibited peaks of biotite, phlogopite, lepidolite, and muscovite (Figure 1) that were more prominent than those in the above results [38,39], and a fluorine concentration of 2647 mg/kg was measured (the fluorine concentration in the soil sample with a size of 2 mm or less from the same soil was 1078 mg/kg). This finding indicated that fluorine was possibly present in the soil sample (with a size of 2 mm or less) as result of the fragmentation of mica minerals.

3.3.2. Sequential Extraction Results

Table 6 shows the fluorine sequential extraction results for the target soil sample. The residual fraction (F5) amounted to 99.2–99.6%, thus confirming that most of the fluorine was present in soil in a chemically stable form. When the sample, obtained by manually collecting mica minerals with a size of 2 mm or higher and pulverizing them, was subjected to fluorine sequential extraction using the same method, the residual fraction (F5) was found to be 99.8%, which was similar to the value obtained for the soil sample.

As confirmed in Section 3.3.1, the target soil sample represents a case in which fluorine of a natural origin (mica) is present and has accumulated above the soil environmental criterion. In such a case, fluorine is present in a chemically stable form because it exists as minerals. This appears to have contributed to the significantly low efficiencies for the chemical washing procedures tested.

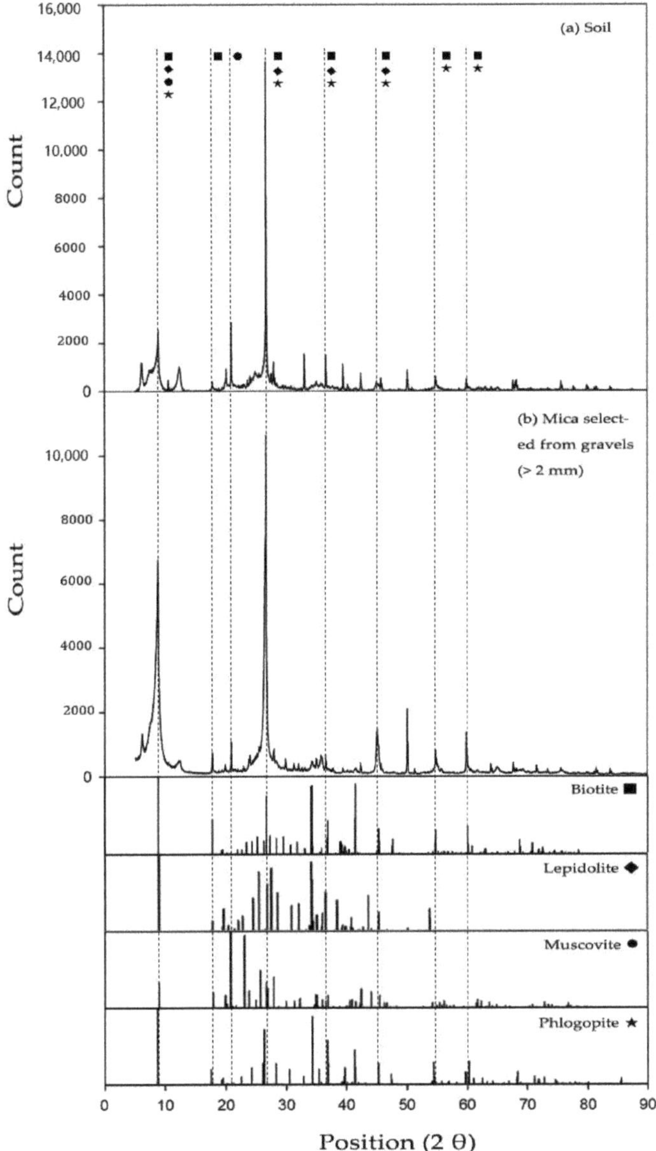

Figure 1. X-ray diffraction analysis results for (**a**) the soil sample and (**b**) mica selected manually from the gravel (>2 mm).

Table 6. Sequential extraction results for the soil sample and mica selected from gravel.

Fraction of Fluorine in Soil	Soil Sample (<2 mm)		Sample of Crushed Mica Selected by Hand from Gravel (>2 mm)	
	Extracted Fluorine Concentration (mg/kg)	Extraction Efficiency (%)	Extracted Fluorine Concentration (mg/kg)	Extraction Efficiency (%)
Water soluble (F1)	7.19 ± 0.17	0.57	2.31 ± 0.01	0.09
Exchangeable (F2)	0.58 ± 0.04	0.05	0.34 ± 0.34	0.01
Bound to Mn and Fe oxides (F3)	0.09 ± 0.01	0.01	0.08 ± 0.01	0.00
Bound to organic matter (F4)	1.99 ± 0.30	0.16	1.86 ± 0.12	0.07
Residual fraction (F5)	1253 ± 85	99.21	2647 ± 11.55	99.84

4. Conclusions

In this study, the applicability of chemical washing was evaluated for the treatment of a soil containing high levels of fluorine that originated from mica. This soil exceeded the soil environmental criterion for the region. Various conditions (washing solution type, washing solution concentration, solid–liquid ratio, agitation speed, reaction time, aeration, ultrasonic washing, and multi-stage washing), similar to those of the general washing method for soil contaminated with heavy metals and the washing method previously applied to soil contaminated with fluorine of an artificial origin (near a chemical plant, where a 97% washing efficiency was achieved) [17], were tested; however, from all methodologies applied, the maximum washing efficiency achieved in our study for natural fluorine-enriched soil was 6.2%. The sequential extraction results showed that approximately 99% of the fluorine was present as a residual fraction, thus indicating that it occurred in the soil in a chemically stable form, possibly because of the presence of the fragmented mica minerals. This may have contributed to the low washing efficiency. Consequently, treating this soil enriched with fluorine of a natural origin (mica) using general chemical washing methods is not feasible. Therefore, it is recommended that physical separation technology be applied or other approaches be used to manage potential human health and/or ecological risks.

Author Contributions: Conceptualization, M.-Y.J. and J.A.; methodology, J.A.; software, D.-J.B.; validation, D.-J.B., Y.-E.K., and H.-O.Y.; formal analysis, D.-J.B. and Y.-E.K.; investigation, D.-J.B. and Y.-E.K.; resources, H.-O.Y.; data curation, D.-J.B.; writing—original draft preparation, D.-J.B. and J.A.; writing—review and editing, J.A.; visualization, D.-J.B.; supervision, J.A.; project administration, M.-Y.J. and J.A.; funding acquisition, J.A. All authors have read and agreed to the published version of the manuscript.

Funding: This work was supported by the Korea Ministry of Environment (MOE) as Waste to Energy-Recycling Human Resource Development Project, and the Geo–Advanced Innovation Action (GAIA) project (Grant No. 2013000530001) of the Korea Environmental Industry & Technology Institute (KEITI).

Data Availability Statement: Data is contained within the article and Appendix A.

Conflicts of Interest: The authors declare no conflict of interest.

Appendix A

Table A1. Composition of the major elements in the soil samples determined using an X-ray fluorescence spectrometer.

Major Element	Composition (%)
SiO_2	52.85
Al_2O_3	24.63
Fe_2O_3	11.03
K_2O	4.25
MgO	2.41
TiO_2	1.85
CaO	1.55
Na_2O	0.44
P_2O_5	0.43
SO_3	0.14

Table A2. Characteristics of the soil samples used in this study.

pH	Organic Matter Content (%)	CEC (cmol/kg)	Fe Oxides (mg/kg)	Al Oxides (mg/kg)	Mn Oxides (mg/kg)
7.4	0.6	11.8	26,655	2584	374

References

1. An, J.; Kim, J.-A.; Yoon, H.-O. A review on the analytical techniques for the determination of fluorine contents in soil and solid phase samples. *J. Soil Groundw. Environ.* **2013**, *18*, 112–122. [CrossRef]
2. Kabir, H.; Gupta, A.K.; Tripathy, S. Fluoride and human health: Systematic appraisal of sources, exposures, metabolism, and toxicity. *Crit. Rev. Environ. Sci. Technol.* **2020**, *50*, 1116–1193. [CrossRef]
3. Luo, K.; Ren, D.; Xu, L.; Dai, S.; Cao, D.; Feng, F.; Tan, J.A. Fluorine content and distribution pattern in Chinese coals. *Int. J. Coal Geol.* **2004**, *57*, 143–149. [CrossRef]
4. Mukherjee, I.; Singh, U.K. Groundwater fluoride contamination, probable release, and containment mechanisms: A review on Indian context. *Environ. Geochem. Health* **2018**, *60*, 2259–2301. [CrossRef]
5. Deshmukh, A.N.; Wadaskar, P.M.; Malpd, D.B. Fluorine in environment: A review. *Gondwana Geol. Mag.* **1995**, *9*, 1–20.
6. Fuge, R.; Andrews, M.J. Fluorine in the UK environment. *Environ. Geochem. Health* **1988**, *10*, 96–104. [CrossRef] [PubMed]
7. Berger, T.; Peltola, P.; Drake, H.; Åström, M. Impact of a fluorine-rich granite intrusion on levels and distribution of fluoride in a small boreal catchment. *J. Aquat. Geochem.* **2012**, *18*, 77–94. [CrossRef]
8. Lee, J.-H.; Jeong, J.-O.; Kim, K.-K.; Lee, S.-W.; Kim, S.-O. Geochemical study on the naturally originating fluorine distributed in the area of Yongyudo and Sammokdo, Incheon. *Econ. Environ. Geol.* **2019**, *52*, 275–290. [CrossRef]
9. Noda, T.; Roy, R. OH-F exchange in fluorine phlogopite. *Am. Mineral.* **1956**, *41*, 929–932.
10. Oh, H.-J.; Lee, J.-Y. A Study on the characteristical evaluation of metals and fluorine concentrations in the southern part of Seoul. *J. Soil Groundw. Environ.* **2003**, *8*, 68–73.
11. Rao, C.R.N. Fluoride and Environment—A Review. In Proceedings of the Third International Conference on Environment and Health, Chennai, India, 15–17 December 2003; pp. 386–399.
12. Sen Gupta, J.G. Determination of fluorine in silicate and phosphate rocks, micas and stony meteorites. *Anal. Chim. Acta.* **1968**, *42*, 119–125. [CrossRef]
13. Czerwinski, E.; Nowak, J.; Dabrowska, D.; Skolarczyk, A.; Kita, B.; Ksiezyk, M. Bone and joint pathology in fluoride-exposed workers. *Arch. Environ. Health* **1988**, *43*, 340–343. [CrossRef] [PubMed]
14. Hodge, H.C.; Smith, F.A. Occupational fluoride exposure. *J. Occup. Med.* **1977**, *19*, 12–39. [CrossRef] [PubMed]
15. Grandjean, P. Classical syndromes in occupational medicine occupational fluorosis through 50 years: Clinical and epidemiological experiences. *Am. J. Ind. Med.* **1982**, *3*, 227–236. [CrossRef] [PubMed]
16. WHO. Preventing Disease through Healthy Environments: Inadequate or Excess Fluoride: A Major Public Health Concern. License: CC BY-NC-SA 3.0 IGO World Health Organization. 2019. Available online: https://apps.who.int/iris/handle/10665/329484 (accessed on 11 November 2020).
17. Moon, D.H.; Jo, R.; Koutsospyros, A.; Cheong, K.H.; Park, J.-H. Soil washing of fluorine contaminated soil using various washing solutions. *Bull. Environ. Contam. Toxicol.* **2015**, *94*, 334–339. [CrossRef]
18. Korean Ministry of Environment. *The Korean Standard Test (KST) Methods for Soils*; Korean Ministry of Environment: Gwachun-si, Korea, 2010.

19. Reed, B.E.; Carriere, P.C.; Moore, R. Flushing of a Pb(II) contaminated soil using HCl and CaCl$_2$. *J. Environ. Eng.* **1996**, *122*, 48–50. [CrossRef]

20. Moutsatsou, A.; Gregou, M.; Matsas, D.; Protonotarios, V. Washing as a remediation technology applicable in soils heavily polluted by mining–metallurgical activities. *Chemosphere* **2006**, *63*, 1632–1640. [CrossRef]

21. Tokunaga, S.; Hakuta, T. Acid washing and stabilization of an artificial arsenic-contaminated soil. *Chemosphere* **2002**, *46*, 31–38. [CrossRef]

22. Abumaizar, R.J.; Smith, E.H. Heavy metal contaminants removal by soil washing. *J. Hazard. Mater.* **1999**, *70*, 71–86. [CrossRef]

23. Alaboudi, K.A.; Ahmed, B.; Brodie, G. Soil washing technology for removing heavy metals from a contaminated soil: A case study. *Pol. J. Environ. Stud.* **2020**, *29*, 1029–1036. [CrossRef]

24. Ye, M.; Sun, M.M.; Kengara, F.O.; Wang, J.T.; Ni, N.; Wang, L.; Song, Y.; Yang, X.L.; Li, H.X.; Hu, F.; et al. Evaluation of soil washing process with carboxymethyl-β-cyclodextrin and carboxymethyl chitosan for recovery of PAHs/heavy metals/fluorine from metallurgic plant site. *J. Environ. Sci.* **2014**, *26*, 1661–1672. [CrossRef] [PubMed]

25. Ahn, Y.; Pandi, K.; Cho, D.-W.; Choi, J. Feasibility of soil washing agents to remove fluoride and risk assessment of fluoride-contaminated soils. *J. Soils Sediments* **2020**, 1–8. [CrossRef]

26. Kitsopoulos, K.P. Cation-exchange capacity (CEC) of zeolitic volcaniclastic materials: Applicability of the ammonium acetate saturation (AMAS) method. *Clays Clay Miner.* **1999**, *47*, 688–696. [CrossRef]

27. Robinson, G.W. A new method for the mechanical analysis of soils and other dispersions. *J. Agric. Sci.* **1922**, *12*, 306–321. [CrossRef]

28. Sparks, D.L.; Page, A.L.; Helmke, P.A.; Loeppert, R.H.; Soltanpour, P.N.; Tabatabai, M.A.; Johnson, C.T.; Sumner, M.E. *Methods of Soil Analysis, Part 3*; Chemical Methods ASA-SSSA: Madison, WI, USA, 1996.

29. Walkley, A.; Black, I.A. An examination of the Degtjareff method for determining soil organic matter, and a proposed modification of the chromic acid titration method. *Soil Sci.* **1934**, *37*, 29–37. [CrossRef]

30. McQuaker, N.R.; Gurney, M. Determination of total fluoride in soil and vegetation using an alkali fusion-selective ion electrode technique. *Anal. Chem.* **1977**, *49*, 53–56. [CrossRef]

31. Gao, H.; Zhang, Z.; Wan, X. Influences of charcoal and bamboo charcoal amendment on soil-fluoride fractions and bioaccumulation of fluoride in tea plants. *Environ. Geochem. Health.* **2012**, *34*, 551–562. [CrossRef]

32. Yi, X.; Qiao, S.; Ma, L.; Wang, J.; Ruan, J. Soil fluoride fractions and their bioavailability to tea plants (*Camellia sinensis* L.). *Environ. Geochem. Health.* **2017**, *39*, 1005–1016. [CrossRef]

33. Wenzel, W.W.; Kirchbaumer, N.; Prohaska, T.; Stingeder, G.; Lombi, E.; Adriano, D.C. Arsenic fractionation in soils using an improved sequential extraction procedure. *Anal. Chim. Acta.* **2001**, *436*, 309–323. [CrossRef]

34. Davison, C.M.; Thomas, R.P.; McVey, S.E.; Perala, R.; Littlejohn, D.; Ure, A.M. Evaluation of a sequential extraction procedure for the speciation of heavy metals in sediments. *Anal. Chim. Acta.* **1994**, *291*, 277–286. [CrossRef]

35. Chen, W.; Pang, X.-F.; LI, J.-H.; Hang, X. Effect of oxalic acid and humic acid on the species distribution and activity of fluoride in soil. *Chem. Asian. J.* **2013**, *25*, 469–474. [CrossRef]

36. FÖrstner, U. Trace metal analysis of polluted sediments, Part 1. Assessment of sources and intensities. *Environ. Technol. Lett.* **1980**, *1*, 494–505. [CrossRef]

37. Haque, M.A.; Subramanian, V. Cu, Pb and Zn pollution of soil environment. *CRC Crit. Rev. Environ. Contr.* **1982**, *12*, 13–90. [CrossRef]

38. Bassett, W.A. Role of hydroxyl orientation in mica alteration. *Geol. Soc. Am. Bull.* **1960**, *71*, 449–456. [CrossRef]

39. Kalinowski, B.E.; Schweda, P. Kinetics of muscovite, phlogopite, and biotite dissolution and alteration at pH 1–4, room temperature. *Geochim. Cosmochim. Acta* **1995**, *60*, 367–385. [CrossRef]

 *minerals*

Article

Arsenic Fixation in Polluted Soils by Peat Applications

Antonio Aguilar-Garrido [1],*, Ana Romero-Freire [2], Minerva García-Carmona [3], Francisco J. Martín Peinado [1], Manuel Sierra Aragón [1] and Francisco J. Martínez Garzón [1]

[1] Departamento de Edafología y Química Agrícola, Facultad de Ciencias, Universidad de Granada, 18071 Granada, Spain; fjmartin@ugr.es (F.J.M.P.); msierra@ugr.es (M.S.A.); fjgarzon@ugr.es (F.J.M.G.)

[2] Instituto de Investigaciones Marinas (IIM-CSIC), 36208 Vigo, Pontevedra, Spain; anaromero@iim.csic.es

[3] Departamento de Agroquímica y Medio Ambiente, Universidad Miguel Hernández, 03202 Elche, Alicante, Spain; minerva.garciac@umh.es

* Correspondence: antonioag@ugr.es; Tel.: +34-695-406-897

Received: 10 September 2020; Accepted: 27 October 2020; Published: 29 October 2020

Abstract: Soil arsenic (As) pollution is still a major concern due to its high toxicity and carcinogenicity, thus, the study of decontamination techniques, as the organic amendment applications, keeps upgrading. This research evaluates the potential remediation of peat in different As-polluted soils, by assessing the decrease of As solubility and its toxicity through bioassays. Obtained reduction in As solubility by peat addition was strongly related to the increase of humic substances, providing colloids that allow the complexation of As compounds. Calcareous soils have been the least effective at buffering As pollution, with higher As concentrations and worse biological response (lower soil respiration and inhibition of lettuce germination). Non-calcareous soils showed lower As concentrations due to the higher iron content, which promotes As fixation. Although in both cases, peat addition improves the biological response, it also showed negative effects, hypothetically due to peat containing toxic polyphenolic compounds, which in the presence of carbonates appears to be concealed. Both peat dose tested (2% and 5%) decreased drastically As mobility; however, for calcareous soils, as there is no phytotoxic effect, the 5% dose is the most recommended; while for non-calcareous soils the efficient peat dose for As decontamination could be lower.

Keywords: soil remediation; toxicity bioassays; humic substances; calcium carbonate; iron oxides; polyphenolic compounds

1. Introduction

Soil pollution by potentially harmful elements (PHEs) is a worldwide problem with a diverse origin, e.g., waste generated by industrial, mining or smelting activities, intensive use of agrochemicals or wastewater irrigation [1,2]. Reducing the concentration of PHEs in soils is essential to minimize the current and future impacts produced on surrounding ecosystems. However, due to the high capacity of soil to retain contaminants, the cost and complexity of the existing techniques is a challenge not yet solved [3]. Therefore, intensive research is still necessary for the development of profitable, efficient, and environmentally responsible techniques. As Mirsal [1] states, the properties of the pollutants, the degree of pollution, and the natural processes that will take place in situ should be considered for selecting the remediation techniques. Some authors argue that remediation treatments must meet broad objectives, such as waste volume reduction efficiency, pollutant toxicity reduction efficiency, and profitability and environmental compliance [4].

Arsenic (As) is a metalloid found in highly variable concentrations in different environments [5]. By its hazard and presence worldwide, As highlights among the PHEs. Currently, its presence is associated with both natural and anthropogenic sources, and in many areas, its environmental levels in the water, sediments, or soils are of major concern due to its high toxicity that often threatens human health and ecosystems [6]. Besides, arsenic distribution also occurs in several biological species by direct uptake, thus, it can easily enter the food chain [7]. For all these reasons, many studies about arsenic pollution have been carried out, and pollution problems have been located in different regions of the world (hotspots), such as that found in soils of mining areas in China [8]; industrial complexes in, e.g., Korea [9], Spain [10], and Canada [11]; abandoned mine soils in Australia [12], Brazil [13], and Spain [14]; agroecosystems in Germany and France [15]; rice fields in major rice-growing countries (India, Bangladesh, China, Pakistan, Malaysia, Thailand, Japan, and Australia) [16]; and riverside areas of North East England [17]; among others.

Arsenic concentration in soils varies widely because it depends on the initial concentration (background) in the parent material, the natural geochemical cycles, and the type of soil [18]. According to the literature, the mean As values in natural soils varies from 0.1 to 80 $mg \cdot kg^{-1}$ worldwide [19–22]. As far as government information is concerned, the guideline values proposed for soil protection vary greatly among countries and legislations. For example, the United States Environmental Protection Agency (EPA) establishes the permissible limit in 24 $mg \cdot kg^{-1}$ [23]. Whereas in Andalusia (Spain), the Generic Reference Level (NGR) applicable to soils polluted with As is 36 $mg \cdot kg^{-1}$ [24]. In Europe, other examples of As guideline values are 50 $mg \cdot kg^{-1}$ [25], 76 $mg \cdot kg^{-1}$ [26], and 43 $mg \cdot kg^{-1}$ [27], for Germany, Netherlands, and United Kingdom, respectively.

Due to the elevated environmental levels of arsenic observed and the risks that it poses for human health, various technologies, both conventional and more advanced, are being used for arsenic removal from the soil and water systems [23]. In recent years, among them, stands out the soil/water phytoremediation, for being a technique that respects the environment [28,29], and other bioremediation treatments, such as land farming, composting, and biopiles techniques [30]. Another interesting and frequently used methodology for the decontamination of PHEs is the amendments with organic compounds, which is characterized by being a highly viable method of reducing concentrations or toxic effects in polluted soils, and for being economical, efficient, and with good social acceptance [31–34]. The application of organic amendments in soils polluted by PHEs produces important changes in the main soil properties (e.g., increase pH, boost organic carbon content, increment ion exchange capacity, raise soil moisture, and improve soil structure) and modifies the solubility and bioavailability of pollutants [3,35]. Likewise, treating soils contaminated with organic matter facilitates at the same time the activation of the nutrient cycle and the microorganisms involved; thus, that the ecosystemic functions that facilitate the long-term remediation of the soil are recovered. Some examples of this technique are the use of vermicompost [35,36], and, as different researchers point out [37–39], the use of peatland soils, characterized for their high potential to adsorb different metals such as Ni and Sb and the metalloid As present in waters. This high potential is likely due to peat having a large active surface area, a consequence of the high organic matter content, and thus a large number of adsorption sites. Additionally, adsorption is not the only feasible immobilization mechanism of PHEs as processes such as precipitation, coprecipitation, and complex formation are also expected and would aid retention [38,40].

Although the aforementioned studies demonstrate its capacity to adsorb As present in water (natural water and mine wastewater) and atmospheric depositions, there is little research on the use of peat in the remediation of As-polluted soils. For this reason, the present study is focused on the assessment of the remediation capacity of the peat in As-polluted soils. Furthermore, to gain targeted information about its potential use in soil remediation projects, the proposed experiments will be carried out both in anthropogenically polluted soils (areas affected by a mining discharge that present significant residual pollution) [41] and in artificially polluted soils (natural soils that are polluted in the laboratory), selected from previous toxicity studies [42]. The specific aims will be i) to determine the peat doses applicable to soils, depending on their typology, for effective and economical implementation of the remediation technique; ii) the assessment of the As mobility and the evaluation of soil toxicity through the use of quick and economic toxicity bioassays (soil respiration and seed germination/root elongation of *Lactuca sativa* L. tests).

2. Materials and Methods

2.1. Soil Samples

Three soil samples were selected for this study: 1 soil polluted in situ, consisting of the surface layer (uppermost 20 cm) of an eutric Regosol affected by residual pollution after the Aznalcóllar mine spill [43] and located in the Guadiamar Green Corridor (AZN); and 2 ex situ (spiked) soils, 1 sampled from the C horizon of a calcaric Kastanozem (SC), and another from a C horizon of a leptic Regosol (SNC), both samples were selected by their contrasting properties and were described in Romero-Freire et al. [44]. In all cases, soil samples were obtained from composite subsamples of the same horizon and thoroughly homogenized before the analysis. The 2 natural soil samples (SC and SNC) were selected due to the strong differences in the calcium carbonate content to study the role of carbonates in As fixation. AZN samples were taken in an area with high residual pollution, which has partially avoided remediation measures carried out in the area (i.e., the addition of carbonates, clays, and organic wastes) throughout these years [14].

The main parameters analyzed in the soil samples were: pH (soil:water ratio 1:2.5), electrical conductivity (EC) (soil:water ratio 1:5), calcium carbonate content ($CaCO_3$), organic carbon (OC), textural analysis (clay content), cation exchange capacity (CEC), degree of base saturation (BS), bulk density (BD), porosity (Po), and available water (AW). The main physicochemical properties were analyzed according to standard analysis methods [45]. On the other hand, in addition to analyzing total iron content (Fet) by acid digestion in strong acids (HNO_3 + HF), free iron content (Fed) and amorphous iron content (Feo) were also analyzed following the Holmgren [46] and Blakemore [47] techniques, respectively. All analyses were done in triplicate.

2.2. Peat Characterization

The organic compound selected for remediation treatments in the As-polluted soil samples was peat from the Agia's peat bog located in Padul (Granada, Spain), which is an acidic minerotrophic peat bog (or fen type). To have a correct characterization of the peat, the same properties and same analytical techniques as for soils have been applied (Table 1). Besides, the fractionation of the organic carbon (OC) presented in the peat was determined, according to Kononova [48], by the determination of the Total Humic Extract (THE), Humic Acids (HA), and Fulvic Acids (FA). The polyphenolic profile of the peat was determined by high-performance liquid chromatography (HPLC) using an Agilent 1200 series® HPLC-DAD-ESI-MS n (Agilent Technologies, Inc.(R), Santa Clara, CA, USA). All peat analyzes were carried out in triplicate.

Table 1. Main properties of the selected soil samples and peat (mean ± standard deviation).

Properties	AZN	SC	SNC	Peat
pH (H_2O, 1:2.5)	6.77 ± 0.07	8.79 ± 0.02	5.87 ± 0.09	3.50 ± 0.14
EC [2] ($dS \cdot m^{-1}$)	0.40 ± 0.01	0.07 ± 0.01	0.05 ± 0.01	3.10 ± 0.21
$CaCO_3$ [3] (%)	0.53 ± 0.04	92.32 ± 0.86	nd [1]	nd [1]
OC [4] (%)	0.72 ± 0.13	0.38 ± 0.17	0.49 ± 0.02	25.04 ± 0.05
Clay (%)	8.81 ± 0.40	7.70 ± 0.58	8.31 ± 0.12	nd [1]
CEC [5] ($cmol_+ \ kg^{-1}$)	8.46 ± 0.16	2.94 ± 0.13	3.83 ± 0.37	41.77 ± 1.16
BS [6] (%)	97.10 ± 1.12	100.00 ± 0.00	30.70 ± 1.05	66.24 ± 3.35
BD [7] ($g \cdot cm^{-3}$)	1.56 ± 0.01	1.53 ± 0.02	1.57 ± 0.003	0.32 ± 0.006
Po [8] (%)	41.30 ± 0.51	37.75 ± 0.92	40.60 ± 0.10	83.74 ± 0.01
AW [9] (%)	7.18 ± 0.04	5.38 ± 0.06	7.40 ± 0.03	8.51 ± 0.01
As_T [10] ($mg \cdot kg^{-1}$)	120.20 ± 0.14	3.39 ± 0.15	4.39 ± 0.10	11.85 ± 0.10
As_W [11] ($mg \cdot kg^{-1}$)	0.03 ± 0.01	0.01 ± 0.001	0.01 ± 0.001	0.03 ± 0.001
Fet [12] ($g \cdot kg^{-1}$)	68.40 ± 0.31	16.80 ± 0.35	71.20 ± 0.12	14.40 ± 0.06
Fed [13] ($g \cdot kg^{-1}$)	26.30 ± 0.25	3.30 ± 0.03	7.80 ± 0.10	0.97 ± 0.04
Feo [14] ($g \cdot kg^{-1}$)	18.80 ± 0.21	0.01 ± 0.01	1.00 ± 0.06	0.43 ± 0.06
THE [15] (%)	-	-	-	27.26 ± 0.30
HA [16] (%)	-	-	-	22.64 ± 0.20
FA [17] (%)	-	-	-	4.62 ± 0.30

[1] non-detected; [2] electrical conductivity; [3] calcium carbonate content; [4] organic carbon content; [5] cation-exchange capacity; [6] degree of base saturation; [7] bulk density; [8] porosity; [9] available water; [10] total arsenic concentration; [11] water-soluble arsenic concentration; [12] total iron; [13] free iron; [14] amorphous iron; [15] Total Humic Extract; [16] Humic Acid; [17] Fulvic Acid.

2.3. Soil Samples Preparation

The SC and SNC soil samples were spiked with sodium arsenate solutions ($Na_2HAsO_4 \times 7H_2O$) at different concentration ranges (0, 300, and 600 $mg \cdot kg^{-1}$ As) under controlled laboratory conditions. The selected concentrations (well above the permissible limits in soils) were chosen according to the first guideline values proposed for Andalusia in the industrial areas (300 $mg \cdot kg^{-1}$ As), and multiplying the highest concentrations of industrial areas by 2, which had been used in Romero-Freire et al. [42] where As toxicity in relation to soil properties was studied. Once spiked, soil samples were incubated for 4 weeks under controlled humidity and temperature conditions to stabilizing the added pollutant and optimizing the time consumption according to previous observations [41,49,50].

After soil incubation, different peat treatments were applied. Peat doses selected were 0%, 2%, and 5% with respect to the total dry weight of the samples. These doses were in relation to the organic amendments used in the Aznalcóllar soil remediation [51] and other restoration plans of mining areas [52]. After the application of these amendments, samples were incubated again for another 4 weeks, under the same conditions.

After the incubation period, water-soluble extracts (1:1) were done from treated soils. In the obtained extracts pH (pH_W), electrical conductivity (EC_W) were measured with a pH/conductometer 914 Metrohm (Metrohm AG, Herisau, Switzerland) and a Eutech[TM] CON700 (Thermo Fisher Scientific Inc., Waltham, MA, USA) conductivity-meter, respectively, and soluble arsenic concentration (As_W) was measured by inductively coupled plasma–mass spectrometry (ICP-MS) in a spectrometer PerkinElmer ®NexIONTM 300D (Waltham, MA, USA).

In total, 105 treatments were carried out in the 3 different groups of samples: 15 AZN (5 soil samples × 3 peat treatments), 45 SC (5 soil samples × 3 As treatments × 3 peat treatments), and 45 SNC (5 soil samples × 3 As treatments × 3 peat treatments). Each treatment was identified as follows: First, an acronym representing the soil sample considered (AZN, SC, SNC), separated by a hyphen the number that identifies the content in As added (0, 300, 600 $mg \cdot kg^{-1}$) and separated by another hyphen the percentage of peat added (0%, 2%, 5%).

2.4. Determination of As Concentrations

Total arsenic concentration (As_T) was determined from acid digestion in strong acids (HNO_3 + HF), and water-soluble arsenic concentration (As_W) was determined from the soil:water extracts (1:1 ratio). In all cases, As was measured by ICP-MS in a spectrometer PerkinElmer ®NexIONTM 300D (Waltham, MA, USA). Instrumental drift was monitored by regularly running standard element solutions between samples. For calibration, 2 sets of standards containing the analyte of interest at 5 concentrations were prepared using rhodium as an internal standard. Procedural blanks for estimating the detection limits ($3 \times \sigma$; $n = 6$) were <0.21 $\mu g \cdot L^{-1}$ for As. The analytical precision was better than ± 5% in all cases.

2.5. Toxicity Bioassays

To evaluate the soil toxicity after peat treatments, 2 short-term toxicity bioassays were selected:

1. Heterotrophic soil respiration was measured by determining the CO_2 flux from treated soils with a microbiological analyser μ-Trac 4200 SY-LAB model (Neupurkersdorf, Austria) according to ISO 17155 protocol [53]. Soil moisture content was fixed at field capacity and soils were incubated at a constant temperature of 30 °C. The production of CO_2 was determined by absorption in vials with a solution of potash (KOH 0.2%) during 96 h, and related to the mass of soil used to obtain a measure of respiration rate. The results were expressed as the basal respiration rate (BR) in $mg \cdot CO_2 \cdot day^{-1} \cdot kg^{-1} \cdot soil$. This test was done in triplicate in all studied soil treatments and also using only peat samples.

2. Seed germination/root elongation of *Lactuca sativa* L. toxicity test, according to OECD [54] and US EPA [55] recommendations. This test assessed the phytotoxic effects on seed germination and seedling growth in the first days of growth [56]. In Petri dishes, 15 seeds of *Lactuca sativa* L. and 5 mL of soluble extract from the treated soils were placed in an incubator at 25 ± 1 °C, and the number of germinated seeds and the length of the germinated seed roots were measured after 120 h. Two endpoints were calculated: (a) The percentage of germinated seeds (SG) in relation to the control, and (b) the percentage of root elongation (RE) in relation to the control (distilled water). This assay was done in triplicate in all treatments.

2.6. Data Analyses

Normality was checked with the Shapiro–Wilk test and homoscedasticity with the Levene test. As none of these conditions were met, even when transforming the variables, non-parametric Kruskal–Wallis and Mann–Whitney U test ($p < 0.05$) for the analysis of multiple comparisons were applied. Furthermore, in order to analyze the influence of soil properties on the solubility and toxicity of arsenic, Spearman correlation analysis involving soil properties, peat doses added, water-soluble arsenic concentrations, and endpoints of the toxicity bioassays (BR, SG, and RE) were also performed. All these analyses were performed with a confidence level of 95% by using SPSS v.21.0 software (SPSS Inc.[R], Chicago, IL, USA).

3. Results

3.1. Properties of the Soil and Peat Samples

Results observed for the Aznalcóllar polluted field soil (AZN) showed slightly acidic pH, low content of carbonates (<1%), and moderate electrical conductivity. Whereas the As_T level (>100 $mg \cdot kg^{-1}$) and the levels of iron oxides (Fed and Feo) were higher than in the reference soils. The reference soils (SC and SNC) differed from each other by the content of carbonates, pH, degree of base saturation, and the contents in the different forms of Fe, with much higher Fe contents in the SNC samples. The total As (As_T) concentrations were similar, with average content below the considered background level of 29 $mg \cdot kg^{-1}$ [57] (Table 1).

The peat selected had an acidic pH, a high electrical conductivity of 3.10 dS·m^{-1}, and a degree of base saturation of more than 50%; it showed low bulk density (BD) and, consequently, high porosity (Po). The organic carbon content (OC) was 25.04%, of which 27.26% corresponds to Total Humic Extract (THE). This THE was divided into 22.64% Humic Acids (HA) and 4.62% Fulvic Acids (FA). Peat showed a total iron content (Fet) of 14.40 g·kg^{-1}, and As$_T$ of 11.85 mg·kg^{-1}, of which 0.03 mg·kg^{-1} was soluble in water (<1%) (Table 1). The polyphenolic profile of the peat identifies three compounds: 4-hydroxybenzoic-4-glucoside acid (662 mg·kg^{-1}), p-coumaroylquinic acid (1222 mg·kg^{-1}), and lariciresinol-sesquilignan (4783 mg·kg^{-1}), representing 0.66% w/w of the total peat, with the phenolic compound lariciresinol-sesquilignan being the most abundant (Figure S1).

3.2. Arsenic Solubility

The leachate from the AZN samples showed acidic pH$_W$ without statistically significant differences among the peat treatments. Whereas, EC$_W$ increased significantly with additions of 2% and 5% of peat (0.44 and 0.53 dS·m^{-1}, respectively). In SC soils spiked with As, the addition of peat decreased the pH$_W$ significantly. Similar results were obtained in the SNC samples, but with lower pH$_W$ values. In addition, it was observed that the addition of As significantly increased the pHw due to the fact that As had been added in the form of sodium arsenate solutions (Na$_2$HAsO$_4$ × 7H$_2$O). Regarding the EC$_W$, it was observed that in almost all treatments with As and peat, the EC of the leachates increased significantly, except in the SNC-300-2 and SNC-600-2 treatments, where there were no significant differences (Table 2).

Table 2. The pH$_W$, electrical conductivity (EC$_W$) in water-soluble extract (1:1), and water-soluble As content (As$_W$) in the three studied soil samples with the different addition of As and peat (mean ± standard deviation).

Soil	AZN	SC			SNC		
As (mg·kg^{-1})	120.2	0	300	600	0	300	600
Peat (%)				pH$_W$ (1:1)			
0	4.49 ± 0.08 a	7.16 ± 0.23 aA	7.76 ± 0.03 bB	7.91 ± 0.07 cB	3.78 ± 0.15 aA	6.25 ± 0.73 bB	6.45 ± 0.14 bB
2	4.59 ± 0.08 a	7.08 ± 0.02 aA	7.20 ± 0.09 aAB	7.31 ± 0.02 bB	3.72 ± 0.09 aB	5.46 ± 0.09 aA	5.67 ± 0.08 aA
5	4.67 ± 0.08 a	6.90 ± 0.02 aA	7.02 ± 0.09 aAB	7.13 ± 0.02 aB	3.64 ± 0.10 aB	5.32 ± 0.09 aA	5.53 ± 0.07 aA
Peat (%)				EC$_W$ (1:1) (dS·m^{-1})			
0	0.39 ± 0.01 a	0.13 ± 0.01 aA	0.49 ± 0.01 aB	0.89 ± 0.05 aC	0.04 ± 0.01 aA	0.18 ± 0.01 aB	0.33 ± 0.02 aC
2	0.44 ± 0.07 b	0.19 ± 0.01 bA	0.52 ± 0.01 bB	0.95 ± 0.03 abC	0.10 ± 0.01 bA	0.19 ± 0.01 aB	0.33 ± 0.01 aC
5	0.53 ± 0.07 c	0.28 ± 0.02 cA	0.61 ± 0.00 cB	1.01 ± 0.03 bC	0.19 ± 0.01 cA	0.28 ± 0.02 bB	0.42 ± 0.01 bC
Peat (%)				As$_W$ (1:1) (mg·kg^{-1})			
0	0.026 b	0.010 bA	152.513 bB	337.450 bC	0.019 bA	26.427 bB	118.123 bC
2	0.014 a	0.004 aA	0.759 aB	2.588 aC	0.015 abA	0.450 aB	2.592 aC
5	0.013 a	0.006 aA	0.113 aB	0.289 aC	0.004 aA	0.157 aB	0.631 aC

Lowercase letters indicate statistically significant differences for each As treatment with different peat additions; and capital letters among the different treatments with As for the same peat content (Kruskal–Wallis test, $p < 0.05$).

Soluble As (As$_W$) decreased significantly with peat treatments, regardless of the percentage of peat applied and for all treatments with As. In AZN-0, As$_W$ was low (0.026 mg·kg^{-1}); even so, the peat treatments significantly decreased the As solubility, with a reduction of 46.15% in AZN-2 and 50.00%

in AZN-5 compared to AZN-0, respectively. In the reference soils, once polluted, different amounts of As_W were obtained but far from the total As added, showing SC higher As_W than SNC. In SC-300-0 and SC-600-0, the reduction of As_W in relation to the dose of As added was 49.16% and 43.76%, respectively; and in SNC-300-0 and SNC-600-0, the reduction was 91.99% and 80.31%, respectively. The addition of peat significantly enhanced arsenic retention. Peat additions at 2% and 5% decreased As_W with respect to the no addition by more than 98% in all samples. This decrease in As_W was even observed, and significantly, in the reference soil samples when adding peat (Table 2).

3.3. Assessment of Peat Treatments Adequacy from Bioassays. Arsenic Solubility

3.3.1. Basal Soil Respiration

The basal soil respiration test was also done with the peat alone, showing basal respiration (BR) values of 101.84 ± 9.94 mg·CO_2·day^{-1}·kg^{-1}·soil. BR increased significantly as the amount of peat added increased in AZN soils. In calcareous samples (SC), the increase in pollution without peat additions (SC-300-0 and SC-600-0) showed a slight, but not significant, decrease in BR. Peat additions significantly increased BR up to about 100 mg·CO_2·day^{-1}·kg^{-1}·soil. Peat at 2% significantly increased BR compared to the no peat addition. Peat at 5% further increased the respiration rate, with significant differences among the three As treatments. Similar results were obtained in the non-calcareous samples (SNC) with higher levels in BR than in SC. Treatment with 2% of peat considerably increased the BR, being higher in the reference sample and decreasing significantly with As addition. The addition of 5% of peat showed a significantly increases BR with values >100 mg·CO_2·day^{-1}·kg^{-1}·soil. Comparing the results obtained for the same pollution level with increasing peat dose, significant differences appeared between SNC-300-2 and SNC-300-5, and between SNC-600-2 and SNC-600-5 (Figure 1).

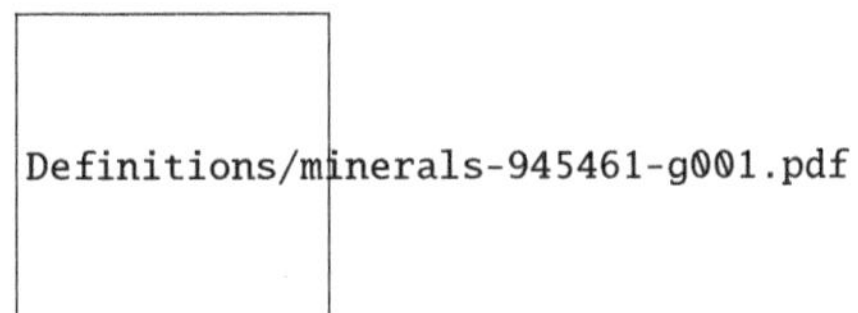

Figure 1. Basal respiration rate (BR) for soil samples with different treatments. Lowercase letters indicate statistically significant differences for each treatment with different peat additions; and capital letters among the different treatments with As for the same peat content (Kruskal-Wallis test, $p < 0.05$).

3.3.2. Germination and Elongation Test with the *Lactuca Sativa* L. Plant

AZN soil showed a low germination rate for AZN-0 and AZN-2, respectively, and even lower for the higher peat treatment (AZN-5), but without significant differences between peat treatments. In SC soils, without As addition, reported germination of 100%, whereas with As addition, it completely inhibits the *Lactuca sativa* L. seed germination (Figure 2). The addition of peat in the polluted SC soil entailed a high germination rate. In non-calcareous soil (SNC), the germination test showed substantially different results compared to calcareous soil (SC). When there was no addition of peat, the germination of *Lactuca sativa* L. occurred in the three As treatments. In both SNC-0-0 and SNC-300-0, the germination was 100%, while in the SNC-600-0, germination decreased significantly by 43%. Besides, peat treatments decreased SG in all treatments without showing significant differences due to the high dispersion in the data (error bars).

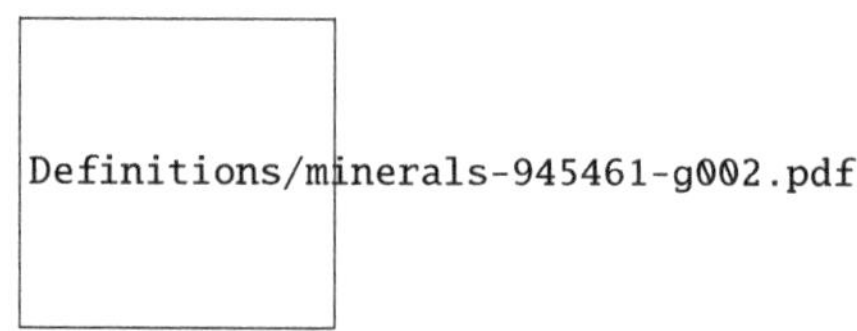

Figure 2. Percentage of germination for *Lactuca sativa* L. seeds (SG%) exposed to the different soils and treatments. Lowercase letters indicate statistically significant differences for each treatment with different peat additions; and capital letters among the different treatments with As for the same peat content (Kruskal-Wallis test, $p < 0.05$).

In AZN soil, a trend to decrease the percentage of RE with the increase of added peat was observed. In calcareous soils (SC), without As addition, RE was 97%. However, samples polluted with As, due to the total inhibition of germination, did not record data on root elongation. However, it is interesting to note that the addition of peat favored RE with values even higher than 100%. Only in the treatment with 600 mg·kg^{-1} of As, RE was lower, but a clear increase was observed with the amount of peat added. For SNC soils, RE was >100% in SNC-0-0, while it decreased significantly with the addition of As, SNC-300-0 (15.7%), and SNC-600-0 (3.3%). Moreover, peat addition to the reference soil SNC caused a significant decrease in RE. In the treatments with 2% and 5% peat addition, there were no statistically significant differences in the RE values. However, when 2% of peat was added, there was an increasing trend in RE; while, when 5% was added, RE was reduced not significantly (Figure 3).

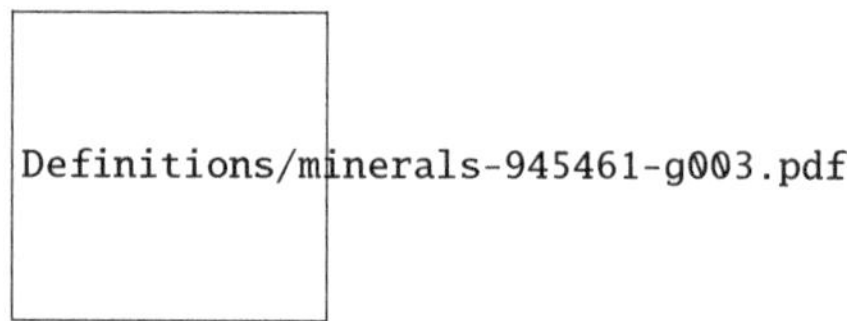

Figure 3. *Lactuca sativa* L. root elongation (RE%) in the soils with different treatments. Lowercase letters indicate statistically significant differences for each treatment with different peat additions; and capital letters among the different treatments with As for the same peat content (Kruskal-Wallis test, $p < 0.05$). Values higher than 100% indicate hormesis.

3.4. Relation between Soil Properties and Their Recovery

In SC samples, higher As solubility (Table 2) is related to soil toxicity by the endpoints used in toxicity bioassays (BR, SG, and RE) ($p < 0.01$). However, the addition of peat improves the three endpoints with a high degree of significance ($p < 0.01$). While, in the SNC and AZN samples, with no or few carbonates, the treatment with peat only shows a direct correlation ($p < 0.01$) with the respiration bioassay (BR). In both soil samples, iron oxides (Fed and Feo) were inversely correlated with water-soluble As concentration ($p < 0.01$), whereas other soil properties such as OC and CEC were directly correlated. Note that in these cases, the behavior of As pollution on the Fed and Feo content was uneven, with no correlation in SNC samples, and high direct correlation ($p < 0.01$) in the case of the AZN samples (Table 3).

Table 3. Spearman correlations between different soil samples, remediation treatment with peat, and the solubility of As in soils.

Variables		Peat	RE [2]	SG [3]	BR [4]	EC [5]	pH [6]	OC [7]	CEC [8]	Fed [9]	Feo [10]	BD [11]
AZN	Peat	-	-	-	0.986 **	0.993 **	−0.761 **	0.971 **	0.988 **	−0.933 **	−0.968 **	−0.999 **
	As$_W$ [1]	−0.787 **	-	-	−0.770 **	−0.803 **	0.661 *	−0.739 *	−0.792 **	0.778 ***	0.766 **	0.801 **
SC	Peat	-	0.662 **	0.583 **	0.839 **	-	−0.733 **	0.944 **	0.994 **	0.916 **	0.987 **	−0.824 **
	As$_W$ [1]	−0.558 **	−0.828 **	−0.913 **	−0.742 **	-	0.896 **	−0.521 **	−0.550 **	−0.346 **	−0.550 **	-
SNC	Peat	-	-	-	0.916 **	0.452 *	-	0.999 **	0.939 **	−0.874 **	−0.918 **	−0.994 **
	As$_W$ [1]	−0.492 **	-	-	−0.559 **	-	0.557 **	−0.490 **	−0.461 **	-	-	0.472 **

[1] water-soluble As concentration; [2] *Lactuca sativa* L. root elongation; [3] *Lactuca sativa* L. seed germination; [4] basal soil respiration rates; [5] soil electrical conductivity (1:5); [6] soil pH (1:2.5); [7] organic carbon content; [8] cation-exchange capacity; [9] free oxides iron; [10] amorphous iron; [11] bulk density. * $p < 0.05$; ** $p < 0.01$.

4. Discussion

According to the different doses of peat added, we observed that the adsorption processes on colloidal fractions of the soils played a key role in the As solubility. Although the cation exchange capacity (CEC) and the organic carbon (OC) content of the analyzed soil samples was low, peat additions increased the soil OC content, as well as the total humic extract and humic and fulvic acids, which provides an important content of reactive colloidal fractions that allow the complexation of the different chemical forms of As (mainly arsenates) [58]. The additions of 2% and 5% of peat enhance arsenic retention above 98% in most cases (except in AZN), thus reducing its solubility. These results are consistent with other studies proposing the use of peatland materials for the decontamination of waters polluted with As and other PHEs, from, e.g., gold mines in Finnish Lapland [38] or Northern England [59], among others. In the present study, soil properties such as pH, iron oxides content (Feo) and OC are essential to As retention probably due to the influence they exert on the control of speciation, bioavailability and solubility of As in soils [60,61]. Soil pH controls strongly the adsorption processes of the arsenic in the soil [62]. Authors such as Jones et al. [63], Simón et al. [64], and Tyler and Olsson [65], among others, agree that within the soil pH range (3.8 to 6.3), arsenic is more easily adsorbed, while arsenic precipitation could decrease from pH > 6.5. These data agree with the obtained results, since the different treatments with As, both in the SC and SNC soil samples, reveal a slight tendency to increase the As solubility when the pH of the solutions increases. For example, in non-calcareous soil samples (SNC) polluted with 300 and 600 mg·kg^{-1} As, without peat additions, the As$_W$ concentration is lower with respect to calcareous soil samples (SC) with a basic character. This, in addition to the action of iron oxides, is because the pH in SNC is below 6.5. However, this increase in solubility is minimal since it is damped by the action of remediation with peat. Thus, when the pH$_W$ of the solution is basic and Ca^{2+} dominates in the cation exchange complex, it plays an important role in the As adsorption promoting binding bridge between the forms of arsenate and the humic substances of the peat, decreasing the As solubility, as it was observed in the SC samples, by the formation of organo-mineral complexes [66]. On the other hand, in the case of SNC samples, the acidic pH of the solution and the higher content of iron oxides, allow the co-precipitation of As with Fe and peat, thereby justifying the decrease in the As solubility, compared to the samples not treated with peat. Coinciding with Mukwaturi [67], the added organic matter is likely to play an important role in the formation of complexes between organic matter, Fe (III) and As, leading to an immobilization of the arsenic. However, without peat additions, the content of As_W was much lower than that added, which shows the greater retention power that iron oxides perform compared to carbonates coinciding with the stated by Kabata-Pendias and Pendias [5] and Pierce and Moore [68]. In fact, in anthropogenically polluted soil samples (AZN), although the As_T concentration was 120.2 mg·kg^{-1}, it showed a low As solubility that could be largely controlled by the high amount of iron oxides present.

The observed decrease in As solubility showed positive results in the microbial activity of the soils, according to Spearman's correlation. However, this increase in microbial activity measured by basal soil respiration appears to be related to the addition of peat rather than the consequent reduction in As solubility; as the addition of peat improves physicochemical properties such as OC content and porosity (Po), and activates soil biota [69]. Kumpiene et al. [70], stating that peat amendments in soils polluted with As and high levels of Fe do not show toxicity to microorganisms, and, therefore, it is the most efficient method of reducing As dissolved in water retained in soil pores and in depth, most likely as a result of low soil density and good air diffusion in the soil, as could be deduced for AZN soil samples. Similarly, Niemeyer [71] obtained a good positive correlation between added organic amendments and basal respiration. However, other researchers maintained that the respiration of certain microorganisms can be reduced by humus-rich peat extract and concluded that they can have a toxic effect [72].

In the *Lactuca sativa* L. phytotoxicity test, the calcareous samples (SC), where the seeds do not germinate at any of the As levels of pollution when they are treated with peat, showed a clear positive response in seed germination (SG) and root elongation (RE), even higher than that obtained in the controls. However, in the samples with no or low carbonates (SNC and AZN), peat addition did not have a positive influence on SG and, consequently, on RE. Conversely, the addition of peat to SNC and AZN on the *Lactuca sativa* L. bioassay showed a negative effect probably caused by the toxic role that polyphenolic compounds such as lariciresinol-sesquilignan, which is found at a high level in the peat, can play. In this sense, Nishiwaki et al. [73] examined the lariciresinol stereoisomers effect on the growth of plants like *Lactuca sativa* L., and observed that most of the diastereomers reduced the growth of *Lactuca sativa* L. at a rate between −20.7% and −1.6%. Likewise, Cutillo et al. [74] observed inhibition of germination of *Lactuca sativa* L. between 50% and 75% at low concentrations of various derivatives of lignans. A toxic effect caused by polyphenolic compounds that appear to be concealed when calcium carbonate is present, as it does not occur in carbonate soil samples.

Finally, if we compare the influence that the peat dose added had in the bioassays, we observed that in calcareous soil (SC) samples, 5% peat slightly improved the respiration rate in soils with high pollution (600 mg·kg^{-1} As) compared to the 2% peat dose. However, this fact is much more prominent in the samples of non-calcareous soil (AZN and SNC) for the proportion of 5% peat in both levels of pollution by As, which agrees with what has been reported by other authors studying the influence of iron oxides rich-amendments, like the peat, in the restoration of As-polluted mine soils [75]. Likewise, the addition of 5% peat is much more effective than the 2% in terms of root elongation for *Lactuca sativa* L. in calcareous soil (SC) since it buffers the phytotoxic effects of the polyphenolic compounds even at the 600 mg·kg^{-1} As level of the SC samples. However, for non-calcareous soils, the addition of a higher dose of peat has a large negative effect on SG and RE that does not compensate for the greater reduction in the As solubility. Taking into consideration all this, and that the reduction in As solubility achieved with the addition of 5% is greater than with the addition of 2%, although, without statistical significance, the dose of peat to be added as the most recommended amendment for calcareous soils is 5%, while for non-calcareous soils it is 2% since it reduces the solubility and the phytotoxic effect of polyphenolic compounds is less severe.

5. Conclusions

The application of peat as an organic amendment in As-polluted soils showed high efficiency in fixing As in both natural and artificially polluted soils, regardless of the presence of calcareous properties in the soils. While soils fix some of the added As, increasingly with higher iron oxide content in the soil, additions of 2% and 5% of peat greatly enhance arsenic retention, with values above 98% in most cases. Peat addition increases OC, total humic extract, and humic and fulvic acids, providing important colloidal reactive fractions that allow the complexation of the As compounds, reducing As solubility. Soil As toxicity differs according to soil properties and peat addition. Biological response to arsenic additions was worse (lower soil respiration and inhibition of *Lactuca sativa* L. seed germination) in calcareous soils than in non-calcareous soils, and the former soils showed the greatest improvement in biological response by lowering As toxicity with peat amendments. In fact, in non-calcareous soils, peat addition has a negative effect on the biological response since peat contains large quantities of polyphenolic compounds such as lariciresinol-sesquiligninan that have a phytotoxic effect, which in the presence of carbonates appears to be concealed; thus additional studies are needed to assess the potential toxicity of the use of peat as an organic amendment and to prevent hazards derived from the environmental management of the As-polluted soils. Since, although both doses of peat tested (2% and 5%) drastically reduced As mobility in calcareous soils, the 5% dose is the most recommended due to the great reduction of As and the absence of phytotoxic effect by polyphenolic compounds; while, in non-calcareous soils, the recommended dose is 2%, although peat amendment should be added carefully in these soils to avoid phytotoxic effects.

Supplementary Materials: The following are available online at http://www.mdpi.com/2075-163X/10/11/968/s1, Figure S1: Chromatogram of peat—Determination of polyphenolic profile of the peat.

Author Contributions: Conceptualization: F.J.M.G., M.S.A., F.J.M.P., A.R.-F.; methodology: F.J.M.G., M.S.A., F.J.M.P., A.R.-F.; validation: A.A.-G., M.G.-C., F.J.M.G., A.R.-F.; formal analysis: F.J.M.G., M.G.-C., A.R.-F., A.A.-G.; investigation: M.G.-C., A.R.-F., A.A.-G., F.J.M.G.; resources: F.J.M.G., M.S.A., F.J.M.P., A.R.-F.; data curation: F.M.G., M.S.A., A.R.-F., M.G.-C.; writing—original draft preparation: F.J.M.G., A.A.-G., M.S.A.; writing—review and editing: A.A.-G., A.R.-F., M.G.-C., M.S.A., F.J.M.P., F.J.M.G.; visualization: F.J.M.G., A.R.-F., M.G.-C., A.A.-G.; supervision: F.J.M.P., F.M.G., M.S.A., A.R.-F.; project administration: F.J.M.G., F.J.M.P., M.S.A.; funding acquisition: F.J.M.P., M.S.A., F.J.M.G., A.A.-G. All authors have read and agreed to the published version of the manuscript.

Funding: This work was supported by the Research Project RTI 2018-094327-B-I00 and the Grant FPU-18/02901 (Spanish Ministry of Science, Innovation and Universities).

Conflicts of Interest: The authors declare no conflict of interest.

References

1. Mirsal, I.A. *Soil Pollution. Origin, Monitoring and Remediation*, 2nd ed.; Springer: Berlin/Heidelberg, Germany, 2008; ISBN 9783540707752.
2. Nagajyoti, P.C.; Lee, K.D.; Sreekanth, T.V.M. Heavy metals, occurrence and toxicity for plants: A review. *Environ. Chem. Lett.* **2010**, *8*, 199–216. [CrossRef]
3. Palansooriya, K.N.; Shaheen, S.M.; Chen, S.S.; Tsang, D.C.W.; Hashimoto, Y.; Hou, D.; Bolan, N.S.; Rinklebe, J.; Ok, Y.S. Soil amendments for immobilization of potentially toxic elements in contaminated soils: A critical review. *Environ. Int.* **2020**, *134*, 105046. [CrossRef]
4. Pavel, L.V.; Gavrilescu, M. Overview of ex situ decontamination techniques for soil cleanup. *Environ. Eng. Manag. J.* **2008**, *7*, 815–834. [CrossRef]
5. Kabata-Pendias, A.; Pendias, H. *Trace Elements in Soils and Plants*, 3rd ed.; CRC Press (Taylor & Francis Group): Boca Raton, FL, USA, 2001; ISBN 0849315751.
6. Nriagu, J.O.; Bhattacharya, P.; Mukherjee, A.B.; Bundschuh, J.; Zevenhoven, R.; Loeppert, R.H. Arsenic in soil and groundwater: An overview. In *Trace Metals and Other Contaminants in the Environment*; Bhattacharya, P., Mukherjee, A.B., Bundschuh, J., Zevenhoven, R., Loeppert, R.H., Eds.; Elsevier Inc.: Amsterdam, The Netherlands, 2007; Volume 9, pp. 3–60. ISBN 9780444518200.
7. Beni, C.; Diana, G.; Marconi, S. Bovine milk chain in Italian farms. I. Arsenic levels in soil, gravitational and clean water, bovine diet, and milk. *Agrochimica* **2008**, *52*, 99–115.
8. Zhong, X.; Chen, Z.; Li, Y.; Ding, K.; Liu, W.; Liu, Y.; Yuan, Y.; Zhang, M.; Baker, A.J.M.; Yang, W.; et al. Factors influencing heavy metal availability and risk assessment of soils at typical metal mines in Eastern China. *J. Hazard. Mater.* **2020**, *400*, 123289. [CrossRef] [PubMed]

9. Kim, H.; Lee, M.; Lee, J.H.; Kim, K.H.; Owens, G.; Kim, K.R. Distribution and extent of heavy metal(loid) contamination in agricultural soils as affected by industrial activity. *Appl. Biol. Chem.* **2020**, *63*, 1–8. [CrossRef]

10. Fernández-Caliani, J.C. Risk-based assessment of multimetallic soil pollution in the industrialized peri-urban area of Huelva, Spain. *Environ. Geochem. Health* **2012**, *34*, 123–139. [CrossRef]

11. Wang, S.; Mulligan, C.N. Occurrence of arsenic contamination in Canada: Sources, behavior and distribution. *Sci. Total Environ.* **2006**, *366*, 701–721. [CrossRef] [PubMed]

12. Fazle Bari, A.S.M.; Lamb, D.; Choppala, G.; Bolan, N.; Seshadri, B.; Rahman, M.A.; Rahman, M.M. Geochemical fractionation and mineralogy of metal(loid)s in abandoned mine soils: Insights into arsenic behaviour and implications to remediation. *J. Hazard. Mater.* **2020**, *399*, 123029. [CrossRef]

13. De Souza Neto, H.F.; da Silveira Pereira, W.V.; Dias, Y.N.; de Souza, E.S.; Teixeira, R.A.; de Lima, M.W.; Ramos, S.J.; do Amarante, C.B.; Fernandes, A.R. Environmental and human health risks of arsenic in gold mining areas in the eastern Amazon. *Environ. Pollut.* **2020**, *265*, 114969. [CrossRef]

14. García-Carmona, M.; García-Robles, H.; Turpín Torrano, C.; Fernández Ondoño, E.; Lorite Moreno, J.; Sierra Aragón, M.; Martín Peinado, F.J. Residual pollution and vegetation distribution in amended soils 20 years after a pyrite mine tailings spill (Aznalcóllar, Spain). *Sci. Total Environ.* **2019**, *650*, 933–940. [CrossRef] [PubMed]

15. Tarvainen, T.; Reichel, S.; Müller, I.; Jordan, I.; Hube, D.; Eurola, M.; Loukola-Ruskeeniemi, K. Arsenic in agro-ecosystems under anthropogenic pressure in Germany and France compared to a geogenic As region in Finland. *J. Geochem. Explor.* **2020**, *217*, 106606. [CrossRef]

16. Ali, W.; Mao, K.; Zhang, H.; Junaid, M.; Xu, N.; Rasool, A.; Feng, X.; Yang, Z. Comprehensive review of the basic chemical behaviours, sources, processes, and endpoints of trace element contamination in paddy soil-rice systems in rice-growing countries. *J. Hazard. Mater.* **2020**, *397*, 122720. [CrossRef] [PubMed]

17. Neal, C.; Robson, A.J. A summary of river water quality data collected within the Land-Ocean Interaction Study: Core data for eastern UK rivers draining to the North Sea. *Sci. Total Environ.* **2000**, *251–252*, 585–665. [CrossRef]

18. Díez, M.; Simón, M.; Dorronsoro, C.; García, I.; Martín, F. Background arsenic concentrations in Southeastern Spanish soils. *Sci. Total Environ.* **2007**, *378*, 5–12. [CrossRef]

19. O'Neill, P. Arsenic. In *Heavy Metals in Soils*; Alloway, B.J., Ed.; Springer Science & Business Media: New York, NY, USA, 1995; pp. 105–121. ISBN 9789400744691.

20. Adriano, D.C. Arsenic. In *Trace Elements in Terrestrial Environments. Biogeochemistry, Bioavailability and Risks of Metals*; Adriano, D.C., Ed.; Springer Science & Business Media: New York, NY, USA, 2001; pp. 219–261. ISBN 9781468495058.

21. Bohn, H.L.; McNeal, B.L.; O'Connor, G.A. *Soil Chemistry*, 3rd ed.; John Wiley & Sons: New York, NY, USA, 2001; ISBN 978-1-118-62923-9.

22. Kabata-Pendias, A.; Mukherjee, A.B. *Trace Elements from Soil to Human*; Springer Science & Business Media: Berlin, Germany, 2007; ISBN 9783540327134.

23. Singh, R.; Singh, S.; Parihar, P.; Singh, V.P.; Prasad, S.M. Arsenic contamination, consequences and remediation techniques: A review. *Ecotoxicol. Environ. Saf.* **2015**, *112*, 247–270. [CrossRef]

24. BOJA. Boletín Oficial de la Junta de Andalucía. Decreto 18/2015, de 27 de enero, por el que se aprueba el reglamento que regula el régimen aplicable a los suelos contaminados. *BOJA* **2015**, *38*, 28–64. Available online: http://www.juntadeandalucia.es/medioambiente/web/2012_provisional/2015/reglamento_suelos_contaminados.pdf (accessed on 3 October 2020).

25. German Federal Government. Federal Soil Protection Act of 17 March 1998 (Federal Law Gazette I p. 502). 1998. Available online: http://www.bmu.de/files/pdfs/allgemein/application/pdf/soilprotectionact.pdf (accessed on 3 October 2020).

26. VROM. *Circular on Target and Intervention Values for Soil Remediation, Reference DBO/1999226863*; Ministry of Housing Spatial Planning and the Environment: Bilthoven, The Netherlands, 2000. Available online: https://www.esdat.net/Environmental%20Standards/Dutch/annexS_I2000Dutch%20Environmental%20Standards.pdf (accessed on 3 October 2020).

27. Environmental Agency. Government of the United Kingdom. In *Human Health Toxicological Assessment of Contaminants in Soil. Science Report—Final SC050021/SR2*; Environmental Agency: Bristol, UK, 2009; ISBN

9781844328581. Available online: https://www.claire.co.uk/information-centre/water-and-land-library-wall/44-risk-assessment/178-soil-guideline-values?showall=1 (accessed on 3 October 2020).

28. Dickinson, N.M.; Baker, A.J.M.; Doronila, A.; Laidlaw, S.; Reeves, R.D. Phytoremediation of inorganics: Realism and synergies. *Int. J. Phytoremediat.* **2009**, *11*, 97–114. [CrossRef]

29. Ali, H.; Khan, E.; Sajad, M.A. Phytoremediation of heavy metals—Concepts and applications. *Chemosphere* **2013**, *91*, 869–881. [CrossRef]

30. García-Carmona, M.; Romero-Freire, A.; Sierra Aragón, M.; Martínez Garzón, F.J.; Martín Peinado, F.J. Evaluation of remediation techniques in soils affected by residual contamination with heavy metals and arsenic. *J. Environ. Manage.* **2017**, *191*, 228–236. [CrossRef]

31. Aguilar, J.; Dorronsoro, C.; Fernández, E.; Fernández, J.; García, I.; Martín, F.; Sierra, M.; Simón, M. Arsenic contamination in soils affected by a pyrite-mine spill (Aznalcóllar, SW Spain). *Water. Air. Soil Pollut.* **2007**, *180*, 271–281. [CrossRef]

32. Shaheen, S.M.; Shams, M.S.; Khalifa, M.R.; El-Dali, M.A.; Rinklebe, J. Various soil amendments and environmental wastes affect the (im)mobilization and phytoavailability of potentially toxic elements in a sewage effluent irrigated sandy soil. *Ecotoxicol. Environ. Saf.* **2017**, *142*, 375–387. [CrossRef] [PubMed]

33. Park, J.H.; Lamb, D.; Paneerselvam, P.; Choppala, G.; Bolan, N.; Chung, J.W. Role of organic amendments on enhanced bioremediation of heavy metal(loid) contaminated soils. *J. Hazard. Mater.* **2011**, *185*, 549–574. [CrossRef] [PubMed]

34. Kumpiene, J.; Lagerkvist, A.; Maurice, C. Stabilization of As, Cr, Cu, Pb and Zn in soil using amendments—A review. *Waste Manag.* **2008**, *28*, 215–225. [CrossRef]

35. Sierra Aragón, M.; Nakamaru, Y.M.; García-Carmona, M.; Martínez Garzón, F.J.; Martín Peinado, F.J. The role of organic amendment in soils affected by residual pollution of potentially harmful elements. *Chemosphere* **2019**, *237*, 1–9. [CrossRef]

36. Nakamaru, Y.M.; Martín Peinado, F.J. Effect of soil organic matter on antimony bioavailability after the remediation process. *Environ. Pollut.* **2017**, *228*, 425–432. [CrossRef]

37. González, Z.I.; Krachler, M.; Cheburkin, A.K.; Shotyk, W. Spatial distribution of natural enrichments of arsenic, selenium and uranium in a minerotrophic peatland, Gola di Lago, Canton Ticino, Switzerland. *Environ. Sci. Technol.* **2006**, *40*, 6568–6574. [CrossRef]

38. Palmer, K.; Ronkanen, A.K.; Kløve, B. Efficient removal of arsenic, antimony and nickel from mine wastewaters in Northern treatment peatlands and potential risks in their long-term use. *Ecol. Eng.* **2015**, *75*, 350–364. [CrossRef]

39. Cloy, J.M.; Farmer, J.G.; Graham, M.C.; Mackenzie, A.B. Retention of As and Sb in ombrotrophic peat bogs: Records of As, Sb, and Pb deposition at four Scottish sites. *Environ. Sci. Technol.* **2009**, *43*, 1756–1762. [CrossRef]

40. Rothwell, J.J.; Taylor, K.G.; Ander, E.L.; Evans, M.G.; Daniels, S.M.; Allott, T.E.H. Arsenic retention and release in ombrotrophic peatlands. *Sci. Total Environ.* **2009**, *407*, 1405–1417. [CrossRef]

41. Martín, F.; Simon, M.; Arco, E.; Romero, A.; Dorronsoro, C. Arsenic behaviour in polluted soils after remediation activities. In *Soil Health and Land Use Management*; Hernandez Soriano, M.C., Ed.; InTech: Shangai, China, 2012; pp. 201–216. ISBN 978-953-307-614-0.

42. Romero-Freire, A.; Sierra-Aragón, M.; Ortiz-Bernad, I.; Martín-Peinado, F.J. Toxicity of arsenic in relation to soil properties: Implications to regulatory purposes. *J. Soils Sediments* **2014**, *14*, 968–979. [CrossRef]

43. Simón, M.; Ortiz, I.; García, I.; Fernández, E.; Fernández, J.; Dorronsoro, C.; Aguilar, J. El desastre ecológico de doñana. *Edafología* **1998**, *5*, 153–161.

44. Romero-Freire, A.; Martin Peinado, F.J.; van Gestel, C.A.M. Effect of soil properties on the toxicity of Pb: Assessment of the appropriateness of guideline values. *J. Hazard. Mater.* **2015**, *289*, 46–53. [CrossRef]

45. Ministerio de Agricultura Pesca y Alimentación. *Métodos Oficiales de Análisis. Tomo III. Plantas, Productos Orgánicos Fertilizantes, Suelos, Agua, Productos Fitosanitarios y Fertilizantes Inorgánicos*; Publicaciones del Ministerio de Agricultura Pesca y Alimentación: Madrid, Spain, 1986.

46. Holmgren, G.G.S. A Rapid Citrate-Dithionite Extractable Iron Procedure. *Soil Sci. Soc. Am. J.* **1967**, *31*, 210–211. [CrossRef]

47. Blakemore, L.C. Exchange complex dominated by amorphous material (ECDAM). In *The Andisol Proposal*; Smith, G.D., Ed.; Soil Bureau (Department of Scientific and Industrial Research): Wellington, New Zealand, 1978; Volume 21.

48. Kononova, M.M. *Materia Orgánica del Suelo: Su Naturaleza, Propiedades y Métodos de Investigación*, 1st ed.; OIKOS-TAU Ediciones: Barcelona, Spain, 1982; ISBN 8428104964.

49. Fendorf, S.; La Force, M.J.; Li, G. Heavy Metals in the Environment Temporal Changes in Soil Partitioning and Bioaccessibility. *J. Environ. Qual.* **2004**, *33*, 2049–2055. [CrossRef] [PubMed]

50. Tang, X.Y.; Zhu, Y.G.; Cui, Y.S.; Duan, J.; Tang, L. The effect of ageing on the bioaccessibility and fractionation of cadmium in some typical soils of China. *Environ. Int.* **2006**, *32*, 682–689. [CrossRef]

51. Aguilar, J.; Dorronsoro, C.; Fernández, E.; Fernández, J.; García, I.; Martín, F.; Sierra, M.; Simón, M. Remediation of As-contaminated soils in the Guadiamar river basin (SW, Spain). *Water. Air. Soil Pollut.* **2007**, *180*, 109–118. [CrossRef]

52. Abad-Valle, P.; Iglesias-Jiménez, E.; Álvarez-Ayuso, E. A comparative study on the influence of different organic amendments on trace element mobility and microbial functionality of a polluted mine soil. *J. Environ. Manage.* **2017**, *188*, 287–296. [CrossRef]

53. ISO 17155. *Soil Quality. Determination Abundance Activity Soil Microflora Using Respiration Curves*; International Standard ISO No.17155; International Organization for Standardization: Geneva, Switzerland, 2002; Available online: https://www.iso.org/standard/53529.html (accessed on 24 July 2020).

54. OECD. *Guideline for the Testing of Chemicals. Proposal for Updating Guideline 208. Terrestrial Plant. Test.: 208: Seedling Emergence and Seedling Growth Test*; Organization for Economic Cooperation and Development: Paris, France, 2003; Available online: https://www.oecd.org/chemicalsafety/testing/33653757.pdf (accessed on 24 July 2020).

55. US EPA. *Ecological Effects Test. Guidelines. Seed Germination/Root Elongation Toxicity Test. OPPTS 850.4200*; United States Environmental Protection Agency: Washington DC, USA, 1996; 2p. Available online: https://nepis.epa.gov/Exe/tiff2png.cgi/P100RF5I.PNG?-r+75+-g+7+D%3A%5CZYFILES% 5CINDEX%20DATA%5C95THRU99%5CTIFF%5C00003181%5CP100RF5I.TIF (accessed on 24 July 2020).

56. Torres, M.T.R. Empleo de los ensayos con plantas en el control de contaminantes tóxicos ambientales. *Rev. Cubana Hig. Epidemiol.* **2003**, *41*, 2–3.

57. Reimann, C.; De Caritat, P. *Chemical Elements in the Environment: Factsheets for the Geochemist and Environmental Scientist*; Springer Science & Business Media: Berlin/Heidelberg, Germany, 2012; ISBN 3642720161.

58. Zhang, D.-r.; Chen, H.r.; Xia, J.-l.; Nie, Z.y.; Fan, X.l.; Liu, H.-c.; Zheng, L.; Zhang, L.-j.; Yang, H.-y. Humic acid promotes arsenopyrite bio-oxidation and arsenic immobilization. *J. Hazard. Mater.* **2020**, *384*, 121359. [CrossRef]

59. Rothwell, J.J.; Taylor, K.G.; Evans, M.G.; Allott, T.E.H. Contrasting controls on arsenic and lead budgets for a degraded peatland catchment in Northern England. *Environ. Pollut.* **2011**, *159*, 3129–3133. [CrossRef] [PubMed]

60. Bissen, M.; Frimmel, F.H. Arsenic—A review. Part I: Occurrence, toxicity, speciation, mobility. *Acta Hydrochim. Hydrobiol.* **2003**, *31*, 9–18. [CrossRef]

61. Juhasz, A.L.; Naidu, R.; Zhu, Y.G.; Wang, L.S.; Jiang, J.Y.; Cao, Z.H. Toxicity issues associated with geogenic arsenic in the groundwater-soil-plant-human continuum. *Bull. Environ. Contam. Toxicol.* **2003**, *71*, 1100–1107. [CrossRef]

62. Roberts, D.; Nachtegaal, M.; Sparks, D.L. Speciation of metals in soils. In *Chemical Processes in Soils*; Tabatabai, M.A., Sparks, D.L., Eds.; Soil Science Society of America: Washington, DC, USA, 2005; pp. 619–654. ISBN 9780891188926.

63. Jones, C.A.; Inskeep, W.P.; Neuman, D.R. Arsenic transport in contaminated mine tailings following liming. *J. Environ. Qual.* **1997**, *26*, 433–439. [CrossRef]

64. Simón, M.; Martín, F.; García, I.; Bouza, P.; Dorronsoro, C.; Aguilar, J. Interaction of limestone grains and acidic solutions from the oxidation of pyrite tailings. *Environ. Pollut.* **2005**, *135*, 65–72. [CrossRef] [PubMed]

65. Tyler, G.; Olsson, T. Concentrations of 60 elements in the soil solution as related to the soil acidity. *Eur. J. Soil Sci.* **2001**, *52*, 151–165. [CrossRef]

66. Kong, Y.; Kang, J.; Shen, J.; Chen, Z.; Fan, L. Influence of humic acid on the removal of arsenate and arsenic by ferric chloride: Effects of pH, As/Fe ratio, initial As concentration, and co-existing solutes. *Environ. Sci. Pollut. Res.* **2017**, *24*, 2381–2393. [CrossRef]

67. Mukwaturi, M.; Lin, C. Mobilization of heavy metals from urban contaminated soils under water inundation conditions. *J. Hazard. Mater.* **2015**, *285*, 445–452. [CrossRef]

68. Pierce, M.L.; Moore, C.B. Adsorption of arsenite and arsenate on amorphous iron hydroxide. *Water Res.* **1982**, *16*, 1247–1253. [CrossRef]

69. Six, J.; Bossuyt, H.; Degryze, S.; Denef, K. A history of research on the link between (micro)aggregates, soil biota, and soil organic matter dynamics. *Soil Tillage Res.* **2004**, *79*, 7–31. [CrossRef]

70. Kumpiene, J.; Desogus, P.; Schulenburg, S.; Arenella, M.; Renella, G.; Brännvall, E.; Lagerkvist, A.; Andreas, L.; Sjöblom, R. Utilisation of chemically stabilized arsenic-contaminated soil in a landfill cover. *Environ. Sci. Pollut. Res.* **2013**, *20*, 8649–8662. [CrossRef] [PubMed]

71. Niemeyer, J.C.; Lolata, G.B.; de Carvalho, G.M.; Da Silva, E.M.; Sousa, J.P.; Nogueira, M.A. Microbial indicators of soil health as tools for ecological risk assessment of a metal contaminated site in Brazil. *Appl. Soil Ecol.* **2012**, *59*, 96–105. [CrossRef]

72. Minderlein, S.; Blodau, C. Humic-rich peat extracts inhibit sulfate reduction, methanogenesis, and anaerobic respiration but not acetogenesis in peat soils of a temperate bog. *Soil Biol. Biochem.* **2010**, *42*, 2078–2086. [CrossRef]

73. Nishiwaki, H.; Kumamoto, M.; Shuto, Y.; Yamauchi, S. Stereoselective syntheses of all stereoisomers of lariciresinol and their plant growth inhibitory activities. *J. Agric. Food Chem.* **2011**, *59*, 13089–13095. [CrossRef]

74. Cutillo, F.; D'Abrosca, B.; DellaGreca, M.; Fiorentino, A.; Zarrelli, A. Lignans and neolignans from Brassica fruticulosa: Effects on seed germination and plant growth. *J. Agric. Food Chem.* **2003**, *51*, 6165–6172. [CrossRef]

75. Abad-Valle, P.; Álvarez-Ayuso, E.; Murciego, A. Evaluation of ferrihydrite as amendment to restore an arsenic-polluted mine soil. *Environ. Sci. Pollut. Res.* **2015**, *22*, 6778–6788. [CrossRef]

Publisher's Note: MDPI stays neutral with regard to jurisdictional claims in published maps and institutional affiliations.

Article

Effect of the Long-Term Application of Sewage Sludge to A Calcareous Soil on Its Total and Bioavailable Content in Trace Elements, and Their Transfer to the Crop

Armelle Zaragüeta [1,2], Alberto Enrique [1,*], Iñigo Virto [1], Rodrigo Antón [1], Henar Urmeneta [3] and Luis Orcaray [2]

1   Departamento Ciencias, IS-FOOD, Universidad Pública de Navarra, 31006 Pamplona, Spain; azaragueta@intiasa.es (A.Z.); inigo.virto@unavarra.es (I.V.); rodrigo.anton@unavarra.es (R.A.)
2   Área de Innovación, Sección de Sistemas Sostenibles, Instituto Navarro de Tecnologías e Infraestructuras Agroalimentarias, 31610 Villava, Spain; lorcaray@intiasa.es
3   Departamento Estadística, Informática y Matemáticas, Universidad Pública de Navarra, 31006 Pamplona, Spain; henar@unavarra.es
*   Correspondence: alberto.enrique@unavarra.es

Abstract: Sewage sludge (SS) can be used as an organic amendment in agricultural soils, provided they comply with the relevant legislation. This use can incorporate traces of metals into the soil, which can cause environmental or human health problems. In the study period between 1992 and 2018 (26 years), it was observed that the use of SS as an organic fertilizer significantly increased the total concentration of Zn, Cu, Cr, Ni and Hg of this study between 55.6% (Hg) and 7.0% (Ni). The concentration of Zn, Cu, Pb, Ni and Cd extracted with DTPA, also increased between 122.2% (Zn) and 11.3% (Cd). In contrast, the Mn concentrations extracted with Diethylene Triamine Pentaacetic Acid (DTPA)were 6.5% higher in the treatments without SS. These changes in the soil had an impact on the crop, which showed a significant increase in the concentration of Zn, Cu and Cr in the grain, between 15.0% (Cr) and 4.4% (Cu), and a decrease in the concentration of Mn, Cr and Ni in the barley straw when SS was added to the soil between 32.2% (Mn) and 29.6% (Ni). However, the limits established by current legislation on soil protection and food were not exceeded. This limited transfer to the crop, is likely due to the high content of carbonates and organic matter in the soil, which limit the bioavailability of most of the trace metals (TM) in the soil. As a conclusion, we observe that the use of SS as an organic amendment increased the concentration of some TM in the soil, in its bioavailable forms, and in the crop.

Keywords: trace metals; sewage sludge; calcareous soil; extraction DTPA; crop; transfer; long time

Citation: Zaragüeta, A.; Enrique, A.; Virto, I.; Antón, R.; Urmeneta, H.; Orcaray, L. Effect of the Long-Term Application of Sewage Sludge to A Calcareous Soil on Its Total and Bioavailable Content in Trace Elements, and Their Transfer to the Crop. *Minerals* **2021**, *11*, 356. https://doi.org/10.3390/min11040356

Academic Editor: Ana Romero-Freire and Hao Qiu

Received: 6 March 2021
Accepted: 26 March 2021
Published: 30 March 2021

Publisher's Note: MDPI stays neutral with regard to jurisdictional claims in published maps and institutional affiliations.

1. Introduction

As a consequence of the urbanization and industrialization of urban environments, there is a growing generation of wastewater of domestic and industrial origin. This water must be treated to ensure an environmentally safe and economically viable final destination. This wastewater treatment is generally carried out in treatment plants, and involves a generation of sludge (sewage sludge, SS). As in many other countries, in Spain, SS is considered waste, in accordance with the provisions of Law 22/2011 [1] on waste and contaminated soils, and must be managed in accordance. Depending on its final composition, SS can have three possible destinations within the Spanish legislative framework. By order of priority, these destinations are: (i) the application on agricultural soils, (ii) incineration or co-incineration, and (iii) deposition in landfills. In Spain, the annual generation of SS accounts for approximately 1,000,000 tons of dry matter, remaining stable in recent years, approximately 80% of which is used in agricultural soils [2].

The agricultural use of SS as an organic amendment has been reported to provide several benefits to agrosystems [3–6]. They include an increase in macro and micro nutrients in soil [7,8] and an increase in the content of organic matter [9–12]. These inputs normally result in higher crop yields [10,13,14].

However, this application involves a series of environmental and human health risks. Some of these risks are the possible transfer of genes encoding resistance to antibiotics [11], and the accumulation of persistent organic pollutants [4] or of trace metals (TM) in soil. Trace metals can accumulate in the surface layers [12,15–17] and/or leach out of the agrosystem [18–20], causing environmental contamination [4,21], ecotoxicity problems [14,22,23] or becoming part of the trophic chain due to their translocation in crops [21,24].

To control these risks, different strategies have been developed in order to minimize the impact of TM on soils where SS is applied [25]. On the one hand, there are legislative tools addressing soil protection [25–27] and animal and human food protection [28,29]. On the other hand, SS purification and stabilization techniques have been improved to obtain higher quality biosolids with a lower concentration in TM [30–32].

Along with these strategies, long-term studies are of great importance, as they allow observing the actual evolution of TM in soil, and therefore, the assessment of their true risk of accumulation and possible transfer to food, in the particular conditions of each soil and crop [16,17,32–34]. In this sense, the literature shows variable results, with some studies showing enrichments of TM in crops [16,33], and other studies reporting no accumulation [24,35]. This reflects the possible influence of edaphoclimatic conditions and SS composition on the consequences of their application, and highlights the need to consider these conditions in its study.

In this framework, a relevant factor in the assessment of the consequences of SS application to soil is to determine the proportion of TM present in the soil that are available to be adsorbed by crops. This proportion constitutes the so-called bioavailable fraction, and is influenced by the nature of the element, their interaction with soil components, soil properties, and the contact time of soil with the metal [36]. Some studies have shown that this concentration increases faster than the total concentration in soil as a consequence of the application of SS [37]. A series of methodologies have been developed to determine the concentration of bioavailable TMs in soils. A widely extended and accepted one is the use of the chelating agent DTPA (Diethylene Triamine Pentaacetic Acid) as an extractant [16,37–41].

Among the aspects related to the composition of soil that can modulate the bioavailability of TM, the amount and type of organic matter received with SS is determinant, as it can determine the formation of chelates, exchange complexes or organometallic complexes [12,42,43]. These phenomena can lengthen the retention time of trace metals in soil [44], limiting their bioavailability and their introduction into the trophic chain. On the other hand, in calcareous soils, the bioavailability of TMs in relation to their total content is assumed to be lower than in other types of soils due to the presence of carbonates. Carbonates maintain a high soil pH and are associated with sorption and precipitation processes of TM [15,45].

The general objective of this study was to assess the effect of long-term application of SS on the content of TM and their analytical bioavailability, and actual transfer to the most frequent crop in the rotation used in the area of study, in a cultivated calcareous soil. To this end, we studied the effect of the accumulated dose of SS on the total and DTPA-extracted content of TM in the soil after 26 years of SS application at different doses and frequencies in an experimental field. Then, the actual absorption by the crop was assessed by studying the concentration in TM of grains and leaves of barley grown after 26 years of experimentation.

2. Materials and Methods

2.1. Site Description

The study was carried out in an experimental field located in the municipality of Arazuri, in the region of Navarre, in Northern Spain. The field belongs to the local municipal wastewater treatment company (Commonwealth of Pamplona, Navarre, Spain).

The experimental field was setup in 1992, in order to assess the long-term effect of the direct application of SS as an agricultural amendment both in the soil and crops.

The experimental field is located on a Calcaric Cambisol [46], with a loam-clay-silty texture and more than 16% carbonates in the upper horizon (0–30 cm). The main soil characteristics at this depth are shown in Table 1. It is a well-drained soil, without problems of effective depth and under rainfed agricultural use for decades. Climate in this region is temperate sub-humid Mediterranean, with an average annual temperature of 12.9 °C, an average annual rainfall of 771 mm, and an average potential evapotranspiration according to Thornthwaite [47] of 696.7 mm between 1981 and 2010. [48].

Table 1. Average values of the physicochemical properties of the tilled horizon (0–30 cm) of the study soil and analytical methods used. Control soil and soil amended with sewage sludge for 26 years. Values are given as the mean $\pm$ standard deviation ($n = 4$).

Soil Physical-Chemical Properties	Control	After 26 Years	Analysis Methods
pH	8.67 ± 0.03	8.45 ± 0.11	Soil pH in water 1:2 [49]
Electric Conductivity (μs cm^{-3})	169 ± 10	233 ± 11	Diluted soil:water Extract 1:2.5 [50]
Bulk density (g cm^{-3})	1.59 ± 0.08	1.54 ± 0.01	Core Method [51]
Carbonates (%)	16.0 ± 2.1	16.7 ± 1.4	Bernard calcimeter [52]
Organic Carbon (%)	1.35 ± 0.02	1.56 ± 0.03	Wet Oxidation-Redox Titration Method [53]
Total Phosphorus (mg kg^{-1})	603 ± 52	945 ± 38	Microwave digestion + ICP-MS [54,55]
Available Phosphorus (mg kg^{-1})	32.2 ± 1.5	99.4 ± 8.4	Sodium Bicarbonate-Extractable P at pH 8.5 [56]
Total Potassiun (mg kg^{-1})	9294 ± 550	9709 ± 185	Microwave digestion + ICP-MS [54,55]
Available Potassiun (mg kg^{-1})	109 ± 15.0	110 ± 9.00	Ammonium Acetate Method at pH 7.0 [57]

2.2. Experimental Design

The experimental field was setup to study the effect of SS in the soil and crops, by verifying which doses were the most adequate ones to avoid risks of contamination by TM, on a conventional crop rotation in this area. The design is block factorial, with four repetitions ($n = 4$), and with combinations of two application rates (40 Mg ha^{-1} and 80 Mg ha^{-1}) of SS, and three frequencies (1, 2 and 4 years). Therefore, the treatments applied are 40 Mg ha^{-1} annually, 40 Mg ha^{-1} every two years, 40 Mg ha^{-1} every four years, 80 Mg ha^{-1} annually, 80 Mg ha^{-1} every two years and 80 Mg ha^{-1} every four years. There is also a control treatment in which SS or other fertilizers are not applied denominated "Control", and a treatment that receives mineral fertilization (46% urea and Ammonium Sulfate) at the commercially recommended dose denominated "MinFer". The individual plots per replicate, corresponding to the different treatments, had a surface area of 35 m^2 (10 m $\times$ 3.5 m).

The doses were calculated in 1992 based on the restrictive regulation of 250 N fertilizer units (NFU) per hectare and year, without taking into account the N yield for the year of application. As such, the treatment using 40 Mg ha^{-1} per year was set as the one corresponding to this dose, and those receiving less or more SS were used to follow the evolution of soil and crops below and above the recommended dose at the time. The frequencies of two and four years were established according to the common practices of farmers in the area, and the recommendations of the entity that supplied the SS. The N equivalence of the added doses has varied (Table 2) due to the SS composition changes over time (Table 3).

Table 2. N fertilizer units (NFU) provided by the study doses according to the N concentration in SS as used at the beginning of the study (1992), and at the end of the study (2018), without taking into account the annual yield of N for the dose provided.

Doses Sewage Sludge	NFU 1992	NFU 2018
40 Mg ha^{-1} every year	250	480
80 Mg ha^{-1} every year	500	960

Due to the interest of the cumulative study of TM, from the total doses provided in each treatment during the 26 years of the study, 5 ranges of accumulated SS were established: (i) 0 Mg ha^{-1} of SS (MinFer and Control), (ii) 250 Mg ha^{-1} of SS (40 Mg ha^{-1} every 4 years), called 250MgSS, (iii) correspondent 520 Mg ha^{-1} of SS and 560 Mg ha^{-1} of SS (40 Mg ha^{-1} every 2 years and 80 Mg ha^{-1} every 4 years) called 500MgSS, (iv) 1040 Mg ha^{-1} of SS (40 Mg ha^{-1} every year and 80 Mg ha^{-1} every 2 years) called 1000MgSS and (v) 2080 Mg ha^{-1} of SS (80 Mg ha^{-1} every year) called 2000MgSS. The crops used corresponded to a rainfed conditions in a rotation of 3 years (barley–barley–sunflower), managed in a conventional way (annual tillage with a 30 cm deep moldboard plough, and application of phytosanitary products according to the needs of the crops each year). The 2018 yield values expressed in percent 12% moisture were provided as supplementary material, firstly, for each fertilization treatment (Table S1) and, secondly, an average value was provided for treatments that had received SS and those that had not received SS (Table S2).

2.3. Sewage Sludge Characteristics and Application

The application of SS in each campaign was carried out in September, coinciding with the work of preparing the soil for cultivation. This application was made using a 3.5-m wide spreader trailer. Once the amount of SS corresponding to each treatment was spread, it was mixed with a moldboard plough down to 30 cm. This was conducted annually between three and four weeks before seeding.

The SS used in this study comes from the wastewater treatment plant corresponding to the city of the Pamplona region and its metropolitan area, with approximately 335,000 inhabitants [58]. Wastewater undergoes a primary treatment, followed by a biological treatment and a nitrification and denitrification treatment. The resulting sludge undergoes a mesophilic and dehydration treatment, obtaining biologically stable SS, with the physical-chemical characteristics detailed in Table 3. These characteristics comply with the Spanish national regulations on the maximum concentration of TM in SS used as organic amendments in Agriculture [26].

Table 3. Physical-chemical properties of the sewage sludge (SS) used in this study in 1992 (start of the experiment) and 2018 (sampling) together with the analytical methods used. Mean annual values ± standard deviation. The maximum legal limits of TM allowed in SS intended for agricultural use are indicated for each TM.

Sewage Sludge Physical-Chemical Properties	1992	2018	Legal Limit	Analysis Method
		General parameters		
pH	8.01	8.16 ± 0.03	NA	Soil pH in water 1:5 [49]
Electric Conductivity (μs cm^{-3})	NA	1795 ± 28	NA	Diluted Extracts 1:5 [50]
Dry material (%)	28.8	18.1 ± 0.4	NA	Direct calcination at 540 °C [59]
Volatile matter (*% of dry substance*)	NA	62.8 ± 1.9	NA	Direct calcination at 540 °C [59]
C/N	10	5.35 ± 0.08	NA	
		Fertilizing elements (% of dry substance)		
N total	2.18	5.85 ± 0.13	NA	Kjeldahl digestion and distillation [60]
N ammonium	0.18	0.75 ± 0.02	NA	Kjeldahl digestion and distillation [60]
Phosphorus (P$_2$O$_5$)	3.62	5.59 ± 0.22	NA	Microwave digestion + ICP-MS [54,55]
Potassium (K$_2$O)	0.51	0.62 ± 0.05	NA	Microwave digestion + ICP-MS [54,55]
Iron (Fe)	1.48	1.68 ± 0.04	NA	Microwave digestion + ICP-MS [54,55]
Calcium (CaO)	NA	7.98 ± 0.29	NA	Microwave digestion + ICP-MS [54,55]
Magnesium (MgO)	NA	1.10 ± 0.05	NA	Microwave digestion + ICP-MS [54,55]
		Trace metals (mg Kg^{-1} dry weight)		
Cadmium (Cd)	<10	0.88 ± 0.09	40	Microwave digestion + ICP-MS [54,55]
Copper (Cu)	302	187 ± 11	1750	Microwave digestion + ICP-MS [54,55]
Nickel (Ni)	75	32.1 ± 0.77	400	Microwave digestion + ICP-MS [54,55]
Lead (Pb)	191	39.0 ± 1.2	1200	Microwave digestion + ICP-MS [54,55]
Zinc (Zn)	1230	874 ± 38	4000	Microwave digestion + ICP-MS [54,55]
Mercury (Hg)	NA	0.003 ± 0.003	25	Microwave digestion + ICP-MS [54,55]
Chromium (Cr)	112	58.3 ± 3.2	1500	Microwave digestion + ICP-MS [54,55]
Manganese (Mn)	NA	NA	NA	Microwave digestion + ICP-MS [54,55]

NA: Not analyzed.

2.4. Soil and Crops Sampling

After 26 years from the beginning of the trial, a specific sampling was carried out to evaluate the concentration and availability of trace metals in the soil, as well as in the aerial parts (grain and straw) of barley.

Soil sampling was carried out in September, after the previous crop cycle ended, and before the application of SS for the following campaign. Samples of each treatment and replicate were collected at 0–30 cm depth, with an auger. A sample composed of three subsamples was collected in each plot. Samples were immediately transferred to the laboratory in polyethylene bags protected from sunlight, were dried at room temperature and ground to 2 mm.

Crop sampling was carried out at complete physiological maturity, 5 days before harvest. A sample of the aerial part of the crop was taken from each treatment and repetition, avoiding the edges of each plot. Crop samples were transferred to the laboratory in paper bags, and oven-dried at 50 °C for 7 days. Once dry, they were shelled with a 6mm sieve to separate the grain from the straw, and each part was ground separately with an agate ball mill.

2.5. Analytical Methods

2.5.1. Total Concentration in TM in Soil and Crop Samples

Soil, grain and straw samples for the determination of Cr, Cu, Mn, Ni, Zn, Cd, Pb and Hg were microwave-digested. Following the methodology described in EPA 3051 [55], 0.3 g of soil, 6 mL of 67–69% TraceMetal™ HNO_3, 2 mL 34–37% TraceMetal™ HCl and 2 mL of Milli-Q water were used for total Soil TM determinations. For the analysis of both parts of the crop (straw and grain), the methodology described in EPA 3052 [61] was followed; 0.25 g of sample, 4 mL of 67–69% TraceMetal™ HNO_3, 2 mL of H_2O_2 at 30% and 4 mL of Milli-Q water were used.

The mixing of the samples with the reagents was carried out directly in teflon tubes (PTFE) and then closed. They were treated in an ETHOS UP, Microwave digestion system (Milestone, Sorisole, Italy). For soil and crop samples, microwaves were emitted for 50 min and 1 h, respectively. To obtain a complete digestion, soil samples reached 200 °C for 20 min, and crop samples, 220 °C for 30 min. The temperature was controlled by a system of probes that constantly measure the process temperature. Once the digested samples had cooled within the system, each sample was filtered and transferred pre-tared 50 mL polypropylene containers, which were made up to their capacity with Milli-Q water. Between samples, the PTFE was rinsed with 5% HNO_3, five times and subsequently washed with Milli-Q water. Between different matrices, to avoid contamination, a blank digestion was carried out at 170 °C and, subsequently rinsed with Milli-Q water.

External calibration standards, and internal standards, were included in the sample set by means of a T system in the equipment's sample introduction system.

2.5.2. Extraction with DTPA of Soil Samples

To evaluate the bioavailability of TM in the soil, the DTPA-extractable concentrations of Cr, Cu, Mn, Ni, Zn, Cd and Pb were determined, using the extraction procedure described in the UNE 77315:2001 standard [62], equivalent to the international standard ISO 14870:2001 [63]. This extraction procedure aims to detect the TM concentrations in the most labile fractions of soil, and therefore potentially absorbable by plants [64].

The extraction solution was prepared by mixing, first, 0.735 g of $CaCl_2 \cdot 2H_2O$, 0.984 g of diethylene triamine pentaacetic acid—DTPA ($C_{14}H_{23}N_3O_{10}$) and 7.46 g of Triethanolamine—TEA ($C_6H_{15}NO_3$) in a beaker. The mixture was then diluted with 800 mL of deionized water and the pH was adjusted to 7.3 with HCl. The solution was finally transferred to a 1000 mL flask, made up to the mark and homogenized. The solution was stored at 20 °C until used. Subsequently, in a 100 mL wide-mouth polypropylene container, 20 g of soil sample and 40 mL of the solution were mixed. The recipient was hermetically closed and stirred for 2 h at 20 °C on a reciprocating shaker at 30 rpm. Then a fraction of the extract was decanted, placed in a centrifuge tube, and centrifuged for 10 min at 6000 rpm. The supernatant was filtered with a membrane filter with a pore size of 0.45 μm, collected in a polyethylene bottle, and stored in the refrigerator. The blank extractions for each batch of analysis were carried out in the same way but without soil samples.

All glass and plastic material used in this procedure were previously washed in 5% HNO_3, and in deionized water, 3 times. Then they were rinsed with distilled water and with the extracting solution, 3 and 1 times, respectively.

2.5.3. Trace Metals Determination

The determination of TM concentrations in digested soil and crop samples was carried out by ICP-MS in a 7700x analyzer (Agilent Technologies, Santa Clara, CA, USA), following the UNE-EN 17053 standard [54]. The DTPA extracts were analyzed before 48 h from their preparation, using the same analysis method.

2.6. Statistical Analysis

A statistical analysis was carried out for each MT, with a significance level of 5%, to contrast the difference in means depending on the fertilizer dose received, understand this factor in six different levels associated with the ranges of accumulated contribution of SS, and distinguish the control of mineral fertilization. Thus, it was analyzed by one-factor ANOVA, carried out with R (R Core Team 2019, Vienna, Austria), with six levels: control (Control), mineral fertilizer (MinFer), 250 Mg per ha of SS (250MgSS), 520–560 Mg per ha of SS (500MgSS), 1040 Mg per ha of SS (1000MgSS) and 2080 Mg per ha of SS (2000MgSS). We will refer to the factor as "Fertilizer Treatment".

The ANOVA results are presented accompanied by BoxPlot graphs with whiskers where indicated, when the existence of any difference was confirmed, the groupings of factor levels (a, b, c...) after a study of multiple comparison of means two to two using the Tukey test.

In a complementary way, in order to contrast the effect of the application of SS per se, a study was carried out for each TM for the difference in means depending on whether the treatment had received SS (regardless of the quantity) called "With SS" or had not received SS (control and MinFer) called "Without SS", at 5% significance. This statistical analyses were carried out with R (R Core Team 2019, Vienna, Austria).

3. Results
3.1. Total Concentration of Trace Metals in Soil

The ANOVA showed significant differences in the total concentration of Zn, Cu, Cr, Ni and Hg in soil between the different Fertilizer Treatments (Figure 1).

The total concentrations of Cu, Cr and Hg showed significant differences between the control treatment (either SS or mineral fertilization) and the treatment with the highest accumulated SS dose (2000MgSS), which displayed the lowest and highest concentrations, respectively.

The total concentration of Ni in soil was significantly different between the treatments that had not received SS (control and MinFer) and those that had received SS.

In the case of Zn, significant differences were found between the treatments that had not received SS and those that had received SS. Moreover, in treatment 2000MgSS, the concentration increased.

No significant differences were found in the total concentration in Pb, Cd, and Mn as a function of the different fertilizer treatments studied.

The complementary study based on the contrast between treatments with SS and without SS (Table 4) confirmed the results obtained from ANOVA. This analysis, revealed higher values overall for the total concentration of Zn, Cu, Cr, Ni and Hg in the treatments that had received SS than in those that had not received SS. The average increase with SS was of 55.6% in Hg, 29.3% in Zn, 21.5% in Cu, 7.7% in Cr and 7.0% in Ni.

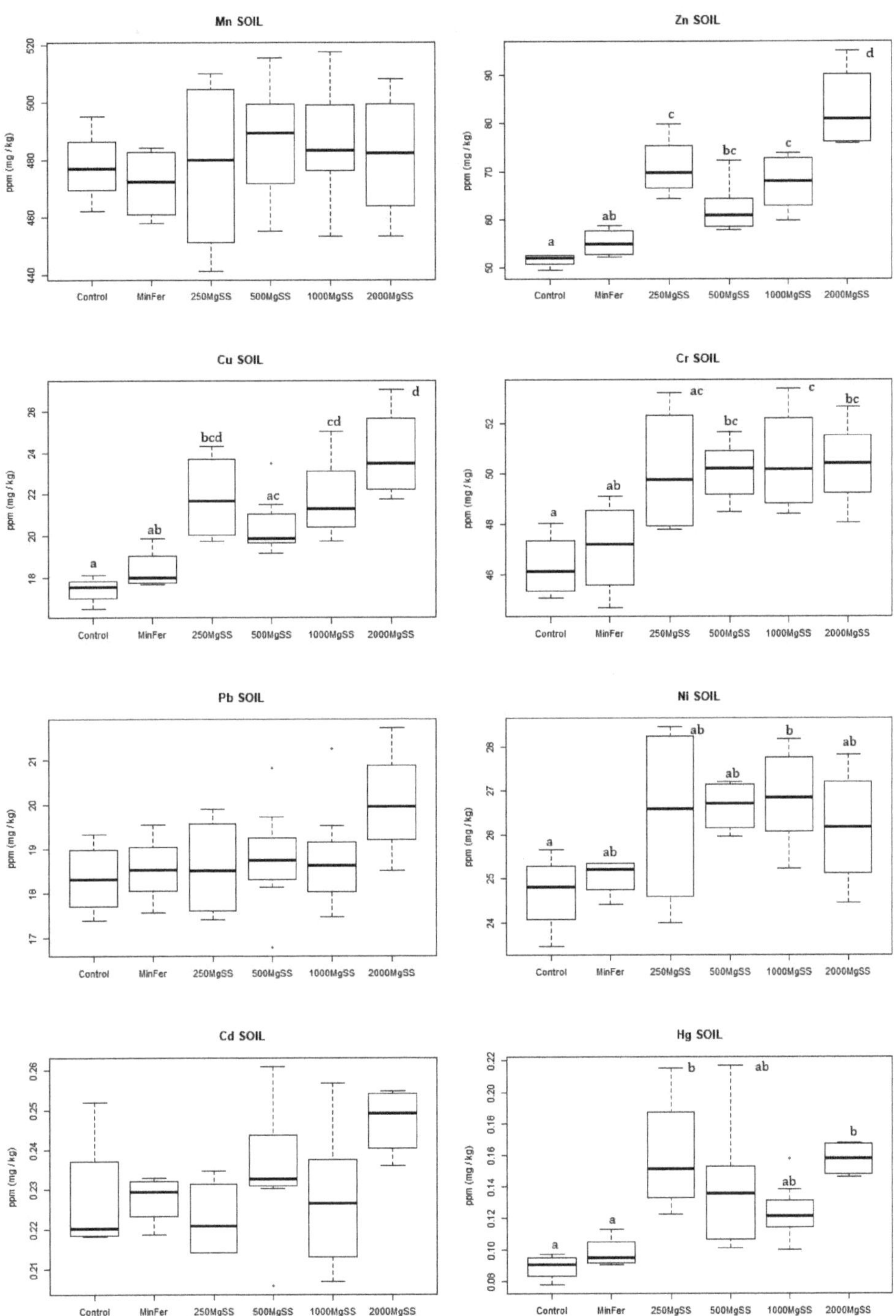

Figure 1. Box-and-whisker plot indicating different homogeneous groups in the multiple comparison Scheme (0–30 cm). Values showing the same letter belong to the same homogeneous group according to Tukey's test ($p < 0.05$).

Table 4. Mean ± standard deviation of the total concentration of Mn, Zn, Cu, Cr, Pb, Ni, Cd and Hg in soil, depending on whether they have received SS or not, together with their *p*-value.

Treatment	TM$_{\text{SOIL TOTAL}}$ (mg kg^{-1})							
	Mn	Zn	Cu	Cr	Pb	Ni	Cd	Hg
With SS	484 ± 21	69.0 ± 9.2	21.7 ± 2.1	50.3 ± 1.7	19.0 ± 1.3	26.6 ± 1.2	0.23 ± 0.02	0.14 ± 0.03
Without SS	475 ± 13	53.4 ± 2.3	17.9 ± 1.0	46.7 ± 1.6	18.4 ± 0.8	24.9 ± 0.7	0.23 ± 0.01	0.09 ± 0.01
p-value	0.13	0.00 *	0.00 *	0.00 *	0.14	0.00 *	0.22	0.00 *

* Significant difference.

3.2. Concentration of DTPA-Extracted Trace Metals in Soil

The results showed that the concentration in DTPA-extracted Zn, Cu, Pb, Ni and Cd, increased with the use and dose of SS (Figure 2).

In all TM except for Mn, DTPA-extracted concentrations were found to be significantly lower in the control than in the treatment 2000MgSS. Furthermore for Pb, a significant difference was observed between the treatments that did not receive SS and those that received SS. For these TMs, the minimum amount of SS applied represented a significant increase in the extractable concentration by DTPA.

In the case of Zn extracted with DTPA (Figure 2), a significant difference was observed between the treatments that did not receive SS and those that received SS. At treatment 2000MgSS, the bioavailable concentration in soil increased.

Cu showed three statistically different groups, with the treatments corresponding to 0 Mg per ha of SS accumulated (control and MinFer) grouped together with the lowest concentrations, the treatments comprised between 250MgSS to 1000MgSS in another homogeneous group with intermediate values, and the treatment 2000MgSS with the highest observed value.

Finally, it was not possible to quantify the concentrations of Cr extracted by DTPA, as it was below the detection limit of 0.5939 ppb.

The complementary statistical study found that the average concentration in Mn extracted with DTPA was 6.5% higher in the plots that had not received any dose of SS, compared to those that had received SS (Table 5). This same analysis confirmed the results of the ANOVA analysis, revealing higher values in the treatments that had received SS applications compared to the treatments without SS for DTPA-extractable Zn, Cu, Pb, Ni and Cd. The average gains were of 122.2% in Zn, 61.1% in Cu, 27.3% in Ni, 14.9% in Pb and 11.3% in Cd. Finally, only Cu and Zn showed a clear correlation in their DTPA-extracted concentrations and those found in bulk soil (Figure S1).

Table 5. Mean ± standard deviation of the DTPA extractable concentration of Mn, Zn, Cu, Pb, Ni and Cd in soil, depending on whether they have received SS or not, together with their *p*-value.

Treatment	TM$_{\text{SOIL DTPA}}$ (mg kg^{-1})					
	Mn	Zn	Cu	Pb	Ni	Cd
With SS	8.85 ± 0.86	4.20 ± 1.27	2.82 ± 0.53	2.00 ± 0.11	0.28 ± 0.04	0.07 ± 0.00
Without SS	9.47 ± 0.28	1.89 ± 0.36	1.75 ± 0.16	1.74 ± 0.07	0.22 ± 0.02	0.06 ± 0.00
p-value	0.01 *	0.00 *	0.00 *	0.00 *	0.00 *	0.00 *

* Significant difference.

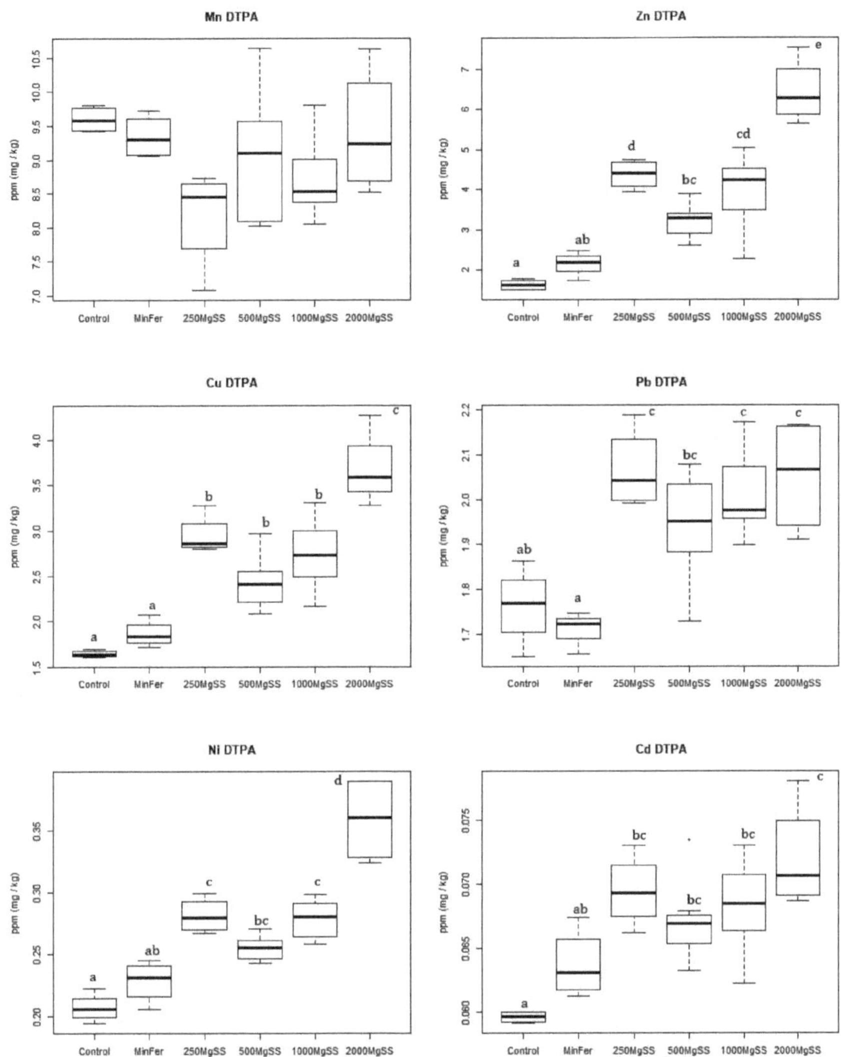

Figure 2. Box-and-whisker plots that indicate different homogeneous groups in the multiple comparison of means two to two of the concentration extracted by DTPA of the trace metal Mn, Zn, Cu, Pb, Ni and Cd existing in the soil according to the fertilizer treatment received in the study horizon (0–30 cm). Values showing the same letter belong to the same homogeneous group according to Tukey's test ($p < 0.05$).

3.3. Concentration of Trace Metals in Barley Grain and Straw

In relation to grain, no significant differences were detected in the concentration of TM in grain as a function of the different treatments, with the exception of Zn (Table 6). Zn concentration showed the highest concentration in grain in the fertilizer treatment corresponding to 2000MgSS. The result of the complementary analysis, confirmed the significant difference in Zn, and revealed differences in Cu and Cr (Table 7). For these TM, the concentrations in barley grains were significantly higher in the treatments that had received SS than in those without SS application. The average increases were of 15.0% in Cr, 8.3% in Zn and 4.4% in Cu.

Table 6. Different homogeneous groups in the multiple comparison of two to two means of the concentration of the trace metal Mn, Zn, Cu, Cr, Pb, Ni, Cd and Hg in grain and straw according to the fertilizer treatment received in the study horizon (0–30 cm). Mean ± standard deviations. Values in the same column showing the same letter belong to the same homogeneous group according to Tukey's test ($p < 0.05$).

Fertilizer Treatment	TM Crop (mg kg^{-1})															
	Mn		Zn		Cu		Cr		Pb		Ni		Cd		Hg	
	Grain	Straw	Grain	Straw	Grain	Straw	Grain	Straw	Grain	Straw	Grain	Straw	Grain	Straw	Grain	Straw
250MgSS	17.3 ± 0.6	16.8 [ab] ± 2.4	34.3 [ab] ± 2.4	8.16 ± 1.00	5.94 ± 0.39	2.88 ± 0.28	1.35 ± 0.23	1.71 ± 0.44	0.10 ± 0.00	0.24 ± 0.01	0.83 ± 0.11	0.95 ± 0.25	0.01 ± 0.00	0.04 ± 0.01	0.01 ± 0.00	0.01 ± 0.00
500MgSS	16.8 ± 1.6	17.4 [a] ± 2.2	31.0 [a] ± 2.6	7.31 ± 0.96	5.69 ± 0.32	2.50 ± 0.41	1.23 ± 0.18	1.58 ± 0.19	0.10 ± 0.01	0.25 ± 0.07	0.81 ± 0.09	1.01 ± 0.21	0.01 ± 0.00	0.03 ± 0.00	0.01 ± 0.00	0.01 ± 0.00
1000MgSS	17.1 ± 1.8	15.6 [a] ± 4.6	35.2 [ab] ± 3.2	6.98 ± 1.68	5.67 ± 0.32	2.67 ± 0.63	1.31 ± 0.3	1.84 ± 0.71	0.10 ± 0.01	0.22 ± 0.09	0.83 ± 0.13	1.02 ± 0.32	0.01 ± 0.00	0.03 ± 0.00	0.01 ± 0.00	0.01 ± 0.00
2000MgSS	17.8 ± 1.2	17.7 [ac] ± 3.1	37.1 [b] ± 4.1	7.00 ± 0.37	5.63 ± 0.44	2.78 ± 0.53	1.41 ± 0.08	1.54 ± 0.82	0.11 ± 0.01	0.27 ± 0.14	0.90 ± 0.02	0.96 ± 0.48	0.01 ± 0.00	0.03 ± 0.01	0.00 ± 0.00	0.01 ± 0.00
MinFer	18.4 ± 1.1	24.4 [bc] ± 3.4	30.4 [ab] ± 2.6	6.91 ± 1.76	5.66 ± 0.21	2.82 ± 0.55	1.03 ± 0.06	2.26 ± 1.01	0.10 ± 0.01	0.31 ± 0.10	0.73 ± 0.05	1.51 ± 0.57	0.01 ± 0.00	0.04 ± 0.01	0.01 ± 0.00	0.01 ± 0.00
Control	17.3 ± 1.8	25.1 [c] ± 5.3	32.1 [ab] ± 2.6	9.26 ± 1.60	5.27 ± 0.17	2.25 ± 0.39	1.26 ± 0.47	2.55 ± 0.45	0.12 ± 0.03	0.23 ± 0.04	0.78 ± 0.25	1.34 ± 0.40	0.01 ± 0.00	0.03 ± 0.01	0.00 ± 0.00	0.01 ± 0.00

Table 7. Mean ± standard deviation of the concentration of Mn, Zn, Cu, Cr, Pb, Ni, Cd and Hg in the crop (grain and straw), depending on whether or not they have received SS, together with their *p*-value.

Treatment	TM$_{CROP}$ (mg kg^{-1})							
	Mn	Zn	Cu	Cr	Pb	Ni	Cd	Hg
GRAIN With SS	17.1 ± 1.4	34.0 ± 3.7	5.71 ± 0.35	1.30 ± 0.22	0.10 ± 0.01	0.83 ± 0.10	0.01 ± 0.00	0.01 ± 0.00
GRAIN Without SS	17.8 ± 1.6	31.4 ± 2.6	5.47 ± 0.27	1.13 ± 0.30	0.11 ± 0.03	0.75 ± 0.15	0.01 ± 0.00	0.01 ± 0.00
p-value	0.32	0.045 *	0.038 *	0.047 *	0.64	0.062	0.25	0.59
STRAW With SS	16.8 ± 3.3	7.29 ± 1.21	2.66 ± 0.48	1.68 ± 0.53	0.24 ± 0.09	1.00 ± 0.29	0.03 ± 0.01	0.01 ± 0.00
STRAW Without SS	24.8 ± 4.1	8.08 ± 2.00	2.53 ± 0.54	2.40 ± 0.74	0.27 ± 0.08	1.42 ± 0.47	0.03 ± 0.01	0.01 ± 0.00
p-value	0.00 *	0.19	0.26	0.01 *	0.40	0.00 *	0.71	0.46

* Significant difference.

In straw, no significant differences were found in the concentration of TM analyzed as a function of the fertilizer treatment, with the exception of Mn (Table 6). The Mn concentration in straw in the treatments without SS was significantly higher than in those that had received SS. The result of complementary analysis confirmed the significant difference in Mn and revealed differences in Cr and Ni (Table 7). In the three TM, the concentrations in the straw of the treatments that had not received SS were higher than those that had received SS. The average decrease was 32.2% in Mn, 30.0% in Cr and 29.6% in Ni.

4. Discussion

4.1. Influence of Sewage Sludge in Total and DTPA-Extracted Concentrations of Trace Metals in Soil

4.1.1. Total TM Concentrations in Soil

In general, the TM concentration found in soil, in descending order was Mn> Zn> Cr> Ni> Cu> Pb> Cd> Hg. The application of SS significantly increased the total concentration of Zn, Cu, Cr, Ni and Hg in the soil. These results are in line with those found in other locations and experimental sites. For instance, in a study carried out on a Mediterranean agricultural soil, Reference [16] obtained similar results, where the concentrations of Cu, Zn, Ag, Sb, Hg and Pb in the soil increased with the application of SS over 15 years, compared to the control and to mineral fertilization. In studies conducted on other soil types such as Cumulic Haplustolls and an Udic Calciustoll [65], it was also concluded that Cr, Cu, Cd, Pb and Zn accumulated in the most superficial layers of the soil after the application of SS over several years.

The different behavior of the TM analyzed can be explained by both their concentration in SS, and their particular behavior in soil. As it can be seen in Figure 1, the concentrations of Zn, Cu, Cr and Hg were significantly higher in the treatments receiving the highest dose of SS, 2000MgSS (corresponding 80 Mg ha^{-1} of SS every year for 26 years), compared to the control plots. This dose is possible only within the framework of research. Although the annual amounts of TM that can be applied to agricultural soils comply with the current legislation RD 1310/1190 [26], the regional regulations on the use of SS in agriculture in Navarra (Orden Foral 359/2010 [66]) is more strict. This regulation prohibits exceeding the dose of 250 NFU annually in general, and of 170 NFU in areas designated as vulnerable to nitrate contamination according to EU Directive 91/676/CEE [67]. The dose of 80 Mg ha^{-1} of SS every year is therefore well above the one allowed at the local level (Table 2), which corresponds to 20 Mg ha^{-1} every four years in rainfed crops. To this, it has to be added that current regulations on organic fertilization also impose limits that make this dose above those allowed [68].

In the particular case of Zn, the significant enrichment observed in the soil with increasing SS application can be related to the high concentration of Zn in SS (874 mg Kg^{-1}, Table 3). These results were also observed in a previous study carried out in this experimental field [11].

In Ni, a significant difference was observed between the treatments that had received SS compared to those that had not received SS (Figure 1). This was not observed in studies previously conducted under similar conditions [39], and in the same experimental field [11]. This suggests, that although the concentration of Ni in SS was low (32.1 mg Kg^{-1}, Table 3) it was sufficient to generate an increase in concentration in the plots that received SS at any dose after 26 years of continuous application. The application time and the accumulated dose seem therefore to have consequences in the accumulation of this element in soils like the one studied here. A significant increase in total Ni was also observed in a study carried out in two soils in India, with neutral pH, warm steppe climate, sunflower and wheat cultivation, fewer applications and lower doses of SS than those used in this study [41].

In Mn, Pb and Cd there were no significant differences in their total concentration in the soil between any fertilizer treatment, whether they received SS or not, as previously found in this experimental site [11], and in other studies with similar conditions but lower doses of SS, for Mn and Cd [16]. Despite amending with lower doses, Iglesias [16] found a significant increase in Pb, probably due to the concentration of this element in the SS used (112 ± 21 mg ka^{-1}), while in our study, the concentration of Pb was lower (39 ± 1.2 mg ka^{-1}, Table 3). In another study in Sweden, since 1997, in a silty clay soil (42% clay), cereal crop but acid pH, Pb did not suffer significant increases and Cd increased only when SS with metal salts was added [10]. It is important to note that the doses applied were lower than in our case study.

It has to be noted that, despite the observed cumulative effect of SS application on total Zn, Cu, Cr, Ni and Hg concentrations, these concentrations were always below the legal limit for soils with a pH greater than 7, as indicated by the Spanish national legislation RD 1310/1190 [26], and the European Directive 86/278/EEC [27]. Previous studies carried out under similar conditions coincide with this observation [12,16,17]. Our results indicate, however, that the use of SS for fertilization purposes, in the conditions used here can have a relevant and cumulative effect on the total concentration of TM in the soil, which should be monitored in the long term.

4.1.2. DTPA-Extractable TM Concentrations in Soil

In general, the order of concentration of TM in the DTPA extractable fraction was Mn > Zn > Cu > Pb > Ni > Cd. It was observed that for Zn, Cu, Pb, Ni and Cd, the concentration extracted with DTPA was significantly higher in the treatments that received SS than in those without SS, as already observed in many other studies [22,35]. Assuming that these concentrations represent the most readily available forms of TMs in soil, they would be affected by three factors: the source or origin of these TM (in the case, the SS), the characteristics of the soil and soil conditions associated with their application, and the particular behavior of each TM. In relation to the source, anthropogenic TMs are known to be more bioavailable than pedogenic metals [42,69,70].

In relation to soil parameters that might influence TMs bioavailability in the experimental site of this study, the most relevant ones are pH, the presence of carbonates, and the possible modifications in the retention of cations associated with the presence of organic matter. The high pH (8.5) associated with a high content of carbonates in the depth of study, hinders the mobility of TM in the soil to phases easily absorbed by the crops [15,45]. To this fact we must add that increasing the content of organic matter by continuous addition of SS can favor the formation of organo-metallic complexes [12,42,43], and increase the cation-exchange capacity of soil [42,71,72], which would also reduce the mobility and availability of TMs. In addition, the use of phosphate compounds has been proven to be efficient for the chemical immobilization of heavy metals in soils [73,74]. In this study, SS contributed a high concentration of phosphorus (P) to the soil, due to the affinity of this element with the solid fraction of SS during the wastewater treatment process [6]. It has been indeed observed in this experimental field that the application of SS increased the total and available content of P in soil [11], which could increase the immobilization of different TMs [71]. The behavior of the different TM analyzed was however not homogeneous. In Ni and Cd extracted by DTPA, only the treatment 2000MgSS caused a significant increase in the concentration with respect to the control (Figure 2). Both TMs have similar mobility in SS [39]. Cd tends to be found in a greater proportion strongly absorbed to organic matter and/or easily reducible Mn oxides and amorphous iron oxides, while Ni can also be strongly bound or incorporated into organic matter or into other oxidizable species [39]. When SS is incorporated into the soil, these TM undergo a minimum redistribution, remaining above all strongly adsorbed to organic matter and carbonates [39,75,76]. This mobility in SS and soil, together with the low concentrations of both TMs in SS (Table 3), can explain the observed increase in bioavailability only in the treatments corresponding to the highest SS doses, as well as the lack of a clear relationship between total and DTPA-extractable concentrations (Figure S1). These results coincide with those described by Navarro [77], who determined, by means of a column experiment, that the Cd and Ni extracted by DTPA increased only after the use of large doses of SS in calcareous soils. In the case of Cd, it has however been observed that in soils modified with biosolids greater mobility can be observed as exchangeable Cd associated to the greater exchange capacity of some organic components added to the soils [76]. It is important to take this into account since Cd is a non-essential TM.

For Pb extracted by DTPA, a significantly higher concentration was observed in the treatments with SS compared to those that had not received SS. In this case, the contributions through SS were low (39.0 mg Kg^{-1}, Table 3), and considering its affinity with organic ligands [78], it can be thought that a relevant part of its concentration was associated to organic matter in SS. The mineralization of organic matter would thus imply a gradual release of the Pb attached to it. In the case of calcareous soils, this release would be associated with the formation of forms that are not readily bioavailable for crops, mostly in the form of carbonates [76,79,80]. As such, the increasing concentration of Pb with increasing accumulation of SS-derived organic matter would be outbalanced by this behavior of Pb in calcareous soils that would imply its progressive incorporation into stable forms as it is released by organic matter mineralization, explaining also the low correlation between total and DTPA-extractable Pb (Figure S1).

In relation to Zn extracted by DTPA (Figure 2), the response was similar to that observed for total concentration in soil (Figure 1 and Figure S1). This is probably due to the high concentration of Zn in SS (874 mg Kg^{-1}, Table 3) and the mobility of this TM. According to Morera [39], in SS such as the one used, Zn is mainly found together to organic matter and Fe and Mn oxides. However, about 10% can be in easily mobile forms. Some studies have determined that Zn may be biologically available and mobile in soils that have been treated with SS for long periods [81]. Kabata Pendias [82] already mentioned that the long-term addition of biosolids to the soil can increase the fraction of bioavailable Zn. This mobility has been observed to increase with time of SS addition to soil since, as it can form soluble chelates with humified and non-humidified forms of organic matter, which in turn can be easily degraded by soil microorganisms [83,84].

DTPA-extractable Cu showed significant differences between the different doses of accumulated SS, with an increasing concentration with the increase of cumulative SS dose (Figure 2 and Figure S1). This behavior can be related to the great affinity that Cu has with organic matter [78,85]. It is likely that the Cu present in SS was mostly bound to organic matter. When added to the soil would hardly undergo any redistribution, as it can be retained to the organic fraction of the soil through stable organic-mineral complexes [40]. At the same time, the increment in the density and activity of the soil microbial biomass associated to the addition of SS with a low concentration in TM to the soil [86], can contribute to gradually incorporate this Cu into more available forms. Some studies have shown that the amount of SS that actually affects the increase in Cu availability is that accumulated mainly on the last 4 years of application [87,88]. This seems to support the observation that the fertilizer treatment with 2000MgSS, corresponding to an annual application of 80 Mg ha^{-1} of SS, was the treatment with the highest concentration of Cu extracted with DTPA. This observed correlation of DTPA-extracted Zn and Cu and their total concentration in soil has been observed in fact in other studies including SS additions to calcareous soils [38].

In the Mn extracted with DTPA, a higher concentration was observed in the treatments that had not received SS. This may be due to the affinity that Mn has with carbonates in SS and in soil, coupled with the low Mn contribution normally associated with SS [33]. Mn in SS is usually bound to carbonates, Fe and Mn oxides, and/or bound to the residual fraction [89,90]. When incorporated into calcareous soils, Mn hardly undergoes a redistribution [80], remaining bound to carbonates. This could explain the low efficiency of SS in incorporating Mn into the soil and a higher efficiency of mineral amendments. Previous results where the highest concentration of Mn extractable by DTPA was obtained in the control amended with mineral fertilization, support this hypothesis [40].

4.2. Concentration of TM in Barley

The study of TM concentrations in crops, as a reliable method to assess the actual bioavailability of TMs in the studied experimental field, detected some differences and trends between TMs, and for the different parts of the crop (grains and straw).

As a general trend, it was observed how the essential TM were adsorbed in higher concentrations than the non-essential ones, being the order of concentration from highest to lowest Zn > Mn > Cu > Cr > Ni > Pb > Cd > Hg, and Mn > Zn > Cu > Cr > Ni > Pb > Cd = Hg, for grain and straw respectively. However, significant differences were observed for Mn, Zn, Cu, Cr and Ni accumulation in barley grains and straw. In grain, Zn, Cu and Cr showed higher concentrations in the treatments that had received SS than in those without SS (Table 7). On the contrary, in straw, Mn, Cr and Ni showed higher concentrations in the treatments that had not received SS (Table 7).

In the case of Zn, the highest concentration in grain coincided with the highest concentration extracted with DTPA, corresponding to 2000MgSS, in line with the general correspondence between the total concentration of this TM and of Cu in the soil and that found in plants in soils amended with SS [3,38]. However, the differences observed for soil Cu concentrations between doses were not observed in grain (Table 6). This could be explained by what was observed by Montaghian [91], who found that in calcareous soils amended with SS, the concentration of bioavailable Cu was lower in the soil of the rhizosphere than in the rest of the soil. They observed a redistribution of Cu between the different soil fractions, increasing the proportion linked to organic matter as a consequence of an increase in dissolved organic carbon released by the roots. This suggests a possible antagonistic effect, hindering the accumulation of Cu in grain (the organ that normally accumulates the most Cu in mature plants [92]) beyond a certain level of accumulation of Cu in the soil of this study. In the case of Mn, a tendency to a greater accumulation was observed in the grains of barley plants from the treatment MinFer. This highlights the low efficiency of SS to supply Mn to crops.

In straw, the treatments without SS contained higher concentrations of Mn, Ni and Cr than the treatments that had received any dose of SS. This can be related to the "dilution effect" for these elements caused differences in yield between treatments. As SS produced an increase in yields (Supplementary Tables S1 and S2) also observed in other studies [14,38,93] and the applied and available concentrations of these TM were low, a dilution of these TM in the crop can be expected. Similar results have been reported before for Mn in long-term field [94] and pot experiments for Mn and Ni [71].

Finally, in the case of Cr, for which the DTPA-extracted concentration was below the method's detection limit, it was not possible to relate this concentration and that observed in the crop aerial parts. However, our results seem to indicate a different efficiency in the use of Cr depending on the plant organ and SS dose received. In a study carried out in pots, a high correlation was observed between the Cr concentration in the grain of different varieties of wheat and that extracted from soil with EDTA [35].

The study of TM concentrations in aerial crop parts allowed to contrast them with the limits and thresholds defined in the legislation. The European regulation referring to food products intended for human [29] and animal consumption [28] only contemplates Pb, Cd and Hg as risk elements. In our study, in general terms, it was found that the concentrations found in barley that complied with current regulations, coinciding with other studies [16,94]. In the case of grain, the concentrations were below the limit allowed for both norms. In the case of straw, however Pb concentrations were above that allowed by the legislation for human consumption. Despite the fact that straw is not intended for this type of use, it should be noted that even the control was found above this limit, which indicates that SS was not the main reason for Pb exceeding the value allowed by the legislation.

5. Conclusions

The application of SS as a long-term organic amendment in the conditions of soil, climate, management and type of sludge studied here, caused an accumulation of total Zn, Cu, Cr, Ni and Hg in soil, of Zn, Cu, Pb, Ni and Cd in their bioavailable form extracted by DTPA and of Zn, Cu and Cr in grain. This suggests the existence of potential environmental risks and supports the need for long-term monitoring of this type of practice. On the other hand, a higher concentration of Mn extracted with DTPA, as well as a higher concentration of Mn, Ni and Cr in the straw was observed, in the treatments without SS amendment.

Therefore, it was observed that Zn, Cu and probably Cr, suffered an increase in their concentration in soil and extractable by DTPA, that was visualized in grain. Other TM, such as Hg, increased their total concentration in soil and others, such as Cd and Pb, increase their concentration extracted by DTPA. In the case of Ni, both the total and DTPA-extracted concentration increased without any influence in the grain content.

Despite the existence of an increase in the total and bioavailable concentration of some TMs, the resulting crop complied with the current regulations imposed by the European Union for human and animal nutrition.

Supplementary Materials: The following are available online at https://www.mdpi.com/article/10.3390/min11040356/s1, Table S1: Yield expressed in percentage 12% humidity in 2018 of each fertilizer treatment (mean ± standard deviation); Table S2: Yield expressed in percentage 12% humidity in 2018 of the treatments that had received SS and those that had not received SS (mean ± standard deviation); Figure S1: Total TM concentration (TMSoil) vs Extractable TM concentration by DTPA (TMBio).

Author Contributions: Conceptualization, A.Z., A.E., I.V. and L.O.; methodology, A.Z., A.E., I.V., H.U. and L.O.; software, H.U.; validation, A.Z., A.E., I.V., H.U. and L.O.; formal analysis, A.Z., A.E., I.V., H.U. and L.O.; investigation, A.Z., A.E. and I.V.; writing—original draft preparation, A.Z., A.E., I.V., H.U. and L.O.; writing—review and editing, A.Z., A.E., I.V., H.U., R.A. and L.O.; supervision, A.Z., A.E., I.V. and H.U.; project administration, L.O. All authors have read and agreed to the published version of the manuscript.

Funding: This study has been financed and promoted by the Sectoral Program for Agricultural and Food Research and Development of the Ministry of Agriculture, Fisheries and Food (M.AP.A.) from its beginning in 1992 to 1993 within "Other Projects" and since 1993 until 1997 through Project SC94-025. On the other hand, the National Institute for Agricultural and Food Research and Technology (INIA) granted aid for the hiring of teaching and research staff through the RTA2017-00088-C03-01 project from 2018 to 2022.

Data Availability Statement: The data presented in this study are available on request from the corresponding author. The data are not publicly available due to their belonging to a multi stakeholder project.

Acknowledgments: We are grateful for the services and disposition of the Commonwealth of Pamplona, which contributed the experimental plot and the sludge from the Arazuri treatment plant. We also thank the team of the Soil Science Laboratory of the Public University of Navarra, and the Environment and Soil Sciences team, and the Technical Scientific Service for Elemental Analysis of the University of Lleida for their help and advice.

Conflicts of Interest: The authors declare no conflict of interest.

References

1. Ministerio de la Presidencia, Relaciones con las Cortes y Memoria Democrática. *Ley 22/2011, de 28 de Julio, de Residuos y Suelos Contaminados*; Agencia Estatal Boletín Oficial del Estado: Madrid, España, 2011; Número *181*; pp. 85650–85705. (In Spanish)
2. Ministerio de agricultura pesca y alimentación. *Resolución de 8 de Octubre de 2015*; Agencia Estatal Boletín Oficial del Estado: Madrid, España, 2015; Número *253*; pp. 99107–99116. (In Spanish)
3. McBride, M.B. Toxic metals in sewage sludge-amended soils: Has promotion of beneficial use discounted the risks? *Adv. Environ. Res.* **2003**, *8*, 5–19. [CrossRef]
4. Singh, R.P.; Agrawal, M. Potential benefits and risks of land application of sewage sludge. *Waste Manag.* **2008**, *28*, 347–358. [CrossRef]

5. Roig, N.; Sierra, J.; Martí, E.; Nadal, M.; Schuhmacher, M.; Domingo, J.L. Long-term amendment of Spanish soils with sewage sludge: Effects on soil functioning. *Agric. Ecosyst. Environ.* **2012**, *158*, 41–48. [CrossRef]

6. Alvarenga, P.; Mourinha, C.; Farto, M.; Santos, T.; Palma, P.; Sengo, J.; Morais, M.C.; Cunha-Queda, C. Sewage sludge, compost and other representative organic wastes as agricultural soil amendments: Benefits versus limiting factors. *Waste Manag.* **2015**, *40*, 44–52. [CrossRef] [PubMed]

7. Latare, A.M.; Kumar, O.; Singh, S.K.; Gupta, A. Direct and residual effect of sewage sludge on yield, heavy metals content and soil fertility under rice-wheat system. *Ecol. Eng.* **2014**, *69*, 17–24. [CrossRef]

8. Lloret, E.; Pascual, J.A.; Brodie, E.L.; Bouskill, N.J.; Insam, H.; Juárez, M.F.D.; Goberna, M. Sewage sludge addition modifies soil microbial communities and plant performance depending on the sludge stabilization process. *Appl. Soil Ecol.* **2016**, *101*, 37–46. [CrossRef]

9. Parat, C.; Chaussod, R.; Lévêque, J.; Andreux, F. Long-term effects of metal-containing farmyard manure and sewage sludge on soil organic matter in a fluvisol. *Soil Biol. Biochem.* **2005**, *37*, 673–679. [CrossRef]

10. Börjesson, G.; Kirchmann, H.; Kätterer, T. Four Swedish long-term field experiments with sewage sludge reveal a limited effect on soil microbes and on metal uptake by crops. *J. Soils Sediments* **2014**, *14*, 164–177. [CrossRef]

11. Urra, J.; Alkorta, I.; Mijangos, I.; Epelde, L.; Garbisu, C. Application of sewage sludge to agricultural soil increases the abundance of antibiotic resistance genes without altering the composition of prokaryotic communities. *Sci. Total Environ.* **2019**, *647*, 1410–1420. [CrossRef]

12. Hamdi, H.; Hechmi, S.; Khelil, M.N.; Zoghlami, I.R.; Benzarti, S.; Mokni-Tlili, S.; Hassen, A.; Jedidi, N. Repetitive land application of urban sewage sludge: Effect of amendment rates and soil texture on fertility and degradation parameters. *Catena* **2019**, *172*, 11–20. [CrossRef]

13. Chiba, M.K.; Mattiazzo, M.E.; Oliveira, F.C. Sugarcane cultivation in a sewage-sludge treated ultisol. I—Soil nitrogen availability and plant yield. *Rev. Bras. Cienc. Solo* **2008**, *32*, 643–652. [CrossRef]

14. López-Rayo, S.; Laursen, K.H.; Lekfeldt, J.D.S.; Delle Grazie, F.; Magid, J. Long-term amendment of urban and animal wastes equivalent to more than 100 years of application had minimal effect on plant uptake of potentially toxic elements. *Agric. Ecosyst. Environ.* **2016**, *231*, 44–53. [CrossRef]

15. Alloway, B.J.; Jackson, A.P. The behaviour of heavy metals in sewage sludge-amended soils. *Sci. Total Environ.* **1991**, *100*, 151–176. [CrossRef]

16. Iglesias, M.; Marguí, E.; Camps, F.; Hidalgo, M. Extractability and crop transfer of potentially toxic elements from mediterranean agricultural soils following long-term sewage sludge applications as a fertilizer replacement to barley and maize crops. *Waste Manag.* **2018**, *75*, 312–318. [CrossRef] [PubMed]

17. Marguí, E.; Iglesias, M.; Camps, F.; Sala, L.; Hidalgo, M. Long-term use of biosolids as organic fertilizers in agricultural soils: Potentially toxic elements occurrence and mobility. *Environ. Sci. Pollut. Res.* **2016**, *23*, 4454–4464. [CrossRef] [PubMed]

18. Moolenaar, S.W.; Beltrami, P. Heavy metal balances of an Italian soil as affected by sewage sludge and bordeaux mixture applications. *J. Environ. Qual.* **1998**, *27*, 828–835. [CrossRef]

19. Gascó, G.; Lobo, M.C.; Guerrero, F. Land application of sewage sludge: A soil columns study. *Water SA* **2005**, *31*, 309–318. [CrossRef]

20. Toribio, M.; Romanyà, J. Leaching of heavy metals (Cu, Ni and Zn) and organic matter after sewage sludge application to Mediterranean forest soils. *Sci. Total Environ.* **2006**, *363*, 11–21. [CrossRef]

21. Cameron, K.C.; Di, H.J.; McLaren, R.G. Is soil an appropriate dumping ground for our wastes? *Aust. J. Soil Res.* **1997**, *35*, 995–1035. [CrossRef]

22. Smith, S.R. A critical review of the bioavailability and impacts of heavy metals in municipal solid waste composts compared to sewage sludge. *Environ. Int.* **2009**, *35*, 142–156. [CrossRef] [PubMed]

23. Yoshida, H.; ten Hoeve, M.; Christensen, T.H.; Bruun, S.; Jensen, L.S.; Scheutz, C. Life cycle assessment of sewage sludge management options including long-term impacts after land application. *J. Clean. Prod.* **2018**, *174*, 538–547. [CrossRef]

24. Sharma, S.; Sharma, P.; Bhattacharyya, A.K. Accumulation of Heavy Metals in Wheat (*Triticum aestivum* L.) Seedlings Grown in Soils Amended with Electroplating Industrial Sludge. *Commun. Soil Sci. Plant Anal.* **2010**, *41*, 2505–2516. [CrossRef]

25. Collivignarelli, M.; Abbà, A.; Frattarola, A.; Carnevale Miino, M.; Padovani, S.; Katsoyiannis, I.; Torretta, V. Legislation for the Reuse of Biosolids on Agricultural Land in Europe: Overview. *Sustainability* **2019**, *11*, 6015. [CrossRef]

26. Ministerio de agricultura pesca y alimentación. *Real Decreto 1310/1990U, de 29 de Octubre, por el que se Regula la Utilización de los Lodos de Depuración en el Sector Agrario*; Agencia Estatal Boletín Oficial del Estado: Madrid, España, 1990; Número 262; pp. 32339–32340. (In Spanish)

27. European Commission. *Council Directive of 12 June 1986 on The Protection of The Environment, and in Particular of The Soil, When Sewage Sludge is Used in Agriculture*; Official Journal of the European Communities: Aberdeen, UK, 1986; Volume 181, pp. 6–12.

28. European Commission. *Directive 2002/32/EC of the European Parliament and of the Council of 7 May 2002 on Undesirable Substances in Animal Feed*; Official Journal of the European Communities: Aberdeen, UK, 2002; Volume 140, pp. 10–22.

29. European Commission. *Commission Regulation (EC) No. 1881/2006 of 19 December 2006 Setting Maximum Levels for Certain Contaminants in Foodstuffs*; Official Journal of the European Communities: Aberdeen, UK, 2006; Volume 364, pp. 5–24.

30. Kacprzak, M.; Neczaj, E.; Fijałkowski, K.; Grobelak, A.; Grosser, A.; Worwag, M.; Rorat, A.; Brattebo, H.; Almås, Å.; Singh, B.R. Sewage sludge disposal strategies for sustainable development. *Environ. Res.* **2017**, *156*, 39–46. [CrossRef] [PubMed]

31. Campbell, H.W. Sludge management—Future issues and trends. *Water Sci. Technol.* **2000**, *41*, 1–8. [CrossRef]
32. Börjesson, G.; Menichetti, L.; Kirchmann, H.; Kätterer, T. Soil microbial community structure affected by 53 years of nitrogen fertilisation and different organic amendments. *Biol. Fertil. Soils* **2012**, *48*, 245–257. [CrossRef]
33. Hamnér, K.; Kirchmann, H. Trace element concentrations in cereal grain of long-term field trials with organic fertilizer in Sweden. *Nutr. Cycl. Agroecosyst.* **2015**, *103*, 347–358. [CrossRef]
34. Cambier, P.; Michaud, A.; Paradelo, R.; Germain, M.; Mercier, V.; Guérin-Lebourg, A.; Revallier, A.; Houot, S. Trace metal availability in soil horizons amended with various urban waste composts during 17 years—Monitoring and modelling. *Sci. Total Environ.* **2019**, *651*, 2961–2974. [CrossRef]
35. Jamali, M.K.; Kazi, T.G.; Arain, M.B.; Afridi, H.I.; Jalbani, N.; Kandhro, G.A.; Shah, A.Q.; Baig, J.A. Heavy metal accumulation in different varieties of wheat (Triticum aestivum L.) grown in soil amended with domestic sewage sludge. *J. Hazard. Mater.* **2009**, *164*, 1386–1391. [CrossRef]
36. Naidu, R.; Rogers, S.; Gupta, V.V.S.R.; Kookana, R.S.; Bolan, N.S.; Adriano, D. Bioavailability of metals in the soil-plant environment and its potential role in risk assessment. In *Bioavailability, Toxicity, and Risk Relationship in Ecosystems*; Naidu, R., Rogers, S., Gupta, V.V.S.R., Kookana, R.S., Bolan, N.S., Adriano, D., Eds.; Science Publishers: Enfield, NH, USA, 2003; pp. 21–59.
37. Madrid, F.; López, R.; Cabrera, F. Metal accumulation in soil after application of municipal solid waste compost under intensive farming conditions. *Agric. Ecosyst. Environ.* **2007**, *119*, 249–256. [CrossRef]
38. Karami, M.; Afyuni, M.; Rezainejad, Y.; Schulin, R. Heavy metal uptake by wheat from a sewage sludge-amended calcareous soil. *Nutr. Cycl. Agroecosystems* **2009**, *83*, 51–61. [CrossRef]
39. Morera, M.T.; Echeverría, J.C.; Garrido, J.J. Mobility of heavy metals in soils amended with sewage sludge. *Can. J. Soil Sci.* **2001**, *81*, 405–414. [CrossRef]
40. Nogueirol, R.C.; De Melo, W.J.; Alleoni, L.R.F. Testing extractants for Cu, Fe, Mn, and Zn in tropical soils treated with sewage sludge for 13 consecutive years. *Water Air Soil Pollut.* **2013**, *224*, 1–13. [CrossRef]
41. Dhanker, R.; Chaudhary, S.; Bhatia, T.; Goyal, S. Impact of Long Term Application of Municipal Solid Waste on Physicochemical and Microbial Parameters and Heavy Metal Distribution in Soils in Accordance to Its Agricultural Uses. *Int. J. Agric. Biosyst. Eng.* **2017**, *11*, 351–359.
42. Naidu, R.; Kookana, R.S.; Sumner, M.E.; Harter, R.D.; Tiller, K.G. Cadmium Sorption and Transport in Variable Charge Soils: A Review. *J. Environ. Qual.* **1997**, *26*, 602–617. [CrossRef]
43. Achiba, W.B.; Lakhdar, A.; Gabteni, N.; Du Laing, G.; Verloo, M.; Boeckx, P.; Van Cleemput, O.; Jedidi, N.; Gallali, T. Accumulation and fractionation of trace metals in a Tunisian calcareous soil amended with farmyard manure and municipal solid waste compost. *J. Hazard. Mater.* **2010**, *176*, 99–108. [CrossRef]
44. Raikhy, N.P.; Takkar, P.N. Zinc and Copper Adsorption by a Soil with and without Removal of Carbonates. *J. Indian Soc. Soil Sci.* **1983**, *31*, 611–614.
45. Smith, S.R. Effect of soil pH on availability to crops of metals in sewage sludge-treated soils. I. Nickel, copper and zinc uptake and toxicity to ryegrass. *Environ. Pollut.* **1994**, *85*, 321–327. [CrossRef]
46. FAO-UNESCO Soil Map of the World. Available online: http://www.fao.org/soils-portal/data-hub/soil-maps-and-databases/faounesco-soil-map-of-the-world/en/.# (accessed on 24 November 2020).
47. Thornthwaite, C.W. An approach towards a rational classification of climate. *Geogr. Rev.* **1948**, *38*, 55–94. [CrossRef]
48. Gobierno de Navarra Mapa de Estaciones. Available online: http://meteo.navarra.es/estaciones/mapadeestaciones.cfm# (accessed on 12 December 2020).
49. Hendershot, W.H.; Lalande, H. Chapter 16. Soil Reactionand Exchangeable Acidity. In *Soil Sampling and Methods of Analysis*; Carter, M.R., Ed.; CRC Press LLC: Boca Raton, FL, USA, 1993; pp. 141–142.
50. Pansu, M.; Gautheyrou, J. Chapter 18. Soluble Salts. In *Handbook of Soil Analysis. Mineralogical, Organic and Inorganic Methods*; Springer Netherlands: Dordrecht, The Netherlands, 2003; pp. 608–609.
51. Culley, J.L.B. Chapter 50. Density and Compressibility. In *Soil Sampling and Methods of Analysis*; Carter, M., Ed.; CRC Press LLC: Boca Raton, FL, USA, 1993; pp. 530–533.
52. Pansu, M.; Gautheyrou, J. Chapter 17. Carbonates. In *Handbook of Soil Analysis. Mineralogical, Organic and Inorganic Methods*; Springer Netherlands: Dordrecht, The Netherlands, 2003; pp. 593–601.
53. Tiessen, H.; Moir, J.O. Chapter 21. Total and Organic Carbon. In *Soil Sampling and Methods of Analysis*; Carter, M., Ed.; CRC Press LLC: Boca Raton, FL, USA, 1993; pp. 187–191.
54. AENOR. *Alimentos para animales. Métodos de Muestreo y Análisis. Determinación de Elementos Traza, Metales Pesados y Otros Elementos en los Alimentos Para Animales por ICP-MS (UNE-EN 17053)*; Asociación Española de Normalización: Génova, Madrid, 2018. (In Spanish)
55. U.S. EPA. *Method 3051A (SW-846): Microwave Assisted Acid Digestion of Sediments, Sludges, and Oils, Revision 1*; United States Environmental Protection Agency: Washington, DC, USA, 2007.
56. Pansu, M.; Gautheyrou, J. Chapter 29. Phosphorus. In *Handbook of Soil Analysis. Mineralogical, Organic and Inorganic Methods*; Springer Netherlands: Dordrecht, The Netherlands, 2003; p. 809.
57. Pansu, M.; Gautheyrou, J. Chapter 26. Cation Exchange Capacity. In *Handbook of Soil Analysis. Mineralogical, Organic and Inorganic Methods*; Springer Netherlands: Dordrecht, The Netherlands, 2003; pp. 732–738.
58. Instituto Nacional de Estadística INE. Available online: https://www.ine.es/ (accessed on 12 December 2020).

59. Ministerio de la Presidencia, Relaciones con las Cortes y Memoria Democrática. *Orden de 17 de Septiembre de 1981 por la que se Establecen Métodos Oficiales de Análisis de Aceites y Grasas, Aguas, Carnes y Productos Cárnicos, Fertilizantes, Productos Fitosanitarios, Leche y Productos Lácteos, Piensos y sus Primeras Materias, Producto*; Agencia Estatal Boletín Oficial del Estado: Madrid, España, 1981; Número 246; pp. 24003–24034. (In Spanish)

60. McGill, W.B.; Figueiredo, C.T. Chapter 22. Total Nitrogen. In *Soil Sampling and Methods of Analysis*; Carter, M., Ed.; CRC Press LLC: Boca Raton, FL, USA, 1993; pp. 201–207.

61. U.S. EPA. *Method 3052 (SW-846): Microwave Assisted Acid Digestion of Siliceous and Organically Based Matrices*; United States Environmental Protection Agency: Washington, DC, USA, 1996.

62. AENOR. *Calidad del suelo. Extracción de Oligoelementos con una Disolución Tampón de ADTP (UNE 77315:2001)*; Asociación Española de Normalización: Génova, Madrid, 2001. (In Spanish)

63. ISO. *Soil Quality—Extraction of Trace Elements by Buffered DTPA Solution*; ISO: Geneva, Switzerland, 2001.

64. González, D.; Almendros, P.; Álvarez, J.M. Metodos de analisis de elementos en suelos: Disponibilidad y fraccionamiento. *An. Real Soc. Española Química* **2009**, *105*, 205–212.

65. Yang, Y.; Wang, Y.; Westerhoff, P.; Hristovski, K.; Jin, V.L.; Johnson, M.V.V.; Arnold, J.G. Metal and nanoparticle occurrence in biosolid-amended soils. *Sci. Total Environ.* **2014**, *485–486*, 441–449. [CrossRef]

66. Gobierno de Navarra. *Orden Foral 359/2010, de 26 de Julio, de la Consejera de Desarrollo Rural y Medio Ambiente, por la que se Regula la Utilización de Lodos de Depuración en la Agricultura de la Comunidad Foral de Navarra*; Boletín Oficial de Navarra: Navarra, España, 2010; Número 16 de Agosto de 2010; pp. 1–11. (In Spanish)

67. European Commission. *Council Directive 91/676/CEE of 12 December 1991 Concerning the Protection of Waters Against Pollution Caused by Nitrates from Agricultural Sources*; Official Journal of the European Communities: Aberdeen, UK, 1991; Volume 375, pp. 1–8.

68. The European Parliament and the Council of the European Union. *Regulation (Eu) 2019/1009 of The European Parliament and of the Council of 5 June 2019 Laying Down Rules on the Making Available on the Market of EU Fertilising Products and Amending Regulations (EC) No 1069/2009 and (EC) No 1107/2009 and Repealing Regula*; Official Journal of the European Communities: Aberdeen, UK, 2019; Volume 170, pp. 1–114.

69. Merrington, G.; Oliver, I.; Smernik, R.J.; McLaughlin, M.J. The influence of sewage sludge properties on sludge-borne metal availability. *Adv. Environ. Res.* **2003**, *8*, 21–36. [CrossRef]

70. Reyes, Y.; Vergara, I.; Torres, O.E.; Díaz, M.; González, E. Heavy metals contamination: Implications for health and food safety Yulieth. *Rev. Ing. Investig. Desarro.* **2016**, *16*, 66–77. [CrossRef]

71. Kidd, P.S.; Domínguez-Rodríguez, M.J.; Díez, J.; Monterroso, C. Bioavailability and plant accumulation of heavy metals and phosphorus in agricultural soils amended by long-term application of sewage sludge. *Chemosphere* **2007**, *66*, 1458–1467. [CrossRef]

72. Alcantara, S.; Pérez, D.V.; Almeida, M.R.A.; Silva, G.M.; Polidoro, J.C.; Bettiol, W. Chemical Changes and Heavy Metal Partitioning in an Oxisol Cultivated with Maize (*Zea mays*, L.) after 5 Years Disposal of a Domestic and an Industrial Sewage Sludge. *Water. Air. Soil Pollut.* **2009**, *203*, 3–16. [CrossRef]

73. Bolan, N.S.; Adriano, D.C.; Naidu, R. Role of phosphorus in (im)mobilization and bioavailability of heavy metals in the soil-plant system. *Rev. Environ. Contam. Toxicol.* **2003**, *177*, 1–44. [CrossRef]

74. Naidu, R.; Bolan, N.S.; Megharaj, M.; Juhasz, A.L.; Gupta, S.K.; Clothier, B.E.; Schulin, R. Chapter 1. Chemical bioavailability in terrestrial environments. In *Chemical Bioavailability in Terrestrial Environment*; Hartemink, A.E., McBratney, A.B., Naidu, R.B.T.-D., Eds.; Elsevier: Amsterdam, The Netherlands, 2008; Volume 32, pp. 1–6. ISBN 0166-2481.

75. de Melo, W.J.; de Stéfani Aguiar, P.; Maurício Peruca de Melo, G.; Peruca de Melo, V. Nickel in a tropical soil treated with sewage sludge and cropped with maize in a long-term field study. *Soil Biol. Biochem.* **2007**, *39*, 1341–1347. [CrossRef]

76. Pardo, F.; Jordán, M.M.; Sanfeliu, T.; Pina, S. Distribution of Cd, Ni, Cr, and Pb in amended soils from alicante province (SE, Spain). *Water. Air. Soil Pollut.* **2011**, *217*, 535–543. [CrossRef]

77. Navarro-Pedreño, J.; Almendro-Candel, M.B.; Jordán-Vidal, M.M.; Mataix-Solera, J.; García-Sánchez, E. Mobility of cadmium, chromium, and nickel through the profile of a calcisol treated with sewage sludge in the southeast of Spain. *Environ. Geol.* **2003**, *44*, 545–553. [CrossRef]

78. Taylor, R.W.; Xiu, H.; Mehadi, A.A.; Shuford, J.W.; Tadesse, W. Fractionation of residual cadmium, copper, nickel, lead, and zinc in previously sludge-amended soil. *Commun. Soil Sci. Plant Anal.* **1995**, *26*, 2193–2204. [CrossRef]

79. Cabral, A.R.; Lefebvre, G. Use of sequential extraction in the study of heavy metal retention by silty soils. *Water. Air. Soil Pollut.* **1998**, *102*, 329–344. [CrossRef]

80. Moral, R.; Gilkes, R.J.; Jordán, M.M. Distribution of heavy metals in calcareous and non-calcareous soils in Spain. *Water. Air. Soil Pollut.* **2005**, *162*, 127–142. [CrossRef]

81. Kabala, C.; Karczewska, A.; Szopka, K.; Wilk, J. Copper, zinc, and lead fractions in soils long-term irrigated with municipal wastewater. *Commun. Soil Sci. Plant Anal.* **2011**, *42*, 905–919. [CrossRef]

82. Kabata-Pendias, A. Zinc. In *Trace Elements in Soil and Plants*; CRC Press LLC: Boca Raton, FL, USA, 2000; pp. 146–158.

83. Kiekens, L. Zinc. In *Heavy Metals in Soils*; Alloway, B.J., Ed.; Blackie Academic and Professional: Glasgow, Scotland, 1995; pp. 284–305.

84.	Gonzalez Flores, E.; Tornero Campante, M.A.; Sandoval Castro, E.; Pérez Magaña, A.; Gordillo Martínez, A.J. Biodisponibilidad y Fraccionamiento de Metales Pesados en Suelos Agrícolas Enmendados con Biosólidos de Origen Municipal. Available online: http://www.scielo.org.mx/scielo.php?script=sci_arttext&pid=S0188-49992011000400002 (accessed on 23 December 2020).
85.	Bolan, N.; Adriano, D.; Mani, S.; Khan, A. Adsorption, complexation, and phytoavailability of copper as influenced by organic manure. *Environ. Toxicol. Chem.* **2003**, *22*, 450–456. [CrossRef]
86.	Bünemann, E.K.; Smernik, R.J.; Marschner, P.; McNeill, A.M. Microbial synthesis of organic and condensed forms of phosphorus in acid and calcareous soils. *Soil Biol. Biochem.* **2008**, *40*, 932–946. [CrossRef]
87.	Rundle, H.; Calcroft, H.; Holt, C. Agricultural disposal of sludges on a historic sludge disposal site. *Water Pollut. Control* **1982**, *81*, 619–632.
88.	McGrath, S.P.; Zhao, F.J.; Dunham, S.J.; Crosland, A.R.; Coleman, K. Long-Term Changes in the Extractability and Bioavailability of Zinc and Cadmium after Sludge Application. *J. Environ. Qual.* **2000**, *29*, 875–883. [CrossRef]
89.	Nkinahamira, F.; Suanon, F.; Chi, Q.; Li, Y.; Feng, M.; Huang, X.; Yu, C.P.; Sun, Q. Occurrence, geochemical fractionation, and environmental risk assessment of major and trace elements in sewage sludge. *J. Environ. Manag.* **2019**, *249*. [CrossRef]
90.	Zhu, N.; Li, Q.; Guo, X.J.; Zhang, H.; Deng, Y. Sequential extraction of anaerobic digestate sludge for the determination of partitioning of heavy metals. *Ecotoxicol. Environ. Saf.* **2014**, *102*, 18–24. [CrossRef] [PubMed]
91.	Motaghian, H.R.; Hosseinpur, A.R. Rhizosphere effects on Cu availability and fractionation in sewage sludge-amended calcareous soils. *J. Plant Nutr. Soil Sci.* **2015**, *178*, 713–721. [CrossRef]
92.	Garnett, T.P.; Graham, R.D. Distribution and Remobilization of Iron and Copper in Wheat. *Ann. Bot.* **2005**, *95*, 817–826. [CrossRef] [PubMed]
93.	Morera, M.T.; Echeverría, J.; Garrido, J. Bioavailability of heavy metals in soils amended with sewage sludge. *Can. J. Soil Sci.* **2002**, *82*, 433–438. [CrossRef]
94.	Codling, E.E. Long-Term Effects of Biosolid-Amended Soils on Phosphorus, Copper, Manganese, and Zinc Uptake by Wheat. *Soil Sci.* **2014**, *179*, 21–27. [CrossRef]

Article

Adsorption of *p*-Arsanilic Acid on Iron (Hydr)oxides and Its Implications for Contamination in Soils

Yifan Yang [1,2,3], **Shiyong Tao** [1,2,3], **Zhichun Dong** [1,2,3], **Jing Xu** [1,2,3], **Xiang Zhang** [1,2,3,*] **and Guoyan Pan** [1,2,3,*]

[1]   State Key Laboratory of Water Resources and Hydropower Engineering Science, Wuhan University, Wuhan 430072, China; yangyfian@whu.edu.cn (Y.Y.); taoshiyong@whu.edu.cn (S.T.); zcdong@whu.edu.cn (Z.D.); jingxu0506@whu.edu.cn (J.X.)

[2]   Hubei Key Laboratory of Water System Science for Sponge City Construction, Wuhan University, Wuhan 430072, China

[3]   Hubei Provincial Collaborative Innovation Center for Water Resources Security, Wuhan 430072, China

*   Correspondence: zhangxiang@whu.edu.cn (X.Z.); guoyan.pan@whu.edu.cn (G.P.)

Abstract: Because of the diversification of industries in developing cities, the phenomenon of the simultaneous contamination of various kinds of pollutants is becoming common, and the environmental process of pollutants in multi-contaminated environmental mediums has attracted attention in recent years. In this study, *p*-arsanilic acid (ASA), a kind of organic arsenic feed additive that contains the arsenic group in a chemical structure, is used as a typical contaminant to investigate its adsorption on iron oxides and its implication for contaminated soils. The adsorption kinetics on all solids can be fitted to the pseudo-second-order kinetic model well. At the same mass dosage conditions, the adsorption amount per unit surface area on iron oxides follows the order α-FeOOH > γ-Fe$_2$O$_3$ > α-Fe$_2$O$_3$, which is significantly higher than that for actual soil, because of the lower content of iron oxides in actual soil. Lower pH conditions favor ASA adsorption, while higher pH conditions inhibit its adsorption as a result of the electrostatic repulsion and weakened hydrophobic interaction. The presence of phosphate also inhibits ASA adsorption because of the competitive effect. Correlations between the amount of ASA adsorption in actual soil and the Fe$_2$O$_3$ content, total phosphorus content, arsenic content, and organic matter content of actual soil are also investigated in this work, and a moderate positive correlation (R^2 = 0.630), strong negative correlation (R^2 = 0.734), insignificant positive correlation (R^2 = 0.099), and no correlation (R^2 = 0.006) are found, respectively. These findings would help evaluate the potential hazard of the usage of organic arsenic feed additives, as well as further the understanding of the geochemical processes of contaminants in complicated mediums.

Keywords: *p*-arsanilic acid; iron oxides; soil contamination; adsorption

Citation: Yang, Y.; Tao, S.; Dong, Z.; Xu, J.; Zhang, X.; Pan, G. Adsorption of *p*-Arsanilic Acid on Iron (Hydr)oxides and Its Implications for Contamination in Soils. *Minerals* **2021**, *11*, 105. https://doi.org/10.3390/min11020105

Received: 5 December 2020

Accepted: 19 January 2021

Published: 22 January 2021

1. Introduction

Industries in developing cities are diversified, including agriculture, manufacturing, mining industry, etc. This diversification could lead to a multitude of sources and types of pollution to the environment. The simultaneous contamination of nutrients, heavy metals, toxic organics, Pharmaceutical and Personal Care Products, etc., in water and surface soil has been widely reported [1–3]. Such simultaneous contamination is the joint result of domestic sewage discharge, industrial wastewater discharge, livestock and poultry industrial wastewater discharge, and agricultural non-point source pollution.

The geochemical behavior of pollutants in a multi-contaminated environment has attracted attention in recent years. Both competitive and cooperative effects on contaminants' adsorption in the complex system are reported [4–10]. The competitive effect, which is mainly caused by limited surface sites for adsorption, is reported to be relatively more, while the cooperative effect, which is mainly caused by bridge ions or compounds, is relatively less. Because of the intricacy of contamination in the urban environment, studying the transportation behavior of pollutants on actual environmental mediums is of great importance.

Organic arsenic feed additives, 4-aminobenzenearsenic acid (*p*-arsanilic acid (ASA)) and 4-hydorxy-3-nitrobenzenearsenic acid (roxarsone, ROX), have been used in poultry production for decades. They are types of antibiotics that can promote protein synthesis and animal growth, and can prevent the growth of parasites and microorganisms [11,12]. Organic arsenic feed additives tend to be excreted by animals with no significant chemical structural change, and the waste of these animals is often used as a fertilizer in nearby farms, which could lead to the potential contamination of agricultural fields [13]. The occurrence of organoarsenicals in the surface water, soil, and sediment surrounding swine farms has been reported [14]. These organoarsenicals are made up of the total arsenic in the environmental mediums. Although these two organoarsenicals are less toxic than inorganic arsenic, they can turn into inorganic arsenic through the biological or chemical process during their long-term existence in the environment, which will eventually cause damage to the soil and harm human health [15–18].

Studies about the adsorption of arsenic species on solids have been done for decades, the majority of which focus on the behavior of inorganic arsenic species, because of the higher percentage in the determined total arsenic species [19]. These studies have reported the adsorption behavior of inorganic arsenic on pure iron oxides and collected actual soil or sediment, both in the presence or absence of other contaminants. The results show that the adsorption kinetics and adsorption amount of inorganic arsenic were highly affected by the presence of other contaminants, especially phosphate, due to its similar chemical structure to arsenic [20–22]. Tofan-Lazar and Al-Abadleh compared the adsorption kinetics of phosphate on the surface of iron oxides at various conditions, and found that the adsorption rate was the fastest on freshly prepared iron (oxyhydr)oxide and slowest on arsenate-covered iron (oxyhydr)oxide [23]. Although aromatic organoarsenicals are also important as arsenic pollution sources, their adsorption behavior was relatively less studied previously, and has only received attention in recent years. Those works concerning organoarsenical adsorption mainly report on the behavior and mechanisms of organoarsenicals on iron oxides because of the abundance of iron species in soil [24–27], while the information of their adsorption on actual soil, especially contaminated soil, remains unclear.

Because of the complexity of the environmental medium in urban areas, investigations into the sorption pattern of pollutants on actual soil are important in order to understand the geochemical behavior of pollutants. In this work, the adsorption of ASA on both pure iron oxides (α-Fe$_2$O$_3$, γ-Fe$_2$O$_3$, and α-FeOOH) and collected actual surface soil is investigated. In pure iron oxide systems, the adsorption kinetics are investigated at various dosage conditions, and the effects of pH and the presence of phosphate on the ASA adsorption amount were also studied. The actual soil samples were collected from a multi-industrial city (including the agriculture and mining industries) in Hubei Province, China, in order to illustrate the relationship between the adsorption amount/rate and the chemical properties of solids. Describing the environmental transportation behavior of organoarsenicals would help with evaluating the potential hazards associated with the usage of organic arsenic feed additives, and further the understanding of the geochemical behavior of pollutants in multi-contaminated mediums.

2. Materials and Methods

2.1. Chemicals

The ASA (98%) was purchased from Aladdin Co. (Shanghai, China) and was used without further purification. α-Fe$_2$O$_3$ and γ-Fe$_2$O$_3$ were purchased from Aladdin Co. (Shanghai, China), and α-FeOOH was purchased from Sigma-Aldrich Co. (St. Louis, MO, USA). All of the other reagents were analytically pure and were purchased from Sinopharm Chemical Reagent Co., Ltd. Ultrapure water (18.2 MΩ, obtained through a water purification system, Ming-Che 24UV, Millipore, France) was used for the reagent preparation and experiments. All of the prepared solutions were stored in polypropylene plastic bottles (Nalgene, Rochester, NY, USA) avoiding light.

2.2. Soil Collection and Pretreatment

The soils were collected from Jingmen City (Hubei Province, China)—an important agricultural and industrial area with many farms and mining factories. Both the wastewater and waste generated from farms and factories in Jingmen City are extensive and could lead to contamination of the surrounding surface soils. The ecological and environmental protection in the Yangtze River Basin has attracted attention in recent years, and the government of Hubei Province has launched a pollution survey project to control the contamination situation and annual pollution loads, including for Jingmen City, for the purpose of the environmental management and restoration. The sampling sites in this work are shown in Figure 1. The soil samples were collected 10–15 cm below the surface, in order to avoid collecting anthropogenic impurities. The fresh soil samples were preserved at a low temperature and kept out of light before pretreatment.

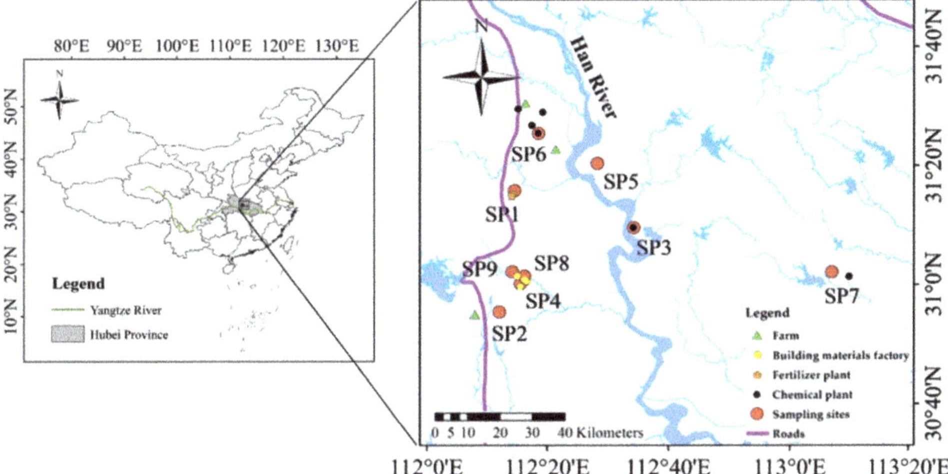

Figure 1. Map showing the locations of the sampling sites.

The collected soil samples were dried at room temperature and were protected from light. Then, the dried solids were ground and sieved using a 100-mesh sieve. The sieved samples were transferred into sample bags and preserved at 4 °C.

2.3. Adsorption Experiments

The adsorption experiments were conducted in a Nalgene bottle with continuous stirring using a magnetic stirrer at a speed of 750 r/min. The mixed solution, containing 20 μM ASA and 10 mM NaCl, was prepared and adjusted to the desired pH value with diluted NaOH/HCl before adding the solids. Samples were taken at interval times and were filtered with a 0.22 μm polyethersulfone filter membrane for further analysis. In the kinetic adsorption experiments, the solution volume was 200 mL, and the pH value was controlled by re-adjusting several times throughout the entire adsorption period. In the pH effect experiments, the solution volume was 100 mL, the pH value was also re-adjusted several times, and the precise pH value was recorded. In the phosphate competition experiments, the solution volume was also 100 mL, a certain amount of phosphate was also added to the mixed solution, and the rest procedures were the same as described above. Here, >90% of ASA was be desorbed by the co-presence of phosphate and alkaline (pH > 12, 2 mM PO_4^{3-}), confirming that no degradation occurred during adsorption.

2.4. Analytical Methods

The concentration of ASA was analyzed as described in our previous works [12,18,28]. High-performance liquid chromatography (HPLC; a 20ADVP pump, a DAD-20AVP detector, Shimadzu Instrument Co. Ltd., Kyoto, Japan) with a C18 column (Supelco Discovery, 4.6 mm × 250 mm, 5 μm) was used for the analysis. The mobile phase was a mixture of a 2.5% formic acid and methanol solution (95:5, *v/v*). The flow rate was set to 1 mL/min and the detection wavelength was set to 254 nm.

The BET surface area of the iron oxide solids was analyzed using a surface area and porosimetry analyzer (V-Sorb 2800P, Gold APP Instrument Co., Beijing, China). The total phosphorus content of the collected soil samples was analyzed according to the national standard using an ICAP6300 Plasma Emission spectrometer (Thermo Fisher, Waltham, MA, USA). The element content (Fe_2O_3 and As) of the soils was analyzed using an X-ray fluorescence spectroscopy (XRF, Explorer 9000, Jiangsu Skyray Instrument Co., Ltd., Kunshan, China). The content of organic matter (OM) was estimated by measuring the ignition loss. The pH of the solutions was determined using a pH meter (F2-Meter, Mettler Toledo, Greifensee, Switzerland).

2.5. Statistical Analysis

The amount of ASA (q_t, $\mu mol \cdot m^{-2}$) adsorbed was calculated using the difference between their concentrations at the initial time and at time t. A pseudo-second-order kinetic model was used to fit the kinetic data for ASA, which can be expressed as follows:

$$\frac{t}{q_t} = \frac{1}{q_e^2 * k_2} + \frac{1}{q_e}t \tag{1}$$

where k_2 ($m^2 \cdot \mu mol^{-1} \cdot h^{-1}$) is the pseudo-second-order rate constant and q_e is the amount of adsorption at equilibrium time. k_2 and q_e can be obtained from the slope and y-intercept of the plots of t/q_t vs. t. The adsorption percentage was calculated by the difference between the concentrations at the initial and ending times, which can be expressed as follows:

$$adsorption\ percentage\ (\%) = \left(1 - \frac{C_t}{C_0}\right) \times 100\% \tag{2}$$

where C_0 and C_t are the concentrations of ASA at the initial and ending times, respectively.

3. Results and Discussion

3.1. Adsorption Kinetics of ASA on Iron Oxides

The adsorption kinetics of ASA on three iron oxides is investigated at various iron oxide dosages at pH 5. The adsorption kinetics and equilibrium time of ASA on three iron oxides were close. As shown in Figure 2a–c, after adding iron oxides into the mixture solution, the concentrations of ASA decreased significantly in the early stage (within 4 h), while the changes gradually became slow in the later stage, and reached equilibrium at around 24 h. Among the three iron oxides, α-Fe_2O_3 had the fastest adsorptive removal efficiency; more than 95% of ASA was be removed from the solution within 30 min at 1 g·L^{-1} dosage, whereare α-FeOOH was the slowest, with the removal efficiency decreasing to only 49% at the same condition. The BET surface areas of α-Fe_2O_3, γ-Fe_2O_3, and α-FeOOH were 125.04, 53.39, and 9.50 m$^2 \cdot$g^{-1}, respectively. The significant difference in adsorptive removal efficiency among each of the iron oxides could be caused by the highly different surface areas. The calculated adsorption amount of ASA, normalized to the surface area, is shown in Figure 2d–f. The obtained parameters for the pseudo-second-order kinetic model are shown in Table 1, and the obtained calculated kinetics from the parameters are also given in Figure 2d–f.

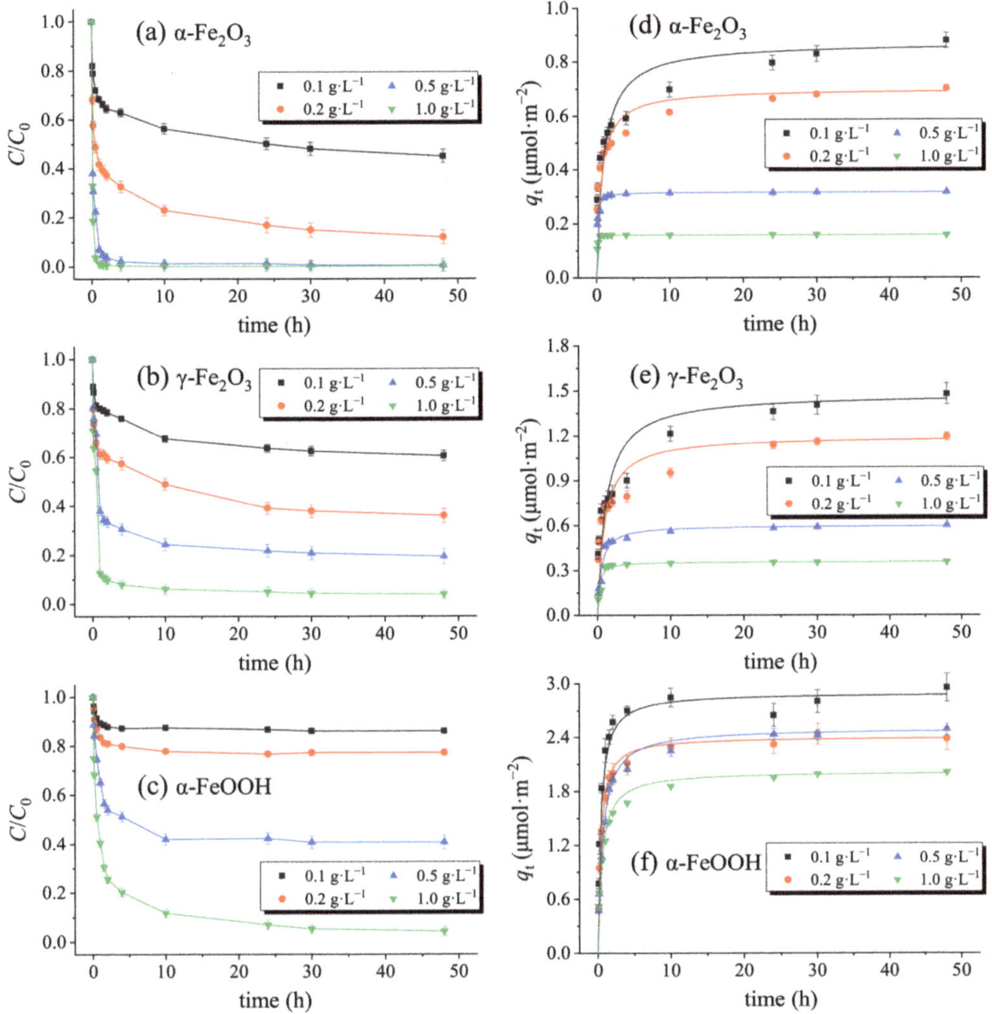

Figure 2. Adsorption kinetics of *p*-arsanilic acid (ASA) on α-Fe$_2$O$_3$, γ-Fe$_2$O$_3$, and α-FeOOH. The solid points are experimental data. The polylines in (**a**–**c**) are for the visual guides. The curves in (**d**–**f**) are the fitted results according to the pseudo-second-order kinetic model. ASA = 20 μM, NaCl = 10 mM, and pH = 5.

Table 1. Fitted parameters for ASA adsorption on iron oxides.

Parameters	Dosage	α-Fe$_2$O$_3$	γ-Fe$_2$O$_3$	α-FeOOH
	0.1 g·L^{-1}	0.87	1.48	2.90
q_e	0.2 g·L^{-1}	0.70	1.20	2.41
(μmol·m^{-2})	0.5 g·L^{-1}	0.32	0.60	2.50
	1.0 g·L^{-1}	0.16	0.36	2.02
	0.1 g·L^{-1}	1.22	0.60	0.94
k_2	0.2 g L^{-1}	2.32	0.92	1.20
(m^2·μmol^{-1}·h^{-1})	0.5 g·L^{-1}	42.44	3.69	0.67
	1.0 g·L^{-1}	602.02	13.30	0.97

As can be seen from Table 1, q_e and k_2 show an opposite trend to each other. With q_e for α-Fe$_2$O$_3$, γ-Fe$_2$O$_3$ decreases dramatically when increasing the solid dosages (i.e., total surface area) by an order of magnitude, and the difference can be as high as 5.5 times for α-Fe$_2$O$_3$. However, the change is less significant for α-FeOOH at the same condition. Such results are thought to be caused by the valid adsorption sites on iron oxide surface [29]. At limited surface area conditions, the adsorption sites on the surface of the iron oxides could be nearly fully occupied by ASA, while at abundant surface area conditions, the adsorption sites are also abundant, and thus would not be completely occupied. At the same solid dosage conditions, the surface areas of α-Fe$_2$O$_3$ and γ-Fe$_2$O$_3$ were much larger than that of α-FeOOH; therefore, the two former iron oxides had more vacancy sites. Although the dosage of α-FeOOH increased from 0.1 g·L^{-1} to 1.0 g·L^{-1}, the amount of adsorption sites were still limited compared with the dosed ASA, thus the change in q_e was less significant than that of α-Fe$_2$O$_3$ and γ-Fe$_2$O$_3$. In order to avoid the extreme adsorption circumstances and to obtain a better observation, the dosage of iron oxides in the following experiment was 0.2 g·L^{-1}.

3.2. Effect of pH

Previous works have reported the obvious effect of pH on organic compounds' adsorption behavior, which is caused by the joint effect of compounds and solids [30,31]. Here, experiments were therefore conducted at various pH conditions in order to investigate the adsorption behavior of ASA on the three iron oxides. First, 10 mM NaCl was introduced to the mixed solution in order to eliminate the ionic strength effect caused by the pH. As can be seen from Figure 3a, among the three iron oxides, α-Fe$_2$O$_3$ shows a relatively higher adsorptive removal percentage for the entire investigated pH range, and that of α-FeOOH is the lowest at the same mass dosage. This trend was reversed when calculating the adsorption amount normalized to the surface area (Figure 3b): α-FeOOH showed the highest q_e for the overall investigated pH conditions. For all three oxides, ASA adsorption showed a decrease trend with an increase of pH. The adsorption percentage of ASA on α-Fe$_2$O$_3$ showed a sharp decrease from 93% to less than 10% when the pH increased from 4.08 to 11.88, and that for α-FeOOH also showed a gradual decrease from 31% to less than 10% when the pH increased from 4.00 to 11.48.

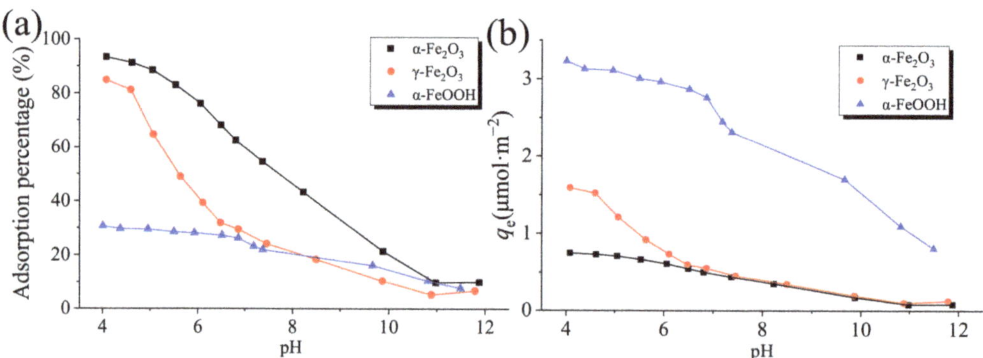

Figure 3. Effect of pH on ASA adsorption on iron oxides. (**a**) Data shown as adsorption percentage (%) and (**b**) data shown as adScheme 2. Solid points are experimental data; polylines are for the visual guide. ASA = 20 μM, iron oxdies = 0.2 g·L^{-1}, and NaCl = 10 mM.

Such a trend is similar to the previous reported works. Bell-shaped adsorption curves with pH have been widely reported for the adsorption of organic compounds on (hydr)oxides [5,31,32], and both bell-shaped and cliff-shaped curves have been found for the adsorption of inorganic arsenic species on oxides or minerals [10,33,34]. The adsorption behavior of ASA is also highly affected by its arsenic group in its chemical structure.

ASA has 3 acidity coefficients—2.00, 4.02, and 8.92. Therefore, at neutral and basic conditions, ASA exists in anion form [18,35]. In the meantime, the point of zero charge for α-Fe_2O_3, γ-Fe_2O_3, and α-$FeOOH$ were at a circumneutral pH [36–38], also showing a negative charge at high pH conditions. Therefore, the higher pH would exacerbate the electrostatic repulsion and weaken the hydrophobic interaction between the oxyanion compounds and solids.

3.3. Competing Ion (Phosphate) Effect on Equilibrium Adsorption

Because of the similarity in the chemical structure between the phosphate and arsenate groups, the effect of phosphate on the chemical behavior of arsenic is often considered when investigating the adsorption of arsenic species. Cheng et al. studied the effect of various anions (CO_3^{2-}, SiO_3^{2-}, Cl^-, F^-, SO_4^{2-}, NO_3^-, and HPO_4^{2-}) on the adsorption behavior of As(III) in iron-containing materials, and found that HPO_4^{2-} has the highest inhibition effect [39]. The removal efficiency by adsorption decreased from >90% to only about 60% in the presence of HPO_4^{2-}. Wang et al. reported the obvious competitive effect between the inorganic arsenic adsorption and phosphorus release by sediments [21]. Lin et al. proposed that the use of phosphorus fertilizers could enhance the mobility of arsenic towards groundwater in arsenic-contaminated aquifers [40]. The inhibition effect of phosphate on organic arsenic adsorption has previously been reported [25,41].

In this work, the effect of phosphate on the adsorption of ASA on α-Fe_2O_3, γ-Fe_2O_3, and α-$FeOOH$ were investigated at a wide phosphate concentration range. The inhibition effect could be observed over the whole investigated pH range (4–12) for all the three iron oxides (Figure 4). A significantly higher inhibition effect was observed when the concentration of phosphate increased from 0.02 mM to 2 mM. This strong competitive effect reduced the adsorption ability of ASA on phosphorus abundant soil, thus leading to an enhancement in ASA mobility. Despite this, the high amount of phosphate did not show the complete inhibition of ASA, especially in acid conditions. Liu et al. used the concentrated phosphate solution (0.5 M H_3PO_4) to extract arsenic species from contaminated soils, and the extraction efficiency for ASA was relatively low (67%) with a high error bar, while that for As (V) was close to 100% [14]. They suggested that the degradation of ASA might have occurred during the long time extraction process (16 h), which, in fact, might be also caused by an improper acid extraction condition (pH < 2 for 0.5 M H_3PO_4).

3.4. Actual Soil Adsorption of ASA

The content of actual soil is complex, and contains metal oxides, nutrients, and inorganic and organic contaminants. Such compounds might all have implications on the contaminant adsorption behavior and mobility, resulting in a positive or negative effect. Table 2 shows the contents of Fe_2O_3, total phosphorus (TP), As, and organic matter (OM) of the nine collected soils. The BET surface area and soil pH are also represented in Table 2. The correlation of each of the parameters is investigated. As our parameters and compound contents were moderately/strongly skewed, we computed their log-transformation. Interestingly, a slight correlation between these parameters could be observed. Figure 5a shows a slight positive correlation between Fe_2O_3 and As ($R^2 = 0.278$). These results are not surprising, because the adsorption of arsenic species is highly related to the presence of iron oxides [42]. As the sampling area has many mining activities, it could lead to the contamination of heavy metals, including arsenic species. The higher content of iron oxides in actual soil could therefore immobilize more heavy metals on the surface soils. Nevertheless, the content of As is a jointly affected by the surrounding human activities and the adsorption ability of the soil. Therefore, the current As content/contamination level did not show a strong positive correlation with the iron content. Hafeznezami et al. also reported an increased trend in the adsorption ability of As at a higher content of amorphous Fe, although the statistical correlation was not significant [43]. Surprisingly, opposite to As, a negative correlation between Fe_2O_3 and TP was observed (Figure 5b, $R^2 = 0.505$), which was unexpected, as the presence of iron oxides should enhance the retention ability of phosphorus in soil [44,45]. This phenomenon might be attributed to the proportion of amorphous species. There was no correlation between As content and TP content ($R^2 = 0.010$, results not shown), indicating that the available adsorption sites were still abundant for the pollutant sorption, and the competitive effect between As and TP was insignificant for the current contamination situation. The contributions of each composition to the surface area of the soil samples were also investigated. A slight positive correlation between Fe_2O_3 content and the surface area can be observed in Figure 5c ($R^2 = 0.329$), indicating a relatively important contribution. TP showed a moderate negative correlation with the surface area ($R^2 = 0.602$, data not shown), which might be caused by its relationship with the Fe_2O_3 content. The content of OM and As showed unimportant contributions to the surface area ($R^2 = 0.016$ for OM and $R^2 = 0.006$ for As).

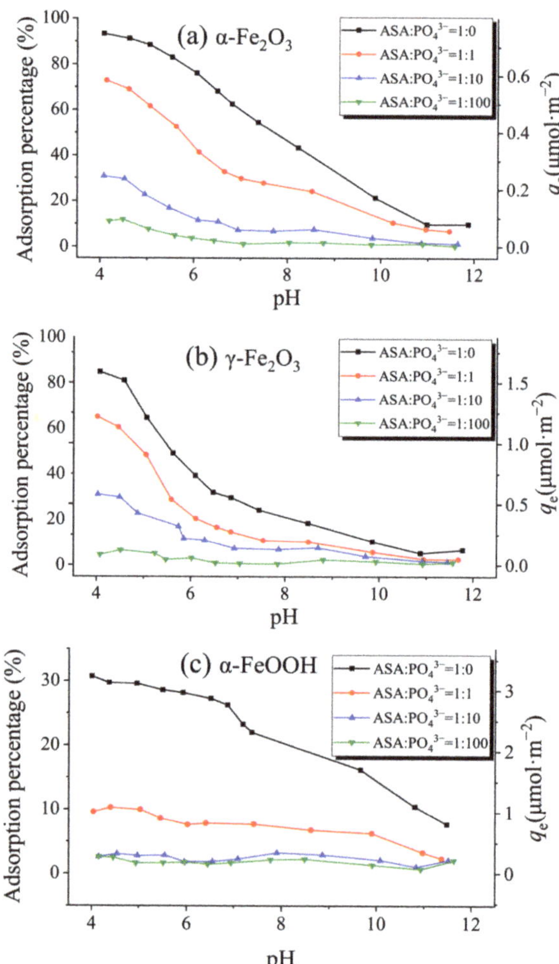

Figure 4. Effect of phosphate on ASA adsorption. Solid points are experimental data; polylines are for the visual guide. ASA = 20 μM, iron oxdies = 0.2 g·L^{-1}, and NaCl = 10 mM. (**a**) α-Fe$_2$O$_3$, (**b**) γ-Fe$_2$O$_3$, (**c**) α-FeOOH.

Table 2. Properties of the collected actual soil.

Characters	SP1	SP2	SP3	SP4	SP5	SP6	SP7	SP8	SP9
Fe_2O_3%	4.284	4.917	5.276	5.921	6.031	6.518	6.717	7.509	8.228
TP%	0.44	0.08	0.06	0.02	0.03	0.02	0.03	0.04	0.05
As% $\times 10^4$	18.5	8.9	13.1	14.3	14.8	17.8	20.7	33.3	20.0
OM%	6.77	3.54	2.12	6.96	3.63	3.79	6.92	4.22	3.82
BET surface area ($m^2 \cdot g^{-1}$)	13.60	42.55	15.54	34.72	46.36	61.82	56.18	54.75	29.33
pH *	7.21	7.42	7.17	7.37	7.05	7.14	7.11	7.42	7.44

* determined after the ASA adsorption reaction.

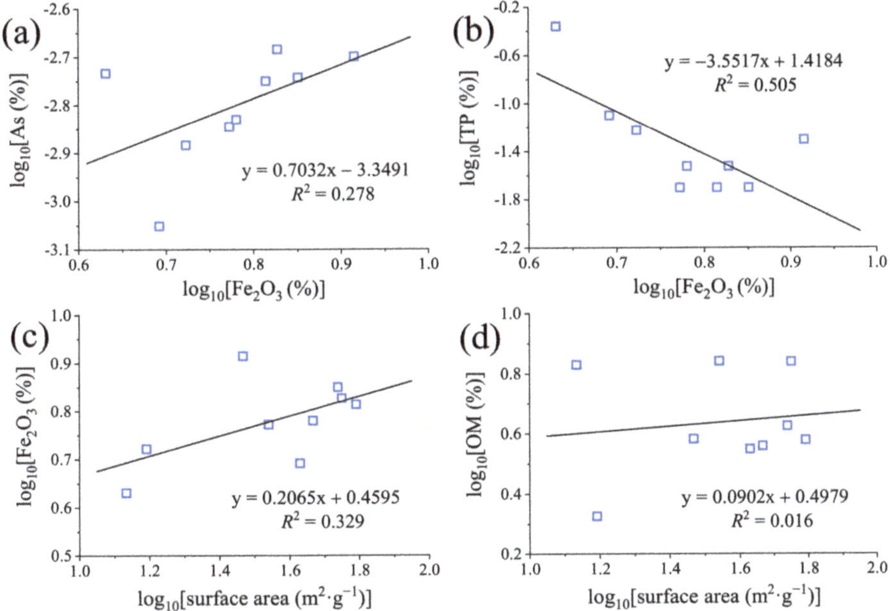

Figure 5. Correlation between the (**a**) Fe_2O_3 content and As content, (**b**) Fe_2O_3 content and total phosphorus (TP) content, (**c**) Fe_2O_3 content and surface area, and (**d**) organic matter (OM) content and surface area for the collected actual soil.

The adsorption kinetics of ASA for the nine collected soil samples were then investigated. The strong interactions between arsenic species and iron oxides have been widely reported, while the interactions between arsenic species and pure silica (which is the main composition of soil) seem to be less important for the adsorption efficiency of arsenic species. Although the surface area of the soil samples was close to pure iron oxides (γ-Fe_2O_3 and α-FeOOH), the valid adsorption sites on the solid surface were be much lower. In order to ensure experimental accuracy (i.e., a relatively obvious adsorption percentage), a high solid dosage (5 g·L^{-1}) was used in this section. No pH adjustment was done before or during the adsorption reaction, as the high dosage of soil could form a buffer system, although the final pH was determined. The pseudo-second-order kinetic model was also used to fit the adsorption kinetic results. The experimental results and fitted results are shown in Figure 6, and the calculated parameters for the pseudo-second-order kinetic model are shown in Table 3. The obtained q_e varied significantly with the different soil samples, with SP8 showing the highest result (0.0309 μmol·m^{-2}), which was ~8.4 times as high as the lowest (SP1, 0.0037 μmol·m^{-2}), while all of them were obviously lower than that of the pure iron oxide surface (α-Fe_2O_3, γ-Fe_2O_3, and α-FeOOH), indicating invalid adsorption sites on the solid surface. Such a big difference would be mainly caused by the content of iron oxides. The adsorption rate constants were mainly opposite to the trend of the adsorption amount.

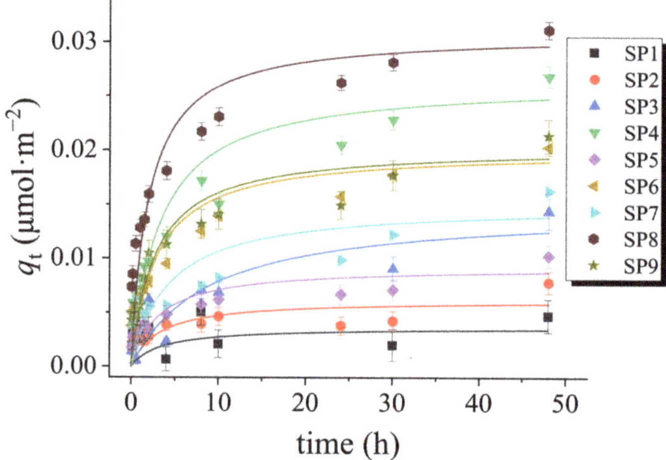

Figure 6. Adsorption kinetics of ASA on the collected actual soil. Solid points are experimental data, and curves are fitted results by pseudo-second-order kinetic model. ASA = 20 μM, NaCl = 10 mM, and soil dosage = 5 g·L^{-1}.

Table 3. Fitted parameters for ASA adsorption of soil by a pseudo-second order kinetic model.

Parameters	SP1	SP2	SP3	SP4	SP5	SP6	SP7	SP8	SP9
$q_e \times 10^2$ (μmol m^{-2})	0.37	0.62	1.47	2.64	0.93	2.01	1.50	3.09	2.04
k_2 (m^2·μmol^{-1}·h^{-1})	80.44	52.10	7.98	11.91	35.74	17.66	16.32	16.85	18.38

Previous works have reported that the adsorption of arsenic species is positively correlated with the metal oxide content and is negatively correlated with the phosphorus content [46,47]. The correlations of the ASA adsorption parameters with the Fe$_2$O$_3$ and TP content in the actual soils were fitted in this work. As can be seen from Figure 7a, a moderate positive correlation between the q_e and Fe$_2$O$_3$ content can be observed, as expected. As iron oxides have strong interactions with ASA, the higher Fe$_2$O$_3$ content would obviously facilitate the adsorption of ASA. However, the positive correlation was not very strong in the limited soil samples, indicating the effect of other physical and chemical parameters on ASA adsorption, which was similar to the situation of the arsenate adsorption on sandy sediments [43]. In contrast, the correlation between q_e and TP content is strongly negative (see Figure 7b). The strong negative correlation between the ASA adsorption and TP content would be caused by the competitive effect, as discussed in Section 3.3, which therefore decreased the statistical correlation between the q_e and Fe$_2$O$_3$ content. The correlation between the q_e and As content was also investigated, and an insignificant positive correlation was found, indicating potential relevance between ASA adsorption ability and the current As contamination level. Such results also reveal abundant available adsorption sites on the surface of the soil, although these soil samples have been contaminated by As to a certain degree. It seems that the OM content did not affect q_e (Figure 7d), probably because of the joint reason of its low content and its weak sorption ability. The final pH after 48 h of the adsorption reaction was very close between the nine samples; therefore, the statistic correlation between q_e and pH was insignificant ($R^2 = 0.005$, data not shown). In general, the adsorption results on the collected actual soils indicate that predicting the adsorption behavior was not possible because of single or very few physical and chemical parameters.

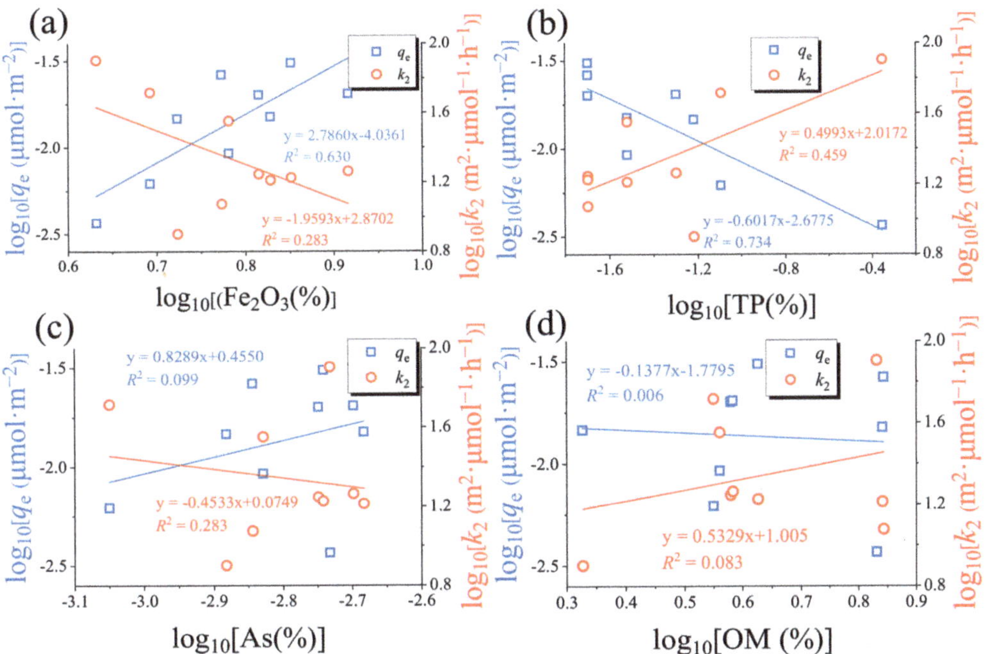

Figure 7. Correlation fitting between ASA adsorption amount (q_e) and (**a**) Fe$_2$O$_3$ content, (**b**) TP content, (**c**) As content, and (**d**) OM content.

4. Conclusions

The adsorption behavior of ASA on three kinds of pure iron oxides and nine collected actual soil samples are studied in this work. The adsorption kinetics on all solids are well fitted to the pseudo-second-order kinetic model. The dosage of iron oxides, pH conditions, and the concentration of co-present phosphate could all affect the adsorption amount of ASA on iron oxides. The parameters of the actual soil that can affect ASA adsorption are more complicated than those of pure iron oxides. Although the adsorption amount shows a moderate positive correlation with the Fe$_2$O$_3$ content, strong negative correlation with the TP content, insignificant positive correlation with the As content, and no correlation with the OM content, it is still difficult to obtain an accurate prediction model because of the complexity of the physical and chemical properties of those soils. Investigating the transportation of contaminants in multi-contaminated environmental mediums helps to further understand their geochemical processes and helps with formulating a remediation strategy for the contaminated area.

Author Contributions: Conceptualization, X.Z.; data curation, Z.D.; formal analysis, Y.Y. and G.P.; funding acquisition, J.X.; investigation, Y.Y., S.T., and Z.D.; methodology, G.P.; resources, J.X. and X.Z.; supervision, J.X.; validation, Z.D.; visualization, Y.Y. and S.T.; writing—original draft, Y.Y.; writing—review and editing, S.T., X.Z., and G.P. All authors have read and agreed to the published version of the manuscript.

Funding: National Natural Science Foundation of China: 21707106. National Natural Science Foundation of China: 42077350. Strategic Priority Research Program of the Chinese Academy of Sciences: XDA23040300. China Postdoctoral Science Foundation: 2016M602358.

Institutional Review Board Statement: Not applicable.

Informed Consent Statement: Not applicable.

Data Availability Statement: Data is contained within the article.

Acknowledgments: We thank T. Chen and T. Luo for their help in the actual soil sampling. Comments from the anonymous reviewers are also appreciated.

Conflicts of Interest: The authors declare no conflict of interest.

References

1. Khatri, N.; Tyagi, S. Influences of natural and anthropogenic factors on surface and groundwater quality in rural and urban areas. *Front. Life Sci.* **2015**, *8*, 23–39. [CrossRef]
2. Li, Q.; Zhang, Y.; Lu, Y.; Wang, P.; Suriyanarayanan, S.; Meng, J.; Zhou, Y.; Liang, R.; Khan, K. Risk ranking of environmental contaminants in Xiaoqing River, a heavily polluted river along urbanizing Bohai Rim. *Chemosphere* **2018**, *204*, 28–35. [CrossRef] [PubMed]
3. Wu, Y.; Yu, C.P.; Yue, M.; Liu, S.P.; Yang, X.Y. Occurrence of selected PPCPs and sulfonamide resistance genes associated with heavy metals pollution in surface sediments from Chao Lake, China. *Environ. Earth Sci.* **2016**, *75*, 1–8. [CrossRef]
4. Xu, J.; Marsac, R.; Wei, C.; Wu, F.; Boily, J.F.; Hanna, K. Cobinding of Pharmaceutical Compounds at Mineral Surfaces: Mechanistic Modeling of Binding and Cobinding of Nalidixic Acid and Niflumic Acid at Goethite Surfaces. *Environ. Sci. Technol.* **2017**, *51*, 11617–11624. [CrossRef]
5. Xu, J.; Marsac, R.; Costa, D.; Cheng, W.; Wu, F.; Boily, J.F.; Hanna, K. Co-Binding of Pharmaceutical Compounds at Mineral Surfaces: Molecular Investigations of Dimer Formation at Goethite/Water Interfaces. *Environ. Sci. Technol.* **2017**, *51*, 8343–8349. [CrossRef]
6. Li, S.; Zhang, C.; Wang, M.; Li, Y. Adsorption of multi-heavy metals Zn and Cu onto surficial sediments: Modeling and adsorption capacity analysis. *Environ. Sci. Pollut. Res.* **2014**, *21*, 399–406. [CrossRef]
7. Wang, Y.J.; Jia, D.A.; Sun, R.J.; Zhu, H.W.; Zhou, D.M. Adsorption and Cosorption of Tetracycline and Copper(II) on Montmorillonite as Affected by Solution pH. *Environ. Sci. Technol.* **2008**, *42*, 3254–3259. [CrossRef]
8. Pei, Z.; Shan, X.Q.; Kong, J.; Wen, B.; Owens, G. Coadsorption of ciprofloxacin and Cu(II) on montmorillonite and kaolinite as affected by solution pH. *Environ. Sci. Technol.* **2010**, *44*, 915–920. [CrossRef]
9. Neupane, G.; Donahoe, R.J.; Arai, Y. Kinetics of competitive adsorption/desorption of arsenate and phosphate at the ferrihydrite-water interface. *Chem. Geol.* **2014**, *368*, 31–38. [CrossRef]
10. Goldberg, S. Competitive Adsorption of Arsenate and Arsenite on Oxides and Clay Minerals. *Soil Sci. Soc. Am. J.* **2002**, *66*, 413–421. [CrossRef]
11. Frost, D.V.; Overby, L.R.; Spruth, H.C. Studies with Arsanilic Acid and Related Compounds. *J. Agric. Food Chem.* **1955**, *3*, 235–243. [CrossRef]
12. Shen, X.; Xu, J.; Pozdnyakov, I.P.; Liu, Z. Photooxidation of *p*-arsanilic acid in aqueous solution by UV/persulfate process. *Appl. Sci.* **2018**, *8*, 615. [CrossRef]
13. Mangalgiri, K.P.; Adak, A.; Blaney, L. Organoarsenicals in poultry litter: Detection, fate, and toxicity. *Environ. Int.* **2015**, *75*, 68–80. [CrossRef] [PubMed]
14. Liu, X.; Zhang, W.; Hu, Y.; Cheng, H. Extraction and detection of organoarsenic feed additives and common arsenic species in environmental matrices by HPLC-ICP-MS. *Microchem. J.* **2013**, *108*, 38–45. [CrossRef]
15. Cortinas, I.; Field, J.A.; Kopplin, M.; Garbarino, J.R.; Gandolfi, A.J.; Sierra-Alvarez, R. Anaerobic biotransformation of roxarsone and related N-substituted phenylarsonic acids. *Environ. Sci. Technol.* **2006**, *40*, 2951–2957. [CrossRef] [PubMed]
16. Zhu, X.D.; Wang, Y.J.; Liu, C.; Qin, W.X.; Zhou, D.M. Kinetics, intermediates and acute toxicity of arsanilic acid photolysis. *Chemosphere* **2014**, *107*, 274–281. [CrossRef]
17. Xie, X.; Hu, Y.; Cheng, H. Mechanism, kinetics, and pathways of self-sensitized sunlight photodegradation of phenylarsonic compounds. *Water Res.* **2016**, *96*, 136–147. [CrossRef]
18. Xu, J.; Shen, X.; Wang, D.; Zhao, C.; Liu, Z.; Pozdnyakov, I.P.; Wu, F.; Xia, J. Kinetics and mechanisms of pH-dependent direct photolysis of *p*-arsanilic acid under UV-C light. *Chem. Eng. J.* **2018**, *336*, 334–341. [CrossRef]
19. Nicomel, N.R.; Leus, K.; Folens, K.; Van Der Voort, P.; Du Laing, G. Technologies for arsenic removal from water: Current status and future perspectives. *Int. J. Environ. Res. Public Health* **2015**, *13*, 1–24. [CrossRef]
20. Zeng, H.; Fisher, B.; Giammar, D.E. Individual and competitive adsorption of arsenate and phosphate to a high-surface-area iron oxide-based sorbent. *Environ. Sci. Technol.* **2008**, *42*, 147–152. [CrossRef]
21. Wang, J.; Xu, J.; Xia, J.; Wu, F.; Zhang, Y. A kinetic study of concurrent arsenic adsorption and phosphorus release during sediment resuspension. *Chem. Geol.* **2018**, *495*, 67–75. [CrossRef]
22. Leus, K.; Folens, K.; Nicomel, N.R.; Perez, J.P.H.; Filippousi, M.; Meledina, M.; Dîrtu, M.M.; Turner, S.; Van Tendeloo, G.; Garcia, Y.; et al. Removal of arsenic and mercury species from water by covalent triazine framework encapsulated γ-Fe$_2$O$_3$ nanoparticles. *J. Hazard. Mater.* **2018**, *353*, 312–319. [CrossRef] [PubMed]

23. Tofan-Lazar, J.; Al-Abadleh, H.A. Kinetic ATR-FTIR studies on phosphate adsorption on iron (Oxyhydr)oxides in the absence and presence of surface arsenic: Molecular-level insights into the Ligand exchange mechanism. *J. Phys. Chem. A* **2012**, *116*, 10143–10149. [CrossRef] [PubMed]
24. Mitchell, W.; Goldberg, S.; Al-Abadleh, H.A. In situ ATR-FTIR and surface complexation modeling studies on the adsorption of dimethylarsinic acid and *p*-arsanilic acid on iron-(oxyhydr)oxides. *J. Colloid Interface Sci.* **2011**, *358*, 534–540. [CrossRef] [PubMed]
25. Chen, W.R.; Huang, C.H. Surface adsorption of organoarsenic roxarsone and arsanilic acid on iron and aluminum oxides. *J. Hazard. Mater.* **2012**, *227*, 378–385. [CrossRef]
26. Depalma, S.; Cowen, S.; Hoang, T.; Al-Abadleh, H.A. Adsorption thermodynamics of *p*-arsanilic acid on iron (oxyhydr)oxides: In-situ ATR-FTIR studies. *Environ. Sci. Technol.* **2008**, *42*, 1922–1927. [CrossRef] [PubMed]
27. Folens, K.; Leus, K.; Nicomel, N.R.; Meledina, M.; Turner, S.; Van Tendeloo, G.; Du Laing, G.; Van Der Voort, P. Fe3O4@MIL-101 – A Selective and Regenerable Adsorbent for the Removal of As Species from Water. *Eur. J. Inorg. Chem.* **2016**, *2016*, 4395–4401. [CrossRef]
28. Li, S.; Xu, J.; Chen, W.; Yu, Y.; Liu, Z.; Li, J.; Wu, F. Multiple transformation pathways of *p*-arsanilic acid to inorganic arsenic species in water during UV disinfection. *J. Environ. Sci.* **2016**, *47*, 39–48. [CrossRef] [PubMed]
29. Wang, Z.M.; Giammar, D.É. Mass Action Expressions for Bidentate Adsorption in Surface Complexation Modeling: Theory and Practice. *Environ. Sci. Technol.* **2013**, *47*, 3982–3996. [CrossRef]
30. Marsac, R.; Martin, S.; Boily, J.-F.; Hanna, K. Oxolinic acid binding at goethite and akaganéite surfaces: Experimental Study and Modeling. *Environ. Sci. Technol.* **2016**, *50*, 660–668. [CrossRef]
31. Zhang, H.; Huang, C.H. Adsorption and oxidation of fluoroquinolone antibacterial agents and structurally related amines with goethite. *Chemosphere* **2007**, *66*, 1502–1512. [CrossRef] [PubMed]
32. Wu, Q.; Li, Z.; Hong, H. Adsorption of the quinolone antibiotic nalidixic acid onto montmorillonite and kaolinite. *Appl. Clay Sci.* **2013**, *74*, 66–73. [CrossRef]
33. Manning, B.A.; Fendorf, S.E.; Goldberg, S. Surface structures and stability of arsenic(III) on goethite: Spectroscopic evidence for inner-sphere complexes. *Environ. Sci. Technol.* **1998**, *32*, 2383–2388. [CrossRef]
34. Ghosh, M.M.; Yuan, J.R. Adsorption of inorganic arsenic and organoarsenicals on hydrous oxides. *Environ. Prog.* **1987**, *6*, 150–157. [CrossRef]
35. Jaafar, J. Separation of Phenylarsonic Compounds by Ion Pairing–Reversed Phase–High Performance Liquid Chromatography. *J. Teknol.* **2001**, *35*, 71–79. [CrossRef]
36. Mustafa, S.; Tasleem, S.; Naeem, A. Surface charge properties of Fe_2O_3 in aqueous and alcoholic mixed solvents. *J. Colloid Interface Sci.* **2004**, *275*, 523–529. [CrossRef]
37. Watanabe, H.; Seto, J. The Point of Zero Charge and the Isoelectric Point of γ-Fe_2O_3 and α-Fe_2O_3. *Bull. Chem. Soc. Jpn.* **1986**, *59*, 2683–2687. [CrossRef]
38. Tripathy, S.S.; Kanungo, S.B. Adsorption of Co^{2+}, Ni^{2+}, Cu^{2+} and Zn^{2+} from 0.5 M NaCl and major ion sea water on a mixture of δ-MnO_2 and amorphous FeOOH. *J. Colloid Interface Sci.* **2005**, *284*, 30–38. [CrossRef]
39. Cheng, W.; Xu, J.; Wang, Y.; Wu, F.; Xu, X.; Li, J. Dispersion-precipitation synthesis of nanosized magnetic iron oxide for efficient removal of arsenite in water. *J. Colloid Interface Sci.* **2015**, *445*, 93–101. [CrossRef]
40. Lin, T.Y.; Wei, C.C.; Huang, C.W.; Chang, C.H.; Hsu, F.L.; Liao, V.H.C. Both Phosphorus Fertilizers and Indigenous Bacteria Enhance Arsenic Release into Groundwater in Arsenic-Contaminated Aquifers. *J. Agric. Food Chem.* **2016**, *64*, 2214–2222. [CrossRef]
41. Chabot, M.; Hoang, T.; Al-Abadleh, H.A. ATR-FTIR studies on the nature of surface complexes and desorption efficiency of *p*-arsanilic acid on iron (oxyhydr)oxides. *Environ. Sci. Technol.* **2009**, *43*, 3142–3147. [CrossRef]
42. Thakur, J.K.; Thakur, R.K.; Ramanathan, A.; Kumar, M.; Singh, S.K. Arsenic Contamination of Groundwater in Nepal—An Overview. *Water* **2010**, *3*, 1–20. [CrossRef]
43. Hafeznezami, S.; Zimmer-Faust, A.G.; Dunne, A.; Tran, T.; Yang, C.; Lam, J.R.; Reynolds, M.D.; Davis, J.A.; Jay, J.A. Adsorption and desorption of arsenate on sandy sediments from contaminated and uncontaminated saturated zones: Kinetic and equilibrium modeling. *Environ. Pollut.* **2016**, *215*, 290–301. [CrossRef] [PubMed]
44. Bolland, M.D.A.; Allen, D.G. Spatial variation of soil test phosphorus and potassium, oxalate-extractable iron and aluminum, phosphorus-retention index, and organic carbon content in soils of Western Australia. *Commun. Soil Sci. Plant Anal.* **1998**, *29*, 381–392. [CrossRef]
45. Kuo, S.; Mikkelsen, D.S. Distribution of iron and phosphorus in flooded and unflooded soil profiles and their relation to phosphorus adsorption. *Soil Sci.* **1979**, *127*, 18–25. [CrossRef]
46. Smedley, P.L.; Kinniburgh, D.G. A review of the source, behaviour and distribution of arsenic in natural waters. *Appl. Geochem.* **2002**, *17*, 517–568. [CrossRef]
47. Lambkin, D.C.; Alloway, B.J. Arsenate-induced phosphate release from soils and its effect on plant phosphorus. *Water. Air. Soil Pollut.* **2003**, *144*, 41–56. [CrossRef]

Article

Describing Phosphorus Sorption Processes on Volcanic Soil in the Presence of Copper or Silver Engineered Nanoparticles

Jonathan Suazo-Hernández [1,2] , Erwin Klumpp [3], Nicolás Arancibia-Miranda [4,5], Patricia Poblete-Grant [2] , Alejandra Jara [2,6], Roland Bol [3] and María de La Luz Mora [2,6,*]

1   Doctoral Program in Science of Natural Resources, Universidad de La Frontera, Av. Francisco Salazar 01145, P.O. Box 54-D, Temuco, Chile; j.suazo06@ufromail.cl

2   Center of Plant, Soil Interaction and Natural Resources Biotechnology, Scientific and Biotechnological Bioresource Nucleus (BIOREN-UFRO), Universidad de La Frontera, Avenida Francisco Salazar, 01145 Temuco, Chile; patty.grant87@gmail.com (P.P.-G.); alejandra.jara@ufrontera.cl (A.J.)

3   Institute of Bio- and Geosciences, Agrosphere (IBG–3), Forschungszentrum Jülich, Wilhelm Johnen Str., 52425 Jülich, Germany; e.klumpp@fz-juelich.de (E.K.); r.bol@fz-juelich.de (R.B.)

4   Faculty of Chemistry and Biology, Universidad de Santiago de Chile, Av. B. O'Higgins, 3363 Santiago, Chile; nicolas.arancibia@usach.cl

5   Center for the Development of Nanoscience and Nanotechnology, CEDENNA, 9170124 Santiago, Chile

6   Department of Chemical Sciences and Natural Resources, Universidad de La Frontera, Av. Francisco Salazar 01145, P.O. Box 54-D, Temuco, Chile

*   Correspondence: mariluz.mora@ufrontera.cl; Tel.: +56-4574-4240; Fax: +56-4532-5053

Citation: Suazo-Hernández, J.; Klumpp, E.; Arancibia-Miranda, N.; Poblete-Grant, P.; Jara, A.; Bol, R.; de La Luz Mora, M. Describing Phosphorus Sorption Processes on Volcanic Soil in the Presence of Copper or Silver Engineered Nanoparticles. *Minerals* **2021**, *11*, 373. https://doi.org/10.3390/min11040373

Academic Editors: Ana Romero-Freire and Hao Qiu

Received: 5 March 2021
Accepted: 29 March 2021
Published: 1 April 2021

Publisher's Note: MDPI stays neutral with regard to jurisdictional claims in published maps and institutional affiliations.

Abstract: Engineered nanoparticles (ENPs) present in consumer products are being released into the agricultural systems. There is little information about the direct effect of ENPs on phosphorus (P) availability, which is an essential nutrient for crop growth naturally occurring in agricultural soils. The present study examined the effect of 1, 3, and 5% doses of Cu^0 or Ag^0 ENPs stabilized with L-ascorbic acid (suspension pH 2–3) on P ad- and desorption in an agricultural Andisol with total organic matter (T-OM) and with partial removal of organic matter (R-OM) by performing batch experiments. Our results showed that the adsorption kinetics data of $H_2PO_4^-$ on T-OM and R-OM soil samples with and without ENPs were adequately described by the pseudo-second-order (PSO) and Elovich models. The adsorption isotherm data of $H_2PO_4^-$ from T-OM and R-OM soil samples following ENPs addition were better fitted by the Langmuir model than the Freundlich model. When the Cu^0 or Ag^0 ENPs doses were increased, the pH value decreased and $H_2PO_4^-$ adsorption increased on T-OM and R-OM. The $H_2PO_4^-$ desorption (%) was lower with Cu^0 ENPs than Ag^0 ENPs. Overall, the incorporation of ENPs into Andisols generated an increase in P retention, which may affect agricultural crop production.

Keywords: adsorption; engineered nanoparticles; organic matter; phosphorus; nutrients; pollution; volcanic soil

1. Introduction

In the past decade, the incorporation of engineered nanoparticles (ENPs) into consumer products [1,2] has led to a significant increase in their turnover from $250 billion in 2009 to $3 trillion in 2020 [3]. Two of the most widely used ENPs in consumer products are metallic copper (Cu^0) and silver (Ag^0), due to their antibacterial properties. Cu^0 ENPs are added to biocides, electronics, paints, cosmetics, agrochemicals, ceramics, and film [1,3,4], whereas Ag^0 ENPs are used in textiles, air filters, bandages, paints, food storage containers, agrochemicals, deodorants, toothpaste, and household appliances [5]. Thus, as a consequence of extensive and diverse commercial applications, these ENPs can be released into the environment. Soil is the main sink of disposal for most of the released ENPs [6]. Adverse effects on human health and ecosystems may be expected, making it necessary to improve our current understanding of environmental risks, fate, transformations and aggregation behaviors of metallic ENPs [7].

The geochemistry of metallic Cu^0 and Ag^0 ENPs in soils is complex, due to their chemical transformation between Cu^0, Cu^+ and Cu^{2+} as well as between Ag^0 and Ag^+, respectively [1,4], also due to their strong binding capacity to various soil components like clay minerals, organic matter, microorganisms, among others. Transformations of metallic ENPs in soil include oxidation, dissolution, and sulfidation. Over time, Cu^0 ENPs can be oxidized in the soil to form CuO (tenorite) and Cu_2O (cuprite) nanoparticles with a core-shell structure. Any of these, both forms of copper oxide nanoparticles, can dissolve and release cuprous and/or cupric ions into solution [8]. Meanwhile, the Ag^0 ENPs show a slow oxidation process, which can be promoted in acid soils. The metallic ENPs oxidation in soils can be diminished when organic molecules are used as stabilizing agents [9]. Transformation on metallic ENPs is an important consideration to developing risk assessments of ENPs [4,9].

Several studies have intended to determine the effects caused by ENPs on soil properties. In these studies, it has been shown than due to metallic Cu^0 and Ag^0, ENPs are characterized by a high surface area and chemical reactivity, variable surface charge and chemical transformation [10]. Once in contact with soil, ENPs may therefore modify their structural and physico-chemical properties such as pH, electric conductivity, redox potential, porosity, and hydraulic conductivity [10–12]. This could affect reactions and processes of elements in soil, such as precipitation, dissolution, co-precipitation, complexation, oxidation/reduction, plant uptake, and ad- and desorption. Particularly, ad- and desorption are important because they control the availability and mobility of contaminants and nutrients [10]. In this context, Taghipour and Jalali [13] reported that metal oxide ENPs (Al_2O_3 and TiO_2) caused immobilization of phosphorus (P) in calcareous soils from Hamadan, Western Iran, and reduced the bioavailability of P.

In volcanic soils (Andisol and Ultisol), P is an essential crop macronutrient and this soil contains between 1000 and 3500 $mg \cdot kg^{-1}$ [14]. However, P availability for plant growth is limited because it can form inner-sphere complexes by ligand exchange with surface -OH and $-OH_2^+$ groups of soil components like ferrihydrite, imogolite, allophane, and Al(Fe)-humus complexes [15–17]. Numerous studies have focused on P availability in volcanic soils considering the effects on soils of fertilizers [18], liming [19], microorganisms [20,21], enzymes [22], inorganic/organic ligands [23], specific surface area [24], surface charge [25], organic matter content [26], and pH and mineralogy [27].

In relation to effects caused by ENPs in volcanic soils, no studies have assessed the influence of metallic ENPs on the adsorption of nutrients. In this context, the aim of this research was to evaluate the effect of Cu^0 or Ag^0 ENPs on phosphorus sorption processes in volcanic soils and its relationship with organic matter content. Overall, the results provide new information about the implication of ENPs for nutrient availability in soils.

2. Materials and Methods

2.1. Chemicals Used

The reagents used were $CuCl_2 \cdot 2H_2O$, $AgNO_3$, L-ascorbic acid, KH_2PO_4, KCl, HCl, and KOH (analytical grade, Merck) and double-distilled water. The pH electrode (Orion Star A211 pH Benchtop Meter, Thermo Fischer Scientific Beverly, Waltham, MA, USA) was calibrated using standard buffers of 4.01, 7.01, and 10.01 (Hanna, Woonsocket, RI, USA).

2.2. Synthesis of Cu^0 and Ag^0 ENPs

$CuCl_2 \cdot 2H_2O$ and $AgNO_3$ were used for the formation of Cu^0, and Ag^0 ENPs, respectively, and L-ascorbic acid was added as a reducing and capping agent [28]. Cu^0 ENPs (or Ag^0 ENPs) was synthesized by mixing 10.0 $mmol \cdot L^{-1}$ $CuCl_2 \cdot 2H_2O$ (or 10.0 $mmol \cdot L^{-1} AgNO_3$) in 50 mL double-distilled water. An Erlenmeyer flask (100 mL), containing the $CuCl_2 \cdot 2H_2O$ (or $AgNO_3$) solution, was heated in a water bath at 80 °C with magnetic stirring; 50 mL of L-ascorbic acid (1.0 $mol \cdot L^{-1}$) was added dropwise into the flask while stirring. The aqueous dispersion of stabilized Cu^0 ENPs (or Ag^0 ENPs) obtained was kept at 80 °C for 24 h and it was finally saved to ambient conditions for later research.

2.3. Soil Samples

The soil used was an Andisol belonging to Santa Barbara series from Southern Chile (36°50′ S; 71°55′ W). The soil was collected from the top 20 cm depth of the soil horizon. The soil was passed through a <2 mm mesh sieve and freeze-dried (total organic matter soil sample = T-OM). For partial removal of organic matter (OM), the T-OM soil sample was treated several times with H_2O_2 until adding did not result anymore in air bubbles emanating from the aqueous solution and maintained at 40 °C in a thermoregulated bath [29]. The resulting sample was then washed four times with double-distilled water (partial removal of OM soil sample = R-OM). Finally, both soil samples were freeze-dried and stored at 4 °C.

2.4. Characterization of Ag^0 and Cu^0 ENPs

The synthetized Cu^0 and Ag^0 ENPs were characterized using transmission electron microscopy (TEM) on a Hitachi model HT7700 (Hitachi, Tokyo, Japan) with Olympus camera (Veleta 2000 × 2000) using high resolution mode at 120 kV. The TEM images obtained were analyzed manually to calculate the particle size with the ImageJ program (version 1.50i, Wayne Rasband, National Institute of Health, Bethesda, MD, USA). The ultraviolet-visible (UV–Vis) spectra was recorded with a double-beam Rayleigh UV-2601 spectrophotometer (BRAIC Co. Ltd., Beijing, China) using 1 cm path length glass cell. The zeta potential (ZP) of Cu^0 and Ag^0 ENPs (25 mg) was measured in the presence of 10 mL KCl 0.01 M using a Nano ZS apparatus (Malvern Instruments, Worcestershire, UK) at 20 °C and the isoelectric point (IEP) was obtained from graphs of ZP versus pH. The Fourier-transform infrared spectroscopy (FT-IR) were recorded with a 1 mL of ENPs suspension. FT-IR analysis was realized using a Cary 630 spectrometer (Agilent Technologies, Santa Clara, CA, USA). The transmission spectrum was acquired with 4 cm^{-1} resolution and the operating range was 600 cm^{-1} to 4000 cm^{-1} at atmospheric pressure and 20 °C. The pH of the suspensions of ENPs was measured with 10 mL using a pH Meter.

2.5. Characterization of Soil Samples

The morphological characteristics of both soil samples were obtained by scanning electron microscopy with a STEM SU-3500 transmission module (Hitachi, Tokyo, Japan) and the QUANTAX 100 energy-dispersive X-ray spectrometer detector (EDX), (Bruker, Berlin, Germany) was used for the semi-quantitative analysis of the elemental composition (Al, Si, and Fe content). 20 mg of each soil sample were deposited onto 300-mesh Formvar/carbon-coated grids and were inspected under a high-vacuum. Confocal analysis was performed by laser scanning confocal microscopy (LSCM) using the Olympus Fluoview1000 (Olympus Optical Co., Melville, New York, NY, USA). 50 μL of the suspensions were collocated on a microscope slide with a micropipette and the sample was dried on a stove at 40 °C. The total organic carbon (TOC) of T-OM and R-OM soil samples was calculated using a Shimadzu TOC-V CPH instrument (Shimadzu, Tokyo, Japan). The TOC was transformed into soil organic matter content using the conversion factor of 1.72 [30]. The specific surface area of R-OM and T-OM soils was obtained using the Brunauer, Emmett and Teller (BET) theory. Approximately 200 mg of soil sample was degassed for 2 h at 105 °C and then was conducted using N_2 gas at −196 °C in the relative pressure range (P/P0) of 0.05–0.4. Surface area measurements were made with a Quantachrome Nova 1000e analyzer (Quantachrome Instruments, Boynton Beach, FL, USA). The average pore volume and size were obtained using the Barrett-Joyner-Halenda (BJH) model. For the FT-IR absorption spectrum, soil samples were dried at 50 °C for 12 h to eliminate the interference produced by the absorption of the water molecules. To determine the functional groups in both soil samples, the analysis was performed under similar conditions to the ENPs. Soil pH was determined in 1:2.5 soil: double-distilled water ratio after 5 min shaking and 120 min resting, using the same pH Meter used for ENPs determination. Total P was extracted from the soil samples by alkaline oxidation with sodium hypobromite (NaBrO) [31]. After each extraction, the supernatant was filtered (5C, Advantec) and then the concentration of total P in the supernatant was determined using a spectrophotometer Rayleigh UV-2601 with a wavelength of 880 nm [32]. Exchangeable Al was extracted with KCl (1 M) and measured using a Unicam model Solaar 969 atomic absorption spectrophotometer (AAS) (Unicam Ltd, Cambridge, UK). Exchangeable base cations (Na, K, Mg and Ca) in soils were extracted using NH_4Ac (1 M, pH 7.0) and were measured by AAS [33]. Effective cation exchange capacity (ECEC) was calculated as the sum of exchangeable Al plus the exchangeable base cations [33].

The ZP and IEP of the soil samples were determined pre- and post-adsorption of $H_2PO_4^-$ on T-OM and R-OM soil samples in the absence and presence of 5% Cu^0 or Ag^0 ENPs using the high point adsorption isotherms similar to the procedure followed by ENPs.

2.6. Adsorption Experiments

Batch experiments were conducted to investigate the adsorption of phosphate (indicated as $H_2PO_4^-$) on T-OM and R-OM soil samples in the absence and presence of 0, 1, 3, and 5% Cu^0 or Ag^0 ENPs doses (% *w/w*). Cu^0 or Ag^0 ENPs doses were added to 0.5 g (dry weight) of soil samples in polyethylene tubes and mixed with 20 mL $H_2PO_4^-$ solution. The adsorbed amounts of $H_2PO_4^-$ (q_t, mmol·kg^{-1}) were determined as the difference between initial concentration and final concentration of $H_2PO_4^-$ in the solution (Equation (1)).

$$q_t = \frac{(C_0 - C_t)V}{w} \tag{1}$$

where, C_o is the initial concentrations of $H_2PO_4^-$ and C_t is the concentrations of $H_2PO_4^-$ at time t or the equilibrium concentration (mmol·L^{-1}), w the weight (kg) of the soil and V is the volume (L).

To evaluate the pH effect on the adsorption of $H_2PO_4^-$ onto T-OM and R-OM soil samples, stock solutions of 6.47 mmol·L^{-1} of $H_2PO_4^-$ were prepared with double-distilled water at pH ranging from 4.5 to 8.5 by adding 0.1 M HCl or KOH and ionic strength 0.01 M KCl (background electrolyte). The $H_2PO_4^-$ solutions were added to soil samples with and without ENPs and were stirred at 200 rpm for 24 h at 20 ± 2 °C.

For the kinetic study, the initial solution of 6.47 mmol·L^{-1} of $H_2PO_4^-$ was adjusted to pH 5.5 ± 0.2 by adding 0.1 M HCl or KOH at ionic strength 0.01 M KCl and 20 ± 2 °C. Samples were taken from the suspension at 2.5, 5, 10, 30, 45, 60, 120, 180, 360, 720, and 1440 min, and $H_2PO_4^-$ was determined in solution. Furthermore, the initial pH (pH$_i$) and the final pH (pH$_f$) were measured after $H_2PO_4^-$ solution was added to soil samples (time 0 min) and after $H_2PO_4^-$ adsorption (1440 min), respectively.

Adsorption isotherms were obtained by varying the initial $H_2PO_4^-$ concentrations from 0.016 to 9.71 mmol·L^{-1} and were initially adjusted to pH 5.5 ± 0.2 and ionic strength 0.01 M KCl. The suspensions were stirred at 200 rpm in an orbital shaker at 20 ± 2 C for 24 h. To determine the effect of copper (Cu^{2+}) or silver cations (Ag^+) or L-ascorbic acid on $H_2PO_4^-$ adsorption onto T-OM and R-OM soil samples, adsorption isotherms were made in the presence of 3% Cu^{2+} or Ag^+ or L-ascorbic acid (% *w/w*) under the aforementioned experimental conditions.

The desorption experiment was performed once the adsorption isotherm procedure had ended by adding 20 mL of double-distilled water three times, and the samples were then stirred at 200 rpm in an orbital shaker at 20 ± 2 °C for 24 h. The desorption percentages (%) were calculated by the equation used by Silva-Yumi et al. [34] All the samples of the adsorption experiments were first centrifuged at 10,000 rpm for 10 min, using a centrifuge RC-5B Plus (Sorvall, Newtown, CT, USA) and then filtered through 0.22 μm syringe filters. In all experiments, the concentration of $H_2PO_4^-$ in the supernatant was determined according to the procedure followed for total P. To minimize manipulation errors in the analysis, the adsorption experiments were performed in triplicate.

2.7. Data Analysis

The kinetics adsorption (e.g., pseudo-first-order, pseudo-second-order, and Elovich) and isotherm (e.g., Langmuir and Freundlich) models used in this study are presented in Tables 1 and 2, respectively.

Table 1. The kinetic models used for the description of phosphate adsorption.

Kinetic Equations	Expression Formula	Parameters	References
Pseudo-first-order (PFO)	$q_t = q_e\left(1 - e^{-k_1 t}\right)$	q_t = amount of anion adsorbed at any time (mmol·kg^{-1}). q_e = amount of anion adsorbed at equilibrium (mmol·kg^{-1}). k_1 = PFO rate constant (min^{-1}).	
Pseudo-second-order (PSO) *	$q_t = \frac{k_2 q_e^2 t}{1 + k_2 q_e t}$	k_2 = PSO rate constant (kg·mmol^{-1}·min^{-1}). t = time (min)	[35,36]
Elovich	$q_t = \frac{1}{\beta} \ln(1 + \alpha\beta t)$	α = initial rate constant (mmol·kg$_-$·min^{-1}). β = number of sites available for the sorption and desorption constant (mmol·kg^{-1}).	

* From PSO initial adsorption rate (h), can be calculated by multiplying $k_2 q_t^2$ (mmol·kg^{-1}·min^{-1}).

Table 2. The isotherm models used for the description of phosphate adsorption.

Isotherm Equations	Expression Formula	Parameters	References
Langmuir	$q_e = \frac{q_m K_L C_e}{1 + K_L C_e}$	q_e = amount of adsorbed anion per unit mass of the adsorbent at equilibrium (mmol·kg^{-1}). q_{max} = maximum adsorption capacity (mmol·kg^{-1}). C_e = concentration of anion at equilibrium in the solution (mmol·L^{-1}). K_L = constant related to the affinity (L·mmol^{-1}). K_F = freundlich adsorption coefficient ((mmol·kg^{-1}) (L·kg^{-1})$^{1/n}$).	[34,37]
Freundlich	$q_e = K_F C_e^{\frac{1}{n}}$	n = adsorption intensity (1 < n < 10).	

The data were evaluated through the Chi-square (χ^2), adding the coefficient of determination (r^2) (Equations (2) and (3)). The lowest χ^2 and highest r^2 values were used as the best fit [37]. The statistical analysis of the adsorption data was conducted using Origin Pro 8.0.

$$\chi^2 = \sum \frac{\left(q_{e,exp} - q_{e,cal}\right)^2}{q_{e,cal}} \tag{2}$$

$$r^2 = \sum \frac{\left(q_{e,mean} - q_{e,cal}\right)^2}{\left(q_{e,cal} - q_{e,mean}\right)^2 + \left(q_{e,cal} - q_{e,exp}\right)^2} \tag{3}$$

where, $q_{e,mean}$ is the average value of experimental adsorption capacity (mmol·kg^{-1}), $q_{e,cal}$ is the equilibrium capacity from a model (mmol·kg^{-1}) and $q_{e,exp}$ is the experimental adsorption capacity.

3. Results

3.1. Characterization of Cu^0 and Ag^0 ENPs and Soils

The size, morphology, surface charge and the presence of functional groups on the surface of prepared ENPs were determined by TEM images, UV-Vis, ZP and FT-IR analyses. TEM images showed that both ENPs had spherical morphology (Figure S1a,b in Supplementary Materials). Cu^0 ENPs had a diameter between 8 and 29 nm, whereas Ag^0 ENPs showed a diameter between 7 and 27 nm (Figure S2a,b). The UV-Vis spectra of Cu^0 and Ag^0 ENPs showed an extended peak in the range of 342–512 and 337–474 nm, respectively (Figure S3). The FT-IR spectra for pure L-ascorbic acid showed a band corresponding to a stretching vibration carbon–carbon double bond at 1674 cm^{-1} and the peak of enol hydroxyl at 1322 cm^{-1} (Figure S4a). After the reduction of Cu^{2+} and Ag$^+$ by L-ascorbic acid, the peaks disappeared and new peaks at 3481 cm^{-1} and 1636 cm^{-1} were observed (Figure S4b,c), which were associated with the conjugated hydroxyl and carbonyl groups, respectively. The pH of Cu^0 and Ag^0 ENPs suspension was 2.46 and 2.35, respectively. The IEP of Cu^0 ENPs was 2.7, whereas Ag^0 ENPs had a negatively charged surface in the studied pH range (Figure S5).

Physico-chemical properties of the soils untreated (T-OM) and treated with H_2O_2 (R-OM) are shown in Table 3. The T-OM and R-OM were a typical Andisol exhibiting acidic characteristics showing pH values of 5.40 (strongly acidic) for T-OM and 6.20 (slightly acidic) for R-OM. Total P and OM in T-OM were 1.8 and 3.1 times higher as compared to R-OM, whereas the Al and Fe contents for R-OM were 1.2 and 1.4 times higher than T-OM. The SEM images revealed a decreased number of aggregates in R-OM compared to T-OM (Figure 1a,b). The contrasting OM content was also indicated in confocal images (Figure 1c,d) by a higher green fluorescence intensity for T-OM as compared to R-OM images. The IEP for T-OM was 3.2, while it was 5.7 for R-OM. Furthermore, the BET-specific surface area and pore volume increased 1.4 and 11.5 times for R-OM in comparison to T-OM. The FT-IR analysis (Figure S6) showed that R-OM had bands at 1003 cm^{-1} and 913 cm^{-1} corresponding to alumina and silica-rich allophane, respectively, while T-OM only showed the band at 1003 cm^{-1} [29]. T-OM had more effective cation exchange capacity (ECEC) than R-OM (Table 3).

Table 3. Physico-chemical properties of soil with total organic matter (T-OM) and with partial removal of matter (R-OM).

Parameter	T-OM	R-OM
pH (H_2O)	5.4 ± 0.0	6.2 ± 0.0
Total P (mg·kg^{-1})	1766.4 ± 27.0	996.6 ± 15.0
Si (%)	15.9 ± 3.5	16.3 ± 2.9
Al (%)	11.7 ± 1.1	14.1 ± 1.5
Fe (%)	7.5 ± 0.6	10.5 ± 1.2
OM (%)	14.1 ± 0.1	4.6 ± 0.1
ECEC (cmol(+) kg^{-1}) *	8.8 ± 0.5	7.8 ± 0.0
Isoelectric point	3.2	5.7
BET- specific surface area (m^2·g^{-1})	17.4	24.4
Average pore volume (cm^3·g^{-1})	0.002	0.023
Average pore diameter (A)	10.7	10.4

* ECEC: Effective cation exchange capacity.

3.2. $H_2PO_4^-$ Adsorption on Soils with and without Cu^0 or Ag^0 ENPs

3.2.1. Effect of pH Solution

Figure 2 shows the effect of the $H_2PO_4^-$ pH solution between 4.5–8.5 on $H_2PO_4^-$ adsorption on T-OM and R-OM soil samples in the absence and presence of ENPs. The $H_2PO_4^-$ adsorbed on T-OM decreased slightly with increasing pH without and with ENPs. When Cu^0 or Ag^0 ENPs content increased, the $H_2PO_4^-$ adsorption on T-OM was 1.4–1.8 times higher than without ENPs (Figure 2a,c). In addition, the $H_2PO_4^-$ adsorption on R-OM increased with increased Cu^0 ENPs doses, but with Ag^0 ENPs showed no changes (Figure 2b,d).

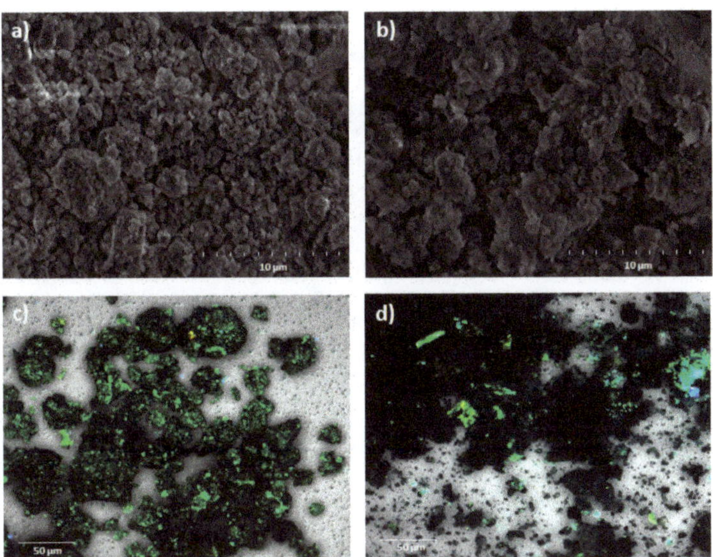

Figure 1. SEM analysis to soil with (**a**) total organic matter (T-OM) and (**b**) partial removal of organic matter (R-OM) and confocal images to soil with (**c**) T-OM and (**d**) R-OM.

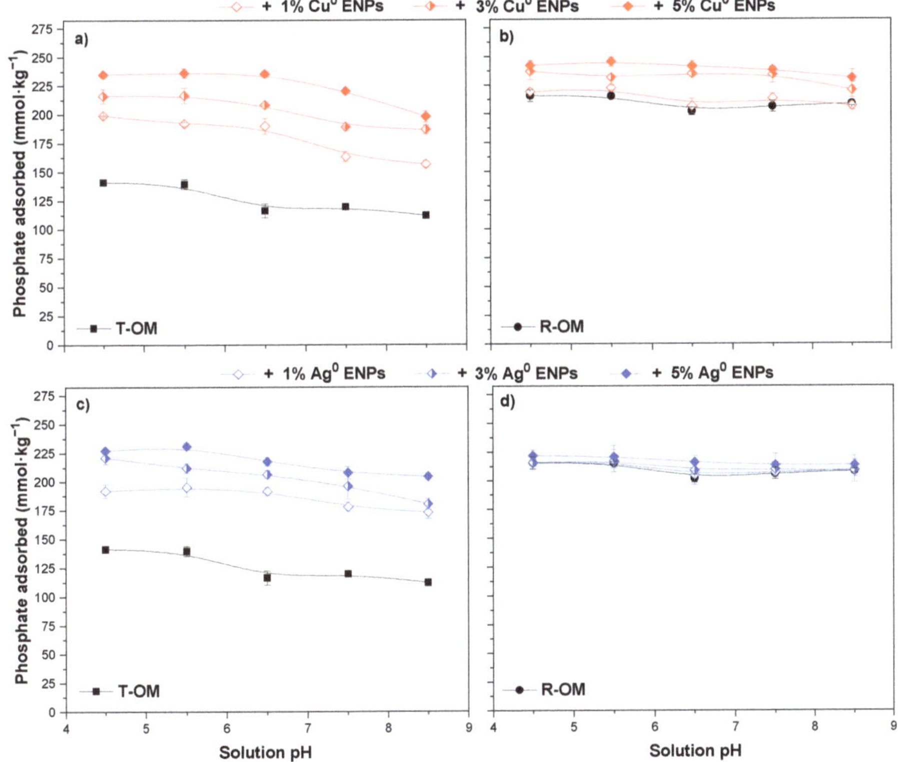

Figure 2. Initial pH effect of the solution on the adsorption of $H_2PO_4^-$ in the presence of Cu^0 ENPs on soil with (**a**) total organic matter (T-OM) and (**b**) partial removal of organic matter (R-OM) and Ag^0 ENPs on soil with (**c**) T-OM and (**d**) R-OM.

3.2.2. Adsorption Kinetics

The kinetic studies are shown in Figure 3. We observed that increasing in contact time at pH 5.5 as well as in the presence of Cu^0 or Ag^0 ENPs there was a subsequent increase in the adsorption of $H_2PO_4^-$ in T-OM and R-OM soil samples. It was also shown that adsorption comprised a fast initial phase at 45 min, followed by a slower rate stage until equilibrium was reached at 360 min for T-OM and at 720 min for R-OM, whereas in the presence of ENPs for most systems it was reached at 720 min. Based on the Table 4, in the absence of ENPs after $H_2PO_4^-$ adsorption on T-OM and R-OM soil samples, the final pH (pH_f) showed an increase in relation to the initial pH (pH_i). A similar tendency was obtained with increasing the ENPs doses and the pH_i and pH_f values were lower compared with systems without ENPs.

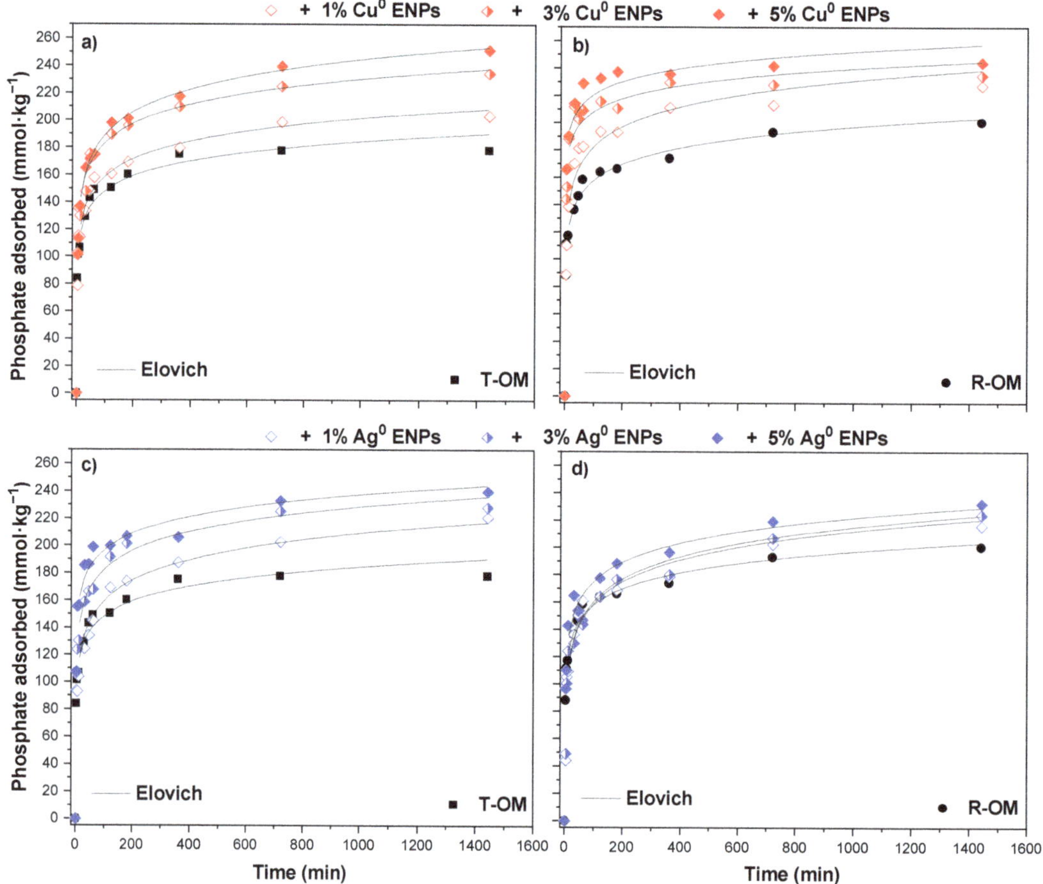

Figure 3. Phosphate adsorption kinetics at pH 5.5 ± 0.2 of the solution in the presence of Cu^0 ENPs on soil with (**a**) total organic matter (T-OM) and (**b**) partial removal of organic matter (R-OM) and Ag^0 ENPs on soil with (**c**) T-OM and (**d**) R-OM modelled by the Elovich model.

Table 4. pH changes associated to $H_2PO_4^-$ adsorption in the absence and presence of different doses of Cu^0 or Ag^0 ENPs and two levels of soil organic matter content (total organic matter, T-OM and partial removal of organic matter, R-OM). Experimental conditions: 6.47 mmol·L^{-1} $H_2PO_4^-$ solution at pH 5.5, 0.01 M KCl at 20 ± 2 °C. Initial pH (pH_i) and final pH (pH_f) were measured after $H_2PO_4^-$ solution added to soil samples (time 0 min) and after $H_2PO_4^-$ adsorption (1440 min), respectively.

| | Cu^0 ENPs | | | | | | | | Ag^0 ENPs | | | | | |
| | 0% | | 1% | | 3% | | 5% | | 1% | | 3% | | 5% | |
	pH_i	pH_f	pH_i	pH_f	pH_i	pH_f	pH_i	pH_f	pH_i	pH_f	pH_i	pH_f	pH_i	pH_f
T-OM	5.4 ± 0.1	5.7 ± 0.0	5.0 ± 0.1	5.4 ± 0.0	4.3 ± 0.1	4.7 ± 0.0	4.0 ± 0.0	4.2 ± 0.0	5.0 ± 0.1	5.6 ± 0.2	4.5 ± 0.1	4.9 ± 0.0	4.2 ± 0.1	4.7 ± 0.0
R-OM	5.2 ± 0.0	5.6 ± 0.1	4.7 ± 0.1	5.3 ± 0.1	4.2 ± 0.0	4.4 ± 0.0	3.8 ± 0.0	3.9 ± 0.0	4.6 ± 0.1	5.0 ± 0.1	4.2 ± 0.0	4.4 ± 0.0	3.9 ± 0.0	4.0 ± 0.0

To determine the kinetic constants and understand the adsorption mechanisms, the experimental kinetics data were modeled by the pseudo-second-order (PSO) Elovich (Table 5) and pseudo-first-order (PFO) (Table S1) models. PSO and PFO models describe the kinetics of the adsorbate on an adsorbent based on chemical-adsorption and physical-adsorption, respectively, with respect to the adsorbent capacity [36]. On the other hand, the Elovich model describes the sorption of adsorbate onto a heterogeneous surface [38,39].

Based on the higher r^2 and the lower χ^2 values, the PSO model fitted to the adsorption kinetics data better than the PFO model. According to the PSO model, the amount of $H_2PO_4^-$ adsorbed at equilibrium ($q_{e,cal}$) in T-OM and R-OM soil samples increased with ENPs contents and it was higher in R-OM than T-OM, except for 3 and 5% Ag^0 ENPs doses. The kinetic rate (k_2) did not show a clear trend at low ENPs contents. However, it increased in T-OM with 5% Ag^0 ENPs and with 3 and 5% Cu^0 ENPs in R-OM as compared to the soils without ENPs. Similar behavior was observed for the initial adsorption rate (h) in the presence of ENPs leading to increases by adding 3% Cu^0 and 5% Ag^0 ENPs for T-OM and R-OM soil samples and 5% Cu^0 for R-OM and 3% Ag^0 ENPs for T-OM.

Experimental kinetic data at pH 5.5 in T-OM and R-OM soil samples without and with increasing Cu^0 or Ag^0 ENPs content also adequately fitted the Elovich model (r^2 = 0.927 − 0.998 and χ^2 = 9 − 279). This means that the $H_2PO_4^-$ adsorption happened on a heterogeneous substrate [38]. The initial rate (α) and the surface coverage (β) obtained from this model showed a similar tendency to h and k_2, respectively, calculated from the PSO model. Thus, both PSO and Elovich models were capable of describing the kinetics of $H_2PO_4^-$ adsorption on volcanic soils. Similar results have been obtained by other researchers for an acid soil [40] and for adsorbents such as biochar [38] and chitosan [41].

3.2.3. Adsorption Isotherms

The isotherm adsorptions at pH 5.5 (Figure 4) showed that the amount of $H_2PO_4^-$ adsorbed was slightly higher in R-OM than T-OM and $H_2PO_4^-$ adsorption increased with increasing ENPs contents. In general, all adsorption isotherm described curves type L [42]. This means that a high affinity of $H_2PO_4^-$ anions exist in both soils. In particular, in T-OM samples, the curve reached a strict asymptotic plateau, while in R-OM the curve did not reach it. This difference indicated that the number of adsorption sites in the T-OM sample for $H_2PO_4^-$ is limited; on the contrary, the R-OM sample had a greater number of adsorption sites for $H_2PO_4^-$. At the same time, by increasing the Cu^0 or Ag^0 ENPs content, the curves showed a much less strict plateau for both soil samples, suggesting that the number of available adsorption sites for $H_2PO_4^-$ increased [42,43].

Table 5. Pseudo-second-order and Elovich parameters ($\pm$ standard error) obtained from $H_2PO_4^-$ adsorption kinetics at pH 5.5 ± 0.2 for the soil with total organic matter (T-OM) and with partial removal of organic matter (R-OM) in the absence and presence of different doses of Cu^0 or Ag^0 ENPs.

	ENPs Doses (%)	Pseudo-Second-Order						Elovich			
		$q_{e,exp}$ (mmol·kg^{-1})	$q_{e,cal}$ (mmol·kg^{-1})	k_2 ($\times 10^{-4}$ kg·mmol^{-1}·min^{-1})	h (mmol·kg^{-1}·min^{-1})	r^2	χ^2	α ($\times 10^3$ mmol·kg^{-1}·min^{-1})	β ($\times 10^{-2}$ kg·mmol^{-1})	r^2	χ^2
T-OM	0	177.9 ± 2.4	164.3 ± 5.1	16.6 ± 1.9	45.6 ± 0.0	0.940	157	1.9 ± 0.3	6.4 ± 0.4	0.985	40
R-OM		200.4 ± 4.2	174.4 ± 6.1	16.9 ± 2.5	50.5 ± 0.0	0.924	226	2.0 ± 0.3	5.9 ± 0.3	0.993	21
					Cu^0						
T-OM	1	203.1 ± 5.1	177.4 ± 6.8	14.0 ± 4.0	44.0 ± 0.0	0.915	271	1.0 ± 0.2	5.4 ± 0.3	0.986	43
R-OM		227.1 ± 2.8	206.5 ± 4.5	10.9 ± 1.7	46.5 ± 0.0	0.973	114	1.3 ± 0.4	4.8 ± 0.4	0.972	118
T-OM	3	234.2 ± 3.9	202.4 ± 8.4	14.0 ± 4.4	57.4 ± 0.0	0.900	430	1.7 ± 0.4	4.9 ± 0.3	0.989	45
R-OM		234.5 ± 1.5	221.4 ± 3.5	26.4 ± 4.0	129.4 ± 0.0	0.980	86	590.5 ± 15.2	7.3 ± 0.5	0.970	123
T-OM	5	251.1 ± 7.2	214.9 ± 9.0	9.4 ± 2.8	43.4 ± 0.0	0.910	454	0.7 ± 0.1	4.2 ± 0.0	0.998	9
R-OM		244.0 ± 4.3	232.9 ± 4.5	28.8 ± 5.6	156.2 ± 0.0	0.970	139	1518.7 ± 101.6	7.4 ± 0.8	0.980	91
					Ag^0						
T-OM	1	220.9 ± 4.8	183.4 ± 6.5	9.0 ± 3.5	30.1 ± 0.0	0.941	280	0.4 ± 0.1	4.6 ± 0.2	0.988	43
R-OM		215.9 ± 6.0	185.8 ± 7.7	8.1 ± 2.3	28.1 ± 0.0	0.933	306	0.3 ± 0.2	4.6 ± 0.5	0.927	279
T-OM	3	228.1 ± 4.1	201.7 ± 7.9	13.6 ± 4.0	55.3 ± 0.0	0.927	373	1.7 ± 0.3	5.0 ± 0.2	0.995	21
R-OM		224.0 ± 5.2	187.9 ± 8.7	7.9 ± 2.5	27.9 ± 0.0	0.905	386	0.2 ± 0.1	4.2 ± 0.3	0.968	132
T-OM	5	239.6 ± 3.6	213.2 ± 5.6	18.1 ± 3.9	82.3 ± 0.0	0.952	186	11.2 ± 1.0	6.6 ± 0.4	0.974	112
R-OM		232.0 ± 4.5	194.2 ± 8.8	13.7 ± 4.7	51.7 ± 0.0	0.942	263	1.1 ± 0.2	4.9 ± 0.4	0.979	84

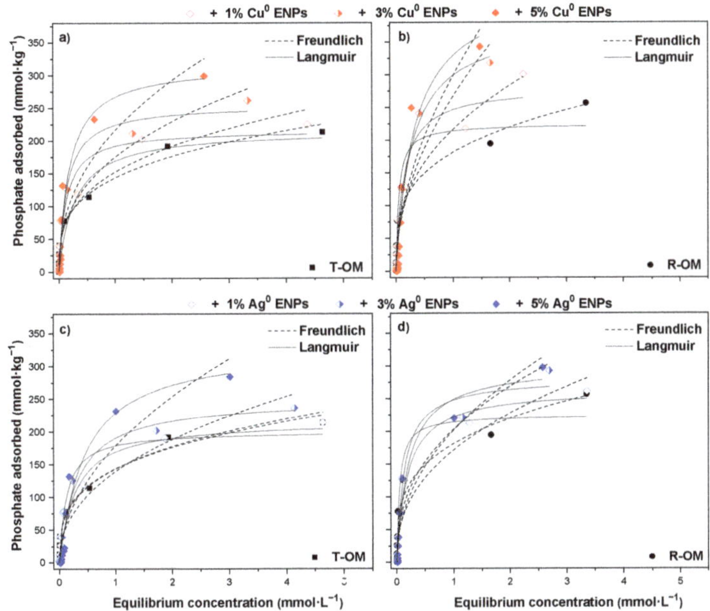

Figure 4. Phosphate adsorption isotherms at pH 5.5 $\pm$ 0.2 of the solutions in the presence of Cu^0 ENPs on soil with (**a**) total organic matter (T-OM) and (**b**) partial removal of organic matter (R-OM) and Ag^0 ENPs on soil with (**c**) T-OM and (**d**) R-OM modelled by the Langmuir and Freundlich models.

The adsorption isotherm data were fitted by Langmuir and Freundlich models (Table 6), which have been frequently used to explain $H_2PO_4^-$ adsorption on different soils [44,45].

Table 6. Langmuir and Freundlich parameters ($\pm$ standard error) obtained from $H_2PO_4^-$ adsorption isotherms at pH 5.5 $\pm$ 0.2 and desorption (%) for the soil with total organic matter (T-OM) and with partial removal of organic matter (R-OM) in the absence and presence of different doses of Cu^0 or Ag^0 ENPs.

	ENPs Doses (%)	Langmuir					Freundlich				Desorption
		$q_{e,exp}$ (mmol·kg^{-1})	K_L (L·mmol^{-1})	q_{max} (mmol·kg^{-1})	r^2	χ^2	K_F ((mmol·kg^{-1})(L·kg^{-1})$^{1/n}$)	n	r^2	χ^2	(%)
T-OM	0	213.6 $\pm$ 0.6	3.6 $\pm$ 1.5	216.1 $\pm$ 20.0	0.934	396	143.5 $\pm$ 5.0	3.4 $\pm$ 0.3	0.982	111	31.6
R-OM		256.4 $\pm$ 1.4	20.1 $\pm$ 7.3	224.7 $\pm$ 15.6	0.944	437	181.0 $\pm$ 9.0	3.6 $\pm$ 0.4	0.961	299	9.7
						Cu^0					
T-OM	1	225.0 $\pm$ 1.1	7.8 $\pm$ 3.2	216.2 $\pm$ 16.6	0.943	388	152.1 $\pm$ 10.6	3.0 $\pm$ 0.4	0.928	488	24.0
R-OM		301.0 $\pm$ 0.7	7.3 $\pm$ 2.8	280.4 $\pm$ 23.2	0.942	600	216.3 $\pm$ 10.2	2.6 $\pm$ 0.2	0.966	352	9.5
T-OM	3	262.0 $\pm$ 0.2	7.2 $\pm$ 1.6	255.4 $\pm$ 14.3	0.968	274	175.0 $\pm$ 13.6	2.6 $\pm$ 0.3	0.923	738	10.7
R-OM		318.4 $\pm$ 1.4	3.6 $\pm$ 0.8	382.7 $\pm$ 31.5	0.963	434	266.8 $\pm$ 25.8	1.9 $\pm$ 0.3	0.870	1550	6.4
T-OM	5	299.2 $\pm$ 0.5	5.4 $\pm$ 1.5	316.4 $\pm$ 26.1	0.945	604	216.0 $\pm$ 22.0	2.3 $\pm$ 0.4	0.854	1590	9.5
R-OM		343.3 $\pm$ 2.4	3.0 $\pm$ 0.9	440.2 $\pm$ 55.3	0.926	1000	298.2 $\pm$ 33.5	1.9 $\pm$ 0.3	0.836	2208	4.5
						Ag^0					
T-OM	1	213.8 $\pm$ 3.0	8.1 $\pm$ 2.7	200.5 $\pm$ 13.3	0.953	290	143.4 $\pm$ 7.0	3.3 $\pm$ 0.3	0.964	219	31.5
R-OM		261.3 $\pm$ 0.3	5.2 $\pm$ 0.8	264.0 $\pm$ 11.4	0.982	151	171.3 $\pm$ 13.3	2.5 $\pm$ 0.3	0.921	724	16.8
T-OM	3	236.2 $\pm$ 1.9	3.2 $\pm$ 0.8	249.9 $\pm$ 19.2	0.948	361	139.9 $\pm$ 14.2	2.3 $\pm$ 0.4	0.870	946	29.4
R-OM		292.7 $\pm$ 2.1	7.3 $\pm$ 1.5	281.4 $\pm$ 15.5	0.970	302	202.9 $\pm$ 13.0	2.5 $\pm$ 0.3	0.940	604	21.0
T-OM	5	284.1 $\pm$ 2.8	2.2 $\pm$ 0.6	332.8 $\pm$ 30.8	0.951	500	178.6 $\pm$ 18.1	2.0 $\pm$ 0.3	0.871	1327	22.2
R-OM		297.5 $\pm$ 1.4	4.7 $\pm$ 1.2	301.4 $\pm$ 21.4	0.958	434	201.5 $\pm$ 15.6	2.2 $\pm$ 0.3	0.924	881	22.1

The Freundlich model fitted the experimental data of T-OM and R-OM soil samples better than the Langmuir model (Table 6). However, in the presence of ENPs in T-OM and R-OM soil samples, the Langmuir model, except for R-OM—1% Cu^0 ENPs and T-OM—1% Ag^0 ENPs systems, showed a better fit to the experimental data ($r^2 = 0.926 - 0.982$ and $\chi^2 = 151 - 1000$). According to the Langmuir model, the maximum $H_2PO_4^-$ adsorption capacity (q_{max}) in R-OM and T-OM soils increased with ENPs contents, and it was higher on R-OM than T-OM, except for 5% Ag^0 ENPs dose, in contrast to the affinity coefficient (K_L).

3.2.4. Desorption

The desorption (%) depends on the chemical nature and energy of the bonds between soil components and phosphate [46]. In this sense, after the soil samples were treated with double-distilled water repeatedly (three times), $H_2PO_4^-$ desorption was about 3.2 times higher from T-OM than R-OM (Table 6). In the presence of ENPs, the desorption from R-OM and T-OM soils decreased with increasing Cu^0 ENPs doses as well as from T-OM with 3 and 5% Ag^0 ENPs. In contrast, with increasing Ag^0 ENPs content, desorption from R-OM was greater than without ENPs.

4. Discussion

4.1. Characterization of Cu^0 and Ag^0 ENPs and Soil Samples Studied

The particle size average of Cu^0 (19 nm) and Ag^0 ENPs (17 nm) was low due to L-ascorbic acid coating, which provides colloidal stability to the nanoparticles by electrostatic repulsion. The stability effect of the L-ascorbic acid coating could be attributed to the presence of a polyhydroxyl structure on the surface of both nanoparticles [28]. This was supported by the high negative values of ZP, which is normally related to the negatively charged groups of the capping agents [28,47]. Similar results using organic molecules as reducing and capping agents for the preparation of ENPs have been reported previously [28,47–50].

The organic matter in volcanic soils is highly stabilized [51], whereby after repeated treatment with H_2O_2, only a part of the OM was removed from soil, accounting a 14.1% of OM (T-OM), obtaining a soil sample with 4.6% of OM (R-OM) (Table 3). The partial removal of OM significantly changed the aggregate structure of the soil because OM acts as a binding agent [52]. In addition, T-OM had more aggregates, a higher P concentration and an effective cation exchange capacity (ECEC) as compared to R-OM. In this sense, it is knowing that the functional groups of OM such as carboxyl, alcoholic hydroxyl, and phenolic hydroxyl contribute to the aggregation of soil particles, formation of humic (organic matter)-Al (Fe)-phosphate complexes and cations adsorption [52,53]. Likewise, R-OM samples had a higher IEP and BET-specific surface area than T-OM. This can be explained by the exposure of $\equiv$Fe-OH and $\equiv$Al-OH active sites from amorphous components of the soil, which decreased the negative charges of the surface and increased BET-specific surface area [34]. In general, allophane and ferrihydrite minerals can interact with negatively charged ENPs through attraction (Van der Waals) forces contributing to their retention in the soil [54].

4.2. Ad- and Desorption of Phosphate on Soils

The phosphate adsorption isotherms on T-OM and R-OM soil samples in the absence of ENPs were best fitted to the Freundlich model (Table 6), which reflected the heterogenic nature of soil components. The intensity of adsorption (n) and relative adsorption capacity (K_F) for R-OM were higher than T-OM. The difference between K_F and n for two soil samples may be due to the higher OM content of T-OM, since OM could block adsorption-specific sites leading to a lower availability of surface-reactive sites and weak interaction with $H_2PO_4^-$ [55]. The OM can act by preventing the irreversible retention of $H_2PO_4^-$ and increasing the nutrient recovery. We found that, after partial OM removal, the $H_2PO_4^-$ desorption from R-OM was lower than from T-OM (Table 6), indicating a strong interaction between the phosphate and mineral components of R-OM [15,16,23]. These results are supported by the higher BET-specific surface area and lower negative surface charge of R-OM as compared to T-OM. Similar results were obtained by Zeng et al. [56] for $H_2PO_4^-$ desorption in volcanic soils exhibiting contrasting OM contents. However, these findings were in contrast to the results reported by Debicka et al. [57] by removing the OM from sandy soil resulted in decreases of K_F and n values. Contrasting results could be attributed to the particularly components in each soil. According to the FAO-WRB soil classification, sandy soils such as Brunic Arenosols are mainly characterized by minerals such as hematite, goethite, and maghemite [57,58]. On the contrary, Santa Barbara soil is formed by minerals such as allophane (>50%), followed by 1–5% halloysite and vermiculite [59]. In this context, Parfitt [60] found that phosphate was adsorbed in the order hematite ~ goethite < ferrihydrite < allophane. Moreover, $H_2PO_4^-$ can be rapidly and strongly adsorbed on the most reactive aluminol ($\equiv$Al-OH) groups of the allophane by ligand exchange forming monodentate or/and binuclear complexes.

According to the PSO model, the higher $H_2PO_4^-$ adsorption ($q_{e,cal}$) was in the R-OM as compared to T-OM (Table 5), which could due to the destruction of OM in T-OM, leading to a larger pore volume and BET-surface area. In addition, R-OM improved the accessibility to active sites for $H_2PO_4^-$ according to the higher values of α and h obtained for R-OM (Table 5) [57]. The h parameter can be associated to the chemical and/or hydrogen bonding interaction between $H_2PO_4^-$ and surface hydroxyls in soil samples at the initial adsorption process [16]. Moreover, considering the Elovich model and increase in pH_f values after $H_2PO_4^-$ adsorption with respect to pH_i (Table 4), we might suggest that $H_2PO_4^-$ adsorption in T-OM and R-OM soil samples was performed mainly through ligand exchange (chemi-adsorption) onto Fe/Al (hydr)oxides forming monodentate or bidentate complexes. The pH changes were consistent with the studies carried out by Vistoso et al. [24], who reported that $H_2PO_4^-$ was adsorbed through ligand exchange mechanism in volcanic soils with contrasting properties.

The $H_2PO_4^-$ adsorption on T-OM was pH-dependent in contrast to R-OM (Figure 2). In this context, the IEP of T-OM was 3.2 whereas it was 5.7 for R-OM. Therefore, in acidic pH $H_2PO_4^-$ solution the surface hydroxyl (–OH) groups in R-OM were more protonated than in T-OM, causing a favorable effect on electrostatic interaction and ligand exchange [61]. However, at alkaline pH $H_2PO_4^-$ solution, mainly for T-OM, there was a decrease in the ligand exchange and an increase in electrostatic repulsion due to deprotonation from soil superficial groups. Likewise, at a higher pH, the competition between OH^- and $H_2PO_4^-$ on the T-OM surface would also reduce the $H_2PO_4^-$ adsorption [62].

4.3. Ad- and Desorption of Phosphate on Soils in the Presence of Cu^0 or Ag^0 ENPs

The increasing phosphate adsorption with increasing ENPs content in soils indicated that in the presence of ENPs, the number of adsorption sites increased. Although, there was a decrease in the initial adsorption rate (h) with 1% ENPs content, which implied that during the first few minutes ENPs compete with $H_2PO_4^-$ for the adsorption sites of the soil surface. Additionally, h strongly increased with 3 and 5% ENPs content, suggesting that ENPs also contributed to new adsorption sites for $H_2PO_4^-$ [63,64]. Accordingly, Duncan and Owens [63] indicated that CeO_2 ENPs can be adsorbed on soil adsorption sites before Pb(II) and Sun et al. [64] determined a similar trend for h with increasing carbon nanotubes (CNTs) content after studying the effects of CNTs with outer diameter of 25 nm and inner diameter of 5 nm on Cd(II) adsorption in sediments.

The adsorption isotherms of $H_2PO_4^-$ on T-OM and R-OM following Cu^0 or Ag^0 ENPs addition fitted to the Langmuir model (Table 6). Similarly, Sun et al. [64] found that in the presence of CNTs the isotherms for Cd(II) on sediment showed a better fit to the Langmuir than the Freundlich model; however due to the adsorption sites of sediments with CNTs are heterogeneous, they used the Freundlich to describe their results. Therefore, the fit of adsorption data to the Langmuir model in the presence of ENPs should be more studied.

Adsorption enhancement was larger through Cu^0 than Ag^0 ENPs. According to Afshinnia and Baalousha [65], the decrease in the zeta potential after $H_2PO_4^-$ adsorption on T-OM and R-OM soil samples with ENPs could be associated with $H_2PO_4^-$ adsorption/complexation onto the ENPs surface (Figure S7). In this context, Niaura et al. [66] indicated that $H_2PO_4^-$ was adsorbed through monodentate surface coordination on Cu^0 ENPs, while on Ag^0 ENPs it was performed through hydrogen bonding [66,67]. Although both coated ENPs had a low rate of oxidation and dissolution [68], it was probable that these processes could be favored by an acidic soil pH as well as a consequence of the ionic exchange between $H_2PO_4^-$ and L-ascorbic acid on the surface of the ENPs, being similar to the mechanism observed for citric acid [50]. Under such conditions, Cu^0 could be oxidized to Cu^{2+} ($E°_{Cu2+/Cu0}$ = 0.337 V) and the amount of phosphate adsorbed in T-OM and R-OM soil samples increased (Figure S8) because Cu^{2+} could be linked to $H_2PO_4^-$ and hydroxyl groups of OM via a cation bridge [69]. Furthermore, this could be attributed to the formation of complexes between Cu^{2+} and $H_2PO_4^-$ and the precipitate of $Cu_3(PO_4)_2$ (Ksp = 2.07×10^{-33}) [70]. Meanwhile, in the case that Ag^+ ions were released from Ag^0 ENPs into solution ($E°_{Ag+/Ag0}$ = 0.799 V), the formation of AgCl precipitate was more favorable (Ksp = 1.77×10^{-10}) than a Ag_3PO_4 formation (Ksp = 8.89×10^{-17}) [71,72].

On the other hand, the presence of L-ascorbic acid free in soil solution slightly competes with $H_2PO_4^-$ for available adsorption sites, decreasing $H_2PO_4^-$ adsorption on T-OM and R-OM soil samples (Figure S8). However, as a consequence of the addition of Cu^0 or Ag^0 ENPs suspensions to soil samples, the pH_i values decreased, being less acidic in T-OM as compared to R-OM (Table 4), which was consistent with the buffering capacity of OM [73]. An acid pH can be associated with a decrease in the electrostatic repulsion between $H_2PO_4^-$ and the negatively charged surface of the organic matter (-COOH, -OH) due to a decrease in the number of deprotonated surface groups [74]. Furthermore, the protonation of surface hydroxyl groups of Fe/Al (hydr)oxides might be favored by acid pH values, promoting the $H_2PO_4^-$ adsorption through a ligand exchange [24,75,76]. In the same way, it has been reported that below 4.5 of pH values the mineral dissolution is favored, promoting the precipitation reactions between $H_2PO_4^-$ and cations in solution (Al^{3+} and Fe^{3+}) [77], and to form $H_2PO_4^-$-cation-organic matter complexes [53].

The increase of the $H_2PO_4^-$ adsorption at a low pH has been demonstrated on pillared bentonites [75], AgNPs-tea activated carbon [76], sediments [78] and in Andisol soils [24]. Future research should be addressed to corroborate whether, in the presence of both ENPs, one of these mechanisms was prevalent for $H_2PO_4^-$ adsorption on T-OM and R-OM soil samples, or whether several mechanisms acted together.

The $H_2PO_4^-$ adsorption in the presence of ENPs through chemical interactions onto a heterogeneous surface was indicated by the adequate fits of the kinetic data to the PSO and Elovich models (Table 5). In addition, the desorption behavior supported the adsorption mechanisms proposed in the presence of ENPs. With Cu^0 ENPs, the desorption of $H_2PO_4^-$ from T-OM and R-OM soil samples was smaller than Ag^0 ENPs. These results can be supported by a chemisorption-like interaction between $H_2PO_4^-$ and Cu^0 ENPs. Similarly, desorption studies of U(VI) on the soil in the presence of nano-crystalline goethite showed that U(VI) was more resistant to released due to an increase in the inner-sphere complexes on the soil surface [79]. In addition, Elkhatib et al. [80] revealed that sorption of Hg(II) on arid soils in the presence of water treatment residual nanoparticles occurred mainly through inner-sphere complexes, which enhanced Hg immobilization in the arid soils. The high desorption of $H_2PO_4^-$ in R-OM following Ag^0 ENPs addition needs further investigation. One possible explanation for this is that the Ag^0 ENPs were attached to the potential $H_2PO_4^-$ adsorption sites, such as allophane and Fe oxides, leading to a blocking effect for $H_2PO_4^-$ on this soil with lower levels of OM. Then, the $H_2PO_4^-$ physi-adsorbed (through hydrogen bonding) on the surface of the attached Ag^0 ENPs was more desorbable.

5. Conclusions

Our study demonstrated that the phosphate adsorption process in the presence of ENPs was dependent on the amount of ENPs and soil organic matter content. The addition of Cu^0 caused a higher increase in phophate adsorpion on T-OM and R-OM as compared to the Ag^0 ENPs. The Elovich and pseudo-second-order (PSO) models correctly described the kinetic adsorption of phosphate on T-OM and R-OM soil samples without and with ENPs.

The phosphate adsorption with both ENPs was better described by the Langmuir isotherm model than the Freundlich model. According to the Langmuir model, by increasing the ENPs content from 0 to 5%, the maximum adsorption capacity (q_{max}) of $H_2PO_4^-$ for T-OM ranged from 216.1 to 316.4 mmol·kg^{-1} following the Cu^0 ENPs addition and to 332.8 mmol·kg^{-1} using Ag^0 ENPs. Meanwhile, with the increase from 0 to 5% of ENPs, the q_{max} of $H_2PO_4^-$ for R-OM ranged from 224.7 to 440.2 mmol·kg^{-1} with Cu^0 ENPs and to 301.4 mmol·kg^{-1} with Ag^0 ENPs. Phosphate desorption in T-OM and R-OM soils following Cu^0 ENPs addition was lower than Ag^0 ENPs. In the future, more attention should be pointed globally to management agriculture practices based on nanotechnology, because the incorporation of ENPs into the soil have the potential to reduce the already limited crop phosphorus availability.

Supplementary Materials: The following are available online at https://www.mdpi.com/article/10.3390/min11040373/s1, Figure S1: TEM images L-ascorbic acid-stabilized (a) Cu^0 and (b) Ag^0 ENPs, Figure S2: Histograms with the corresponding particle size distribution for L-ascorbic acid-stabilized (a) Cu^0 and (b) Ag^0 ENPs, Figure S3: UV-Vis absorption spectra for L-ascorbic acid-stabilized Cu^0 and Ag^0 ENPs, Figure S4: FT-IR spectra of (a) Pure L-ascorbic acid, (b) L-ascorbic acid-stabilized Cu^0 ENPs and (c) L-ascorbic acid-stabilized Ag^0 ENPs, Figure S5: Zeta potential of L-ascorbic acid-stabilized Cu^0 and Ag^0 ENPs in 0.01 M KCl, Figure S6: FT-IR spectrum for soil samples with (a) total organic matter (T-OM) and (b) partial removal of organic matter (R-OM), Figure S7: Zeta potential curves in the presence of 9.71 mmol·L^{-1} $H_2PO_4^-$ and 5% Cu^0 or 5% Ag^0 ENPs at constant ionic strength (0.01 M KCl) for soil with (a) total organic matter (R-OM) and (b) partial removal of organic matter (R-OM), Figure S8: Adsorption isotherm curves of $H_2PO_4^-$ on (a) total organic matter (T-OM) and /9b) partial removal of organic matter (R-OM) in the presence of 3% L-ascorbic acid and Cu^{2+} and Ag^+. Reaction conditions: Concentrations from 0.016 to 9.71 mmol·L^{-1} $H_2PO_4^-$ on 0.5 g soil in 0.01 M KCl at 20 ± 2 °C and pH 5.5, Table S1: Pseudo-first-order parameters (± standard error) obtained from $H_2PO_4^-$ adsorption kinetics in the absence and presence of different doses of Cu^0 and Ag^0 ENPs at pH 5.5 ± 0.2 for soil with total organic matter (T-OM) and with partial removal of organic matter (R-OM).

Author Contributions: Conceptualization, E.K., M.d.L.L.M. and A.J.; methodology, J.S.-H.; software, J.S.-H.; validation, M.d.L.L.M., E.K., N.A.-M. and R.B.; formal analysis, J.S.-H.; P.P.-G. and A.J.; investigation, J.S.-H. and P.P.-G.; resources, M.d.L.L.M.; data curation, J.S.-H.; writing—original draft preparation, J.S.-H., N.A.-M. and E.K.; writing—review and editing, E.K., R.B., N.A.-M. and A.J.; visualization, J.S.-H. and R.B.; supervision, M.d.L.L.M. and N.A.-M.; project administration, M.d.L.L.M.; funding acquisition, M.d.L.L.M. and J.S.-H. All authors have read and agreed to the published version of the manuscript.

Funding: This research was funded by the Fondo Nacional de Desarrollo Científico y Tecnológico (FONDECYT) projects N° 1181050 and 1191018 and by the Agencia Nacional de Investigación y Desarrollo (ANID) Ph.D. scholarships N° 21171685.

Data Availability Statement: Data are contained within this article.

Acknowledgments: Jonathan Suazo-Hernández acknowledges to Daniela Vergara, the FONDECYT project N° 3210228, the Technological Bioresource Nucleus (BIOREN-UFRO) and the Soil and Plant Laboratory.

Conflicts of Interest: The authors declare no conflict of interests.

References

1. Shah, V.; Luxton, T.P.; Walker, V.K.; Brumfield, T.; Yost, J.; Shah, S.; Wilkinson, J.E.; Kambhampati, M. Fate and impact of zero-valent copper nanoparticles on geographically-distinct soils. *Sci. Total Environ.* **2016**, *573*, 661–670. [CrossRef]
2. Ramyadevi, J.; Jeyasubramanian, K.; Marikani, A.; Rajakumar, G.; Rahuman, A.A. Synthesis and antimicrobial activity of copper nanoparticles. *Mater. Lett.* **2012**, *71*, 114–116. [CrossRef]
3. Roco, M.C.; Mirkin, C.A.; Hersam, M.C. Nanotechnology research directions for societal needs in 2020: Summary of international study. *J. Nanoparticle Res.* **2011**, *13*, 897–919. [CrossRef]

4. Li, M.; Wang, P.; Dang, F.; Zhou, D.M. The transformation and fate of silver nanoparticles in paddy soil: Effects of soil organic matter and redox conditions. *Environ. Sci. Nano* **2017**, *4*, 919–928. [CrossRef]

5. Baskar, V.; Venkatesh, J.; Park, S.W. Impact of biologically synthesized silver nanoparticles on the growth and physiological responses in Brassica rapa ssp. pekinensis. *Environ. Sci. Pollut. Res.* **2015**, *22*, 17672–17682. [CrossRef]

6. Boxall, A.B.A.; Tiede, K.; Chaudhry, Q. Engineered nanomaterials in soils and water: How do they behave and could they pose a risk to human health? *Nanomedicine* **2007**, *2*, 919–927. [CrossRef] [PubMed]

7. Conway, J.R.; Adeleye, A.S.; Gardea-Torresdey, J.; Keller, A.A. Aggregation, dissolution, and transformation of copper nanoparticles in natural waters. *Environ. Sci. Technol.* **2015**, *49*, 2749–2756. [CrossRef] [PubMed]

8. Kent, R.D.; Vikesland, P.J. Dissolution and Persistence of Copper-Based Nanomaterials in Undersaturated Solutions with Respect to Cupric Solid Phases. *Environ. Sci. Technol.* **2016**, *50*, 6772–6781. [CrossRef]

9. Liu, J.; Hurt, R.H. Ion release kinetics and particle persistence in aqueous nano-silver colloids. *Environ. Sci. Technol.* **2010**, *44*, 2169–2175. [CrossRef]

10. Ben-Moshe, T.; Frenk, S.; Dror, I.; Minz, D.; Berkowitz, B. Effects of metal oxide nanoparticles on soil properties. *Chemosphere* **2013**, *90*, 640–646. [CrossRef] [PubMed]

11. Bayat, H.; Kolahchi, Z.; Valaey, S.; Rastgou, M.; Mahdavi, S. Iron and magnesium nano-oxide effects on some physical and mechanical properties of a loamy Hypocalcic Cambisol. *Geoderma* **2019**, *335*, 57–68. [CrossRef]

12. Torrent, L.; Marguí, E.; Queralt, I.; Hidalgo, M.; Iglesias, M. Interaction of silver nanoparticles with mediterranean agricultural soils: Lab-controlled adsorption and desorption studies. *J. Environ. Sci.* **2019**, *83*, 205–216. [CrossRef] [PubMed]

13. Taghipour, M.; Jalali, M. Effect of nanoparticles on kinetics release and fractionation of phosphorus. *J. Hazard. Mater.* **2015**, *283*, 359–370. [CrossRef] [PubMed]

14. Escudey, M.; Galindo, G.; Förster, J.E.; Briceño, M.; Diaz, P.; Chang, A. Chemical Forms of Phosphorus of Volcanic Ash-Derived Soils in Chile. *Commun. Soil Sci. Plant Anal.* **2001**, *32*, 601–616. [CrossRef]

15. Mora, M.L.; Galindo, G.; Escudey, M. The role of iron oxides and organic matter on phosphate adsorption in model allophanic synthetic soils. *Chil. J. Agric. Res.* **1992**, *52*, 416–421.

16. Wang, H.; Zhu, J.; Fu, Q.L.; Xiong, J.W.; Hong, C.; Hu, H.Q.; Violante, A. Adsorption of phosphate onto ferrihydrite and ferrihydrite-humic acid complexes. *Pedosphere* **2015**, *25*, 405–414. [CrossRef]

17. Borie, F.; Aguilera, P.; Castillo, C.; Valentine, A.; Seguel, A.; Barea, J.M.; Cornejo, P. Revisiting the Nature of Phosphorus Pools in Chilean Volcanic Soils as a Basis for Arbuscular Mycorrhizal Management in Plant P Acquisition. *J. Soil Sci. Plant Nutr.* **2019**, *19*, 390–401. [CrossRef]

18. Paredes, C.; Menezes-Blackburn, D.; Cartes, P.; Gianfreda, L.; Mora, M.L. Phosphorus and nitrogen fertilization effect on phosphorus uptake and phosphatase activity in ryegrass and tall fescue grown in a Chilean Andisol. *Soil Sci.* **2011**, *176*, 245–251. [CrossRef]

19. Mora, M.L.; Cartes, P.; Demanet, R.; Cornforth, I.S. Effects of lime and gypsum on pasture growth and composition on an acid Andisol in Chile, South America. *Commun. Soil Sci. Plant Anal.* **2002**, *33*, 2069–2081. [CrossRef]

20. Jorquera, M.A.; Hernández, M.T.; Rengel, Z.; Marschner, P.; Mora, M.L. Isolation of culturable phosphobacteria with both phytate-mineralization and phosphate-solubilization activity from the rhizosphere of plants grown in a volcanic soil. *Biol. Fertil. Soils* **2008**, *44*, 1025–1034. [CrossRef]

21. Osorio, N.W.; Habte, M. Soil Phosphate Desorption Induced by a Phosphate-Solubilizing Fungus. *Commun. Soil Sci. Plant Anal.* **2014**, *45*, 451–460. [CrossRef]

22. Calabi-Floody, M.; Velásquez, G.; Gianfreda, L.; Saggar, S.; Bolan, N.; Rumpel, C.; Mora, M.L. Improving bioavailability of phosphorous from cattle dung by using phosphatase immobilized on natural clay and nanoclay. *Chemosphere* **2012**, *89*, 648–655. [CrossRef] [PubMed]

23. Jara, A.A.; Violante, A.; Pigna, M.; Mora, M.L. Mutual Interactions of Sulfate, Oxalate, Citrate, and Phosphate on Synthetic and Natural Allophanes. *Soil Sci. Soc. Am. J.* **2006**, *70*, 337–346. [CrossRef]

24. Vistoso, E.; Theng, B.K.G.; Bolan, N.S.; Parfitt, R.L.; Mora, M.L. Competitive sorption of molybdate and phosphate in Andisols. *J. Soil Sci. Plant Nutr.* **2012**, *12*, 59–72. [CrossRef]

25. Cartes, P.; Cea, M.; Violante, A.; Mora, M.L.; Jara, A. Description of mutual interactions between silicon and phosphorus in Andisols by mathematical and mechanistic models. *Chemosphere* **2015**, *131*, 117–164. [CrossRef] [PubMed]

26. Vistoso, E.M.; Bolán, N.S.; Theng, B.K.G.; Mora, M.L. Kinetics of Molybdate and Phosphate Sorption by Some Chilean Andisols. *Rev. Cienc. Suelo Nutr. Veg.* **2009**, *9*, 55–68. [CrossRef]

27. Pigna, M.; Jara, A.A.; Mora, M.L.; Violante, A. Effect Of pH, Phosphate and/or Malate on Sulfate Sorption on Andisols. *Rev. Cienc. Suelo Nutr. Veg.* **2007**, *7*, 62–73. [CrossRef]

28. Xiong, J.; Wang, Y.; Xue, Q.; Wu, X. Synthesis of highly stable dispersions of nanosized copper particles using L-ascorbic acid. *Green Chem.* **2011**, *13*, 900–904. [CrossRef]

29. Siéwé, J.M.; Djoufac Woumfo, E.; Djomgoue, P.; Njopwouo, D. Activation of clay surface sites of Bambouto's Andosol (Cameroon) with phosphate ions: Application for copper fixation in aqueous solution. *Appl. Clay Sci.* **2015**, *114*, 31–39. [CrossRef]

30. Khatoon, H.; Solanki, P.; Narayan, M.; Tewari, L. Role of microbes in organic carbon decomposition and maintenance of soil ecosystem. *Int. J. Chem. Stud.* **2017**, *5*, 1648–1656.

31. Dick, W.A.; Tabatabai, M.A. An Alkaline Oxidation Method for Determination of Total Phosphorus in Soils. *Am. Soc. Agron.* **1976**, *41*, 501–514. [CrossRef]
32. Murphy, J.; Riley, J.P. A modified single solution method for the determination of phosphate in natural waters. *Anal. Chim. Acta* **1962**, *27*, 31–36. [CrossRef]
33. Sadzawka, R.A.; Carrasco, R.M.A.; Grez, Z.R.; Mora, M.L.; Flores, P.H.; Neaman, A. *Métodos de Análisis Recomendados Para Suelos Chilenos*; Comisión de Normalización y Acreditación (CNA), Sociedad Chilena de la Ciencia del Suelo: Santiago, Chile, 2006.
34. Silva-Yumi, J.; Escudey, M.; Gacitua, M.; Pizarro, C. Kinetics, adsorption and desorption of Cd (II) and Cu (II) on natural allophane: Effect of iron oxide coating. *Geoderma* **2018**, *319*, 70–79. [CrossRef]
35. Lin, J.; Wang, L. Comparison between linear and non-linear forms of pseudo-first-order and pseudo-second-order adsorption kinetic models for the removal of methylene blue by activated carbon. *Front. Environ. Sci. Eng. China* **2009**, *3*, 320–324. [CrossRef]
36. Febrianto, J.; Kosasih, A.N.; Sunarso, J.; Ju, Y.H.; Indraswati, N.; Ismadji, S. Equilibrium and kinetic studies in adsorption of heavy metals using biosorbent: A summary of recent studies. *J. Hazard. Mater.* **2009**, *162*, 616–645. [CrossRef]
37. Wang, J.; Guo, X. Adsorption isotherm models: Classification, physical meaning, application and solving method. *Chemosphere* **2020**, *258*, 127279. [CrossRef]
38. Eduah, J.O.; Nartey, E.K.; Abekoe, M.K.; Weck Henriksen, S.; Neumann Andersen, M. Mechanism of orthophosphate (PO4-P) adsorption onto different biochars. *Environ. Technol. Innov.* **2019**, *17*, 100572–100583. [CrossRef]
39. Rawajfih, Z.; Nsour, N. Adsorption of γ-picoline onto acid-activated bentonite from aqueous solution. *Appl. Clay Sci.* **2010**, *47*, 421–427. [CrossRef]
40. Ghodszad, L.; Reyhanitabar, A.; Oustan, S. Biochar effects on phosphorus sorption-desorption kinetics in soils with dissimilar acidity. *Arab. J. Geosci.* **2021**, *14*, 366–383. [CrossRef]
41. Zhang, B.; Chen, N.; Feng, C.; Zhang, Z. Adsorption for phosphate by crosslinked/non-crosslinked-chitosan-Fe (III) complex sorbents: Characteristic and mechanism. *Chem. Eng. J.* **2018**, *353*, 361–372. [CrossRef]
42. Giles, C.H.; Macewan, T.H.; Nakhwa, S.N.; Smit, D. 786. Studies in adsorption. Part XI. A System of Classi$cation of Solution Adsorption Isotherms, and its Use in Diagnosis of Adsorption Mechanisms and in Measurement of Specific Surface Areas of Solids. *J. Chem. Soc.* **1960**, *846*, 3973–3993. [CrossRef]
43. Limousin, G.; Gaudet, J.P.; Charlet, L.; Szenknect, S.; Barthès, V.; Krimissa, M. Sorption isotherms: A review on physical bases, modeling and measurement. *Appl. Geochem.* **2007**, *22*, 249–275. [CrossRef]
44. Mermoz, S.J.; Emmanuel, D.W.; Dieudonne, B.; Francois, F.; Paul, D.; Daniel, N.; Tamfuh, A.P. Andosols of the Bambouto Mountains (West Cameroon): Characteristics, Superficial Properties—Study of the Phosphate Ions Adsorption. *Open Inorg. Chem. J.* **2008**, *2*, 106–115. [CrossRef]
45. Yang, X.; Chen, X.; Yang, X. Effect of organic matter on phosphorus adsorption and desorption in a black soil from Northeast China. *Soil Tillage Res.* **2019**, *187*, 85–91. [CrossRef]
46. Fink, J.R.; Inda, A.V.; Bavaresco, J.; Barrón, V.; Torrent, J.; Bayer, C. Adsorption and desorption of phosphorus in subtropical soils as affected by management system and mineralogy. *Soil Tillage Res.* **2016**, *155*, 62–68. [CrossRef]
47. Zain, N.M.; Stapley, A.G.F.; Shama, G. Green synthesis of silver and copper nanoparticles using ascorbic acid and chitosan for antimicrobial applications. *Carbohydr. Polym.* **2014**, *112*, 195–202. [CrossRef]
48. Kobayashi, Y.; Ishida, S.; Ihara, K.; Yasuda, Y.; Morita, T.; Yamada, S. Synthesis of metallic copper nanoparticles coated with polypyrrole. *Colloid Polym. Sci.* **2009**, *287*, 877–880. [CrossRef]
49. Kobayashi, Y.; Sakuraba, T. Silica-coating of metallic copper nanoparticles in aqueous solution. *Colloids Surfaces a Physicochem. Eng. Asp.* **2008**, *317*, 756–759. [CrossRef]
50. Njoki, P.N. Transformation of Silver Nanoparticles in Phosphate Anions: An Experiment for High School Students. *J. Chem. Educ.* **2019**, *96*, 546–552. [CrossRef]
51. Calabi-Floody, M.; Bendall, J.S.; Jara, A.A.; Welland, M.E.; Theng, B.K.G.; Rumpel, C.; Mora, M.L. Nanoclays from an Andisol: Extraction, properties and carbon stabilization. *Geoderma* **2011**, *161*, 159–167. [CrossRef]
52. Krause, L.; Rodionov, A.; Schweizer, S.A.; Siebers, N.; Lehndorff, E.; Klumpp, E.; Amelung, W. Microaggregate stability and storage of organic carbon is affected by clay content in arable Luvisols. *Soil Tillage Res.* **2018**, *182*, 123–129. [CrossRef]
53. Gerke, J. Humic (organic matter)-Al (Fe)-phosphate complexes: An underestimated phosphate form in soils and source of plant-available phosphate. *Soil Sci.* **2010**, *175*, 417–425. [CrossRef]
54. Hoppe, M.; Mikutta, R.; Kaufhold, S.; Utermann, J.; Duijnisveld, W.; Wargenau, E.; Fries, E.; Guggenberger, G. Retention of sterically and electrosterically stabilized silver nanoparticles by soil minerals. *Eur. J. Soil Sci.* **2016**, *67*, 573–582. [CrossRef]
55. Nafiu, A. Effects of soil properties on the kinetics of desorption of phosphate from Alfisols by anion-exchange resins. *J. Plant Nutr. Soil Sci.* **2009**, *172*, 101–107. [CrossRef]
56. Zeng, L.; Johnson, R.L.; Li, X.; Liu, J. Phosphorus removal from aqueous solutions by sorption on two volcanic soils. *Can. J. Soil Sci.* **2011**, *83*, 547–556. [CrossRef]
57. Debicka, M.; Kocowicz, A.; Weber, J.; Jamroz, E. Organic matter effects on phosphorus sorption in sandy soils. *Arch. Agron. Soil Sci.* **2015**, *62*, 840–855. [CrossRef]
58. Hirsch, F.; Bonhage, A.; Bauriegel, A.; Schneider, A.; Raab, T.; Raab, A.; Gypser, S. The occurrence, soil parameters and genesis of rubified soils ('Fuchserden') of northeastern Germany. *Catena* **2019**, *175*, 77–92. [CrossRef]

59. Cáceres-Jensen, L.; Rodríguez-Becerra, J.; Parra-Rivero, J.; Escudey, M.; Barrientos, L.; Castro-Castillo, V. Sorption kinetics of diuron on volcanic ash derived soils. *J. Hazard. Mater.* **2013**, *261*, 602–613. [CrossRef]
60. Parfitt, R.L. Phosphate reactions with natural allophane, ferrihydrite and goethite. *J. Soil Sci.* **1989**, *40*, 359–369. [CrossRef]
61. Zhou, A.; Tang, H.; Wang, D. Phosphorus adsorption on natural sediments: Modeling and effects of pH and sediment composition. *Water Res.* **2005**, *39*, 1245–1254. [CrossRef]
62. Li, R.; Wang, J.J.; Zhou, B.; Awasthi, M.K.; Ali, A.; Zhang, Z.; Gaston, L.A.; Lahori, A.H.; Mahar, A. Enhancing phosphate adsorption by Mg/Al layered double hydroxide functionalized biochar with different Mg/Al ratios. *Sci. Total Environ.* **2016**, *559*, 121–129. [CrossRef] [PubMed]
63. Duncan, E.; Owens, G. Metal oxide nanomaterials used to remediate heavy metal contaminated soils have strong effects on nutrient and trace element phytoavailability. *Sci. Total Environ.* **2019**, *678*, 430–437. [CrossRef]
64. Sun, W.; Jiang, B.; Wang, F.; Xu, N. Effect of carbon nanotubes on Cd (II) adsorption by sediments. *Chem. Eng. J.* **2015**, *264*, 645–653. [CrossRef]
65. Afshinnia, K.; Baalousha, M. Effect of phosphate buffer on aggregation kinetics of citrate-coated silver nanoparticles induced by monovalent and divalent electrolytes. *Sci. Total Environ.* **2017**, *581–582*, 268–276. [CrossRef] [PubMed]
66. Niaura, G.; Gaigalas, A.K.; Vilker, V.L. Surface-enhanced Raman spectroscopy of phosphate anions: Adsorption on silver, gold, and copper electrodes. *J. Phys. Chem. B* **1997**, *101*, 9250–9262. [CrossRef]
67. White, P.; Hjortkjaer, J. Preparation and characterisation of a stable silver colloid for SER(R)S spectroscopy. *J. Raman Spectrosc.* **2014**, *45*, 32–40. [CrossRef]
68. Cornelis, G.; Doolette Madeleine Thomas, C.; McLaughlin, M.J.; Kirby, J.K.; Beak, D.G.; Chittleborough, D. Retention and Dissolution of Engineered Silver Nanoparticles in Natural Soils. *Soil Sci. Soc. Am. J.* **2012**, *76*, 891–902. [CrossRef]
69. Pérez-Novo, C.; Fernández-Calviño, D.; Bermúdez-Couso, A.; López-Periago, J.E.; Arias-Estévez, M. Influence of phosphorus on Cu sorption kinetics: Stirred flow chamber experiments. *J. Hazard. Mater.* **2011**, *185*, 220–226. [CrossRef]
70. Liu, R.; Zhao, D. In situ immobilization of Cu (II) in soils using a new class of iron phosphate nanoparticles. *Chemosphere* **2007**, *68*, 1867–1876. [CrossRef]
71. Zhang, H.; Wang, J.N.; Zhu, Y.G.; Zhang, X. Research and application of analytical technique on δ18Opof inorganic phosphate in soil. *Chin. J. Anal. Chem.* **2015**, *43*, 187–192. [CrossRef]
72. Zhou, W.; Liu, Y.-L.; Stallworth, A.M.; Ye, C.; Lenhart, J.J. Effects of pH, Electrolyte, Humic Acid, and Light Exposure on the Long-Term Fate of Silver Nanoparticles. *Environ. Sci. Technol.* **2016**, *50*, 12214–12224. [CrossRef] [PubMed]
73. Funakawa, S.; Hirooka, K.; Yonebayashi, K. Temporary storage of soil organic matter and acid neutralizing capacity during the process of pedogenetic acidification of forest soils in Kinki District, Japan. *Soil Sci. Plant Nutr.* **2008**, *54*, 434–448. [CrossRef]
74. Poggere, G.C.; Melo, V.F.; Serrat, B.M.; Mangrich, A.S.; França, A.A.; Corrêa, R.S.; Barbosa, J.Z. Clay mineralogy affects the efficiency of sewage sludge in reducing lead retention of soils. *J. Environ. Sci.* **2019**, *80*, 45–57. [CrossRef] [PubMed]
75. Yan, L.G.; Xu, Y.Y.; Yu, H.Q.; Xin, X.D.; Wei, Q.; Du, B. Adsorption of phosphate from aqueous solution by hydroxy-aluminum, hydroxy-iron and hydroxy-iron-aluminum pillared bentonites. *J. Hazard. Mater.* **2010**, *179*, 244–250. [CrossRef]
76. Trinh, V.T.; Nguyen, T.M.P.; Van, H.T.; Hoang, L.P.; Nguyen, T.V.; Ha, L.T.; Vu, X.H.; Pham, T.T.; Nguyen, T.N.; Quang, N.V.; et al. Phosphate Adsorption by Silver Nanoparticles-Loaded Activated Carbon derived from Tea Residue. *Sci. Rep.* **2020**, *10*, 1–13. [CrossRef] [PubMed]
77. Bulmer, D.; Hamilton, J.; Kar, G.; Dhillon, G.; Si, B.C.; Peak, D. Effects of Citrate on the Rates and Mechanisms of Phosphate Adsorption and Desorption on a Calcareous Soil. *Soil Sci. Soc. Am. J.* **2019**, *83*, 332–338. [CrossRef]
78. Liu, Z.; Zhang, Y.; Han, F.; Yan, P.; Liu, B.; Zhou, Q.; Min, F.; He, F.; Wu, Z. Investigation on the adsorption of phosphorus in all fractions from sediment by modified maifanite. *Sci. Rep.* **2018**, *8*, 1–13. [CrossRef]
79. Jung, H.B.; Xu, H.; Konishi, H.; Roden, E.E. Role of nano-goethite in controlling U(VI) sorption-desorption in subsurface soil. *J. Geochem. Explor.* **2016**, *169*, 80–88. [CrossRef]
80. Elkhatib, E.; Moharem, M.; Mahdy, A.; Mesalem, M. Sorption, Release and Forms of Mercury in Contaminated Soils Stabilized with Water Treatment Residual Nanoparticles. *Land Degrad. Dev.* **2017**, *28*, 752–761. [CrossRef]

Article

Analysis of Spatial Variability of River Bottom Sediment Pollution with Heavy Metals and Assessment of Potential Ecological Hazard for the Warta River, Poland

Joanna Jaskuła *, Mariusz Sojka, Michał Fiedler and Rafał Wróżyński

Department of Land Improvement, Environmental Development and Spatial Management, Faculty of Environmental Engineering and Mechanical Engineering, Poznań University of Life Sciences, Piątkowska 94, 60-649 Poznań, Poland; mariusz.sojka@up.poznan.pl (M.S.); michal.fiedler@up.poznan.pl (M.F.); rafal.wrozynski@up.poznan.pl (R.W.)
* Correspondence: joanna.jaskula@up.poznan.pl

Abstract: Pollution of river bottom sediments with heavy metals (HMs) has emerged as a main environmental issue related to intensive anthropopressure on the water environment. In this context, the risk of harmful effects of the HMs presence in the bottom sediments of the Warta River, the third longest river in Poland, has been assessed. The concentrations of Cr, Ni, Cu, Zn, Cd, and Pb in the river bottom sediments collected at 24 sample collection stations along the whole river length have been measured and analyzed. Moreover, in the GIS environment, a method predicting variation of HMs concentrations along the whole river length, not at particular sites, has been proposed. Analysis of the Warta River bottom sediment pollution with heavy metals in terms of the indices: the Geoaccumulation Index (Igeo), Enrichment Factor (EF), Pollution Load Index (PLI), and Metal Pollution Index (MPI), has proved that, in 2016, the pollution was heavier than in 2017. Assessment of the potential toxic effects of HMs accumulated in bottom sediments, made on the basis of Threshold Effect Concentration (TEC), Midpoint Effect Concentration (MEC), and Probable Effect Concentration (PEC) values, and the Toxic Risk Index (TRI), has shown that the ecological hazard in 2017 was much lower. Cluster analysis revealed two main groups of sample collection stations at which bottom sediments showed similar chemical properties. Changes in classification of particular sample collection stations into the two groups analyzed over a period of two subsequent years indicated that the main impact on the concentrations of HMs could have their point sources in urbanized areas and river fluvial process.

Keywords: heavy metals; pollution; bottom sediments; river

Citation: Jaskuła, J.; Sojka, M.; Fiedler, M.; Wróżyński, R. Analysis of Spatial Variability of River Bottom Sediment Pollution with Heavy Metals and Assessment of Potential Ecological Hazard for the Warta River, Poland. *Minerals* **2021**, *11*, 327. https://doi.org/10.3390/min11030327

Academic Editors: Ana Romero-Freire and Hao Qiu

Received: 21 February 2021
Accepted: 19 March 2021
Published: 21 March 2021

1. Introduction

Surface water accumulated in the rivers, lakes and water reservoirs makes invaluable resources of fresh water vital for all kinds of use, e.g., for household users and industry, agriculture purposes, production of renewable energy, water transport, and recreation purposes. The possibility of using water resources is closely related to a number of factors modified by the type of natural or anthropogenic sources and climate changes. These factors include the quality of water [1–5], the quality of bottom sediments [6–8], processes of eutrophication and overgrowth [9,10], hydrological processes, and extreme natural phenomena [11,12].

One of the greatest hazards to the water environment is the pollution of water and bottom sediments with heavy metals. According to Liu et al. [13] and Jain and Sharma [14], over 97% of the mass transport of heavy metals to oceans involves river bottom sediments. Water pollution with heavy metals is a common problem, so their concentrations, distribution and sources have been of global interest [15–18]. The presence of heavy metal pollutants in the water environment has been of great concern because of their established negative effect on the health of humans and state of ecosystems [19–21]. Moreover,

Wu et al. [22] have suggested that future climate changes will increase the harmful effect of heavy metals in water environment because of their release from bottom sediments to water.

Pollution with heavy metals is considered as a serious threat to the water environment because of its chronic character, stability, toxicity, heavy metal ability to enter the food chain and bioaccumulation [23–25]. After the release, heavy metals can be distributed through different elements of the river environment, they can accumulate in water, bottom sediments, as well as the river fauna and flora. As follows from the hitherto studies, some small amount of heavy metals remains in the water column, while their majority gets accumulated in bottom sediments [26]. It has been documented that the river bottom sediments are of key importance in sorption and transport of heavy metals in the water environment [27].

Heavy metals combine with the bottom sediments in the hyporheic zone as a result of many processes, including coprecipitation, particle surface adsorption, ionic exchange, hydrolysis, and deposition in organic matter [23,28]. The most abundant sources of heavy metals are anthropogenic point and area sources related to agricultural and industrial activity [29,30]. After introduction to the water ecosystem, most of heavy metal pollutants attach to solid particulate matter [6,31] and gets deposited in bottom sediments leading to harmful biological effects, even if the water quality meets the criteria specified in a certain regulations [32]. The content of heavy metals in bottom sediments also depends on hydrodynamic conditions [33]. According to the hitherto studies, the content of heavy metals in river bottom sediments is usually lower than in the water reservoirs supplied by the rivers [23].

Accumulation of heavy metals in bottom sediments poses a persistent threat to the uniform parts of water and other elements of the natural environment. When released from bottom sediments heavy metals can pose a threat to the river flora and agricultural crops as they may accumulate in the plants with water from irrigation installations, along with causing harmful effects to the fauna [34]. Some of the metals accumulated in bottom sediments have been proved to be dangerous even when present in small amounts, depending on their origin [30]. In view of the above, it is of great importance to define and analyze the indices describing the quality of bottom sediments in order to establish the risk of contamination and toxicity related to the presence of heavy metals in the water environment. So far, a number of such indices have been defined, including the Geoaccumulation Index (Igeo), Enrichment Factor (EF), Pollution Load Index (PLI), and Metal Pollution Index (MPI), and some regulations concerning the quality of bottom sediments have been formulated [26,35,36].

This study was undertaken to perform spatial analysis of variability of heavy metal concentration in the bottom sediments of the Warta River, the third longest river in Poland. The analysis was performed on the basis of data collected at 24 sample collection stations localized along the river course. In the samples collected at these sites the concentrations of the metals: Cr, Ni, Cu, Zn, Cd, and Pb were measured. The overriding aim was realized through fulfilment of the following objectives: (1) analysis of the concentrations of heavy metals in the river bottom sediments; (2) determination of the temporal and spatial variability of the heavy metals concentration; (3) analysis of contamination of the river bottom sediments; (4) evaluation of ecological hazard; and (5) identification of potential sources and factors determining the content and spatial distribution of heavy metals in the Warta River.

2. Materials and Methods

2.1. Study Site Description

The Warta River of 808 km in length is the third longest river in Poland. Its catchment area is 54.5.103 km^2, which makes about 17.4% of the area of Poland. The river flows in the western part of Poland and is the right tributary of the river Odra. The main tributaries of the Warta River are: Ner, Prosna, Obra, and Noteć (Figure 1). The source of the Warta River is at the altitude of about 380 m a.s.l., while its mouth is at 12 m a.s.l. The parent rocks of the soil in the river catchment area are postglacial formations. The near-surface layer is dominated with sand and clay formations. The dominant form of land use in the catchment area of the Warta River is agriculture (arable land), covering about 60% of the whole area. The total annual precipitation in the area varies from over 650 mm in the upper course (the Krakowsko-Częstochowska Upland) to nearly 500 mm in the middle course (e.g., catchment area of the upper course of the river Noteć), for the sake of comparison the mean annual precipitation for Poland is 600 mm.

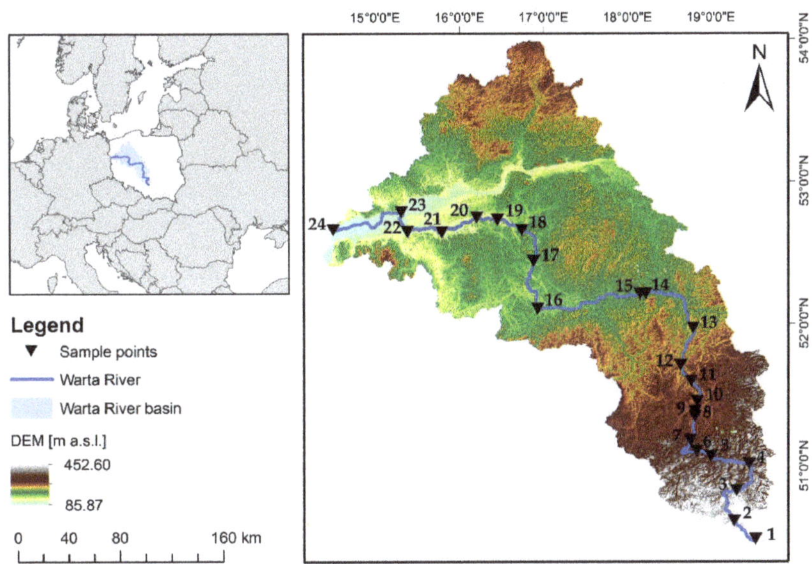

Figure 1. Study site location.

2.2. Materials

The analysis was made on the basis of the measurements of the concentrations of the metals: Cr, Ni, Cu, Zn, Cd, and Pb in the bottom sediments of the Warta River, performed within the program of State Monitoring of the Natural Environment from the period of 2016–2017. Samples of the river bottom sediments were collected in 2016 and 2017 at 24 sample collection stations localized along the course of the Warta River (Figure 1). Sample collection stations were located at the borders of the catchment area, at the mouths of tributaries and below large cities or towns where industrial plants are located. Chemical analysis was performed for a 5 cm thick surface layer of the bottom sediment collected at, from 4 to 5 sites over a section of 50 m for each of the 24 sample collection stations. The samples were mixed and subjected to chemical analysis. Measurements were made by the inductively coupled plasma atomic emission spectroscopy (ICP–OES) according to PN-EN ISO 11885:2009 PB/I/13/D:04.2013.

The boundaries of the Warta River basin area were delimited on the basis of the Raster Hydrographical Map of Poland (MPHP) developed by the Section of Hydrography and Morphology of Riverbeds of the Institute of Meteorology and Water Management (IMGW). The map shows a complete hydrography of Poland in vector format at a scale of 1:50,000, in the PUWG-92 coordinate system. The type of landscape sculpture was characterized on the basis of the Digital Terrain Model (DTM) of the mesh size of at least 100 m, provided by the Head Office of Geodesy and Cartography (GUGiK). The structure of land use over the river basin area was characterized on the basis of the digital database Corine Land Cover (CLC) obtained from the Chief Inspectorate of Environmental Protection.

2.3. Methods

Analysis of the content of heavy metals in the Warta River bottom sediments was performed in four stages. At first, a general characterization of the contents of heavy metals in the bottom sediments in 2016 and 2017 was made using the basic tools of statistical analysis. The second stage was devoted to assessment of contamination of the bottom sediments with heavy metals based on the calculation of Geoaccumulation Index (Igeo), Enrichment Factor (EF), Pollution Load Index (PLI), and Metal Pollution Index (MPI). At the third stage analysis of the possible hazardous effects of the heavy metals present in the river bottom sediments on the water flora and fauna was made using the procedure proposed by MacDonald et al. [37] based on the Threshold Effect Concentration (TEC), Probable Effect Concentration (PEC), and Midpoint Effect Concentration (MEC). Another parameter taken into account was the Toxic Risk Index (TRI), which permitted identification of the sample collection stations running the highest risk of toxic effect of heavy metals on water organisms. At the last stage, the multidimensional statistical methods were used to identify the river sections of similar contents of heavy metals in bottom sediments and to expose potential sources of contamination. Moreover, the content of heavy metals in bottom sediments was analyzed in relation to the hydrological conditions in the years 2016 and 2017, as well as the structure of land use in the land zone adjacent to the river, above the sampling site. It was expected to show if the content of heavy metals was increasing down the river.

2.3.1. General Characterization of the Content of Heavy Metals in the Warta River Bottom Sediments

To characterize the contents of the metals Cr, Ni, Cu, Zn, Cd, and Pb in the Warta River bottom sediments, the mean concentrations of the metals, their minimum and maximum values and median were calculated. The calculations were performed for 2016 and 2017 separately and jointly for the two years. For each metal, the percentiles were found (1%, 5%, 10%, 25%, 50%, 75%, 90%, 95%, and 99%). The data on the concentrations of Cr, Ni, Cu, Zn, Cd, and Pb were analyzed to establish the type of the values distribution, the presence of divergent results and the character of their variation. The distributions of Cr, Ni, Cu, Zn, Cd, and Pb concentration were characterized by performing one-sample Kolmogorov–Smirnov test (K–S). The divergent results in the analyzed data set were detected with the Grubbs–Beck (G–B) test. As the distributions of concentrations of all heavy metals considered differed from the normal distribution, in further analysis the nonparametric methods were applied. The Kruskal–Wallis (K–W) test and Dunn's test, as a post-hoc procedure, were used to find out the possible differences in the median values of concentrations of particular metals between 2016 and 2017. The variations in the concentrations of Cr, Ni, Cu, Zn, Cd, and Pb in the years 2016 and 2017 were characterized on the basis of the range, interquartile range (IQR), median absolute deviation (MAD), and quartile coefficient of dispersion (QCD). The correlations between the concentrations of particular heavy metals in the bottom sediments were checked using the Spearman's rank correlation coefficient. The Grubbs–Beck (G–B) test, Kruskal–Wallis (K–W), and Dunn's test, Kolmogorov–Smirnov test (K–S) and Spearman's correlation analysis were carried out using the STATISTICA software version 13.1 (Statistica).

2.3.2. Geoaccumulation Index

The Geoaccumulation Index (Igeo) was first developed by Muller [35] and is used to assess different pollution levels in bottom sediments and soils. The value of Igeo was calculated from the following, Equation (1):

$$I_{geo} = log_2[C_i/1.5B_i], \qquad (1)$$

where C_i is the concentration of each heavy metal (HM) in the sediment (mg·kg^{-1}) and B_i is reference geochemical background value of each HM.

The following geochemical background values were adopted in this study: Cd—0.5 mg·kg^{-1}, Cu—6 mg·kg^{-1}, Cr—5 mg·kg^{-1}, Fe—10,000 mg·kg^{-1}, Ni—5 mg·kg^{-1}, Pb—10 mg·kg^{-1} and Zn—48 mg·kg^{-1} [38]. The factor 1.5 is used for correction of the possible variability of the background data due to lithological conditions. On the basis of the Igeo value, the level of pollution of the bottom sediments can be classified into six different classes (Table 1) [39].

Table 1. Geoaccumulation Index (Igeo) classification.

Class	Value	Sediment Pollution
0	Igeo $\leq$ 0	no enrichment
1	0 < Igeo $\leq$ 1	minor enrichment
2	1 < Igeo $\leq$ 2	moderate enrichment
3	2 < Igeo $\leq$ 3	moderately severe enrichment
4	3 < Igeo $\leq$ 4	severe enrichment
5	4 < Igeo $\leq$ 5	very severe enrichment
6	5 < Igeo	extremely severe enrichment

2.3.3. Enrichment Factor

Calculation of the Enrichment Factor (EF) allows differentiation of the sources of metals, which can be natural or anthropogenic. EF is mostly used for quantification of the human impact on the concentration of each HM in sediments [40]. The Enrichment Factor describes the stabilization of sediments relative to reference elements, e.g., aluminum (Al), iron (Fe), manganese (Mn), scandium (Sc), or titanium (Ti) [26,41]. In this study, anthropogenic metal enrichment was measured by using iron (Fe) as a reference element. The EF values were calculated as Equation (2):

$$EF_i = (C_i/C_{Fe}) / (B_i/B_{Fe}), \tag{2}$$

where C_i is the concentration of each HM in the sediment (mg·kg^{-1}), C_{Fe} is the concentration of iron (Fe), B_i is the reference geochemical background value of each HM, B_{Fe} is the reference geochemical background value of iron (Fe). On the basis of the obtained EF values, bottom sediments can be classified into different classes presented in Table 2 [42].

Table 2. Enrichment Factor (EF) classification.

Class	Value	Sediment Pollution
0	EF $\leq$ 1	no enrichment
1	1 < EF $\leq$ 3	minor enrichment
2	3 < EF $\leq$ 5	moderate enrichment
3	5 < EF $\leq$ 10	moderately severe enrichment
4	10 < EF $\leq$ 25	severe enrichment
5	25 < EF $\leq$ 50	very severe enrichment
6	50 < EF	extremely severe enrichment

2.3.4. Pollution Load Index

The Pollution Load Index (PLI) was originally developed by Tommilson et al. [36] to assess the extent of pollution status of the summarized heavy metals (HMs) in bottom sediments. The PLI is calculated as follows Equation (3):

$$PLI = (CF_{i1} \times CF_{i2} \times \ldots \times CF_{in})^{\frac{1}{n}}, \tag{3}$$

where n is the number of heavy metals and CF is a contamination factor defined for each studied heavy metal, which is one of the most recognized and effective tools in monitoring the heavy metal concentrations. The value of CF was calculated from the following Equation (4) [43]:

$$CF_i = C_i / B_i, \tag{4}$$

where C_i is the concentration of each HM in the sediment (mg·kg^{-1}) and B_i is the reference geochemical background value of each HM.

On the basis of the PLI, the Warta River bottom sediments can be classified into two classes: no pollution (PLI < 1) or pollution (PLI $\geq$ 1).

2.3.5. Metal Pollution Index

The Metal Pollution Index (MPI) is a presentation of all metal concentrations in sediments as a single value, overcoming the difficulties with application and understanding statistical analysis [44]. To compare the total content of all heavy metals at different sample collection stations, the MPI was calculated from Equation (5):

$$MPI = \left(C_{i1} \times C_{i2} \times \ldots \times C_{in}\right)^{\frac{1}{n}}, \tag{5}$$

where C_i is the concentration of each HM in the sediment (mg·kg^{-1}) and n is the number of heavy metals considered.

2.3.6. Ecological Risk Assessment

In order to assess the ecological risk of the impact of heavy metals on aquatic organisms the ecotoxicological criteria were applied. Assessment of the potential toxic effects of heavy metals accumulated in bottom sediments was made on the basis of TEC, MEC and PEC values [37]. The Threshold Effect Concentration (TEC) value is defined as the limit below which no harmful effects on aquatic organisms are expected, the Probable Effect Concentration (PEC) marks the limit above which a toxic effect on aquatic organisms can be expected. The Midpoint Effects Concentrations (MEC) is the mean of TEC and PEC limits. On the basis of the method proposed by MacDonald et al. [37], four levels of bottom sediment pollution and their impact on organisms have been distinguished (Table 3). According to the adopted methodology, the sediments classified as level I, II, III ($\leq$PEC) may cause sporadic harmful effects on organisms, while the sediments at level IV (>PEC) often have harmful effect on living organisms. It is assumed that the sediments are classified as harmful to living organisms when PEC calculated for at least one analyzed heavy metal in the sediments exceeds the limit value.

Table 3. Ecotoxicological criteria for assessing the quality of bottom sediments using the Threshold Effect Concentration (TEC), Midpoint Effect Concentration (MEC), and Probable Effect Concentration (PEC) values (mg·kg^{-1}).

HMs	Level I ($\leq$ TEC)	Level II (>TEC $\leq$ MEC)	Level III (>MEC $\leq$ PEC)	Level IV (>PEC)
Cd	$\leq$0.99	0.99–3.0	3.0–5.0	>5.0
Cr	$\leq$43	43–76.5	76.5–110	>110
Cu	$\leq$32	32–91	91–150	>150
Ni	$\leq$23	23–36	36–49	>49
Pb	$\leq$36	36–83	83–130	>130
Zn	$\leq$120	120–290	290–460	>460

2.3.7. Toxic Risk Index

Toxic Risk Index (TRI) was introduced by Zhang et al. [45] to assess potential integrated toxic risk caused by heavy metals to aquatic organisms. The TRI values calculated on the basis of the threshold (TEC) and probable (PEC) effect concentrations of heavy metals are presented in Table 2. The TEC and PEC values have been successfully used to assess ecological risk in many previous studies concerning rivers, lakes and reservoirs sediments [23,45,46]. The Toxic Risk Index is calculated according to the following Equation (6):

$$TRI = \sum_{i=1}^{n} TRI_i = \{[(C_i/TEC_i)^2 + (C_i/PEC_i)^2]/2\}^{\frac{1}{2}},$$ (6)

where n is the number of HMs, C_i is the concentration of each HM in the sediment (mg·kg^{-1}), TEC_i is the threshold effect concentration of each HM, PEC_i is the probable effect concentration of each HM, and TRI_i is the toxic risk index of each HM. The toxic risks are classified into five categories based on the obtained TRI values [47] (Table 4).

Table 4. Toxic Risk Index (TRI) classification.

Class	Value	Toxic Risk
0	TRI $\leq$ 5	no toxic risk
1	5 < TRI $\leq$ 10	low
2	10 < TRI $\leq$ 15	moderate
3	15 < TRI $\leq$ 20	considerable
4	20 < TRI	very high

2.3.8. Analysis of Spatial Variability of the Heavy Metal Content in the Warta River Bottom Sediments

Cluster analysis (CA) was applied to distinguish the groups of the sample collection stations at which the bottom sediments samples revealed similar contents of heavy metals. The grouping was performed by the Ward method assuming the square of Euclidean distance as a measure of similarity. The division of sample collection stations into groups and subgroups was made assuming the limits of $D_{link} \cdot D_{lnik.max.} 100\%$ at 66% and 25% [48]. For the groups and subgroups the mean heavy metal concentrations and median were calculated. The differences in heavy metal concentrations between the groups and subgroups were analyzed by the nonparametric Kruskal–Wallis (K–W) test and Dunn's tests as post hoc procedures. The cluster analysis (CA) was carried out using STATISTICA software (version 13.1, StatSoft, Kraków, Poland).

The types of land use were established for the zone adjacent to the river of 600 m in width. Taking into account great differences in spatial concentration of heavy metals, independent of the km of the river course, the land use character was analyzed separately at the distances of 1, 2, 3, 4, and 5 km from the sample collection station (Figure 2). The structure of land cover in the zone adjacent to the river was analyzed using the ArcGIS software and Buffer function.

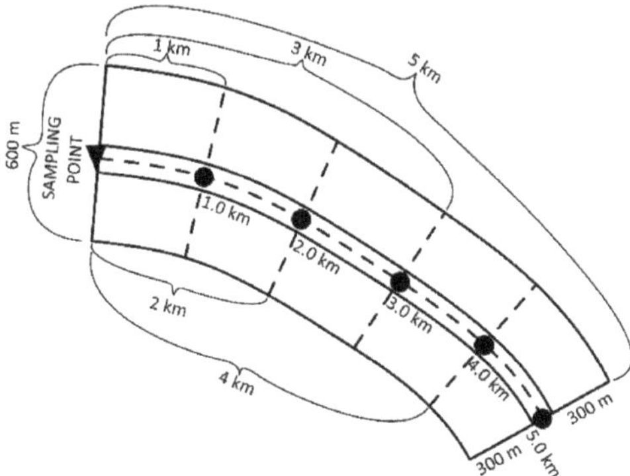

Figure 2. Scheme of land cover analysis.

2.3.9. Linear Interpolation

Point representation of the data on the contamination with heavy metals does not permit presentation of the changes taking place along the river course in a continuous way. However, on the basis of the point data a linear interpolation was performed to make a graphic presentation of the distribution of heavy metal contamination in the form of raster graphics of pixel resolution of 100 m. The linear interpolation can be described by Equation (7):

$$C_i = ((C_x - C_y) / p_{xy}) \cdot p_{xi} + C_x, \tag{7}$$

where C_i is the concentration of each HM in the sediment (mg·kg^{-1}) in the cell, C_x—the concentration of each HM in the sediment (mg·kg^{-1}) at the point of measurement below a reference localization along the river course, p_{xy}—the number of pixels of a certain spatial resolution between the points of measurements, p_{xi}—the number of pixels between the point of measurement below along the river course and the reference point (Figure 3). The linear interpolation was performed in the ArcGIS software (10.7 version, ESRI, Redlands, California, CA, USA).

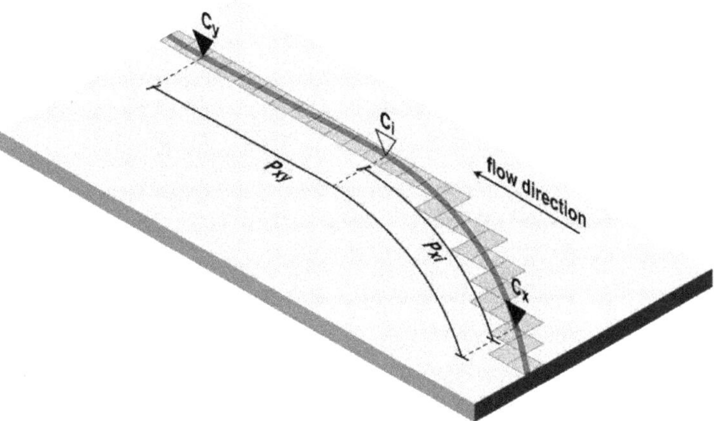

Figure 3. Scheme of linear interpolation developed in this study.

3. Results

3.1. General Characterization of Heavy Metal Concentrations

The concentrations of Cd, Cr, Cu, Ni, Pb, and Zn in the samples of the Warta River bottom sediments measured in 2016 and 2017 at particular sample collection stations are presented in Table 5, with reference to the values of TEC, MEC, and PEC. The concentrations of individual metals in the river bottom sediments varied in the following ranges: Cd: 0.03–14.50 mg·kg^{-1}, Cr: 0.78–193.0 mg·kg^{-1}, Cu: 0.40–116.0 mg·kg^{-1}, Ni: 0.56–36.7 mg·kg^{-1}, Pb: 1.0–144.0 mg·kg^{-1} and Zn: 0.50–519.0 mg·kg^{-1} (Tables 5 and 6). Analysis of the data set with the Grubbs–Beck (G–B) test revealed a number of deviating results. In 2016 the deviating observations included the concentrations of Cd at site numbers 19, 20, and 23, and the concentrations of Cu and Pb at site 23. In 2017, the deviating observations were the concentration of Cr at site 1 and that of Pb at sites 1 and 24. Analysis of the concentrations of Cd, Cr, Cu, Ni, Pb, and Zn proved that they showed mostly a logarithmic-normal distribution, only the concentrations of Ni in 2016 showed a normal distribution.

Table 5. Concentrations of heavy metals in the bottom sediments of Warta River (mg·kg^{-1}) in 2016 and 2017 on the background of ecotoxicological criteria.

No.	Cd		Cr		Cu		Ni		Pb		Zn	
	2016	2017	2016	2017	2016	2017	2016	2017	2016	2017	2016	2017
1	0.37	0.53	193.00	152.00 *	23.20	36.98	22.20	16.80	27.00	106.00 *	168.00	234.00
2	0.52	0.47	4.32	0.78	0.40	2.93	2.84	0.77	12.00	5.99	200.00	168.50
3	0.13	0.03	17.40	6.26	0.40	6.12	10.30	5.99	16.00	6.32	132.00	55.79
4	2.62	1.04	41.40	25.05	39.10	20.07	28.60	16.66	69.00	41.66	440.00	207.50
5	0.05	0.31	37.00	4.12	6.50	14.30	13.10	3.21	15.00	27.90	58.90	31.20
6	0.05	0.17	12.60	6.57	1.16	5.12	15.90	10.11	6.00	6.26	73.10	48.48
7	0.05	0.12	5.75	4.20	0.91	4.49	12.40	2.47	1.00	3.32	11.10	13.52
8	0.05	0.19	39.30	7.33	3.09	7.48	14.40	6.23	16.00	7.72	0.50	29.84
9	0.05	0.03	19.60	3.46	4.55	1.42	9.96	3.76	12.00	1.91	40.80	24.31
10	0.05	0.03	52.20	2.90	8.28	2.40	21.90	2.47	34.00	2.13	64.60	19.62
11	0.21	0.06	39.40	3.46	7.74	3.79	18.60	2.58	27.00	4.35	22.80	21.11
12	0.61	0.31	64.50	13.79	11.20	6.91	19.70	6.36	43.00	10.30	84.20	55.35
13	0.05	0.54	33.30	1.41	2.54	7.30	13.80	1.09	15.00	2.94	0.50	6.09
14	2.16	0.44	86.30	15.23	23.80	8.69	16.60	3.88	27.00	9.12	170.00	46.45
15	0.96	0.29	63.10	7.11	21.00	3.52	18.40	1.94	26.00	6.90	153.00	25.67
16	0.05	1.15	6.60	3.17	0.40	11.92	1.48	1.65	2.00	3.90	19.80	9.05
17	1.56	0.05	100.00	4.68	83.80	4.82	20.20	1.93	85.00	5.95	327.00	14.44
18	0.47	2.00	27.40	24.40	10.20	24.34	8.92	7.49	17.00	21.38	78.60	97.10
19	4.22 *	3.23	69.60	35.89	52.60	21.77	23.20	6.86	45.00	21.39	296.00	95.61
20	6.46 *	3.52	81.30	40.26	58.30	39.00	21.60	16.60	52.00	31.81	283.00	173.50
21	0.71	0.67	56.90	10.66	12.30	10.17	24.90	4.58	34.00	10.18	74.70	38.67
22	1.70	0.12	27.00	2.12	38.20	9.60	7.36	0.56	19.00	4.12	88.90	19.02
23	14.50 *	1.17	184.00 *	21.33	116.00	16.11	36.70	6.61	144.00 *	16.22	519.00	67.95
24	0.05	0.12	3.20	6.56	9.00	13.55	1.30	1.91	10.00	90.47 *	11.30	95.45

*-deviating values, Ecotoxicological criteria for assessment of the lakes quality using TEC, MEC, and PEC values: X—level 1, X—level 2, X—level 3, X—level 4.

Table 6. The characteristic concentrations of heavy metals (mg.kg^{-1}) in the Warta River bottom sediments in 2016 and 2017.

Period		2016						2017						2016/2017					
HMs		Cd	Cr	Cu	Ni	Pb	Zn	Cd	Cr	Cu	Ni	Pb	Zn	Cd	Cr	Cu	Ni	Pb	Zn
Minimum		0.05	3.20	0.40	1.30	1.00	0.50	0.03	0.78	1.42	0.56	1.91	6.09	0.03	0.78	0.40	0.56	1.00	0.50
Mean		1.57	52.7	22.3	16.0	31.4	138	0.69	16.8	11.8	5.52	18.7	66.6	1.13	34.7	17.0	10.8	25.0	102
Maximum		14.5	193	116	36.7	144	519	3.5	152	39.0	16.8	106	234	14.5	193	116	36.7	144	519
Percentile	1%	0.05	3.20	0.40	1.30	1.00	0.50	0.03	0.78	1.42	0.56	1.91	6.09	0.03	0.78	0.40	0.56	1.00	0.50
	5%	0.05	4.32	0.40	1.48	2.00	0.50	0.03	1.41	2.40	0.77	2.13	9.05	0.03	2.12	0.40	1.09	2.00	6.09
	10%	0.05	5.75	0.40	2.84	6.00	11.1	0.03	2.12	2.93	1.09	2.94	13.5	0.05	3.17	1.16	1.48	2.94	11.1
	25%	0.05	18.5	2.82	10.1	13.5	31.8	0.12	3.46	4.66	1.94	4.24	20.4	0.05	5.22	4.14	2.71	6.13	22.0
	50%	0.42	39.3	9.60	16.3	22.5	81.4	0.31	6.57	8.09	3.82	7.31	42.6	0.34	18.5	8.85	8.21	15.5	61.7
	75%	1.63	67.1	31.0	21.8	38.5	185	0.86	18.3	15.2	6.74	21.4	95.5	1.10	40.8	21.4	16.7	29.9	161
	90%	4.22	100	58.3	24.9	69.0	327	2.00	35.9	24.3	16.6	41.7	174	3.23	86.3	39.1	22.2	69.0	283
	95%	6.46	184	83.8	28.6	85.0	440	3.23	40.3	37.0	16.7	90.5	208	4.22	152	58.3	24.9	90.5	327
	99%	14.5	193	116	36.7	144	519	3.52	152	39.0	16.8	106	234	14.5	193	116	36.7	144	519
Range		14.5	190	116	35.4	143	519	3.5	151	37.6	16.2	104	228	14.5	192	116	36.1	143	519
IQR		1.58	48.6	28.2	11.6	25.0	153	0.74	14.8	10.6	4.80	17.2	75.2	1.05	35.6	17.2	14.0	23.7	139
MAD		0.37	24.5	8.95	5.80	11.0	70.2	0.24	3.88	4.43	2.29	3.70	24.5	0.29	15.0	5.84	6.27	11.3	42.0
QCD		0.94	0.57	0.83	0.36	0.48	0.71	0.75	0.68	0.53	0.55	0.67	0.65	0.91	0.77	0.68	0.72	0.66	0.76

On the basis of the measured concentrations of individual metals, their minimum, maximum mean values and median were calculated. Moreover, for each metal considered, the percentile values were found (Table 6). The calculations were made separately for 2016 and 2017 as well as for the two years together. The mean concentrations of heavy metals in the Warta River bottom sediments in 2016 and 2017 can be ordered as Zn > Cr > Pb > Cu > Ni > Cd and Zn > Pb > Cr > Cu > Ni > Cd, respectively. The mean concentrations of all heavy metals considered were higher in 2016 than in 2017. Analysis of the differences between the medians of heavy metal concentrations in the bottom sediments in 2016 and 2017 performed by the nonparametric Kruskal–Wallis test proved that the differences were statistically significant for Cr, Ni, and Pb at the level of 0.05, while for Zn – at the level of 0.10. For Cd and Cu, the differences were statistically insignificant.

In 2016 the concentrations of Cr showed the highest variation, while in 2017—the concentrations of Pb. When considering the two years together, the highest variations were noted for the concentrations of Cd and Zn, while the lowest was for Ni. The variations of Cd, Cr, Cu, Ni, Pb, and Zn concentrations in 2016 and 2017 are presented in the form of violin plots (Figure 4), which clearly reveals the considerable variation of the concentrations of the analyzed metals as well as the fact that in 2016 the heavy metal concentrations showed greater variation than in 2017. Moreover, the mean concentrations of individual heavy metals (marked in Figure 4 as a square) were higher than the corresponding medians (marked as a cross). These results confirm the earlier results proving the right skewed distributions of Cd, Cr, Cu, Ni, Pb, and Zn in the Warta River bottom sediments. The exceptions are the concentrations of Ni in 2016, whose mean values and medians are close, indicating the normal distribution of concentrations.

3.2. Assessment of the Contamination of the Warta River Bottom Sediment

As follows from the calculated Igeo index values, the river bottom sediments were more contaminated in 2016 than in 2017. The highest contribution to the bottom sediments contamination with heavy metals in 2016 brought chromium; from among 24 samples, for six, the values of the Igeo index for Cr indicated heavy contamination, while for two samples—very heavy contamination. For two samples, the values of Igeo for Cd and Cu were in the range of heavy contamination, while for one sample Igeo for Pb indicated heavy contamination. At sample collection stations 14 and 24, the values of Igeo for all analyzed heavy metals were lower than zero, which points to no contamination of the bottom sediments, (Figure 5a). The highest level of contamination was detected for the bottom sediments at sample collection station 23, with Igeo for Cd, Cr, Cu, and Pb indicating heavy and very heavy contamination. In 2017, the general level of contamination with heavy metals was lower. Only at station 1, the Igeo value for Cr indicated very heavy contamination. For three samples collected at stations 19, 20, and 24 in the lower course of the Warta River, the Igeo values were higher than 2, which corresponds to moderate contamination of the bottom sediments (Figure 5b).

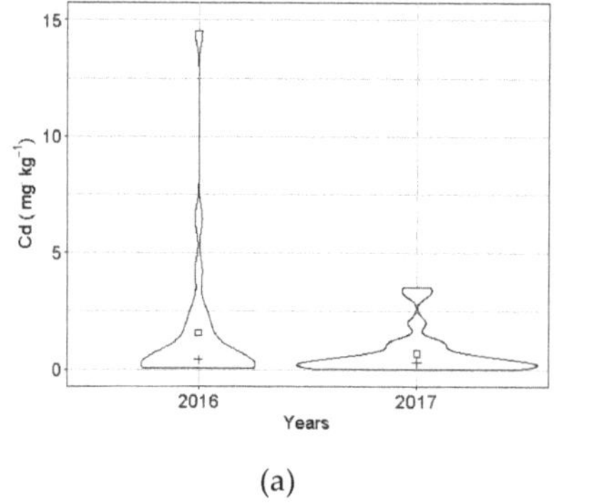

(a)

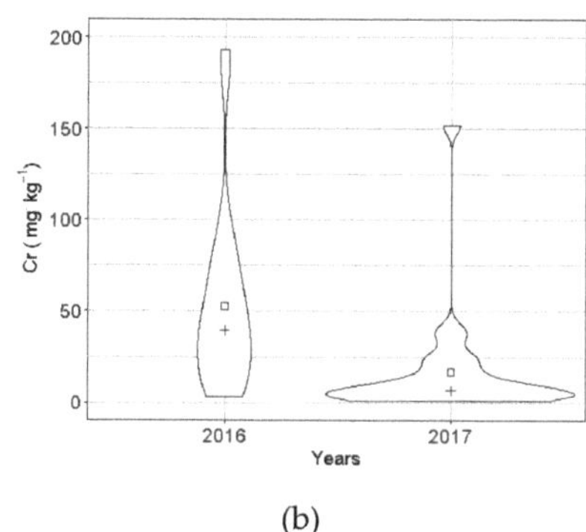

(b)

Figure 4. *Cont.*

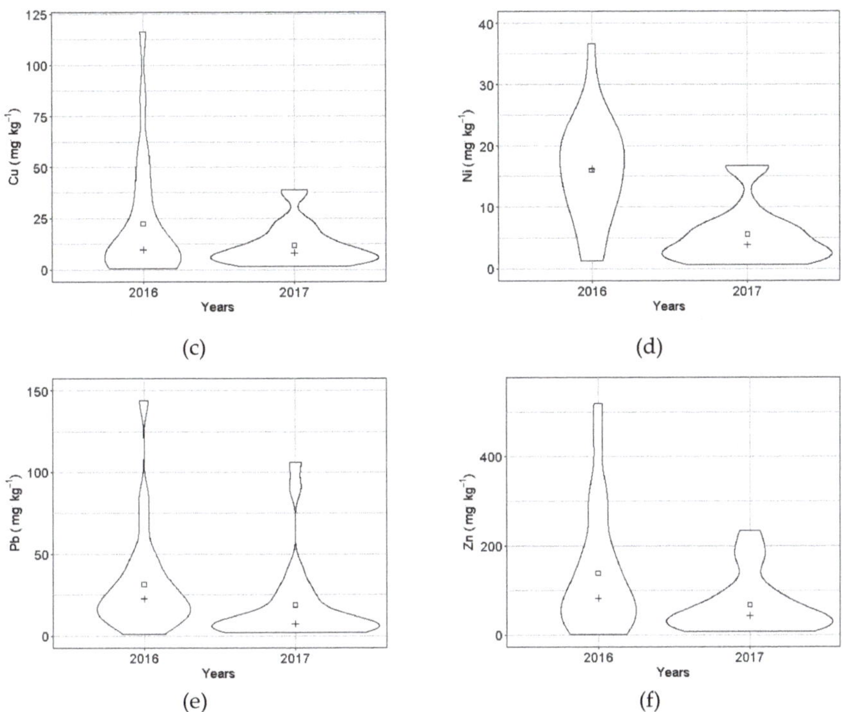

Figure 4. Variation of the concentrations of (a) Cd, (b) Cr, (c) Cu, (d) Ni, (e) Pb, and (f) Zn in the Warta River bottom sediments in 2016 and 2017.

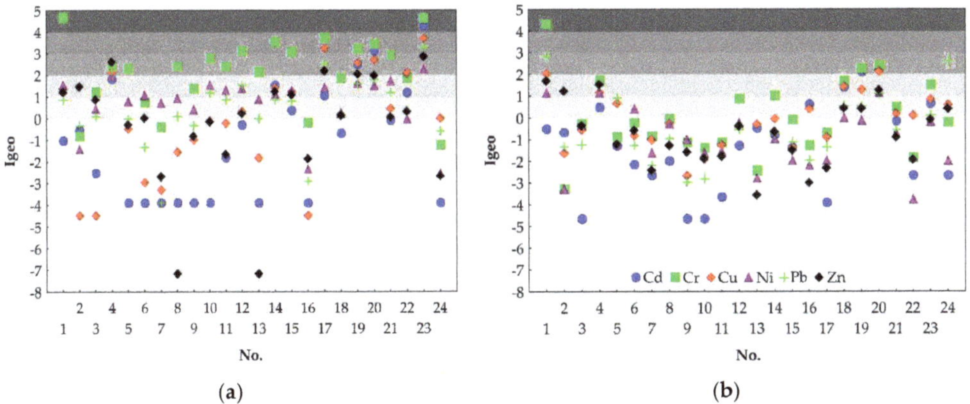

Figure 5. Geoaccumulation Index (Igeo) in bottom sediments of the Warta River in 2016 (a) and 2017 (b).

Analysis of the EF index data also revealed the heaviest contamination of the Warta River bottom sediments with Cr. In 2016, the values of EF calculated for Cr indicated higher than average contamination of the river bottom sediments (Figure 6).

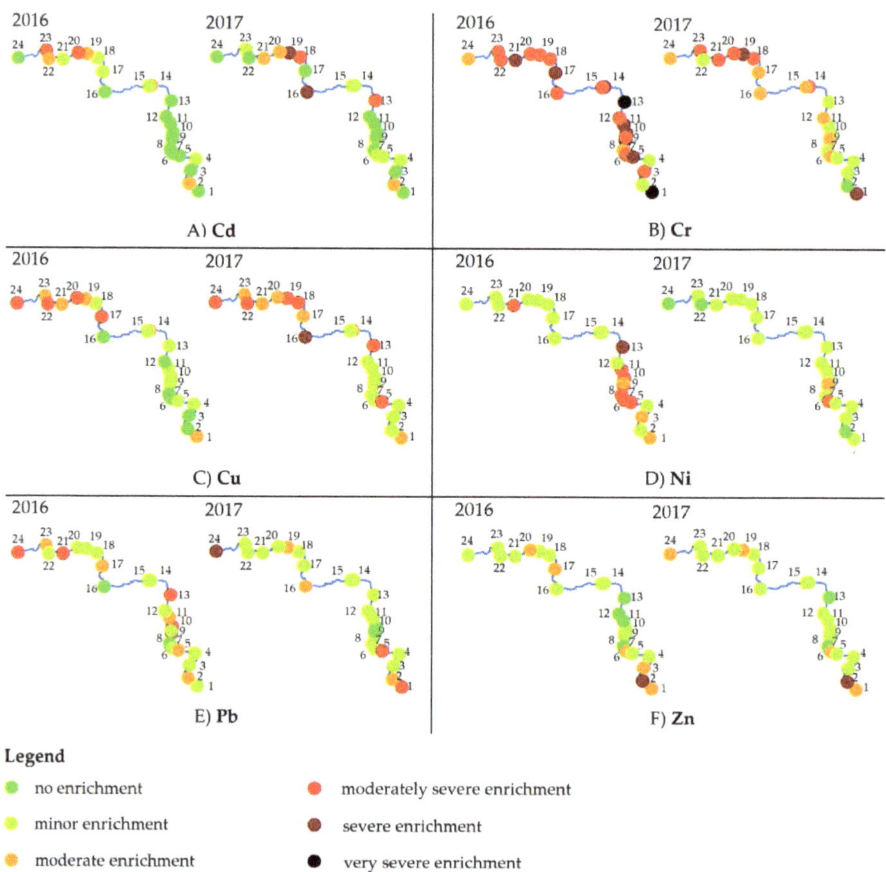

Figure 6. Enrichment Factor (EF) in bottom sediments of Warta River in 2016 and 2017.

The EF values calculated for Cd and Cu signaled the heavier contamination of the bottom sediments in the lower course of the Warta River. In 2016 for Ni the highest EF were obtained for the bottom sediments from the middle course of the river, while for Pb a similar level of contamination was noted along the whole course of the river. According to EF data, the poorest contamination of the bottom sediment was observed with Zn; only in the upper course of the river at a few sample collection stations elevated EF values were calculated (Figure 6). The elevated EF values found for almost all heavy metals considered along the whole river course imply that the contaminants come mainly from local point sources. In 2017, the EF values for Ni, Pb, and Zn were at a level indicating poor contamination of bottom sediments with these elements, however, the EF values for Cd, Cr, and Cu were higher, indicating above average contamination with these metals.

The calculated PLI values in the period 2016–2017 varied from 0.25 to 16.93. The mean PLI values calculated for the years 1016 and 2017 were 3.32 and 1.46, respectively. According to the PLI values analysis for the concentrations of all heavy metals jointly, the river bottom sediments were highly contaminated at 15 of the 24 sample collection stations in 2016 (Figure 7a), and at 11 of the 24 stations in 2017 (Figure 7b). At 17 sample collection stations, the PLI values decreased in 2017 relative to those in 2016, while only at seven stations did the PLI values increase in 2017. Particular attention should be given to the increase in the value of PLI at station 24 in 2017, relative to the value determined in 2016. In general, PLI values indicate the heaviest contamination of the bottom sediments in the low course of the river. In 2017, in the upper and middle course of the river, the data indicate the scattered character of contamination, suggesting its origin from point sources.

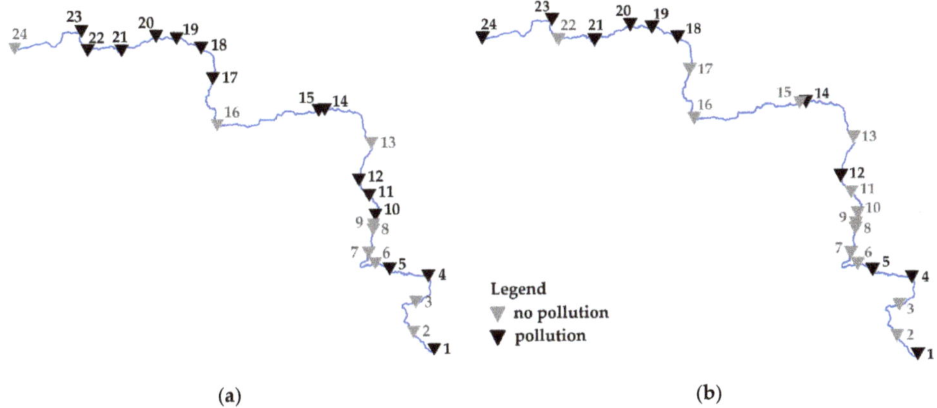

Figure 7. Classes of Pollution Load Index (PLI) in bottom sediments of the Warta River in 2016 (**a**) and 2017 (**b**).

A similar picture of contamination of the Warta River bottom sediments with heavy metals emerges from the analysis of MPI values, pointing generally to a reduced level of contamination in 2017. The mean value of MPI in 2016 was 10.7 higher than that obtained in 2017. According to MPI analysis, in 2017 (MPI > 15) the highest contamination was noted at stations 1, 4, 18, 19, and 20. In 2016, the MPI values were higher than 15 at 11 of the 24 stations: 1, 4, 10, 12, 14, 15, 17, 19, 20, 21, 22, and 23 (Figure 8).

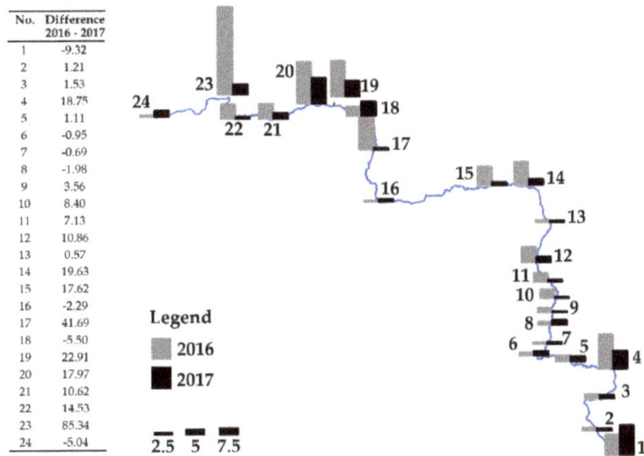

Figure 8. Comparison of Metal Pollution Index (MPI) in bottom sediments of Warta River in 2016 and 2017.

3.3. Toxic Effect of Heavy Metals

Analysis of the potential effect of heavy metals accumulated in the Warta River bottom sediments on aquatic organisms was performed according to the procedure proposed by MacDonald et al. [37]. It showed that in 2016 the PEC limits were exceeded at two sample collection stations: station 1 for Cr and station 23 for Cd, Cr, Pb, and Zn. In the vicinity of these stations, a harmful effect on living aquatic organisms is predicted (Figure 9). In 10 of the 24 stations, numbers 5, 6, 7, 8, 9, 11, 13, 16, 18, and 24, the concentrations of all heavy metals considered were lower than the TEC limit, which points to no harmful effect related to the presence of these elements in the river bottom sediments. At the other stations, the content of heavy metals in the river bottom sediments indicates sporadic toxic effect on aquatic organisms, at seven stations the heavy metal concentrations were above the TEC limit but below the MEC limit, while at five stations they were below the MEC limit. In 2017, only at station no. 1, the concentration of Cr was higher than the PEC limit, whereas at 15 stations, the concentrations of Cd, Cr, Cu, Ni, Pb, and Zn were lower than the TEC limit. At five sample collection stations, the river bottom sediments were contaminated with heavy metals at level II, according to the classification introduced by MacDonald et al. [37], while at three stations—at level III, which implied sporadic toxic effect on aquatic organisms. In general, the highest toxic threat to aquatic life had the high concentration of Cr and at some sites the concentrations of Cd, Pb, and Zn.

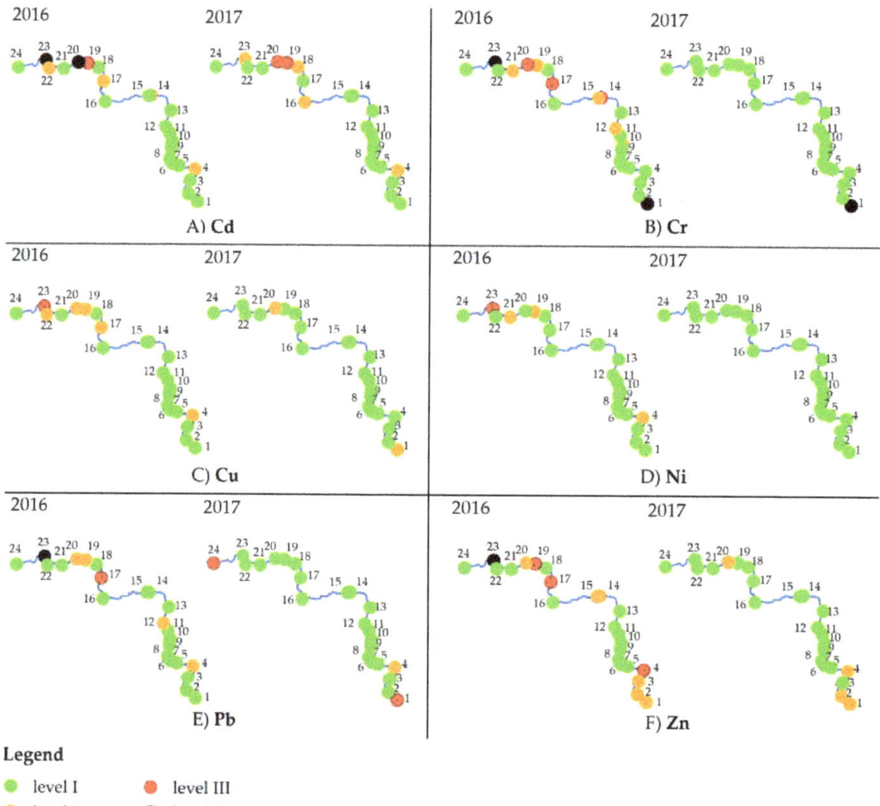

Figure 9. Ecotoxicological criteria for assessing the lakes quality using TEC, MEC, and PEC values in bottom sediments of Warta River in 2016 and 2017.

The values of TRI in 2016 varied from 0.37 to 23.77, with the mean value of 4.60. In 2017, TRI varied from 0.37 to 8.06, with the mean value of 2.04. Assessment of the potential toxic effect of heavy metals accumulated in the Warta River bottom sediments in 2016 indicated a small effect on aquatic life at the sample collection stations 1, 4, 14, 17, and 19. According to TRI, only at stations 20 and 23 a possible moderate and very strong effects, respectively, on aquatic organisms could occur (Figure 10a). In 2017, only at two stations, 1 and 20, were the values of TRI 8.06 and 6.42—so high enough to indicate a possible small toxic effect on aquatic life (Figure 10b).

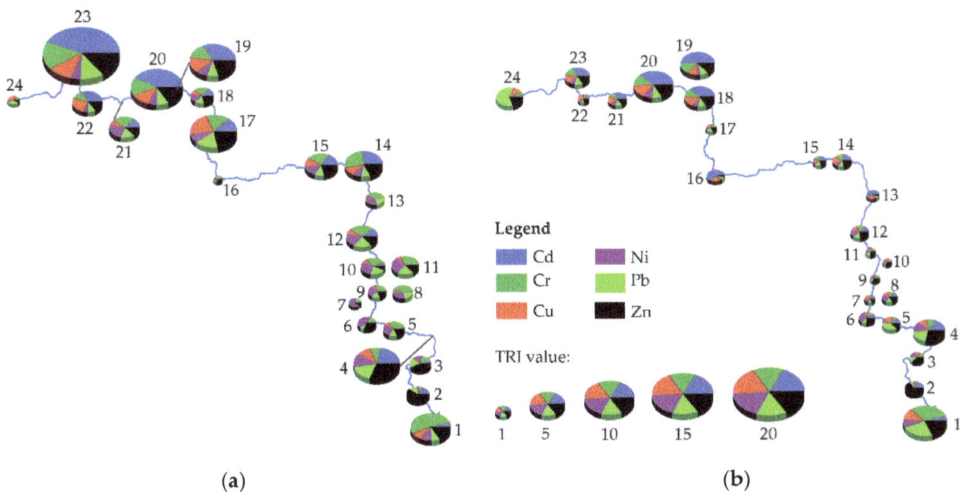

Figure 10. Toxic Risk Index (TRI) in bottom sediments of Warta River in 2016 (**a**) and 2017 (**b**).

3.4. Spatial Analysis of Contamination of the Warta River Bottom Sediments with Heavy Metals

Cluster analysis was performed separately for the data for 2016 and 2017 (Figure 11a,b). The cluster analysis of the data from 2016 revealed two groups of stations at which the bottom sediment samples were characterized by similar concentrations of Cd, Cr, Cu, Pb, Ni, and Zn. Group 1 included 19 sample collection stations; the bottom sediment samples collected at them showed lower concentrations of Cd, Cr, Cu, Pb, Ni, and Zn than the samples collected at group 2 of stations. The differences between the medians of concentrations of Cd, Cr, Cu, Pb, Ni, and Zn calculated for groups 1 and 2 were statistically significant at the level of 0.05. The stations from group 1 were subdivided into two subgroups 1–1 and 1–2. The samples collected at the stations 1–1 were characterized by higher concentrations of all heavy metals considered than those collected at the stations 1–2. The differences were statistically significant at the level of 0.05. Group 2 included only five stations, of which four were localized in the low course of the Warta River, below the city of Poznań, the largest city on the river. On the basis of the data for 2017, the sample collection stations were also divided into two groups. Group 1 included seven stations, at which the samples of bottom sediments contained higher concentrations of all heavy metals considered than those collected at the stations from group 2. The differences were statistically significant at the level of 0.05. In 2017, the stations of group 1 included four stations, numbers 4, 19, 20, and 23, classified to group 2 in 2016. Moreover, three stations (no. 1, 18, and 24) were classified as those at which the bottom sediment samples showed the strongest contamination with the heavy metals studied. Similarly as in 2016, the stations at which the bottom sediment samples contain the greatest amounts of heavy metals were in the low course of the river, except stations 1 and 4, at which probably the effect of local anthropogenic sources of contamination was reflected. It should be mentioned that the samples collected at the stations from group 2 were characterized by greater variation of the heavy metals concentrations than those collected at the stations from group 1; however, the highest concentrations of heavy metals were found in the samples collected at stations 1 and 24.

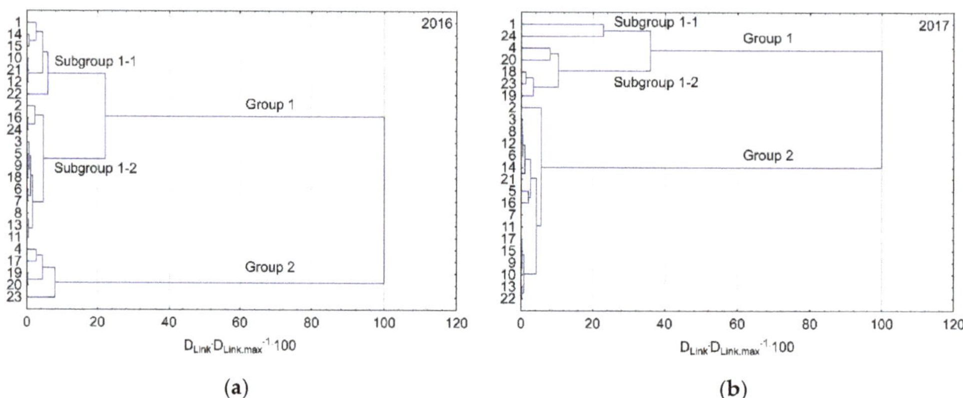

(a)   (b)

Figure 11. Division of sample collection stations into groups as a result of cluster analysis: in 2016 (**a**) and 2017 (**b**).

3.5. Linear Interpolation

The main limitation of point measurements is the lack of possibility of presentation of changes in the heavy metals concentrations over the whole length of the river. On the basis of the concentrations determined at the stations, a linear interpolation was performed to obtain a graphical presentation of the contamination with heavy metals along the whole course of the river. Figure 12 presents the variation in the concentration of lead (Pb) along the river section between stations 22 and 24. Considering the results of the interpolation over this section it is apparent that the concentrations of lead in 2016 and 2017 were significantly different along the whole section, which implies that their negative effects can be significant not only near the sample collection stations, but also along the sections not subjected to monitoring. Linear interpolation of the data along the course of the river reveals important information both for identification of new sources of contamination and for making decisions on activities aimed at limitation of a given type contamination on the environment.

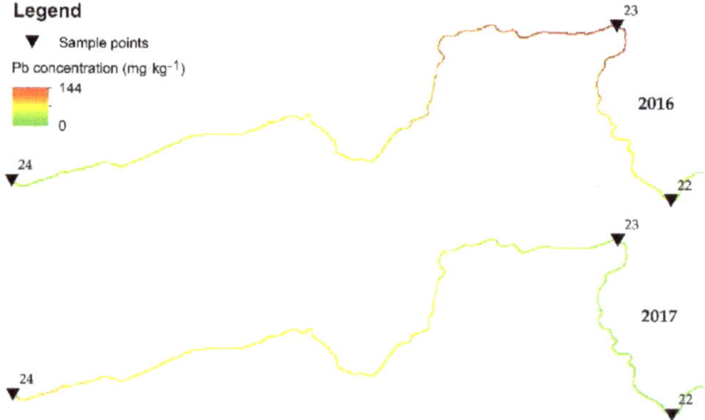

Figure 12. Linear interpolation of Pb concentration between sample collection stations 22–24.

4. Discussion

According to the hitherto studies, the content of heavy metals in river bottom sediments is related to the terrain relief of the catchment area, hydrological conditions and anthropogenic activity in the catchment area, including the type of land use. Analysis of the data from a two-year period of monitoring the content of heavy metals (Cd, Cu, Ni, Zn, Fe, Pb, Cr, and Mn) in the mouths of the rivers revealed no seasonal changes and no relations to tides. The observed high content of heavy metals in the form of particles was probably a result of anthropogenic pressure [49]. Zhou et al. [50] determined the content of main and trace elements in bottom sediments of 26 large rivers in south China and on the basis of EF data analysis they concluded that high concentrations of As, Cd, Hg, Mn, Mo, Pb, and Zn in the majority of the rivers studied were strongly related to the natural factors, in particular to geological conditions. Le Gall et al. [51] analyzed the contamination of bottom sediments of the river Seine on the basis of determination of stable isotopes of lead and EF index. Their conclusion was that the contamination of the river bottom sediments below Paris was mainly caused by point influx from urbanized areas [51]. According to the results presented in our paper, the concentrations of some heavy metals in the Warta River bottom sediments show high variation over the period of the two years studied, 2016 and 2017. This character of results collected in two subsequent years may indicate a significant effect of point sources of contamination from urbanized areas on the heavy metal contents as in such a short period of time the type of land use was unchanged (Figure 13).

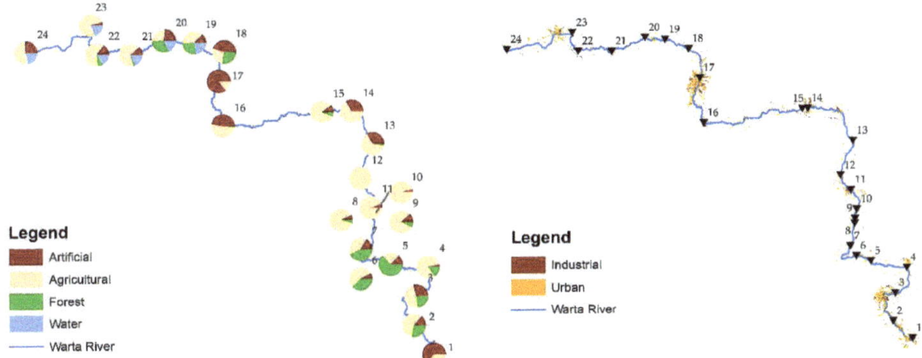

Figure 13. Land cover analysis.

The results presented in this paper corroborate the conclusions drawn by Li et al. [52], who have identified the industrial activities as the largest contributor (48.0%) of heavy metals to the river–lake sediments, followed by the agricultural activities (27.3%) and mix sources (24.7%). Zeng et al. [46] have shown that heavy metal concentrations in the central city in a given area were generally higher than those in other areas. Heavy metals in sediments pose a considerable ecological risk. The main contributors to the contamination with heavy metals were Hg and Cd whose presence posed the highest potential ecological risk in the central city. According to Custodio et al. [53], the rivers of the Mantaro River watershed are exposed to contamination by heavy metals and metalloids from natural and anthropogenic sources; among the latter, the mining metallurgical industry, agricultural activity, and manufacturing industry are the main sources. One of the key variables affecting the heavy metals release from sediments are also hydrometeorological conditions [54]. River fluvial process depends on flow rate value, so flash floods and low flows also have impact on HMs concentration. According to Li et al. [54]. The flow rate significantly affected the release amount of Zn, Pb, and Cr, while it slightly affected the concentration of Cu and Cd.

Sojka et al. [23] have shown that the high values of Igeo, EF, PLI, and MPI in the samples collected at particular stations may indicate the anthropogenic origin of the heavy metal contamination and the influx of heavy metals from point sources. Bing et al. [47] have suggested that local anthropogenic activity increases the concentration of certain heavy metals in the river bottom sediments, in particular in the upper courses of the rivers. Saleem et al. [55] have reported a relatively high level of heavy metals at the sites close to large cities and towns. The results presented in this paper confirm the earlier observations indicating a significant effect of urbanized areas on the level of contamination with heavy metals in the river bottom sediments. This conclusion implies the necessity of looking for new solutions supporting monitoring and aimed at protection of water resources along the whole course of the river, e.g., with the use of GIS tools. New approach to classical interpolation method which that takes into account the curvature of the river was never presented regarding to HMs contamination. This interpolation method allows to receive accurate information on the status of the whole courses of the rivers not only at single sample points. The solution proposed in this paper permits analysis of the changes in the river bottom sediments along the whole course of the river, which is essential for identification of potential sources of contamination and for taking up actions aimed at limitation of the effect of such sources. The main limitation of the method proposed is the fact that it is based only on point measurements, disregarding a variable related to the land use, e.g., the effect of urbanized areas.

5. Conclusions

The above presented results and their analyses permit drawing the following conclusions:

- As shown by the results of analysis of Igeo, EF, PLI, and MPI values, the level of contamination of the Warta River bottom sediments with heavy metals was higher in 2016 than in 2017.
- According to the assessment of the potential toxic effects of heavy metals accumulated in bottom sediments made on the basis of TEC, MEC, PEC, and TRI, the ecological risk related to the presence of heavy metals in the river bottom sediments was much lower in 2017 than in 2016.
- Cluster analysis permitted distinction of two groups of the sample collection stations at which bottom sediments showed similar chemical character. Changes in the classification of particular stations to particular groups indicated that the concentration of heavy metals in the Warta river bottom sediments is mainly related to the point sources of contamination in urbanized areas and connected with river fluvial process.
- In view of the necessity of taking up measures aimed at protection of water resources, it is vital to find methods providing the possibly most accurate information on the status of the whole courses of the rivers not only at their certain points. The solution proposed in this paper permits analysis of the changes in the river bottom sediments along the whole course of the river, which is essential for identification of potential sources of contamination and for taking up actions aimed at limitation of the effect of such sources. The main drawback of the method proposed is the fact that it is based on point measurements, disregarding a variable related to the land use, e.g., the effect of urbanized areas.

Author Contributions: Conceptualization, J.J. and M.S.; methodology J.J. and M.S.; software, J.J. and M.S.; validation, J.J. and M.S.; investigation, J.J.; data curation, J.J., M.S., and M.F.; writing—original draft preparation, J.J. and M.S.; writing—review and editing, J.J; visualization, J.J., R.W.; project administration, J.J. All authors have read and agreed to the published version of the manuscript.

Funding: The publication was co-financed within the framework of the Ministry of Science and Higher Education programme as "Regional Initiative Excellence" in the years 2019–2022, Project No. 005/RID/2018/19.

Conflicts of Interest: The authors declare no conflict of interest.

References

1. Dąbrowska, J.; Dąbek, P.B.; Lejcuś, I. A GIS based approach for the mitigation of surface runoff to a shallow lowland reservoir. *Ecohydrol. Hydrobiol.* **2018**, *18*, 420–430. [CrossRef]
2. Dąbrowska, J.; Pawęska, K.; Dąbek, P.B.; Stodolak, R. The implications of economic development, climate change and European water policy on surface water quality threats. *Acta Sci. Pol. Form. Circumiectus* **2017**, *16*, 111. [CrossRef]
3. Tung, T.M.; Yaseen, Z.M. A survey on river water quality modelling using artificial intelligence models: 2000–2020. *J. Hydrol.* **2020**, *585*, 124670.

4. Sojka, M.; Jaskuła, J.; Wicher-Dysarz, J. Assessment of Biogenic Compounds elution from the Główna River catchment in the Years 1996–2009. *Annu. Set Environ. Prot.* **2016**, *18*, 815–830.

5. Kuriata-Potasznik, A.; Szymczyk, S.; Skwierawski, A. Influence of cascading river–lake systems on the dynamics of nutrient circulation in catchment areas. *Water* **2020**, *12*, 1144. [CrossRef]

6. Sojka, M.; Siepak, M.; Jaskuła, J.; Wicher-Dysarz, J. Heavy metal transport in a river-reservoir system: A case study from central Poland. *Pol. J. Environ. Stud.* **2018**, *27*, 1725–1734. [CrossRef]

7. Kuriata-Potasznik, A.; Szymczyk, S.; Skwierawski, A.; Glińska-Lewczuk, K.; Cymes, I. Heavy metal contamination in the surface layer of bottom sediments in a flow-through lake: A case study of Lake Symsar in Northern Poland. *Water* **2016**, *8*, 358. [CrossRef]

8. Deng, Q.; Wei, Y.; Yin, J.; Chen, L.; Peng, C.; Wang, X.; Zhu, K. Ecological risk of human health posed by sediments in a karstic river basin with high longevity population. *Environ. Pollut.* **2020**, *265*, 114418. [CrossRef]

9. Jaskuła, J.; Sojka, M.; Wicher-Dysarz, J. Analysis of the vegetation process in a two-stage reservoir on the basis of satellite imagery—a case study: Radzyny Reservoir on the Sama River. *Annu. Set Environ. Prot.* **2018**, *20*, 203–220.

10. Jaskuła, J.; Sojka, M. Assessment of spectral indices for detection of vegetative overgrowth of reservoirs. *Pol. J. Environ. Stud.* **2019**, *28*, 4199–4211. [CrossRef]

11. Dysarz, T.; Wicher-Dysarz, J.; Sojka, M.; Jaskuła, J. Analysis of extreme flow uncertainty impact on size of flood hazard zones for the Wronki gauge station in the Warta river. *Acta Geophys.* **2019**, *67*, 661–676. [CrossRef]

12. Wicher-Dysarz, J.; Szałkiewicz, E.; Jaskuła, J.; Dysarz, T.; Rybacki, M. Possibilities of controlling the river outlets by weirs on the example of Noteć Bystra River. *Sustainability* **2020**, *12*, 2369. [CrossRef]

13. Liu, M.; Chen, J.; Sun, X.; Hu, Z.; Fan, D. Accumulation and transformation of heavy metals in surface sediments from the Yangtze River estuary to the East China Sea shelf. *Environ. Pollut.* **2019**, *245*, 111–121. [CrossRef]

14. Jain, C.K.; Sharma, M.K. Distribution of trace metals in the Hindon River system, India. *J. Hydrol.* **2001**, *253*, 81–90. [CrossRef]

15. Nawrot, N.; Wojciechowska, E.; Rezania, S.; Walkusz-Miotk, J.; Pazdro, K. The effects of urban vehicle traffic on heavy metal contamination in road sweeping waste and bottom sediments of retention tanks. *Sci. Total Environ.* **2020**, *749*, 141511. [CrossRef]

16. Mandeng, E.P.B.; Bidjeck, L.M.B.; Bessa, A.Z.E.; Ntomb, Y.D.; Wadjou, J.W.; Doumo, E.P.E.; Dieudonné, L.B. Contamination and risk assessment of heavy metals, and uranium of sediments in two watersheds in Abiete-Toko gold district, Southern Cameroon. *Heliyon* **2019**, *5*, e02591. [CrossRef]

17. Sun, X.; Fan, D.; Liu, M.; Tian, Y.; Pang, Y.; Liao, H. Source identification, geochemical normalization and influence factors of heavy metals in Yangtze River Estuary sediment. *Environ. Pollut.* **2018**, *241*, 938–949. [CrossRef]

18. Duodu, G.O.; Goonetilleke, A.; Ayoko, G.A. Comparison of pollution indices for the assessment of heavy metal in Brisbane River sediment. *Environ. Pollut.* **2016**, *219*, 1077–1091. [CrossRef]

19. Stamatis, N.; Kamidis, N.; Pigada, P.; Sylaios, G.; Koutrakis, E. Quality Indicators and Possible Ecological Risks of Heavy Metals in the Sediments of three Semi-closed East Mediterranean Gulfs. *Toxics* **2019**, *7*, 30. [CrossRef]

20. He, Z.; Li, F.; Dominech, S.; Wen, X.; Yang, S. Heavy metals of surface sediments in the Changjiang (Yangtze River) Estuary: Distribution, speciation and environmental risks. *J. Geochem. Explor.* **2019**, *198*, 18–28. [CrossRef]

21. Brady, J.P.; Ayoko, G.A.; Martens, W.N.; Goonetilleke, A. Enrichment, distribution and sources of heavy metals in the sediments of Deception Bay, Queensland, Australia. *Mar. Pollut. Bull.* **2014**, *81*, 248–255. [CrossRef] [PubMed]

22. Wu, Q.; Qi, J.; Xia, X. Long-term variations in sediment heavy metals of a reservoir with changing trophic states: Implications for the impact of climate change. *Sci. Total Environ.* **2017**, *609*, 242–250. [CrossRef]

23. Sojka, M.; Jaskuła, J.; Siepak, M. Heavy metals in bottom sediments of reservoirs in the lowland area of western Poland: Concentrations, distribution, sources and ecological risk. *Water* **2019**, *11*, 56. [CrossRef]

24. Doležalová Weissmannová, H.; Mihočová, S.; Chovanec, P.; Pavlovský, J. Potential ecological risk and human health risk assessment of heavy metal pollution in industrial affected soils by coal mining and metallurgy in Ostrava, Czech Republic. *Int. J. Environ. Res. Public Health* **2019**, *16*, 4495. [CrossRef] [PubMed]

25. Gao, L.; Wang, Z.; Shan, J.; Chen, J.; Tang, C.; Yi, M.; Zhao, X. Distribution characteristics and sources of trace metals in sediment cores from a trans-boundary watercourse: An example from the Shima River, Pearl River Delta. *Ecotox. Environ. Safe* **2016**, *134*, 186–195. [CrossRef]

26. Ahamad, M.I.; Song, J.; Sun, H.; Wang, X.; Mehmood, M.S.; Sajid, M.; Khan, A.J. Contamination level, ecological risk, and source identification of heavy metals in the hyporheic zone of the Weihe River, China. *Int. J. Environ. Res. Public Health* **2020**, *17*, 1070. [CrossRef]

27. Malvandi, H. Preliminary evaluation of heavy metal contamination in the Zarrin-Gol River sediments, Iran. *Mar. Pollut. Bull.* **2017**, *117*, 547–553. [CrossRef]

28. Dong, X.; Wang, C.; Li, H.; Wu, M.; Liao, S.; Zhang, D.; Pan, B. The sorption of heavy metals on thermally treated sediments with high organic matter content. *Bioresour. Technol.* **2014**, *160*, 123–128. [CrossRef]

29. Cymes, I.; Glińska-Lewczuk, K.; Szymczyk, S.; Sidoruk, M.; Potasznik, A. Distribution and potential risk assessment of heavy metals and arsenic in sediments of a dam reservoir: A case study of the Łoje retention reservoir, NE Poland. *J. Elementol.* **2017**, *22*. [CrossRef]

30. Nguyen, B.T.; Do, D.D.; Nguyen, T.X.; Nguyen, V.N.; Nguyen, D.T.P.; Nguyen, M.H.; Bach, Q.V. Seasonal, spatial variation, and pollution sources of heavy metals in the sediment of the Saigon River, Vietnam. *Environ. Pollut.* **2020**, *256*, 113412. [CrossRef]

31. Lin, Q.; Liu, E.; Zhang, E.; Li, K.; Shen, J. Spatial distribution, contamination and ecological risk assessment of heavy metals in surface sediments of Erhai Lake, a large eutrophic plateau lake in southwest China. *Catena* **2016**, *145*, 193–203. [CrossRef]

32. Bibi, M.H.; Ahmed, F.; Ishiga, H. Assessment of metal concentrations in lake sediments of southwest Japan based on sediment quality guidelines. *Environ. Geol.* **2007**, *52*, 625–639. [CrossRef]

33. Noronha-D'Mello, C.A.; Nayak, G.N. Assessment of metal enrichment and their bioavailability in sediment and bioaccumulation by mangrove plant pneumatophores in a tropical (Zuari) estuary, west coast of India. *Mar. Pollut. Bull.* **2016**, *110*, 221–230. [CrossRef] [PubMed]

34. Lee, A.S.; Huang, J.J.S.; Burr, G.; Kao, L.C.; Wei, K.Y.; Liou, S.Y.H. High resolution record of heavy metals from estuary sediments of Nankan River (Taiwan) assessed by rigorous multivariate statistical analysis. *Quat. Int.* **2019**, *527*, 44–51. [CrossRef]

35. Muller, G. Index of geoaccumulation in sediments of the Rhine River. *Geol. J.* **1969**, *2*, 108–118.

36. Tomlinson, D.L.; Wilson, J.G.; Harris, C.R.; Jeffrey, D.W. Problems in the assessment of heavy-metal levels in estuaries and the formation of a pollution index. *Helgol. Mar. Res.* **1980**, *33*, 566–575. [CrossRef]

37. MacDonald, D.D.; Ingersoll, C.G.; Berger, T.A. Development and evaluation of consensus-based sediment quality guidelines for freshwater ecosystems. *Arch. Environ. Contam. Toxicol.* **2000**, *39*, 20–31. [CrossRef]

38. Bojakowska, I.; Sokołowska, G. Geochemiczne klasy czystości osadów wodnych. *Przegląd Geol.* **1998**, *46*, 49–54.

39. Sakan, S.M.; Đorđević, D.S.; Manojlović, D.D.; Predrag, P.S. Assessment of heavy metal pollutants accumulation in the Tisza river sediments. *J. Environ. Manag.* **2009**, *90*, 3382–3390. [CrossRef]

40. Selvaraj, K.; Ram Mohan, V.; Szefer, P. Evaluation of metal contamination in coastal sediments of the Bay of Bengal, India: Geochemical and statistical approaches. *Mar. Pollut. Bull.* **2004**, *49*, 174–185. [CrossRef]

41. Salati, S.; Moore, F. Assessment of heavy metal concentration in the Khoshk River water and sediment, Shiraz, Southwest Iran. *Environ. Monit. Assess.* **2010**, *164*, 677–689. [CrossRef] [PubMed]

42. Al Rashdi, S.; Arabi, A.A.; Howari, F.M.; Siad, A. Distribution of heavy metals in the coastal area of Abu Dhabi in the United Arab Emirates. *Mar. Pollut. Bull.* **2015**, *97*, 494–498. [CrossRef] [PubMed]

43. Martin, J.M.; Meybeck, M. Elemental mass-balance of material carried by major world rivers. *Mar. Chem.* **1979**, *7*, 173–206. [CrossRef]

44. Singovszka, E.; Balintova, M.; Demcak, S.; Pavlikova, P. Metal pollution indices of bottom sediment and surface water affected by acid mine drainage. *Metals* **2017**, *7*, 284. [CrossRef]

45. Zhang, G.; Bai, J.; Zhao, Q.; Lu, Q.; Jia, J.; Wen, X. Heavy metals in wetland soils along a wetland-forming chronosequence in the Yellow River Delta of China: Levels, sources and toxic risks. *Ecol. Indic.* **2016**, *69*, 331–339. [CrossRef]

46. Zeng, Y.; Bi, C.; Jia, J.; Deng, L.; Chen, Z. Impact of intensive land use on heavy metal concentrations and ecological risks in an urbanized river network of Shanghai. *Ecol. Indic.* **2020**, *116*, 106501. [CrossRef]

47. Bing, H.; Wu, Y.; Zhou, J.; Sun, H.; Wang, X.; Zhu, H. Spatial variation of heavy metal contamination in the riparian sediments after two-year flow regulation in the Three Gorges Reservoir, China. *Sci. Total Environ.* **2019**, *649*, 1004–1016. [CrossRef]

48. Ptak, M.; Sojka, M.; Choiński, A.; Nowak, B. Effect of environmental conditions and morphometric parameters on surface water temperature in Polish lakes. *Water* **2018**, *10*, 580. [CrossRef]

49. Beltrame, M.O.; De Marco, S.G.; Marcovecchio, J.E. Dissolved and particulate heavy metals distribution in coastal lagoons. A case study from Mar Chiquita Lagoon, Argentina. *Estuar. Coast. Shelf Sci.* **2009**, *85*, 45–56. [CrossRef]

50. Zhou, G.; Sun, B.; Zeng, D.; Wei, H.; Liu, Z.; Zhang, B. Vertical distribution of trace elements in the sediment cores from major rivers in east China and its implication on geochemical background and anthropogenic effects. *J. Geochem. Explor.* **2014**, *139*, 53–67. [CrossRef]

51. Le Gall, M.; Ayrault, S.; Evrard, O.; Laceby, J.P.; Gateuille, D.; Lefevre, I.; Mouchel, J.M.; Meybeck, M. Investigating the metal contamination of sediment transported by the 2016 Seine River flood (Paris, France). *Environ. Pollut.* **2018**, *240*, 125–139. [CrossRef] [PubMed]

52. Li, Y.; Chen, H.; Teng, Y. Source apportionment and source-oriented risk assessment of heavy metals in the sediments of an urban river-lake system. *Sci. Total Environ.* **2020**, *737*, 140310. [CrossRef] [PubMed]

53. Custodio, M.; Cuadrado, W.; Peñaloza, R.; Montalvo, R.; Ochoa, S.; Quispe, J. Human risk from exposure to heavy metals and arsenic in water from rivers with mining influence in the Central Andes of Peru. *Water* **2020**, *12*, 1946. [CrossRef]

54. Li, H.; Shi, A.; Li, M.; Zhang, X. Effect of pH, temperature, dissolved oxygen, and flow rate of overlying water on heavy metals release from storm sewer sediments. *J. Chem.* **2013**, *2013*, 1–11. [CrossRef]

55. Saleem, M.; Iqbal, J.; Akhter, G.; Shah, M.H. Fractionation, bioavailability, contamination and environmental risk of heavy metals in the sediments from a freshwater reservoir, Pakistan. *J. Geochem. Explor.* **2018**, *184*, 199–208. [CrossRef]

Article

Natural and Anthropogenic Origin of Metals in Lacustrine Sediments; Assessment and Consequences—A Case Study of Wigry Lake (Poland)

Anna Kostka [1,*] and **Andrzej Leśniak** [2]

1 Department of Environmental Protection, Faculty of Geology, Geophysics and Environmental Protection, AGH University of Science and Technology, Al. Mickiewicza 30, 30-059 Kraków, Poland

2 Department of Geoinformatics and Applied Computer Science, Faculty of Geology, Geophysics and Environmental Protection, AGH University of Science and Technology, Al. Mickiewicza 30, 30-059 Kraków, Poland; lesniak@agh.edu.pl

* Correspondence: kostka@agh.edu.pl

Abstract: The contamination of aquatic sediments by metals is a worldwide phenomenon and its assessment is a fairly complex issue, as numerous factors affect the distribution of particular contaminants in the environment, as well as their bioavailability. Wigry Lake, as the object of this study, is almost a perfect water body for such considerations. It has been well investigated and densely sampled (up to 459 sediment samples). The quantities of seven metals were determined using the atomic absorption spectrometry (AAS) or inductively coupled plasma (ICP)-MS methods, following previous extraction in a microwave oven. The levels of concentration of the examined elements were as follows (min–max (mg·kg^{-1})): Cd—0.003–3.060; Cr—0.20–22.61; Cu—0.02–59.70; Fe—80–32,857; Mn—18–1698; Pb—7.0–107.5; Zn—3.1–632.1. Significant differences were also registered in terms of particular metal concentrations in different sediment types found at the lake bottom. Five different geochemical backgrounds and sediment quality guidelines implemented in the study enabled a very scrupulous contamination assessment of the lake sediments' condition, as well as the evaluation of the natural and anthropogenic contribution to the enrichment of examined sediments in metals. Although Wigry Lake is situated in a pristine region, it is still subject to anthropopressure, which seems to be the lowest in respect to Cr and Mn, while the highest in the case of Pb. The chemoecological state of the lake was ultimately assessed as good. The study highlighted the necessity of an integrated approach to the assessment of contamination or pollution in the course of an environmental research.

Keywords: lacustrine sediments; metals; geochemical mapping; spatial distribution; contamination assessment; environmental risk assessments; sediment quality guidelines

Citation: Kostka, A.; Leśniak, A. Natural and Anthropogenic Origin of Metals in Lacustrine Sediments; Assessment and Consequences—A Case Study of Wigry Lake (Poland). *Minerals* **2021**, *11*, 158. https://doi.org/10.3390/min11020158

Academic Editor: Ana Romero-Freire

Received: 30 December 2020

Accepted: 31 January 2021

Published: 3 February 2021

Publisher's Note: MDPI stays neutral with regard to jurisdictional claims in published maps and institutional affiliations.

1. Introduction

Metals are native elements, which can be found in the earth's crust, but environmental contamination or pollution is generally connected with anthropogenic activity, such as mining and smelting, agriculture, industry (especially energy production and distribution, plastic, textile, microelectronic, and wood industries) or sewage effluents [1]. Some metals (such as Co, Cu, Cr, Fe, Mg, Mn, Mo, Ni, Se, Zn) are essential for living organisms, playing the role of microelements, but become toxic while applied in higher doses [2,3]. Others seem to be neutral, or their biological role is unknown, while some elements (e.g., Hg, Pb) are considered biologically useless and harmful in all concentrations [3–5]. It is also worth noting that the biological role of some metals, previously considered only harmful, has now been recognized. An example of such an element is cadmium [6].

The term "heavy metals", although commonly used in science, is ambiguous, as it is defined in terms of different properties, such as density, atomic number, or relative atomic mass [7–10]. In environmental sciences, heavy metals are identified as toxic, ecotoxic, or hazardous elements, including some metalloids as well, e.g., arsenic. This term is also often used synonymously with "trace metals" or "trace elements" [8,9,11–13]. Ali and Khan [10] recently proposed a new definition of the term "heavy metals" as 'naturally occurring metals having an atomic number (Z) greater than 20 and an elemental density greater than 5 g·cm^{-3}'; however, this "semantic problem" still seems to be unresolved. The authors therefore decided to avoid the term "heavy metals", using instead some more neutral forms, such as metals, elements, or simply their names, as suggested by Chapman and Holzmann [14].

Soil and water environments are particularly susceptible to the accumulation of metals [3]. In aquatic environments, metals dissolved in water (as the most assimilable for living organisms) are usually quickly removed through sorption (physical, chemical, or ion exchange-based) on solid particles of suspended matter, or by precipitation and bonding in the structure of minerals. Finally, such elements are accumulated in sediments, thus becoming much less hazardous, but they may be, however, released again in the event of unfavorable environmental conditions [15–19]. The accumulation of metals in aquatic sediments depends on many variables, e.g., the geological character of the catchment and its development, or morphometric and hydrologic characteristics of a particular water body. One of the key factors is the chemical and granular composition of deposits [15], wherein the most important contributors, which facilitate the accumulation of metals in aquatic deposits, are: organic matter, clays, Fe/Mn oxides, sulfides [20–24], and the small size of sediment particles [25–28]. All of these factors also affect the bioavailability of hazardous substances bonded in the sediment matrix, including metals. Bioavailability is further dependent on physiochemical conditions (pH, redox gradient, salinity, water hardness, temperature), chemical and physical speciation of metals, some processes that may take place within the sediment (resuspension, deposition, pore water convection), route of exposure, and finally on the properties of living organisms exposed to harmful chemicals [29–35]. The most important factors, which influence the bioavailability of metals, are the potential of hydrogen (pH) and the redox potential (Eh) [35].

As the presence of metals in aquatic sediments may be considered quasi-persistent, research on their concentration in that component may be a good indicator of the chemoe-cological state of the particular water body, as well as of the historical contamination or pollution. There are many indicators and criteria for the assessment of the condition of sediments, which are used around the world. One of the first, and nowadays widely used indicators, is the geoaccumulation index (I_{geo}), defined by Müller [36]. Some other subsequently proposed indicators include, e.g., the contamination factor (C_f) and the degree of contamination (C_d) [37] or enrichment factor (EF), primarily used in order to speculate on the origin of some elements in the atmosphere, e.g., [38,39], but further adopted for soils, e.g., [40,41], sediments, e.g., [42,43], rainwater, e.g., [44,45], bioindicators, e.g., [46], and other environmental components [47]. Most of the indicators used in contamination or pollution assessment refer to some background values; therefore, the establishment and appropriate definition of the geochemical background seems to be a crucial, but also a complicated issue [48]. Background values, understood as concentrations of particular chemicals or elements naturally present in the environment, may be determined in many ways [28,47–55]. Some researchers tend to compare their data with some "globally" defined geochemical background, proposed, e.g., by Turekian and Wedephol [56], Håkanson [37], or Taylor and McLennan [57], while others consider that any geochemical background values should be determined locally. The main role of a geochemical background in environmental research is to determine whether the studied area is anthropogenically affected and whether we are dealing with the effect of contamination or pollution. Contamination of a particular environment, by some chemical, means that concentration of that substance is elevated when compared to the background value, whereas when adverse effects on biota occur, the environment is considered polluted [58]. The geochemical background does not provide any information on the influence of a given substance or element on living organisms. Therefore, sediment quality guidelines (SQGs), which define certain critical threshold concentration values of selected chemicals associated with specifically determined effects of biota, were provided. These include: threshold effect level (TEL; concentration below which adverse effects are expected to occur only rarely) and probable effect level (PEL; the concentration above which adverse effects are expected to occur frequently) proposed by Smith et al. [59]. Those limit values were further adjusted by MacDonald et al. [60] and proposed as threshold effect concentration (TEC) and probable effect concentration (PEC).

Considering all of the above, Wigry Lake seems to be a perfect water body for the analysis of environmental assessment issues, as it has been very well investigated in many aspects, densely sampled and geochemically examined in terms of 7 metals: Cd, Cr, Cu, Fe, Mn, Pb, and Zn. This article summarizes more than a decade of research conducted on this lake and focuses on the concentration of metals in sediments. The main issues of the article are as follows: (1) the description of the spatial distribution of metals in Wigry Lake sediments in a geochemical and environmental context; and (2) the assessment of contamination and evaluation of natural and anthropogenic contribution to the enrichment of recent Wigry sediments in metals.

2. Materials and Methods

2.1. Study Area

Wigry (located in north-eastern Poland, at the area of Suwałki Lakeland—Figure 1) is one of the most valuable Polish lakes, discovered for science only in the second half of the 19th century [61]. Nevertheless, the lake, as well as its vicinity, quickly became an object of interest for researchers from many fields of science and many papers were published during the next century [62–73], including even Slavonic studies [74]. In 1998, Wigry Lake, as the first in the world, was "adopted" by the International Association of Theoretical and Applied Limnology [75]. More "contemporary" research on Wigry Lake started in 1997, as a result of the passion and commitment of Professor Jacek Rutkowski, professionally associated with the Polish Geological Society, the Polish Limnological Society, and the AGH University of Science and Technology [76], in cooperation with the authorities and personnel of the Wigry National Park, as well as other scientists. During more than two decades that followed, numerous studies were published in many fields, such as hydrobiology [77–86], paleoclimatology, paleolimnology, isotopic investigations and dating [87–94], limnology and sedimentology [95–102], and finally geochemistry and environmental contamination, also with metals [103–114], which is further discussed in more detail.

Wigry is a postglacial lake, typical of northern Poland, but it differs from other water bodies in this region due to its interesting and complicated morphology with a diversified coastline pattern, numerous islands, and shallows. The lake was finally formed during Weichselian (Vistulian, Baltic) glaciation and may be divided into five main parts, characterized by diverse bathymetry and a slightly different origin (of a furrow or morainic type), which are usually separated by distinct shallows and narrowings of the lake basin (Figure 1b). Wigry is also one of the biggest (water area: 21.2 km^2; islands area: 0.68 km^2; capacity: 336.7 mln m^3) and the deepest (maximum depth: 73 m; mean depth: 15.8 m) lakes in Poland. Wigry is fed and drained mainly by waters of the Czarna Hańcza River, which flows through the northern part of the lake. Other elements of the Wigry Lake water system include several smaller lakes nearby, streams, springs, and precipitations [94,101,111]. Sediments of Wigry Lake have been extensively investigated and represent several types of deposits, but with the vast majority being carbonate sediments. Profundal (60–75% of the lake's bottom) sediments, as the most homogeneous in this regard, are represented by carbonate gyttja (CaCO$_3$: 54–87%; organic matter: 8–30%; silt fraction: 95–99%), while littoral sediments are more diverse. The most typical deposit present in the shallow parts of the lake is lacustrine chalk (CaCO$_3$: 52–98%; organic matter: 2–7%; silt fraction: 41–72%). Other littoral deposits, which are found only locally, include clastic sediment, i.e., sands and gravels (CaCO$_3$: 7–16%; organic matter: <1%; silt fraction: <1%), and organic gyttja, a specific variety of which is fluvial-lacustrine sediment present in the mouth of the Czarna Hańcza River ($CaCO_3$: 3–14%; organic matter: 10–51%; silt fraction: not determined) [95,97,98,101–103,107,110,114].

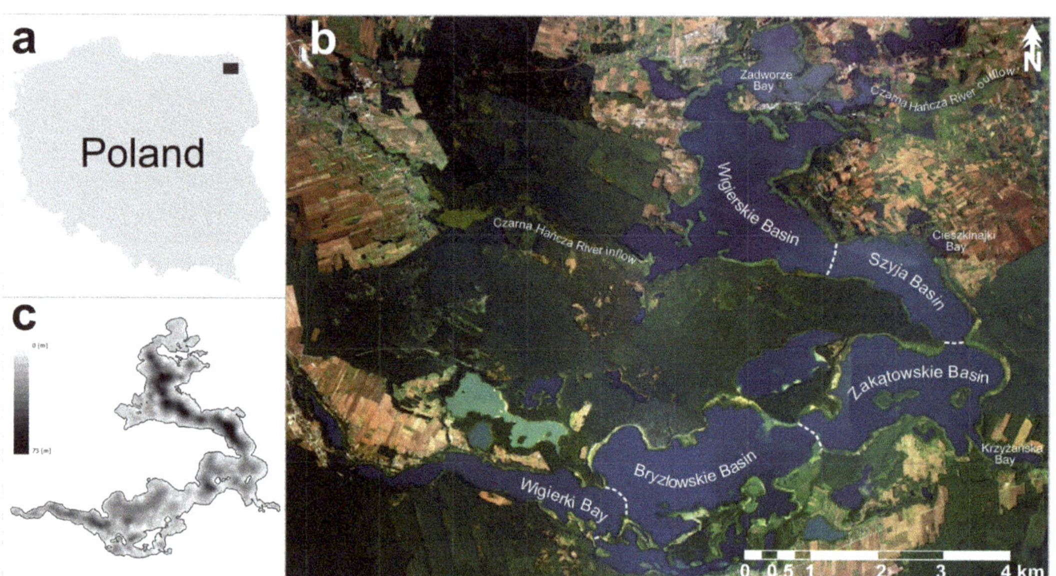

Figure 1. Study area: (**a**) location of the study area (gray rectangle) on the map of Poland; (**b**) Wigry Lake photomap (obtained courtesy of the Wigry National Park authorities) with the designation of five main parts of the lake (Wigierskie Basin, Szyja Basin, Zakątowskie Basin, Bryzglowskie Basin, and Wigierki Bay) and other important zones (Czarna Hańcza River inflow, Czarna Hańcza River outflow, Zadworze Bay, Cieszkinajki Bay, and Krzyżańska Bay); (**c**) bathymetry sketch of the Wigry Lake bottom.

2.2. Sampling and Laboratory Research

Wigry Lake sediments were sampled during summer stagnation periods during more than ten successive years (starting from 1997), ultimately resulting in the collection of over 1200 sediment cores of up to 1.5 m in length. Samples were collected from a motorboat using a gravitational sampler, designed and constructed specifically for this research [115]. Sampling point locations were determined by GPS and their depths by echo-sounder (model 381/382 by FURUNO Electronic Company, Nagasaki, Japan) or using a multi-parameter water quality monitor (model 6920 by YSI Inc., Yellow Springs, OH, USA), sometimes also using a calibrated pole (in the shallows). Adequately secured cores were then transported to the laboratory of the Wigry National Park (Krzywe, Poland), sliced into subsamples and subject to further analysis, including lithological and geochemical studies. For the purposes of metal concentration analysis, the top 0–5 cm of the cores was separated and dried and further analysis was conducted in the laboratory of the AGH University of Science and Technology (Kraków, Poland). Homogenized samples were treated with a mixture of 10 mL of 65% HNO_3 (POCH, Poland) and 2 mL of 30% H_2O_2 (Chempur, Poland) and digested in a microwave oven (model MDS 2000 by CEM Corporation, Charlotte, NC, USA). The extracted metals were determined using atomic absorption spectrometry (AAS—model SP9 by PYE UNICAM, Cambridge, UK; Fe: limit of quantification (LOQ)— 60 $\mu g \cdot L^{-1}$, Mn: LOQ—27 $\mu g \cdot L^{-1}$, Pb: LOQ—10 $\mu g \cdot L^{-1}$ and Zn: LOQ—1 $\mu g \cdot L^{-1}$) or using an inductively coupled plasma induced mass spectrometer (ICP-MS—model HP 4500 by Hewlett-Packard, Palo Alto, CA, USA; Cd: LOQ—0.16 $\mu g \cdot L^{-1}$, Cr: LOQ—0.21 $\mu g \cdot L^{-1}$ and Cu: LOQ—0.23 $\mu g \cdot L^{-1}$). In order to ensure high analytical quality, reagents were of the highest purity, reagent blanks were used, and reference material (LKSD-4) was analyzed, additionally, using inductively coupled plasma atomic emission spectrometry (ICP-AES—model Plasma 40 by Perkin-Elmer, Wellesley, MA, USA) (Table 1). The quality of the results was evaluated through the analysis of variance by the Robust statistics method [116,117] and estimated as very good.

Table 1. Concentrations of selected metals in the LKSD-4 reference material, determined using different instrumental methods [113,114].

Metal	LKSD-4 *	ICP-MS	ICP-AES	AAS
	$mg·kg^{-1}$			
Cd	1.9	2.2 ± 0.2	2.0 ± 0.2	2.0 ± 0.3
Cu	31.0	29.8 ± 1.7	28.5 ± 1.6	30.5 ± 2.6
Mn	420	412 ± 7.9	-	405 ± 12
Pb	91	101 ± 3.8	99 ± 4.5	81.8 ± 5.1
Zn	195	202 ± 9.6	200 ± 4.0	190 ± 11

* The LKSD-4 reference material is a mixture of sediments from the Big Gull Lake in Ontario and the Key and Sea Horse Lakes in Saskatchewan and its main components are: SiO_2 (41.6%), organic matter (as LOI, 500 °C) (40.8%), Al_2O_3 (5.9%), Fe_2O_3 (4.1%), and CaO (1.8%) [118].

2.3. Data Processing

Metal concentrations were determined ultimately for 162 (in case of Pb) to 459 (in case of Fe) samples (Table 2), which gave a sampling density from about 8 to 22 per km^2, respectively. All samples were identified in terms of belonging to one of five deposit types that can be found at the Wigry Lake bottom, i.e., lacustrine chalk, carbonate gyttja, fluvial-lacustrine sediment, organic gyttja, and clastic sediment (characterized briefly above). Basic statistical parameters were calculated with the use of STATISTICA software (ver. 13.3, TIBCO Software Inc., Palo Alto, CA, USA) both for particular sediment types and for all samples collectively (Table 2). The data was also statistically analyzed using principal component analysis (PCA), in order to investigate the multivariate relationship between the concentrations of metals and the type of sediments and the depth. Lead was excluded from analysis, as its concentration values were in the majority determined on a separate samples set from other metals. The PCA was therefore carried out on the basis of samples containing information on the type of sediment, depth and the concentration of other elements, i.e., Cd, Cr, Cu, Fe, Mn, and Zn, excluding those with incomplete data. This ultimately resulted in a set of 177 samples taken for PCA, which was carried out with a use of MATLAB software (ver. 8.2.0.701 (R2013b), The MathWorks Inc., Natick, MA, USA) (Figure 2).

All maps were prepared with the use of the ArcGIS Pro software (ver. 2.6.3, ESRI Inc., Redlands, CA, USA). The bathymetry map (Figure 1c) and the maps of metal spatial distribution (Figures 3a–9a) were created using an ordinary kriging procedure [119], which gave the best interpolation results for the presented data and which has been described in more detail previously [113,114]. The bathymetry map was based on depth data received during "contemporary" Wigry Lake research (after 1997), while collecting sediment samples, and gathered during the performance of seismoacoustic studies [96,99,100]. Similarly, maps of spatial distribution of metals were based on data gathered for over a decade, starting from 1997. Maps presenting Wigry Lake sediments mapped on the basis of geochemical backgrounds (Figures 3b–9b) and sediment quality guidelines (Figures 5c–9c) were based on particular metal spatial distribution maps (Figures 3a–9a) and appropriate threshold values (Table 3). In the case of GB3, upper range limits were considered for analysis.

3. Results and Discussion

Characteristic of the Wigry Lake is the fact that particular types of its sediments differ significantly in terms of their metal concentrations, and these differences (as mean values) can even be several dozen-fold (Table 2). The least enriched with the analyzed elements is lacustrine chalk (for Cd, Cr, Cu, and Fe) or clastic sediment (for Mn, Pb, and Zn), while the richest in metals is fluvial-lacustrine sediment. This observation has been emphasized in previous studies concerning the Wigry Lake [103,106,111,112], but gathered and summarized in detail only recently [113,114], which was possible due to the availability of an abundant data set.

Table 2. Basic statistics for metal concentrations in different types of Wigry Lake sediments [113,114].

Metal		Lacustrine Chalk	Carbonate Gyttja	Fluvial-lacustrine Sediment	Organic Gyttja	Clastic Sediment	All Sediments
				$mg \cdot kg^{-1}$			
Cd	n	98	149	17	4	3	271
	Min	0.003	0.010	0.160	0.120	0.006	0.003
	Max	0.557	0.870	3.060	0.620	0.633	3.060
	Mean	0.133	0.340	1.077	0.338	0.236	0.310
Cr	n	88	136	7	4	4	239
	Min	0.20	0.42	4.07	0.82	3.39	0.20
	Max	4.31	12.25	22.61	7.78	4.15	22.61
	Mean	1.48	3.69	16.17	3.63	3.73	3.24
Cu	n	102	144	7	4	4	261
	Min	0.02	0.20	12.27	3.07	0.08	0.02
	Max	8.97	26.27	59.70	10.30	6.55	59.70
	Mean	2.24	6.78	41.56	6.68	2.34	5.87
Fe	n	200	217	21	9	12	459
	Min	80	484	5863	2542	581	80
	Max	5588	10,654	32,857	15,876	3181	32,857
	Mean	983	3670	18,377	7529	1496	3191
Mn	n	200	216	21	9	12	458
	Min	18	56	142	86	28	18
	Max	206	1698	1373	372	85	1698
	Mean	94	354	518	230	51	238
Pb	n	72	64	14	5	7	162
	Min	36.2	46.8	35.6	62.3	7.0	7.0
	Max	88.1	84.7	107.5	79.4	63.8	107.5
	Mean	71.6	60.6	78.2	71.4	21.8	65.7
Zn	n	200	213	21	9	12	455
	Min	4.6	7.1	84.5	6.2	3.1	3.1
	Max	103.4	119.3	632.1	105.9	60.2	632.1
	Mean	17.8	44.6	339.2	62.7	14.7	46.0

Two types of sediments covering almost the whole bottom of the lake, i.e., carbonate gyttja and lacustrine chalk, represent nearly perfect examples of the influence of two important contributors favoring metals accumulation in aquatic environments, i.e., granular composition of sediment particles [15,25–28] and chemical composition of deposits [15,20–24]. Both of them, carbonate gyttja and lacustrine chalk, are composed mainly of silt and clay fraction, but in the case of chalk sandy and even gravel fraction may also be observed [95]; this type of deposit is then relatively coarse-grained. That is the main reason why metals are less likely to accumulate in lacustrine chalk, which has been confirmed also by some specific, detailed studies. For example, calcium carbonate, which makes up lacustrine chalk, is partly precipitated with the participation of Charophyta (a group of algae), which leads to the formation of calcite tubes of even more than 1 cm in length, and it was found that those tubes are made of almost pure $CaCO_3$ [98]. Moreover, research on pore water mineralization of Wigry Lake sediments has revealed that it is relatively low in the case of lacustrine chalk (544.5 mg/L) when compared to other types of deposits (e.g., carbonate gyttja—717.7 mg/L; fluvial-lacustrine sediment—1670.9 mg/L) [120]. Consequently, apart from lacustrine chalk, clastic sediment, as the most coarse-grained, is also the least enriched with metals. The granularity of sediment particles also explains the correlation of concentration of elements with the depth of the lake, which has been found significant for Fe (Figures 1c and 3a), Zn (Figures 1c and 8a), and especially for Mn (Figures 1c and 4a) [111,113]. The reason for that is the tendency of fine-grained sediments to accumulate in deeper areas of the lake bottom [15].

In the context of the chemical composition of Wigry Lake sediments, it has been found that the accumulation of metals is mostly facilitated by the presence of organic matter, as it acts as a sorbent for metals primarily via the formation of complexes [20,34]. Organic matter in aquatic environments comes mostly from decay of plant and animal material and is usually present in the form of humic substances. As to the sorption of metals, the most important are fine particles (of a silt and clay fraction), due to the relatively larger interaction area of a unit amount of material made up of particles with a smaller diameter. This kind of organic material comes mainly from benthic invertebrate fecal material, biofilms, and decay of aquatic macrophytes [21,34,121–124]. Profundal sediment—carbonate gyttja—is more organic-rich and metals-rich than littoral lacustrine chalk or clastic sediment. Other deposits found on low depths, such as organic gyttja, and especially fluvial-lacustrine sediment, are rich in metals when compared to lacustrine chalk or clastic sediment (Table 2), which is mainly due to their relatively high organic matter content. Metal concentrations (except for Fe) in organic gyttja are comparable with those recorded for carbonate gyttja, which is related to the specific conditions in which organic gyttja is formed. This type of sediment can be found in Cieszkinajki Bay and Krzyżańska Bay (Figure 1b), which are quite isolated from the main lake basin. This is a factor that restricts water circulation and favors eutrophication, and a symptom of that phenomenon is the reduction of $CaCO_3$ precipitation and the displacement of carbonate sediments by organic ones [98]. Finally, the most organic-rich fluvial-lacustrine sediment found at the mouth of the Czarna Hańcza River (Figure 1b) is the richest in all examined metals (Table 2, Figures 3a, 4a, 5a, 6a, 7a, 8a and 9a).

All of the above was also confirmed by the results of the principal component analysis (PCA), which revealed that the two largest variances in the data set, related to the PC1 and PC2 axes, together explain 87.5% of the variability of data set. PC1 explains 64.5% of the data variability (Figure 2a) and it coincides well with the type of sediment. PC2 explains 23.0% of the variability (Figure 2b) and is correlated to the depth of sediments. Most of the fluvial-lacustrine sediment samples form a separate cluster with high values on the PC1 axis. The remaining data show the greatest variability in the intermediate direction between PC1 and PC2, and form clusters, but less apparent. The highest values in this direction are found in the case of carbonate gyttja samples, while the smallest ones are in the case of lacustrine chalk (Figure 2a). In Figure 2b, characteristic data variability in a direction close to the PC2 axis may be observed. Samples taken from the shallows have the lowest values, while samples taken from the depths—the highest. Manganese stands out here from other metals used in PCA analysis (i.e., Cd, Cr, Cu, Fe, and Zn), as its relationship with PC1 and PC2 axes is comparable, which means that this element is the least related to the sediment type (in comparison to other metals), but the most related to the depth, which is further discussed.

The assessment of the chemoecological state of the Wigry Lake is a very interesting issue, as very extensively investigated lake sediments, abundant data set, and as many as five different geochemical background values, which could be applied in this study, allowed for a very scrupulous analysis. In this paper, the authors decided not to demonstrate some of the very commonly used indices, e.g., contamination factor (C_f), degree of contamination (C_d) [37], or the pollution load index (PLI) [125], because they have been calculated and discussed in detail before [114], for five of the seven metals presented here (Cd, Cr, Cu, Pb, and Zn), and values calculated additionally for Fe and Mn would not add anything new. The other widely used indicator—the geoaccumulation index (I_{geo}) [36]—could not be used in this study, as the index formula refers to metal concentrations in pelitic sediment fraction [126], while bulk samples have been examined in the case of Wigry Lake. Moreover, all mentioned indices are based on geochemical backgrounds; therefore, their values are strongly affected by the background values used, and those might be diversified. This was shown before [114], where the obtained assessment results for the condition of Wigry Lake sediments ranged from "uncontaminated" to "extremely polluted", depending on background values adopted for calculations. In this paper, therefore, we focus only on background values themselves. Three of them are local, while the other two are more globally defined.

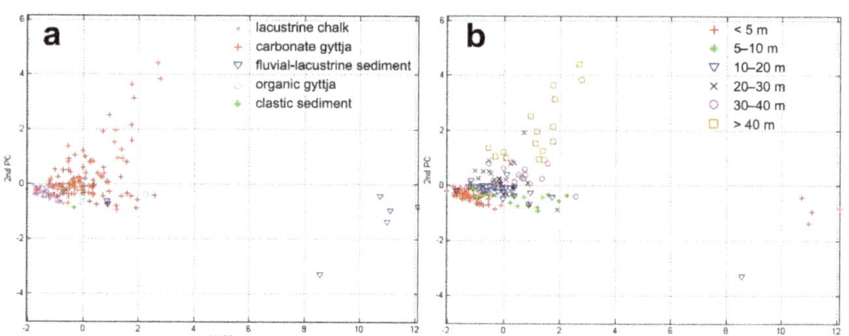

Figure 2. Principal component analysis (PCA) carried out for 177 Wigry Lake sediment samples, visualized according to sediment type (**a**) and depth (**b**).

The first of the local geochemical backgrounds (GB1) was established on the basis of results obtained for five lacustrine chalk samples, taken from the vicinity of the contemporary Wigry Lake and underlying peat dated with ^{14}C method at 7970 ± 70 years BP [103,127]. The second one (GB2) was implemented as a range of metal concentration values obtained for the bottom layer of five core samples taken from the basin of the Wigry Lake. The cores were dated and their bottom layers are known to originate from the preindustrial period [104]. The third local geochemical background (GB3) is a fairly new idea and was implemented for the first time in the case of Wigry Lake only recently [114], in accordance with the concept of Matschullat et al. [48]. The method proposed by authors assumes that the concentrations of a particulate substance naturally present in the environment should demonstrate normal distribution, and any abnormal values (usually associated with contamination or pollution) lead to the right-skewness of the distribution function. A threshold value calculated using the method called a "calculated distribution function" is defined as the median plus 2σ, where σ is the standard deviation of a normal distribution estimated on the basis of data lower than the median. GB3, defined previously for Cd, Cr, Cu, Pb, and Zn [114], could be calculated only for carbonate gyttja samples, as in case of other sediment types, data sets were too small or the method assumptions were not met. Then, consequently, GB3 for Fe and Mn was calculated here also on the basis of carbonate gyttja samples.

The next geochemical background (GB4) was adopted from the Geochemical Atlas of Poland [128], as mean values of metal concentrations in 993 sediment samples taken from various aquatic environments (e.g., rivers, streams, channels, lakes, reservoirs, ponds) in Poland. The last one—the world geochemical background (GB5)—was defined as mean concentrations of elements in carbonate rocks, according to Turekian and Wedephol [56], as carbonate sedimentation predominates in the Wigry Lake basin. All presented background values (gathered in Table 3) have advantages and disadvantages, which have been listed before [114]. Briefly, GB1 and GB2 were calculated on quite a small number of samples. Moreover, GB1 refers to lacustrine chalk, covering only up to 30% of the current lake bottom, but on the other hand, this type of sediment was dominant in the paleo-lake [93,101]. GB3 in turn refers only to carbonate gyttja, but this type of sediment covers the majority of the current lake bottom. GB4 (as based on aquatic samples of different origin and chemical and granular composition) and GB5 (as based on carbonate rocks of different origin and not being aquatic sediments) seem to be too "general". It is also worth noting that GB1–GB5 values are very divergent, differing even by several orders of magnitude. Generally, the lowest values for all metals are in the case of GB1, while the highest in the case of GB3 (for Cd, Cu, Pb, and Zn), GB2 (for Fe and Mn), or GB5 (for Cr). Nevertheless, implementing so many different geochemical backgrounds, along with the critical approach to their informative value, gives a fairly reasonable base for assessment of the chemoecological state of Wigry Lake sediments, which has been visualized on the maps (Figures 3b, 4b, 5b, 6b, 7b, 8b and 9b).

The quality of Wigry Lake sediments was also based on sediment quality guidelines, which define certain critical concentrations associated with specifically defined effects on biota, proposed by Smith et al. [59]. The guidelines determine two threshold values: the threshold effect level (TEL) and the probable effect level (PEL) adjusted by MacDonald et al. [60] and renamed to threshold effect concentration (TEC) and probable effect concentration (PEC). These two threshold values give three ranges of possible effects on biota. Below the TEL/TEC (minimal effect range)—adverse effects are expected to occur only rarely; between the TEL/TEC and PEL/PEC (possible effect range)—adverse effects may occasionally occur; above the PEL/PEC (probable effect range)—adverse effects occur frequently. It is notable that both the TEL and TEC, as well as PEL and PEC values, do not differ significantly, and all of these threshold concentrations (listed in Table 3) were implemented in the course of the present study. SQGs were used previously to assess the quality of Wigry Lake sediments [114], however here they are visualized in the form of maps for Cd, Cr, Cu, Pb, and Zn (Figures 5c, 6c, 7c, 8c and 9c); there are no values of SQGs for Fe and Mn. It should also be remembered that SQGs have some limitations too. For example, false positive assessment results (if SQGs indicate that a sediment is toxic when in fact it is not) and false negative assessment results (when SQGs suggest that a sediment is nontoxic, but, in fact, it is toxic) may occur. Moreover, SQGs do not take into account the bioavailable fraction of contaminants [58], which is known to be affected by many factors [29–35]. Nevertheless, the application of SQGs values, along with five different geochemical backgrounds, provides a solid basis for the assessment of the chemoecological state of Wigry Lake sediments. Particular metals, which display similar characteristics or significance, are analyzed further together, in order to make the discussion more clear and concise.

Table 3. Threshold values of geochemical backgrounds (GB1–GB5) and sediment quality guidelines (TEL, TEC, PEL, PEC) implemented for the assessment of the chemoecological state of Wigry Lake.

Metal	GB1 [1]	GB2 [2]	GB3 [3]	GB4 [4]	GB5 [5]	TEL [6]	TEC [7]	PEL [8]	PEC [9]
	mg·kg^{-1}								
Cd	0.003 *	-	0.631	0.500	0.035	0.596	0.99	3.53	4.98
Cr	0.99	-	6.08	5.00	11.00	37.3	43.4	90	111
Cu	0.5	<1.0–4.0	11.1	6.0	4.0	35.7	31.6	197	149
Fe	344	100–13,800	5640	1000	3800	-	-	-	-
Mn	82	53–2046	493	500	1100	-	-	-	-
Pb	0.2	4.0–8.0	70.9	10.0	9.0	35	35.8	91.3	128
Zn	4.0	2.0–36.0	73.5	48.0	20.0	123	121	315	459

* below the limit of quantification; [1] GB1—local geochemical background "a" [103,127]; [2] GB2—local geochemical background "b" [104]; [3] GB3—local geochemical background "c" [48]; [4] GB4—Polish geochemical background for aquatic sediments [128]; [5] GB5—world geochemical background for carbonates [56]; [6] TEL—threshold effect level [59]; [7] TEC—threshold effect concentration [60]; [8] PEL—probable effect level [59]; [9] PEC—probable effect concentration [60].

Spatial distribution of iron (Figure 3a) and manganese (Figure 4a) in Wigry Lake sediments, as well as the results of principal component analysis (discussed above) (Figure 2), and the coefficients of correlation between the concentrations of these metals, and the depth of the lake (0.77 and 0.60, respectively) [113] indicate that both elements have fairly similar geochemistry and, likely, origin (at least to some extent). Manganese and iron are both redox-sensitive elements [129,130] and tend to coexist in the form of oxides and hydroxides [131]. Their correlation with depth may be explained by sediment diagenesis, during which Fe and Mn undergo selective dissolution and migration in pore waters in an upwards direction [15]. This leads to the constant enrichment of the upper layer of sediments with these metals, which is the highest in the deepest parts of the lake, with the thickest layer of deposits, as has been discussed more closely before [113]. The mean values of concentrations of both of these metals are the highest in fluvial-lacustrine sediment. However, maximum amounts of Mn have been observed in the case of carbonate gyttja samples located in the deepest part of the lake—the Szyja Basin (Figure 1b,c and Figure 4a), while maximum concentrations of Fe can be found in the mouth of the Czarna Hańcza River (Figures 1b and 3a). This phenomenon may be explained by specific conditions present in the area of the mixing of Czarna Hańcza River and Wigry Lake waters, where water flow parameters and physio-chemical parameters are subject to rapid changes, which favors the precipitation of any contaminants carried out by the river [108]. A similar phenomenon was observed for example in the case of other Polish lake [27]. This observation is in line with the study on the ecological state of the Wigry National Park water system [104], as well as with the research on phosphorus concentrations and precipitation in the lake environment [132,133]. Moreover, it is known that iron and phosphorus have a tendency to coprecipitate [134]. Manganese, on the other hand, is oxidized more slowly [135,136] and can be transported over longer distances. This is in line with results obtained from the PCA analysis, which revealed that among all examined metals, Mn was the most strongly related to the PC2 axis, corresponding with the lake depth. Both metals may originate from the weathering of post-glacial rock fragments (rich in these elements), mainly igneous and metamorphic. In the petrographic composition of gravels found in the Wigry Lake and its vicinity, crystalline rocks (originating mainly from Scandinavia) constitute about 45%, which is typical for northern Poland [98,137].

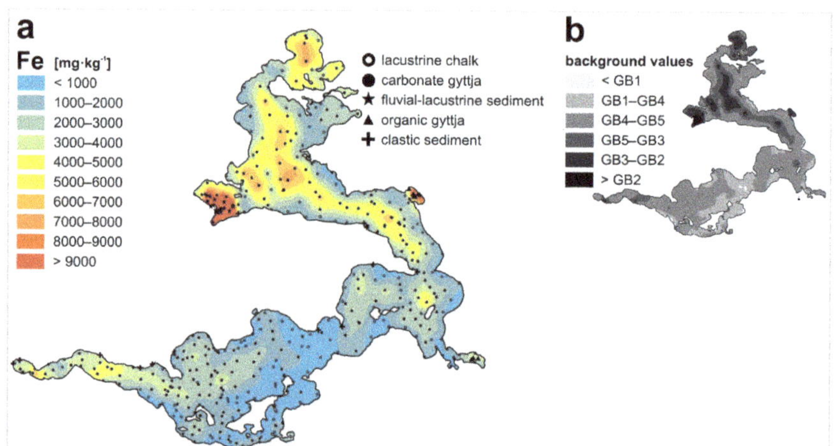

Figure 3. Iron in recent bottom sediments of the Wigry Lake: (**a**) spatial distribution of Fe; (**b**) Fe mapped according to different geochemical background values implemented in the study (Table 3).

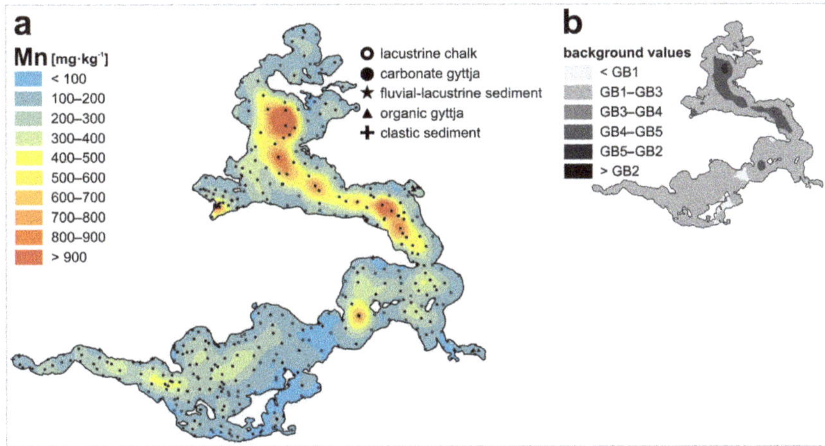

Figure 4. Manganese in recent bottom sediments of the Wigry Lake: (**a**) spatial distribution of Mn; (**b**) Mn mapped according to different geochemical background values implemented in the study (Table 3).

High concentrations of both Fe and Mn in ground waters in the area of the Wigry National Park [73,111] also suggest a mostly natural origin of these metals in lake sediments. The geochemical background values are also notable. Figures presenting Wigry Lake sediments mapped according to five different GB values (Figures 3b and 4b) show that, in both cases, deposits with metal concentrations below the GB1 value (the lowest one) can be found. In the case of manganese, most of the lake area is covered by sediments with Mn concentrations between GB1 and GB3 (the second lowest) and there are no deposits in which the content of this element exceeds the GB2 value (the highest one). It is also notable that the local geochemical background (GB2) [104] expressed in the form of a range (Table 3), and the range of Fe and Mn concentrations in recent Wigry Lake sediments (Table 2), are quite similar, which further supports the conclusion that the origin of these two elements is mostly natural, especially in the case of Mn. Moreover, GB2, which is based on fossil sediments, is much higher (taking into account the upper edge of the range) than GB3, which is based on recent sediments. It should be noted, however, that both values were established in different ways. On the other hand, a significant portion of iron seems to be delivered to the lake by its tributaries, mainly by the Czarna Hańcza River, but also by the Wiatrołuża River, which flows into the lake from the north, in the Zadworze Bay (Figures 1b and 3a). That may indicate a natural, as well as anthropogenic contribution to the enrichment of Wigry Lake sediments in Fe, as this element may originate from the natural leaching of the catchment built of postglacial rocks rich in this metal, as well as from human-related contamination of the water system in the studied region.

Spatial distribution maps of Cd (Figure 5a), Cr (Figure 6a), Cu (Figure 7a), and Zn (Figure 8a) demonstrate similarities and although they differ in details, some general accuracies may be observed. The area that is the most enriched with these metals is the area of the Czarna Hańcza River mouth (Figure 1b) and the corresponding fluvial-lacustrine sediment (Table 2). The lowest concentration of elements can be observed in the central part of the lake (Zakątowskie Basin and Bryzglowskie Basin), while the northern part (Wigierskie Basin, Szyja Basin) and, to a lesser extent, the south-west part of the lake (Wigierki Bay; Figure 1b), are more enriched with metals. It seems that similar phenomena affect the spatial distribution of these metals and their origin is similar as well, which was confirmed by correlation coefficients between particular pairs of metals, ranging from 0.55 (for Cd–Zn) to 0.85 (for Cr–Cu) [114]. These phenomena seem to be also quite consistent with those observed in the case of Fe and Mn, as spatial models of their distribution show many similarities. This observation is further supported by correlation coefficients between metal concentrations and the depth of the lake, which are lower than in the case of Fe and Mn, but still significant: Cd-depth—0.34, Cr-depth—0.46, Cu-depth—0.49, and Zn-depth—0.58 [114], as well as by the results of PCA analysis (Figure 2b).

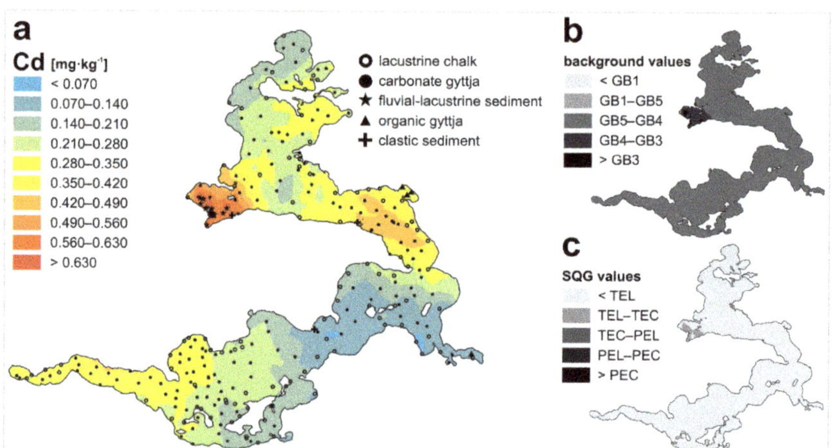

Figure 5. Cadmium in recent bottom sediments of the Wigry Lake: (**a**) spatial distribution of Cd; (**b**) Cd mapped according to different geochemical background values implemented in the study (Table 3); (**c**) Cd mapped according to sediment quality guideline values (TEL—threshold effect level, TEC—threshold effect concentration, PEL—probable effect level, PEC—probable effect concentration [59,60]; Table 3).

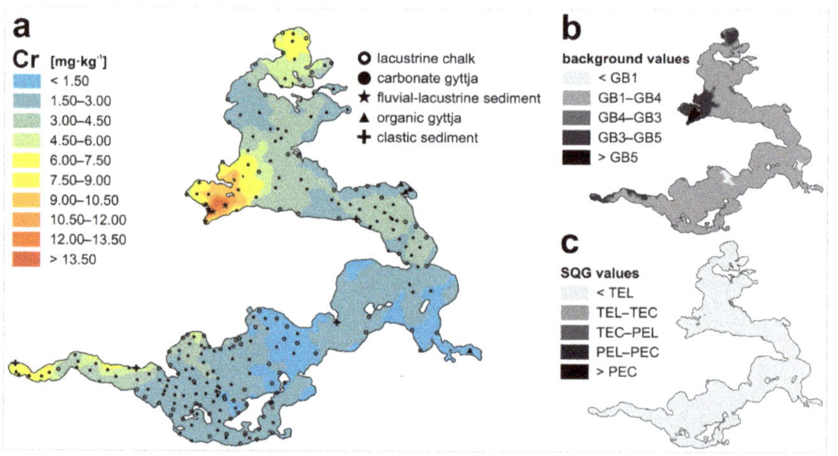

Figure 6. Chromium in recent bottom sediments of the Wigry Lake: (**a**) spatial distribution of Cr; (**b**) Cr mapped according to different geochemical background values implemented in the study (Table 3); (**c**) Cr mapped according to sediment quality guideline values (TEL—threshold effect level, TEC—threshold effect concentration, PEL—probable effect level, PEC—probable effect concentration [59,60]; Table 3).

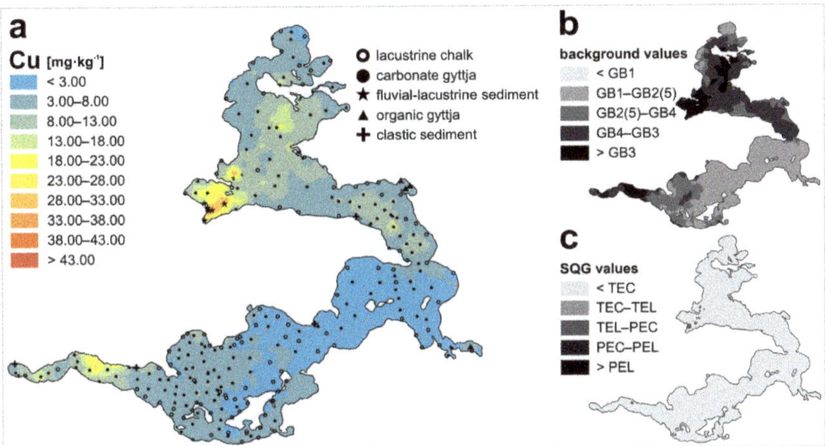

Figure 7. Copper in recent bottom sediments of the Wigry Lake: (**a**) spatial distribution of Cu; (**b**) Cu mapped according to different geochemical background values implemented in the study (Table 3); (**c**) Cu mapped according to sediment quality guideline values (TEL—threshold effect level, TEC—threshold effect concentration, PEL—probable effect level, PEC—probable effect concentration [59,60]; Table 3).

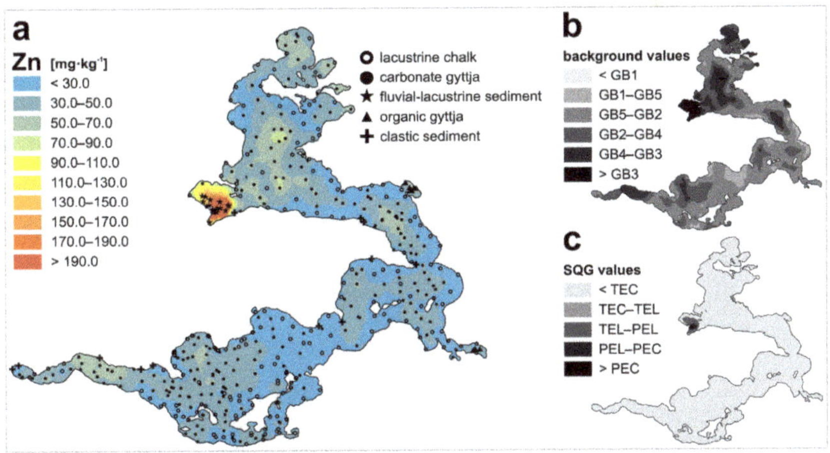

Figure 8. Zinc in recent bottom sediments of the Wigry Lake: (**a**) spatial distribution of Zn; (**b**) Zn mapped according to different geochemical background values implemented in the study (Table 3); (**c**) Zn mapped according to sediment quality guideline values (TEL—threshold effect level, TEC—threshold effect concentration, PEL—probable effect level, PEC—probable effect concentration [59,60]; Table 3).

What differentiates these four metals from Fe and Mn is their origin, which in the case of Cd, Cr, Cu, and Zn seems to be more anthropogenically affected. GB2 values (established on a basis of fossil Wigry Lake sediments [104]) for Cd and Cr were not available, however in the case Cu and Zn those values were significantly lower than the ranges of the concentration of metals measured for recent Wigry Lake sediments (Tables 2 and 3), and lower than GB3 [48]. This indicates that deposits of the contemporary Wigry Lake are significantly enriched with these metals, which was also confirmed by the studies of the sediment cores [101,104,107,110]. On the other hand, the comparison of Wigry

Lake sediments with different geochemical background values (Figures 5b, 6b, 7b and 8b) revealed that only small areas of the current lake bottom are covered by deposits with metal concentrations exceeding the highest GB values, i.e., GB3 (in case of Cd, Cu, and Zn) or GB5 (in case of Cr). This means that, although the analyzed sediments are enriched with metals, the observed contamination does not pose a serious threat to the Wigry Lake environment. This conclusion is also supported by the comparison of the sediment state with the sediment quality guidelines (Figures 5c, 6c, 7c and 8c), which showed that concentrations of Cd, Cr, Cu, and Zn in the vast majority of the area of examined lake deposits do not exceed threshold values of TEL/TEC, which means that there is no threat to biota, or this threat is very low, as those values represent the upper end of the concentrations of the minimal effect range (adverse effects are expected to occur only rarely) [59,60]. In the case of Cr, not even a single sample result was higher than TEL/TEC and sediment areas of the concentration of elements below GB1 (the lowest one) were observed. This indicates that, of those four metals, the concentrations of Cr in Wigry Lake sediments are the least hazardous and the least anthropogenically affected.

The spatial distribution model of lead (Figure 9a) differs significantly from the maps for other metals. Although the Czarna Hańcza River estuary and the northern part of the Wigry Lake are enriched with this metal, as with other elements, significantly elevated concentrations of Pb are observed also within the Bryzglowskie Basin (Figure 1b)—a part of the lake that is rather uncontaminated in terms of other metals. Mean lead concentrations in five different types of sediments demonstrate relatively little variation (Table 2), which is quite unusual in the case of Wigry Lake as well. It is also difficult to indicate any particular factor, which affects the spatial distribution of this element in sediments, and it was ultimately concluded that the main source of Pb in the Wigry Lake environment is precipitation and agriculture [111,114]. Atmospheric deposition was also identified as the main source of Pb in lake sediments, e.g., in Norway [138] and in the USA [139].

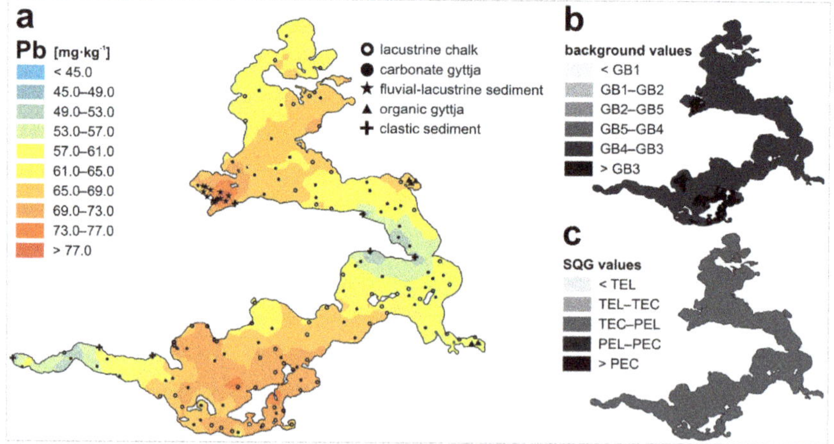

Figure 9. Lead in recent bottom sediments of the Wigry Lake: (**a**) spatial distribution of Pb; (**b**) Pb mapped according to different geochemical background values implemented in the study (Table 3); (c) Pb mapped according to sediment quality guideline values (TEL—threshold effect level, TEC—threshold effect concentration, PEL—probable effect level, PEC—probable effect concentration [59,60]; Table 3).

Due to the specific geochemical characteristics of lead, its concentrations do not correlate with other metals (though, in some cases, such correlations could not be calculated as Pb and other elements were analyzed mostly on separate data sets) [114]. What is also untypical in the case of the Wigry Lake is that lacustrine chalk is on average more enriched with Pb than carbonate gyttja (Table 2). This may be caused by the tendency of lead to be accumulated and immobilized in carbonates [104,112,135,140]. Figures presenting Wigry Lake sediment mapped according to five different geochemical backgrounds (Figure 9b) and sediment quality guidelines (Figure 9c) demonstrate that of all metals examined, lead is the most anthropogenically affected. The majority of the lake bottom is covered by sediments with lead concentrations between GB4 and GB3 (the highest one) and all deposits fall within the range between TEC and PEL values (concentrations occasionally associated with adverse biological effects) (Table 3), which indicates a moderate threat to biota [59,60]. It is notable, however, that SQGs do not take into account the real bioavailability of metals trapped in sediments, and in the case of the Wigry Lake, it has been evaluated as low [114] due to the alkaline environment of the lake sediments [104], which favors metal immobilization and the reduction of their bioavailability [141].

The enrichment of Wigry Lake sediments with metals is the effect of progressive eutrophication of its environment. Although the lake is located in one of the most pristine regions of Europe [104], agricultural activity, development of the nearby Suwałki city, the development of tourism [89] and a significant nutrient load supplied to the lake [132,133], have significantly accelerated natural processes, observed from the formation of the lake [91]. Anthropopressure was the most significant in the 1960–1990s [83,101], while nowadays, constant improvement of the Wigry Lake environment can be observed, which is mainly due to the construction of a sewage treatment plant in Suwałki in 1986 and its modernization in the 1990s.

The chemoecological state of Wigry Lake sediments was additionally compared to other lakes in north-central Poland and northern Europe. This region was chosen due to a fairly comparable climate, age, and origin of the lakes (post-glacial). It was also noted whether the compared lakes were located within relatively pristine areas. It should be remembered, however, that each environment is unique in terms of, e.g., sedimentation conditions, nature, or development of the catchment, morphometry, depth of lake basin, etc. Moreover, results are often obtained by various research methods, which has been previously pointed out [114]; therefore, such comparison may only be vague. The lakes selected in Poland included 23 lakes of the Suwałki Lakeland and Mazury Lakeland [26], Symsar Lake [27], 11 lakes of the Wielkopolski National Park [142,143], and Gopło Lake [144], while north-European lakes included 49 lakes in Latvia [145], 33 lakes in Norway [146], Lake Rõuge Liinjärv in Estonia [147], and Lake Lehmilampi in Finland [148]. The average concentrations of examined metals were lower or higher in other lakes, but generally at levels comparable to those in the Wigry Lake. It is notable, however, that in the case of the Wigry Lake, the ranges of results were much higher, which is probably due to significantly larger set of samples analyzed in the case of Wigry Lake. Moreover, when fluvial-lacustrine sediment (which, after all, occupies only a small area of the lake) was not taken into account, Wigry turned out to be nearly the most pristine of all of the compared lakes.

4. Conclusions

The presented paper summarizes a long-term multidisciplinary research conducted on Wigry Lake and its surroundings. An unusually abundant data set and a meticulous investigation of the lake environment, as well as the application of various study methods, made it possible to highlight and re-draw attention to some environmental issues, which, although commonly known, still pose significant challenges. For example, it is known that the spatial distribution of metals in lacustrine sediment is affected by numerous factors, such as chemical and granular composition of the lake sediments, local physiochemical conditions, or the character of the catchment. Thus, an extensive investigation of the deposits of a particular water body under analysis is very important and sediment samples that are taken for analysis should fairly represent the different sediment types and different depths. However, a precise investigation of a particular environment is time- and cost-consuming, which often makes it difficult or even impossible. The other important issue highlighted by the present, as well as previous studies on the Wigry Lake, is the Environmental Risk

Assessment. Although there are many ways (indices, indexes, geochemical backgrounds) that may be used to assess the chemoecological condition of aquatic sediments, each of them has its own disadvantages. Most of them are based on geochemical background values; therefore, these should be very carefully chosen. Wigry Lake is unique, also, in terms of the number of geochemical backgrounds, because as many as five different values could be applied. However, it still failed to provide a clear answer on the chemoecological state of Wigry Lake sediments, mainly due to the large variability in GB values.

Nevertheless, Wigry Lake sediments may be ultimately assessed as uncontaminated or slightly contaminated and the risk to biota can be defined as low. Spatial distribution of metals, the implementation of five different geochemical background values and the sediment quality guidelines led us to conclude that all examined elements originated from natural as well as anthropogenic sources, while their contribution was variable. Mn was found to be mostly of natural origin and was the most correlated with the depth. Concentrations of Cd, Cr, Cu, Fe, and Zn in sediments were more anthropogenically affected, but the lowest enrichment with these metals was found in the case of Cr. Pb was recognized as an element in the case of which human activity was the most significant. The natural source of the examined metals has been identified as mainly the leaching of the Wigry Lake catchment, while anthropogenic sources include agricultural activity, sewage effluents from the nearby Suwałki city and adjacent villages, and tourism. In the case of Pb, atmospheric deposition should additionally be taken into account. The chemoecological state of Wigry Lake sediments does not differ significantly from other pristine lakes in Poland or Northern Europe.

Author Contributions: Conceptualization, A.K. and A.L.; methodology, A.K. and A.L.; software, A.L.; validation, A.K; formal analysis, A.K. and A.L.; investigation, A.K.; resources, A.K.; data curation, A.K.; writing—original draft preparation, A.K.; writing—review and editing, A.K.; visualization, A.K. and A.L.; supervision, A.K.; project administration, A.K.; funding acquisition, A.L. All authors have read and agreed to the published version of the manuscript.

Funding: This research was funded by the AGH University of Science and Technology, Statutory Research grant number 16.16.140.315. The APC was funded by the AGH University of Science and Technology.

Data Availability Statement: Data is contained within the article.

Conflicts of Interest: The authors declare no conflict of interest.

References

1. Tchounwou, P.B.; Yedjou, C.G.; Patlolla, A.K.; Sutton, D.J. Heavy Metal Toxicity and the Environment. In *Molecular, Clinical and Environmental Toxicology. Experientia Supplementum*; Luch, A., Ed.; Springer: Basel, Switzerland, 2012; Volume 101, pp. 133–164. [CrossRef]
2. World Health Organization. *Trace Elements in Human Nutrition and Health*; World Health Organization: Geneva, Switzerland, 1996; Available online: https://apps.who.int/iris/handle/10665/37931 (accessed on 19 January 2021).
3. Sharma, R.K.; Agrawal, M. Biological effects of heavy metals: An overview. *J. Environ. Biol.* **2005**, *26*, 301–313.
4. Bernhoft, R.A. Mercury Toxicity and Treatment: A Review of the Literature. *J. Environ. Public Health* **2012**, 460508. [CrossRef] [PubMed]
5. Assi, M.A.; Hezmee, M.N.M.; Haron, A.W.; Sabri, M.Y.M.; Rajion, M.A. The detrimental effects of lead on human and animal health. *Vet. World* **2016**, *9*, 660–671. [CrossRef] [PubMed]
6. Dmytryk, A.; Tuhy, Ł.; Samoraj, M.; Chojnacka, K. Biological Functions of Cadmium, Nickel, Vanadium, and Tungsten. In *Recent Advances in Trace Elements*; Chojnacka, K., Saeid, A., Eds.; John Wiley & Sons Ltd.: Hoboken, NJ, USA, 2018; pp. 219–234. [CrossRef]
7. Hawkes, S.J. What Is a "Heavy Metal"? *J. Chem. Educ.* **1997**, *74*, 1374. [CrossRef]
8. Duffus, J.H. "Heavy metals"—A meaningless term? (IUPAC Technical Report). *Pure Appl. Chem.* **2002**, *74*, 793–807. [CrossRef]
9. Hübner, R.; Astin, K.B.; Herbert, R.J.H. 'Heavy metal'—time to move on from semantics to pragmatics? *J. Environ. Monit.* **2010**, *12*, 1511–1514. [CrossRef]
10. Ali, H.; Khan, E. What are heavy metals? Long-standing controversy over the scientific use of the term 'heavy metals'—proposal of a comprehensive definition. *Toxicol. Environ. Chem.* **2018**, *100*, 6–19. [CrossRef]
11. Goldsmith, R.H. Metalloids. *J. Chem. Educ.* **1982**, *59*, 526–527. [CrossRef]

12. Baldwin, D.R.; Marshall, W.J. Heavy metal poisoning and its laboratory investigation. *Ann. Clin. Biochem.* **1999**, *36*, 267–300. [CrossRef]
13. Vernon, R.E. Which Elements Are Metalloids? *J. Chem. Educ.* **2013**, *90*, 1703–1707. [CrossRef]
14. Chapman, P.M.; Holtzmann, M. Heavy metal—music, not science. *Environ. Sci. Technol.* **2007**, *41*, 6. [CrossRef]
15. Salomons, W.; Förstner, U. *Metals in the Hydrocycle*; Springer: Berlin/Heidelberg, Germany, 1984.
16. Schindler, P.W. The regulation of heavy metal concentrations in natural aquatic systems. In *Heavy Metals in the Environment*; Vernet, J.-P., Ed.; Elsevier: Amsterdam, The Netherlands, 1991; pp. 95–123.
17. Belzunce Segarra, M.J.; Helios-Rybicka, E.; Prego, R. Distribution of heavy metals in the River Ulla and its estuary (North-West Spain). *Oceanol. Stud.* **1997**, *26*, 139–152.
18. Tao, F.; Jiantong, L.; Bangding, X.; Xiaoguo, C.; Xiaoqing, X. Mobilization Potential of Heavy Metals: A Comparison between River and Lake Sediments. *Water Air Soil Pollut.* **2005**, *161*, 209–225. [CrossRef]
19. Bjerregaard, P.; Andersen, C.B.I.; Andersen, O. Ecotoxicology of Metals—Sources, Transport, and Effects on the Ecosystem. In *Handbook on the Toxicology of Metals*, 4th ed.; Gunnar, F.N., Fowler, B.A., Nordberg, M., Eds.; Academic Press: Burlington, MA, USA, 2015; pp. 425–459. [CrossRef]
20. Coquery, M.; Welbourn, P.M. The relationship between metal concentration and organic matter in sediments and metal concentration in the aquatic macrophyte *Eriocaulon septangulare*. *Water Res.* **1995**, *29*, 2094–2102. [CrossRef]
21. El Bilali, L.; Rasmussen, P.E.; Hall, G.E.M.; Fortin, D. Role of sediments composition in trace metal distribution in lake sediments. *Appl. Geochem.* **2002**, *17*, 1171–1181. [CrossRef]
22. Glasby, G.P.; Szefer, P.; Geldon, J.; Warzocha, J. Heavy-metal pollution of sediments from Szczecin Lagoon and the Gdansk Basin, Poland. *Sci. Total Environ.* **2004**, *330*, 249–269. [CrossRef] [PubMed]
23. Lim, W.Y.; Aris, A.Z.; Zakaria, M.P. Spatial Variability of Metals in Surface Water and Sediment in the Langat River and Geochemical Factors That Influence Their Water-Sediment Interactions. *Sci. World J.* **2012**, 652150. [CrossRef]
24. Salam, M.A.; Paul, S.C.; Shaari, F.I.; Rak, A.E.; Ahmad, R.B.; Kadir, W.R. Geostatistical Distribution and Contamination Status of Heavy Metals in the Sediment of Perak River, Malaysia. *Hydrology* **2019**, *6*, 30. [CrossRef]
25. Selvaraj, K.; Mohan, V.R.; Szefer, P. Evaluation of metal contamination in coastal sediments of the Bay of Bengal, India: Geochemical and statistical approaches. *Mar. Pollut. Bull.* **2004**, *49*, 174–185. [CrossRef]
26. Tylmann, W.; Łysek, K.; Kinder, M.; Pempkowiak, J. Regional Pattern of Heavy Metal Content in Lake Sediments in Northeastern Poland. *Water Air Soil Pollut.* **2011**, *216*, 217–228. [CrossRef]
27. Kuriata-Potasznik, A.; Szymczyk, S.; Skwierawski, A.; Glińska-Lewczuk, K.; Cymes, I. Heavy Metal Contamination in the Surface Layer of Bottom Sediments in a Flow-Through Lake: A Case Study of Lake Symsar in Northern Poland. *Water* **2016**, *8*, 358. [CrossRef]
28. Matys Grygar, T.; Popelka, J. Revisiting geochemical methods of distinguishing natural concentrations and pollution by risk elements in fluvial sediments. *J. Geochem. Explor.* **2016**, *170*, 39–57. [CrossRef]
29. Burton, G.A., Jr. Assessing the toxicity of freshwater sediments. *Environ. Toxicol. Chem.* **1991**, *10*, 1585–1627. [CrossRef]
30. Riba, I.; García-Luque, E.; Blasco, J.; DelValls, T.A. Bioavailability of heavy metals bound to estuarine sediments as a function of pH and salinity values. *Chem. Speciat. Bioavailab.* **2003**, *15*, 101–114. [CrossRef]
31. Ahlf, W.; Drost, W.; Heise, S. Incorporation of metal bioavailability into regulatory frameworks—metal exposure in water and sediment. *J. Soils Sediments* **2009**, *9*, 411–419. [CrossRef]
32. De Jonge, M.; Teuchies, J.; Meire, P.; Blust, R.; Bervoets, L. The impact of increased oxygen conditions on metal-contaminated sediments part I: Effects on redox status, sediment geochemistry and metal bioavailability. *Water Res.* **2012**, *46*, 2205–2214. [CrossRef]
33. Li, H.; Shi, A.; Li, M.; Zhang., X. Effect of pH, Temperature, Dissolved Oxygen, and Flow Rate of Overlying Water on Heavy Metals Release from Storm Sewer Sediments. *J. Chem.* **2013**, 434012. [CrossRef]
34. Zhang, C.; Yu, Z.; Zeng, G.; Jiang, M.; Yang, Z.; Cui, F.; Zhu, M.; Shen, L.; Hu, L. Effects of sediment geochemical properties on heavy metal bioavailability. *Environ. Int.* **2014**, *73*, 270–281. [CrossRef]
35. de Paiva Magalhães, D.; da Costa Marques, M.R.; Baptista, D.F.; Buss, D.F. Metal bioavailability and toxicity in freshwaters. *Environ. Chem. Lett.* **2015**, *13*, 69–87. [CrossRef]
36. Müller, G. Index of geoaccumulation in sediments of the Rhine River. *Geojournal* **1969**, *2*, 108–118.
37. Håkanson, L. An Ecological Risk Index for Aquatic Pollution Control. A Sedimentological Approach. *Water Res.* **1980**, *14*, 975–1001. [CrossRef]
38. Chester, R.; Stoner, J.H. Pb in Particulates from the Lower Atmosphere of the Eastern Atlantic. *Nature* **1973**, *245*, 27–28. [CrossRef]
39. Zoller, W.H.; Gladney, E.S.; Duce, R.A. Atmospheric Concentrations and Sources of Trace Metals at the South Pole. *Science* **1974**, *183*, 198–200. [CrossRef] [PubMed]
40. Barbieri, M. The Importance of Enrichment Factor (EF) and Geoaccumulation Index (Igeo) to Evaluate the Soil Contamination. *J. Geol. Geophys.* **2016**, *5*, 237. [CrossRef]
41. Jimoh, A.; Agbaji, E.B.; Ajibola, V.O.; Funtua, M.A. Application of Pollution Load Indices, Enrichment Factors, Contamination Factor and Health Risk Assessment of Heavy Metals Pollution of Soils of Welding Workshops at Old Panteka Market, Kaduna-Nigeria. *Open J. Anal. Bioanal. Chem.* **2020**, *4*, 011–019. [CrossRef]
42. Loska, K.; Cebula, J.; Pelczar, J.; Wiechuła, D.; Kwapiliński, J. Use of Enrichment, and Contamination Factors Together with Geoaccumulation Indexes to Evaluate the Content of Cd, Cu, and Ni in the Rybnik Water Reservoir in Poland. *Water Air Soil Pollut.* **1997**, *93*, 347–365. [CrossRef]

43. Malsiu, A.; Shehu, I.; Stafilov, T.; Faiku, F. Assessment of Heavy Metal Concentrations with Fractionation Method in Sediments and Waters of the Badovci Lake (Kosovo). *J. Environ. Public Health* **2020**, 3098594. [CrossRef]
44. Keresztesi, Á.; Nita, I.-A.; Birsan, M.-V.; Bodor, Z.; Pernyeszi, T.; Micheu, M.M.; Szép, R. Assessing the variations in the chemical composition of rainwater and air masses using the zonal and meridional index. *Atmos. Res.* **2020**, *237*, 104846. [CrossRef]
45. Strzebońska, M.; Gruszecka-Kosowska, A.; Kostka, A. Chemistry and Microbiology of Urban Roof Runoff in Kraków, Poland with Ecological and Health Risk Implications. *Appl. Sci.* **2020**, *10*, 8554. [CrossRef]
46. Kłos, A.; Rajfur, M.; Wacławek, M. Application of enrichment factor (EF) to the interpretation of results from the biomonitoring studies. *Ecol. Chem. Eng. S* **2011**, *18*, 171–183.
47. Reimann, C.; de Caritat, P. Distinguishing between natural and anthropogenic sources for elements in the environment: Regional geochemical survey versus enrichment factors. *Sci. Total Environ.* **2005**, *337*, 91–107. [CrossRef] [PubMed]
48. Matschullat, J.; Ottenstein, R.; Reimann, C. Geochemical background—can we calculate it? *Environ. Geol.* **2000**, *39*, 990–1000. [CrossRef]
49. Cobelo-García, A.; Prego, R. Heavy metal sedimentary record in a Galician Ria (NW Spain): Background values and recent contamination. *Mar. Pollut. Bull.* **2003**, *46*, 1253–1262. [CrossRef]
50. Reimann, C.; Filzmoser, P.; Garret, R.G. Background and threshold: Critical comparison of methods of determination. *Sci. Total Environ.* **2005**, *346*, 1–16. [CrossRef] [PubMed]
51. Gałuszka, A. Different Approaches in Using and Understanding the Term "Geochemical Background"—Practical Implications for Environmental Studies. *Pol. J. Environ. Stud.* **2007**, *16*, 389–395.
52. Gałuszka, A.; Migaszewski, Z.M. Geochemical background—an environmental perspective. *Mineralogia* **2011**, *42*, 7–17. [CrossRef]
53. Dung, T.T.T.; Cappuyns, V.; Swennen, R.; Phung, N.K. From geochemical background determination to pollution assessment of heavy metals in sediments and soils. *Rev. Environ. Sci. Biotechnol.* **2013**, *12*, 335–353. [CrossRef]
54. Bábek, O.; Matys Grygar, T.; Faměra, M.; Hron, K.; Nováková, T.; Sedláček, J. Geochemical background in polluted river sediments: How to separate the effects of sediment provenance and grain size with statistical rigour? *Catena* **2015**, *135*, 240–253. [CrossRef]
55. Xu, F.; Liu, Z.; Yuan, S.; Zhang, X.; Sun, Z.; Xu, F.; Jiang, Z.; Li, A.; Yin, X. Environmental background values of trace elements in sediments from the Jiaozhou Bay catchment, Qingdao, China. *Mar. Pollut. Bull.* **2017**, *121*, 367–371. [CrossRef]
56. Turekian, K.K.; Wedephol, K.H. Distribution of the Elements in Some Major Units of the Earth's Crust. *Geol. Soc. Am. Bull.* **1961**, *72*, 175–192. [CrossRef]
57. Taylor, S.R.; McLennan, S.M. The geochemical evolution of the continental crust. *Rev. Geophys.* **1995**, *33*, 241–265. [CrossRef]
58. Burton, G.A., Jr. Sediment quality criteria in use around the world. *Limnology* **2002**, *3*, 65–75. [CrossRef]
59. Smith, S.L.; MacDonald, D.D.; Keenleyside, K.A.; Ingersoll, C.G.; Field, L.J. A Preliminary Evaluation of Sediment Quality Assessment Values for Freshwater Ecosystems. *J. Great Lakes Res.* **1996**, *22*, 624–638. [CrossRef]
60. MacDonald, D.D.; Ingersoll, C.G.; Berger, T.A. Development and Evaluation of Consensus-Based Sediment Quality Guidelines for Freshwater Ecosystems. *Arch. Environ. Contam. Toxicol.* **2000**, *39*, 20–31. [CrossRef] [PubMed]
61. Połujański, A. *Wandering around the Augustów Governorate for Research Purposes*; Drukarnia Gazety Codziennej: Warszawa, Poland, 1859. (In Polish)
62. Kulwieć, K. Lake Wigry Physiography Materials. *Pamiętnik Fizyograficzny* **1904**, *18*, 1–42. (In Polish)
63. Lityński, A. General data on the Wigry lakes complex. *Sprawozdania Stacji Hydrobiologicznej na Wigrach* **1922**, *1*, 11–14. (In Polish)
64. Lityński, A. Limnological studies on Wigry Lake. I. Limnographic part. *Archiwum Hydrobiologii i Rybactwa* **1926**, *1*, 1–78. (In Polish)
65. Dembowska, S.; Dembowski, J. Morphometric measurements of Wigry lakes complex. 1. Uklejowa Bay and Białe Lake. *Archiwum Hydrobiologii i Rybactwa* **1922**, *1*, 15–21. (In Polish)
66. Dembowska, S.; Dembowski, J. Morphometric measurements of Wigry lakes complex. 2. Wigierki Bay. *Archiwum Hydrobiologii i Rybactwa* **1924**, *1*, 7–9. (In Polish)
67. Dembowska, S.; Dembowski, J. Morphometric measurements of Wigry lakes complex. 3. Eastern part of Wigierki Bay. *Archiwum Hydrobiologii i Rybactwa* **1927**, *2*, 160–165. (In Polish)
68. Stangenberg, M. *Chemical Composition of Deep Sediments in the lakes of Suwałki Region*; Rozprawy i Sprawozdania, Instytut Badawczy Lasów Państwowych: Warszawa, Poland, 1938. (In Polish)
69. Stangenberg, M. Nitrogen and carbon in the bottom-deposits of lakes and in the soils under carp-ponds. *Verh. Int. Ver. Limnol.* **1949**, *10*, 422–437. [CrossRef]
70. Rybak, J.I. Bottom sediments of the lakes of various trophic type. *Ekol. Pol. A* **1969**, *17*, 611–662.
71. Czeczuga, B.; Gołębiewski, Z. Ecological changes in Wigry Lake in the post-glacial period. Part I. Chemical investigations. *Polskie Archiwum Hydrobiologii* **1976**, *23*, 189–205.
72. Czeczuga, B.; Kossacka, W. Ecological changes in Wigry Lake in the post-glacial period. Part II. Investigations of the Cladoceran stratigraphy. *Polskie Archiwum Hydrobiologii* **1977**, *24*, 259–277.
73. Woroniecka-Stasiak, A. Chemical composition of interstitial waters in bottom sediments of some Polish lakes of the Wigry group (Northern Poland). *Acta Hydrobiol.* **1980**, *22*, 347–360.
74. Falk, K.O. *Wigry and Huta Waters–Toponomastic Study*; Almqvist & Wiksell: Uppsala, Sweden, 1941. (In Polish)

75. Kamiński, M. Lake Wigry, the lake "adopted" by International Association of Theoretical and Applied Limnology (SIL "Lake Adoption" Project). *Pol. J. Ecol.* **1999**, *47*, 215–224.
76. Król, K. Obituary—Prof. Jacek Rutkowski (1934–2016). *Geochronometria* **2017**, *44*, 1-1.
77. Niewolak, S. The Evaluation of the Contamination Degree and the Sanitary and Bacteriological State of the Waters in the Czarna Hańcza River in the Region of Suwałki and Wigry National Park. *Pol. J. Environ. Stud.* **1998**, *7*, 229–241.
78. Niewolak, S. Evaluation of Pollution and the Sanitary-Bacteriological State of Lake Wigry, Poland. Part I. Pelagic Waters of Lake Wigry. *Pol. J. Environ. Stud.* **1999**, *8*, 89–100.
79. Niewolak, S. Evaluation of Pollution and the Sanitary-Bacteriological State of Lake Wigry, Poland. Part II. Near-shore Waters of Lake Wigry. *Pol. J. Environ. Stud.* **1999**, *8*, 169–177.
80. Niewolak, S. Bacteriological Monitoring of Lake Water in Wigry National Park in the Summer. *Pol. J. Environ. Stud.* **1999**, *8*, 231–249.
81. Niewolak, S. Bacteriological Monitoring of River Water Quality in the North Area of Wigry National Park. *Pol. J. Environ. Stud.* **2000**, *9*, 291–299.
82. Niewolak, S.; Opieka, A. Potentially Pathogenic Microorganisms in Water and Bottom Sediments in the Czarna Hańcza River. *Pol. J. Environ. Stud.* **2000**, *9*, 183–194.
83. Niewolak, S. The Evaluation of the Degree of Pollution and Sanitary-Bacteriological State of Surface Water in Wigry Lake, North-East Poland. Part III. Waters of Hanczanska Bay and the Areas Adjoing Wigry Lake. *Pol. J. Environ. Stud.* **2001**, *10*, 167–174.
84. Alexandrowicz, W.P. The malacofauna of the upper Holocene lacustrine sediments of Wigry Lake (N Poland). *Folia Malacol.* **2000**, *8*, 141–149. [CrossRef]
85. Brzeziński, T.; Kołodziejczyk, A. Distribution of *Potamopyrgus antipodarum* (Gray, 1843) in waters of the Wigry National Park (NE Poland) and the effect of selected habitat factors on its occurrence. *Folia Malacol.* **2001**, *9*, 125–135. [CrossRef]
86. Czeczuga, B.; Kozłowska, M. Fertility of *Eudiaptomus*, *Bosmina* and *Daphnia* (Crustacea) Representatives in Lake of Varied Trophic States in the Suwałki District. *Pol. J. Environ. Stud.* **2002**, *11*, 23–31.
87. Pawlyta, J.; Pazdur, A.; Piotrowska, N.; Poręba, G.; Sikorski, J.; Szczepanek, M.; Król, K.; Rutkowski, J.; Hałas, S. Isotopic investigations of uppermost sediments from Lake Wigry (NE Poland) and its environment. *Geochronometria* **2004**, *23*, 71–78.
88. Kupryjanowicz, M. Postglacial Development of Vegetation in the Vicinity of the Wigry Lake. *Geochronometria* **2007**, *27*, 53–66. [CrossRef]
89. Paprocka, A. Stable Carbon and Oxygen Isotopes in Recent Sediments of Lake Wigry, NE Poland: Implications for Lake Morphometry and Environmental Changes. *Terrestrial Ecol.* **2007**, *1*, 267–281. [CrossRef]
90. Piotrowska, N.; Hajdas, I.; Bonani, G. Construction of the Calendar Timescale for Lake Wigry (NE Poland) Sediments on the Basis of Radiocarbon Dating. *Radiocarbon* **2007**, *49*, 1133–1143. [CrossRef]
91. Zawisza, E.; Szeroczyńska, K. The Development History of Wigry Lake as Shown by Subfossil Cladocera. *Geochronometria* **2007**, *27*, 67–74. [CrossRef]
92. Drzymulska, D.; Żurek, S. Subfossil vegetation of near-shores mires from the vicinity of the Wigry Lake (NE Poland). In Proceedings of the 12th International Palynological Congress (IPC-12), Bonn, Germany, 30 August–5 September 2008; pp. 70–71.
93. Rutkowski, J.; Aleksander-Kwaterczak, U.; Prosowicz, D.; Żurek, S.; Piotrowska, N.; Krzysztofiak, L. On paleo-Wigry Lake (NE Poland) and its sediments covered contemporary by peats. In *Anthropogenic and Natural Transformations of Lakes*; Bajkiewicz-Grabowska, E., Borowiak, D., Eds.; Department of Limnology, University of Gdańsk, Polish Limnological Society: Gdańsk, Poland, 2008; Volume 2, pp. 167–169.
94. Drzymulska, D.; Fiłoc, M.; Kupryjanowicz, M. Reconstruction of landscape paleohydrology using the sediment archives of three dystrophic lakes in northeastern Poland. *J. Paleolimnol.* **2014**, *51*, 45–62. [CrossRef]
95. Rutkowski, J.; Król, K.; Krzysztofiak, L.; Prosowicz, D. Recent sediments of the Wigry Lake (Bryzgiel Basin). *Limnol. Rev.* **2002**, *2*, 353–362.
96. Rutkowski, J.; Rudowski, S.; Pietsch, K.; Król, K.; Krzysztofiak, L. Sediments of Lake Wigry (NE Poland) in the light of high-resolution seismic (seismoacoustic) survey. *Limnol. Rev.* **2002**, *2*, 363–371.
97. Rutkowski, J.; Król, K.; Krzysztofiak, L.; Prosowicz, D. Recent sediments of Wigry Lake (Szyja Basin), NE Poland. *Limnol. Rev.* **2003**, *3*, 197–203.
98. Rutkowski, J. Sediments of Wigry Lake. *Rocznik Augustowsko-Suwalski* **2004**, *IX*, 19–37. (In Polish)
99. Rutkowski, J.; Pietsch, K.; Król, K.; Rudowski, S.; Krzysztofiak, L. High-resolution seismic survey in the Wigry Lake (NE Poland). *Peribalticum* **2005**, *9*, 147–162.
100. Osadczuk, A.; Rutkowski, J.; Krzysztofiak, L. Variability of southern part of Lake Wigry bottom (NE Poland) in the light of survey using the Rox-Ann acoustic system. *Prace Komisji Paleogeografii Czwartorzędu PAU* **2006**, *III*, 179–185. (In Polish)
101. Rutkowski, J.; Król, K.; Szczepańska, J. Lithology of the Profundal Sediments in Słupiańska Bay (Wigry Lake, NE Poland)—Introduction to Interdisciplinary Study. *Geochronometria* **2007**, *27*, 47–52. [CrossRef]
102. Rutkowski, J.; Prosowicz, D.; Aleksander-Kwaterczak, U.; Krzysztofiak, L. Profundal sediments of Wigry Lake (NE Poland). In *Anthropogenic and Natural Transformations of Lakes*; Bajkiewicz-Grabowska, E., Borowiak, D., Eds.; Department of Limnology, University of Gdańsk, Polish Limnological Society: Gdańsk, Poland, 2008; Volume 2, pp. 171–173.
103. Prosowicz, D.; Helios-Rybicka, E. Trace metals in recent bottom sediments of Lake Wigry (Bryzgiel Basin). *Limnol. Rev.* **2002**, *2*, 323–332.

104. Migaszewski, Z.M.; Gałuszka, A.; Pasławski, P. Baseline versus background concentrations of trace elements in sediments of Lake Wigry, NE Poland. *Limnol. Rev.* **2003**, *3*, 165–172.

105. Migaszewski, Z.M.; Gałuszka, A.; Pasławski, P. The use of the barbell cluster ANOVA design for the assessment of environmental pollution: A case study, Wigierski National Park, NE Poland. *Environ. Pollut.* **2005**, *133*, 213–223. [CrossRef] [PubMed]

106. Prosowicz, D. Trace Metals and Their Chemical Forms of Bonding in Bottom Sediments of Lake Wigry. Ph.D. Thesis, AGH University of Science and Technology, Kraków, Poland, 2006. (In Polish).

107. Aleksander-Kwaterczak, U.; Prosowicz, D. Distribution of Cd and Pb in the lake sediments cores from the Hańczańska Bay (Wigry Lake, NE Poland). *Limnol. Rev.* **2007**, *7*, 215–219.

108. Helios-Rybicka, E.; Kostka, A. Distribution of Fe, Mn and Zn in Bottom Sediments of Wigierskie Basin—Wigry Lake (NE Poland). *Pol. J. Environ. Stud.* **2007**, *16*, 162–167.

109. Kostka, A.; Prosowicz, D.; Helios-Rybicka, E. Use of GIS for Spatial Analysis of Heavy Metals Distribution in the Bottom Sediments of Wigry Lake (NE Poland). *Pol. J. Environ. Stud.* **2008**, *17*, 307–312.

110. Aleksander-Kwaterczak, U.; Prosowicz, D.; Rutkowski, J.; Szczepańska, J. Changes of Selected Micro-Pollution Concentrations in the Long Carbonate Sediment Cores of the Southern Part of Wigry Lake (NE Poland). *Pol. J. Environ. Stud.* **2009**, *18*, 51–55.

111. Kostka, A. Assessment of Spatial Variability of Trace Metals in Bottom Sediments of Lake Wigry Using GIS Techniques. Ph.D. Thesis, AGH University of Science and Technology, Kraków, Poland, 2009. (In Polish).

112. Aleksander-Kwaterczak, U.; Kostka, A. Lead in the environment of Lake Wigry (NE Poland). *Limnol. Rev.* **2011**, *11*, 59–68. [CrossRef]

113. Kostka, A.; Leśniak, A. Spatial and geochemical aspects of heavy metal distribution in lacustrine sediments using the example of Lake Wigry (Poland). *Chemosphere* **2020**, *240*, 124879. [CrossRef]

114. Aleksander-Kwaterczak, U.; Kostka, A.; Leśniak, A. Multiparameter assessment of select metal distribution in lacustrine sediments. *J. Soils Sediments* **2021**, *21*, 512–529. [CrossRef]

115. Rutkowski, J. Simple gravity sampler for taking cores from lake sediments. In *Anthropogenic and Natural Transformation of Lakes*; Kubiak, J., Bajkiewicz-Grabowska, E., Eds.; Agricultural University of Szczecin, Polish Limnological Society: Szczecin, Poland, 2007; Volume 1, pp. 108–109.

116. Ramsey, M.H.; Thompson, M.; Hale, M. Objective evaluation of precision requirements for geochemical analysis using robust analysis of variance. *J. Geochem. Explor.* **1992**, *44*, 23–36. [CrossRef]

117. Ramsey, M.H. Sampling and analytical quality control (SAX) for improved error estimation in the measurement of Pb in the environment using robust analysis of variance. *Appl. Geochem.* **1993**, *8*, 149–153. [CrossRef]

118. *LKSD-1 to LKSD-4 Lake Sediment Samples*; CCRMP, CANMET Mining and Mineral Sciences Laboratories: Ottawa, ON, Canada, 2003. Available online: https://www.nrcan.gc.ca/sites/www.nrcan.gc.ca/files/mineralsmetals/pdf/mms-smm/tect-tech/ccrmp/cer-cer/lksd-1-eng.pdf (accessed on 19 January 2021).

119. Burrough, P.A.; McDonnell, R.A. *Principles of Geographical Information Systems*; Oxford University Press: Oxford, UK, 1998.

120. Aleksander-Kwaterczak, U.; Zdechlik, R. Hydrogeochemical characteristics of interstitial water and overlying water in the lacustrine environment. *Environ. Earth Sci.* **2016**, *75*, 1352. [CrossRef]

121. Stumm, W.; Baccini, P. Man-Made Chemical Perturbation of Lakes. In *Lakes: Chemistry, Geology and Physics*; Lerman, A., Ed.; Springer: New York, NY, USA, 1978; pp. 91–126.

122. Helios-Rybicka, E. Phase-specific bonding of heavy metals in sediments of the Vistula River, Poland. *Appl. Geochem.* **1993**, *2*, 45–48. [CrossRef]

123. Helios-Rybicka, E.; Jędrzejczyk, B. Preliminary studies on mobilisation of copper and lead from contaminated soils and readsorption on competing sorbents. *Appl. Clay Sci.* **1995**, *10*, 259–268. [CrossRef]

124. Prego, R.; Bezlunce Segarra, M.J.; Helios-Rybicka, E.; Barciela, M.C. Cadmium, manganese, nickel and lead contents in surface sediments of the lower Ulla River and its estuary (northwest Spain). *Bol. Inst. Esp. Oceanogr.* **1999**, *15*, 495–500.

125. Tomlinson, D.L.; Wilson, J.G.; Harris, C.R.; Jeffery, D.W. Problems in the assessment of heavy-metals levels in estuaries and the formation of a pollution index. *Helgol. Meeresunters.* **1980**, *33*, 566–575. [CrossRef]

126. Förstner, U.; Ahlf, W.; Calmano, W.; Kersten, M. Sediment criteria development. Contributions from environmental geochemistry to water quality management. In *Sediments and Environmental Geochemistry: Selected Aspects and Case Histories*; Heling, D., Rothe, P., Förstner, U., Stoffers, P., Eds.; Springer: Berlin/Heidelberg, Germany, 1990; pp. 331–338. [CrossRef]

127. Król, K. Preliminary results of lacustrine chalk survey from Lake Wigry. *Sprawozdania z Czynności i Posiedzeń PAU* **1998**, *LXI*, 112–114. (In Polish)

128. Lis, J.; Pasieczna, A. *Geochemical Atlas of Poland on a scale 1:2,500,000*; Państwowy Instytut Geologiczny: Warszawa, Poland, 1995. (In Polish)

129. Balistrieri, L.S.; Murray, J.W.; Paul, B. The cycling of iron and manganese in the water column of Lake Sammamish, Washington. *Limnol. Oceanogr.* **1992**, *37*, 510–528. [CrossRef]

130. Schaller, T.; Wehrli, B. Geochemical-Focusing of Manganese in Lake Sediments—An Indicator of Deep-Water Oxygen Conditions. *Aquat. Geochem.* **1997**, *2*, 359–378. [CrossRef]

131. Azzam, A.M.; Elatrash, A.M.; Ghattas, N.K. The Co-Precipitation of Manganese by Iron(III) Hydroxide. *J. Radioanal. Chem.* **1969**, *2*, 255–262. [CrossRef]

132. Stawecki, K.; Zdanowski, B.; Dunalska, J. Seasonal changes in phosphorous concentrations in the waters of Lake Wigry. *Limnol. Rev.* **2003**, *3*, 217–222.

133. Zdanowski, B. Precipitation of phosphorus in the zone of river and lake water mixing: R. Hańcza and lake Wigry (North-East Poland). *Pol. J. Ecol.* **2003**, *51*, 143–154.
134. Santoro, V.; Martin, M.; Persson, P.; Lerda, C.; Said-Pullicino, D.; Magnacca, G.; Celi, L. Inorganic and organic P retention by coprecipitation during ferrous iron oxidation. *Geoderma* **2019**, *348*, 168–180. [CrossRef]
135. Hamilton-Taylor, J.; Davison, W. Redox-driven cycling of trace elements in lakes. In *Physics and Chemistry of Lakes*; Lerman, A., Imboden, D., Gat, J., Eds.; Springer: Berlin/Heidelberg, Germany, 1995; pp. 217–263.
136. Łącka, B.; Starnawska, E.; Kuźniarski, M.; Chróst, L. Mineralogy and geochemistry of the Lake Gościąż Holocene sediments. In *Lake Gościąż, Central Poland. A Monographic Study*; Ralska-Jasiewiczowa, M., Goslar, T., Madeyska, T., Starkel, L., Eds.; W. Szafer Institute of Botany, Polish Academy of Sciences: Kraków, Poland, 1998; pp. 196–202.
137. Ber, A. Stratigraphy of the Quaternary of the Suwałki Lakeland and its substrate based on recent data. *Kwart. Geol.* **1989**, *33*, 463–478.
138. Fjeld, E.; Rognerud, S.; Steinnes, E. Influence of Environmental Factors on Heavy Metal Concentration in Lake Sediments in Southern Norway Indicated by Path Analysis. *Can. J. Fish. Aquat. Sci.* **1994**, *51*, 1708–1720. [CrossRef]
139. Graney, J.R.; Halliday, A.N.; Keeler, G.J.; Nriagu, J.O.; Robbins, J.A.; Norton, S.A. Isotopic record of lead pollution in lake sediments from the northeastern United States. *Geochim. Cosmochim. Acta* **1995**, *59*, 1715–1728. [CrossRef]
140. Bojakowska, I.; Krasuska, J.; Retka, J.; Wiłkomirski, B. Lead concentration and the content of selected macroelements in lake sediments in Poland. *J. Elem.* **2014**, *3*, 627–636. [CrossRef]
141. Gäbler, H.-E. Mobility of heavy metals as a function of pH of samples from an overbank sediment profile contaminated by mining activities. *J. Geochem. Explor.* **1997**, *58*, 185–194. [CrossRef]
142. Zerbe, J.; Sobczyński, T.; Elbanowska, H.; Siepak, J. Speciation of Heavy Metals in Bottom Sediments of Lakes. *Pol. J. Environ. Stud.* **1999**, *8*, 331–339.
143. Sobczyński, T.; Siepak, J. Speciation of Heavy Metals in Bottom Sediments of Lakes in the Area of Wielkopolski National Park. *Pol. J. Environ. Stud.* **2001**, *10*, 463–474.
144. Juśkiewicz, W.W.; Marszelewski, W.; Tylmann, W. Differentiation of the concentration of heavy metals and persistent organic pollutants in lake sediments depending on the catchment management (Lake Gopło case study). *Bull. Geogr. Phys. Geogr.* **2015**, *8*, 71–80.
145. Kļaviņš, M.; Briede, A.; Kļaviņa, I.; Rodinov, V. Metals in sediments of lakes in Latvia. *Environ. Int.* **1995**, *21*, 451–458. [CrossRef]
146. Rognerud, S.; Hongve, D.; Fjeld, E.; Ottesen, R.T. Trace metal concentrations in lake and overbank sediments in southern Norway. *Environ. Geol.* **2000**, *39*, 723–732. [CrossRef]
147. Lepane, V.; Varvas, M.; Viitak, A.; Alliksaar, T.; Heinsalu, A. Sedimentary record of heavy metals in Lake Rõuge Liinjärv, southern Estonia. *Estonian J. Earth Sci.* **2007**, *56*, 221–232. [CrossRef]
148. Augustsson, A.; Peltola, P.; Bergbäck, B.; Saarinen, T.; Haltia-Hovi, E. Trace metal and geochemical variability during 5500 years in the sediment of Lake Lehmilampi, Finland. *J. Paleolimnol.* **2010**, *44*, 1025–1038. [CrossRef]

Article

Environmental Assessment of Trace Metals in San Simon Bay Sediments (NW Iberian Peninsula)

A.M. Ramírez-Pérez [1],*, M.A. Álvarez-Vázquez [2] , E. De Uña-Álvarez [2] and E. de Blas [1,3]

[1]   Departamento Bioloxía Vexetal e Ciencia do Solo, Universidade de Vigo, 36310 Vigo, Spain; eblas@uvigo.es
[2]   GEAAT Group (Group of Studies of Archeology, Antiquity and Territory), Department of History, Art and Geography, Area of Physical Geography, University of Vigo, 36310 Vigo, Spain; mianalva@uvigo.es (M.A.Á.-V.); edeuna@uvigo.es (E.D.U.-Á.)
[3]   CITACA (Agri-Food Research and Transfer Cluster, Campus Auga), University of Vigo, 36310 Vigo, Spain
*   Correspondence: alexandra@uvigo.es

Received: 5 August 2020; Accepted: 17 September 2020; Published: 19 September 2020

Abstract: A gravity core (220 cm depth) was collected to investigate the geochemistry, enrichment, and pollution of trace metals in anoxic sediments from San Simon Bay, an ecosystem of high biological productivity in the northwest of Spain. A five-step sequential extraction procedure was used. The Cu, Pb, and Zn contents decreased with depth, with maximum values in the top layers. Ni and Zn were bound to pyrite fractions, while Cd and Pb were associated with the most mobile fractions. The analyzed metals were associated with the fractions bound to organic matter, mainly with the strongly bound to organic matter fraction. High Cd and Cu values were observed. The fractionation showed a high mobility for Cd (28.3–100%) and Pb (54.0–70.2%). Moreover, the pollution factor and the geoaccumulation index reflected a high contamination for Pb and a moderate contamination for Cu and Zn in the superficial layers, pointing to a possible ecotoxicological risk to organisms in San Simon Bay.

Keywords: marine sediments; trace metals; speciation; contamination; San Simon Bay

---

1. Introduction

Trace metals occur naturally in the environment, differing in contents along the Earth's crust [1,2], but presenting a surprisingly consistent background in sediments [3]. Human activities can drastically increase their contents, mainly from agricultural, urban, and industrial land use [4,5]. Industrial and mining activities contribute to 48% of the total release of contaminants in Europe [6]. The human delivery of some trace metals to the coastal environment, putting coastal ecosystems under stress, represents a current issue of concern. This is due to their persistence, potential toxicity, mobility, and ability to be incorporated and accumulated in food chains, with important implications for risk assessment, public health, and contamination management [7–11]. Trace metal pollution is related to a deterioration of water quality and/or accumulation in plants and animals [12–14], being incorporated into sediments as one of their reservoirs [15,16].

The distribution of trace metals in the different sediment fractions, their capacity for complexation, and the magnitude of the contamination-related risks are conditioned by the intrinsic properties of both metals and sediments, as well as their, mobilization in interstitial water and other factors (salinity, bioturbation, redox or pH). Metal accumulation in sediments and the determination of their natural and anthropogenic sources makes sedimentary records good environmental indicators and a powerful tool by which evaluate geochemical changes, reconstructing the human–nature relationship over time, particularly in coastal areas [17–19]. With the purpose of assessing trace metals contents in sediments and their contamination degree, this study considers the distribution of Cadmium (Cd), Copper (Cu),

Nickel (Ni), Lead (Pb), and Zinc (Zn) in a gravity core (220 cm depth) retrieved from San Simon Bay (NW of the Iberian Peninsula). The bay is a Special Area of Conservation (SAC; included in the Natura 2000 Network), with high natural values affected by intense human pressure. It is a suitable site to study natural and anthropogenic dynamics from pre-industrial society to today.

Numerous studies have focused on determining the contamination of the Galician Rias (e.g., [20–24]), mainly in San Simon Bay (e.g., [25–27]). Some studies have analyzed the metal fractionation [21,25,28,29]. However, these studies cover only the inner or middle zones of the Ria of Vigo, analyzing the uppermost part of the sedimentary sequence (<100 cm deep). The objective of this work was to analyze the trace metal contents in the San Simon Bay sediments in order to conduct an environmental risk assessment to provide relevant information for environmental protection and future management.

2. Materials and Methods

2.1. Study Area

The San Simon Bay, with a length of 8 km and a maximum width of 3.6 km, has a surface area of 19.4 km^2 with an average water depth < 5 m (Figure 1). The San Simon Bay can be considered as an estuary, according to its hydrology and sedimentology [30]. The orientation and morphology of this area cause low-energy hydrodynamics, which lead to the development of intertidal flats and marshes [24,26]. The sedimentary environments in the bay are controlled by tidal processes, being in the medium tidal range of 2.2 m. The upwelling process provides cold waters rich in nutrients, causing San Simon Bay (as the Ria of Vigo) to have a high productivity. According to [31], the primary production values range between 0.05 and 2.8 g C/m^2 day. The high productivity increase the flow of organic matter to the seabed, having sediments with total organic carbon (TOC) average values from 7% to 10% [32]. The TOC/TN (total nitrogen) ratio is from 11 to 21 [33], indicating that the organic matter is of a continental origin [34]. The sediments are fine-grained and the pH values are close to neutral or slightly alkaline. The sedimentation rates ranged between 3 and 6 mm y^{-1} [27]. The high sedimentation rate, together with the high biological productivity, favor the development of anoxic conditions in the sediment and biogenic methane production [33,35]. Shallow gas accumulations have been observed in the Holocene sedimentary record [36], delimiting the sulfate-methane transition zone (SMTZ) between an 80 and 100 cm depth in the inner zone of Ria of Vigo [35].

Intertidal areas are usually sinks for trace metals. Anoxic conditions, pH values near to neutrality, and high TOC contents in the San Simon Bay sediment could determine the speciation of trace metals and the behavior in these environments.

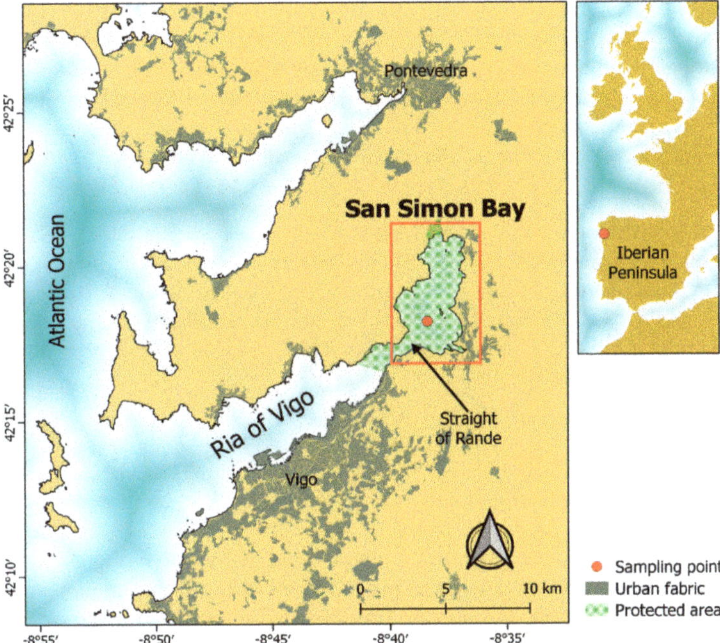

Figure 1. Location of San Simon Bay in the Ria of Vigo (NW Iberian Peninsula) and the position of the sampling point. Map elaborated with QGIS 3.4 with layers from the Spanish National Geographic Institute (©IGN) and the Directorate General of Nature Conservation of the Ministry of the Environment, Territory and Infrastructures s (©Xunta de Galicia).

2.2. Sampling and Analysis

In order to characterize the behavior and trace metals distribution in sediments with high organic matter contents, a gravity core (220 cm depth) was retrieved in the subtidal area of San Simon Bay (Latitude 42.302972°; Longitude −8.638778°, Figure 1) in November 2012, on board the R/V Mytillus. This core showed Eh values <−174 mV, and the pH ranged between 6.8 and 7.9. The sediments are fine-grained (48% and 46% on average of silt and clay, respectively). The TOC values varied from 4% to 15% [37].

The core was promptly sampled in the ship's laboratory in order to minimize the oxidation of the sampled material. The core was sliced into 2.5 cm segments and frozen at −18 °C until chemical analysis. For the analysis of this work, 1 out of 4 samples (0–2.5 cm; 10–12.5 cm, and so on) were used.

The samples were analyzed in duplicate after the separation of pore water by centrifugation for 45 min at 4000 rpm. The percent wetness of the samples was calculated in order to express the final results in dry weight. The partitioning of the solid-phase trace metal contents was performed using the sequential extraction procedure proposed by [38]. This procedure allows separate five phases according to the nomenclature [37,39]:

- Metals present in the ion-exchangeable form and bound to carbonate (F1), were obtained by adding 40mL of 1M ammonium acetate ($NH_4CH_3CO_2$) adjusted to pH 5 with acetic acid to a 0.5 g sample under stirring for 24 h at room temperature.
- Metals present in the reductive phase bound to manganese-iron oxides (F2), obtained with 20 mL of 1M hydroxylammonium chloride ($HONH_2 \cdot HCl$) and 20 mL of acetic acid (CH_3COOH, 25%) after 24 h of shaking at room temperature.

- Metals weakly bound to organic matter (F3), obtained by adding 20 mL of 0.1 M HCl and stirring for 24 h at room temperature.
- Metals strongly bound to organic matter (F4), obtained by adding 20 mL of 0.5 M NaOH under stirring for 24 h at room temperature. For sediment samples with high organic carbon content, this treatment should be repeated until a clear solution is obtained. All the solutions obtained from the solids were dried at 55 °C and digested using 8 mL of HNO_3 (65%) at 115 °C for 30 min.
- Metals bound to sulfide (F5), was obtained with 20 mL of 8 M HNO_3 and digested for 3 h at 85 °C.

All the residues were washed with 5 mL of distilled water and the washings were added to the solution. The trace metals contents were measured by inductively coupled plasma optical emission spectrometry (ICP-OES) with a Perkin Elmer Optima 4300 DV instrument. The reagents for analysis were ISO grade and ACS. Calibration solutions were prepared in deionized water from stock solution. The detection limits were: Cd: 0.005 mg L^{-1}, Cu: 0.010 mg L^{-1}, Ni: 0.010 mg L^{-1}, Pb: 0.020 mg L^{-1}, and Zn: 0.005 mg L^{-1}.

2.3. Assessment Indices

In the present study, the mobility factor (MF) and three others indices were selected for the environmental assessment—i.e., the contamination factor (CF), geo-accumulation index (Igeo), and risk assessment code (RAC).

The mobility factor (MF, [40]) describes the potential of a metal to become mobile in the environment and to be bioavailable. This factor strongly depends on the specific chemical forms of association of each metal [Me], and it is calculated following Equation (1), where Fn is the metal content (mg kg^{-1}) obtained for each extracted fraction (from F1 to F5):

$$MF_{Me} = ((F1 + F2)/(F1 + F2 + F3 + F4 + F5)) \times 100. \qquad (1)$$

The Contamination Factor (CF, [41]) is calculated (Equation (2)) as the quotient between the content of a single metal in the sediments $[Me]_i$ and the background content of the same metal $[Me]_{BG}$. The criteria adopted to determine the extent of the contamination were as follows: no/low contamination (CF < 2), moderate ($2 \leq CF < 3$), high ($3 \leq CF < 6$), and very high ($CF \geq 6$).

$$CF = [Me]_i/[Me]_{BG} \qquad (2)$$

The geoaccumulation index (Igeo; [42]) is used to quantify possible metal pollution in marine sediments by comparing the current contents with preindustrial levels. The mathematical expression of the Igeo is presented in Equation (3); a 1.5 factor is included to take into consideration the natural variations caused by factors such as the lithology.

$$Igeo = \log_2 ([Me]_i/1.5 [Me]_{BG}) \qquad (3)$$

Based on [42], a sediment classification into seven grades can be established—i.e., grade 0: uncontaminated ($Igeo \leq 0$); grade 1: slightly contaminated ($0 < Igeo \leq 1$); grade 2: moderately contaminated ($1 < Igeo \leq 2$); grade 3: moderately/severely contaminated ($2 < Igeo \leq 3$); grade 4: severely contaminated ($3 < Igeo \leq 4$); grade 5: severely/extremely contaminated ($4 < Igeo \leq 5$); grade 6: extremely contaminated ($Igeo > 5$).

In this study, the background/preindustrial contents $[Me]_{BG}$ were established by the identification of uncontaminated sediments within the core ("low down-hole metal profile" [3]) with a common procedure found in the literature (e.g., [43–45]). This approach considers the local complexities, more appropriate than the use of general or regional references [46], because small variations in the geographic location may result in noticeable differences in the metal content (e.g., [47]).

The risk assessment code (RAC) is defined as the percentage of the exchangeable fraction (F1) in the total content of the selected metal. A five-level classification consists of no risk (<1%), low risk (1–10%), medium risk (11–30%), high risk (31–50%), and very high risk (>50%) [48].

3. Results and Discussion

Assuming a rough sedimentation rate between 3 and 6 mm y^{-1} [27], the bottom of the core (220 cm) could respond to sediments deposited 300–700 years ago. However, these sediments might be older according to [49], who dated the sediments in San Simon Bay deposited at a 200–220 cm depth to have sunk 8905 ± 305 cal. aBP. Anyway, the depth of the core supports the assumption that it has reached a pre-industrial section.

The copper, Pb, and Zn total extractable contents can serve to identify the industrial boundary (when the industrialization is clearly noticeable) in the 30 cm depth (Figure 2). Below this depth, the contents remained nearly constant (RSD < 15%, except for Cd). In consequence, sediments below a 30 cm depth can be considered as a pre-industrial record, assimilating them to the background contents [3,44]. The sum of the five fractions, or metal total extractable contents (hereinafter denoted as Tex), in the preindustrial layers of the core ranged between 18 and 30 mg Cu kg^{-1}, 24–45 mgNi kg^{-1}, 17–29 mgPb kg^{-1}, and 85–135 mgZn kg^{-1}. These preindustrial contents were coherent with previous works in the Ria of Vigo (see Table 1), pointing to a minor contribution of the residual phase (not measured in this work); therefore, Tex will be operatively considered as the total content. However, it is necessary to acknowledge that a certain portion of the elements would be missing in the residual portion of the sediments.

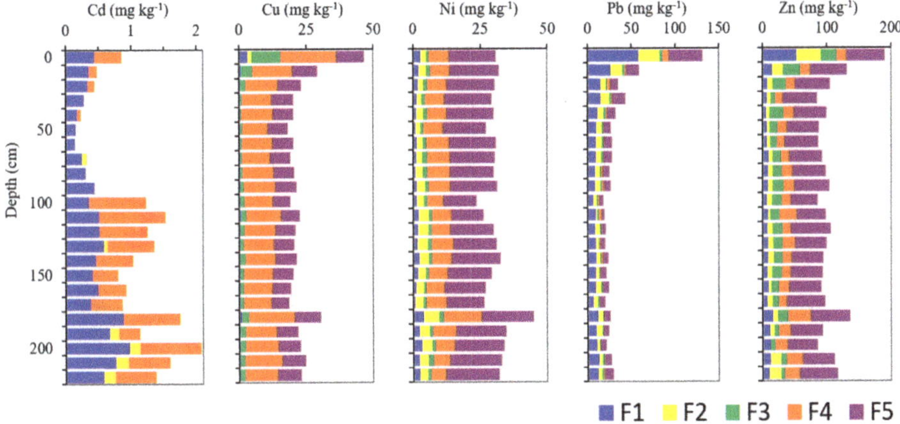

Figure 2. Vertical distribution of the metal fractions (Cd, Cu, Ni, Pb, and Zn) in the sediments of San Simon Bay. F1, metals present in exchangeable form and bound to carbonate; F2, metals present in the reductive phase bound to Mn-Fe oxides; F3, metals weakly bound to organic matter; F4, metals strongly bound to organic matter; F5, metals bound to sulfide.

Table 1. Local background $[Me]_{BG}$ levels for trace elements (mg kg^{-1}).

Scale	Cd	Cu	Ni	Pb	Zn	Reference
Ria of Vigo	–	39	26	66	106	[21]
Ria of Vigo	–	25	30	25	100	[50]
Ria of Vigo	–	22.5	32.2	73.3	133.3	[51]
Ria of Vigo	–	29.4	30.3	51.3	105.3	[24]
Ria of Vigo	–	20	30.3	25	105.3	[46]
San Simon Bay	–	21	33	51	110	[52]
San Simon Bay	–	19	33	54	111	[28]
San Simon Bay	–	30	35	34	104	[28]
San Simon Bay (Tex)	1.0*	21.2	30.6	23.3	98.2	This study

* Indicative, the real background value was not possible to estimate due to the diagenetic mobilization of the element. Tex: total extractable content.

The Tex content for Cd in the preindustrial section (0.1–2.1 mg kg^{-1}) cannot be established as a reference due to the high mobility of this element in sediments, affected by intense post-depositional migrations [53]. Nevertheless, the Cd average Tex content in the preindustrial section (1.0 mg kg^{-1}) is similar to the top limit for the Cd background values previously established in the nearby Ria of Arousa—i.e., 0.99–0.92 mg kg^{-1}—respectively in [54,55]. This value is presented as indicative in Table 1, but it needs to be considered carefully, and this level is probably an overestimation, as will be discussed further.

The different distributions of the five trace metals under study among the five chemically defined fractions are shown in Figure 2.

3.1. Cadmium

The total extractable contents of Cd varied between 0.1 and 2.1 mg kg^{-1}, coherent with the ranges of total contents previously determined in San Simon Bay (i.e., 0.2–5.5 mg kg^{-1}; [56]. This trace element is present along the core in the labile fraction (particularly F1) and bound to organic matter (mainly F4), and was not detected in the sulfide fraction. Organic matter seems to be the major source of variation. After simple regression, considering the Cd Tex content (mg kg^{-1}) as the dependent variable and the total organic carbon (%TOC) as the independent variable, the resulting function ($[Cd]_{Tex} = 4.71e^{0.434}[TOC]$) showed a moderately strong relationship between the variables ($r^2 = 0.56$; p-value < 0.05). Cadmium increased largely below a 90 cm depth, mainly in the strongly bound to organic matter fraction (F4). This increase happened below the sulfate-methane transition zone (SMTZ; see [35]). This enrichment of Cd may be related to the bacteria-mediated processes of solid sulfide mineral production [53] or with the formation of organic-Cd complexes [57].

Cadmium was the most mobile element of those studied. The calculated MF (Figure 3a) showed an average value of 66% ± 24%. Its mobility increased from the surface (51%) to a 30 cm depth (100%). Below 90 cm, the values of Cd bound to the F4 fraction reduced its mobility drastically (28% at a 100 cm depth), increasing again towards the bottom, but never passing the 72% mark. Similar to the MF, the Risk Assessment Code (RAC, Figure 3b) indicated that Cd was at the level of very high risk, on average at $RAC_{0-220\,cm}$ = 66%, but the RAC is again irregularly distributed. Higher values were found in the uppermost layer (average $RAC_{0-90\,cm}$ = 88%). However, the total contents were low (average$_{0-90\,cm}$, 0.35 ± 0.21mg kg^{-1}), probably without environmental meaning, and so under the threshold effects of the sediment quality guidelines that can be found in the literature (e.g., 1.2 mg kg^{-1}, [58] freshwater ecosystems; [0.05] 0.5–1.5 mg kg^{-1} [59]; 0.68 mg kg^{-1} [60]). In the bottom core (100–200 cm depth), the average $RAC_{100-220\,cm}$ was 49% and the total content average was 1.29 ± 0.37 mg kg^{-1}, indicating that this core section acted as a sink of Cd and was a risk if the sediments were dredged, remobilized, or resuspended. Contents of up to 2.1 mg kg^{-1} are above the available threshold effects and close to the probable effect levels (e.g., 2.6 mg kg^{-1} [58]; 1–9.6 mg kg^{-1} [59]; 4.21 mg kg^{-1} [60]).

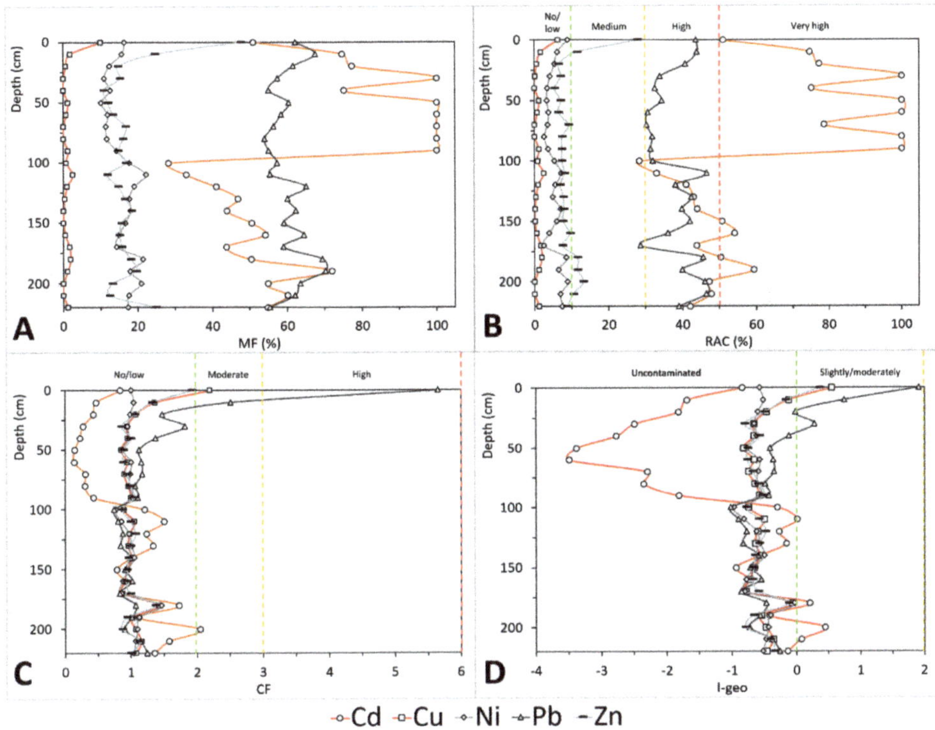

Figure 3. Vertical distribution of the tested indices for the studied metals (Cd, Cu, Ni, Pb, and Zn) in the sediments of San Simon Bay. (**A**) Mobility Factor; (**B**) Risk Assessment Code; (**C**) Contamination Factor; (**D**) Geoaccumulation Index. Vertical broken lines in B, C, and D represent the thresholds used in the assessment of each index.

The CF (Figure 3c) was not a good measure of anthropogenic pressure for Cd due to its mobility. However, using as indicative value of 1.0 mgCd kg^{-1} (Table 1), the 90 cm depth boundary is revealed. Sediments above 90 cm seemed to be depleted in Cd (average CF = 0.35), and the sediments below were probably enriched by post-depositional migration (average CF = 1.28). The use of the Igeo (Figure 3d), more restrictive than CF and taking into consideration a factor for natural variability, showed equivalent results below the 90 cm boundary. Although Igeo cannot be used to measure contamination in this case, its high negative values pointed to a possible overestimation of the indicative value used as background, well above the upper continental crust reference (i.e., 0.09 mg kg^{-1}; [61]) or the average in regional (Galicia) continental sedimentary deposits (i.e., 0.10 ± 0.06 mg kg^{-1}; [57]).

3.2. Copper

The copper Tex contents were the highest at the surface—i.e., 46.4 mg kg^{-1}—decreasing until being relatively constant below the 30 cm depth—i.e., 21.1 ± 2.8 (RSD = 13%). Only on the surface (0–20 cm) the Cu Tex contents were slightly higher than the previous background values established in the Ria of Vigo and in San Simon Bay (Table 1). They were regularly distributed within the five fractions along the core, with the major carrier being organic matter (F3 + F4 ~ 63%), particularly in the strongly bound to organic matter fraction (F4, 54.2 ± 3.0%). The second reservoir was bound to sulfide (F5, 36.1 ± 3.9%), being the labile forms (F1 + F2 ~ 1%) of lesser importance. In the uppermost layers (0–30 cm), we observed a general exponential increase (r^2 ≥ 0.9) towards the top core in all the five fractions, mainly in the organic fractions (F3 + F4). Comparing the bottom core trend with the uppermost layer enrichment, the mass balance showed the following sequence from higher to lower: F3 (from 1.6 ± 0.6 mg kg^{-1} $_{40-220\,cm}$ to 10.9 mg kg^{-1} $_{0\,cm}$), F4 (from 11.60 ± 1.6 mg kg^{-1} $_{40-220\,cm}$ to 20.4 mg kg^{-1} $_{0\,cm}$), F1 (from 0.2 ± 0.2 mg kg^{-1} $_{40-220\,cm}$ to 3.0 mg kg^{-1} $_{0\,cm}$), F5 (from 7.7 ± 0.8 mg kg^{-1} $_{40-220\,cm}$ to 10.5 mg kg^{-1} $_{0\,cm}$), and F2 (from negligible mg kg^{-1} $_{40-220\,cm}$ to 1.7 mg kg^{-1} $_{0\,cm}$). Accordingly, the organic matter (both F3 and F4) seemed to be the major carrier of the uppermost core enrichment in Cu.

Between the five studied elements, Cu is the least mobile. Its MF (Figure 3a) showed values below 3% in the whole core (range$_{10-220\,cm}$ from negligible to 2.5%), except in the surficial layer (0 cm), where the MF increased up to 10%. Cu was not detected in the reductive phase bound to manganese-iron oxides (F2), except in the 0–10 cm section. The results calculated for the RAC (Figure 3b) were equal to the MF below 10 cm, ranging in the core from 0% to 6.4%. The qualitative assessment for the RAC placed Cu in the ranges between no risk and low risk. Unlike what was observed for Cd, although the Cu contents in labile forms (F1 and F2) were scarce, the Tex contents were within the values between the threshold effect (e.g., 16 mg kg^{-1} [58]; 16–70 mg kg^{-1} [59]; 18.7 mg kg^{-1} [60]) and the probable effect levels (e.g., 34 mg kg^{-1} [58]; 36–390 mg kg^{-1} [59]; 108 mg kg^{-1} [60]), suggesting any kind of potential risk.

The contaminant factor (CF, Figure 3c) is relatively constant in the sediment core (average CF$_{40-220\,cm}$ = 1.0 ± 0.1), except in the superficial layer, where the CF increased from 30 cm (0.9) to the surface (2.2), reaching values in the range of moderate contamination. The Igeo (Figure 3d) showed similar results (-0.6 $\pm$ 0.2 below 30 cm), increasing towards the surface (up to 0.6) to reach a value classified as slightly contaminated in the uppermost layer. Although the inflexion in the trend is observed above a 30 cm depth, only the surficial sediments showed indexes indicative of low contamination, with Tex contents slightly above the local background values (Table 1), and also over the range for unpolluted marine sediments (i.e., 30–35 mg kg^{-1}), provided by [62] for the regional and global scale.

3.3. Lead

The lead Tex contents ranged between 17.3 and 131.7 mg kg^{-1}, showing the highest values at the surface, over the local background values previously reported in the area (Table 1). The Tex contents were relatively constant below a 40 cm depth, of 23.3 ± 3.4 mg kg^{-1} (RSD = 15%). The enrichment in the uppermost 40 cm, ranging from 32.0 to 131.7 mg kg^{-1}, increased towards the surface adjusted to a power function ($r^2 \sim 0.95$). Lead was mainly present in labile forms throughout the core (F1 + F2 gathers 60%, RSD < 10%), especially in the ion-exchangeable form, and was bound to carbonate (F1, 38% $\pm$ 6%). The second fraction in importance was the bound to sulfide (F5, 31% $\pm$ 4%), followed by the Pb bound to organic matter (9% $\pm$ 5%), both weakly (F3, 6% $\pm$ 2%) and strongly bound (F4, 3% $\pm$ 4%). The standard deviations provided (or the relative standard deviations), except in the more variable and less Pb-containing fractions (F3 and F4), showed a low variability. In consequence, all the fractions increased in Pb content in the uppermost 40 cm, adjusting to a power function towards the surface ($r^2 > 0.8$, except for F3 -0.6- and F2 -0.7-).

The mobility factor (MF, Figure 3a) was high along the core, varying from 54% to 70%. Its average mobility (65% $\pm$ 5%) was close to that of Cd but much less variable, not presenting any apparent trend in its fluctuations. The RAC (Figure 3b, on average 38% $\pm$ 6%) showed three well differentiated zones: (1) an uppermost zone (0–40 cm depth), ranging from 33% to 44%, with increasing values towards the surface; (2) an intermediate zone (50–90 cm), ranging from 30% to 35%, with values more or less constant; and (3) a downcore trend, varying from 29% to 47%, with higher fluctuations. This assessment, based on the mobility of Pb (MF and RAC), could be relevant in the up-core section (0–30 cm), where the Tex contents of Pb (34.4–131.7 mg kg^{-1}) were higher than the threshold effect levels (e.g., 31 mg kg^{-1} [58]; [0.1, 0.01] 28–85 mg kg^{-1} [59]; 30.2 mg kg^{-1} [60]) and even above the probable effect levels (e.g., 68 mg kg^{-1} [58]; [0.3] 38–530 mg kg^{-1} [59]; 112 mg kg^{-1} [60]).

The CF (Figure 3c) showed a considerable variation (1.3 $\pm$ 1.0; RSD = 76%). The fluctuations in the CF values again determined two zones: a down-core constant trend in the range of no/low contamination (CF = 1.0 ± 0.2), with progressively increasing values above the 50 cm depth, from 1.1 to 5.7 at the surface, reaching the classifications of moderate (10 cm) and high contamination (0 cm). Similar results were provided by the Igeo (Figure 3d), below the inflexion at a 50 cm depth, with the Igeo averages of −0.6 ± 0.2 being classified as uncontaminated. The increasing values above this depth (50 cm)—i.e., from −0.4 to 1.9—classify the sediments as slightly (30, 10 cm) and moderately contaminated (0 cm). In coherence, the superficial sediments (0 cm) overpass the background values at the local (Table 1), regional, and global scales (e.g., 5–78 mgPb kg^{-1} [62]).

3.4. Nickel

The nickel Tex contents were relatively constant along the core (RSD = 13%), averaging 30.6 ± 4.1 mg kg^{-1}, being in accordance with the previously established background values presented in Table 1. The nickel fractionation showed that Ni bound to sulfide was the most abundant fraction (F5, 56% ± 4%), followed by the organic fractions (F3 + F4, 29% ± 3%), and particularly the strongly bound to organic matter fraction (F4, 24 ± 3). The labile fractions were the minor drivers (F1, 6% ± 2%; F2, 10% ± 2%). The variability within the fractions was low (except in F1, RSD = 46%), so the proportions were relatively constant in the sedimentary column.

The mobility factor (MF, Figure 3a) allowed distinguishing some differences along the core. This index presented an average of 16% ± 4%, the lowest between the studied metals after Cu. Two sections were differentiated in the core, occurring at the boundary with the SMTZ (90 cm depth, see [35]). Above this depth, the MF slightly increased towards the surface (an average: 13%±2%), while below 90 cm the average mobility is higher, at 18% ± 3%. The RAC values (Figure 3b) followed a similar dynamic to those for the MF. From the SMTZ (90 cm depth) towards the surface the values increased regularly, from 2.7% (80 cm) to 9.0% (0 cm), averaging 4.7% ± 2.0%. Below 90 cm, the RAC was more variable, showing slightly higher values (on average 6.1% ± 1.9%). The RAC values in the core were indicative of low risk (<10%). However, the Ni contents (23-45) were similar to or slightly higher than the threshold effect levels (e.g., 7.5 mg kg^{-1} [58]; 16–40 mg kg^{-1} [59]; 15.9 mg kg^{-1} [60]), but below or close to the lower values for the probable effect levels (e.g., 19 mg kg^{-1} [58]; 36–130 mg kg^{-1} [59]; 42.8 mg kg^{-1} [60]).

Both the CF and Igeo indices for Ni were indicative of uncontaminated sediments along the core (Figure 3c,d). The contamination factor average was 1.0 ± 0.1 and the mean Igeo was −0.6 ± 0.2. Although a slight difference can be observed over (CF average 1.0 ± 0.1, and Igeo average 0.6 ± 0.1) and below (CF 1.0 ± 0.2 and Igeo 0.6 ± 0.2) the 90 cm depth, the variations were not significantly different. This lack of evidence of contamination was coherent with the similarity of the values to those in the local (Table 1) and broader references (e.g., 7–38 mgNi kg^{-1}; [62]).

3.5. Zinc

The variations in Zn Tex in the sedimentary column were similar to those of Cu and Pb. Contents averaged 103 ± 23 mg kg^{-1}, being relatively constant below a 30 cm depth (98 ± 12 mg kg^{-1}, RSD = 12%) and increasing towards the surface, from 84 mg kg^{-1} (30 cm) to 189 mg kg^{-1} (0 cm). Below 30 cm, the Zn Tex contents were slightly lower than the background values found in the literature, but the top-core contents were well above the references (Table 1). Zinc bound to sulfide (F5) was the most abundant fraction (52 ± 7%), presenting a low variation (RSD = 13%). The organic fractions (F3 + F4) showed 31%±7% of Zn, with similar contents in the weakly (F3, 14% ± 5%) and strongly (F4, 17% ± 5%) bound to organic matter fractions. The labile fractions (F1 and F2) were responsible, in average, for the 18% ± 8% of the Zn content. Both fractions—metals present in an ion-exchangeable form and bound to carbonate (F1, 9% ± 4%) and metals present in the reductive phase bound to manganese-iron oxides (F2, 9% ± 4%)—increased above 30 cm depth, reaching 48% of the Zn content at the core surface (F1, 28%; F2, 20%).

The MF (Figure 3a), average 17%±8%, clearly identified a bottom, nearly constant, low-mobility trend (16% ± 3%) below 30 cm, and a top-core with a higher mobile Zn above 30 cm (26% ± 15%). This is a consequence of the top-core Zn enrichment on the most labile fractions (F1 and F2). The RAC values (Figure 3b) were nearly constant along the core, being on average (9% ± 4%) within the limit for low risk. It reached the medium risk classification above the 20 cm depth (12%), being maximum at the surface (28%), presenting in this point a value below but close to the high risk level. It was also highlightable a bottom-core section (180–210 cm) presenting RAC values (10–13%) which are in the low range of medium risk. Only at the surface, the Tex content (189 mg kg^{-1}) was within or slightly over the threshold effect levels reported in the literature (e.g., 163 mg kg^{-1} [58]; 80–200 mg kg^{-1} [59]; 124 mg kg^{-1} [60]), but the risk seems to be low or irrelevant because the content is well below the probable effect levels (e.g., 305 mg kg^{-1} [58]; 100-700 mg kg^{-1} [59]; 271 mg kg^{-1} [60]).

Below a 30 cm depth, both indices (CF and Igeo) did not show evidence of reliable contamination (Figure 3c,d). Their values were in this major section of the core in the range of uncontaminated (CF$_{30-220cm}$ = 1.0 ± 0.1 and Igeo$_{30-220cm}$ = −0.6 ± 0.2). The contamination indices also increased towards the surface in the uppermost 30 cm, but only on the most superficial layer was a legible contamination observed. The Igeo and CF values (0.4 and 1.9, respectively) at the surface indicated low contamination according to the assumed criteria, and showed that the low contamination at the surface was above the local references (Table 1), and also over those in a world context (e.g., 20–136 mgZn kg^{-1}; [62]).

3.6. Environmental Assessment

The San Simon Bay sediments have a clay-silty texture and high TOC contents (4–15%), resulting from a mixture of terrestrial and marine inputs [37]. The high specific surface and surficial charge of clay minerals together with the organic-rich fine-grained sediments favor the trace metals accumulation [63]. Organic matter and oxides act as principal adsorbents of the trace metals, controlling their mobility and bioavailability [64,65]. The fractionation procedure in this study was useful to evaluate bioavailability and possible risk associated with metals, separated between labile forms (F1 and F2), organic fractions (F3 and F4), and the fraction bound to sulfides (F5).

The order of affinity of metal cations by organic matter (F3 + F4) in the sediments of San Simon Bay was Cu > Cd > Zn > Ni > Pb. These trace metals were bound to strongly bound to the organic matter (Figure 3).

The cluster analysis allowed us to differentiate three behaviors (Figure 4): (1) cadmium, a highly mobile element affected by post-depositional migration being stored below the SMTZ (90 cm depth) and keeping a potential risk in depth sediments if they are dredged or removed; (2) nickel, an stable and conservative element mainly bound to the less mobile fractions (F5 and F4), did not show evidence of important post-depositional migration or legible contamination; (3) copper, Pb. and Zn showed contamination (particularly Pb) in the uppermost core at 30-40 cm. According to the RAC, only Pb presented a high risk in the surface layer.

Several authors have pointed to a variety of contamination sources in the area, related to the urbanization phenomenon coupled with industrial development—i.e., local industries [25,28], wastewater outflows [66], harbor-shipyard activities [67], the use of fossil fuels [25,27], the traffic of vehicles [22,68,69], or the use of fertilizers and pesticides in agriculture [27,52]. The anthropogenic enrichment towards the surface was consistent with [27]. These authors found important metal enrichments (particularly Cu and Pb) in sediments deposited after the 1970s attributed to anthropogenic sources. The industrialization started in the area within the end the 19th century in the City of Vigo and its surrounding area (including San Simon Bay) [70], suffering an important increase in the 1960s, being coherent with the global phenomenon of the "Great Acceleration" of the mid-20th century [71,72].

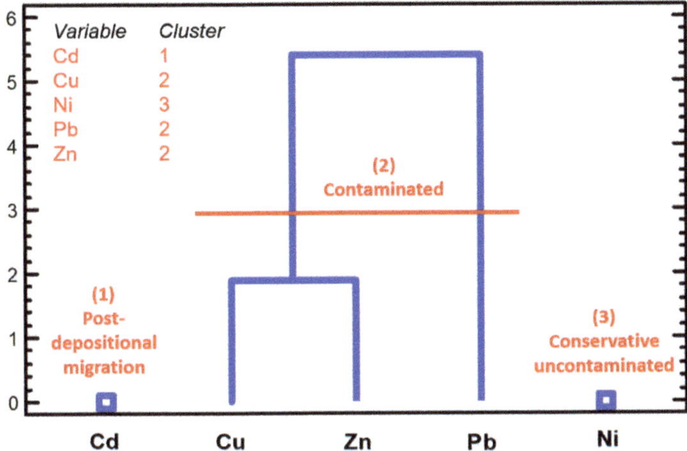

Figure 4. Cluster analysis (median method, squared Eucliean distance) performed over the total extractable contents for each element in the studied core.

4. Conclusions

The sedimentary column allowed us to estimate the background for Cu, Pb, Ni, and Zn. Cadmium was a particularly difficult element in estimating background values due to its high mobility and participation in post-depositional processes, thus only a rough estimation is provided by previous works. The fractionation procedure allowed us to identify the carriers of risk elements within the sediments—i.e., Cd and Cu were driven by their association with organic matter, Pb was mainly in the most mobile form (ion-exchangeable and bound to carbonates), and Ni and Zn were related to the less mobile fraction (bound to sulfide). The use of the Risk Assessment Code index showed some risk in surface sediments, particularly for Pb. It should be mentioned that contamination risks are actual only within those sediment layers which are easily accessible to bottom fauna—i.e., above the depth of 30 cm. In the deeper layers, as is known, most zoobenthic organisms are not found there. The contamination indices showed contamination above 30-40 cm depth, in increasing order, by Zn, Cu, and Pb. The nickel contamination was negligible, and it was not possible to determine for Cd.

This information is a valuable resource for management issues to be taken into account by decision-makers. San Simon Bay sediments might be considered contaminated in the surface mainly due to anthropogenic inputs, especially in the case of Pb, reflecting the enormous human pressure on these ecosystems. Depth sediments showed a sink for Cd due to diagenetic processes. Both Pb and Cd presented contents and mobility factors that can imply concern due to possible environmental risk.

Author Contributions: A.M.R.-P., Conceptualization, Methodology, Writing—original draft, Writing—review & editing, Investigation, Formal analysis; M.A.Á.-V., Conceptualization, Methodology, Writing—original draft, Writing—review & editing, Investigation. E.D.U.-Á., Conceptualization, Methodology, Writing—review & editing; E.D.B., Conceptualization, Methodology, Writing—review & editing, Investigation, Funding acquisition.

Funding: This work was supported by the Spanish Ministry of Economy and Competitiveness and the European Regional Development Fund under research project CGL2012-33584 and by the Xunta de Galicia under research project GRC2015/020. M.A. Álvarez-Vázquez is supported by the Xunta de Galicia through the postdoctoral grant #ED481B-2019-066.

Conflicts of Interest: The authors declare no conflict of interest.

References

1. McLennan, S.M. Relationships between the trace element composition of sedimentary rocks and upper continental crust. *Geochem. Geophys. Geosyst.* **2001**, *2*, 2. [CrossRef]
2. Hu, Z.; Gao, S. Upper crustal abundances of trace elements: A revision and update. *Chem. Geol.* **2008**, *253*, 205–221. [CrossRef]
3. Birch, G.F. Determination of sediment metal background concentrations and enrichment in marine environments—A critical review. *Sci. Total Environ.* **2017**, *580*, 813–831. [CrossRef]
4. Adriano, D.C. *Trace Elements in Terrestrial Environments*; Springer Science and Business Media LLC: Berlin/Heidelberg, Germany, 2001.
5. Kabata-Pendias, A. *Trace Elements in Soils and Plants*; CRC Press: Boca Raton, FL, USA, 2011; p. 548.
6. Panagos, P.; Van Liedekerke, M.; Yigini, Y.; Montanarella, L. Contaminated Sites in Europe: Review of the Current Situation Based on Data Collected through a European Network. *J. Environ. Public Health* **2013**, *2013*, 1–11. [CrossRef]
7. Vareda, J.P.; Valente, A.J.; Durães, L. Assessment of heavy metal pollution from anthropogenic activities and remediation strategies: A review. *J. Environ. Manag.* **2019**, *246*, 101–118. [CrossRef] [PubMed]
8. Nemati, K.; Abu Bakar, N.K.; Abas, M.R.; Sobhanzadeh, E. Speciation of heavy metals by modified BCR sequential extraction procedure in different depths of sediments from Sungai Buloh, Selangor, Malaysia. *J. Hazard. Mater.* **2011**, *192*, 402–410. [CrossRef] [PubMed]
9. Hamelink, J.L.; Landrum, P.F.; Harold, B.L.; William, B.H. *Bioavailability: Physical, Chemical, and Biological Interactions*; CRC Press: Boca Raton, FL, USA, 1994.
10. Voegelin, A.; Barmettler, K.; Kretzschmar, R. Heavy metal release from contaminated soils: Comparison of column leaching and batch extraction results. *J. Environ. Qual.* **2003**, *32*, 865–875. [CrossRef]
11. Fang, T.; Li, X.-D.; Zhang, G. Acid volatile sulfide and simultaneously extracted metals in the sediment cores of the Pearl River Estuary, South China. *Ecotoxicol. Environ. Saf.* **2005**, *61*, 420–431. [CrossRef]
12. Rodgers, K.; McLellan, I.; Peshkur, T.; Williams, R.; Tonner, R.; Knapp, C.W.; Henriquez, F.L.; Hursthouse, A.S. The legacy of industrial pollution in estuarine sediments: Spatial and temporal variability implications for ecosystem stress. *Environ. Geochem. Health* **2019**, *42*, 1057–1068. [CrossRef]
13. Vukovic, D.; Vuković, Ž.; Stanković, S. The impact of the Danube Iron Gate Dam on heavy metal storage and sediment flux within the reservoir. *Catena* **2014**, *113*, 18–23. [CrossRef]
14. Zhong, A.-P.; Guo, S.; Li, F.-M.; Li, G.; Jiang, K.-X. Impact of anions on the heavy metals release from marine sediments. *J. Environ. Sci.* **2006**, *18*, 1216–1220. [CrossRef]
15. Birch, G.; Gunns, T.; Olmos, M. Sediment-bound metals as indicators of anthropogenic change in estuarine environments. *Mar. Pollut. Bull.* **2015**, *101*, 243–257. [CrossRef] [PubMed]
16. De Groot, A.J.; Salomons, W.; Allersma, E. Processes affecting heavy metals in estuarine sediments. *Estuar. Chem.* **1976**, 131–157.
17. Fang, G.-C.; Yang, H.-C. Comparison of heavy metals in marine sediments from coast areas in East and Southeast Asian countries during the years 2000–2010. *Toxicol. Ind. Health* **2011**, *27*, 754–759. [CrossRef]
18. Rubio, B.; Gago, L.; Vilas, F.; Nombela, M.A.; García-Gil, S.; Alejo, I.; Pazos, O. Interpretación de tendencias históricas de contaminación por metales pesados en testigos de sedimentos de la Ría de Pontevedra. *Thalassas* **1996**, *12*, 137–152.
19. Valette-Silver, N.J. The Use of Sediment Cores to Reconstruct Historical Trends in Contamination of Estuarine and Coastal Sediments. *Estuaries* **1993**, *16*, 577–588. [CrossRef]
20. Carballeira, A.; Carral, E.; Puente, X.; Villares, R. Regional-scale monitoring of coastal contamination. Nutrients and heavy metals in estuarine sediments and organisms on the coast of Galicia (northwest Spain). *Int. J. Environ. Pollut.* **2000**, *13*, 534. [CrossRef]
21. Belzunce Segarra, M.J.; Prego, R.; Wilson, M.J.; Bacon, J.; Santos-Echeandía, J. Metal speciation in surface sediments of the Vigo Ria (NW Iberian Peninsula). *Sci. Mar.* **2008**, *72*, 119–126.
22. Marcet, P. *Contribucion al Estudio de la Contaminacion de la ria de Vigo. Contenido y Distribucion de Nutrientes y Metales Pesados en Sedimentos Someros*; Universidad de Vigo: Vigo, Spain, 1994.
23. Nombela, M.A.; Vilas, F.; Evans, G. Sedimentation in the Mesotidal Rías Bajas of Galicia (North-Western Spain): Ensenada De San Simón, Inner Ría De Vigo. *Tidal Signat. Mod. Anc. Sediments* **2009**, 133–149. [CrossRef]

24. Rubio, B.; Nombela, M.A.; Vilas, F. Geochemistry of Major and Trace Elements in Sediments of the Ria de Vigo (NW Spain): An Assessment of Metal Pollution. *Mar. Pollut. Bull.* **2000**, *40*, 968–980. [CrossRef]

25. Alvarez-Iglesias, P.; Rubio, B.; Vilas, F. Pollution in intertidal sediments of San Simón Bay (Inner Ria de Vigo, NW of Spain): Total heavy metal concentrations and speciation. *Mar. Pollut. Bull.* **2003**, *46*, 491–503. [CrossRef]

26. Villares, R.; Puente, X.; Carballeira, A. Heavy Metals in Sandy Sediments of the Rías Baixas (NW Spain). *Environ. Monit. Assess.* **2003**, *83*, 129–144. [CrossRef] [PubMed]

27. Alvarez-Iglesias, P.; Quintana, B.; Rubio, B.; Pérez-Arlucea, M. Sedimentation rates and trace metal input history in intertidal sediments from San Simón Bay (Ría de Vigo, NW Spain) derived from 210Pb and 137Cs chronology. *J. Environ. Radioact.* **2007**, *98*, 229–250. [CrossRef] [PubMed]

28. Álvarez-Iglesias, P.; Rubio, B. Trace Metals in Shallow Marine Sediments from the Ría de Vigo: Sources, Pollution, Speciation and Early Diagenesis. *Geochem. Earth's Syst. Process.* **2012**, 185–210. [CrossRef]

29. Rubio, B.; Alvarez-Iglesias, P.; Vilas, F. Diagenesis and anthropogenesis of metals in the recent Holocene sedimentary record of the Ría de Vigo (NW Spain). *Mar. Pollut. Bull.* **2010**, *60*, 1122–1129. [CrossRef]

30. Evans, G.; Prego, R. Rias, estuaries and incised valleys: Is a ria an estuary? *Mar. Geol.* **2003**, *196*, 171–175. [CrossRef]

31. Fraga, F. Fotosíntesis en la Ría de Vigo. *Invst. Pesq.* **1976**, *40*, 151–167.

32. Vilas, F.; Nombela, M.A.; García-Gil, E.; García-Gil, S.; Alejo, I.; Rubio, B.; Pazos, O. *Cartografía de sedimentos submarinos, Ría de Pontevedra. E: 1:50000*; Xunta de Galicia: Sandtiago de Compostela, Spain, 1996.

33. García-Gil, S.; De Blas, E.; Martínez-Carreño, N.; Iglesias, J.; Rial-Otero, R.; Simal-Gandara, J.; Judd, A. Characterisation and preliminary quantification of the methane reservoir in a coastal sedimentary source: San Simón Bay, Ría de Vigo, NW Spain. *Estuarine Coast. Shelf Sci.* **2011**, *91*, 232–242. [CrossRef]

34. Hedges, J.I.; Keil, R.G. Sedimentary organic matter preservation: An assessment and speculative synthesis. *Mar. Chem.* **1995**, *49*, 81–115. [CrossRef]

35. Ramírez-Pérez, A.; De Blas, E.; García-Gil, S. Redox processes in pore water of anoxic sediments with shallow gas. *Sci. Total. Environ.* **2015**, *538*, 317–326. [CrossRef]

36. Martínez-Carreño, N.; García-Gil, S. The Holocene gas system of the Ría de Vigo (NW Spain): Factors controlling the location of gas accumulations, seeps and pockmarks. *Mar. Geol.* **2013**, *344*, 82–100. [CrossRef]

37. Ramírez-Pérez, A.; De Blas, E.; García-Gil, S. Sulfur, Iron, and Manganese Speciation in Anoxic Sediments with Methane (Ría de Vigo, NW Spain). *CLEAN Soil Air Water* **2017**, *45*, 1600700. [CrossRef]

38. Campanella, L.; D'Orazio, D.; Petronio, B.; Pietrantonio, E. Proposal for a metal speciation study in sediments. *Anal. Chim. Acta* **1995**, *309*, 387–393. [CrossRef]

39. Pagnanelli, F.; Moscardini, E.; Giuliano, V.; Toro, L. Sequential extraction of heavy metals in river sediments of an abandoned pyrite mining area: Pollution detection and affinity series. *Environ. Pollut.* **2004**, *132*, 189–201. [CrossRef]

40. Kabala, C.; Singh, B.R. Fractionation and Mobility of Copper, Lead, and Zinc in Soil Profiles in the Vicinity of a Copper Smelter. *J. Environ. Qual.* **2001**, *30*, 485–492. [CrossRef]

41. Håkanson, L. An ecological risk index for aquatic pollution control.a sedimentological approach. *Water Res.* **1980**, *14*, 975–1001. [CrossRef]

42. Muller, G. Index of Geoaccumulation in sediments of the Rhine River. *Geo J.* **1969**, *2*, 108–118.

43. Mil-Homens, M.; Vale, C.; Naughton, F.; Brito, P.; Drago, T.; Anes, B.; Raimundo, J.; Schmidt, S.; Caetano, M. Footprint of roman and modern mining activities in a sediment core from the southwestern Iberian Atlantic shelf. *Sci. Total. Environ.* **2016**, *571*, 1211–1221. [CrossRef]

44. Álvarez-Vázquez, M.A.; Bendicho, C.; Prego, R. Ultrasonic slurry sampling combined with total reflection X-ray spectrometry for multi-elemental analysis of coastal sediments in a ria system. *Microchem. J.* **2014**, *112*, 172–180. [CrossRef]

45. Birch, G.; Chang, C.-H.; Lee, J.-H.; Churchill, L. The use of vintage surficial sediment data and sedimentary cores to determine past and future trends in estuarine metal contamination (Sydney estuary, Australia). *Sci. Total Environ.* **2013**, *454*, 542–561. [CrossRef]

46. Rubio, B.; Nombela, M.A.; Vilas, F. La contaminación por metales pesados en las Rías Baixas gallegas: Nuevos valores de fondo para la Ría de Vigo (NO de España). *J. Iber. Geol.* **2000**, *26*, 121–149.

47. Álvarez-Vázquez, M.A.; Caetano, M.; Alvarez-Iglesias, P.; Pedrosa-García, M.D.C.; Calvo, S.; De Uña-Álvarez, E.; Quintana, B.; Vale, C.; Prego, R. Natural and Anthropocene fluxes of trace elements in estuarine sediments of Galician Rias. *Estuarine Coast. Shelf Sci.* **2017**, *198*, 329–342. [CrossRef]

48. Perin, G.; Craboledda, L.; Lucchese, M.; Cirillo, R.; Dotta, L.; Zanette, M.L.; Orio, A.A. Heavy metal speciation in the sediments of northern Adriatic Sea. A new approach for environmental toxicity determination. *Heavy Metals Environ.* **1985**, *2*, 454–456.

49. Martínez-Carreño, N.; García-Gil, S. Reinterpretation of the Quaternary sedimentary infill of the Ría de Vigo, NW Iberian Peninsula, as a compound incised valley. *Quat. Sci. Rev.* **2017**, *173*, 124–144. [CrossRef]

50. Barreiro, R. Estudio de metales pesados en medio y organismos de un ecosistema de ría (Pontedeume, A Coruña). Ph.D. Thesis, Universidade de Santiago de Compostela, Santiago de Compostela, Spain, 1991; p. 227.

51. Carral, E.; Puente, X.; Villares, R.; Carballeira, A. Background heavy metal levels in estuarine sediments and organisms in Galicia (northwest Spain) as determined by modal analysis. *Sci. Total Environ.* **1995**, *172*, 175–188. [CrossRef]

52. Álvarez-Iglesias, P.; Rubio, B.; Pérez-Arlucea, M. Reliability of subtidal sediments as "geochemical recorders" of pollution input: San Simón Bay (Ría de Vigo, NW Spain). *Estuarine Coast. Shelf Sci.* **2006**, *70*, 507–521. [CrossRef]

53. Davies-Colley, R.J.; Nelson, P.O.; Williamson, K.J. Sulfide control of cadmium and copper concentrations in anaerobic estuarine sediments. *Mar. Chem.* **1985**, *16*, 173–186. [CrossRef]

54. Barciela-Alonso, M.C.; Pazos-Capeáns, P.; Regueira-Miguens, M.E.; Bermejo-Barrera, A.; Bermejo-Barrera, P. Study of cadmium, lead and tin distribution in surface marine sediment samples from Ria de Arousa (NW of Spain). *Anal. Chim. Acta* **2004**, *524*, 115–120. [CrossRef]

55. Prego, R.; Segarra, M.J.B.; Helios-Rybicka, E.; Barciela, M.C. Cadmium, manganese, nickel and lead contents in surface sediments of the lower Ulla River and its estuary (northwest Spain). *Bol. Inst. Esp. Oceanogr.* **1999**, *15*, 495–500.

56. Howarth, R.J.; Evans, G.; Croudace, I.W.; Cundy, A.B. Sources and timing of anthropogenic pollution in the Ensenada de San Simón (inner Ría de Vigo), Galicia, NW Spain: An application of mixture-modelling and nonlinear optimization to recent sedimentation. *Sci. Total. Environ.* **2005**, *340*, 149–176. [CrossRef]

57. Macías, F.; Calvo de Anta, R. *Niveles Genéricos de Referencia de Metales Pesados y Otros Elementos Traza en Suelos de Galicia*; Consellería de Medio Ambiente e Desenvolvemento Sostible: Santiago de Compostela, Spain, 2008.

58. De Deckere, E.; De Cooman, W.; Leloup, V.; Meire, P.; Schmitt, C.; Von Der Ohe, P.C. Development of sediment quality guidelines for freshwater ecosystems. *J. Soils Sediments* **2011**, *11*, 504–517. [CrossRef]

59. Burton, G.A. Sediment quality criteria in use around the world. *J. Limnol.* **2002**, *3*, 65–76. [CrossRef]

60. Macdonald, D.D.; Carr, R.S.; Calder, F.D.; Long, E.R.; Ingersoll, C.G. Development and evaluation of sediment quality guidelines for Florida coastal waters. *Ecotoxicology* **1996**, *5*, 253–278. [CrossRef] [PubMed]

61. Rudnick, R.; Gao, S. Composition of the Continental Crust. In *Treatise on Geochemistry*; Elsevier: Amsterdam, The Netherlands, 2003; Volume 3.

62. Prego, R.; Cobelo-Garcia, A. Twentieth century overview of heavy metals in the Galician Rias (NW Iberian Peninsula). *Environ. Pollut.* **2003**, *121*, 425–452. [CrossRef]

63. Horowitz, A.J.; Elrick, K.A. The relation of stream sediment surface area, grain size and composition to trace element chemistry. *Appl. Geochem.* **1987**, *2*, 437–451. [CrossRef]

64. Ajala, L.O.; Onwukeme, V.I.; Mgbemena, M.N. Speciation of Some Trace Metals in Floodplain Soil of Eke-Mgbom, Afikpo, Nigeria. *Am. Chem. Sci. J.* **2014**, *4*, 963–974. [CrossRef]

65. McBride, M.B. Toxic Metal Accumulation from Agricultural Use of Sludge: Are USEPA Regulations Protective? *J. Environ. Qual.* **1995**, *24*, 5–18. [CrossRef]

66. Filgueiras, A.V.; Prego, R. Biogeochemical fluxes of iron from rainwater, rivers and sewage to a Galician Ria (NW Iberian Peninsula). Natural versus anthropogenic contributions. *Biogeochemistry* **2007**, *86*, 319–329. [CrossRef]

67. Prego, R.; Filgueiras, A.V.; Santos-Echeandía, J. Temporal and spatial changes of total and labile metal concentration in the surface sediments of the Vigo Ria (NW Iberian Peninsula): Influence of anthropogenic sources. *Mar. Pollut. Bull.* **2008**, *56*, 1031–1042. [CrossRef]

68. Belzunce-Segarra, M.J.; Bacon, J.R.; Prego, R.; Wilson, M.J. Chemical forms of heavy metals in surface sediments of the san Simón inlet, Ría de vigo, galicia. *J. Environ. Sci. Heal. Part A* **1997**, *32*, 1271–1292. [CrossRef]
69. Evans, G.; Howarth, R.J.; Nombela, M.A. Metals in the sediments of Ensenada de San Simón (inner Ría de Vigo), Galicia, NW Spain. *Appl. Geochem.* **2003**, *18*, 973–996. [CrossRef]
70. González-Pérez, J.M.; Pérez González, A. Demographic dynamics and urban planning in Vigo since 1960. The impact of industrialization. *Anal. Geogr. Univ. Complut.* **2003**, *23*, 163–185.
71. Steffen, W.; Crutzen, P.J.; McNeill, J.R. The Anthropocene: Are Humans Now Overwhelming the Great Forces of Nature. *Ambio* **2007**, *36*, 614–621. [CrossRef]
72. Steffen, W.; Grinevald, J.; Crutzen, P.; McNeill, J. The Anthropocene: Conceptual and historical perspectives. *Philos. Trans. R. Soc. A Math. Phys. Eng. Sci.* **2011**, *369*, 842–867. [CrossRef] [PubMed]

Article

Comprehending the Causes of Presence of Copper and Common Heavy Metals in Sediments of Irrigation Canals in Taiwan

Shih-Han Huang [1,2], Tien-Chin Chang [1], Hui-Chen Chien [2], Zih-Sin Wang [2], Yen-Chen Chang [3], Ying-Lin Wang [4] and Hsing-Cheng Hsi [4,*]

[1] Institute of Environmental Engineering and Management, National Taipei University of Technology, Taipei 106, Taiwan; Sh0933891461@gmail.com (S.-H.H.); tcchang@ntut.edu.tw (T.-C.C.)
[2] Soil and Groundwater Remediation Fund Management Board, Environmental Protection Administration, Taipei 100, Taiwan; hcchien@epa.gov.tw (H.-C.C.); zswang@epa.gov.tw (Z.-S.W.)
[3] Apollo Technology Co., Ltd., Taipei 104, Taiwan; ycchang@apollotech.com.tw
[4] Graduate Institute of Environmental Engineering, National Taiwan University, Taipei 106, Taiwan; lynn12783@gmail.com
* Correspondence: hchsi@ntu.edu.tw; Tel.: +88-62-3366-4374

Citation: Huang, S.-H.; Chang, T.-C.; Chien, H.-C.; Wang, Z.-S.; Chang, Y.-C.; Wang, Y.-L.; Hsi, H.-C. Comprehending the Causes of Presence of Copper and Common Heavy Metals in Sediments of Irrigation Canals in Taiwan. *Minerals* **2021**, *11*, 416. https://doi.org/10.3390/min11040416

Academic Editors: Ana Romero-Freire and Hao Qiu

Received: 24 March 2021
Accepted: 12 April 2021
Published: 14 April 2021

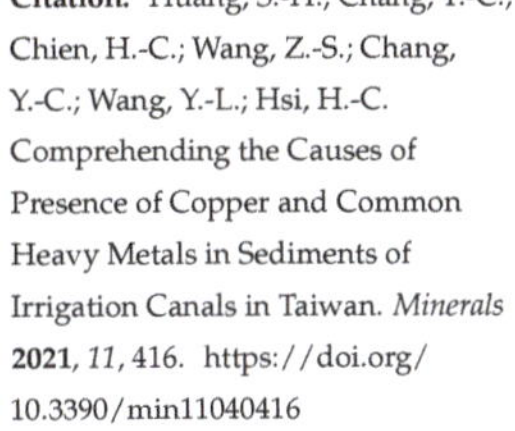

Abstract: In 2019, Taiwan completed its first thorough heavy metal investigation of irrigation canal sediments by this study with the support of Taiwan Environmental Protection Administration. Box-and-whisker plots were used to analyze the sediment distribution and to define metal concentrations. Possible metal pollution sources, the polluted agricultural land, irrigation area, and water sources were also evaluated using spatial analysis to understand the possible causes of sediment pollution. Results showed that the main heavy metal in agricultural land was Cu, found in 77% of contaminated agricultural land sites. Most sites with Cu pollution in sediments were in Taoyuan City and Changhua County. The heavy metals present in the sediment pollution sites in Taoyuan were consistent with those of possible pollution sources upstream, namely, Cr, Cu, and Pb. The main heavy metals in sediment pollution sites in Changhua were Cr, Cu, Ni, Pb, and Zn, whereas those for the polluted agricultural land sites were Cr, Cu, Ni, and Zn, without Pb. The main irrigation water sources in Changhua include drainages and rivers, with some receiving most wastewater pollution mass of release of Changhua, and functions as an irrigation water source with a high release mass in Cr, Cu, Ni and Zn. These findings indicate that the sites of sediment pollution, sites of polluted agricultural land, and the sources of pollution share corresponding heavy metal characteristics. Therefore, in Changhua, the sediments were polluted mainly because (1) the irrigation canals received the highest masses of pollutant releases into drainage wastewater of the county; and (2) the return flow from irrigation and the illegal discharge of wastewater. The preliminary assessment results for sediment pollution in Taoyuan also suggest that the main causes may be irrigation by polluted rivers or drainages and return flows.

Keywords: agricultural land contamination; sediment; irrigation canal; heavy metal; copper

1. Introduction

Agricultural lands contaminated by heavy metals has become a serious problem and is of concern in Taiwan [1]. Although Taiwan has an annual precipitation reaching 95.2 billion m^3, which is approximately 2.6 times the world average, the sum of annual evaporation and runoff to the sea reaches 75.1 billion m^3 because of Taiwan's geographic characteristics of a longer north–south span in comparison to the west–east span and the Central Mountain Range running from the north to the south of the island [2]. This leads to difficulty in water storage and the rainfall that can be used by each person is only 1/6 of world average in Taiwan [3]. Therefore, irrigation canals provide the foundation of rice agriculture in Taiwan. Water storage systems including reservoirs and farm ponds are employed to store a considerable amount of rainwater. Through water transmission systems, such as leading canals, main canals, and lateral canals, water is led to ditches and dispatched to agricultural land for irrigation [4]. The systems enable regions including Changhua, Chiayi, and Tainan Counties to grow rice crops successfully and ensure abundant harvests. However, as of November 2020, the number of agricultural land sites listed as having soil contamination was 7294, of which 5589 contained copper (Cu) pollution, accounting for 77% of the sites. The data thus indicated that Cu is the main pollutant of agricultural land in Taiwan [5].

Excessive trace elements in agricultural soil may inhibit the growth or decrease the yield of food [6], and the food loss caused by the abnormal Cu concentration in the soil may be higher than other causes leading to food safe and food security threats. Cu can cause growth inhibition, oxidative damage, antioxidant response, and disturbances in physiological and biochemical processes in food crops such as wheat, rice, and maize [7–10]. One study found that when the Cu concentration in the culture solution reached 10 μM, the germination rate of rice (*Oryza sativa* L.) was reduced by about 60% compared with the control group (from 93% to 34%), and the root length was reduced by about 40% (from 80.3 mm to 50.1 mm) [11]; when the Cu concentration in the soil is 100 mg/kg, the yield of rice grain decreases by 10%, and when the concentration is between 300 mg/kg and 500 mg/kg, the yield of rice grain is reduced by 50%. When the Cu concentration reaches 1000 mg/kg, the yield of rice grain is reduced by 90%. Cao and Hu [12] investigated that rice grown in paddy soils irrigated by Cu-containing wastewater would lead to the problem of less effective tiller and black roots; in addition, the yield of rice grain was reduced by 18%–25%. The Cu concentration reaching 50 mg/kg also had a significant effect on root weight [13]. Moreover, the Cu concentration in brown rice (15.5 mg/kg) was measured higher than the value of 10 mg/kg of maximum permissible concentration in grains promulgated by Food and Agriculture Organization (FAO) [14] when the Cu concentration in the paddy soil reached 101.2 mg/kg [12]. In Taiwan, the higher bioaccumulation of Cu in the soil–grapevine system in vineyards may also cause a public concern of general population exposure to Cu through grape consumption [15–18].

Rice is the most important food crop in Taiwan; the average annual harvest area in the past 10 years is higher than that of other crops, such as grains, vegetables, and fruits [19]. If the high Cu concentration in farmland soils reduces rice production, it may cause a food crisis in Taiwan. Unfortunately, several agricultural land sites are polluted with Cu and other common heavy metals by irrigation water [20–23] or specifically, suspended solids in irrigation water [24]. Although the strategy of separating irrigation and drainage canals may improve the quality of irrigation water [25], the suspended solids in irrigation water may accumulate at the bottom of canals and form sediments. No large-scale study has been conducted to investigate the sediments of irrigation canals before this study. If sediments are resuspended or scoured because of rainwater flooding into agricultural land, they may pollute the land.

To protect citizen health and secure sediment quality, the Environmental Protection Administration (EPA) of the Executive Yuan in Taiwan incorporated sediment quality management provisions in amendments to the Soil and Groundwater Pollution Remediation Act in 2010 and made subsidiary regulations including "Regulations on the Classification Management and Usage Limitation of Sediment Quality Guidelines" and "Implementation Regulations on Testing Sediments for the Future Reference of Local Industry Competent Authorities" after the implementation of the Act. The regulations were effective as of January 2014. According to Paragraph 5 of Article 6 of the Soil and Groundwater Pollution Remediation Act, the authorities responsible for rivers, irrigation canals, lakes, and reservoirs must test the bottom sediment quality of the listed water bodies and submit the resulting data to the central competent authority.

Therefore, this study is of importance, exploring the distribution of Cu and other common heavy metal pollutants in the sediments of irrigation canals and compared it with that of Cu-polluted agricultural sites. A composite analysis of possible pollution sources, situations of surface waters, and other heavy metal pollution characteristics was conducted to determine the sediment pollution characteristics and possible causes.

2. Materials and Methods

2.1. Data

The total number of water bodies reported on by all industry competent authorities in the initial period of regular reporting (2014–2019) was 557, including 118 rivers, 94 lakes and reservoirs, and 345 irrigation canals. Among the water bodies, 35 rivers and two reservoirs were approved to be exempted from reporting because of gravel on the bottom, and one reservoir was exempted because of ongoing construction. Reports on a total of 519 water bodies were to be completed in the first operation period, consisting of 83 rivers, 91 lakes and reservoirs, and 345 irrigation canals. All competent authorities were to complete all reports on sediment quality by 2019. The sediment samples of one reservoir (under construction) and 63 irrigation canals (without sufficient amounts of sediments) could not be collected. Thus, the actual number of water bodies tested in the first operation period comprised 83 rivers, 90 lakes and reservoirs, and 282 irrigation canals.

2.2. Sampling and Analytical Methods

Relevant sampling, preprocessing, and analytical methods were implemented according to the standard methods provided by the Environmental Analysis Laboratory of the EPA, Executive Yuan. The methods included NIEA S104 (sediment sampling method), NIEA M353 (method for detecting metal in waste and sediments—acid digestion), NIEA M301 (method for detecting metal in waste and sediments—microwave assisted acid digestion), NIEA S310 (method for detecting arsenic in soil and sediments—arsine atomic absorption spectrometry), NIEA M111 (flame atomic absorption spectrometry), NIEA M104 (inductively coupled plasma atomic emission spectrometry), NIEA M317 (method for detecting the total mercury in soil, sediments, and waste—cold vapor atomic absorption spectrometry), and NIEA M318 (method for detecting the total mercury in solid and liquid samples—thermal decomposition, amalgamation, and atomic absorption spectrophotometry).

Hand shovels, sampling spoons, and grab samplers were used to collect samples from sediments in rivers, irrigation canals, lakes, and reservoirs. The samples were air dried (for approximately 7–10 days), oven dried at $30 \pm 4\,°C$, or freeze dried. All dried samples were sifted using 2 mm (10 mesh) standard sieves before they were further ground and sifted through 0.15 mm (100 mesh) standard sieves.

The homogenized samples underwent microwave digestion or acid digestion to complete the digestion process. The microwave digestion method involved adding 1 mg of the sample into 9 ± 0.1 mL concentrated nitric acid and 3 ± 0.1 mL concentrated hydrochloric acid and mixed evenly. Microwave digestion devices were used to heat samples to $175 \pm 5\,°C$ at a rate of approximately $10\,°C/min$, and the temperature was maintained for 10 ± 1 min. Each sample was poured into a volumetric flask and diluted with reagent water. The supernatant digestion solution was collected and analyzed using instruments. The acid digestion method involved adding 0.01 mg of the sample into the digestion vessel. Then, 10 mL of (1:1) nitric acid, 5 mL of concentrated acid, 2 mL of water, and 3 mL of 30% hydrogen peroxide were sequentially added to the vessel, which was heated to $95 \pm 5\,°C$ by vapor collection devices. After each reflux condensation without boiling the sample, 1 mL of 30% hydrogen peroxide was added until the sample was heated and minimum bubbles were present or the sample appearance remained unchanged. Then, 10 mL of concentrated hydrochloric acid was added to the digestion solution, which was then heated to $95 \pm 5\,°C$. The nonboiling sample underwent reflux for 15 min. After the sample was condensed, it was diluted to a constant volume of 100 mL using reagent water. Whatman Grade 40 filter papers (or equivalent products) or centrifugal methods were employed to collect the supernatant digestion solution for instrumental analysis.

Digestion solutions can be determined through flame atomic absorption spectrometry for cadmium (Cd), chromium (Cr), Cu, nickel (Ni), lead (Pb), and zinc (Zn). Graphite furnace atomic absorption spectrometry can be used to determine arsenic (As), Cd, Cr, Cu, Ni, Pb, and Zn content. Inductively coupled plasma atomic emission spectrometry or inductively coupled plasma mass spectrometry can be used to determine As, Cd, Cr, Cu, Hg, Ni, Pb, and Zn.

2.3. Graphic and Spatial Analysis

The sediment data had to be processed using appropriate statistical methods to identify the value distributions and trends. A box-and-whisker plot [26] is a common schematic method for presenting descriptive statistics. The method yields five-number summaries, including the first quartile, median, third quartile, minimum, and maximum. In consideration of the absence of previous irrigation canal baseline data in Taiwan, 1.5× interquartile range can be added to the third quartile to obtain the outlier value as the screening basis for high pollution potential sites.

Bar charts provided concentration distribution trends of pollutants, visualizing pollutants with higher concentrations and concentration trends for different sediment sites with icons. Pie charts and radar charts facilitated the comparisons of different types of data such as the number of sediment pollution sites in different counties and cities and the distribution and number of agricultural land pollution sites.

A geographic information system was adopted to present vector or grid information that provided coordinate information including the sediment sampling site, agricultural land, pollution source, irrigation canal, and surface water. Spatial analysis was conducted to extract the data characteristics of different ranges for spatial distribution patterns.

3. Results and Discussion

3.1. Cu Distribution in Irrigation Canal Sediments

Sediment quality data for irrigation canals, rivers, and reservoirs between 2014 and 2019 were collected. The numbers of sampling sites in irrigation canals, rivers, and reservoirs are 292, 443, and 276, respectively. The Cu concentration distribution is shown in Figure 1 and Table S1 (Supplementary Materials). The median concentrations of the three water body types were similar, with 26.6, 21.6, and 18.3 mg/kg for irrigation canals, rivers, and reservoirs, respectively. The sediment concentration in the third quartile for irrigation canals was approximately twice that of rivers and reservoirs; the sediment concentrations of the three water bodies were 62.5, 31.7, and 26.7 mg/kg, respectively. Compared with rivers and reservoirs, the Cu cumulative concentration in sediments from irrigation canals was notably higher.

The irrigation canal sampling sites were collected in 15 counties and cities. Given the industrial characteristics of counties and cities, different sediment flow and distribution trends may have been present. Therefore, sediment concentration variation in the 15 counties and cities was explored separately. The distribution of the sampling sites of irrigation canals is presented in Figures 2 and 3 and Table S2 (Supplementary Material). The average number of sampling sites per county or city was 17. Changhua had the most sampling sites (32), followed by Hualien (29), whereas Hsinchu City had the fewest (5). Descriptive statistics of sediment concentration data (Figure 3) indicated that the highest average concentration was observed in Changhua, where it reached 340 mg/kg. The concentrations in Taoyuan (105 mg/kg), New Taipei City (105 mg/kg), and Pingtung (92.2 mg/kg) were higher than the overall average concentration (79.2 mg/kg). The median concentrations of Pingtung (80.6 mg/kg) and Taipei City (80.3 mg/kg) were similar and both high. The median concentration of Taoyuan was 75.0 mg/kg. The median concentrations of Changhua (59.5 mg/kg), New Taipei City (56.9 mg/kg), Hsinchu City (41.7 mg/kg), and Yilan (34.8 mg/kg) were higher than the overall median concentration (26.6 mg/kg). In general, compared with the overall irrigation canal sediment concentration distribution pattern in Taiwan (Figure 3), the sediment concentration distribution patterns of Pingtung, Taoyuan, New Taipei City, Hsinchu City, Changhua, and Taipei City were high.

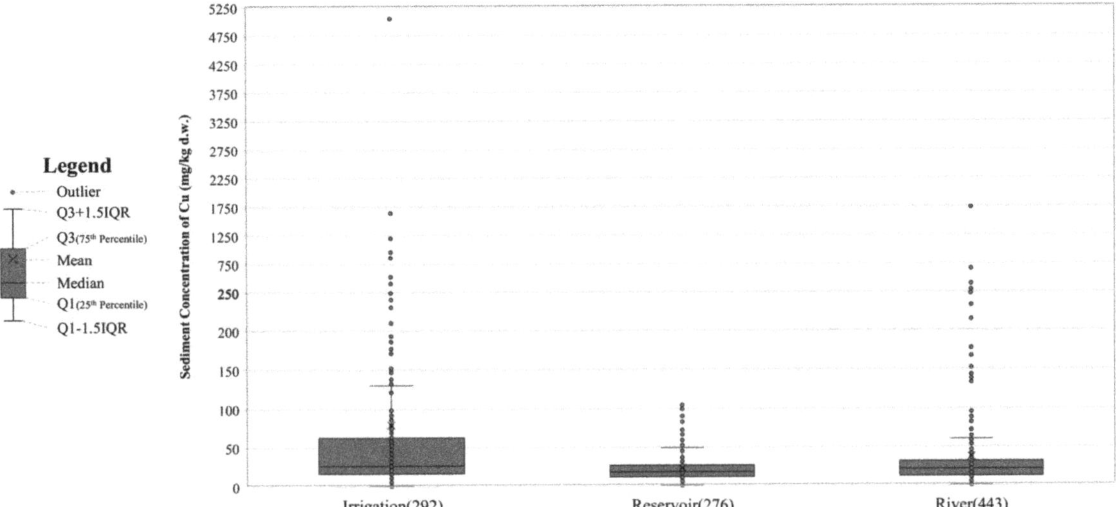

Figure 1. Box-and-whisker plot analysis of sediment quality in irrigation canals, rivers, and reservoirs based on Cu concentration. The number in the parentheses is the number of sampling sites.

3.2. Pollution Characteristics Analysis

3.2.1. Irrigation Canal Sediment Pollution and Agricultural Land Pollution Site Patterns

Taiwan's sediment management policy uses the upper limit value of sediment quality guidelines as the value to commence risk assessment. Table 1 shows the lower and upper limit values of sediment quality guidelines for heavy metals. Based on logistic regression model, the ecological effects of an individual species can be considered to be "rarely observed" ($\leq$25% possibility) if a metal concentration falls below the lower limit; those at upper-limit concentrations are likely to have more apparent ecological effects making their effects more "frequently observed" ($\geq$50%). Regarding the use of risk assessment results as the basis for determining treatment necessity, no control standard to determine pollution level has been stipulated. Therefore, the outlier value exceeding the upper extreme of the box-and-whisker plot was used to determine the pollution of sediments. Moreover, because sediments could resuspend and flow into agricultural land with irrigation water or accumulate by suspended solids in effluent [27], the pollution potential of sediments may be consistent with the characteristics of downstream polluted agricultural land [21]. However, because the sampling time of sediments and soil of polluted agricultural land differed, heavy metal pollutants in sediments across counties and cities and those in control sites of polluted agricultural land soil (hereinafter referred to as the polluted agricultural land site) were compared. Survey results up to the time the current study was reported were used to assess possible pollution intervention patterns.

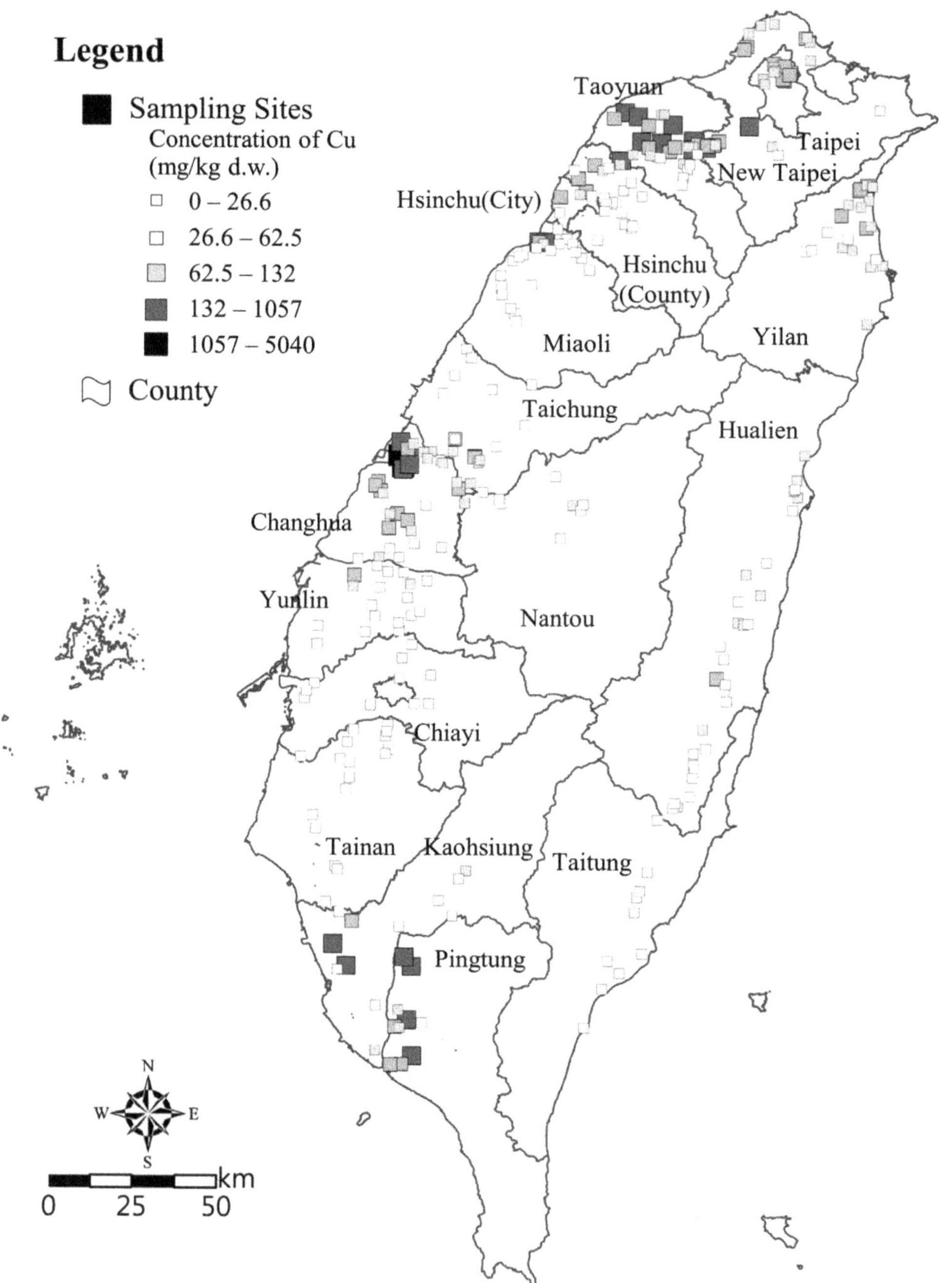

Figure 2. Irrigation canal sediment sampling sites and Cu concentration distribution in Taiwan.

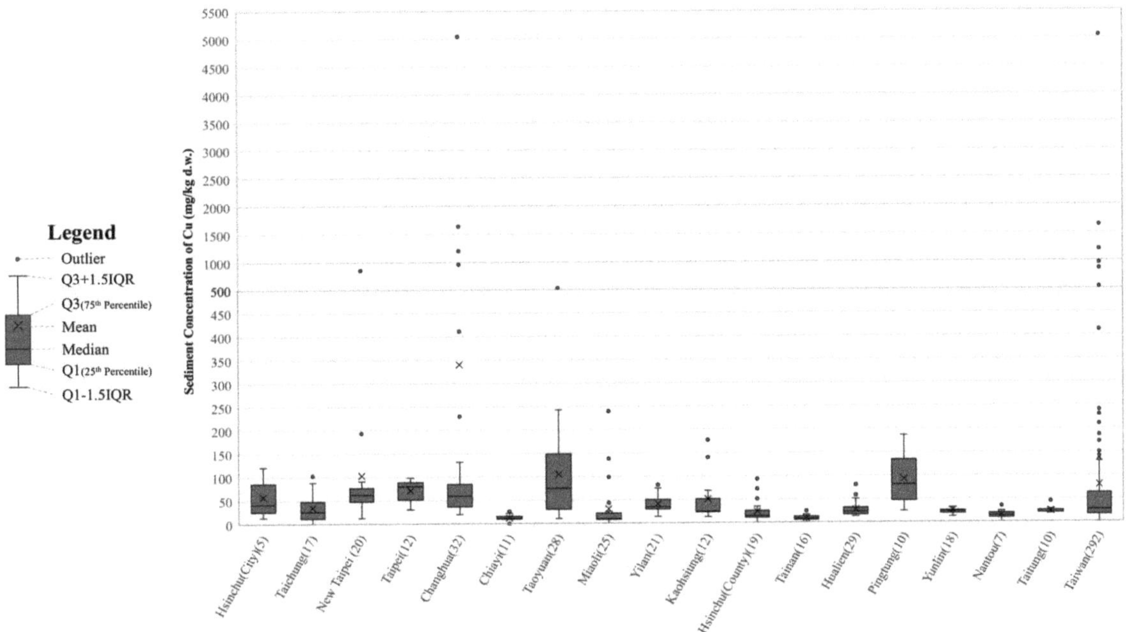

Figure 3. Box-and-whisker plot analysis of Cu concentration in irrigation canal sediment sampling sites in various counties and cities in Taiwan. The number in the parentheses is the number of sampling sites in a city/county.

Table 1. The lower and upper limit values of sediment quality guidelines for heavy metals.

Sediment Quality Indicator	Upper Limit (mg/kg)	Lower Limit (mg/kg)
As	33.0	11.0
Cd	2.49	0.65
Cr	233	76.0
Cu	157	50.0
Hg	0.870	0.230
Ni	80.0	24.0
Pb	161	48.0
Zn	384	140

According to the study hypothesis of pollution determination, the upper extreme of the overall irrigation canal sediment concentration in Taiwan (132 mg/kg) was used as the reference. A total of 25 sites exhibited Cu concentrations higher than the upper extreme in sediments. The distribution of the 25 sediment pollution sites and the sediment pollutants and concentrations are shown in Figure 4. The number of pollution sites was highest in Taoyuan (eight sites), followed by Changhua (seven sites) and Pingtung (four sites). Miaoli, Kaohsiung, and New Taipei City each had two pollution sites. In addition, the overall outliers for As, Cd, Cr, Cu, Hg, Ni, Pb, and Zn in irrigation canals was calculated to understand the pollutants (as detailed in Figure 5 and Table S3 (Supplementary Materials)). Among the 25 sediment pollution sites, most pollution patterns containing Cu concurrently contained Cr, Cu, Ni, Pb, and Zn (eight sites, approximately 32%), followed by the pattern concurrently presenting Cr, Cu, Pb, and Zn (five sites, 20%), and the patterns of Cu + Pb,

Cu + Pb + Zn, and Cu + Zn (each with two sites, 8%). Most other patterns also presented these five heavy metals (Cr, Cu, Ni, Pb, and Zn).

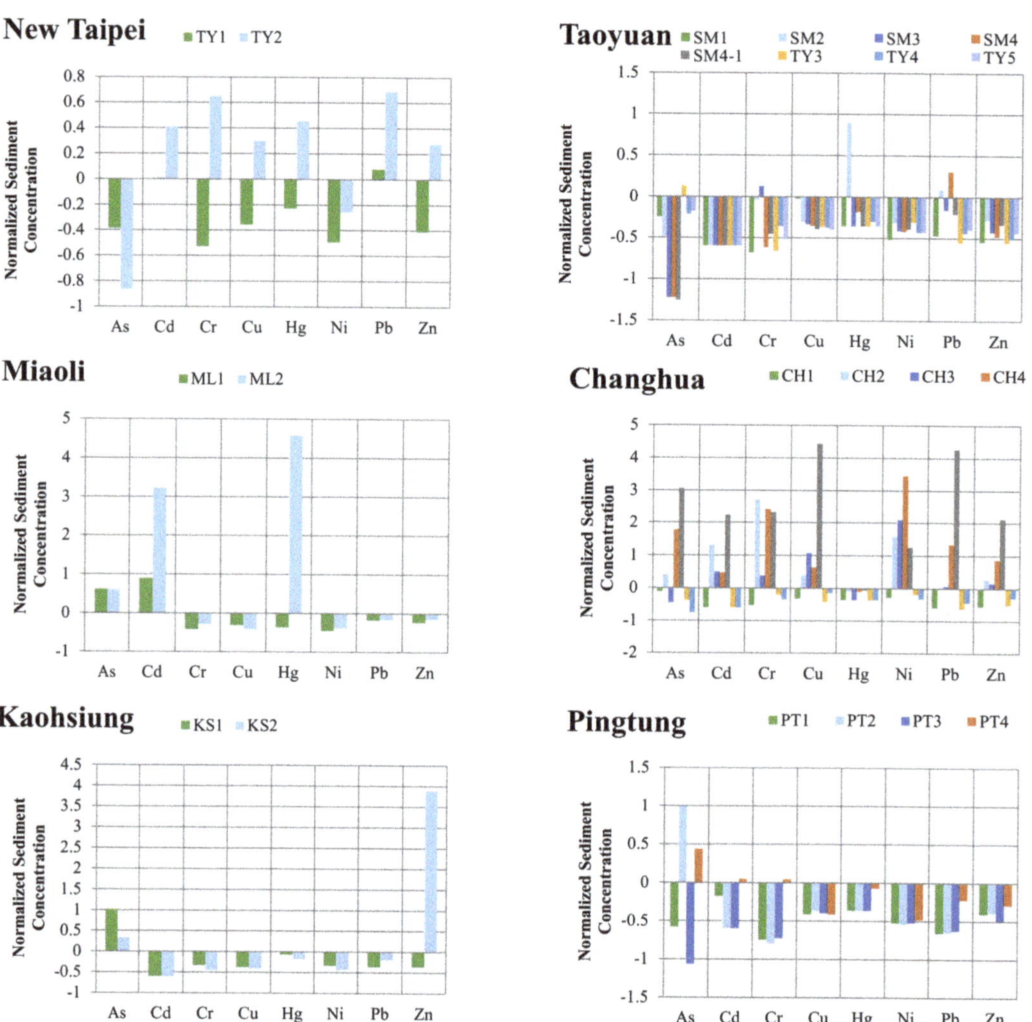

Figure 4. Bar charts of pollutants and concentrations of 25 sediment pollution sites.

Agricultural land pollution site characteristics for each county and city are summarized in Table 2. The agricultural land pollution sites were located mainly in Taoyuan (47%) and Changhua (45%). In the heavy metal pollutant patterns, most sites contained Cu pollution, totaling 3341 sites (59.8%), followed by pollution sites containing Cr, Cu, Ni, and Zn (10.8%) and the pattern of Cu and Zn (7.9%). The aforementioned three patterns accounted for nearly 80% of the agricultural land pollution sites. In addition to Cr, Cu, Ni, and Zn, pollutants including Cd, Pb, As, and Hg were identified in low proportions.

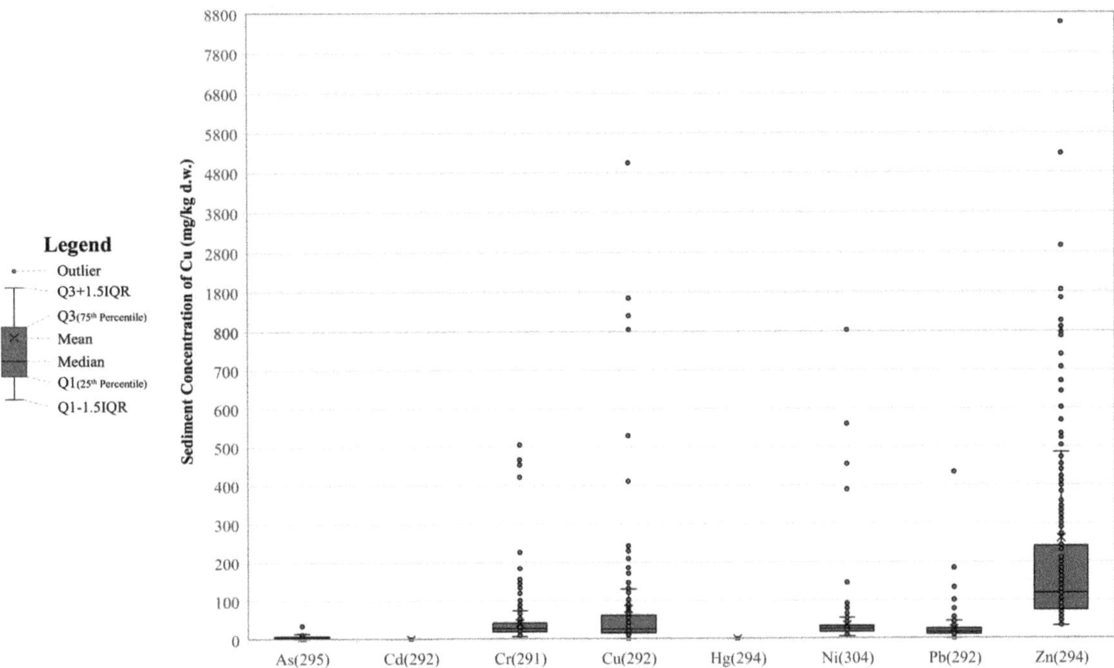

Figure 5. Heavy metal statistics of irrigation canals in Taiwan. The number in the parentheses is the number of metal appearance in the sampled irrigation canals.

Table 2. Summary of pollution characteristics in agricultural land pollution sites containing Cu.

Pollutant (Cu Included)	Yilan	Pingtung	Miaoli	Tao-yuan	Kaohsiung	New Taipei	Hsinchu (City)	Chang-hua	Taichung	Taipei	Tainan	Total	Ratio
As, Cr, Cu, Pb			1									1	0.02%
As, Cr, Cu, Ni, Zn								1				1	0.02%
As, Cu			2					2				4	0.07%
As, Cu, Zn								1				1	0.02%
As, Cd, Cu, Ni, Pb								1				1	0.02%
Cr, Cu				16			4	17	25			62	1.11%
Cr, Cu, Pb, Zn				2	1							3	0.05%
Cr, Cu, Zn				4				10	3		2	19	0.34%
Cr, Cu, Ni	1			1			9	139	89			239	4.28%
Cr, Cu, Ni, Zn				1			1	595	5			602	10.8%
Cu		1	6	2400	1	8	5	850	61	2	7	3341	59.8%

Table 2. *Cont.*

Pollutant (Cu Included)	Yilan	Pingtung	Miaoli	Tao-yuan	Kaohsiung	New Taipei	Hsinchu (City)	Chang-hua	Taichung	Taipei	Tainan	Total	Ratio
Cu, Hg				2				1				3	0.05%
Cu, Pb				1		2						3	0.05%
Cu, Pb, Zn				1	6	1		6			1	15	0.27%
Cu, Zn	1			154		6		266	1		14	442	7.91%
Cu, Ni			1	3			1	285	12		3	305	5.46%
Cu, Ni, Pb								1				1	0.02%
Cu, Ni, Pb, Zn											1	1	0.02%
Cu, Ni, Zn				3				246	5			254	4.54%
Cd, Cr, Cu, Pb, Zn											2	2	0.04%
Cd, Cr, Cu, Ni								12				12	0.21%
Cd, Cr, Cu, Ni, Pb, Zn							154					154	2.76%
Cd, Cr, Cu, Ni, Zn							22	52				74	1.32%
Cd, Cu				13						1		14	0.25%
Cd, Cu, Hg										1		1	0.02%
Cd, Cu, Hg, Ni, Pb, Zn											1	1	0.02%
Cd, Cu, Pb											1	1	0.02%
Cd, Cu, Pb, Zn											6	6	0.11%
Cd, Cu, Zn				4							4	8	0.14%
Cd, Cu, Ni								5				5	0.09%
Cd, Cu, Ni, Pb								3				3	0.05%
Cd, Cu, Ni, Pb, Zn											2	2	0.04%
Cd, Cu, Ni, Zn								8				8	0.14%
Total	2	1	10	2605	7	18	196	2501	201	4	44	5589	100%
Ratio	0.04%	0.02%	0.18%	46.6%	0.13%	0.32%	3.51%	44.75%	3.60%	0.07%	0.79%	100%	

Note: Agricultural land soil pollution was determined according to Soil Pollution Control Standards, in which the criteria for each pollution item are 60 mg/kg for arsenic; 5 mg/kg for cadmium; 250 mg/kg for chromium; 200 mg/kg for copper; 5 mg/kg for mercury; 200 mg/kg for nickel; 500 mg/kg for lead; and 600 mg/kg for zinc.

Again, sediment pollution sites and agricultural land pollution sites were mainly distributed in Taoyuan and Changhua (Figure 6). The agricultural land pollution sites in Taoyuan generally contained Cu, with these totaling 2400 sites and accounting for 92% of the total number of agricultural land pollution sites in Taoyuan. The sediment pollution sites did not present notable patterns. Cr, Cu, Ni, Pb, and Zn were all possible pollutants. The agricultural land pollution sites were mainly distributed in areas near TY4 in northeastern Taoyuan (as detailed in Section 3.2.2). Overall, 71% (5/7) and 26% (648/2501) of sediment and agricultural land pollution sites in Changhua, respectively, exhibited pollution patterns containing Cr, Cu, Ni, and Zn. In addition, the locations of the patterns were distributed closely (as detailed in Section 3.2.3).

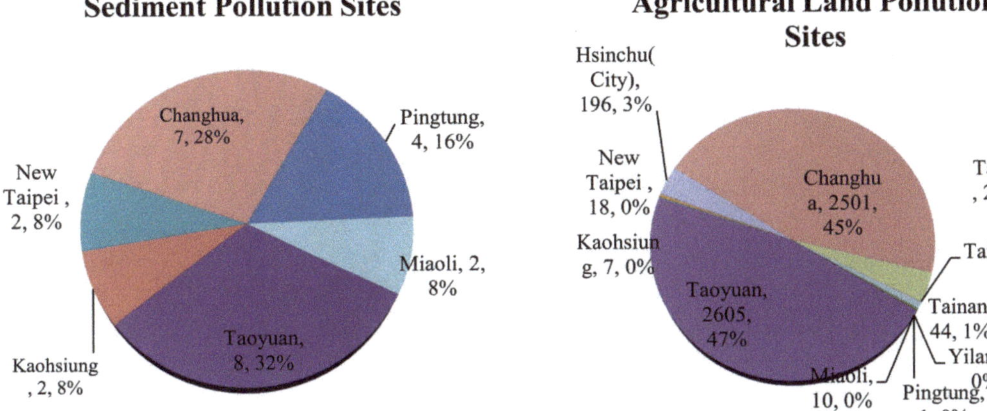

Figure 6. Pie charts of sediment and agricultural land pollution sites containing Cu.

3.2.2. Analysis of Sediment Pollution and Possible Pollution Sources in Taoyuan

Sediment pollution sites, irrigation canals, agricultural land pollution sites, possible pollution sources, and surface waters in Taoyuan are plotted in Figure 7. The areas with both sediment and agricultural land pollution sites were distributed between the trunk lines of Sinjie Drainage and Pusin Drainage. At the west side of the Sinjie Drainage trunk line and the east side of the trunk line of Pusin Drainage, irrigation canals are present. The irrigation area of TY4 was less likely to include irrigation canals supplying irrigation beyond the range between the two drainages. Therefore, the preliminary comparison of the pollution characteristics of sediment and agricultural land pollution sites in the range between these drainages was conducted.

The sediment pollution characteristic items in TY4 indicated mainly Cr, Cu, and Pb. The agricultural land pollution sites were in the area that bordered Sinjie Drainage to the west, Pusin Drainage to the east, Taoyuan Da Zun to the south, and the shoreline to the north. A total of 1503 agricultural pollution sites were present in this area, accounting for 58% of all the agricultural pollution sites in Taoyuan. In this area, 97% of the agricultural land pollution sites were mainly polluted by Cu. Agricultural land containing both Cr and Cu accounted for 0.7%. Two Pb-containing agricultural land pollution sites were 1.3 km from the sediment pollution sites. Given that Pb in agricultural soil must reach 500 mg/kg to be identified as pollution (48.7 mg/kg of Pb in sediment was used as the pollution criterion in this study), a preliminary inference was made that the criterion was possibly related to the low proportion of Pb-containing agricultural land pollution sites. For Cr and Cu, the characteristics of sediment pollution sites corresponded to those of the agricultural land pollution sites. Agricultural land pollution sites with patterns of Cu + Zn pollution (1%), Cd + Cu pollution (0.4%), and Cd + Cu + Zn pollution (0.1%) were located close to the downstream area of drainage trunk lines. The patterns might be related to downstream pollution; Zn and Cd were not found in the sediments.

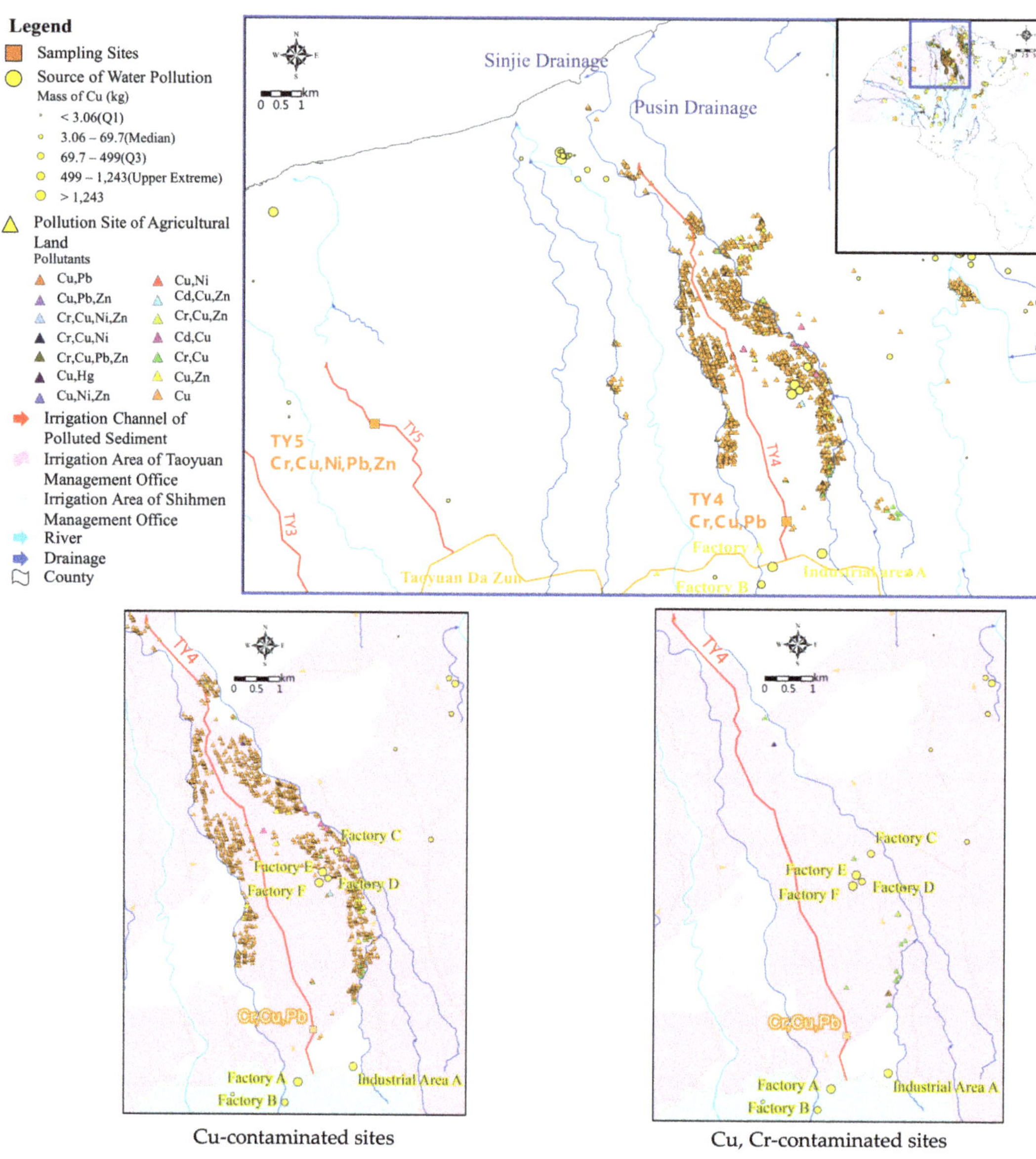

Cu-contaminated sites

Cu, Cr-contaminated sites

Figure 7. *Cont.*

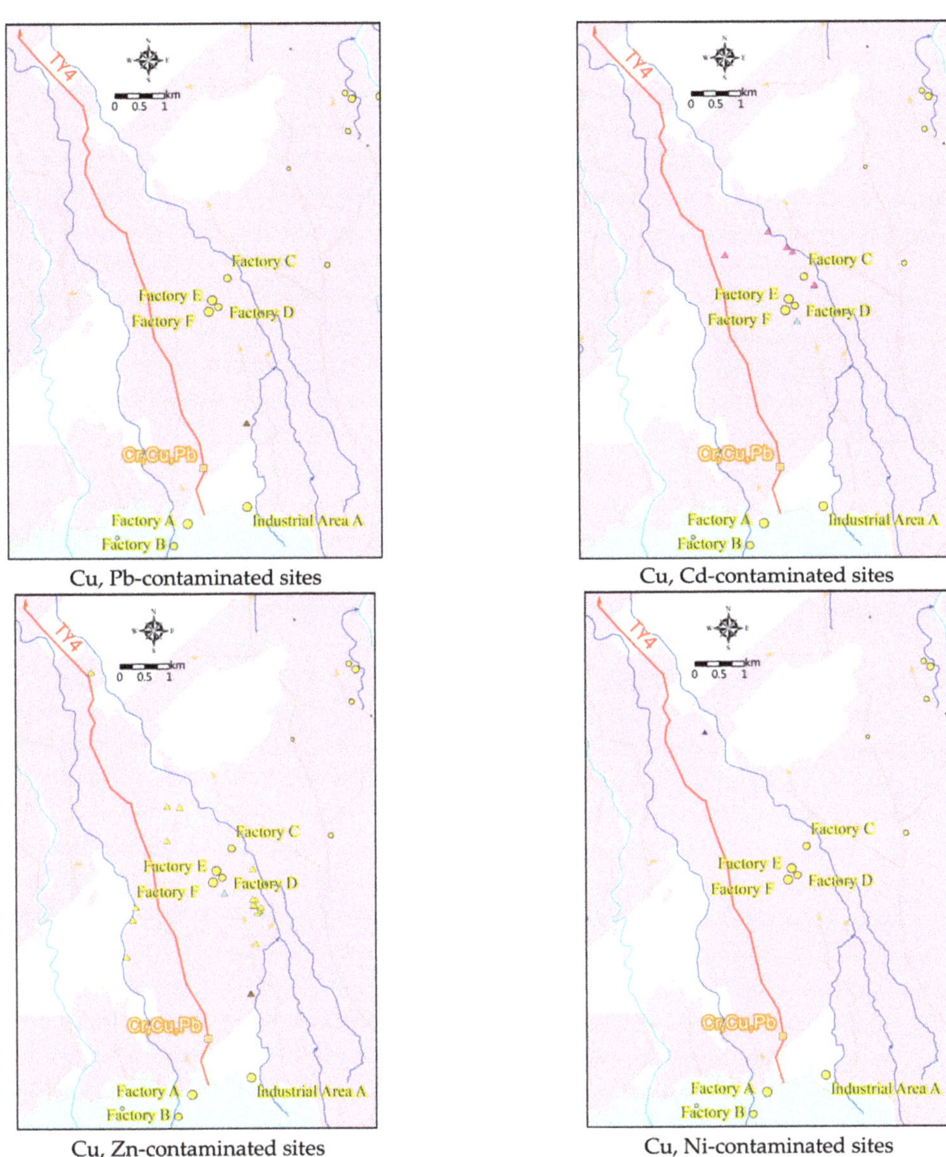

Cu, Pb-contaminated sites

Cu, Cd-contaminated sites

Cu, Zn-contaminated sites

Cu, Ni-contaminated sites

Figure 7. Sediment and agricultural land pollution sites in Taoyuan and the distribution of possible pollution sources.

In terms of the number of potential Cu sources in Taoyuan City, bare printed circuit boards (PCBs) manufacturing has the largest number, accounting for about 43% of the total factories, followed by surface treatments with 15% of the total. The Cu emission amount is also the highest in the PCB manufacturing industry, which accounted for about 78% from 2014 to 2019. A comparison of the pollution sources revealed that two PCB factories (Factory A and B) and an industrial area (Industrial A) were located at the upstream area of sediment pollution sites, two factories operated near the trunk line of Sinjie Drainage, and another 4 PCB factories operated (Factory C, D, E, and F) near the trunk line of Pusin Drainage at the downstream the sediment pollution sites. The heavy metal release mass of the six factories and the industrial area into flowing water during the period 2014–2019 are shown in Figure 8. The two upstream factories were located near the upstream area of the canal with the sediment pollution sites, where the heavy metal masses of releases were mostly Cu (approximately 86%), Ni (approximately 9.3%), and Pb (approximately 3.4%). The release of Cr was small (from only one factory source; approximately 0.1%). The sediments therefore presented pollution characteristics of Cr, Cu, and Pb. However, the level of Ni in the sediments did not attain the level of pollution, possibly because of nonhomogenous samples or different sampling time points or for other reasons. Ni was also not notably present in agricultural land pollution sites. The overall heavy metal mass of release was high in Industrial Area A, with considerably different pollutants. Therefore, most agricultural land pollution sites with Cd, Cr, Ni, Pb, and Zn were located near the Pusin Drainage trunk line. Fewer corresponding characteristics were found in sediment pollution sites.

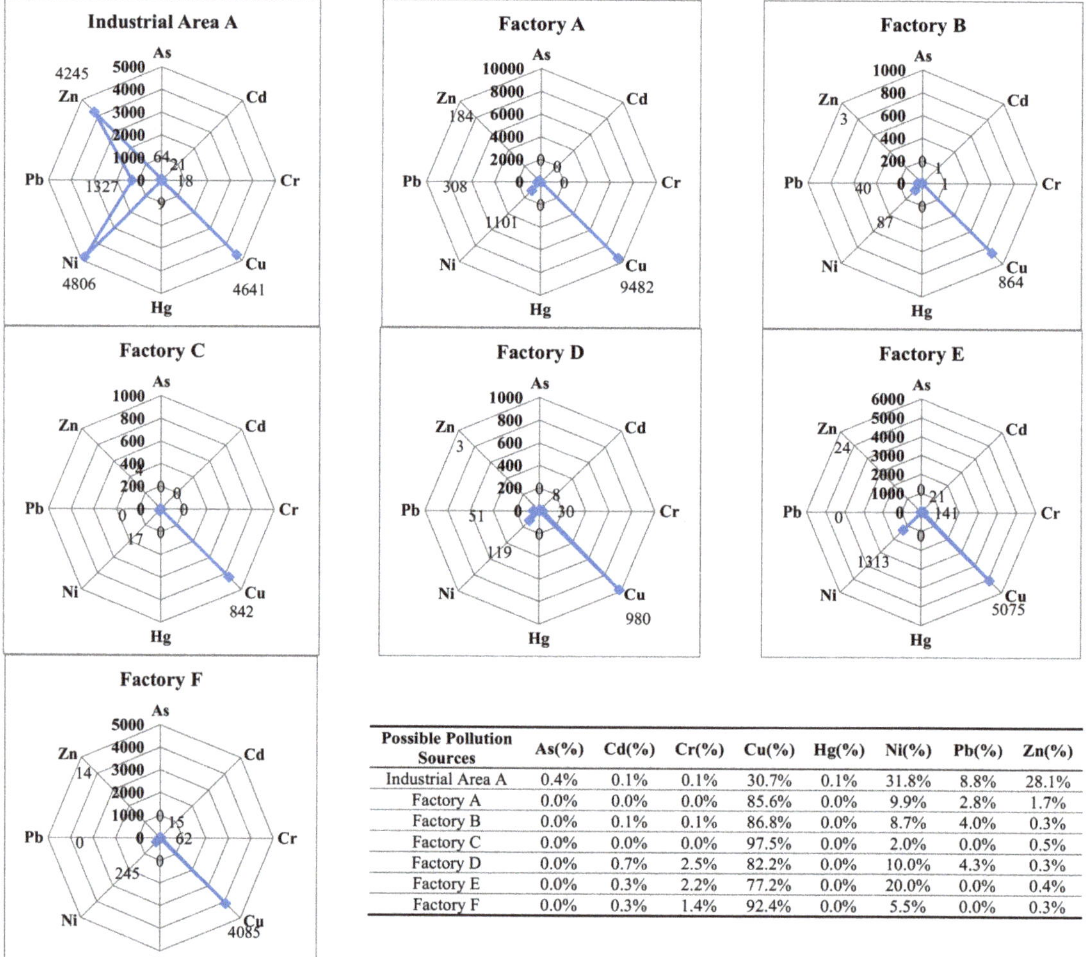

Possible Pollution Sources	As(%)	Cd(%)	Cr(%)	Cu(%)	Hg(%)	Ni(%)	Pb(%)	Zn(%)
Industrial Area A	0.4%	0.1%	0.1%	30.7%	0.1%	31.8%	8.8%	28.1%
Factory A	0.0%	0.0%	0.0%	85.6%	0.0%	9.9%	2.8%	1.7%
Factory B	0.0%	0.1%	0.1%	86.8%	0.0%	8.7%	4.0%	0.3%
Factory C	0.0%	0.0%	0.0%	97.5%	0.0%	2.0%	0.0%	0.5%
Factory D	0.0%	0.7%	2.5%	82.2%	0.0%	10.0%	4.3%	0.3%
Factory E	0.0%	0.3%	2.2%	77.2%	0.0%	20.0%	0.0%	0.4%
Factory F	0.0%	0.3%	1.4%	92.4%	0.0%	5.5%	0.0%	0.3%

Figure 8. Radar charts of heavy metal mass of release at pollution sources near sediment pollution sites in TY4 (Note: Total mass of release of heavy metals during the period 2014–2019, unit: kg).

An investigation of historical survey data on the research area suggested that agricultural land pollution possibly resulted from effluent entering irrigation water sources. Although irrigation water quality might not appear abnormal, the constant accumulation of low concentrations of heavy metal might cause heavy metal buildup in agricultural land soil [28,29]. In addition, the EPA of the Executive Yuan in Taiwan analyzed the water quality and suspended solids at the diversion dams of the Pusin Drainage trunk line and found that most suspended solids were composed of Cu pollutants. The analysis indicated that the water and suspended solids contained 0.103 mg/L and 2890 mg/kg of Cu pollutants, respectively [27]. Suspended solids, with their notably high proportion of Cu pollution, are prone to enter sediments or agricultural land with irrigated water.

Taoyuan's main irrigation source is Shihmen Reservoir, which offers approximately 47% and 48% of the irrigation water consumed in Taoyuan and Shihmen Management Office areas, respectively. For the remaining 53% of irrigation water, supplementation by nearby rivers and drainages is relied upon [30,31]. Between 2014 and 2019, approximately 61% of Cu in industrial wastewater was released into rivers or drainages [32]. To facilitate appropriate water resource use, irrigation water surplus was returned (i.e., return water). Investigation revealed that return water in the irrigation area of Shihmen Management Office accounted for 20% of the overall water resource [30], leading to the possible pollution of agricultural land downstream due to the return of polluted water sources upstream [28]. The summary results of the present study indicated that the pollution characteristics of sediment pollution sites were consistent with those of agricultural land pollution sites (Cr and Cu) and the characteristics of sediment pollution sites were similar to pollution sources (Cr, Cu, and Pb), suggesting similar sources of pollution at the sediment and agricultural land pollution sites. Therefore, the main cause of sediment pollution accumulation in Taoyuan was inferred to be the irrigation canals (rivers or drainages) affected by industrial pollution and return water utilization.

3.2.3. Analysis of Sediment Pollution and Possible Pollution Sources in Changhua

Sediment pollution sites, irrigation canals, areas with different irrigation sources, agricultural land pollution sites, possible pollution sources, and surface waters in Changhua are plotted in Figure 9. Most waterways in Changhua were planned as municipal or business drainage. The north and south sides of the research area were the Wu River and the Zhuoshui River, respectively. In addition to the two rivers, multiple drainages were categorized as irrigation canals in Changhua. Dividing the irrigation area according to water sources renders single water source areas (Wu River, Zhuoshui River, and Jiuzhuoshui River Drainage trunk line), double water source areas (Wu River and Yangzicuo Drainage trunk line, Zhuoshui River and Yuliao River Drainage trunk line, and Jiuzhuoshui River Drainage trunk line and Zhuoshui River), and triple water source areas (Jiuzhuoshui River Drainage trunk line, Zhuoshui River, and Wanxing Drainage trunk line). Sediment and agricultural land pollution sites categorized according to water sources are shown in Table 3. Most sediment pollution sites were irrigation canals with water sources from the Wu River and the Wu River + Yangzicuo Drainage trunk line. The irrigation area of the Wu River contained two sediment pollution sites, and the pollutants from the two sources, respectively, were Cr, Cu, Ni, Pb, and Zn, and Cr, Cu, and Ni. The irrigation area with the Wu River and Yangzicuo Drainage trunk line as the water sources contained five sediment pollution sites, of which three had Cr, Cu, Ni, Pb, and Zn; one had As, Cr, Cu, Ni, and Zn; and one had Cu and Ni. Aside from As, the pollutants in the two irrigation areas were quite similar.

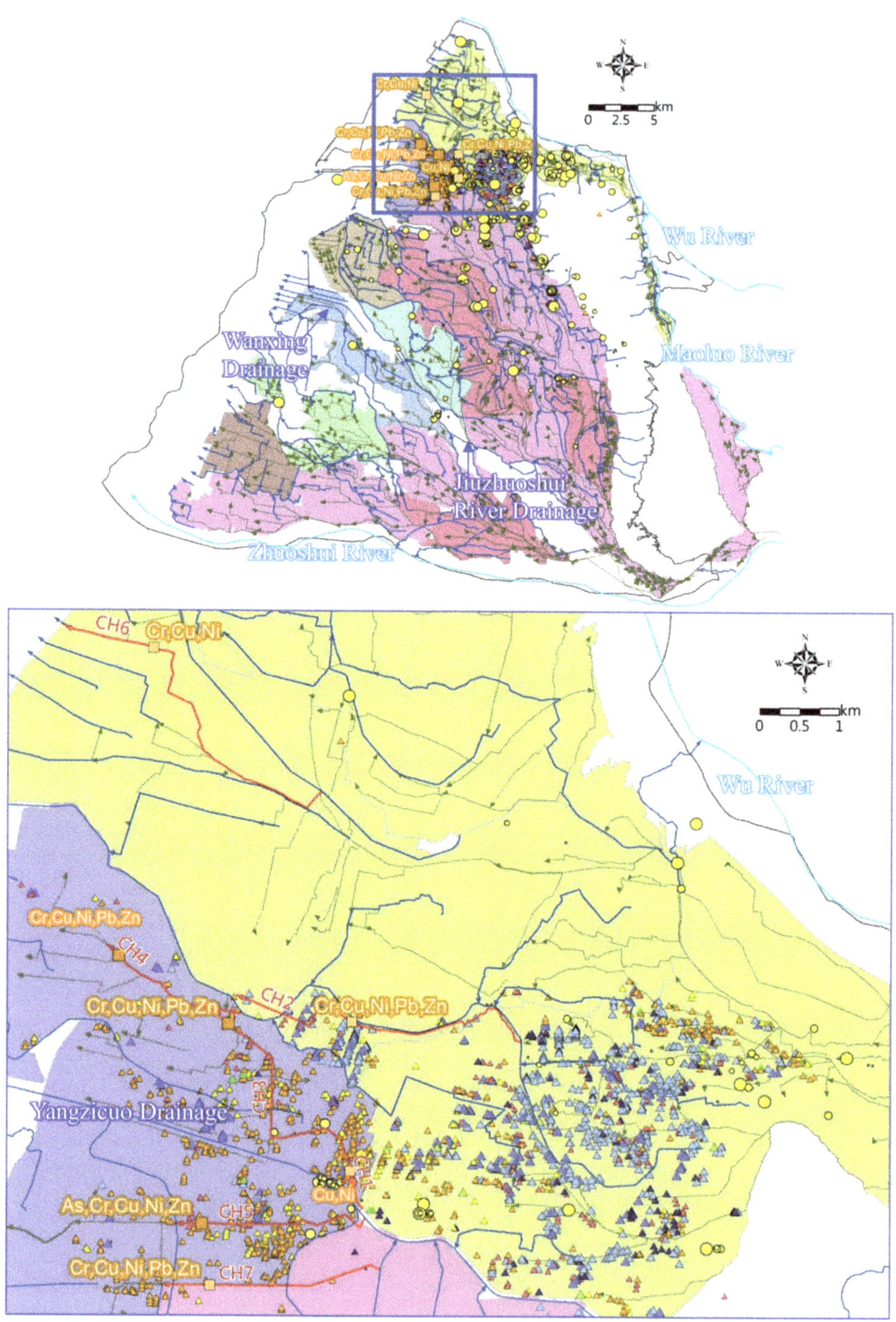

Figure 9. *Cont.*

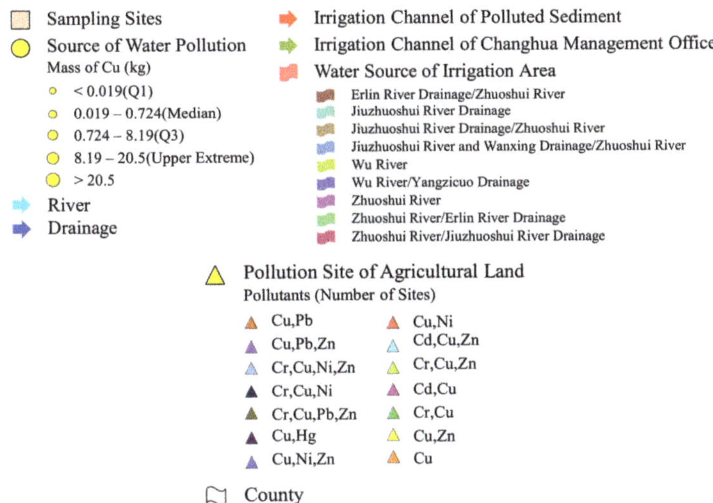

Figure 9. Distribution of the sediment and agricultural land pollution sites as well as possible pollution sources in Changhua County.

Table 3. Summary of the sediment and agricultural land pollution sites of different water sources.

Water Source	Wu River				Wu River/Yangzicuo Drainage Trunk Line				Jiuzhuoshui River Drainage Trunk Line/Zhuoshui River	
	Sediment Pollution Sites Number	Sediment Pollution Sites Ratio	Agricultural Land Pollution Sites Number	Agricultural Land Pollution Sites Ratio	Sediment Pollution Sites Number	Sediment Pollution Sites Ratio	Agricultural Land Pollution Sites Number	Agricultural Land Pollution Sites Ratio	Agricultural land Pollution Sites Number	Agricultural Land Pollution Sites Ratio
Cu	0	0%	287	20%	0	0%	473	63%	0	0%
Cr, Cu, Ni, Zn	1	50%	538	38%	4	80%	22	3%	0	0%
Cu, Ni	0	0%	175	12%	1	20%	60	8%	0	0%
Cu, Zn	0	0%	68	5%	0	0%	129	17%	1	50%
Cu, Ni, Zn	0	0%	159	11%	0	0%	47	6%	0	0%
Cr, Cu, Ni	1	50%	118	8%	0	0%	1	0%	0	0%
Cd, Cr, Cu, Ni, Zn	0	0%	43	3%	0	0%	0	0%	0	0%
Cr, Cu,	0	0%	8	1%	0	0%	7	1%	1	50%
Cd, Cr, Cu, Ni	0	0%	11	1%	0	0%	0	0%	0	0%
Cr, Cu, Zn	0	0%	8	1%	0	0%	0	0%	0	0%
Cd, Cu, Ni, Zn	0	0%	7	0%	0	0%	1	0%	0	0%
Cu, Pb, Zn	0	0%	0	0%	0	0%	6	1%	0	0%
Cd, Cu, Ni	0	0%	4	0%	0	0%	0	0%	0	0%
Cd, Cu, Ni, Pb	0	0%	1	0%	0	0%	0	0%	0	0%
As, Cu	0	0%	1	0%	0	0%	0	0%	0	0%
As, Cr, Cu, Ni, Zn	0	0%	1	0%	0	0%	0	0%	0	0%
As, Cu, Zn	0	0%	0	0%	0	0%	1	0%	0	0%
As, Cd, Cu, Ni, Pb	0	0%	1	0%	0	0%	0	0%	0	0%
Cu, Hg	0	0%	0	0%	0	0%	0	0%	0	0%
Cu, Ni, Pb	0	0%	1	0%	0	0%	0	0%	0	0%
Total	0	100%	1431	100%	0	100%	747	100%	2	100%

Note: An absence of irrigation water source indicates that no agricultural land pollution site was identified in the irrigation area.

In all, 2501 agricultural land pollution sites in Changhua contained Cu. The agricultural land pollution sites in the Wu River irrigation area accounted for 57% (1431 sites) of all the Cu-containing agricultural land pollution sites in Changhua. Many in the irrigation area (38% or 538 agricultural land pollution sites) contained Cr, Cu, Ni, and Zn. Approximately 20% of the sites (287) presented only Cu. Approximately 12% (175 sites) and 11% (159 sites) of the agricultural land pollution sites contained the patterns of Cu + Ni and Cu + Ni + Zn, respectively. The agricultural land pollution sites in the Wu River and Yangzicuo Drainage trunk line irrigation area accounted for approximately 30% (747 sites) of all the Cu-containing agricultural land pollution sites. In this area, most (63%, 473 sites) agricultural land pollution sites contained Cu pollution, followed by those presenting the patterns of Cu + Zn (17%, 129 sites), Cu + Ni (8%, 60 sites), and Cu + Ni + Zn (6%, 47 sites). The pollution patterns of the sediment and agricultural land pollution sites implied that the main pollutants were Cr, Cu, Ni, and Zn. Pb, which exhibited a trend similar to that in Taoyuan, which was found in sediment pollution sites but not in agricultural land pollution sites.

According to the heavy metal mass of the release of various receiving waters between 2014 and 2019 (shown in Table 4), the main irrigation water sources of sediment pollution sites were the Wu River and the Yangzicuo Drainage trunk line. The mass of release of Cu into the Yangzicuo Drainage accounted for 51% of the total amount of mass of release of Cu. Although the mass of release of Cu was only 3% in the Wu River, the receiving waters for 38% of the Cu release were not clearly recorded. Regarding the Cr mass of release, 33% and 7% entered the Yangzicuo Drainage trunk line and the Wu River, respectively. For the Ni mass of release, 9% and 2%, respectively, entered the Yangzicuo Drainage trunk line and the Wu River; whereas the receiving waters for 79% of the Ni release were not recorded. Regarding the Zn mass of release, 22% and 9% entered the Yangzicuo Drainage trunk line and the Wu River, respectively. The results revealed that irrigation water sources were the main receivers of wastewater. A study in 1999 noted that if an annual mean of water conductivity of greater than 750 μmho/cm indicated water pollution, as recorded during the period 1979–1996, then 46%–86.2% of irrigation water was contaminated in Changhua [33]. Figure 9 presents the spatial distribution of possible pollution sources. The distribution suggests that considerable possible pollution sources existed in the irrigation areas of the Wu River and the Zhuoshui River. Historical survey data revealed that along the bank between upstream and downstream areas of the Yangzicuo Drainage trunk line, many unregistered illegal electroplating factories secretly discharged unprocessed waste solutions into the Yangzicuo Drainage trunk line at midnight. Numerous environmental auditing results have revealed that most heavy metal pollution in Changhua was from the illegal discharge of wastewater by legal electroplating factories and unauthorized pipelines installed by illegal factories [34]. In addition, many irrigation canals were discovered to be using return water [34], which further resulted in the polluted water affecting the soil quality of agricultural land downstream. Therefore, considerable irrigation from drainages or rivers affected by industrial pollutants and the presence of both legal and illegal factories in the irrigation area were the main causes of sediment pollution in Changhua.

Table 4. The mass of release, the ratio of heavy metal released in single water source, and the ratio of single heavy metal released among water bodies in Changhua.

River/Drainage Basin	Mass of Release (kg)								
	As	Cd	Cr	Cu	Hg	Ni	Pb	Zn	Total
Yangzicuo Drainage	3	1	928	3988	0	881	35	2421	8257
Others	189	0	573	2964	4	8055	140	6287	18212
Erlin River Drainage	0	0	146	397	0	762	0	504	1810
Wu River	0	0	189	270	0	212	1	969	1640
Yuanlin Drainage	1	1	721	190	0	129	5	238	1283
Wanxing Drainage	0	0	12	21	0	13	0	84	131
Jiuzhuoshui River Drainage	0	0	261	8	0	115	2	299	684
Qingshui River SubsidiaryDrainage	0	0	1	7	0	3	0	38	49
Maoluo River	0	0	0	0	0	0	0	0	0
Total	193	2	2832	7845	4	10,170	182	10,839	32,067

River/Drainage Basin	Ratio of Heavy Metal Released in Single Water Source (%)								
	As	Cd	Cr	Cu	Hg	Ni	Pb	Zn	Total
Yangzicuo Drainage	0%	0%	11%	48%	0%	11%	0%	29%	100%
Others	1%	0%	3%	16%	0%	44%	1%	35%	100%
Erlin River Drainage	0%	0%	8%	22%	0%	42%	0%	28%	100%
Wu River	0%	0%	12%	16%	0%	13%	0%	59%	100%
Yuanlin Drainage	0%	0%	56%	15%	0%	10%	0%	19%	100%
Wanxing Drainage	0%	0%	9%	16%	0%	10%	0%	64%	100%
Jiuzhuoshui River Drainage	0%	0%	38%	1%	0%	17%	0%	44%	100%
Qingshui River SubsidiaryDrainage	0%	0%	3%	14%	0%	6%	0%	77%	100%
Maoluo River	0%	0%	0%	63%	0%	0%	0%	38%	100%
Total	1%	0%	9%	24%	0%	32%	1%	34%	100%

River/Drainage Basin	Ratio of Single Heavy Metal Released among Water Bodies (%)								
	As	Cd	Cr	Cu	Hg	Ni	Pb	Zn	Total
Yangzicuo Drainage	2%	41%	33%	51%	0%	9%	19%	22%	26%
Others	98%	9%	20%	38%	100%	79%	77%	58%	57%
Erlin River Drainage	0%	0%	5%	5%	0%	7%	0%	5%	6%
Wu River	0%	0%	7%	3%	0%	2%	0%	9%	5%
Yuanlin Drainage	0%	49%	25%	2%	0%	1%	3%	2%	4%
Wanxing Drainage	0%	1%	0%	0%	0%	0%	0%	1%	0%
Jiuzhuoshui River Drainage	0%	1%	9%	0%	0%	1%	1%	3%	2%
Qingshui River SubsidiaryDrainage	0%	0%	0%	0%	0%	0%	0%	0%	0%
Maoluo River	0%	0%	0%	0%	0%	0%	0%	0%	0%
Total	100%	100%	100%	100%	100%	100%	100%	100%	100%

4. Conclusions

Taiwan completed its first heavy metal detection survey in sediments in irrigation canals in 2019, during which a total of 292 data were collected. Compared with rivers and reservoirs, sediment heavy metal concentrations in irrigation canals, especially Cu, were substantially higher. A division by counties and cities revealed that most sites with high Cu concentrations in sediments were located in Taoyuan City and Changhua County, where 60% (15/25) of sediment pollution sites were discovered to contain a high Cu content, and 91% (5106/5589) of agricultural land pollution sites were recorded to contain Cu.

The sediment pollution sites, Cu-containing agricultural pollution sites, and pollution sources in Taoyuan and Changhua were compared. Notably, the heavy metal release items of sediment pollution sites and possible pollution sources upstream in Taoyuan were consistent; both contained Cr, Cu, and Pb. In addition, the proportion of Cu mass of release was considerably high (86%). The sediment pollution sites in Changhua were with the main pollutants being Cr, Cu, Ni, Pb, and Zn and the main pollutants in agricultural land pollution sites were Cr, Cu, Ni, and Zn. These results revealed that the pollution characteristics in sediment pollution sites, agricultural land pollution sites, and pollution sources were highly consistent. We also noticed that return water irrigation may compromise downstream irrigation water quality because of polluted water returned upstream. The results indicated that the primary cause of sediment pollution in Taoyuan might be irrigation water sources from polluted rivers or drainages and return water. For Changhua, irrigation sources included drainages that receive the discharge of most wastewater pollution in the county. Incidences of return irrigation and illegal discharge of wastewater were reported in certain regions, which polluted sediments in Changhua. Results from this study establish a valuable linkage of the distribution of Cu and common heavy metal pollutants in the sediments of irrigation canals, polluted agricultural sites, and these potential heavy metal sources to understand the possible causes of sediment pollution. These results also demonstrate the importance of the adequate management of irrigation and drainage systems to prevent the accumulation of Cu and common heavy metal pollutants in sediment and its subsequent pollution to farmland.

Supplementary Materials: The following are available online at https://www.mdpi.com/article/10.3390/min11040416/s1, Table S1: Supporting information of the box-and-whisker plot analysis of sediment quality in irrigation canals, rivers, and reservoirs based on Cu concentration, Table S2: Supporting information of the box-and-whisker plot analysis of Cu concentration in irrigation canal sediment sampling sites in various counties and cities in Taiwan, Table S3: Supporting information of heavy metal statistics of irrigation canals in Taiwan.

Author Contributions: Conceptualization, S.-H.H., T.-C.C. and H.-C.H.; methodology, S.-H.H., H.-C.C. and Z.-S.W.; formal analysis, S.-H.H. and Y.-C.C.; resources, H.-C.C. and Z.-S.W.; writing—original draft preparation, S.-H.H., Y.-C.C., Y.-L.W. and H.-C.H.; writing—review and editing, S.-H.H., T.-C.C., H.-C.C. and H.-C.H. All authors have read and agreed to the published version of the manuscript.

Funding: This research and the APC were funded by Taiwan Environmental Protection Administration, grant number 108A372.

Institutional Review Board Statement: Not applicable.

Data Availability Statement: The authors confirm that the data supporting the findings of this study are available within the article and its Supplementary Materials.

Acknowledgments: This study was funded by the Taiwan Environmental Protection Administration. The views or opinions expressed in this article are those of the writers and should not be con-strued as opinions of the Taiwan Environmental Protection Administration.

Conflicts of Interest: The authors declare no conflict of interest.

References

1. Peng, Y.-P.; Chang, Y.-C.; Chen, K.-F.; Wang, C.-H. A field pilot-scale study on heavy metal-contaminated soil washing by using an environmentally friendly agent-poly-γ-glutamic acid (γ-PGA). *Environ. Sci. Pollut. Res. Int.* **2020**, *27*, 34760–34769. [CrossRef] [PubMed]
2. Liaw, C.-H.; Chiang, Y.-C. Framework for Assessing the Rainwater Harvesting Potential of Residential Buildings at a National Level as an Alternative Water Resource for Domestic Water Supply in Taiwan. *Water* **2014**, *6*, 3224–3246. [CrossRef]
3. Yen, C.-H.; Chen, K.-F.; Sheu, Y.-T.; Lin, C.-C.; Horng, J.-J. Pollution Source Investigation and Water Quality Management in the Carp Lake Watershed, Taiwan. *Clean Soil Air Water* **2012**, *40*, 24–33. [CrossRef]
4. Council of Agriculture, Taiwan. Irrigation Management. Available online: https://www.ia.gov.tw/ (accessed on 6 November 2020).
5. Environmental Protection Administration, Taiwan. Database of Soil and Groundwater Pollution Sites. Available online: https://sgw.epa.gov.tw/ContaminatedSitesMap/Default.aspx (accessed on 13 November 2020).
6. Wuana, R.A.; Okieimen, F.E. Heavy Metals in Contaminated Soils: A Review of Sources, Chemistry, Risks and Best Available Strategies for Remediation. *ISRN Ecol.* **2011**, *2011*, 1–20. [CrossRef]
7. Adrees, M.; Ali, S.; Rizwan, M.; Ibrahim, M.; Abbas, F.; Farid, M.; Zia-Ur-Rehman, M.; Irshad, M.K.; Bharwana, S.A. The effect of excess copper on growth and physiology of important food crops: A review. *Environ. Sci. Pollut. Res. Int.* **2015**, *22*, 8148–8162. [CrossRef] [PubMed]
8. Mourato, M.; Martins, L.; Campos-Andrada, M. Physiological responses of Lupinus luteus to different copper concentrations. *Biol. Plant.* **2009**, *53*, 105–111. [CrossRef]
9. Van Zwieten, L.; Rust, J.; Kingston, T.; Merrington, G.; Morris, S. Influence of copper fungicide residues on occurrence of earthworms in avocado orchard soils. *Sci. Total Environ.* **2004**, *329*, 29–41. [CrossRef] [PubMed]
10. Yang, X.-E.; Long, X.-X.; Ni, W.-Z.; Ye, Z.-Q.; He, Z.-L.; Stoffella, P.J.; Calvert, D.V. Assessing copper thresholds for phytotoxicity and potential dietary toxicity in selected vegetable crops. *J. Environ. Sci. Health B* **2002**, *37*, 625–635. [CrossRef] [PubMed]
11. Mahmood, T.; Islam, K.R.; Muhammad, S. Toxic effects of heavy metals on early growth and tolerance of cereal crops. *Pak. J. Bot.* **2007**, *39*, 451–462.
12. Cao, Z.; Hu, Z. Copper contamination in paddy soils irrigated with wastewater. *Chemosphere* **2000**, *41*, 3–6. [CrossRef]
13. Xu, J.; Yang, L.; Wang, Z.; Dong, G.; Huang, J.; Wang, Y. Toxicity of copper on rice growth and accumulation of copper in rice grain in copper contaminated soil. *Chemosphere* **2006**, *62*, 602–607. [CrossRef] [PubMed]
14. Food and Agriculture Organization. Available online: http://faostat.fao.org/faostat (accessed on 2 March 2021).
15. Lai, H.-Y.; Juang, K.-W.; Chen, B.-C. Copper concentrations in grapevines and vineyard soils in central Taiwan. *Soil Sci. Plant Nutr.* **2010**, *56*, 601–606. [CrossRef]
16. García-Esparza, M.; Capri, E.; Pirzadeh, P.; Trevisan, M. Copper content of grape and wine from Italian farms. *Food Addit. Contam.* **2006**, *23*, 274–280. [CrossRef] [PubMed]
17. Mirlean, N.; Roisenberg, A.; Chies, J. Copper-based fungicide contamination and metal distribution in Brazilian grape products. *Bull. Environ. Contam. Toxicol.* **2005**, *75*, 968–974. [CrossRef] [PubMed]
18. Zheng, N.; Wang, Q.; Zhang, X.; Zheng, D.; Zhang, Z.; Zhang, S. Population health risk due to dietary intake of heavy metals in the industrial area of Huludao city, China. *Sci. Total Environ.* **2007**, *387*, 96–104. [CrossRef] [PubMed]
19. Council of Agriculture, Taiwan. Database of Agriculture. Available online: https://agrstat.coa.gov.tw/sdweb/public/inquiry/InquireAdvance.aspx (accessed on 24 February 2021).
20. Liu, W.H.; Zhao, J.Z.; Ouyang, Z.Y.; Soderlund, L.; Liu, G.H. Impacts of sewage irrigation on heavy metal distribution and contamination in Beijing, China. *Environ. Int.* **2005**, *31*, 805–812. [CrossRef] [PubMed]
21. Cheng, B.-Y.; Fang, W.-T.; Shyu, G.-S.; Chang, T.-K. Distribution of heavy metals in the sediments of agricultural fields adjacent to urban areas in Central Taiwan. *Paddy Water Environ.* **2013**, *11*, 343–351. [CrossRef]
22. Chen, Z.-S. *Relationship between Heavy Metal Concentrations in Soils of Taiwan and Uptake by Crops*; Food and Fertilizer Technology Center: Taiwan, China, 2000.
23. Lin, Y.-P.; Cheng, B.-Y.; Shyu, G.-S.; Chang, T.-K. Combining a finite mixture distribution model with indicator kriging to delineate and map the spatial patterns of soil heavy metal pollution in Chunghua County, central Taiwan. *Environ. Pollut.* **2010**, *158*, 235–244. [CrossRef] [PubMed]
24. LeBlanc, L.A.; Schroeder, R.A. Transport and distribution of trace elements and other selected inorganic constituents by suspended particulates in the Salton Sea Basin, California, 2001. In *The Salton Sea Centennial Symposium*; Springer: Dordrecht, The Netherlands, 2008; pp. 123–135.
25. Council of Agriculture, Taiwan. Protection Program of Agricultural Irrigation Water Quality. Available online: https://law.coa.gov.tw/glrsnewsout/LawContent.aspx?id=GL000497 (accessed on 26 November 2020).
26. Tukey, J.W. Exploratory Data Analysis. Addison-Wesley Publishing Company Incorporation: Boston, MA, USA, 1977.
27. Environmental Protection Administration, Taiwan. Investigation of Contaminated Farmlands in Taoyuan. 2013. Available online: https://epq.epa.gov.tw/Default.aspx (accessed on 18 January 2021).
28. Environmental Protection Administration, Taiwan. Control and Investigation Program for National Agricultural Land with High Potential of Heavy Metal Pollution (II). 2014. Available online: https://epq.epa.gov.tw/Default.aspx (accessed on 10 January 2021).

29. Yadav, R.K.; Goyal, B.; Sharma, R.K.; Dubey, S.K.; Minhas, P.S. Post-irrigation impact of domestic sewage effluent on composition of soils, crops and ground water—A case study. *Environ. Int.* **2002**, *28*, 481–486. [CrossRef]
30. Northern Region Water Resources Office, Water Resources Agency, Ministry of Economic Affairs, Taiwan. Development of Diverse Water Resources—Investigation and Feasibility Assessment of Agricultural Return Flows in the Taoyuan and Hsinchu Areas. 2007. Available online: https://www.wranb.gov.tw/3452/24064/27197/ (accessed on 13 January 2021).
31. Shihmen Management Office, Irrigation Agency, Council of Agriculture, Taiwan. Source of Irrigation Water. Available online: https://www.iasme.nat.gov.tw/view.php?theme=web_structure&id=109 (accessed on 13 January 2021).
32. Environmental Protection Administration, Taiwan. Wastewater and Sewage of Enterprise and Sewage System Management System. Available online: https://wpmis.epa.gov.tw/WPMIS/Login.aspx (accessed on 13 May 2019).
33. Ta-Wei, C.; Wei-Sheng, Y.; Cheng-Chang, H. Assessment of Dilution Water Quality for Polluted Irrigation Water-Chung-hwa Irrigation Area as an Example. *J. Chin. Agric. Eng.* **1999**, *45*.
34. Environmental Protection Administration, Taiwan. Control and Investigation Project for National Agricultural Land with High Potential of Heavy Metal Pollution (IV). 2017. Available online: https://epq.epa.gov.tw/Default.aspx (accessed on 10 January 2021).

Article

Ecological Risk Assessment of Cadmium in Karst Lake Sediments Based on *Daphnia pulex* Ecotoxicology

Faustino Dinis [1], Hongyan Liu [1,2,3,*], Qingdong Liu [1], Xuewen Wang [1] and Meng Xu [2]

1   College of Agriculture, Guizhou University, Guiyang 550025, China; faustinodinisvfm200@gmail.com (F.D.); liuqingwe@gmail.com (Q.L.); wxwzhhaa@gmail.com (X.W.)
2   College of Resources and Environmental Engineering, Guizhou University, Guiyang 550025, China; mengxu192@gmail.com
3   Key laboratory of Karst Georesources and Environment, Guizhou University, Ministry of Education, Guiyang 550025, China
*   Correspondence: hyliu@gzu.edu.cn

Abstract: The background value of cadmium (Cd) in soil and water sediments in the karst area is 0.31 mg kg^{-1}, with a typical high background of cadmium geochemistry. It is well-known that Cd is classified as a highly toxic metal. Therefore, at the Yelang reservoir in Guizhou province, eco-toxicological tests were carried out using *Daphnia pulex*. The Geo-Accumulation Index and Potential Ecological Risk Index were used to assess the environmental risk of Cd in sediments. The Cd contents in the sediments of Yelang reservoir ranged from 2.51 to 5.23 mg kg^{-1}, while the LC$_{50}$ values of the acute toxicity test of *Daphnia pulex* and Cd at 24, 48, 72, and 96 h were 1.17, 0.50, 0.24, and 0.12 mg L^{-1}, respectively, giving a Safe Concentration threshold of Cd of 1.20×10^{-3} mg L^{-1} in the water body. Based on curve fitting the solid–liquid two-phase distribution model of cadmium in Yelang reservoir was Y = $7.59 \times 10^{-9} \times$ X$^{2.58}$ (R^2 = 0.9995). The safety threshold sediment Cd concentration was 103 mg kg^{-1}, and was much higher than the Cd content in the sediment of the Yelang reservoir. The Geo-Accumulation Index (I_{geo} 2.432–3.491) results show that the sediments had reached medium-strong or strong risk levels. The Potential Ecological Risk Index (E_r^i 242.8–505.9) reached a very high or extremely high-risk level. However, due to high concentrations of Ca^{2+} and Mg^{2+}, and the pH being in the neutral–alkaline range of water body in karst areas, the Daphnia ecotoxicology evaluation method showed slight ecological risk, quite different from other assessment results, thus this method could be considered to use in such areas.

Keywords: cadmium; sediment; *Daphnia pulex*; ecotoxicology; LC$_{50}$; ecological risk assessment; karst areas

Citation: Dinis, F.; Liu, H.; Liu, Q.; Wang, X.; Xu, M. Ecological Risk Assessment of Cadmium in Karst Lake Sediments Based on *Daphnia pulex* Ecotoxicology. *Minerals* **2021**, *11*, 650. https://doi.org/10.3390/min11060650

Academic Editors: Ana Romero-Freire and Hao Qiu

Received: 4 May 2021
Accepted: 13 June 2021
Published: 18 June 2021

Publisher's Note: MDPI stays neutral with regard to jurisdictional claims in published maps and institutional affiliations.

1. Introduction

Sediment is an important accumulation site for many natural and anthropogenic heavy metals (HMs) [1,2]. The HMs' concentration within the sediment is perhaps higher in degree than the overlying water [3]. Sediments create very demanding environments for aquatic organisms, as a results harboring pollutants that can directly influence the water quality [2].

A group of freshwater zooplankton commonly referred to as Cladocera's are broadly spread in the aquatic environment, which are available in an extensive range of habitats and are significant links in various food chains [4]. For instance, Daphnia sp. are widely utilized to explore the acute and chronic toxicity of industrial and agricultural chemicals in aquatic ecosystems [5]. This is evident by their relatively short life cycle, less space requirement, adaptability to laboratory conditions, and sensitivity to a wide range of aquatic contaminants [6] and trace quantities of toxic heavy metals in aquatic systems. The trace elements of Cadmium are present in the aquatic environment, which usually leads to significant negative consequences if concentrations are sufficiently raised [7]. Consequently, affecting the growth rate of the population of Cladocerans [8]. Moreover, acute experiments have shown that cadmium is more toxic to daphniids in high temperature compared to low temperature [9]. Interestingly, toxicity may decline if a greater amount of energy can be directed to endure the toxic stress within higher food levels [10].

Numerous indexes have been put forward to assess the environmental risk of HMs in lake sediments based on the total content, bioavailability, and toxicity. These indexes include the enrichment factor, geo-accumulation index, pollution index, potential ecological risk index (RI), and so on [11–13]. However, the different evaluation indexes have varying limitations [1,14]. For instance, the Geo-Accumulation Index (Igeo) method of Müller is the current classical method of assessing the ecological risk of metals in sediments [15]. The Potential Ecological Risk Index (RI) method proposed by Hakanson [16], which is based on the research theory of sedimentology, has also been widely used by researchers both nationally and internationally to assess ecological risks of metals in sediments and soils [11–13]. Both methods are based on the assessment of the entire amount of metals without considering the biological effects of metals in sediments or the interaction between water ions.

Besides the mentioned indexes above, other approaches have been suggested, such as the development of theoretical and empirical single guidelines, evaluations of populations and communities, interstitial water quality, tissue residue, and spiked sediment toxicity, including laboratory and field toxicity testing of single species [1,3]. For instance, the calculation of toxic units is one way to estimate both the concentration and potential toxicity of multiple chemicals in sediments. The concentrations of toxicant in pore-water are divided by the LC_{50} for a reference organism [1,3,17]. These laboratory toxicity tests are of great importance, because they provide information for determining and managing decisions and the consequences of these decisions. Moreover, these tests are critical to the establishment of effective prediction of genuine benthic effect in the environment and the development of appropriate guidance for their application within a regulatory framework [1,3,17,18]. However, very few studies have attempted to compare and combine indexes, including the geo-accumulation index, and potential ecological risk with the LC_{50} *Daphnia Pulex* toxicity test to assess the environmental risk of HMs and detect the specific effects of chemicals on living organisms in lake sediments. These methods will provide adequate information to analyze and understand the ecotoxicology and biotoxic effects of HMs on zooplankton. Thus, an inclusive risk assessment should be given consideration based on a comparison among these indexes.

Guizhou province, located in the southwest of China, is an abnormal geochemical region with respect to cadmium (Cd), representing high background levels of metals [19,20]. For instance, Ling [21] has shown statistically that the geochemical background value of Cd in soils and sediments in Guizhou Province is 0.31 mg kg^{-1}. This is higher than in non-karst regions and is 2.46 times higher than the average value of water sediments in China 0.126 mg kg-1 [22]. Several studies have found that Cd is the leading metal pollutant in various karst lakes, with very high ecological risks [20,22]. The primary objective of this study is to assess the ecological risk of Cd in sediments of Yelang reservoir located in the Guizhou Province based on the *D. pulex* acute toxicity test. A comparative analysis of Geo-Accumulation Index (Igeo) and Potential Ecological Risk Index methods were conducted to provide additional references for environmental and ecological risk assessment.

2. Materials and Methods

2.1. Study Area

Yelang reservoir is located 35 km North of Anshun City, Guizhou Province, in the middle of Sancha Lake, north of Puding County. The reservoir area is about 20 square kilometers, and the water storage is about 420 million cubic meters. Its upstream tributaries into the lake are mainly Sancha Lake and Boyu Lake. The main structures at Yelang reservoir are the middle and upstream comprising Permian and Triassic carbonates, whilst the downstream is generally Cambrian, Ordovician and Silurian carbonate. Moreover, it displays a highly developed karst topography that accounted for 71.5 of the area, rich in mineral resources, mainly coal, iron, ore, copper, and aluminum-zinc. The principal landform is agricultural land, which is practically used for sewage irrigation owing to karst rock desertification [20]. The reservoir has been used for power generation, flood control, tourism, and water supply. Its basic parameters are shown in Table 1.

Table 1. Basic parameters at Yelang reservoir.

Total Reservoir Capacity/km^3	Normal Reservoir Capacity/km^3	Normal Catchment Area/km^2	Flow/km^3	Distance/km	Age/a	Evolution Stage
4.2	2.48	19.25	33.8	238.4	24	Middle level

The above data are from [23].

2.2. Sample Collection

Sediment sampling was done from 15 randomly selected sampling points at the Yelang reservoir in June 2016 (Figure 1). Sediment samples were collected at depths of 0–8 cm from the sites. The collection site was divided into three areas, namely the upstream Sancha lake area, midstream Dachuanbian and Shachong area and downstream Tianfen and Shawan area of the reservoir. The samples were placed in bags, sealed, and taken to the laboratory within 24 h after sampling. Prior to the collection of the sediment samples, surface water samples were collected at the 15 sites for water characterization, major ions, and metal analysis.

Figure 1. Distribution of sampling sites at Yelang reservoir in Guizhou province.

2.3. Sample Determination

The samples were air-dried to remove impurities followed by grinding and later put in a 100-mesh nylon sieve for subsequent analysis. Total cadmium in the sediment was extracted by the HNO_3_HF-$HClO_4$ triacid digestion method and determined by inductively coupled plasma mass spectrometer (X2 ICP-MS, Thermo Fisher Scientific, Waltham, MA). A parallel sample was set up in the experiment to ensure the accuracy of the measured data. The standard reference material GBW07405 (GSS5) was used for quality control.

For each water sample, the water temperature (t), conductivity (EC), pH, and other parameters for the overlying water were measured using a portable multiparametric water quality analyzer (DZB-718, Shanghai Leici Instrument Inc., Shanghai, China). The water samples for calcium and magnesium ions analysis were acidified with high purity nitric acid (HNO_3) to a pH < 2. Moreover, the results were determined by ICP-AES (ACTIVA-M, Horiba, Kyoto, Japan). The final results are shown in Table 2.

Table 2. Water quality parameters and main ionic composition at Yelang reservoir.

pH	EC/μs cm^{-1}	T/°C	DO/mg L^{-1}	Ca^{2+}/mg L^{-1}	Mg^{2+}/mg L^{-1}	HCO$_3{}^{-}$/mg L^{-1}
7.92	535	24.6	8.35	56.44	15.15	151.87

The definitions for the abbreviations within the table are as follows; pH, Water Temperature (t), Electrical Conductivity (EC), Dissolved Oxygen (DO), Calcium Ions (Ca^{2+}), Magnesium Ion (Mg^{2+}), and Bicarbonate Ion ($HCO_3{}^{-}$).

2.4. Biotoxicity Experimental Design

2.4.1. Test Organism and Culture Conditions

Organism experiment and culture conditions followed the procedures of Wu [24]. *D. pulex* were obtained from the Chinese Environmental Science Academy and monocultured in the laboratory of Guizhou University. The culture was maintained at 20.0 ± 0.5 °C with a photoperiod of 14 h light:10 h dark, and light intensity was around 1200 lux. The fresh *Scenedesmus* obliquus (2.0×10^5 cells/mL) was used as a daily feed. Throughout the experiment, half the water in the culture containers was replaced three times weekly. *D. pulex* neonates at 48 h of age were collected for subsequent test.

2.4.2. Test Material and Water Dilution

The culture medium was prepared following [25] with slight modifications. Basically, tap water was oxygenated with an oxygen action machine for more than three days until it reached DO > 8 mg L^{-1}. The Cd^{2+} toxin was prepared by dissolving $CdCl_2$ $2H_2O$ in deionized water, and the final concentration was 100 mg L^{-1}, and kept at 4 °C. The simulated lake water was prepared following the prescribed diluent water formula for Daphnia—China National Standard Formula GB/T13266-1991 [26]: A total of 1 mol L^{-1} high purity hydrochloric acid and 1 mol L^{-1} high purity sodium hydroxide were used and regulated in 7.8 ± 0.2 of pH.

2.4.3. Simulation of Sediment Samples with Different Concentration Gradients

The labeling method of sediment samples of different pollution gradients was simulated based on [27] with slight modification. Eight beakers of capacity 1 L, each numbered 1 to 8, were used to fill the simulated sediment samples. Two hundred grams of dried and sieved sediment was put into each beaker and filled with 800 mL of simulated lake water at room temperature. All beakers' contents were mixed well with a stir bar and then kept for 3 days. After, the overlying water was poured. The 8 beakers were spiked with varying quantities of $CdCl_2$ 0, 50, 100, 200, 600, 800, 1000, and 2000 μg mL^{-1}, respectively. The samples were thoroughly mixed and kept for 7 days. Next, sediments were "washed" with clean simulated lake water every 5 days. The total Cd was measured before each washing. The washing involved draining most of the overlying water, refilling and stirring. The water renewal was repeated until it stabilizes and the process was completed after 40 days. At this point, it was observed that Cd had reached an equilibrium state between the column water and the sediment. Sediment samples were then dried naturally for analysis.

The overlying water samples were filtered through a 0.45 μm filtration membrane before the analysis. Simultaneously, the sediment samples were digested with a mixture of high purity HNO_3, HF, and $HClO_4$ in digested vessels. Lastly, Cd concentrations were analyzed by inductively coupled plasma mass spectrometry (ICP-MS, ACTIVA-M, Horiba, Kyoto, Japan). The conditional distribution coefficient was calculated as follows:

$$Kp = Cs/Cw$$

where Cs is the Cd content in the sediment ($\mu g \, g^{-1}$), and Cw is the Cd concentration in the overlying water ($\mu g \, mL^{-1}$).

2.4.4. Acute Toxicity Test of Cd to *Daphnia pulex* in Water Bodies

Acute toxicity tests were conducted with *Daphnia pulex*. The test solution preparation followed standard guidelines ISO [28]. All steps and details concerned with these tests are reported [25]. Preliminary test was performed to determine the range of the maximum concentration that inhibited the movement of 100% of the *D. pulex*. Based on the set in equal logarithmic spacing, the toxic concentration intervals were determined, setting the Cd^{2+} concentration gradient to 0, 0.1, 0.18, 0.32, 0.56, 1.01, 1.82, and 3.27 mg L^{-1}. The determination of LC_{50} was based on ISO [28], and Zhou and Zhang [29]. *D. pulex* was first rinsed in the simulated lake water 3 times, 5 min each time. Ten neonates were transferred to new test solutions with different concentrations of toxin solution (100 mL). The conditions were the same as the culture conditions except for no feeding during the test. The test was conducted at a constant temperature of 20.0 ± 0.5 °C. The mortality were defined based on sinking to the bottom of the water or does not display movement in the container.

The Safe Concentration (SC) of cadmium was calculated based on the equation given below:

$$SC = 96 \, h - LC_{50} \times AF$$

AF is the application factor, which takes a value of 0.1 or 0.01 [30]. According to the acute toxicity of the different chemical substances, cadmium is a toxic substance with a low decomposition and high accumulation rate, hence the value 0.01 [31,32].

2.4.5. Acute Toxicity Test of Cd to *Daphnia pulex* in Sediments

Before the exposure experiment, two duplications of the simulated sediment, each weighing 45 g. were mixed with 180 mL of the simulated clean lake water. After being placed for 7 days, 40 Daphniids were put in one of the duplications exposing *D. pulex* to the water–sediment system. The other duplication filled with overlying water was drained and 40 Daphniids were added for the overlying water system experiment. *D. pulex* were extracted after 72 h of exposure, and the mortality rate was recorded, according to [27] with slight modification.

2.4.6. Measurement of the Accumulation of *Daphnia pulex* Body

After exposure, 10–15 *D. pulex* from each concentration group were taken and placed into simulated clean lake water for 1–2 h to let the *D. pulex* remove the metal from the internal organs. The *D. pulex* were subsequently removed and washed with distilled water twice. The samples were dried at 80 °C and then digested with concentrated HNO_3 (Superior purity, 68%) at 110 °C till the solution was transparent. After cooling, 2% HNO_3 (high purity) was used to dilute the volume to a measurable range. The Cd concentrations in digested organisms were analyzed by inductively coupled plasma-mass spectrometry (ICP-MS).

2.5. Risk Assessment Method for Heavy Metals in Sediments

2.5.1. Geo-Accumulation Index Method

The Geo-Accumulation Index (I_{geo}) is one of the most widely used methods for the quantitative evaluation of metals in sediments. It reveals the changing characteristics of the distribution of metals and identifies the environmental impact of human activities. It has been used by researchers to evaluate the risk of metals in sediments from rivers [13] and lakes and reservoirs [12]. The calculation formula was based on [13,15] as follows:

$$I_{geo} = \log_2 \frac{C_i}{k \times B_i}$$

where C_i is the measured concentration of the element in the sample, k is the background value change that may be caused by the action of natural rock formation, the overall value is 1.5, and B_i is the geochemical background value of the element in the soil. The classification standards of Müller [15] are shown in Table 3.

Table 3. Geo-Accumulation Index (I_{geo}) and classification of pollution level.

Geoaccumulation Index (I_{geo})	Rank	Pollution Level
$I_{geo} \leq 0$	0	Unpolluted
$0 < I_{geo} \leq 1$	1	Unpolluted to moderately polluted
$1 < I_{geo} \leq 2$	2	Moderately polluted
$2 < I_{geo} \leq 3$	3	Moderately to strongly polluted
$3 < I_{geo} \leq 4$	4	Strongly polluted
$4 < I_{geo} \leq 5$	5	Strongly to extremely polluted
$5 < I_{geo} \leq 10$	6	Extremely polluted

2.5.2. Potential Ecological Risk Index Method

The method is relatively fast, convenient, and straightforward. It not only reveals the impact of various pollutants in sediment environments of a particular area, but it also reveals the combined comprehensive effects of multiple contaminants in the environment. A single metal's Potential Ecological Risk Index were determined by following the formula below [16] and developed by Qi et al. [33]:

$$E_r^i = T_r^i \times C_f^i,$$

$$C_f^i = \frac{C^i}{C_n^i}$$

where E_r^i the is potential ecological risk factors for each heavy metal; T_r^i The toxicity response factor for the given element of "i", the toxicity response factor of Cd is 30 and C_f^i the pollution coefficient of a single element of "i"; C^i is the measured concentration of heavy metal in sediment; C_n^i which is the background value of heavy metal in sediment. The superscript "i" indicates the specific pollutant. The classification standards are shown in Table 4.

Table 4. Criteria for individual potential ecological risk coefficient indices (E_r^i).

Rank	Individual Potential Ecological Risk Index (E_r^i)	Individual Potential Ecological Risk Level
1	<40	Slight
2	40–80	Medium
3	80–160	High
4	160–320	Very High
5	≥320	Extremely High

2.6. Data and Statistical Analysis

The Data Processing System (DPS 2000) was used to carry out statistical analysis of experimental data. The probability value method was applied to obtain the corresponding four times under the LC_{50} [34]. The toxicological concentration and toxic regression equation for Daphnia mortality rate in overlying water was established, and chi-square was tested on regression equations formula [35] using Origin 2019b. Each value signifies the mean of four replicates $\pm$ standard deviation (SD). Levels of statistical significance are shown as * $p < 0.05$ and ** $p < 0.01$.

3. Results and Discussion

3.1. Distribution Coefficient and Fitted Model of Cd in Simulated Sediment and Overlying Water Systems

This crucial stage was carried out by mixing concentrated Cd and sediments. The Cd contamination of the simulated sediment samples were obtained after simulated lake water treatment with varying contents of Cd^{2+}. After repeated washing, Cd reached an equilibrium state between column water and sediment. The concentration of Cd in the simulated sediment and the overlying water system are shown in Figure 2. When the concentration of exogenous Cd increases, the Cd concentration in sediments and overlying water also increase. The ranges of Cd concentration in the sediment was between 2.7 and 3530.0 mg kg^{-1}, while the Cd concentration in the overlying water system was 0.00–10.40 mg kg^{-1}. The distribution coefficients were between 339.42 and 28,787 L kg^{-1}.

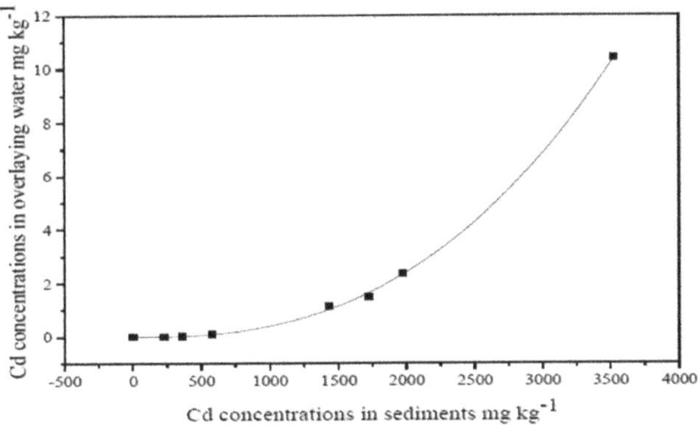

Figure 2. Cadmium concentrations in simulated sediments and overlying water systems (curve fitted by the power function).

The distribution coefficient showed a tendency to decrease because the adsorption sites on the sediment gradually became saturated. When the Cd concentration of the marker increases, the adsorption capacity of the sediment decreases. The Cd concentration relationship between the two systems was fitted by the model $Y = 7.59 \times 10^{-9} \times X^{2.58}$ ($R^2 = 0.9995$). Additionally, based on curve fitting, the solid–liquid two-phase distribution model of cadmium in the Yelang reservoir was $Y = 7.59 \times 10^{-9} \times X^{2.58}$ ($R^2 = 0.9995$). The safety threshold of total cadmium concentration in the sediment was 103 mg kg^{-1}, which was used as a reference value to evaluate the ecological risk of the sediment. The cadmium content in the sediments of Yelang reservoir was at a low level, and the environmental and ecological risks equally low. This was similar to the prior study conducted by Luo et al. [20], in which their finds claimed that the deposition rate of heavy metals in the Yelang Lake sediment is higher than the release rate, and the ecological risk from heavy metals is relatively low. When the release of heavy metals in sediments occurs, it can speedily affect the overlying water in a short time and, consequently, affect the water quality significantly [36].

3.2. Ecological Risk Assessment Based on Daphnia pulex Bio Toxicity Test

In this study, *Daphnia pulex* was selected as test organisms. The choice of *D. pulex* is owed to its sensitivity to toxins and the ability of obtaining precise information. In addition, it is recommended as a standard "Test for Acute Inhibition of Chemical Tritium GB/T 21830-2008)" [37] in carrying out such investigations.

3.2.1. Lethal Effect of Cd^{2+} on *Daphnia pulex*

The sublethal concentration LC_{50} and lethal effect of Cd^{2+} in *D. pulex* at different times are shown in Table 5. Based on the lethal effect of Cd^{2+} on *D. pulex*, the sublethal concentration values and toxicological regression equations were obtained at 24, 48, 72, and 96 h. The results show that, along with time, the sublethal concentration indicated a tendency to decrease gradually. In addition, the chi-square value of the toxicological equation fitted in each period was less than the critical value (χ^2 = 9.49). This illustrates that the toxicological regression equation describing the bio-toxicity of Cd^{2+} on *D. pulex* was feasible. The correlation coefficients for different times were >0.90. According to the Safe Concentration formula, SC = 96 h – LC_{50} × AF, where AF took the value of 0.01 and gives the Safety Concentration SC = 1.20×10^{-3} mg L^{-1}. This value is similar to previously reported values [32], with the sample possessing a limit of second-class water standard of 5×10^{-3} mg L^{-1}. These results are generally less than the recommended ranges of the surface water environmental quality standard GB3838—2002 [38].

Table 5. The lethal effects of Cd2+ on *D. pulex*.

Exposure time/h	LC_{50}/mg L^{-1}	95% Confidence Interval	Toxicology Regression Equation	Coefficient of Correlation R^2	Chi-Square
24	1.17	0.85~1.60	y = 4.89 + 1.66x	0.98	0.01
48	0.50	0.23~1.03	y = 6.38 + 4.47x	0.92	0.68
72	0.24	0.15~0.38	y = 7.16 + 3.46x	0.97	0.23
96	0.12	0.04~0.29	y = 7.70 + 2.80x	0.93	0.3

The fundamental concept in toxicology is the concentration–response relationship. For instance, toxicological evaluation naturally uses estimation points such as LC_{50} to compare species sensitivities [17,39]. Therefore, in this study, the 96 h LC_{50} values of several other aquatic organisms were collected from previous reports and summarized in Table 6. (Due to different experimental designs in the literature, if a species has multiple data, then only the lowest values were collected).

Table 6. The 96 h-LC_{50} values of some common aquatic organisms in other studies.

Species	96 h-LC_{50}/mg L^{-1}	Author(s)	Safe Concentration /mg L^{-1}
Sipunculus nudus	24.328	[40]	0.24
Misgurnus anguillicaudatus	1753.8	[41]	17.54
Argopecten irradiams	3.45	[31]	0.03
Mytilus coruscus	3.1	[42]	0.03
Eriocheir sinensis	40.279	[43]	0.40
Megalobrama terminalis	3.2	[44]	0.03
Carassius auratus gibelio	26.51	[45]	0.27
Gambusia affinis	22.55	[46]	0.23
Tanichthys albonubes	4.447	[47]	0.04
Ctenopharyngodon idella	23.51	[48]	0.24
Tigriopus japonicus	6.31	[32]	0.06
Chironomus javanus	0.06	[49]	6×10^{-4}
Cyathura carinata	37	[50]	0.37
Rasbora sumatrana	0.1	[17]	0.001
Poecilia reticulata	1.06		0.0106
Kryptolebias marmoratus	6.43×10^{-3}	[51]	6.43×10^{-5}
Fundulus heteroclitus	2.94×10^{-3}		2.94×10^{-5}

Table 6. *Cont.*

Species	96 h-LC$_{50}$/mg L^{-1}	Author(s)	Safe Concentration /mg L^{-1}
Macrobrachium lanchesteri	0.007		7×10^{-5}
Stenocypris major	0.013		1.3×10^{-4}
Nais elinguis	0.027		2.7×10^{-4}
Chironomus javanus	0.06	[52]	6×10^{-4}
Rasbora sumatrana	0.1		1×10^{-3}
Poecilia reticulata	0.17		1.7×10^{-3}
Duttaphrynus melanostictus	0.32		3.2×10^{-3}
Melanoides tuberculata	1.49		1.49×10^{-2}
Eurytemora affinis	Male 127.8	[39]	1.28
	Female 90.0		0.09
Orconectes juvenilis	0.06		6×10^{-4}
Orconectes placidus	0.037		3.7×10^{-4}
Orconectes virilis	3.3	[53]	0.03
Procambarus acutus	0.368		3.68×10^{-3}
Procambarus alleni	3.07		0.03
Procambarus clarkii	0.624		6.24×10^{-3}

The 96 h LC$_{50}$ value is found to be 6.43×10^{-3}–1753.8 mg L^{-1}, and its Safe Concentration was evaluated based on the formula SC = 96 h $-$ LC$_{50}$ $\times$ AF (value of 0.01), calculated as 6.43×10^{-5}–17.54, or 0.05–14,615 times the *D. pulex* 96 h LC$_{50}$ (0.12 mg L^{-1}) and the Safe Concentration (1.20×10^{-3} mg L^{-1}). In comparison to the collected aquatic organisms mentioned above, the *D. pulex* is one of the most sensitive aquatic species models in particular for cadmium toxicity [54,55]. Moreover, the 96 h LC$_{50}$ of some organisms can only be obtained in harsh environments, as highlighted in the literature [51], with the lowest value of 6.43×10^{-3} mg L^{-1}. Such a low 96 h LC$_{50}$ requires the ionic concentration in the water body to be extremely low. After adding a certain amount of calcium and magnesium ions, its value is increased significantly. In this study, it was increased from 0.23 to 23.2 mg L^{-1}, which is higher than the 96 h LC$_{50}$ of *D. pulex*. In addition, compared to the results of other freshwater Daphnia studies, Yang [56] measured the 48 h sublethal concentration of Cd^{2+} 0.62 mg L^{-1} in *Daphnia carinata* King (Cladocera—Daphnidae). This is relatively close to this study with a Cd^{2+} concentration of 48 h LC$_{50}$ = 0.50 mg L^{-1}.

Cadmium (Cd) is not an essential element for either plant or animal survival and yet one of the most toxic metals [54,55]. However, similar to a previous report [57], this study also found that the less toxicity for the tested daphniids may very likely be due to the Cd competing with calcium (Ca^{2+}) at enzymatic locations in organisms. The acute toxicity of Cd *D. pulex* is significantly minimized by increasing calcium and magnesium concentrations. The findings of this study is also supported by Clifford et al. [7], who explained that the acute toxicity of CdT to *D. pulex* is significantly influenced by Ca and Mg, which lower the potential risk of cadmium. Additionally, the less toxicity of the cadmium is evident by the dominance of mobile species in well-aerated water bodies with a pH closer to 8 [58,59]. Another possible explanation can be given using the results of researches that focused on mRNA expression as an indicator of the response to metals. For instance, Chen et al. [60] reported that miRNAs and metallothionein (MT) play critical roles in *D. pulex* tolerance to Cd, and confirmed that, when exposed to Cd-polluted environments, aquatic organisms can raise their tolerance to Cd in order to survive. In *Daphnia pulex*, different gene expression and high gene duplication rates have been identified, which enhances its adaptation in adverse environmental conditions [55,61].

3.2.2. Mortality Rate and Cd Accumulation of *Daphnia pulex* after Cd Exposure in Both Systems

Mortality is generally the primary parameter to consider when assessing the impact of contaminants in the environment. Trace metals are predictable by their toxic effects on aquatic organisms [39]. This study investigated the toxic effects of Cd on *Daphnia pulex* in overlaying water and water–sediment systems. It is generally assumed that absorption by organisms will reduce the concentration of metals in water and thus predicted that *D. pulex* only absorbs Cd from overlying water. However, because the change in the value of Cd concentration was small, $\leq 5\%$, the concentration of Cd in the overlying water system is negligible. It was assumed that the concentrations of Cd in the water during the exposure process were constant. The mortality rate of *D. pulex* after exposure to pollutants with different levels of Cd pollution is shown in Figure 3. The mortality rate of *D. pulex* increases with the increase of Cd concentration in sediments. When the Cd concentration in the sediment exceeded 1500 mg kg^{-1}, the mortality rate of *D. pulex* reached 100%. This suggests that Cd in the solid phase (sediment) could also intoxicate the *Daphnia pulex* by sediment mud feeding and other particles. It is supported by Caumette [14], who reported that part of the assessed arsenic signal comes from the sediments ingested by *Daphnia pulex* when the residues were analyzed with soluble arsenic after extraction. Li et al. [3] argued that *D. magna* or daphniids could absorb some heavy metals, and accumulate more Cd by assimilating solid particles. Moreover, Barton et al. [1] defended that benthic organisms exposed to metal-contaminated sediments can accumulate metals by ingesting contaminated sediments and other suspended particles, and by exposure to dissolved metals in the overlying water. The overlying water systems and water–sediment systems showed no significant difference (*t*-test, $p > 0.05$) in the mortality rate of *D. pulex*. The Cd content in the *D. pulex* body after exposure with different levels of Cd pollution is shown in Figure 4. The Cd content in the *D. pulex* body increases with the increasing Cd content in the sediments with significant linear correlation (Overlying water system: $R^2 = 0.936$, $p < 0.01$; water–sediment systems: $R^2 = 0.973$, $p < 0.01$). It is interesting to note that the amount of Cd in the overlying water increased along with Cd concentration in sediment. Notably, when the Cd concentrations in sediment were 2.7 and 3530.0 mg kg^{-1}, the concentrations in the overlying water were 0.00 and 10.40 mg kg^{-1}, respectively. This scenario was also observed in a previous study [3]. Their study hence revealed that when the Cd concentrations in sediment were 344.2 and 742 µg/g, the concentrations in the overlying water were 0.531 and 1.76 mg/L, respectively. The findings of this study are supported by the conclusions of previous researches [3,62]. The increase in mortality and Cd accumulation in the two systems was principally owing to the rise in Cd concentration in the overlying water (see Figure 5).

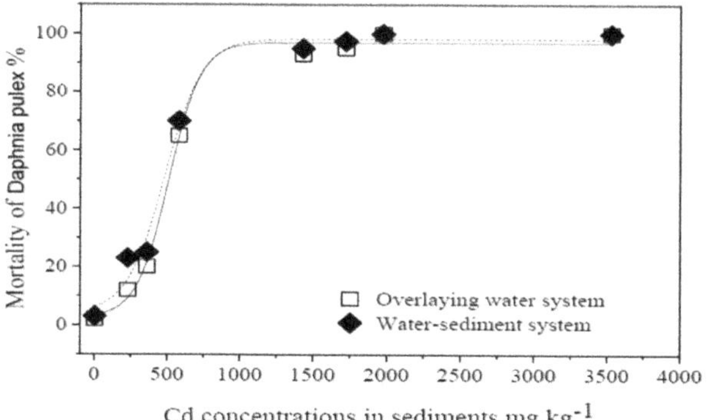

Figure 3. The mortality rate of *Daphnia pulex* under different Cd pollution levels in sediments (curves fitted by the Boltzmann function).

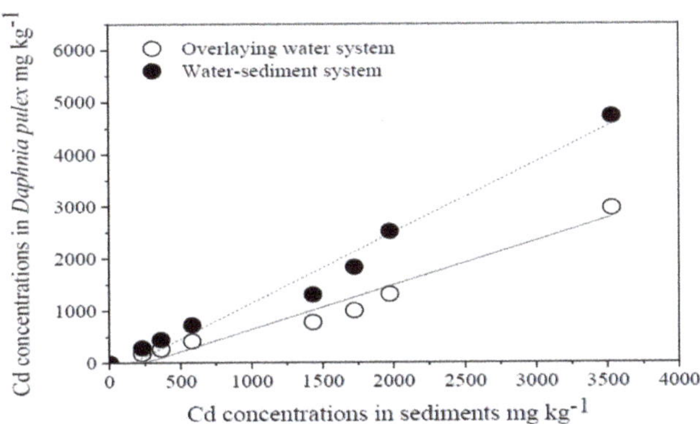

Figure 4. Cadmium concentrations in *Daphnia pulex* under different Cd pollution levels in sediments.

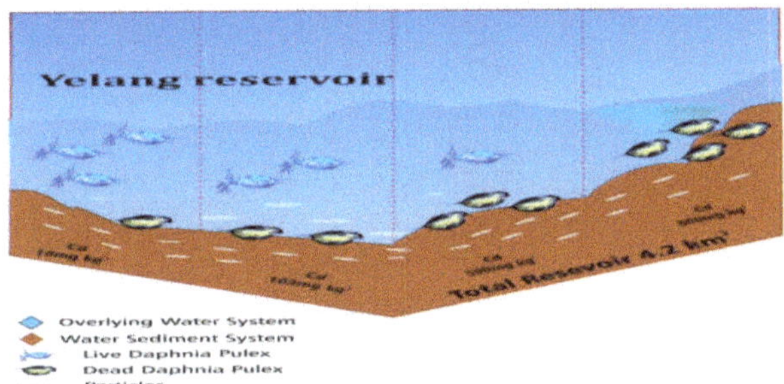

Figure 5. Highlighting the mortality rate of *Daphnia pulex* under different Cd pollution levels in sediments.

3.3. Cadmium Content in Sediments of Yelang Reservoir

The Cd content in the sediments is shown in Table 7. The content ranges from 2.51 to 5.23 mg kg^{-1} and the average value was 3.95 mg kg^{-1}. The obtained values clearly indicated that the samples exceeded the screening value of soil pollution risk on agricultural land and partly greater than the environmental quality control value of 0.6 mg kg^{-1} (GB15618-2018, China) [63]. The Cd content in the middle stream and downstream was smaller than the upstream, which could very likely be due to the larger storage area in the middle stream compared to the upstream and downstream. After an absolute reduction, the risk rates of the downstream and midstream were low. The source of cadmium concentration at the Yelang reservoir is mainly from anthropogenic activities, including domestic, agricultural and mining wastes, sewage discharge, and natural factors such as the influence of carbonate rocks of the karst terrain and its high geological background.

3.4. Ecological Risk Assessment Based on Sedimentology

The evaluation results of Cd content in the sediments of the Yelang reservoir by the Geo-Accumulation Index method are shown in Table 7. The index ranges from 2.432 to 3.491 and the contamination levels reached classes III and IV, suggesting medium–strong and strong risk levels. The degree of Cd pollution in the middle stream is lower than in the upstream and downstream due to the large volume of water which tends to lower potential risk of the Cd contents.

The Potential Ecological Risk Index method evaluation of the Cd content of the sediments of the Yelang reservoir are shown in Table 7. The risk index ranges from 242.8 to 505.9 with an average value of 381.8, suggesting that the risks were between very strong and extremely strong. The results were identical to the evaluation results of the geo-accumulation index method. There were differences in pollution levels at the upstream, middle, and downstream, such that the hazard index was in the order midstream < downstream < upstream.

Table 7. Concentration and evaluation results of Cd in sediments at the Yelang reservoir.

Sampling Site	Concentration of Cd/mg kg^{-1}	I_{geo}	Classification of Geoaccumulation Index	E_r^i	Potential Ecological Risk Level
1	4.13	3.15	4	399.3	Extremely strong
2	3.9	3.068	4	377.4	Extremely strong
3	4.19	3.17	4	405.1	Extremely strong
4	4.18	3.168	4	404.5	Extremely strong
5	4.25	3.191	4	410.9	Extremely strong
6	4.45	3.259	4	430.6	Extremely strong
7	3.69	2.989	3	357.2	Very strong
8	4.43	3.253	4	429.1	Extremely strong
9	2.74	2.557	3	264.8	Very strong
10	4.22	3.18	4	407.9	Extremely strong
11	2.51	2.432	3	242.8	Very strong
12	5.23	3.491	4	505.9	Extremely strong
13	2.8	2.592	3	271.3	Very strong
14	5.04	3.439	4	488.1	Extremely strong
15	3.43	2.884	3	332.1	Extremely strong

The results obtained from the Geo-Accumulation Index method and the Potential Ecological Risk Index method showed that the Cd risk of the Yelang reservoir has reached a high-risk level. The Geo-Accumulation Index method pays more attention to the total amount of metals [64]. The Potential Ecological Risk index method combines metal toxicity, the local metal background value, the general migration and transformation law in the sediments, and evaluation of the regional sensitivity to metal pollution [11,65]. Both evaluation methods were based on the total amount of metals.

A large quantity of Ca, Mg, and other alkaline soil metal ions in carbonate rocks in karst areas enter the water body through erosion, weathering, and transportation, thereby increasing the contents of Ca^{2+} and Mg^{2+}. Furthermore, a significant amount of calcite and dolomite, among others, consume some H^+ in the aqueous solution during the dissolution process, causing the pH of the water body to become weakly alkaline, thereby inhibiting the effectiveness of Cd [51,57]. Many studies [32,51] have confirmed that the bio-toxicity of metals will decrease with increasing pH and hardness of water bodies. Shi [66] and other authors who studied karst areas have found that Ca^{2+} and Mg^{2+} contained in water bodies can antagonize the biological toxicity of metals, increasing the 96 h LC_{50} of the *Daphnia*. Xiong [67] confirmed that, along with the increase in the water body's hardness, the sublethal concentration (96 h LC_{50}) and Safe Concentration of Cd on *Gobiocypris rarus* both increased significantly, showing that the hardness of a water body can effectively reduce the acute toxicity of cadmium in organisms.

3.5. Analysis of the Application of Ecological Risk Assessment Methods for Cadmium in Karst Lake Reservoir Sediments

The sediment quality evaluations were mainly focused on specific chemical analysis (Cd) results, and the results obtained are an accurate reflection of its effects on the aquatic animals at the Yelang reservoir. The sediment metal risk assessment system is a large-scale system that integrates a variety of uncertainties, such as randomness, greyness, uncertainty, and ambiguity [11,68,69]. The application of conventional assessment methods without analysis cannot accurately reveal the real situation of the metal pollution level of sediments [12]. In recent years, some researchers have improved these methods [12,33,70] to make them favorable and adequately applicable. During the assessment of ecological risks in karst areas, the particularity of the region should be taken into account for the evaluation results to be more accurate. However, the biological toxicity test method directly revealed the toxic effect of Cd on *D. pulex*, and did not require considering the particularity of the regional environment. This study only conducted preliminary toxicity tests and did not consider the effects of biomicroscopy. Further studies are required to assess whether further ranks of consumers in the food chain will produce any ecological risk.

Although karst areas are an abnormal geochemical region with respect to cadmium (Cd), due to the high concentration of Ca^{2+} and Mg^{2+} and the pH remains in the neutral–alkaline range of water body, the *Daphnia pulex* ecotoxicology evaluation method showed slight ecological risk, which is quite different from other methods used for environmental risk assessment, thus the method could be considered to use in such areas.

4. Conclusions

The sediment quality evaluations were mainly focused on specific chemical analysis (Cd) results, and the results obtained are an accurate reflection of its effects on the aquatic animals at the Yelang reservoir. The sediment toxicity experiment was carried out by mixing concentrated Cd and sediments. Based on the bio-toxicity experiment, it was deduced that the Safe Concentration threshold of Cd in the sediment was 103 mg kg^{-1}. The cadmium contents in the sediment of the Yelang reservoir ranged from 2.51 to 5.23 mg kg^{-1}. The Geo-Accumulation Index and the Potential Ecological Risk Index showed that the water environment of the Yelang reservoir had strong to extremely strong ecological risks. However, due to the high concentration of Ca^{2+} and Mg^{2+} and that the pH remains in the neutral–alkaline range, the *Daphnia pulex* ecotoxicology evaluation method showed slight ecological risk, which is quite different from the current available for environmental risk assessment, thus the method can be considered for use in the karst area. This study only conducted preliminary toxicity tests and did not consider the effects of biomicroscopy, further studies are required to assess whether further ranks of consumers in the food chain will produce any ecological risk.

Author Contributions: Conceptualization, H.L. and Q.L.; data curation, H.L.; formal analysis, F.D., H.L. and M.X.; funding acquisition, H.L.; investigation, F.D.; methodology, H.L., X.W. and M.X.; project administration, H.L.; resources, H.L.; software, Q.L.; supervision, H.L.; writing—original draft, F.D.; writing—review and editing, F.D. All authors have read and agreed to the published version of the manuscript.

Funding: This project was supported by the Joint National Natural Science Foundation of China and Guizhou Province (U1612442); China National Natural Science Foundation (42067028): The Science and Technology Planning Project of Guizhou Province (Qiankehehoubuzu [2020]3001).

Data Availability Statement: The data that support the findings of this study are available from the corresponding author, upon reasonable request.

Acknowledgments: Special thanks to the Joint National Natural Science Foundation of China and Guizhou Province, China National Natural Science Foundation, the Science and Technology Planning Project of Guizhou Province for the funding of this research. The authors would like to thank Hongyan Liu for the guidance and supervision. Finally, special thanks to all the reviewers for their insightful contributions during the review of the article.

Conflicts of Interest: The authors declare that there are no conflicts of interest.

References

1. Burton, G.A. Metal bioavailability and toxicity in sediments. *Crit. Rev. Environ. Sci. Technol.* **2010**, *40*, 852–907. [CrossRef]
2. Hallare, A.V.; Seiler, T.B.; Hollert, H. The versatile, changing, and advancing roles of fish in sediment toxicity assessment—A review. *J. Soils Sediments* **2011**, *11*, 141–173. [CrossRef]
3. Li, X.; Peng, W.; Jiang, Y.; Duan, Y.; Ren, J.; Liu, Y.; Fan, W. The Daphnia magna role to predict the cadmium toxicity of sediment: Bioaccumlation and biomarker response. *Ecotoxicol. Environ. Saf.* **2017**, *138*, 206–214. [CrossRef] [PubMed]
4. Abe, T.; Saito, H.; Niikura, Y.; Shigeoka, T.; Nakano, Y. Embryonic development assay with Daphnia magna: Application to toxicity of aniline derivatives. *Chemosphere* **2001**, *45*, 487–495. [CrossRef]
5. Hanazato, T. Growth analysis of Daphnia early juvenile stages as an alternative method to test the chronic effect of chemicals. *Chemosphere* **1998**, *36*, 1903–1909. [CrossRef]

6. Emmanuel, E.; Keck, G.; Blanchard, J.M.; Vermande, P.; Perrodin, Y. Toxicological effects of disinfections using sodium hypochlorite on aquatic organisms and its contribution to AOX formation in hospital wastewater. *Environ. Int.* **2004**, *30*, 891–900. [CrossRef] [PubMed]
7. Clifford, M.; McGeer, J.C. Development of a biotic ligand model to predict the acute toxicity of cadmium to *Daphnia pulex*. *Aquat. Toxicol.* **2010**, *98*, 1–7. [CrossRef]
8. Wong, C.K.; Wong, P.K. Life table evaluation of the effects of cadmium exposure on the freshwater cladoceran, Moina macrocopa. *Bull. Environ. Contam. Toxicol.* **1990**, *44*, 135–141. [CrossRef]
9. Heugens, E.H.; Jager, T.; Creyghton, R.; Kraak, M.H.; Hendriks, A.J.; Van Straalen, N.M.; Admiraal, W. Temperature-dependent effects of cadmium on Daphnia magna: Accumulation versus sensitivity. *Environ. Sci. Technol.* **2003**, *37*, 2145–2151. [CrossRef]
10. Heugens, E.H.; Tokkie, L.T.; Kraak, M.H.; Hendriks, A.J.; Van Straalen, N.M.; Admiraal, W. Population growth of Daphnia magna under multiple stress conditions: Joint effects of temperature, food, and cadmium. *Environ. Toxicol. Chem. Int. J.* **2006**, *25*, 1399–1407. [CrossRef]
11. Zhao, Y.; Xu, M.; Liu, Q.; Wang, Z.; Zhao, L.; Chen, Y. Study of heavy metal pollution, ecological risk and source apportionment in the surface water and sediments of the Jiangsu coastal region, China: A case study of the Sheyang Estuary. *Mar. Pollut. Bull.* **2018**, *137*, 601–609. [CrossRef] [PubMed]
12. Yang, J.; Chen, L.; Liu, L.Z.; Shi, W.L.; Meng, X.Z. Comprehensive risk assessment of heavy metals in lake sediment from public parks in Shanghai. *Ecotoxicol. Environ. Saf.* **2014**, *102*, 129–135. [CrossRef]
13. Singh, H.; Yadav, S.; Singh, B.K.; Dubey, B.; Tripathi, K.; Srivastava, V.; Shukla, D.N. Assessment of geochemical environment from study of river sediments in the middle stretch of river Ganga at Ghazipur, Buxar and Ballia Area. *Proc. Natl. Acad. Sci. India Sect. B Biol. Sci.* **2013**, *83*, 371–384. [CrossRef]
14. Caumette, G.; Koch, I.; Moriarty, M.; Reimer, K.J. Arsenic distribution and speciation in daphnia pulex. *Sci. Total Environ.* **2012**, *432*, 243–250. [CrossRef] [PubMed]
15. Müller, G. Schwermetalle in den sedimenten des Rheins-Veränderungen seit 1971. *Umschan* **1979**, *79*, 778–783.
16. Hakanson, L. An ecological risk index for aquatic pollution control. A sedimentological approach. *Water Res.* **1980**, *14*, 975–1001. [CrossRef]
17. Shuhaimi-Othman, M.; Yakub, N.; Ramle, N.A.; Abas, A. Comparative toxicity of eight metals on freshwater fish. *Toxicol. Ind. Health* **2015**, *31*, 773–782. [CrossRef]
18. Moore, D.W.; Farrar, D.; Altman, S.; Bridges, T.S. Comparison of acute and chronic toxicity laboratory bioassay endpoints with benthic community responses in field-exposed contaminated sediments. *Environ. Toxicol. Chem.* **2019**, *38*, 1784–1802. [CrossRef]
19. Peng, Z.Z.; Dan, X.; Yan, H.L.; Rong, H.T.; Ping, L.Y. Adsorption and desorption characteristics of cadmium in yellow soil and limestone soil. *Guizhou Agric. Sci.* **2015**, *436*, 83–86. (In Chinese)
20. Luo, K.; Liu, H.; Yu, E.; Tu, Y.; Gu, X.; Xu, M. Distribution and release mechanism of heavy metals in sediments of Yelang Lake by DGT. *Stoch. Environ. Res. Risk Assess.* **2020**, *34*, 793–805. [CrossRef]
21. Ling, H.S.; Lin, L.C.; Zhong, L.Y.; Guang, H.D.; Sheng, Y.Y.; Wei, Y.S. Geochemistry and environment of the cadmium in the soil and sediments at the surface of Guizhou province. *Guizhou Geol.* **2004**, *4*, 245–250. (In Chinese)
22. Ying, L.Z.; Zhong, C.Z.; Sheng, Y.G.; Ya, C.Z. Concentrations of 39 elements in stream sediment in different landscape zones of china. *Earth Sci. Front.* **2015**, *225*, 226–230. (In Chinese)
23. Fang, Y. Study on the Mass Balance of Mercury in Reservoirs at Different Evolution Stages in the Wujiang River Basin. Ph.D. Thesis, Southwest University, Chongqing, China, 2009. (In Chinese).
24. Wu, Y.G.; Xiong, Y.; Lin, C.X.; Yuan, L. The joint-biotoxicity effect of the different forms of phosphorus on heavy metal (Cu, Zn, Cd). *Acta Sci. Circumstantiae* **2006**, *12*, 2045–2051. (In Chinese)
25. Wu, Y.G.; Huang, J.G.; Yuan, L. Applying phototactic behaviour of Daphnia to monitor water quality. *China Environ. Sci.* **2004**, *243*, 336–339. (In Chinese)
26. Chinese Environmental Protection Administration (CEPA). *Water Quality-Determination of the Acute Toxicity of Substance to Daphnia (Daphnia magna straus) (GB/T 13266-1991)*; China Standards Press: Beijing, China, 1991. (In Chinese)
27. Cairns, M.A.; Nebeker, A.V.; Gakstatter, J.H.; Griffis, W.L. Toxicity of copper-spiked sediments to freshwater invertebrates. *Environ. Toxicol. Chem.* **1984**, *3*, 435–445. [CrossRef]
28. ISO. *Water Quality-Determination of the Inhibition of the Mortality of Daphnia magna Straus (Cladocera: Crustacea)*; Technical Report 6241; International Standards Organisation: Geneva, Switzerland, 1982.
29. Zhou, Y.X.; Zhang, Z.S. *Test Method for Toxicity of Aquatic Organisms*; Agricultural Press: Beijing, China, 1989; pp. 24–145. (In Chinese)
30. Daam, M.A.; Moutinho, M.F.; Espíndola, E.L.G.; Schiesari, L. Lethal toxicity of the herbicides acetochlor, ametryn, glyphosate and metribuzin to tropical frog larvae. *Ecotoxicology* **2019**, *28*, 707–715. [CrossRef]
31. Huan, L.Y.; Hong, L. Acute toxicity study of cadmium on scallops in the bay. *Mar. Fish. Res.* **2006**, *6*, 80–83. (In Chinese)
32. Wei, X.U.; Mu, F.H.; Sun, Y.T.; Cao, Z.Q. Effect of simulated ocean acidification on the acute toxicity of Cu and Cd to Tigriopus japonicus. *Acta Ecol. Sin.* **2014**, *3414*, 3879–3884. (In Chinese)
33. Qi, X.Z.; Jun, N.S.; Guo, T.X.; Jiang, Z.C. Calculation of toxicity coefficients of heavy metals in potential ecological hazard index method. *Environ. Sci. Technol.* **2008**, *2*, 112–115. (In Chinese)

34. Qiang, M.Z.; Jiang, F.; Mi, W. *Environmental Toxicology*; China Environmental Science Press: Beijing, China, 2000; pp. 120–275. (In Chinese)
35. Liang, Y. Study on Biological Toxicity of Toxic Pollutants in Acid Wastewater of Coal Mines in Karst Areas. Ph.D. Thesis, Guizhou University, Guizhou, China, 2009. (In Chinese).
36. Yan, H.-J.; Zhang, H.-Y.; Shi, Y.-J.; Zhou, P.; Li, H.; Wu, D.-L.; Liu, L. Simulation on release of heavy metals Cd and Pb in sediments. *Trans. Nonferrous Met. Soc. China* **2021**, *31*, 277–287. [CrossRef]
37. National Standards of People's Republic of China. *GB/T 21851-2008 (Chemicals-Adsorption-Desorption Using a Batch Equilibrium Method), GB/T 21805-2008 (Chemicals-Alga Growth Inhibition Test), GB/T 21830-2008 (Chemicals-Daphnia sp., Acute Immobilisation Test)*; Standardization Administration of the People's Republic of China: Beijing, China, 2008.
38. MEP. *Environmental Quality Standards for Surface Water*; GB 3838; Ministry of Environmental Protection of the People's Republic of China: Beijing, China, 2002. (In Chinese)
39. Zidour, M.; Boubechiche, Z.; Pan, Y.J.; Bialais, C.; Cudennec, B.; Grard, T.; Drider, D.; Flahaut, C.; Ouddane, B.; Souissi, S. Population response of the estuarine copepod eurytemora affinis to your bioaccumulation of trace metals. *Chemosphere* **2019**, *220*, 505–513. [CrossRef] [PubMed]
40. Jia, L.X.; Hui, P.Y.; Chao, L.C.; Hui, Q.Z. Acute toxicity study of copper and cadmium against Aspergillus. *Fish. Sci.* **2015**, *3402*, 95–99. (In Chinese)
41. Min, Z.Y.; Lei, Z.; Wen, Q.Y.; Hui, Z.B. Bio-toxicity of cadmium-spiked sediments to misgurnus anguillicaudatus. *Asian J. Econ.* **2011**, *601*, 80–86. (In Chinese)
42. Ling, P.; Ning, Z.J.; Zhen, C.Q.; Wei, H.; Ping, D. Effects of cadmium on antioxidant enzyme activity in gill of Mytilus coruscus and acute toxicity of cadmium on Mytilus coruscus. *Environ. Sci. Technol.* **2015**, *38*, 13–24. (In Chinese)
43. Xue, P.; Liang, H.X.; Han, C.S.; Jing, C.S.; Ping, Y.X. Single and joint acute toxicity of Cd+2 and Hg+2 to chinese mitten handed crab eriochersinensis. *Fish. Sci.* **2015**, *34*, 220–226. (In Chinese)
44. Zeng, Y.Y.; Lai, Z.N.; Yang, W.L.; Gao, Y.; Li, Y.F.; Pang, S.X. Toxicities and potential ecological effects of copper and cadmium to natural fish larvae and juveniles from the pearl river. *Asian J. Econ.* **2014**, *91*, 49–55. (In Chinese)
45. Yue, L.Y.; Lin, Z.Z.; Long, D.B.; Yang, F.J.; Mei, L.D. Acute toxicity of cadmium to allogynogenetic crucian carp. *J. Fish.* **2018**, *3106*, 36–39. (In Chinese)
46. Yang, L.; Xi, C.; Hui, Q.J. Acute toxic effect of cadmium on mosquito-eating fish (Gambusia affinis). *J. Saf. Environ.* **2015**, *1503*, 362–366. (In Chinese)
47. Long, W.R.; Zhi, M.G.; Qiang, Z.F. Safety Assessment and Acute toxicity of copper, cadmium and zinc to white clound mountain minnow tanichthys albonubes. *Fish. Sci.* **2006**, *3*, 117–120. (In Chinese)
48. Shou, S.; Lin, S.X.; Yan, W.G. Ecological toxicity of cadmium chloride on grass carp (Ctenopharyngodon idellus). *J. Shenyang Ligong Univ.* **2015**, *34*, 65–68, 72. (In Chinese)
49. Shuhaimi-Othman, M.; Yakub, N.; Umirah, N.S.; Abas, A. Toxicity of eight metals to malaysian freshwater midge larvae chironomus javanus (diptera, chironomidae). *Toxicol. Ind. Health* **2011**, *27*, 879–886. [CrossRef]
50. Martinez-Haro, M.; Moreira-Santos, M.; Marques, J.C.; Ribeiro, R. A short-term laboratory and in situ sediment assay based on the postexposure feeding of the estuarine isopod cyathura carinata. *Environ. Res.* **2014**, *134*, 242–250. [CrossRef]
51. Bielmyer-Fraser, G.K.; Harper, B.; Picariello, C.; Albritton-Ford, A. The influence of salinity and water chemistry on acute toxicity of cadmium to two euryhaline fish species. *Comp. Biochem. Physiol. Part-C Toxicol. Pharmacol.* **2018**, *214*, 23–27. [CrossRef]
52. Shuhaimi-Othman, M.; Nadzifah, Y.; Nur-Amalina, R.; Umirah, N.S. Deriving freshwater quality criteria for copper, cadmium, aluminum and manganese for protection of aquatic life in Malaysia. *Chemosphere* **2013**, *90*, 2631–2636. [CrossRef]
53. Wigginton, A.J.; Birge, W.J. Toxicity of cadmium to six species in two genera of crayfish and the effect of cadmium on molting success. *Environ. Toxicol. Chem.* **2007**, *26*, 548–554. [CrossRef] [PubMed]
54. Shaw, J.R.; Colbourne, J.K.; Glaholt, S.P.; Turner, E.; Folt, C.L.; Chen, C.Y. Dynamics of cadmium acclimation in daphnia pulex: Linking fitness costs, cross-tolerance, an hyper-inaction of metallothionein. *Environ. Sci. Technol.* **2019**, *53*, 14670–14678. [CrossRef]
55. Shaw, J.R.; Colbourne, J.K.; Davey, J.C.; Glaholt, S.P.; Hampton, T.H.; Chen, C.Y.; Folt, C.L.; Hamilton, J.W. Gene response profiles for daphnia pulex exposed to the environmental stressor cadmium reveals novel crustacean metallothioneins. *BMC Genom.* **2007**, *8*, 1–19. [CrossRef] [PubMed]
56. Yang, J.; Yuan, L.; Tang, Y. Monitoring sodium pentachlorophenate and cadmium in aquatic solutions by phototactic behavior of Daphnia carinata clone. *Acta Sci. Circumstantiae* **2006**, *266*, 1011–1015. (In Chinese)
57. Niyogi, S.; Wood, C.M. Biotic ligand model, a flexible tool for developing site-specific water quality guidelines for metals. *Environ. Sci. Technol.* **2004**, *38*, 6177–6192. [CrossRef]
58. Dallinger, R.; Höckner, M. Evolutionary concepts in ecotoxicology: Tracing the genetic background of differential cadmium sensitivities in invertebrate lineages. *Ecotoxicology* **2013**, *22*, 767–778. [CrossRef]
59. Tan, Q.G.; Wang, W.X. Acute toxicity of cadmium in daphnia magna under different calcium and PH conditions: Importance of influx rate. *Environ. Sci. Technol.* **2011**, *45*, 1970–1976. [CrossRef]
60. Chen, S.; Nichols, K.M.; Poynton, H.C.; Sepúlveda, M.S. MicroRNAs are involved in cadmium tolerance in daphnia pulex. *Aquat. Toxicol.* **2016**, *175*, 241–248. [CrossRef] [PubMed]
61. Colbourne, J.K.; Pfrender, M.E.; Gilbert, D.; Thomas, W.K.; Tucker, A.; Oakley, T.H.; Tokishita, S.; Aerts, A.; Arnold, G.J.; Basu, M.K.; et al. The ecoresponsive genome of daphnia pulex. *Science* **2011**, *331*, 555–561. [CrossRef] [PubMed]

62. Wen-Hong, F.; Tang, G.; Zhao, C.M.; Duan, Y.; Zhang, R. Metal accumulation and biomarker responses in daphnia magna following cadmium and zinc exposure. *Environ. Toxicol. Chem.* **2009**, *28*, 305–310.
63. CMEE (China Ministry of Ecology and Environment). *Chinese Soil Environmental Quality: Risk Control Standard for Soil Contamination of Agricultural Land (GB 15618-2018)*; China Environmental Science Press: Beijing, China, 2018. (In Chinese)
64. Vukosav, P.; Mlakar, M.; Cukrov, N.; Kwokal, Ž.; Pižeta, I.; Pavlus, N.; Špoljarić, I.; Vurnek, M.; Brozinčević, A.; Omanović, D. Heavy metal contents in water, sediment and fish in a karst aquatic ecosystem of the plitvice lakes national park (Croatia). *Environ. Sci. Pollut. Res.* **2014**, *21*, 3826–3839. [CrossRef]
65. Liu, P.; Zheng, C.; Wen, M.; Luo, X.; Wu, Z.; Liu, Y.; Chai, S.; Huang, L. Ecological risk assessment and contamination history of heavy metals in the sediments of Chagan Lake, Northeast China. *Water* **2021**, *13*, 894. [CrossRef]
66. Shi, J.B.; Wu, Y.G.; Yan, L.; Liu, F.; Yu, Y.H. The acute biological toxicity of Pb and Zn caused by zinc oxide industrial waste in karst areas in Guizhou. *Guizhou Agric. Sci.* **2009**, *37*, 89–92. (In Chinese)
67. Xiong, X.Q.; Luo, S.; Wu, B.L.; Wu, B.L.; Wang, J.W. Acute toxicity of cadmium and copper to gobiocypris rarus under different water hardness. *Asian J. Econ.* **2016**, *11*, 316–322. (In Chinese)
68. Ali, H.; Khan, E.; Ilahi, I. Environmental chemistry and ecotoxicology of hazardous heavy metals: Environmental persistence, toxicity, and bioaccumulation. *J. Chem.* **2019**, *2019*. [CrossRef]
69. Lim, K.Y.; Zakaria, N.A.; Foo, K.Y. Geochemistry pollution status and ecotoxicological risk assessment of heavy metals in the Pahang river sediment after the high magnitude of flood event. *Hydrol. Res.* **2021**, *52*, 107–124. [CrossRef]
70. Tian, K.; Wu, Q.; Liu, P.; Hu, W.; Huang, B.; Shi, B.; Zhou, Y.; Kwon, B.O.; Choi, K.; Ryu, J.; et al. Ecological risk assessment of heavy metals in sediments and water from the coastal areas of the Bohai Sea and the Yellow Sea. *Environ. Int.* **2020**, *136*, 105512. [CrossRef] [PubMed]

Article

Sequential Extraction and Risk Assessment of Potentially Toxic Elements in River Sediments

Dithobolong L. Matabane [1], Taddese W. Godeto [2,†], Richard M. Mampa [1,*] and Abayneh A. Ambushe [2,*]

1   Department of Chemistry, University of Limpopo, Private Bag X 1106, Sovenga,
Polokwane 0727, South Africa; lavholovia@gmail.com

2   Department of Chemical Sciences, University of Johannesburg, P.O. Box 524, Auckland Park,
Johannesburg 2006, South Africa; Taddese.Godeto@ontario.ca

*   Correspondence: richard.mampa@ul.ac.za (R.M.M.); aambushe@uj.ac.za (A.A.A.);
Tel.: +27-15-268-2334 (R.M.M.); +27-11-559-2329 (A.A.A.); Fax: +27-15-268-1065 (R.M.M.);
+27-11-559-2819 (A.A.A.)

†   Current address: Laboratory Services Branch, Ministry of the Environment, Conservation and Parks,
Toronto, ON M9P 3V6, Canada.

Abstract: In this study, the sequential extraction method was applied to extract selected potentially toxic elements (PTEs) (Cd, Cr, Cu, Fe, Ni, Pb and Zn) in river sediments collected from the Blood River situated in Seshego area, Limpopo Province, South Africa. The study aimed to assess a possible trend of mobilisation of these elements from sediment to water. The accuracy of the sequential extraction method was confirmed by analysing sediment-certified reference material, and quantitative percentage recoveries ranging from 86 to 119%, 81 to 111% and 77 to 119% were achieved for exchangeable, reducible and oxidisable fractions, respectively. The potential risk of the PTEs in sediments was evaluated. The calculated values of contamination factor (CF) as well as risk assessment code (RAC) for Cd, Cu, Ni and Pb revealed the mobility of these elements. The PTEs in river sediments are at a high toxicity-risk level and could therefore cause a threat to organisms dwelling in sediments and humans via consumption of crops irrigated with the polluted river water.

Keywords: contamination factor; potentially toxic elements; risk assessment code; sediment; sequential extraction

Citation: Matabane, D.L.; Godeto, T.W.; Mampa, R.M.; Ambushe, A.A. Sequential Extraction and Risk Assessment of Potentially Toxic Elements in River Sediments. *Minerals* **2021**, *11*, 874. https://doi.org/10.3390/min11080874

Academic Editors: Ana Romero-Freire and Hao Qiu

Received: 27 May 2021
Accepted: 6 July 2021
Published: 12 August 2021

Publisher's Note: MDPI stays neutral with regard to jurisdictional claims in published maps and institutional affiliations.

1. Introduction

Potentially toxic elements (PTEs) are known to be adsorbed directly to the sediments of the water system. However, some of these elements may not be permanently bound by the sediments. The changes in environmental conditions, namely acidification, the redox potential or concentrations of organic ligand can initiate leaching of PTEs from sediment to overlying water, which adversely affects the aquatic environment [1]. The sediments are a sink of pollutants, including PTEs that make sediments as key components of water-quality studies [2]. The risks associated with sediments contaminated by PTEs strongly relate to their specific chemical fractions and binding states. Furthermore, factors such as the sediments' nature, grain size, and properties of adsorbed particles as well as of the element could influence the mobility of PTEs in sediments [3]. The measurement of total elemental concentrations cannot give sufficient information regarding the mobility, bioavailability and toxicity of the PTEs. These properties depend on the chemical forms of the element in the different parts of the sample [4]. Therefore, fractionation of PTEs in sediments by sequential extraction is helpful to understand their source, mode of occurrence, mobility and bioavailability [5].

Sequential extraction methods have been widely used to extract PTEs in solid samples such as soil, sediment, sludge and solid wastes based on their leachability [6–10]. Sequential extraction provides information that helps in identifying the major sites of binding as well as the phase associations of PTEs in river sediment. This could assist us in understanding the processes that govern mobilisation of PTEs and potential risks associated with their mobilisation [1].

In South Africa, rivers, artificial lakes and groundwater are under rising stress due to ongoing pollution caused by industrialisation, urbanisation, mining and agriculture activities as well as electricity generation using coal [11]. Although mining is a substantial contributor to the pollution of water bodies in South Africa [12], contamination from other sources cannot be ruled out. The pollution of the Umtata River in South Africa by PTEs from different sources was reported [13]. These sources include agricultural runoff in the catchment together with contributions from natural and point sources. The coal mine in Lephalale releases dust containing PTEs that could deposit in the Mokolo River as pollutants [14]. This study focused on the Blood River in Seshego, South Africa, which is heavily contaminated by municipal and domestic wastes. Furthermore, runoff from urban areas and wastewater arising from human activities in residential and commercial establishments could be the potential source of PTEs in the Blood River. The aim of this study was to sequentially extract and quantify the levels of selected PTEs (Cd, Cr, Cu, Fe, Ni, Pb and Zn) in river sediments obtained from the Blood River in Limpopo Province, South Africa. Moreover, the contamination factor (CF) and risk assessment code (RAC) were calculated to evaluate the associated risks to the aquatic ecosystem.

2. Materials and Methods

2.1. The Justification of Study Area Selection

The Blood River is located in the Seshego area about 14.5 km away from Polokwane, the capital city of Limpopo Province. The river originates in the west of the Polokwane municipal area. The municipal and domestic wastes, presence of illegal dumping sites close to the river banks, lack of waste removal services and absence of storm water drainage among others contribute to the contamination of the Blood River [15]. During the sampling period, dumping sites near different parts of the Blood River, sand mining activities and direct discharge of effluents into the river were observed and as results, sewage effluents can introduce other chemicals and PTEs into the water system. Therefore, the aforementioned wastes and anthropogenic activities could possibly elevate the levels of PTEs in the Blood River. Sampling sites in the Blood River are shown in Figure 1.

2.2. Reagents, Standard Reference Materials and Standards

Ultrapure deionised water of 18.2 MΩ.cm resistivity, acquired from Milli-Q water purification system (Shanghai Carrex Analytic Instrument Co., Ltd., Shanghai, China) was used for rinsing glassware and diluting solutions. Reagents of 65% HNO_3, 35% H_2O_2 and 37% HCl purchased from Merck (Darmstadt, Germany) were used for mineralisation of river sediments.

A 1000 mg L^{-1} standard solution (Sigma-Aldrich® Chemie GmbH, Buchs, Switzerland) was used for preparation of calibration standards. Certified reference materials SRM 8704 Buffalo River sediments and CRM 280R Lake sediments NIST (Gaithersburg, MD, USA) were used to evaluate the accuracy of the method utilised for quantification of pseudo-total concentrations of PTEs in river sediments. A BCR-701 certified reference material of sediment obtained from NIST (Gaithersburg, MD, USA) was used for the validation of the sequential extraction procedure.

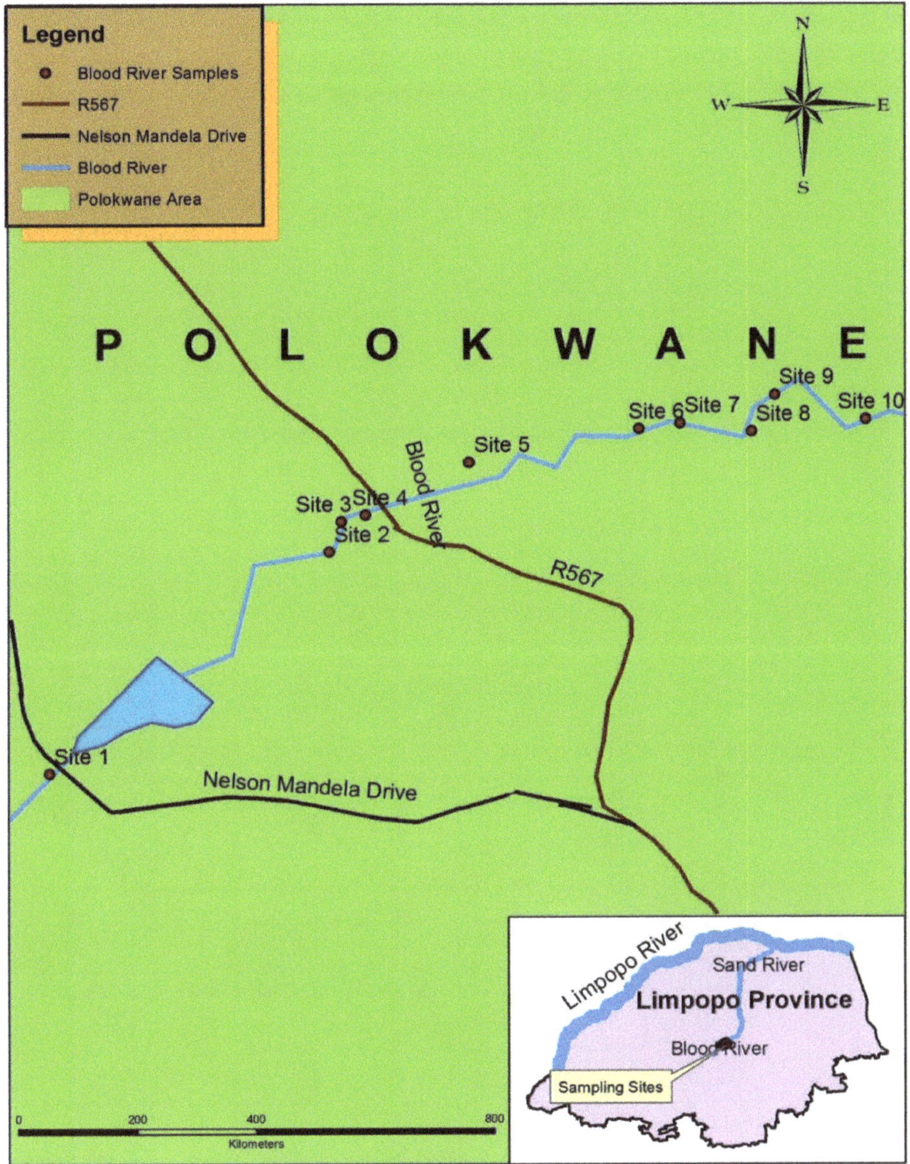

Figure 1. Map of the Blood River showing 10 sampling sites.

2.3. Apparatus and Instrumentation

A mechanical shaker, a centrifuge and an ultrasonic bath purchased from Lenton Furnaces & Scientific Co. (Johannesburg, South Africa) were used to agitate the sediment samples in the centrifuge tubes for extraction of PTEs, for centrifugation of extracts and for agitating the sediment samples along with the supernatant liquid before transferring the liquid to a vial, respectively. A microwave digestion system (CEM Corporation MARS 5, Matthews, NC, USA) was used for mineralisation of the PTEs in river sediments. The measurement of total concentrations of PTEs in river sediments was done by inductively coupled plasma-mass spectrometry (ICP-MS) purchased from PerkinElmer (Elan 6100,

Regensburg, Germany). The sequentially extracted PTEs were quantified using inductively coupled plasma-atomic emission spectrometry (ICP-AES) purchased from Shimadzu (ICPE-9000, Columbia, MD, USA). Total concentration of iron in sediment samples was quantified using flame-atomic absorption spectrometry (F-AAS) purchased from PerkinElmer (PinAAcle 500, Syngistix, Singapore).

2.4. Sample Collection and Preservation

Sediment samples were obtained from about a depth of 20 cm from the surface of the sediment at 10 sampling points along the Blood River. A convenience sampling method was followed based on the accessibility of the sampling point. From each sampling point, about 500 g of sample was collected and placed in polyethylene bottles. The sediment samples from each sampling site were air-dried at room temperature. An agate mortar and pestle were used to grind the dried river sediments. The ground samples were sieved using a 100 µm sieve mesh and kept in polyethylene bottles until extraction and mineralisation of PTEs.

2.5. Sample Preparation

2.5.1. Microwave-Assisted Acid Digestion (Method A)

The fine powdered sediment samples of 0.250 g were mineralised using a microwave-assisted acid digestion system for total elemental determination by ICP-MS. A mixture of reagents containing 2 mL HNO_3, 2 mL H_2O_2 and 2 mL HCl yielded complete digestion of sediment samples. The microwave-based digestion conditions included temperature control with maximum temperature of 150 °C, 20 min ramp, 20 min hold, 800 psi pressure and 1600 W power.

2.5.2. A Sequential Extraction Procedure (Method B)

The BCR sequential extraction procedure reported by Castillo et al. [17] was adopted in this study with slight changes. This method entails three extraction stages and is described as follows. During stage 1 of the BCR protocol, 1.0 g of powdered and sieved sediment sample was weighed into a pre-cleaned 50 mL centrifuge tube. A 40 mL of 0.11 $mol \cdot L^{-1}$ acetic acid was transferred into the centrifuge tube, and the content was shaken for 16 h using the mechanical shaker. The separation of extraction solution from the residue was done by centrifugation at 3000 rpm for 15 min and followed by decantation of the supernatant liquid. This was followed by washing the residue with 4 mL acetic acid (0.11 $mol \cdot L^{-1}$) and shaking for 15 min at 3000 rpm in the centrifuge. Further washing was performed, and the latter was then removed with a pipette and combined with the extracted fraction. The mixture was finally transferred into a 50 mL centrifuge tube topped to the mark with deionised water and kept in a refrigerator at 4 °C till analysis. The residue was used to proceed with stage 2 of the procedure.

In stage 2, 40 mL hydroxylamine hydrochloride of 0.50 $mol \cdot L^{-1}$ was added to the residue from stage 1 and shaken for 16 h in a mechanical shaker. Washing was carried out as in stage 1, and the extracted fraction was stored in a refrigerator until analysis while the residue was used in stage 3 for further extraction.

In stage 3, the residue from stage 2 was mixed with 10 mL hydrogen peroxide of 8.8 $mol \cdot L^{-1}$, and the centrifuge tube was covered and left to react for 1 h at room temperature. Further extraction was done for 1 h at 80 °C in an ultrasonic bath, and the volume was reduced by additional heating of the solution. A 10 mL aliquot of 8.8 $mol \cdot L^{-1}$ hydrogen peroxide was added and was heated to 80 °C for 1 h while the vial was covered. After removing the cover, the volume was reduced to a further $\approx$ 2–3 mL by heating. About 40 mL ammonium acetate of 1.0 $mol \cdot L^{-1}$ was mixed with the cool residue and the PTEs were extracted following stage 1 procedure. A change in the pH of solutions from 2 to 1.5 was implemented after low percent recoveries were observed in the second and third stages of sequential extraction.

An additional fourth stage was added in which the residue was mineralised following the procedure described in method A. The solution was topped to 50 mL with deionised water and stored in a refrigerator until analysis. A flow diagram of the BCR sequential extraction method is shown in Figure 2.

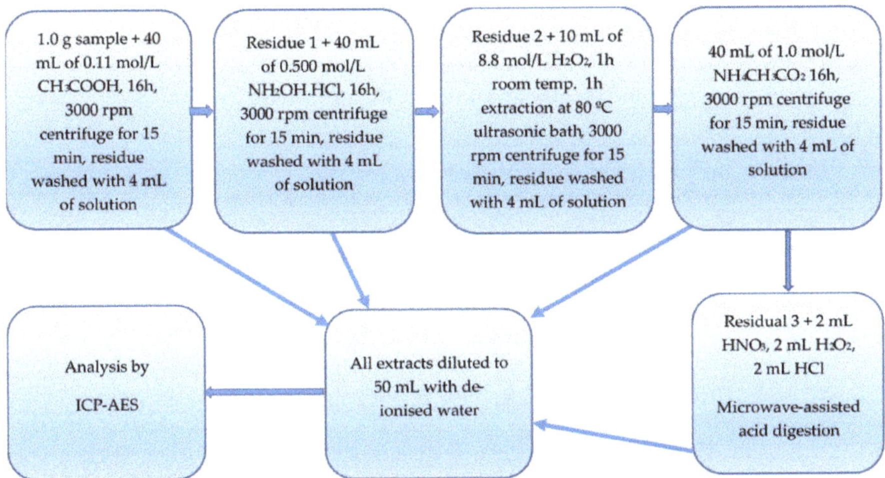

Figure 2. Flow diagram of BCR sequential extraction method.

2.6. Sample Analysis

Sample analyses were conducted using ICP-MS for measurement of total concentrations of PTEs in sediments, ICP-AES for quantification of sequentially extracted PTEs and F-AAS for determination of total concentration of iron in river sediments. The instrumental conditions are summarised in Table 1.

Table 1. Operating conditions of ICP-MS, ICP-AES and F-AAS.

ICP-MS Parameters	ICP-MS Operating Conditions
Lens voltage (V)	10
RF power (W)	1150
Flow rate of plasma gas (L·min^{-1})	12
Flow rate of nebuliser gas (L·min^{-1})	0.95
Flow rate of auxiliary gas (L·min^{-1})	1.2
ICP-AES Parameters	**ICP-AES Operating Conditions**
RF power (W)	1200
Flow rate of plasma gas (L·min^{-1})	10
Flow rate of nebulizer gas (L·min^{-1})	0.7
Flow rate of auxiliary gas (L·min^{-1})	0.60
View direction	Axial
View position	Low
F-AAS Parameters	**F-AAS Operating Conditions**
Flow rate of air (L·min^{-1})	10.0
Flow rate of acetylene (L·min^{-1})	3.30

2.7. Determination of Limit of Detection and Limit of Quantification

The limit of detection (LOD) and limit of quantification (LOQ) values were determined using six blanks, which were prepared following the same procedures of the analytical methods. The values of LOD and LOQ for each analyte were computed based on three and ten times the standard deviation of the six blank measurements, respectively [14].

2.8. Analytical Data Quality Assurance/Quality Control

Standard reference materials (SRM 8704 Buffalo River sediments and BCR 280R Lake sediments) were used to check the accuracy of PTE concentrations in the river sediments. The BCR-701 certified reference material was used to confirm the accuracy of the method applied for determination of PTEs in sequentially extracted solutions.

2.9. Statistical Analysis

The analytical data were evaluated through a *t*-test using SPSS software to examine the statistical significance of the differences in the mean concentrations of Cd, Cr, Cu, Ni, Pb, Zn and Fe, determined in river sediments by method A and method B. Comparison of the means obtained by two methods was done at a 5% level of significance.

3. Results and Discussion

3.1. Performance of the Analytical Method

The calculated LOD and LOQ values are presented in Table 2. In this study, the LODs and LOQs for determination of pseudo-total concentrations of PTEs by ICP-MS ranged from 0.030 to 1.77 $ng·g^{-1}$ and from 0.10 to 5.90 $ng·g^{-1}$, respectively. Iron in sediment samples was quantified by F-AAS, and its LOD and LOQ values were 0.049 $mg·kg^{-1}$ and 0.16 $mg·kg^{-1}$, respectively.

Table 2. Limits of detection and limits of quantification obtained for determination of PTEs in digested sediments by ICP-MS and limit of detection for the sequential extraction procedure obtained by ICP-AES.

	Microwave-Assisted Digestion Procedure		BCR Sequential Extraction Procedure			
Element	LOD ($ng·g^{-1}$)	LOQ ($ng·g^{-1}$)	Stage 1 ($mg·kg^{-1}$)	Stage 2 ($mg·kg^{-1}$)	Stage 3 ($mg·kg^{-1}$)	Residual ($mg·kg^{-1}$)
Cd	0.030	0.10	0.008	0.038	0.0064	0.125
Cr	0.11	0.37	0.218	0.040	0.0064	0.041
Cu	0.48	1.60	0.008	0.013	0.0742	0.022
Fe	0.049 [a]	0.16 [a]	0.087	0.008	0.0573	0.005
Ni	0.62	2.08	0.017	0.004	0.0658	0.003
Pb	0.17	0.57	0.276	0.169	0.503	0.026
Zn	1.77	5.90	0.023	0.004	0.0806	0.041

[a] LOD and LOQ values in $mg·kg^{-1}$ obtained by F-AAS.

Following the same procedure, with slight variation in the volume of the reagents, Melaku et al. [18] reported LOD values of 0.83, 0.59, 0.08, 0.67, 0.64, 0.03 and 0.15 $ng·g^{-1}$ for Cr, Fe, Ni, Cu, Zn, Cd and Pb, respectively. These LOD values were comparable with the LODs, in this study, for Cu, Cd and Pb. The LODs were adequate for quantification of PTEs in the analysed sediment samples.

The LODs for the BCR sequential extraction procedure were determined by ICP-AES and presented in Table 2. For the BCR sequential extraction procedure, the LODs for the exchangeable (stage 1), reducible (stage 2), oxidisable (stage 3) and residual (stage 4) ranged from 0.008 to 0.276, 0.004 to 0.169, 0.0064 to 0.503 and 0.003 to 0.125 $mg·kg^{-1}$, respectively. The LODs reported by Ciceri et al. [19] were 0.066, 0.23, 0.018, 0.41, 0.026 and 0.0059 $mg·kg^{-1}$ in the first stage; 0.16, 0.18, 0.016, 0.11, 0.0069 and 0.0010 $mg·kg^{-1}$ in the second stage; and lastly 0.50, 0.14, 0.023, 0.76, 0.0076 and 0.043 $mg·kg^{-1}$ in the third stage for Cr, Ni, Cu, Zn, Cd and Pb, respectively. These values were found to be higher than the LODs obtained in this study. The low LOD values show that trace levels of these analytes could be measured with confidence.

Two certified reference materials of sediments (SRM 8704 and BCR 280R) were used for checking the accuracy of the method applied for determination of pseudo-total concentrations in the river sediments. The use of two different certified reference materials for the same method of digestion was based on the non-availability of the certified values for some of the analytes in the certified reference material. The certified concentrations available on the SRM 8704 certificate of analysis are for Cd, Cr, Fe, Ni, Pb and Zn, while the available certified values on the BCR 280R certificate of analysis are for Cd, Cu, Cr, Ni and Zn. For this reason, both certified reference materials were analysed by ICP-MS and F-AAS, and the measured concentrations along with the certified values are shown in Table 3. The measured concentrations were in good harmony with the certified concentrations, indicating the accuracy of the method for quantification of PTEs in sediment samples.

Table 3. Analysis of certified reference materials of sediments (BCR 280R and SRM 8704).

Element	BCR-280R			SRM 8704		
	Measured	Certified	Percentage Recovery	Measured	Certified	Percentage Recovery
	Mean $\pm$ SD (mg·kg^{-1})	Mean $\pm$ SD (mg·kg^{-1})	(%)	Mean $\pm$ SD (mg·kg^{-1})	Mean $\pm$ SD (mg·kg^{-1})	(%)
Cd	0.912 $\pm$ 0.002	0.85 $\pm$ 0.10	107	3.33 $\pm$ 0.023	2.94 $\pm$ 0.29	113
Cr	111 $\pm$ 14	126 $\pm$ 7.0	88		na	
Cu	61.4 $\pm$ 0.63	53.0 $\pm$ 6.0	116		NA	
Fe		NA		3.69 [b]	3.97 $\pm$ 0.10	93
Ni	63.3 $\pm$ 2.6	69.0 $\pm$ 5.0	92	38.6 $\pm$ 10	42.9 $\pm$ 3.7	90
Pb		NA		210 $\pm$ 3.8	150 $\pm$ 17	140
Zn	208 $\pm$ 0.91	224 $\pm$ 25	93	402 $\pm$ 2.0	408 $\pm$ 15	98

NA: not available on the SRM certificate; na: not analysed in this study; [b]: analysed with F-AAS and concentration in % (w/w).

The percentage recoveries for the analytes ranged from 88 to 116%, which showed reliability of the applied method, except for Pb. Comparing the measured values with the certified values for the two certified reference materials, quantitative and comparable percentage recoveries were obtained for most analytes. Although low percentage recovery of 88% was observed for Cr, it was still within the acceptable range of 75 to 125% as established by United States Environmental Protection Agency (U.S. EPA) [20].

The evaluation of sequential extraction procedure was carried out by comparing the measured and certified values for the BCR 701 certified reference material. The percentage recoveries obtained by the analysis of the BCR 701 using ICP-AES are presented in Table 4. The BCR sequential extraction protocol yielded the percentage recoveries of 86 to 119%, 81 to 111% and 77 to 119% for exchangeable fraction (F1), reducible fraction (F2) and oxidisable fraction (F3), respectively. These percentage recoveries of analytes were within the acceptable ranges for all three stages [20].

The sequential extraction methods were further validated by comparing the sum of concentrations of PTEs extracted by four stages in sequential extraction procedure including the residual fraction (F4) with the results of the pseudo-total concentrations of PTEs [9,21]. The percentage recovery of the sequential extraction method compared to the pseudo-total concentrations was calculated as follows:

$$\text{Percentage recovery} = \frac{[\text{F1} + \text{F2} + \text{F3} + \text{F4}]}{\text{Pseudo} - \text{total concentration}} \times 100$$

where F1, F2, F3 and F4 stand for the first, second, third and residual fractions, respectively. The results obtained using sequential extraction procedure showed good harmony with the pseudo-total concentrations. For example, for sample 1 the calculated percentage recoveries ranged from 77 to 104% for sequential extraction protocol.

Table 4. Concentrations and percentage recoveries of PTEs in BCR 701 certified reference material obtained by ICP-AES.

Fraction	Element	Certified Value Mean ± SD (mg·kg^{-1})	Measured Value Mean ± SD (mg·kg^{-1})	Percentage Recovery (%)
F1	Cd	7.30 ± 0.40	7.13 ± 0.25	98
	Cr	2.26 ± 0.16	2.69 ± 0.050	119
	Cu	49.3 ± 1.7	52.7 ± 3.0	107
	Ni	15.4 ± 0.9	13.2 ± 0.28	86
	Pb	3.18 ± 0.21	2.87 ± 0.092	90
	Zn	205 ± 6.0	223 ± 3.5	109
F2	Cd	3.77 ± 0.28	3.05 ± 0.36	81
	Cr	45.7 ± 2.0	50.3 ± 1.0	110
	Cu	124 ± 3.0	126 ± 4.0	101
	Ni	26.6 ± 1.3	25.8 ± 2.3	97
	Pb	126 ± 3.0	107 ± 18	85
	Zn	114 ± 5.0	126 ± 9.3	111
F3	Cd	0.27 ± 0.06	0.218 ± 0.010	81
	Cr	143 ± 7.0	110 ± 4.0	77
	Cu	55.0 ± 4.0	63.1 ± 2.1	115
	Ni	15.3 ± 0.9	18.2 ± 0.78	119
	Pb	9.30 ± 2.0	9.25 ± 0.49	99
	Zn	46.0 ± 4.0	46.5 ± 2.7	101

3.2. Fractionation of Potentially Toxic Elements in River Sediments

In this work, sequential extraction was used to evaluate the abundance and distribution of PTEs in sediments by subjection to different reagents under certain operating conditions. The sequential extraction procedure assisted us in determining the distribution of PTEs in the exchangeable, reducible, oxidisable and residual fractions.

The sequential extraction method gives information that could be associated with the physical and chemical factors that affect the mobilisation, bioavailability and transportation of PTEs [7,22]. The concentrations of the fractions represent the pseudo-total concentrations of PTEs in river sediments, thus suitable for comparison studies.

The concentrations of PTEs in different fractions of sediment samples collected from the Blood River were quantified using ICP-AES. The levels of Cd in sequentially extracted sediment samples are given in Figure 3. Furthermore, the total concentrations that are obtained by adding the concentrations of Cd in F1, F2, F3 and F4 are compared with pseudo-total concentrations determined after microwave-assisted acid digestion (Figure 4).

The level of Cd in the sequentially extracted sediments needed to be monitored due to its high toxicity since it has been reported as one of predominantly labile elements [21,23]. The partitioning trend of Cd was similar in all sites (Figure 3). The concentrations detected in the sites were highly favourable in the exchangeable stage. A study carried out by Shozi [24] emphasised Cd as a mineral that is not commonly found in the Earth's crust and associated its existence in water bodies with anthropogenic activities. This could explain the highest level of Cd measured in sediment sample from site 4. Site 4 of the Blood River was found to be adjacent to an illegal dumping site. The domestic wastes could have contributed to the elevated levels of Cd in the river sediment. For most of the sampling sites, Cd in the residual stage was below the LOD value of 0.125 mg·kg^{-1}. The fractionation pattern of Cd followed the order F1 > F2 > F3 > F4. The Cd revealed mobility as its highest portion (44 to 75%) was in the exchangeable fraction in all samples. These findings were in harmony with the studies by Marmolejo-Rodrıguez et al. [23], Díaz-de Alba et al. [25] and Saleem et al. [21], which reported Cd among the most labile elements in their studies. Additionally, Delgado et al. [26] and Kong et al. [27] also reported high percentages of Cd in the first fraction. The highest level of Cd in labile fraction in all sites compared to other fractions might bring risk to the river ecological system and should receive more attention.

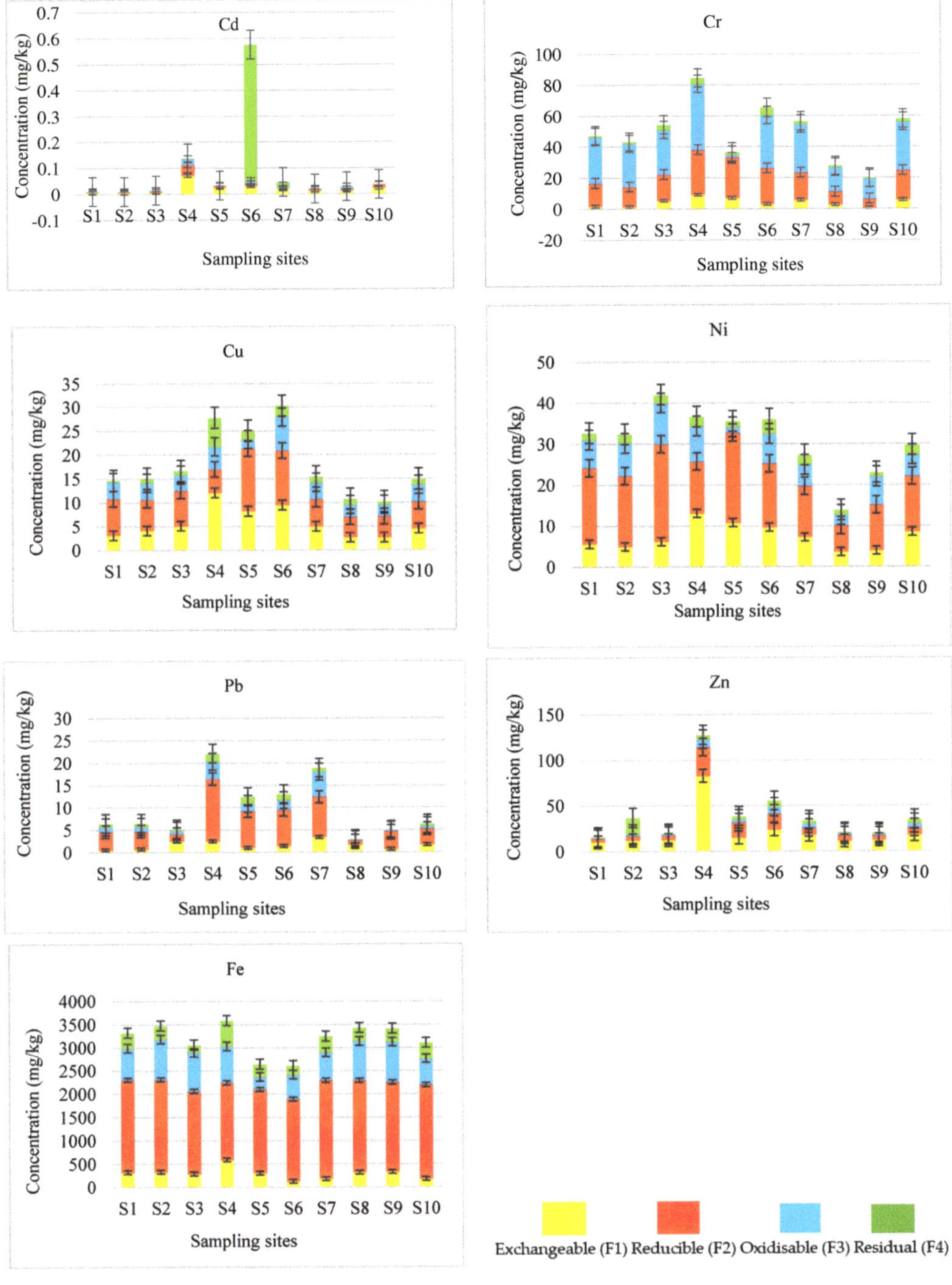

Figure 3. Concentrations of PTEs in different fractions of sediment samples.

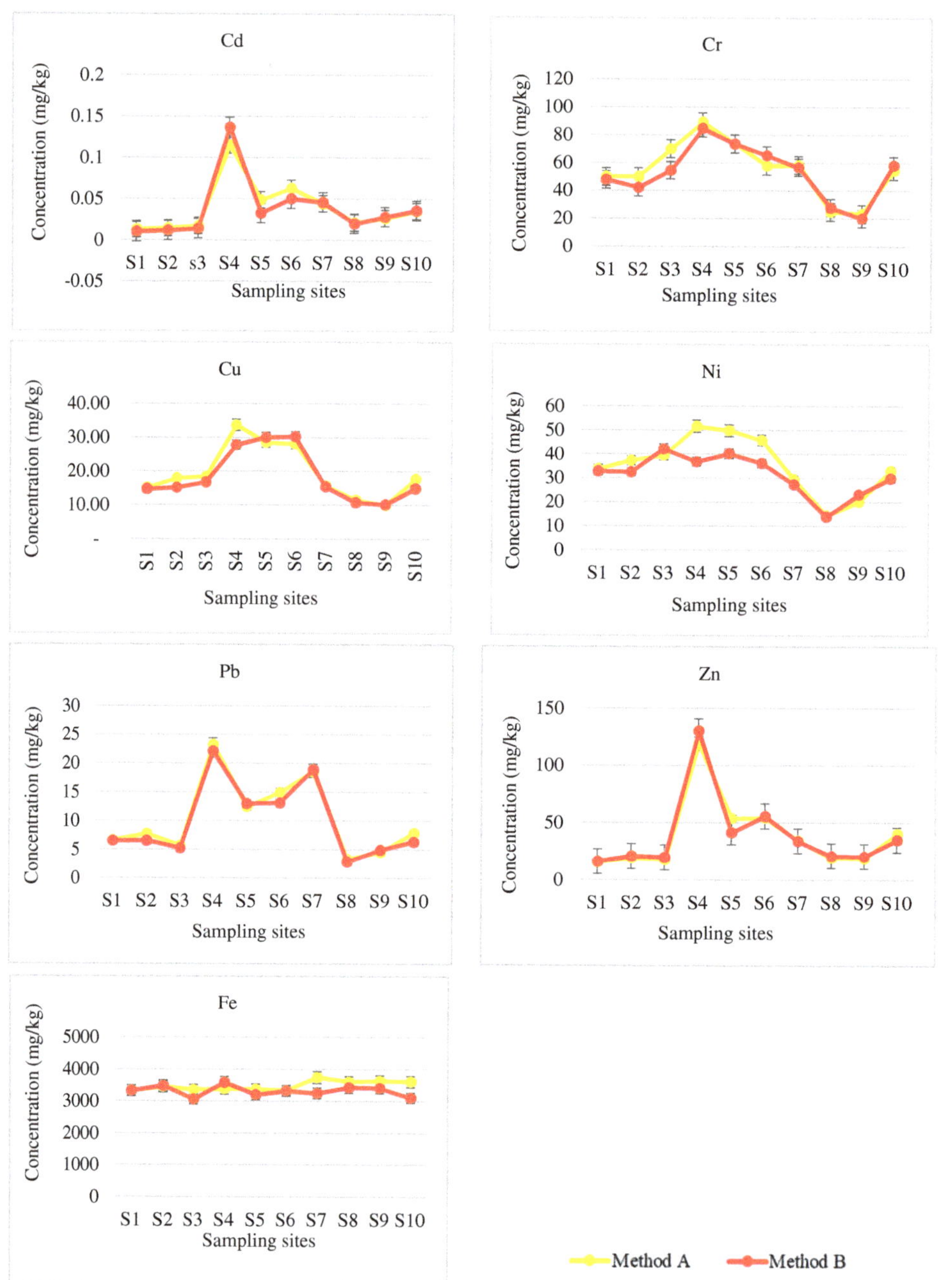

Figure 4. Comparison of total concentrations of PTEs determined by sequential extraction procedure (method B) with pseudo-total concentrations of PTEs determined by microwave-assisted acid digestion (method A).

A noticeable portion of Cd was detected in reducible fraction (20 to 36%) except at site 7 and the oxidisable fraction (17 to 29%) at sites 1 to 9, excluding sites 5 and 10. Relatively low-to-no Cd measured in the residual fraction except at site 7 (Figure 3). A similar trend was reported by Li et al. [28], in which Cd source was attributed to untreated local waste and runoffs from roadside, agricultural and industrial activities. Such activities could have contributed to the presence of the element in the sediments of the Blood River. Other studies reported Cd to be effectively bound to organic matter [29,30].

Figure 4 shows the trend between the total concentrations of sequentially extracted Cd and pseudo-total Cd concentrations in digested sediment samples. The total concentration of Cd obtained from extraction exceeded the pseudo-total concentrations determined by digestion method at site 4. In most sites, the total concentrations and pseudo-total Cd concentrations are in good agreement. The good harmony between method A and method B was confirmed by the *t*-test at a 95% confidence level, which demonstrated the accuracy of sequential extraction method for determination of Cd in river sediments.

Similarly, the concentrations of Cr in sequentially extracted sediment samples are presented in Figure 3. Moreover, the total concentrations of Cr are compared with pseudo-total concentrations determined after microwave-assisted acid digestion (Figure 4).

The Cr revealed a similar partitioning pattern in all the samples with the highest detected level in oxidisable fraction (51 to 65%). Similar findings were reported by Passos et al. [31] and Borgese et al. [32]. A noticeable fraction of Cr was found in F2 (30 to 35%), F1 (3 to 11%) and F4 (2 to 7%). However, the content of Cr in F1, F2 and F4 was below 37% at all sites revealing the soil development source of Cr [32]. It was reported by Chaudhary et al. [33] that Cr levels in the exchangeable fraction were more favourable with acidic pH than alkaline levels. However, this is not the case in this study since the exchangeable Cr was higher in sites 4 and 10 than in all the sites, for the basic pH of 9.32 and 9.29 was recorded at sites 4 and 10, respectively. These high concentrations could be from the metal bound to organic matter in sediments, which may elevate the levels of Cr in the water.

A classical study by Martin and Meybeck [34] introduced the need for the fourth stage and classified it as the residual stage. The authors reported the importance of the residual stage in relation to evaluation of contamination factors. In this study, Cr was detected in residual fraction in all the samples (Figure 3). The pattern of Cr levels in different fractions was F3 > F2 > F1 > F4. The analysis of the BCR 701 sediment certified reference material showed the fractionation pattern of Cr in the order of F3 > F2 > F1. This pattern was consistent with the fractionation pattern of Cr obtained in this study for Blood River sediment samples (F3 > F2 > F1 > F4).

Reliability of the results obtained in the sequential extraction (Method B) for Cr in river sediments was checked by comparison with the pseudo-total digestion concentrations (Method A). There was no significant difference observed between the mean values of Cr in river sediments determined through a *t*-test at a 5% level of significance. A good harmony between the two methods was shown in Figure 4.

The concentrations of Cu in sequentially extracted sediment samples are presented in Figure 3. Furthermore, the total concentrations of Cu in river sediments are compared with pseudo-total concentrations determined after microwave-assisted acid digestion (Figure 4).

The partitioning pattern of Cu in the reducible fraction was highly favourable in all the sites except site 4 (Figure 3). Copper was primarily associated with the reducible fraction (37 to 53%), followed by the exchangeable (21 to 33%), oxidisable (20 to 24%) and lastly, the residual fraction (2 to 11%). A different fractionation pattern was observed in site 4 where Cu was dominant in the exchangeable fraction (44%) followed by the residual fraction (22%), reducible fraction (18%) and finally the oxidisable fraction (17%). Wali et al. [35] and Davutluoglu et al. [36], reported the dominance of Cu in the exchangeable and residual fractions in their studies. Wali et al. [35] reported 39.70% Cu in the exchangeable fraction and 60.30% in the residual fraction.

The sum of Cu concentrations in three fractions plus residual was in good agreement with the pseudo-total concentrations as assessed by a *t*-test at a 95% confidence level and presented in Figure 4. Reliability of the results obtained, in this study, was shown by the partitioning trend of Cu in the BCR 701 sediment certified reference material. The highest Cu concentration was observed in the reducible fraction as it was the case in river sediments. The fractionation pattern followed the order: F2 > F3 > F1 and an additional stage F4 of this study showed the lowest concentration.

The concentrations of Ni in different extraction fractions of river sediments are shown in Figure 3. The sum of Ni concentrations in different fractions plus residual was compared with pseudo-total concentrations determined after microwave-assisted acid digestion in Figure 4. A good agreement was observed between the mean concentrations of Ni measured by the two methods at a 95% level of confidence.

The second fractionation stage represents the elements, which can easily bind to Fe and Mn oxides. These elements may leach from sediments to the water course when subjected to more reductive conditions [9]. The reducible fraction contained the highest Ni content (43 to 57%). A noticeable amount (17 to 29%) was measured in the exchangeable fraction followed by oxidisable fraction (19 to 24%) and residual fraction, which contained the lowest portion (4 to 10%). The fractionation pattern of Ni followed the order F2 > F1 > F3 > F4. Contrary to the findings of this study, Ni was predominantly found in the reducible fraction. Davutluoglu et al. [36] and Favas et al. [37] reported the dominant presence of Ni in the residual fraction compared to the other fractions. The elements predominant in the residual fraction are possibly contained in alumina-silicates and thus are less likely to be released to pore-water [38].

The total concentration obtained from addition of the extraction fractions plus residual was lower than the pseudo-total concentration determined by the digestion method. These differences were elevated in sites 4, 5 and 6. The reliability of the method B was shown by similarity in the partitioning trend of this method for sediment samples with the trend observed for the BCR 701 extraction (F2 > F1 > F3).

The concentrations of Pb in different extraction fractions of river sediments are presented in Figure 3. The sum of Pb concentrations in different extraction fractions plus residual was compared with pseudo-total concentrations determined after microwave-assisted acid digestion in Figure 4.

In normal environmental conditions, Pb that occurs in nature is not very mobile, which shows that the likely sources of Pb are due to human activities [39]. Elevated Pb concentrations were expected in the Fe–Mn oxyhydroxides (reducible) fraction. The oxyhydroxides of Fe–Mn are principal scavengers of the metal in sediments, and Pb is capable of forming stable complexes with oxyhydroxides of Fe–Mn [40,41]. In this study, elevated Pb concentration was measured in reducible fraction in samples from most sites. The highest percentage of Pb was obtained in F2 (47 to 82%), while less than 30%, 27% and 12% was measured in F3, F1 and F4, respectively. The fractionation pattern of Pb followed the order F2 > F3 > F1 > F4, except in sites 3 and 8. Sites 3 and 8 showed a different trend where Pb was dominant in the exchangeable fraction. The highest level of Pb in labile fraction in these sites compared to other fractions might bring high risk to the aquatic biota due to its mobility from this fraction. Zhao et al. [42] reported the highest amount of Pb in the exchangeable and reducible fractions. In their study, they associated the presence of Pb in the exchangeable fraction with anthropogenic activities. The sampling sites in the middle stream sites 4 to 7 are adjacent to a residential area of Seshego Township, and elevated levels of Pb in sites 4 to 7 could be attributed to domestic waste dumped near these sites. The Blood River is mainly influenced by human activities, and Pb primarily originates from municipal wastes, atmospheric inputs and automobile exhaust [21,28]. These activities in Seshego township could be the predominant source of Pb in river sediments.

The reliability of the method B for fractionation of Pb was evaluated by comparing the sum of the concentrations obtained in four stages with the pseudo-total concentrations obtained by method A. In general, the sum of the Pb concentrations in different fractions and pseudo-total concentrations was in good agreement at a 5% level of significance.

The concentrations of Zn in different extraction fractions of sediment samples are given in Figure 3. The sum of concentrations of Zn in sequentially extracted fractions plus residual was compared with pseudo-total concentrations determined after microwave-assisted acid digestion in Figure 4.

Zinc is weakly bound and easily leachable by soluble salts and ions, thus the high levels observed in the exchangeable fraction in this study. The maximum level of Zn in the exchangeable fraction shows that Zn is very mobile. Complexation of Zn with dissolved organic matter has been reported to increase its solubility and mobility [43]. The high levels of Zn at site 4 could have resulted from the release of exchangeable Zn associated with the illegal dumping near the Blood River. The high levels of Zn in exchangeable fraction could be associated with the low-cost houses roofed with Zn corrugated sheets in the township, which may have been leached out from the roofs into the surrounding environment and the river during acidic conditions [44]. The distribution of Zn in the sediments of the Blood River was maximum in the F1 (43 to 65%), followed by F2 (22 to 31%), F3 (7 to 15%) and lastly, F4 (3 to 12%). Zinc partitioning followed the order F1 > F2 > F3 > F4 except at site 5.

Comparison of the sum of Zn concentrations in different fractions (Method B) and pseudo-total concentrations for digested sediments samples (Method A) was done to further check the reliability of the method. In general, the sum of Zn concentrations in different fractions and pseudo-total concentrations of Zn was in good agreement at a 95% confidence level.

The levels of Fe in different extraction fractions of river sediments are presented in Figure 3. The sum of Fe concentrations in sequentially extracted fractions plus residual was compared with pseudo-total concentrations measured after microwave-assisted acid digestion in Figure 4.

In this study, the highest percentage of Fe was obtained in the reducible fraction (46 to 67%) followed by oxidisable fraction (18 to 27%), whereas less than 15% was recorded in the exchangeable and residual fractions (Figure 3). Generally, the concentrations of Fe in the sediments from different sites followed the order F2 > F3 > F4 > F1. The Fe–Mn oxyhydroxide is the major scavenger of all metals. The maximum concentration of Fe in reducible fraction is in good harmony with its association with Fe–Mn oxyhydroxides. Iron is one of abundant metals in the Earth. The continuous increase in the Fe concentration along the sampling sites may be due to the contribution of anthropogenic factors.

Due to lack of the certified value for Fe in the BCR-701 reference material, validation of the analytical results was done by comparing the sum of sequentially extracted concentrations of Fe with the pseudo-total concentrations determined after digestion using a microwave digestion system. The agreement of the results obtained using two methods at a 95% confidence level confirmed the accuracy of the levels of Fe obtained by the sequential extraction method.

3.3. The Implications of Sequential Extracted Concentrations to the Environment

The extent of risks that PTEs could pose to the environment is estimated by determining the contamination factor of PTEs. A high CF of PTEs indicates a high risk to the environment and low retention time [9]. The application of individual contamination factor (ICF) and the global contamination factor (GCF) for assessment of the risks to the environmental and estimation of possible harm to the aquatic organisms and the environment by polluted sediments and water have been reported elsewhere [9,45]. This is based on the PTEs being able to bind to different sediment fractions that determine the bioavailability of the elements from the sediments to the water and the environment [39].

3.3.1. Determination of Contamination Factor

The ICF was determined following the method by Nemati et al. [46]. The ICFs of PTEs in sediments were calculated by dividing the sum of concentrations in the non-residual fractions (F1 + F2 + F3) by the concentrations in residual fractions (F4) of each sample [21].

$$ICF = ([F1 + F2 + F3])/F4$$

The GCF for each site was computed by adding the ICF of all the PTEs determined for the sample [21,47].

$$GCF = \Sigma\, ICF$$

The classifications of ICF and GCF were discussed following the interpretation by Zhao et al. [42], where ICF < 0 and GCF < 6 indicate low contamination, and 1 < ICF < 3 and 6 < GCF < 12 indicate moderate contamination. Additionally, 3 < ICF < 6 and 12 < GCF < 24 indicate considerable contamination. Lastly, ICF > 6 and GCF > 24 are classified as high contamination. The calculated values of ICF and GCF for the ten studied sites are presented in Table 5.

Table 5. Individual contamination and global contamination factors of all the sites in the Blood River obtained employing method B.

Element	ICF										
	S1	S2	S3	S4	S5	S6	S7	S8	S9	S10	Average
Cd	0.0802	0.0724	0.112	108	0.260	0.402	2.600	0.161	0.227	0.287	11.4
Cr	48.9	33.7	15.7	21.4	40.4	12.5	31.4	29.6	34.0	31.1	29.9
Cu	42.4	15.0	18.2	3.6	15.0	11.5	11.7	7.97	14.6	11.2	15.1
Fe	8.98	10.8	16.9	5.5	10.8	12.1	8.3	10.9	10.7	7.9	10.3
Ni	18.1	13.5	20.0	14.8	27.2	9.10	10.4	10.6	21.8	11.4	15.7
Pb	7.05	8.27	16.5	11.3	7.02	8.35	19.9	188	190	13	46.9
Zn	26.6	1.06	30.4	26.5	11.3	7.08	9.05	21.9	12.2	7.0	15.3
GCF	152	82	118	191	112	61	93	269	283	82	144

Based on the calculated values of ICF, there was low contamination by Cd in S1, S2, S3, S5, S6, S8, S9 and S10; S7 showed a moderate contamination by Cd, and site 4 had an ICF value of 108, indicating a high contamination risk of Cd to the environment. This was based on the anthropogenic activities that were dominant at S4. Elevated concentration of Cd in F4 for S6 (Figure 3) and the low ICF index (Table 5) in this sample indicated the natural origin as a likely source of Cd at this site. High concentrations of PTEs in the residual fraction showed that the sediments were relatively unpolluted and the source of PTEs was primarily of geogenic origin [21]. In the residual fraction, PTEs are bound to silicates and as a result, unavailable to the aquatic system [21,38]. The values of ICF for Cr and Cu were above the maximum threshold for high contamination, which shows the risk associated with these elements.

The ICF values of Fe in all the sites were in the high contamination level, except for S3, which was classified as a moderate contaminant. However, with continuous pollution occurring in the area, the Fe level in S3 will likely increase.

The ICF values of Ni in all sites were in the high contamination level. Monitoring the element should be carried out to control possible threats to the aquatic system.

The Pb was classified under high contamination level in all the sites since the ICF values exceeded the ICF of 6 from S1 to S10. The calculated ICF values for Pb in samples from S8 and S9 were also high, which indicated the predominance of anthropogenic origin at these sites. The PTEs from anthropogenic sources are mainly found in the most labile fractions of sediment that are vulnerable to small changes in environmental conditions [21]. Owing to the high toxicity of Pb, the calculated values represent a potential risk of contamination to the water phase, or to the biota, in the sediment samples. Similarly, Zn was also classified as a high contamination risk to the river system based on the ICF values, with an exception in S2, which was under the moderate contamination level.

Overall, the risk assessment showed a high possibility of Cr, Cu, Fe, Ni, Pb, and Zn to be released from the sediments to overlying water with the average ICF values of 29.9, 15.1, 10.3, 15.7, 46.9 and 15.3, respectively. The mean values of ICF followed the order: Pb, Cr, Ni, Zn, Cu, Cd and Fe. Although Cd was under the low classification in most of the sites, it was overall classified under high-risk contamination level based on the average ICF value of 11.2 (Table 5).

The concentration of any element in residual fraction is considered a non-mobile fraction and is an important partition that influences the mobility of the elements, while the non-residual fractions are considered mobile [21]. The results showed Cr, Cu, Fe, Ni, Pb, and Zn have high potential mobility. The mobility of the elements displays the increased potential risk of these elements to the surrounding environment and aquatic organisms.

The GCF values of the PTEs were greater than 24 in all the sites, indicating a high possibility of these elements to be released to water. Thus, the analysed elements in sediments pose high risks to the ecosystem. The GCF values were 152, 82, 118, 191, 112, 61, 93, 269, 283 and 82 from S1 to S10, respectively. The average GCF value was 114, thus classified as high contamination level. The combined effect of the elements along with potential high mobility reveal the likely risks these PTEs could pose to the ecosystem.

3.3.2. The Risk Assessment Code

The risk assessment code (RAC) was first introduced by Perin et al. [48], and was applied to assess the mobility and bioavailability of PTEs in surface sediments [9,49]. The RAC value of each analyte is calculated by the following equation:

$$RAC = [F1/(\Sigma F1 - F4)] \times 100$$

The RAC classifications followed the interpretation by Perin et al. [48] where the RAC value of <1 indicates no risk, between 1 and 10 indicates low risk, 11 to 30 is medium risk, 31 to 50 is high risk and finally, >50 is classified very high risk. Table 6 shows the calculated percentage RAC values of the PTEs in the sediments of the Blood River.

Table 6. Risk assessment code of potentially toxic elements in sediments from the Blood River.

Element	RAC (%)										
	S1	S2	S3	S4	S5	S6	S7	S8	S9	S10	Average
Cd	4.54	3.81	4.42	51.7	89.1	17.5	48.4	8.48	8.15	14.3	25.0
Cr	3.55	3.02	9.70	10.7	9.41	4.72	9.95	9.45	2.98	9.70	7.32
Cu	21.4	27.3	30.6	43.5	27.3	31.2	32.5	24.3	25.6	30.1	29.4
Fe	9.78	9.46	9.44	16.5	9.62	4.90	5.63	9.33	9.66	5.79	9.01
Ni	17.1	15.2	14.6	35.7	27.0	26.9	26.4	25.9	17.1	28.8	23.5
Pb	7.85	10.8	47.5	11.1	7.85	10.8	18.0	58.0	13.9	26.9	21.3
Zn	65.4	33.2	64.2	65.0	36.2	43.0	53.0	54.8	60.1	51.5	52.6

Following the contamination assessment evaluated by ICF and GCF values, the RAC is applied to assess the risk based on the percentage of metal in exchangeable fraction or bound to carbonate [50]. The calculated RAC values in Table 6 showed an average of 52.6% of total Zn in river sediments was either present in exchangeable fraction or bound to carbonate. Thus, it is in the high- to very high-risk category to the aquatic and non-aquatic environment and likely to affect humans via the food chain. The RAC values ranged between 33.2 and 65.45 in the sites, indicating high risk in S2, S5 and S6 and very high risk in S1, S3, S4, S7, S8, S9 and S10.

The RAC values for Cr and Fe revealed that the risks these elements could pose to the aquatic environment are low. The RAC values of Cr ranged from 2.98 to 10.7% with an average value of 7.32%. An average RAC for Fe was 9.01%, ranging from 4.90% in S6 to 16.5% in S4, which falls under the moderate risk category.

Although the concentration of Cd was substantially low in the sediments of the Blood River, the RAC value indicated it as a potential risk to the aquatic and surrounding environment of the river. The average RAC value in the sediments was 25.0% indicating medium risk, while the RAC ranged between 3.81% (low risk) and 89.1% (very high risk). Cadmium in S1, S2, S3, S8 and S9 was classified as high risk and in S6 and S10 as moderate risk.

The average RAC values of Cu (29.4%), Ni (23.4%) and Pb (21.3%) classified the elements as medium risk to the environment. The RAC values of Cu ranged from 21.4% in S1 to 43.5% in S4. Although Cu was classified under the medium-risk category, S3, S4, S6, S7 and S10 were classified under high risk. The presence of Ni in the non-residual fraction categorised its existence in the sediments as medium risk to the environment in the sites with an exception in S4 (35.7%), which was of high risk. Partitioning of Pb in the sediments resulted in the RAC values of Pb ranging from 7.85% in sites 1 and 5 to 58.0% in S8. The presence of this element in S3 and S8 indicates that it poses high risk to the aquatic organisms in the Blood River. Overall, the analyte was classified under the low (S1, S2, S5 and S6) to medium risk (S4, S7, S9 and S10).

These findings indicate potential risks to the human health of residents surrounding the Blood River. These may be caused by direct exposure to the water and indirect exposure via the food chain. The risk assessment of the Blood River followed the decreasing order of Zn > Cu > Cd > Ni > Pb > Fe > Cr. Similar to our findings, Nemati et al. [46] reported Zn and Cd as high-risk elements in river sediments. However, Saleem et al. [21] reported relatively higher mobility of Cd, Ni and Pb, whereas Cu, Fe and Zn were detected predominantly in the residual fractions.

4. Conclusions

Sequential extraction of Cd, Cr, Cu, Fe, Ni, Pb and Zn in river sediments was achieved by employing the BCR sequential extraction procedure. The accuracy of the data was confirmed by analysing the BCR 701 certified reference material of sediment and yielded quantitative percentage recoveries. To assess the retention of PTEs in sediments collected from the Blood River, the contamination factor of each element was calculated. The calculated contamination factors showed the potential of high mobility of PTEs from sediments to the overlaying water. The mobility of the elements shows the increased potential risk of PTEs to the environment and aquatic organisms. Furthermore, the highest concentration of Cd in labile fraction in all the sites compared to other fractions could pose risk to the ecological system of the river. The presence of Cd in water bodies is associated with anthropogenic activities since it is not commonly found in the Earth's crust. The highest level of Zn in the labile fraction could be associated with the low-cost houses roofed with Zn coated sheets in Seshego Township that could have been leached out from the roofs into the river during acidic conditions. In general, the sources of PTEs in the Blood River is predominantly attributed to untreated local waste and runoffs from the roadside, as well as industrial activities. Currently, no regular monitoring of PTEs is done in the Blood River. The measured high concentrations of PTEs in sediments of the Blood River revealed a clear need for continued monitoring of PTEs in the water and sediments. These findings indicate potential risks to the health of human residents surrounding the Blood River. These may be caused by direct exposure to the water and indirect exposure via the food chain. The evaluation of associated risks of PTEs in analysed sediments based on RAC values revealed the risks of Zn > Cu > Cd > Ni > Pb > Fe > Cr.

Author Contributions: Conceptualization, D.L.M. and A.A.A.; methodology, D.L.M. and A.A.A.; software, D.L.M. and A.A.A.; validation, D.L.M. and A.A.A.; formal analysis, D.L.M.; investigation, D.L.M. and A.A.A.; resources, T.W.G., R.M.M. and A.A.A.; data curation, D.L.M. and A.A.A.; writing—original draft preparation, D.L.M. and A.A.A.; writing—review and editing, D.L.M., T.W.G., R.M.M. and A.A.A.; visualization, D.L.M. and A.A.A.; supervision, T.W.G., R.M.M. and A.A.A.; project administration, A.A.A.; funding acquisition, A.A.A. All authors have read and agreed to the published version of the manuscript.

Funding: Dithobolong L. Matabane received financial support from the Sasol Inzalo Foundation and the National Research Foundation (NRF). This study is supported in part by the NRF of South Africa under the Thuthuka Programme Grant Number 117673 and the Water Research Commission (WRC) of South Africa Project Number K5/2515//1. We would like to acknowledge the University of Johannesburg Research Centre for Synthesis and Catalysis and Spectrum for the facility.

Data Availability Statement: Not applicable.

Conflicts of Interest: No conflict of interest.

References

1. Yuan, C.-G.; Shi, J.-B.; He, B.; Liu, J.-F.; Liang, L.-N.; Jiang, G.-B. Speciation of heavy metals in marine sediments from the East China Sea by ICP-MS with sequential extraction. *Environ. Int.* **2004**, *30*, 769–783. [CrossRef] [PubMed]
2. Sarkar, S.K.; Favas, P.J.C.; Rakshit, D.; Satpathy, K.K. Geochemical Speciation and Risk Assessment of Heavy Metals in Soils and Sediments, Environmental Risk Assessment of Soil Contamination, Maria C. Hernandez-Soriano, IntechOpen. 2014. Available online: https://www.intechopen.com/books/environmental-risk-assessment-of-soil-contamination/geochemical-speciation-andrisk-assessment-of-heavy-metals-in-soils-and-sediments (accessed on 29 April 2019).
3. Vetrimurugan, E.; Jonathan, M.; Roy, P.D.; Shruti, V.; Ndwandwe, O. Bioavailable metals in tourist beaches of Richards Bay, Kwazulu-Natal, South Africa. *Mar. Pollut. Bull.* **2016**, *105*, 430–436. [CrossRef]
4. Qiao, Y.; Yang, Y.; Gu, J.; Zhao, J. Distribution and geochemical speciation of heavy metals in sediments from coastal area suffered rapid urbanization, a case study of Shantou Bay, China. *Mar. Pollut. Bull.* **2013**, *68*, 140–146. [CrossRef] [PubMed]
5. Yang, J.; Chen, L.; Liu, L.; Shi, W.; Meng, X. Comprehensive risk assessment of heavy metals in lake sediments from public parks in Shanghai. *Ecotoxicol. Environ. Saf.* **2014**, *102*, 129–135. [CrossRef]
6. Bacon, J.R.; Davidson, C. Is there a future for sequential chemical extraction? *Analyst* **2007**, *133*, 25–46. [CrossRef] [PubMed]
7. Hanay, Ö.; Hasar, H.; Kocer, N.N.; Aslan, S. Evaluation for agricultural usage with speciation of heavy metals in a municipal sewage sludge. *Bull. Environ. Contam. Toxicol.* **2008**, *81*, 42–46. [CrossRef]
8. Nemati, K.; Abu Bakar, N.K.; Sobhanzadeh, E.; Abas, M.R. A modification of the BCR sequential extraction procedure to investigate the potential mobility of copper and zinc in shrimp aquaculture sludge. *Microchem. J.* **2009**, *92*, 165–169. [CrossRef]
9. Nemati, K.; Bakar, A.N.K.; Mhd Abas, R.B.; Sobhanzadeh, E.; Hin Low, K.H. Comparison of unmodified and modified BCR sequential extraction schemes for the fractionation of heavy metals in shrimp aquaculture sludge from Selangor, Malaysia. *Environ. Monit. Assess.* **2011**, *176*, 313–320. [CrossRef]
10. Arenas-Lago, D.; Andrade, M.; Lago-Vila, M.; Rodríguez-Seijo, A.; Vega, F. Sequential extraction of heavy metals in soils from a copper mine: Distribution in geochemical fractions. *Geoderma* **2014**, *230–231*, 108–118. [CrossRef]
11. Oberholster, P.J.; Ashton, P.J. State of the nation report: An overview of the current status of water quality and eutrophication in South African Rivers and Reservoirs. *Counc. Sci. Ind. Res.* **2008**, *3*, 1–7.
12. Bologo, V.; Maree, J.; Carlsson, F. Application of magnesium hydroxide and barium hydroxide for the removal of metals and sulphate from mine water. *Water SA* **2012**, *38*, 23–28. [CrossRef]
13. Fatoki, O.; Lujiza, N.; Ogunfowokan, A.; Fatoki, O.; Lujiza, N.; Ogunfowokan, A. Trace metal pollution in Umtata River. *Water SA* **2002**, *28*, 183–190. [CrossRef]
14. Letsoalo, M.R.; Godeto, T.W.; Magadzu, T.; Ambushe, A.A. Selective speciation of inorganic arsenic in water using nanocomposite based solid-phase extraction followed by inductively coupled plasma-mass spectrometry detection. *J. Environ. Sci. Health Part A* **2019**, *54*, 924–932. [CrossRef] [PubMed]
15. Polokwane Municipality Report. *Integrated Development Planning (IDP) Report for Polokwane Municipality (2012–2013)*; Polokwane Municipal Council: Polowane, South Africa, 2012.
16. Rauret, G.; López Sánchez, J.F.; Sahuquillo, A.; Rubio, R.; Davidson, C.; Ure, A.; Quevauviller, P.H. Improvement of the BCR three step sequential extraction procedure prior to the certification of new sediments and soil reference materials. *J. Environ. Monit.* **1999**, *1*, 57–61. [CrossRef] [PubMed]
17. Castillo, M.L.A.; Alonso, E.V.; Cordero, M.T.S.; Pavon, J.M.C.; De Torres, A.G. Fractionation of heavy metals in sediments by using microwave assisted sequential extraction procedure and determination by inductively coupled plasma mass Spectroscopy. *Microchem. J.* **2011**, *98*, 234–239. [CrossRef]
18. Melaku, S.; Wondimu, T.; Dams, R.; Moens, L. Multi-element analysis of Tinishu river sediments, ethiopia by ICP-MS after microwave assisted digestion. *Can. J. Anal. Sci. Spectrosc.* **2005**, *50*, 31–40.
19. Ciceri, E.; Giussani, B.; Pozzi, A.; Dossi, C.; Recchia, S. Problems in the application of the three-step BCR sequential extraction to low amounts of sediments: An alternative validated route. *Talanta* **2008**, *76*, 621–626. [CrossRef] [PubMed]
20. United States Environmental Protection Agency. *Contract Laboratory Program for National Functional Guidelines for Inorganic Data Review*; United States EPA: Washington, DC, USA, 2010.
21. Saleem, M.; Iqbal, J.; Shah, M.H. Geochemical speciation, anthropogenic contamination, risk assessment and source identification of selected metals in freshwater sediments—A case study from Mangla Lake, Pakistan. *Environ. Nanotechnol. Monit. Manag.* **2015**, *4*, 27–36. [CrossRef]
22. Ure, A.M.; Quevauviller, P.; Muntau, H.; Griepink, B. Speciation of heavy metals in soils and sediments. An account of the improvement and harmonization of extraction techniques undertaken under the auspices of the BCR of the commission of the European communities. *Int. J. Environ. Anal. Chem.* **1993**, *51*, 135–151. [CrossRef]
23. Marmolejo-Rodríguez, A.J.; Prego, R.; Meyer-Willerer, A.; Shumilin, E.; Cobelo-Garcia, A. Total and labile metals in surface sediments of the tropical river-estuary system of Marabasco (Pacific coast of Mexico): Influence of an iron mine. *Mar. Pollut. Bull.* **2007**, *55*, 459–468. [CrossRef]
24. Shozi, M. Assessing the Distribution of Sedimentary Heavy Metals in the Msunduzi River Catchment, Kwazulu-Natal, South Africa. Master's Thesis, University of KwaZulu-Natal, Durban, South Africa, 2015.

25. Diaz-De Alba, M.; Galindo-Riano, M.D.; Casanueva-Marenco, M.J.; Garcia-Vargas, M.; Kosore, C.M. Assessment of the metal pollution, potential toxicity and speciation of sediments from Algeciras Bay (South of Spain) using chemometric tools. *J. Hazard. Mater.* **2011**, *190*, 177–187. [CrossRef] [PubMed]
26. Delgado, J.; Barba-Brioso, C.; Nieto, J.M.; Boski, T. Speciation and ecological risk of toxic elements in estuarine sediments affected by multiple anthropogenic contributions (Guadiana saltmarshes, SW Iberian Peninsula): I. Surficial sediments. *Sci. Total Environ.* **2011**, *409*, 3666–3679. [CrossRef]
27. Kong, M.; Dong, Z.; Chao, J.; Zhang, Y.; Yin, H. Bioavailability and ecological risk assessment of heavy metals in surface sediments of Lake Chaohu. *China Environ. Sci.* **2015**, *35*, 1223–1229.
28. Li, Q.; Wu, Z.; Chu, B.; Zhang, N.; Cai, S.; Fang, J. Heavy metals in coastal wetland sediments of the Pearl River Estuary, China. *Environ. Pollut.* **2007**, *149*, 158–164. [CrossRef]
29. Sundaray, S.K.; Nayak, B.B.; Lin, S.; Bhatta, D. Geochemical speciation and risk assessment of heavy metals in the river estuarine sediments—A case study: Mahanadi basin, India. *J. Hazard. Mater.* **2011**, *186*, 1837–1846. [CrossRef] [PubMed]
30. Zhang, L.; Liao, Q.; Shao, S.; Zhang, N.; Shen, Q.; Liu, C. Heavy metal pollution, fractionation, and potential ecological risks in sediments from Lake Chaohu (Eastern China) and the surrounding rivers. *Int. J. Environ. Res. Public Health* **2015**, *12*, 14115–14131. [CrossRef]
31. Passos, E.A.; Alves, J.C.; Santos, I.S.; Alves, J.P.; Garcia, C.A.B.; Costa, A.C.S. Assessment of trace metals contamination in estuarine sediments using a sequential extraction technique and principal component analysis. *Microchem. J.* **2010**, *96*, 50–57. [CrossRef]
32. Borgese, L.; Federici, S.; Zacco, A.; Gianoncelli, A.; Rizzo, L.; Smith, D.R.; Donna, F.; Lucchini, R.; Depero, L.E.; Bontempi, E. Metal fractionation in soils and assessment of environmental contamination in Vallecamonica, Italy. *Environ. Sci. Pollut. Res.* **2013**, *20*, 5067–5075. [CrossRef]
33. Chaudhary, S.; Banerjee, D.K. Metal phase association of chromium in contaminated soils from an industrial area in Delhi. *Chem. Speciat. Bioavailab.* **2004**, *16*, 145–151. [CrossRef]
34. Martin, J.; Meybeck, M. Elemental mass-balance of material carried by major world rivers. *Mar. Chem.* **1979**, *7*, 178–206. [CrossRef]
35. Wali, A.; Colinet, G.; Ksibi, M. Speciation of heavy metals by modified BCR sequential extraction in soils contaminated by phosphogypsum in Sfax, Tunisia. *Environ. Res. Eng. Manag.* **2015**, *70*, 14–25. [CrossRef]
36. Davutluoglu, O.I.; Seckin, G.; Ersu, C.B.; Yilmaz, T.; Sari, B. Heavy metal content and distribution in surface sediments of the Seyhan River, Turkey. *J. Environ. Manag.* **2011**, *92*, 2250–2259. [CrossRef]
37. Favas, P.J.C.; Pratas, J.; Gomez, M.; Elisa, P.; Cala, V. Selective chemical extraction of heavy metals in tailings and soils contaminate by mining activity: Environmental implications. *J. Geochem. Explor.* **2011**, *111*, 160–171. [CrossRef]
38. Tuzen, M. Determination of trace metals in the river Yesilirmak Sediments in Tokat, Turkey using sequential extraction procedure. *Microchem. J.* **2003**, *74*, 105–110. [CrossRef]
39. Jain, C. Metal fractionation study on bed sediments of River Yamuna, India. *Water Res.* **2004**, *38*, 569–578. [CrossRef] [PubMed]
40. Li, X.; Shen, Z.; Wai, O.; Li, Y. Chemical forms of Pb, Zn and Cu in the sediments profiles of the Peral River Estuary. *Mar. Pollut. Bull.* **2001**, *42*, 215–223. [CrossRef]
41. Wong, C.S.; Wu, S.; Duzgoren-Aydin, N.S.; Aydin, A.; Wong, M.H. Trace metal contamination of sediments in an e-waste processing village in China. *Environ. Pollut.* **2007**, *145*, 434–442. [CrossRef]
42. Zhao, S.; Feng, C.; Yang, Y.; Niu, J.; Shen, Z. Risk assessment of sedimentary metals in the Yangtze Estuary: New evidence of the relationships between two typical index methods. *J. Hazard. Mater.* **2012**, *241–242*, 164–172. [CrossRef]
43. Houben, D.; Sonnet, P. Zinc mineral weathering as affected by plant roots. *Appl. Geochem.* **2012**, *27*, 1587–1592. [CrossRef]
44. Okonkwo, J.O.; Mothiba, M. Physico-chemical characteristics and pollution levels of heavy metals in the rivers in Thohoyandou, South Africa. *J. Hydrol.* **2005**, *308*, 122–127. [CrossRef]
45. Nasr, S.M.; Okbah, M.A.; El Haddad, H.S.; Soliman, N.F. Fractionation profile and mobility pattern of metals in sediments from the Mediterranean Coast, Libya. *Environ. Monit. Assess.* **2015**, *187*, 1–8. [CrossRef] [PubMed]
46. Nemati, K.; Abu Bakar, N.K.; Abas, M.R.; Sobhanzadeh, E. Speciation of heavy metals by modified BCR sequential extraction procedure in different depths of sediments from Sungai Buloh, Selangor, Malaysia. *J. Hazard. Mater.* **2011**, *192*, 402–410. [CrossRef] [PubMed]
47. Ikem, A.; Egiebor, N.O.; Nyavor, K. Trace elements in water, fish and sediment from Tuskegee Lake, Southeastern USA. *Water Air Soil Pollut.* **2003**, *149*, 51–75. [CrossRef]
48. Perin, G.; Craboleda, L.; Lucchese, M.; Cirillo, R.; Dotta, L.; Zanette, M.L.; Orio, A.A. Heavy metal speciation in the sediments of Northern Adriatic Sea- a new approach for environmental toxicity determination. In *Heavy Metal in the Environment*; Lakkas, T.D., Ed.; New Age International: Delhi, India, 1985; pp. 454–456.
49. Mortazavi, S.; Attaeian, B.; Abdolkarimi, S. Risk assessment and environmental geochemistry of Pb, Cu and Fe in surface sediments (case study: Hashilan Wetland, Kermanshah, Iran). *Ecopersia* **2016**, *4*, 1411–1424. [CrossRef]
50. Saeedi, M.; Li, L.Y.; Karbassi, A.; Zanjani, A.J. Sorbed metals fractionation and risk assessment of release in river sediment and particulate matter. *Environ. Monit. Assess.* **2012**, *185*, 1737–1754. [CrossRef] [PubMed]

Review

Trace Metal Contamination of Bottom Sediments: A Review of Assessment Measures and Geochemical Background Determination Methods

Nicole Nawrot [1,*], Ewa Wojciechowska [1], Muhammad Mohsin [2], Suvi Kuittinen [2], Ari Pappinen [2] and Shahabaldin Rezania [3]

[1] Faculty of Civil and Environmental Engineering, Gdańsk University of Technology, 80-233 Gdansk, Poland; esien@pg.edu.pl
[2] School of Forest Sciences, University of Eastern Finland, Yliopistokatu 7, P.O. Box 111, 80100 Joensuu, Finland; muham@uef.fi (M.M.); suvi.kuittinen@uef.fi (S.K.); ari.pappinen@uef.fi (A.P.)
[3] Department of Environment and Energy, Sejong University, Seoul 05006, Korea; shahab.rezania@sejong.ac.kr
* Correspondence: nicnawro@pg.edu.pl

Citation: Nawrot, N.; Wojciechowska, E.; Mohsin, M.; Kuittinen, S.; Pappinen, A.; Rezania, S. Trace Metal Contamination of Bottom Sediments: A Review of Assessment Measures and Geochemical Background Determination Methods. *Minerals* **2021**, *11*, 872. https://doi.org/10.3390/min11080872

Academic Editors: Ana Romero-Freire and Hao Qiu

Received: 16 June 2021
Accepted: 10 August 2021
Published: 12 August 2021

Publisher's Note: MDPI stays neutral with regard to jurisdictional claims in published maps and institutional affiliations.

Abstract: This paper provides an overview of different methods of assessing the trace metal (TM) contamination status of sediments affected by anthropogenic interference. The geochemical background determination methods are also described. A total of 25 papers covering rivers, lakes, and retention tanks sediments in areas subjected to anthropogenic pressure from the last three years (2019, 2020, and 2021) were analysed to support our examination of the assessment measures. Geochemical and ecotoxicological classifications are presented that may prove useful for sediment evaluation. Among the geochemical indices, several individual pollution indices (*CF*, *Igeo*, *EF*, *Pi* (*SPI*), PT_T), complex pollution indices (*PLI*, *Cdeg*, *mCdeg*, Pi_{sum}, PI_{Avg}, PI_{aAvg}, *PIN*, PI_{Prod}, PI_{apProd}, $PI_{vectorM}$, $PI_{Nemerow}$, *IntPI*, *MPI*), and geochemical classifications are compared. The ecotoxicological assessment includes an overview of Sediment Quality Guidelines (SQG) and classifications introduced nationally (as LAWA or modified LAWA). The ecotoxicological indices presented in this review cover individual (ER^i) and complex indices (*CSI*, *SPI*, *RAC*, *PERI*, *MERMQ*). Biomonitoring of contaminated sites based on plant bioindicators is extensively explored as an indirect method for evaluating pollution sites. The most commonly used indices in the reviewed papers were *Igeo*, *EF*, and CF. Many authors referred to ecotoxicological assessment via *SQG*. Moreover, *PERI*, which includes the toxic response index, was just as popular. The most recognised bioindicators include the *Phragmites* and *Salix* species. *Phragmites* can be considered for Fe, Cu, Cd, and Ni bioindication in sites, while *Salix* hybrid cultivars such as Klara may be considered for phytostabilisation and rhizofiltration due to higher Cu, Zn, and Ni accumulation in roots. *Vetiveria zizanoides* demonstrated resistance to As stress and feasibility for the remediation of As. Moreover, bioindicators offer a feasible tool for recovering valuable elements for the development of a circular economy (e.g., rare earth elements).

Keywords: sediments; trace metals; contamination assessment; pollution indices; bioindicators; geochemical background

1. Introduction

Nowadays, the contamination of sediments with trace metals (TMs) is a salient environmental issue [1,2]. TMs delivered to the environment from anthropogenic sources are mostly represented by these elements: Cd, Co, Cr, Cu, Fe, Hg, Mn, Mo, Ni, Pb, Sn, V, and Zn. Supplementation from external sources also includes other potentially toxic elements (PTEs) such as metalloids (As, Sb) and non-metals (Se) [3]. PTEs are not biodegradable, persistent, and can accumulate in the environment in large amounts. TM ions cause health hazards; just to mention a few, these include emphysema (Cd), anaemia (Cd, Zn, Pb), damage to the brain (Pb), kidneys, and bones (Cd), necrosis nephritis (Cr), DNA damage

(Ni), eczema on the hands (Ni), risk of developing cancer (As), central nervous system irritation (Cu), and even death in humans [4]. Due to the toxic nature of PTEs, there is an obvious need to monitor their content. The most significant anthropogenic sources of PTEs are related to industry, urbanisation, transportation, agriculture, and mines [5–8] TMs adhered to particulate matter (PM; including PM10 and PM2.5) could be released into the surrounding environment causing contamination of soils, groundwaters, sediments, and surface waters. PTEs entering the aquatic ecosystem are associated with fine-grained fractions of suspended solids and sediments due to their large surface areas and high sorption capacities [9]. Throughout the hydrological cycle, 99% of PTEs are stored in sediments, which, therefore, act as the main sinks and carriers for contaminants in aquatic environments [10]. Contemporary studies [11–18] have revealed that stormwater run-off carried metals originating from various sources of everyday use associated with tyre wear, corrosion, roof run-off, and fuel combustion products. Zn, Pb, Cu, Cd, Ni, Ce, and As are typically observed in urban and industrial catchments [11,12].

Sediments are the ultimate repository for contaminants and provide valuable information about the environmental status of a catchment area [17]. According to USEPA [19], sediment site management and silting processes in reservoirs present a unique challenge because of the large number of difficult to control contaminant sources, dynamic phenomena influencing movement and sediment stability, dispersion of contaminants, diversified use of areas affecting the sediment environment, as well as difficulties involved in engineering cleanup works.

Sediment contamination may be evaluated in several ways. Geochemical, ecotoxicological, and bioindicator-related approaches can be distinguished. Extensive evaluation includes geochemical and ecotoxicological methods. The geochemical assessment relies on comparing TMs at particular sites to the geochemical background or baseline. The term "geochemical background" (or "natural background") has evolved. In exploration geochemistry, it refers to the non-anomalous concentration of elements in the rocks surrounding the mineral ore body. In environmental sciences, it is used to detect anthropogenic impact on the environment [20]. Establishing the geochemical background raises research concerns because it may determine the overall pollution status of an examined site. Ecotoxicological approaches are based on assessing the effects of metal toxicity on living organisms and determining certain dose limits. Bioindication is an indirect method of verifying the quality of the environment—i.e., by using aquatic plants and hydrophytes able to accumulate a large number of metals, making them suitable for environmental biomonitoring. The distribution and behaviour of aquatic plants are frequently correlated with water and sediment quality [21]. Macrophytes are able to accumulate pollutants at a higher level, irrespective of their content in the surrounding environment [22]. Moreover, the analyses of plant tissues provide time-integrated information about the quality of an examined site, even if contaminant inputs are discontinuous and quickly diluted in water [23]. One of the species widely recognised as featuring correlations between concentration of metals in sediment/soil and roots is *Phragmites australis* [15]. Hydrophytes uptake PTEs through their roots and rhizomes, stems, and leaves [24]. Plant species with a strong ability to accumulate TMs in the aboveground tissues are good candidates for phytoextraction [25].

The most simple way to evaluate TM contamination of sediments is based on total metal content. This approach reflects the geological origin and anthropogenic influence, although it is ineffective in terms of documenting mobility or bioavailability [26]. According to Ure and Davidson [27], a more accurate method taking into account these two features is to measure the "pseudo total" metal content in sediments. Pseudo total content presents the maximum potential soluble or mobile metal content, which can be identified with maximum potential hazard. Sequential extraction (or fractionation) distinguishes several factions—mobile, conditionally mobile, and immobile (depending on the applied method). Therefore, using this approach, potential metal contamination hazards can be more accurately assessed and explored. Combining fractionation with different contamination assessment approaches (geochemical and ecotoxicological) may also offer valuable insight into sediment quality.

Sediment contamination is a well-known problem across the world; however, monitoring programs based on different PTE contamination assessment methods as well as bioindicators have to be implemented to ensure that the assessment is credible and reliable. Various studies use different approaches to evaluate the quality of sediments, which impairs the comparison of results from diverse studies. Even the definition of background values, which is a fundamental step in geochemical assessment, is often controversial. Therefore, there are four aims of this review: (1) to summarise the levels of TM content and methods used to evaluate the contamination status of bottom sediments related to surface run-off receivers from the last 3 years, (2) to assemble and critically evaluate the geochemical and ecotoxicological classifications and contamination indices used to evaluate the PTE-polluted sediment and the subsequent risk, (3) to explain the difference between geochemical background and baseline; and (4) to specify the bioindicators of TMs that have been extensively researched at different sites across the world.

2. Materials and Methods

A literature review was conducted via the Web of Science platform. In order to access a complete spectrum of data on PTEs in the bottom sediments of urban water bodies, 5 specific keywords were chosen: sediments, trace metals, contamination, pollution indices, and urban. The Web of Science Core Collection displayed a total of 519 papers, which are available in the database (starting from the year 1994). From the last 3 years (2019, 2020, 2021), there are 202 publications in total, and a further 145 results in the field of environmental science (as of May 2021). We used 25 of the latest publications from the last 3 years (2019, 2020, and 2021). To explore the last 10 years of updates regarding bioindication species, the search process was based on the following keywords: plant bioindicators, metal indicator, *Salix* or *Willow* as bioindicator, phytoextraction, and bioindicator.

3. PTEs in Bottom Sediments of Urban Water Bodies

Based on 25 of the latest publications [28–52] on the topic of bottom sediments in urban waterbodies, Table A1 in Appendix A presents (1) a list of TMs evaluated under each study, (2) their concentration ranges (in mg/kg d.w.), (3) a basic description of the site and land use, (4) indices, classifications, and an evaluation of TM bioavailability with the use of metal fractionation.

In most of the examined papers, the authors focused on five TMs: Cu, Zn, Pb, Ni, Cr, and a metalloid: As. All of the mentioned elements can be delivered by anthropogenic activities such as metalliferous mining, smelters, metallurgy, rolling, electronic, industry, the dye and paint industry, metal corrosion of materials, plastic industry, chemical industry, combustion of fossil fuels, wood industry, as well as tyre wear and lubricant oils [6]. The scope of analyses in many studies also included Fe, Mn, Co, Hg, as well as rare earth elements, i.e., La and Sc.

Most authors sampled the surface layer of sediments (0–2 cm, 0–5 cm, or 0–10 cm), where reactions occur at the water–sediment interface and where living organisms dwell. The sampling strategy is crucial for the successful collection of sediment samples with undisturbed stratification. With this aim, a grab sampler (for surface sediment extraction), hand (core extraction < 60 cm) and box corer (core extraction < 1 m) devices, or a multiple corer can be used to obtain proper samples. Sediment samples are collected to describe the condition of the sites. The specific reasons for sediment sampling include bioassays (the top 10 cm are usually collected because of biological activities), biosurvey (macroinvertebrate analyses), monitoring, examining non-point pollution, and pollutant sources. Monitoring trends and relationships in sediments requires information on current PTE burden in the uppermost sediment layer (e.g., 0–1 cm). Depending on the receiver (large/small body of water, river, stream, bay), the rate of deposition varies, which means that 1 cm of sediment builds up over different times. Therefore, the EU [53] has recommended sampling the top layer of sediment at a depth of 1–5 cm; however, this depends on the deposition rate. This is also in line with data from the United States, which stated that more than 10% of the volume of bottom sediments (the upper 5 cm of surface sediments) are estimated to have been contaminated. The depth to which sediments are sampled should be one of the key monitoring objectives. Core sample collection is crucial for evaluating the PTE footprint. Sediments from depths greater than 15 cm below the sediment–water interface are usually collected to determine the spatial vertical properties and sedimentation history. Each sampling program requires a compromise between the number of samples, sampling frequency, measurement parameters, and cost effectiveness [54].

The most frequently used indices were Geoaccumulation Index (*Igeo*) (17 studies), Enrichment Factor (*EF*) (13 studies), Contamination Factor (*CF*) (10 studies), Pollution Load Index (*PLI*) (6 studies), Metallic Pollution Index (*MPI*) and Modified Contamination Degree (*mCd*) (each used in 3 studies), Nemerow Pollution Index (*PI*$_{Nemerow}$) (2 studies), *MRI* and Pollution Index (*PI*) (each used in 1 study). Two studies undertook a sediment assessment using the German classification LAWA, while one study refers to Romanian legislation. Reference to the ecotoxicological assessment via Sediment Quality Guidelines (SQG) was carried out in eight studies, while Ecological Risk Index (*Er*^{i}), Risk Index (*RI*), and Potential Ecological Risk Index (*PERI*) (which account for metal toxicity) were used in 17 studies. Moreover, only two studies assessed health risk. Three papers considered the use of chemical sequential extraction for element analyses, which is important to assess the bioavailability and mobility of TMs. Only one study included a Risk Assessment Code (*RAC*) analysis, which takes into account the risk determination based on the metal binding to the exchangeable and carbonate fraction. Health risk—as Hazard Index (*HI*) and Hazard Quotient (*HQ*)—was calculated in two papers. The common use of indices enables the results to be compared between various studies. However, calculations of geochemical indices are strongly dependent on the adopted reference value (geochemical background).

Another important issue, addressed in three studies, is the fractionation analysis of TMs. Fractionation allows the environmental behaviour of trace elements to be assessed [55]. However, the procedure itself is highly labor-intensive and cost-prohibitive. Most works are based on assessing the total metal content of the samples; nevertheless, fractionation is an indispensable method when it comes to exploring the actual metal bioavailability and mobility.

The above characteristic offers an overview of the most often used indices for geochemical and ecotoxicological evaluation of sediments. The next part of this review presents a "conventional" division of tools for assessment of sediment quality, taking into account geochemical and ecotoxicological approaches.

4. Assessment Techniques for Sediments Contaminated with HMs

The most effective approach in assessing sediment contamination with PTEs relies on the use of several assessment tools that can be grouped according to geochemical and ecotoxicological evaluation, as shown in Figure 1. Another method for indirect assessment of TM sediment contamination is the use of bioindicators.

Geochemical evaluation is usually based on pollution indices (PIs), while geochemical classifications are usually implemented by national governments. Both approaches rely on the establishment of a geochemical background. Among ecotoxicological methods, descriptive classifications (usually based on the ecotoxicological effect on benthic organisms—Sediment Quality Guidelines (SQG) as well as those based on toxicological, chemical, and ecological tests) and ecotoxicological indices (EIs) are used. Since the preference for adopting a sediment assessment method is usually on the side of the lawmakers, many classification and indices have been introduced since 1980.

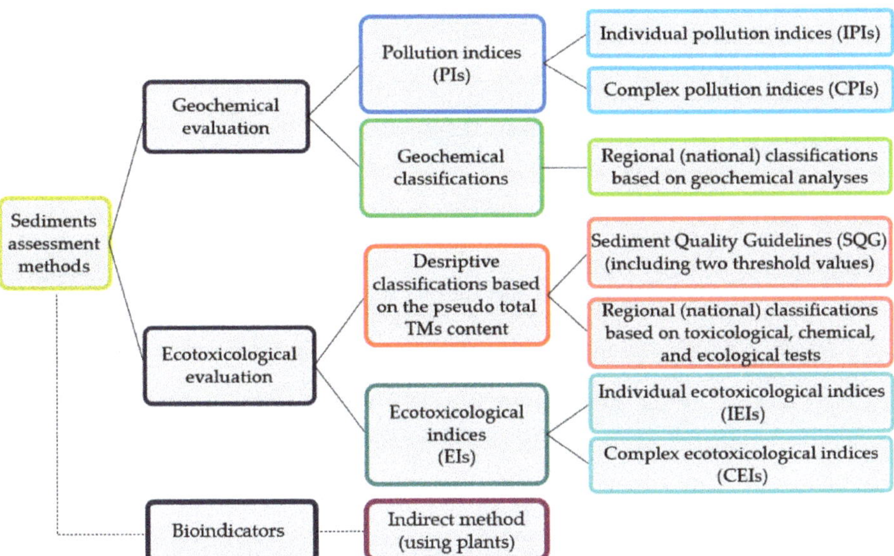

Figure 1. Division of the methods of assessment of sediment contamination with PTEs.

4.1. Geochemical Evaluation

4.1.1. Geochemical Background and Geochemical Baseline

There are two controversial issues within contemporary geochemistry and environmental biochemistry: the geochemical background and the geochemical baseline of an element or chemical compound. The geochemical baseline is easier to determine—it is commonly used for assessing the environmental status before embarking on a specific investment (e.g., constructing a power plants). In exploratory geochemistry, geochemical background means the natural concentration of an element in the environment, i.e., a non-anomalous concentration of elements that corresponds to Clarke's characteristic for their geochemical environment [20]. There are several methods to determine the accurate geochemical background. Once this value is determined, the pollution status can more precisely be assessed. These methods rely on statistical (indirect, theoretical) and empirical (direct, geochemical) determinations. An integrated approach combining both methods is also in use. Geochemical methods involve analysing samples unaffected by anthropogenic activities, which are also referred to as preindustrial samples. Usually, deep core samples and/or samples collected at a certain distance from pollution sources are used to establish geochemical background values. Statistical methods involve a determination and elimination of outliers, which are identified by standard deviation, regression analysis, or by using the Tukey boxplot method [56].

The geochemical background is not a constant value over time due to natural processes taking place at the interface of water and sediment. Additionally, the geochemical background value is local or regional rather than global. The adopted geochemical background affects the overall assessment environment quality. Sometimes, the reference level follows the shale standard (from 1961), i.e., the average content of elements in the earth's crust [57], although this approach is incorrect.

4.1.2. Pollution Indices

According to Weissmannová and Pavlovský [6], pollution indices (PIs) can be divided into two categories including Individual (single) PIs (IPIs) and Complex (total) PIs (CPIs). In such a division, IPIs are calculated based on the contents of each metal separately in particular sediments. IPIs can be used to classify sediments into several classes according to the degree of pollution presented by a single metal. CPIs are calculated as multielement indices based on IPIs. Moreover, CPIs can be composed of IPIs separately. PIs are of most use when they meet several criteria: (1) they are easy to calculate, (2) they make reference to the geochemical background, and (3) they are widely used, which makes it possible to compare sediments from different sites.

Individual Pollution Indices (IPIs)

The IPIs used for sediment assessment are presented in Table A2 in Appendix B. In this group, five indices were identified: Contamination Factor (CF) [58], Geoaccumulation Index (I_{geo}) [59], Enrichment Factor (EF) [60], Pollution Index (P_i) [15,54], and Threshold Pollution Index (PI_T) [28]. Each of these indices compares the concentration of a given metal in a sample ($C_{mSample}$) to its comparative concentration—represented by C_{mPre} (preindustrial concentration of TM) or $C_{mBackground}$ (geochemical background concentration of TM) [40,58,61–64].

CF is originally defined as a ratio of mean TM content averaged from at least five samples ($C_{mSample}$) and metal concentration in preindustrial sediments [58,61,62]. According to Dung et al. [20] the ratio can also refer to background TM content. This approach is included in P_i. However, usually both indices (CF and P_i) are treated by authors interchangeably. The difference lies in the classification of the contamination levels; it also depends on whether the values of C_{mPre} and $C_{mBackground}$ are consistent with each other. CF and P_i are easy to calculate, although both omit natural variability, grain-size, and reference elements ratios. Moreover, they do not take into account metal fractionation even though the formulas of the equations themselves make it possible.

I_{geo} proposed by Müller [59] is a commonly used index, which uses a completely different approach from other IPIs, since it defines seven classes of TM sediment contamination. The TM content in a sample is referred to as $C_{mBackground}$. The formula includes a constant value of 1.5, which is used to offset the natural TM fluctuations in the environment and detect minor anthropogenic impact [6].

EF is an effective tool for TM content comparison [13]. In comparison to other IPIs, EF requires the most extensive data input, such as geochemical background concentration of TM ($C_{mBackground}$) and reference element concentration in a sample (C_{ref}) as well as in the reference environment (B_{ref}). The reference element has a low occurrence variability. The reference element—and this could be Sc, Mn, Ti, Al, Fe, and Ca—has a low occurrence variability. Frequently used reference metals are Mn, Al, and Fe [12]. Usually, Fe has a relatively high natural concentration in comparison to other metals and is therefore not expected to be enriched from anthropogenic sources [64]. The use of EF seeks to reduce the metal variability associated with the granulometric composition (mud/sand ratios). The EF could be also expressed as a percentage.

PI_T is based on the single Pollution Index (P_i), which refers the TM content in a sample to the tolerance levels of metal concentration (C_{TL}). C_{TL} could be established by national guidelines or criteria regarding health-threatening TM content [63]. The determination of the threshold parameter (C_{TL}) is freeform and can be adjusted to the state of the environment and defined individually by national regulations. Therefore, this index may be an effective tool in the hands of environmental managers.

All of the described IPIs omit the metal fractionation and mobility/bioavailability. IPIs are usually calculated in accordance with total metal content; however, they can be easily applied in calculations that include fractionation (as described in Section 4.3).

Complex Pollution Indices (CPIs)

A list of CPIs is presented in Appendix C. CPIs mostly rely on CF or P_i values for TMs under analysis. Thus, the starting point for a comprehensive metal contamination assessment is IPI analysis. Table A3 in Appendix C presents calculation formulas with classifications and descriptions (pros and cons considered by the authors) for the following CPIs: Pollution Load Index (PLI) [62], Contamination Degree (C_{Deg}) [58], Modified Contamination Degree (mC_{Deg}) [64], Sum of Pollution index (PI_{sum}) [63], Average of Pollution Index (PI_{Avg}) [63], Weighted average of Pollution Index (PI_{wAvg}) [58,59,63], Background Enrichment Factor = New Pollution Index (PIN) [65], Product of Pollution Index (PI_{Prod}) [63], Weighted power product of Pollution Index (PI_{wpProd}) [63], Vector modulus of Pollution Index ($PI_{vectorM}$) [63], Nemerow Pollution Index ($PI_{Nemerow}$) [66], Integrated Pollution Index ($IntPI$) [67], and Metallic Pollution Index (MPI) [68,69].

PLI is a very easy-to-use index based on CF values specified for a selected number of metals. Moreover, the deterioration of site quality could be straightforwardly specified because the unity value of PLI (PLI = 1) is the boundary differentiating the adverse impact on the quality of sediments from non-polluted sites [62]. The overarching merit is a comprehensive assessment of sediments, while the disadvantages are related to the shortcomings of the CF index.

C_{Deg} and mC_{Deg} are also based on CF values for analysed TMs. The difference lies in the classification of sediment contamination status. C_{Deg} defines sediment contamination referring to four categories, while mC_{Deg} distinguishes seven ranges of sediment quality [58,64].

The equations for calculating PI_{sum} and PI_{Avg} are similar to those for C_{Deg} and mC_{Deg}. The difference lies in the use of individual Pi ($Pi = SPI = PI$) factors related to $C_{mBackground}$ (not to C_{mPre}) [63]. In order to evaluate the sediments with PI_{sum}, it is necessary to specify the classification scale for each individual case. The defined scale should refer to the number of selected TMs used to assess contamination. The pollution evaluation with PI_{Avg} is easy to apply due to the reference to unity (1). PI_{wAvg} application when the TM weight is equal to unity ($w_i = 1$) is a special form of PI_{sum}.

PIN is based on Pi ($Pi = SPI = PI$) and requires the TM class to be determined (from 1 to 5 in reference to Pi). This very precise scale classifies sediments into five categories [68].

PI_{Prod} is a Cartesian product of Pi applied for a selected number of metals. The classification should be powered by n (count of TMs) to determine the contamination of sediments [63]. PI_{wpProd} uses the weight of Pi in an equation, which is similar to those used for calculating PI_{Prod} [63]. $PI_{vectorM}$ was introduced by Gong et al. [63] and depends on Pi and the number of TMs selected for investigation. $PI_{Nemerow}$ determination emphasises the maximum value of a single Pollution Index (Pi) for all TMs; however, it does not include the weight of Pi [66]. $IntPI$ is an easy-to-calculate index based on the Pi mean value. The classification based on $IntPI$ divides sediment contamination into low, moderate, and strong [67,70].

MPI is equal to PLI. The general difference lies in $C_{mBackground}$ used in the equation for the Pi calculations [71].

In relation to CPIs, one major advantage is the possibility to comprehensibly evaluate the effects of several metals simultaneously. The authors of this review believe that the the Pi weight, which represents the impact of an individual metal on the overall sediment pollution rating, plays a significant role in the calculations. This approach is used by PI_{wAvg} and PI_{wpProd}. Moreover, it would seem to be more accurate to include the maximum Pi as a reference in the calculations of $PI_{Nemerow}$. Despite the mentioned advantages of PI_{wAvg}, PI_{wpProd}, and $PI_{Nemerow}$, it is still desirable to use simpler calculations, which are included by PLI, C_{Deg}, mC_{Deg}, PI_{sum}, etc.—due to their simplicity and frequent use in different studies.

4.1.3. Geochemical Classifications Referring to the Geochemical Background

Extensive environmental monitoring is required to develop geochemical classification and a database. Often, such studies end by mapping a given area to present geochemical data as accurately as possible [72]. Since determining the geochemical background is usually a fundamental step while assessing the environmental status of a given site, such classifications have a temporary value [20].

In Poland, the first attempts to classify lake and river sediments contaminated with TMs were made by Bojakowska and Sokołowska [73]. This so-called "geochemical classification of aquatic sediments" approach involves standardising samples, taking into account a grain size smaller than 0.2 mm. Moreover, this classification applies to samples digested with hydrochloric acid, nitric acid, or aqua regia. The geochemical criteria distinguish four geochemical quality classes presented and described in Table 1. This classification was developed in 1998, so the geochemical background could have changed since then. However, it is still used in Poland in the absence of other regulations. In Table 1, the geochemical quality classes are collated with geochemical background values established using different approaches. Depending on the goals, TM contents in analysed sediments can be compared with preindustrial concentrations [58], trace metal content in world soils [74], or shale standard [57].

Table 1. The concentration of various trace metals (TMs) in mg/kg d.w. for each class of sediments—geochemical classification of aquatic sediments [57,58,73–75].

Trace Metal	Geochemical Background [in mg/kg d.w.]					Geochemical Quality Classes			
	Preindustrial	World Soils	Shale Standard	Gdansk Region Sediments (Poland)	Sediments (Poland)	I	II	III	IV
	[58]	[74]	[57]	[75]	[73]	[73]	[73]	[73]	[73]
As	-	1.5	13	-	<5	<10	<20	<50	$\geq$50
Cd	1.0	0.62	0.3	0.5	1	<1	<3.5	<6	$\geq$6
Cr	90	84	90	7	5	<50	<100	<400	$\geq$400
Cu	50	25.8	45	5	6	<40	<100	<200	$\geq$200
Pb	70	29.2	20	11	10	<50	<200	<500	$\geq$500
Hg	0.25	-	0.4	-	$\leq$0.05	<0.1	<0.5	<1.0	$\geq$1.0
Ni	-	33.7	68	4	5	<30	<50	<100	$\geq$100
Zn	175	59.8	95	41	48	<200	<1000	<2000	$\geq$2000

Bold—geochemical quality classes description according to Bojakowska and Sokołowska: Class I—uncontaminated sediments (boundary value of elements concentration determined as from 5 to 10 times higher than the geochemical background); Class II—moderately contaminated sediments (TMs content from 10 to 20 times higher than the geochemical background); Class III—contaminated sediments (TMs content from 20 to 100 times higher than the geochemical background); Class IV—highly contaminated sediments [73].

4.2. Ecotoxicological Evaluation

In order to assess the level of contamination that is still safe for living organisms, it is necessary to apply ecotoxicological criteria or use indicators that take into account the level of metal toxicity.

4.2.1. Descriptive Classifications Based on Total TM Content

Internationally, a range of sediment quality guidelines (SQGs) have been proposed for contaminants. Originally, they provided the information on SQGs compared with a reference or background concentration of analysed substances [76]. After the 1980s, SQGs were developed to evaluate sediment quality where harm to aquatic organisms and the environment was observed [77]. In general, SQG approaches that use different criteria and factors are divided into two main categories [76]:

- Empirically based SQGs—relying on the empirical relationships needed to determine the sediment contamination level at which a toxic response occurs; these are frequently used for TMs and As.
- Theoretically based SQGs—relying on the parameters that describe the bioavailability of contaminants (equilibrium partitioning, EqP); these are mainly used for organic compounds (less often for TMs).

Both types of SQG predict adverse ecological effects caused by sediment contamination by the response of benthic organisms. SQGs feature two concentration thresholds: the first below which adverse effects are not expected to occur (threshold effect concentration—TEC) and the second above which adverse effects are expected to occur more often than not (probable effect concentration—PEC). The limitation of this approach is that it leaves significant uncertainty in the "grey" area of contaminant concentrations that lie between TEC and PEC. Therefore, it may be necessary to perform a site-specific analysis. Another relevant issue in the case of SQGs is that these descriptions apply to only one type of contaminant, which may lead to inaccurate conclusions in the contamination description, as the sediments are a sink for various type of contaminants that can contribute to their overall toxicity. Table 2 outlines the most popular empirical SQGs used in practice. All of them are based on differently described terms of TEC and PEC (e.g., TEL/PEL, ERL/ERM, or LEL/SEL) [78–80]. Moreover, we presented the average TEC and PEC values calculated based on the literature (TEC_{lit} and PEC_{lit}).

Table 2. Sediment quality guidelines (SEQs) for TMs and metalloid in mg/kg d.w. that reflect threshold effect concentration (TEC) and probable effect concentration (PEC) [77–82].

| Element | Empirical SQGs with TEC and PEC in mg/kg·d.w. | | | | | | | | | | | |
| | Smith et al. [78] | | Long and Morgan [79] | | Persaud et al. [80] | | MENVIQ [81] | | ANZECC/ARMCANZ [82] | | Average Values Based on the Literature | |
	TEL	PEL	ERL	ERM	LEL	SEL	MET	TET	SQGV	SQGV-High	TEC_{lit}	PEC_{lit}
As	5.9	17	33	85	6	33	7	17	20	70	14	44
Cd	0.596	3.53	5	9	0.6	10	0.9	3	1.5	10	1.72	7
Cr	37.3	90	80	145	26	110	55	100	80	370	56	163
Cu	35.7	197	70	390	16	110	28	86	65	270	43	211
Pb	35	91.3	35	110	31	250	42	170	50	220	39	168
Hg	0.174	0.486	0.15	1.3	0.2	2	0.2	1	0.15	1.0	0.17	1
Ni	18	36	30	50	16	75	35	61	21	52	24	55
Zn	123	315	120	270	120	820	150	540	200	410	143	471
Limitations for empirical SQGs:												
Advantages	(1) Allows for thresholds prediction at which a toxic response of benthic organisms is probable (2) A useful tool that allows making the first guess on the sediment contamination problem (3) Quite simple to use with a large database of lab and field tests											
Disadvantages	(1) The assessment of risk and possible negative outcomes associated with toxic concentration levels, which require a separate risk evaluation (2) "Grey" area between thresholds test errors (3) Difficult to apply in a context of contaminant mixtures (4) Relatively expensive research and development of new SQGs (5) The availability of SQGs guidelines criteria and documentation is limited											

TEC—Threshold-Effects Concentration; PEC—Probable-Effects Concentration; TEL—Threshold-Effects Level; PEL—Probable-Effects Level; ERL—Effects Range: Low; ERM—Effects Range: Median; MET—Minimal Effect Threshold; TET—Toxic Effect Threshold; SQGV—Sediment Quality Guideline Value; ANZECC/ARMCANZ [82]—currently used classification in Australia and New Zealand.

In parallel with the intense work undertaken on SQGs in North America and Australia, several approaches and descriptive classifications and recommendations were also developed in Europe. It is worth noting that the authorities and legislators usually consider water quality as a priority, paying less attention to the state of sediments. It should be emphasised that contaminants accumulated for decades in the bottom sediments of water bodies could be released into the water column and cause its recontamination, thereby directly affecting water quality. Despite the large-scale processes that can affect the stability of sediments, there are different inceptors of secondary water recontamination, such as pH fluctuation, or the disturbance of sediment structure during dredging operations or boat movement.

In Germany, the Länder-Arbeitsgemeinschaft Wasser classification (LAWA) [83] is used to assess Hg, Cd, Zn, Pb, Cu, Cr, and Ni content in water, suspended solids, and sediments. The LAWA classification is constrained to the methodology of digestion with the use of aqua regia. This approach does not allow silicates to dissolve; therefore, the term "pseudo total" concentration of TMs is used in the literature. However, in terms of the actual contamination status of sediments and their impact on the aquatic environment, the mobile and unstable forms of TMs are most relevant [75]. In the light of this, digestion by using aqua regia or diluted acids (e.g., hydrochloric acid, nitric acid) may determine the bioavailable fraction of TMs that take part in environmental processes, while it remains impossible to determine the content of TMs bound in minerals resistant to weathering processes. LAWA classification divides sediments into seven classes of contamination, depending on the TM contents in aquatic ecosystems. The boundary values for each class are presented in Table 3.

Table 3. The concentration of various trace metals (TMs) in mg/kg d.w. for each class of sediments—classification LAWA and modified LAWA [83,84].

Classification	Trace Metal [in mg/kg d.w.]	Contamination Class						
		I	I–II	II *	II–III	III	III–IV	IV
LAWA [83]	Cd	≤0.3	≤0.6	≤1.2	≤2.4	≤4.8	≤9.6	>9.6
	Cr	≤80	≤90	≤100	≤200	≤400	≤800	>800
	Cu	≤20	≤40	≤60	≤120	≤240	≤480	>480
	Pb	≤25	≤50	≤100	≤200	≤400	≤800	>800
	Hg	≤0.2	≤0.4	≤0.8	≤1.6	≤3.2	≤6.4	>6.4
	Ni	≤30	≤40	≤50	≤100	≤200	≤400	>400
	Zn	≤100	≤150	≤200	≤400	≤800	≤1600	>1600
Modified LAWA [84]	As	3–5	<10	<20	<40	< 70	<100	>100
	Cd	0.2–0.4	<0.5	<1.2	<5	< 10	<25	>25
	Cr	60–80	<90	<100	<150	< 250	<500	>500
	Cu	20–30	<40	<60	<150	<250	<500	>500
	Pb	25–30	<50	<100	<150	<250	<500	>500
	Hg	0.2–0.4	<0.5	<0.8	<5	<10	<25	>25
	Ni	10–30	<40	<50	<150	<250	<500	>500
	Zn	90–110	<150	<200	<500	<1000	<2000	>2000

* Quality goal—Class II. Class I refers to uncontaminated sediment; Class I–II—sediments unpolluted or with very small anthropogenic interference; Class II corresponds to moderately polluted sediments, and it is a reference for the other levels of pollution; Class II–III—twice as much as Class II—defines moderate to significant contamination of sediments; Class III—four times more than Class II—defines significant pollution; Class III–IV—eight times more than Class II—refers to very strong pollution; Class IV—means more than eight times exceeding Class II—ultimate pollution. Comparative value is the average contents of TM expressed in mg/kg [83,84].

Following the LAWA classification, Ahlf et al. [84] introduced guidelines for evaluating sediment quality based on integrated toxicological, chemical, and ecological approaches (hereafter referred to as modified LAWA). This approach is based on a series of three different tests that should be performed step-by-step, starting from an ecotoxicological assessment. If the ecotoxicological test confirms sediment contamination, the second step is introduced, which focuses on chemical analyses. The final assessment stage involves an additional and thorough analysis of factors that may harm the environment. The classification of TM sediment contamination applied by Ahlf et al. [84] (Table 3) combines temporary valuation standards of river mud and ATV standards concerning sewage and garbage. The final evaluation is based on the concentration limits for class II, which constitutes a point of reference.

In accordance with the Water Framework Directive, water and sediment quality assessment considers the extent of deviation from reference conditions with no or with very minor anthropogenic influence. The general conclusion from analyses of sediments guidelines is that they have advantages as well as limitations for their application in sediment quality assessment [85].

4.2.2. Ecotoxicological Indices (EIs)

EIs as indicators of ecotoxicology comprise numerous indices, such as Ecological Risk Factor ($ER = Er^i$ = Risk Index (RI)) [58]—the only IEI, Contamination Severity Index (CSI) [86], Sediment Pollution Index (SPI) [58,64], Risk Assessment Code (RAC) [35], Risk Index ($PERI$) [87,88], and the probability of Toxicity Index ($MERMQ$) [86]. The list of IEIs and CEIs are presented in Table A4 Appendix D with classification and a summary of strengths and weaknesses.

ER describes the ecological risk caused by TM. T_r^i in the ER formula represents the toxic-response factor for a given metal, and so this index ties the risk assessment with the metal content in a sample [6]. SPI also refers to T_r^i of TMs and additionally takes into account the average shale concentration of TMs. Moreover, SPI is tailored for the comprehensive evaluation of several elements at the same time. Another difference between ER and SPI lies in diverse points of reference—ER classification refers to ecological risk, and SPI refers to the sediment pollution level. $PERI$ is a complex index based on a single

ER. The toxic response factors for TMs according to Håkanson [58] are as follows: Hg—40, Cd—30, As—10, Cu—5, Pb—5, Cr—2, and Zn—1. T_r^i in the limnic system was discussed and established by Håkanson according to the "abundance principle" of elements in different matrixes. The element with the highest mean concentration must be ranked as 1.0; according to this assumption, analysed elements were marked with the following abundance number: Zn = 1 < Cu = 3.4 < Pb = 13 < Cr = 110 < As = 140 < Cd = 230 < Hg = 1160. Note that the abundance number is not equal to the toxic factor. Håkanson corrected the abundance number to reflect the "sink-effect". This effect is related to different footprints caused by substances (e.g., Cr has the lowest sink-factor = 2, which implies that this element leaves the heaviest footprint in the sediment, while Hg has the highest sink-factor =320, which means that much more Hg can be found in water compared to sediments). Taking into account the dimension problem, Håkanson corrected the abundance numbers that correspond to T_r^i. Moreover, according to Kabata-Pendias and Pendias [58], the highest T_r^i corresponds to strong chemical and biological activity in the case of Hg and high solubility in acidic environments and mobility for Cd. Quite low T_r^i for Pb is related to its low mobility. $T_r^i = 2$ for Cr is related to the principle that only Cr(VI) is toxic; at the same time, Cr(VI) is easily reduced to poorly soluble Cr(III), which is generally difficult for plants and benthic organisms to absorb [58].

CSI requires the use of computed weight of each TM according to Pejman et al. [86]: (Cu—0.075, Zn—0.075, Cr—0.134, Ni—0.215, Pb—0.251, Cd—0.25). The weight of TMs may result from statistical analyses (e.g., Principal Component Analyses—PCAs and/or Factor Analyses—FAs). Moreover, *CSI* requires the application of ERL and ERM, and at the same time, it is highly accurate and could provide a precise evaluation of sediments.

RAC depends on exchangeable and carbonate fractions determination (determined with CH_3COOH 0.11 M solution) [35]. The classification refers to the percentage of mobile fractions in relation to the total concentration of bound TMs in all fractions together. It is worth noting that *RAC* is the sole index that combines the speciation analyses of TMs. Therefore, it is very valuable for evaluating the mobility and bioavailability of metals.

MERMQ is also referred to in the literature as a combined effect of Toxic Metals (*mPECQ)*. This complex EI is based on TM content in the sample and the second threshold level of effects (e.g., ERM for *MERMQ* and PEC for *mPECQ)* [86].

4.3. Extended Sediments Quality Ranking

4.3.1. Integrated Geoaccumulation Index—$Igeo_{integrated}$

The approach described by Von Tümpling et al. in 2013 [89] relies on *Igeo* introduced by Müller [59], and this index is combined with metal speciation analysis based on BCR (Community Bureau of Reference) three-stage sequential extraction—this approach is referred to as "extended use of *Igeo* in combination with BCR fractionation". To compare the results obtained with $Igeo_{integrated}$ between fractions, an integrated *Igeo* was introduced ($\Sigma Igeo_{integrated}$). The sum of $Igeo_{integrated}$ offers a general overview of the contamination degree of the considered elements in one fraction in comparison to $\Sigma Igeo_{integrated}$ of other fractions. The $\Sigma Igeo_{integrated}$ equation is as follows:

$$\sum Igeo_{integrated} = \sum_{j=1}^{m} \sum_{k=1}^{n} m_{j,k}$$

where m is the quality class, j is the sampling point (SP), and k is the element. The geoaccumulation classes (described in Table A2 Appendix B) of all the examined TMs are added for each fraction. If there is no contamination or no accumulation of an investigated element, the resulting sum is 0. The maximum class is 6.

The $Igeo_{integrated}$ approach can completely change the result of the sediment assessment based on the total TM content performed by *Igeo*. The results reported by Von Tümpling et al. [89] for Klinke River in Germany indicated that F3, F2, and F1 had the greatest $\Sigma Igeo_{integrated}$. At the same time, the "good chemical status" classification in accordance with total TM content was overturned by the sequential extraction method, which demonstrated that changes in element speciation can lead to higher TM mobility, which may increase the hazardous potential of sediments.

4.3.2. Fractionation of TMs

In order to establish how TMs behave in the sediment–water or soil–water environments, the presence of physical and chemical forms of metals—i.e., fractionation, also called speciation—must be determined. Therefore, speciation is an identification of various physicochemical forms of elements in the analysed material and/or mutual proportions of these forms. This approach helps to understand the mobility of metals in the environment, although the sample processing is time consuming [90].

The analytical methods for the chemical section of TMs in the solid phase involve single reagent leaching, sequential extractions, and ion exchange resins [55]. In general, the processes involving the separation of different species are based on selective chemical reactions and/or extraction procedures (e.g., liquid–liquid, solid-phase extraction). The final result of element contents is presented in mg/kg and or mg/L (ppm—parts per million) or ng/kg and/or ng/L (ppb—parts per billion). Therefore, the overall process is complex, arduous, and susceptible to critical errors that are sometimes neglected. In the field of geochemistry, sequential extraction has proven its value. Nevertheless, accuracy at every stage of sample processing and analysis is crucial to guarantee an acceptable level for detecting (LOD—limit of detection) and quantifying the analytes [91]. TMs may be determined by commonly used instruments such as flame (FAAS), hydride generation (HGAAS), and electrometry (ETAAS) atomic absorption spectrometry, as well as inductively coupled plasma mass spectrometry (ICP-MS) and emission spectrometry (ICP-OES). In comparison with procedures in which total TM contents are determined in a direct manner, the extraction methods generate more uncertainties. This inconvenience is caused by difficulties in isolating compounds from substrates, upsetting the equilibrium between different chemical species, insufficient analytical sensitivity, or a lack of certified reference materials [92].

Many selective extraction methods have been developed, among others, by Tessier et al. [93], Boszke [94], Miller et al. [95], and Fiedler et al. [21]. One method that seeks to minimise errors during sample processing and analysis was proposed by the European Community Bureau of Reference—the BCR method [96].

The BCR three-step extraction procedure is similar to those proposed by Tessier et al. [90]. The key difference involves combining the first fraction of the procedure (instead of evaluating the exchangeable and carbonate fractions separately, the BCR procedure combines both); therefore, BCR distinguishes four factions (F1—exchangeable, F2—reducible, F3—oxidisable, and F4—residual). The description of subsequent steps of the BCR extraction procedure is presented in Table 4.

Table 4. The BCR three-steps sequential extraction scheme [90].

Extraction Step	Fraction	Operational Definition	Chemical Reagents/Conditions
1	F1	Acid extractable (water and acid soluble)	Acetic acid: CH_3COOH (0.11 mol/L); pH 2.85
2	F2	Reducible (Fe/Mn oxides)	Hydroxyloammonium chloride: $NH_2OH \cdot HCl$ (0.1 mol/L), pH 2.0
3	F3	Oxidisable (organic substance and sulphides)	Hydrogen peroxide: H_2O_2 (8.8 mol/L) followed by ammonium acetate: CH_3COONH_4 (1.0 mol/L), pH 2.0
4	F4	Residual (remaining, non-silicate bound metals)	Acid digestion (e.g., Aqua regia: 3 HCl + HNO_3 or HF + HNO_3)

4.4. Bioindicators

In recent decades, the interest in using bioindicators as monitoring tools to examine environmental pollution caused by TMs has been growing steadily. Numerous plants are used as bioindicators to construct assumptions regarding environmental conditions. In this context, plants including macrophytes can be used as bioindicators for TM-related environmental pollution due to their ability to absorb metals from air, water, sediment, soil, and the food chain [97]. In general, bioindicators are characterised by the following abilities: they are easy to grow and cultivate, can withstand polluted environments, and can accumulate high pollutant concentration in their biomass; they are ecologically and economically important for society, and they should be abundant and widespread. In addition, bioindicators not only provide sediment, soil, and water remediation but are also a feasible tool for the recovery of valuable elements to develop a circular economy [98]. In practice, aquatic and related fauna, microbial systems, fungi, animals, and plants are popular targets of TM monitoring programs used as bioindicators to formulate conclusions regarding the environmental contamination status. Biological indicators provide information on the long-term impact of TM contamination [97]. There are many reasons to use a given species as bioindicators, while the key issue lies in the correlation of analysed toxic compounds/elements between species and contaminated matrix (as waters, sediments, or soils) [99].

In this review, different bioindicator plant species were identified from different studies based on their survival, highest metal accumulation, and biomass production in different polluted environments (soil and water) (Table 5). Moreover, metals such as Zn, Cd, Cu, and Ni were predominantly investigated in most regions. For instance, Salinitro et al. [100] depicted the linear relation of Ni in soils and in *Poa annua* shoots ($R^2 = 0.78$) and *Senecio vulgaris* ($R^2 = 0.88$). Therefore, these species have been demonstrated to be reliable indicators of both total and bioavailable Ni fractions in an anthropogenic environment. Extensive research on bioindicators in an estuarine and coastal environment was performed by Farias et al. [98]. In this study, *Ulva australis*, *Zostera muelleri*, and *Ruppia megacarpa* were selected for bioindicator identification due to their great abundance and the highest content of TMs inside their tissues. These species were verified as potential bioindicators of TM (As, Cu, Pb, and Zn) pollution in estuary sediments.

Some authors also focused on highly toxic elements such as Hg and As. These non-essential elements are uptaken passively by plants [24]. Many plant species presented tolerance to As contamination, as presented by Singh et al. [101] based on the example of *Vetiveria zizanoides*, which demonstrated resistance to As stress and feasibility for revegetation/remediation of As-contaminated soils. Although Hg in plants has no metabolic function, it could be easily translocated from roots and rhizomes to shoots [102].

High resistance to changes in salinity, fertility, textures, and pH is presented by an extensively distributed species Phragmites australis [103]. Ganjali et al. [104] demonstrated that Phragmites roots can be used as bioindicators of Fe, Cu, Cd, and Ni.

Phragmites and *Salix* species were studied widely for the phytoremediation of TMs in temperate and boreal climates. *Salix schwerinii* and *S. myrsinifolia* are known as bioindicators in boreal climates. In addition, hybrid cultivars of *Salix*, such as Klara (*Salix viminalis* × *S. schwerinii* × *S. dasyclados*) and/or Karin (*S. viminalis* × *S. schwerinii*) are in use in bioindication. Salam et al. [105] proved that Klara (*Salix viminalis* × *S. schwerinii* × *S. dasyclados*) accumulated high concentrations of Cu, Zn, and Ni in leaves, shoots, and roots from soils. At the same time, Klara may be considered for phytostabilisation and rhizofiltration due to the highest TM accumulation in the roots. *Phragmites* is an emergent plant worldwide that has demonstrated high TMs phytostabilisation properties [106,107]. In three reviewed papers, the authors focused on rare earth elements (lanthanide) such as Pr, Nd, Tb, Sm, Dy, and Er. Mikołajczyk et al. [108] have thoroughly documented the potential of *Artemisia vulgaris*, *Papaver rhoeas*, and *Taraxacum officinale* for phytoextraction of rare earth elements whose occurrence is increasing due to the development of modern technologies.

Table 5. A list of plant species tested as candidates for bioindication across the world.

Plant Species/Cultivars	Climate/Region	Analysed Pollutants	Trial	Source	Reference
Phragmites australis	Temperate/China	Zn and Cd	Pot experiment	Hydroponic	[109]
Atraclytis seratuloides, Lygeum spartum, and *Gymnocarpos decander*	Mediterranean/Tunisia	Cr, Co, Zn, Pb	Field	Soil, cement site	[110]
Phragmites australis	Temperate/Poland	Cr, Ni, Cu, Zn, Cd, Pb	Pot experiment	Sediment	[107]
Vetiveria zizanoides	Tropical/India	As	Field	Soil	[101]
Alopecurus pratensis, Elytrigia repens, Poa angustifolia, Holcus lanatus, Arrhenatherum elatius, Bromus inermis Leyss, Artemisia vulgaris, Urtica dioica, Achillea millefolium, Galium mollugo, Stellaria holostea, and *Silene vulgaris*	Temperate/Germany	Cr, Ni, Cu, Zn, Cd, Pb, As	Pot experiment (cold-house)	Soil	[3]
Boehmeria nivea		Sc, Y, La, Ce, Pr, Nd, Sm, Eu, Gd, Tb, Dy, Ho, Er, Tm, Yb, and Lu	Growth chamber	Hydroponic	[111]
Morus alba, Acacia nilotica, Acacia ampliceps, and *Azadirachta indica*	Tropical/Pakistan	Pb	Greenhouse	Municipal and industrial wastewater	[112]
Salix schwerinii	Boreal/Finland	Ni, Cu, and Zn	Growth chamber	Soil	[113]
Senecio vulgaris, Polygonum aviculare, and *Poa annua*	Mediterranean/Italy	Cr, Cu, Zn, Cd, and Pb	Field	Soil	[100]
Klara (*S. viminalis* × *S. schwerinii* × *S. dasyclados*)	Boreal/Finland	Ni, Cu, and Zn	Pot experiment (greenhouse)	Soil	[105]
Artemisia vulgaris *Phalaris arundinacea* *Heracleum sphondylium* *Bistorta officinalis*	Temperate/UK	Mn, Zn, and As Mn and Ni Cr and Zn Mn and Zn	Field	Soil	[114]
Conocarpus lancifolius	Temperate/Pakistan	Zn, Cd, and Pb	Pot experiment (greenhouse)	Soil	[115]
Salix subfragilis	Temperate/Korea	Mn, Cu, Zn, Cd, and Pb	Field trial	Sediment (wetland)	[116]
Ruppia megacarpa, Zostera muelleri, and *Ulva australis*	Temperate-desert/Australia	Cu, Zn, Cd, Pb, As, and Se	Field trial	Estuary site	[98]
Achillea millefolium, Artemisia vulgaris, Papaver rhoeas, Tripleurospermum inodorum, and *Taraxacum officinale*	Temperate/Poland	Sc, Y, La, Ce, Pr, Nd, Sm, Eu, Gd, Tb, Dy, and Lu	Field trial	Soil	[108]
Salix schwerinii	Boreal/Finland	Cr, Ni, Cu, and Zn	Pot experiment (greenhouse)	Soil (landfill)	[117]
Amaranthus viridis, Bassia indica, Conyza bonariensis Cronquist, Portulaca oleracea, Rumex dentatus, Solanum nigrum, Lycopersicon esculentum, Phragmites australis, and *Pluchea dioscoridis*	Semi-desert/Egypt	Cr, Mn, Fe, Co, Ni, Cu, Zn, Cd, and Fe	Field trial	Soil (sewage sludge dump site)	[118]
Salix schwerinii, Klara, Karin and *Salix myrsinifolia*	Boreal/Finland	Cu and Zn	Pot experiment (greenhouse)	Soil	[119]
Phalaris arundinacea	Temperate/Czech Republic	Cr, Cd, and Hg	Field trial	Constructed wetland	[120]
Asclepias syriaca, Desmodium canadense, Panicum virgatum, Raphanus sativus, and *Solanum lycopersicum*	Boreal/Canada	Pr, Nd, Sm, Tb, Dy, and Er	Growth chamber	Soil	[121]
Oenanthe sp., *Juncus* sp., *Typha* sp., *Callitriche* sp.1, and *Callitriche* sp.2	Temperate/France	Cr, Ni, Cu, Zn, Cd, Pb, and As	Field trial	Urban stormwater	[122]
Phragmites australis	Tropical/Iran	Fe, Ni, Cu, and Cd	Field trial	Sediments	[104]

Bioindicator use in biomonitoring programs may be an effective approach, although each time, it requires extended research to recognise how metals in a potential bioindicator correlate with the evaluated matrix (water, sediments, or soils). Species properties regarding metal uptake, the antagonistic and synergistic properties between the elements as well as whether the metal is essential or non-essential play a key role in evaluating bioindication technology [23]. Nevertheless, this is a useful assessment tool, offering the additional potential for recovering valuable elements, and so it is expected to develop relatively soon.

5. Conclusions

By applying geochemical and ecotoxicological assessment and classification indices, a numeric evaluation may be obtained regarding the extent of sediment contamination. An analysis of the results yielded by recent studies using the Web of Science Core Collection (based on five keywords: sediments, trace metals, contamination, pollution indices, urban) shows that the following occur most often: (1) the researchers studied metal concentration in the surface layer of sediments (0–2 cm, 0–5 cm, 0–10 cm); (2) the most frequently used geochemical indices were *Igeo*, *EF*, and *CF*; (3) ecotoxicological evaluation was mainly performed by using *SQG* and *PERI*, which includes the toxic response index. The above-presented indices and classifications summarise the geochemical and ecotoxicological approaches of sediment assessment. The geoaccumulation indices help to determine whether the accumulation of TMs is due to natural processes or anthropogenic interference. Ecotoxicological assessment supports the overall contamination assessment of sediments and its harm for living organisms. The final results can be calculated for individual elements (*CF*, *Igeo*, *EF*, Pi (*SPI*), PT_T) and by using complex pollution indices (*PLI*, *Cdeg*, *mCdeg*, Pi_{sum}, PI_{Avg}, PI_{aAvg}, *PIN*, PI_{Prod}, PI_{apProd}, $PI_{vectorM}$, $PI_{Nemerow}$, *IntPI*, and *MPI*) for a larger set of TMs. Similar approaches can be applied for ecotoxicological indices (individual: ER^i and complex: *CSI*, *SPI*, *RAC*, *PERI*, and *MERMQ*).

Geochemical indicators largely refer to the total metal content in the sample, although the inclusion of mobile and bioavailable fractions would appear to be indispensable for geochemical and ecotoxicological assessment. The use of $Igeo_{integrated}$ may help comprehensively verify the sediment quality. This approach includes contamination analysis, taking into account metal fractionation. A "good quality assessment" of sediments via *Igeo* may turn out to be worse after checking TM fractionation and referring to $Igeo_{integrated}$ (especially when a large proportion of TMs are bound in mobile or water-soluble fractions). Conversely, when TMs are mainly bound to an immobile and/or residual fraction, the *Igeo* assessment may be mendacious, resulting in an overestimation of the pollution level.

Bioindication is an indirect method of environmental assessment, which can support environmental quality assessment monitoring programs. Nowadays, the simultaneous use of different classifications and indices reflects a more accurate assessment of TM pollution. The most recognised bioindicators include *Phragmites* and *Salix* for TM phytoremediation, particularly Zn, Cd, Cu, and Ni in a boreal and temperate climate. Hybrid cultivars of *Salix* such as Klara (*Salix viminalis* × *S. schwerinii* × *S. dasyclados*) and Karin (*S. viminalis* × *S. schwerinii*) are also applied in biomonitoring and considered for phytostabilisation and rhizofiltration. *Vetiveria zizanoides* demonstrated resistance to As. Rare earth elements (lanthanides) such as Pr, Nd, Tb, Sm, Dy, and Er constitute a separate group of analysed metals investigated in plant research. This issue is of serious concern for the recovery of valuable elements in the development of a circular economy. *Artemisia vulgaris*, *Papaver rhoeas*, and *Taraxacum officinale* present lanthanide phytoextraction potential.

In order to be properly managed or remediated, sediments must be comprehensively evaluated. Monitoring programs for sediments in urban retention systems are crucial in this regard. A promising evaluation method is the use of bioindicators, which also have the potential to recover valuable elements. Therefore, future research should focus on developing this tool while evaluating the quality of sediments.

Author Contributions: Conceptualisation, N.N. and E.W.; methodology, N.N.; investigation, N.N., M.M., S.K.; resources, N.N.; data curation, N.N.; writing—original draft preparation, N.N. and E.W.; writing—review and editing, N.N. and E.W.; visualisation, N.N.; supervision, E.W., S.R., A.P.; funding acquisition, N.N. and E.W. All authors have read and agreed to the published version of the manuscript.

Funding: The authors acknowledge the support provided by InterPhD2 program funded by the European Union (Project No. POWR.03.02.00-IP.08–00-DOK/16).

Conflicts of Interest: The authors declare no conflict of interest.

Appendix A

Table A1. Comparison of trace elements investigated in bottom sediments of diverse urban water bodies (river, lakes, receivers, retention tanks, etc.) with concentration ranges in mg/kg d.w., indication of indices used for pollution evaluation, classifications, and verification of health risks and an indication of the use of metal fractionation.

Authors	Location	Use	Analysed HMs	TMs Ranges [mg/kg d.w.]	Sampling	Indices	SQG	Heath Risk	Chemical Speciation Analysis
Sun et al., 2018 [28]	Songhua River, Jilin City (China)	Urban area with petrochemical industries	Five elements: Cr, Ni, Cu, Zn, Pb	Total HMs content: 123.98–346.34	Surface sediment (0–5 cm)	*Igeo, PERI, RAC*			+
Vieira et al., 2019 [29]	Arthur Thomas Lake, Londrina (Brazil)	Urban lake; strongly agroindustrial economy (with coffee production)	27 elements: Na, K, Mg, Ca, Sr, Ba, Sc, La, Th, Ti, V, Cr, Mo, Mn, Fe, Co, Ni, Ag, Au, Al, Ga, P, As, Sb, Bi, S, Se	Na (1–3), K (10–30), Mg (50–290), Ca (90–570), Sr (5.00–32.0), Ba (37–180), Sc (19.4–57.1), La (14.0–35.0), Th (3.20–5.90), Ti (220–950), V (323–1331), Cr (74.0–185), Mo (0.30–1.00), Mn (640–174.-3), Fe (8940–16830), Co (19.70–62.30), Ni (18.5–63.6), Ag (0.01–3.20), Au (0.50–247.1), Ga (17.0–28.0), P (50–170), As (1.30–3.50), Sb (0.10–0.70), Bi (0.00–0.60), S (10–180), Se (0.01–1.40)	Sediment cores (0–90 cm)	*EF, Igeo*	+		
Kumar et al., 2020 [30]	Yamuna River, Chambal River, Gulf of Mannar, Ganges River, Betwa River, Ken River, Beas River, Gomti River, and Gangotri River (India)	Review of different TMs from sediment samples from Indian rivers'	10 elements: Cr, Mn, Fe, Co, Ni, Cu, Zn, Cd, Pb, As	Cr (0.075–2628.3), Mn (0.097–2436.5), Fe (4.23–2312.44), Co (1.10–624), Ni (0.01–1813), Cu (0.019–5214), Zn (0.13–2759), Cd (0.015–272), Pb (0.17–1297), As (0.12–197)	No data	*CF, EF, Igeo, PERI (=MRI), RI*		*HI, HQ*	
Cui et al., 2019 [31]	Harbin City, Song (China)	Urban and rural rivers:Majiagou River (urban section, industrial zone) Yunliang River (rural section)	Six elements: Cr, Ni, Cu, Zn, Cd, Pb	Urban river: Cr (75.12–203.15), Ni (7.91–30.38), Cu (4.00–82.54), Zn (128.17–1416.71), Cd (0.08–4.08), Pb (8.86–57.49) Rural river: Cr (53.65–81.92), Ni (BDL *.-13.11), Cu (15.75–22.29), Zn (113.23–2474.05), Cd (BDL–4.29), Pb (9.31–114.42)	Surface sediment	*CF (=Pi), PI$_{Nemerow}$, RI*			
Liu et al., 2020 [32]	Pearl River Estuary, Xixiang River, Gongle Chung, Gushu Chung, Nanchang Chung, Tiegang reservoir flood discharge river, and Southern Airport drainage river (China)	Urban area	Six elements: Cr, Ni, Cu, Zn, Hg, Pb	Cr (35–510), Ni (15.0–194), Cu (38,0–1600), Zn (105–2600), Hg (0.009–0.85), Pb (22.9–160)	Surface sediment	*Igeo, RI*			

Table A1. *Cont.*

Authors	Location	Use	Analysed HMs	TMs Ranges [mg/kg d.w.]	Sampling	Indices	SQG	Heath Risk	Chemical Speciation Analysis
Chassiot et al., 2019 [33]	Saint-Charles River, Lake St. Charles (Canada)	Upstream an urban reservoir; urban area with diverse industries	14 samples: V, Cr, Mo, Mn, Co, Ni, Cu, Ag, Zn, Cd, Hg, Sn, Pb, As	No data of ranges	Sediment cores and surface sediment	*EF, Igeo, MPI*			
Hanfi et al., 2020 [34]	Sediment samples around the world: Europe [Europe], Asia [Asia], Africa [Africa], North America [North America]	Review of different TMs from sediment samples around the world (including 41 research paper reported between 1980 and 2018)	Six elements: Cr, Ni, Cu, Zn, Cd, Pb	Cr (29–196[Europe]; 0.85–144[Asia]; 1.4–85.7[Africa]; 17.1–125[North America]), Ni (35–128.45[Europe]; 12–126[Asia]; 1.9–67[Africa]; 18.1–26.5[North America]), Cu (73–466.9[Europe]; 10.99–269[Asia]; 11.3–243[Africa]; 15–356[North America]), Zn (125–1166[Europe]; 50.6–2377[Asia]; 13.1–1840[Africa]; 59–1811[North America]), Cd (0.2–4.6[Europe]; 0.12–72[Asia]; 0.33–6.9[Africa]; 0.1–8[North America]), Pb (48–1880[Europe]; 13.3–2582.5[Asia]; 11.2–737[Africa]; 10.9–2583[North America])	Surface sediment	*CF (=Pi), Igeo*			
Xia et al., 2020 [35]	Caohai Wetland (China)	The natural area of black-necked crane habitats in the Caohai wetland	Nine elements: Be, V, Cr, Ni, Cu, Zn, Cd, Hg, Pb	Be (0.83–2.19), V (49.1–103.1), Cr (83.5–145.9), Ni (40.3–65.7), Cu (13.5–30.9), Zn (108.9–365.4), Cd (0.5–7.34), Hg (0.30–1.34), Pb (37.3–76)	Surface sediment (0–10 cm)	*EF, Igeo, PERI, RAC*			
Dhamodharan et at., 2019 [36]	Cooum River, Chennai, (India)	Urban area	10 elements: Cr, Mn, Fe, Ni, Cu, Zn, Cd, Hg, Pb, As	Cr (7.12–155), Mn (139–2167), Fe (17,389–49,568), Ni (3.54–53.1), Cu (12.3–59.39), Zn (19.7–438), Cd (0.7–24.4), Hg (0.01–0.79), Pb (0–30.6), As (45–497)	Surface sediment	*Igeo, EF, CF, PLI, PERI*			
Siddiqui and Pandey, 2019 [37]	Ganga River (China)	Urban area	Eight elements: Cr, Mn, Fe, Ni, Cu, Zn, Cd, Pb	Cr (7.12–155), Mn (139–2167), Fe (17,389–49,568), Ni (3.54–53.1), Cu (2.1–73.98), Zn (6.3–104.3), Cd (0.21–3.6), Pb (2.1–36.5)	Surface sediment (0–10 cm)	*CF, EF, Er^i, Igeo, mCd, MPI, PERI, PI*	+		
Hafijur Rahaman Khan et al., 2020 [38]	Ganges-Brahmaputra-Meghna (Bangladesh)	Bengal Basin river system	19 elements: Sc, Th, U, Hf, Nb, Ta, W, Cr, Mo, Co, Ni, Cu, Cd, Ga, In, Tl, Ge, Pb, Bi	Sc (7.93–16.79), Th (13.28–29.51), U (2.5–4.71), Hf (3.31– 12.50), Nb (11.75– 17.68), Ta (1.08–1.58), W (1.46–2.90), Cr (43.48–120.61), Mo (0.12–0.72), Co (9.99–19.81), Ni (19.20–85.80), Cu (11.70–48.96), Cd (0.02–0.17), Ga (13.48–23.48), In (0.01–0.09), Tl (0.12–1.04), Ge (1.33–1.63), Pb (19.63–28.78), Bi (0.14–0.88)	Surface sediment	*CF, EF, Igeo, PLI*			

Table A1. *Cont.*

Authors	Location	Use	Analysed HMs	TMs Ranges [mg/kg d.w.]	Sampling	Indices	SQG	Heath Risk	Chemical Speciation Analysis
Dević et al., 2020 [39]	Belgrade (Serbia)	Urban area of New Belgrade; Sava River and reservoirs for diesel fuel and mazut; high traffic zone	10 elements: V, Cr, Mn, Fe, Co, Ni, Cu, Zn, Cd, Pb	V (49.9–299.9), Cr (37–150.9), Mn (395–925), Fe (17,400–32,400), Co (15.99–36.99), Ni (50–139.9), Cu (5.5–30.9), Zn (1–615), Cd (1–4), Pb (20–190)	Sediment cores and surface sediment	*Igeo, PERI, PLI, $PI_{Nemerow}$*			
Wang et al., 2019 [40]	Mid-channel of the Wen-Rui Tang River and its tributaries (China)	Rural–urban area	Five elements: Cr, Cu, Zn, Cd, Pb	Average: Cr (248 ± 131), Cu (995 ± 2011), Zn (2345 ± 2901), Cd (62 ± 125), Pb (217 ± 226)	Surface sediment (0–10 cm)	$E_r{}^i$, *PERI, RAC*	+		
Cui et al., 2020 [41]	Dongfenggou River, Miaotaigou River, Huaijiagou River, Harbin (China)	Suburban rivers	Six elements: Cr, Ni, Cu, Zn, Cd, Pb	Cr (64.0–180.6), Ni (11.8–106.4), Cu (14.6–182.5), Zn (175.8–1198.8), Cd (0.3–3.8), Pb (16.8–150.6)	Surface sediments	*PLI, RI*			
Barhoumi et al., 2019 [42]	Someşu Mic River (Romania)	Different human activities around, e.g., industries, urban, and agriculture	Eight elements: Cr, Mn, Fe, Ni, Cu, Zn, Cd, Pb	Cr (9.39–43.15), Mn (159.90–4707.21), Fe (11,359–35,661.28), Ni (14.73–47.69), Cu (7.22–65.56), Zn (42.12–236.82), Cd (0.04–0.35), Pb (12.27–131.39)	Surface sediment (0–20 cm)	*EF, Igeo,* Romanian legislation	+		
Nodefarahani et al., 2020 [43]	Namak Lake (Iran)	Seasonal lake nourished by surface run-off and groundwater resources	Nine elements: V, Cr, Mn, Fe, Ni, Cu, Zn, Al, Pb	V (0.55–3.19), Cr (0.507–1.83), Mn (5.1–34.1), Fe (195.6–1117), Ni (0.239–1.209), Cu (0.57–1.3), Zn (0.346–1.225), Al (1.140–1.903), Pb (0.15–1.16)	Surface sediment	*Igeo, EF, mPECQ*	+		
Nargis et al., 2018 [44]	River Buriganga (Bangladesh)	Most of the industries and/or factories, such as tanneries, metal goods manufacturing, electroplating, batteries, shipyard, are located on the banks of the river	15 elements: Ba, U, V, Cr, Mo, Mn, Ni, Cu, Zn, Cd, Hg, As, Bi, Se, Pb	Ba (20.80 ± 2.21[M] **; 23.09 ± 2.63[W] ***), U (0.45 ± 0.09[M]; 0.50 ± 0.11[W]), V (7.51 ± 2.25[M]; 8.66 ± 2.77[W]), Cr (39.70 ± 18.84[M]; 41.45 ± 15.88[W]), Mo (0.40 ± 0.09[M]; 0.44 ± 0.11[W]), Mn (37.58 ± 3.13[M]; 39.06 ± 2.72[W]), Ni (6.39 ± 0.96[M]; 7.14 ± 1.11[W]), Cu (14.07 ± 15.93[M]; 15.93 ± 18.38[W]), Zn (36.73 ± 34.38[M]; 40.71 ± 37.33[W]), Cd (0.21 ± 0.02[M]; 0.23 ± 0.03[W]), Hg (0.016 ± 0.001[M]; 0.018 ± 0.001[W]), As (0.18 ± 0.12[M]; 0.21 ± 0.13[W]), Bi (0.33 ± 0.02[M]; 0.36 ± 0.02[W]), Se (1.07 ± 0.05[M]; 1.19 ± 0.05[W]), Pb (10.41 ± 13.61[M]; 11.40 ± 15.09[W])	Surface sediment (0–5 cm)	C_{deg}, *CF*, $E_r{}^i$, *PLI, PERI*			

Table A1. *Cont.*

Authors	Location	Use	Analysed HMs	TMs Ranges [mg/kg d.w.]	Sampling	Indices	SQG	Heath Risk	Chemical Speciation Analysis
Xia et al., 2020 [45]	Wuhan (China)	20 lakes along a rural to the urban gradient in central China	11 elements: Cr, Mn, Fe, Co, Ni, Cu, Zn, Cd, Pb, Al, As	Cr (111.35 ± 43.93[RRG] ****; 95.50 ± 4.89[RCFG] *****; 111.29 ± 36.18[UPG] ******; 99.78 ± 8.47[UCFG] *******), Mn (970 ± 640[RRG]; 420 ± 140[RCFG]; 590 ± 100[UPG]; 710 ± 160 [UCFG]), Fe (37,570 ± 12,660[RRG]; 38,640 ± 3110[RCFG]; 38,260 ± 3010[UPG]; 39,900 ± 3080 [UCFG]), Co (19.81 ± 7.54[RRG]; 15.93 ± 1.40[RCFG]; 14.12 ± 1.37[UPG]; 16.63 ± 1.09 [UCFG]), Ni (32.10 ± 9.24[RRG]; 43.67 ± 4.25[RCFG]; 40.47 ± 3.30[UPG]; 44.61 ± 3.50 [UCFG]), Cu (105.45 ± 113.98[RRG]; 37.37 ± 5.47[RCFG]; 93.72 ± 74.94[UPG]; 62.65 ± 35.21 [UCFG]), Zn (105.75 ± 43.16[RRG]; 95.97 ± 13.15[RCFG]; 166.71 ± 4.71[UPG]; 134.41 ± 37.27 [UCFG]), Cd (0.50 ± 0.16[RRG]; 0.32 ± 0.06[RCFG]; 0.61 ± 0.16[UPG]; 0.45 ± 0.11 [UCFG]), Pb (32.59 ± 6.55[RRG]; 29.51 ± 2.15[RCFG]; 62.87 ± 27.83[UPG]; 39.06 ± 11.88 [UCFG]), Al (25,770 ± 15,580[RRG]; 23,600 ± 9400[RCFG]; 45,500 ± 3290[UPG]; 33,760 ± 16,500 [UCFG]), As (6.53 ± 1.89[RRG]; 10.04 ± 1.05[RCFG]; 10.69 ± 0.85[UPG]; 11.15 ± 1.49 [UCFG])	Surface sediment (0–5 cm)	*EF, RI*			
Wojciechowska et al., 2019 [46]	Gdansk (Poland)	Bottom sediments of urban retention tanks	Six elements: Cr, Ni, Cu, Zn, Cd, Pb	Cr (2.43–25.8), Ni (3.76–11.03), Cu (3.04–1133), Zn (17.9–362), Cd (0.088–0.60), Pb (5.77–162)	Surface sediment (0–5 cm)	*CF, Igeo, LAWA, PLI, RI,*		HQ	
Nawrot et al., 2020 [47]	Gdansk (Poland)	Bottom sediments of urban retention tanks	Six elements: Cr, Fe, Ni, Cu, Zn, Cd, Pb	Cr (2.45–74.5), Fe (3993–63,817), Ni (1.57–25.8), Cu (3.24–119), Zn (12.5–584), Cd (0.003–0.716), Pb (4.91–309)	Sediment cores	*AF, EF, mC$_{deg}$*			+

Table A1. *Cont.*

Authors	Location	Use	Analysed HMs	TMs Ranges [mg/kg d.w.]	Sampling	Indices	SQG	Heath Risk	Chemical Speciation Analysis
Jaskuła et al., 2021 [48]	Warta River (Poland)	Bottom sediments of the third longest river in Poland	Six elements: Cr, Ni, Cu, Zn, Cd, Pb	Cr (0.78–193), Ni (0.56–36.7), Cu (0.40–116), Zn (0.50–519), Cd (0.03–14.5), Pb (1.0–144)	Surface sediment (0–5 cm)	*EF, Igeo, MPI, PLI*	+		
Kostka and Leśniak, 2021 [49]	Wigry Lake (Poland) * WL	Bottom sediments	Seven elements: Cr, Mn, Fe, Cu, Zn, Cd, Pb	Cr (0.20–22.61), Mn (18–1698), Fe (80–32,857), Cu (0.02–59.7), Zn (3.1–632.1), Cd (0.003–3.060), Pb (7.0–107.5)	Surface sediment (0–5 cm)	SQG	+		
Ribbe et al., 2021 [50]	Lake Victoria, Ugandan part (Uganda)	Bottom sediments of the largest tropical lake in the world	Seven elements: Cr, Ni, Cu, Zn, Cd, Pb, As	Cr (29–100), Ni (19–56), Cu (21–121), Zn (49–103), Cd (0.06–0.26), Pb (10–25), As (2.9–6.6)	Surface sediment (the upper 15 cm)	*Igeo*, LAWA			+
Xiao et al., 2021 [51]	Lijiang River, Guilin City (China)	Analysis of a 160 km section of the river	10 elements: Cr, Mn, Co, Ni, Cu, Zn, Cd, Hg, Pb, As	Cr (24.38–95.38), Mn (176.25–1572.50), Co (4.50–15.38), Ni (11.63–37.13), Cu (9.38–102.75), Zn (53.63–258.0), Cd (0.16–4.41), Hg (0.08–2.13), Pb (17.88–171.75), As (9.97–36.44)	Surface sediment (0–5 cm)	*Igeo, mC$_{deg}$, RI*			
Castro et al., 2021 [52]	San Luis River (Argentina)	Bottom sediments	Nine elements: Cr, Mn, Co, Ni, Cu, Zn, Cd, Pb, As	Cr (0.5–32), Mn (10–420), Co (0.5–14), Ni (1–19), Cu (0.5–70), Zn (0–600), Cd (0–1,5), Pb (0–45), As (0.5–18)	Surface sediment (0–2 cm)	*CF, EF, Igeo*	+		

Explanations: * BDL—below detection limit; ** M—in monsoon; *** W—in winter; **** RCFG—rural commercial fishing group; ***** RRG—rural reservoir group; ****** UPG—urban park group; ******* UCFG—urban commercial fishing group * WL Wigry Lake in Poland is located in Wigry National Park (non urban area)—just to comparison with other studies.

Appendix B

Table A2. A list of Individual (single) Pollution Indices (IPIs) given in the literature with classification and a basic description of strengths and weaknesses.

Index	Formula	Classification	Description (Pros "+" and Cons "–")	References
Contamination Factor (*CF*)	$CF = \frac{C_{mSample}}{C_{mPre}}$ $C_{mSample}$—TM content * C_{mPre}—preindustrial concentration of TM	CF < 1—low, 1 ≤ CF < 3—moderate, 3 ≤ CF < 6—considerable, CF ≥ 6—very high	+ dedicated to individual metal + simple in use (easy to calculate) + comparing the sample to the reference contamination status of analysed element (pre-industrial) − omits the TM availability and mobility in the environment − omits the natural variability process, grain-size, and reference elements ratios	[58]

285

Table A2. *Cont.*

Index	Formula	Classification	Description (Pros "+" and Cons "–")	References
Geoaccumulation Index (I_{geo})	$I_{geo} = \log_2 \left[\dfrac{C_{mSample}}{1.5 \times C_{mBackground}} \right]$ $C_{mSample}$—TM content $C_{mBackground}$—geochemical background concentration of TM	Class 0: Igeo $\leq$ 0—uncontaminated, Class 1: 0 < Igeo $\leq$ 1—uncontaminated to moderately contaminated, Class 2: 1 < Igeo $\leq$ 2—moderately contaminated, Class 3: 2 < Igeo $\leq$ 3—moderately to strongly contaminated, Class 4: 3 < Igeo $\leq$ 4—strongly contaminated, Class 5: 4 < Igeo $\leq$ 5—strongly to extremely contaminated, Class 6: Igeo > 5—extremely contaminated	+ dedicated to individual metal + simple in use (easy to calculate) + comparing the sample to the background contamination status of analysed element + 1.5 multiplication factor reduces the possible variation of lithogenic effects + precise scale + widely used – omits the TM availability and mobility in the environment – omits the natural variability process, grain-size, and reference elements – incorrect selection of *Background* leads to mistaken results	[59]
Enrichment Factor (*EF*)	$EF = \dfrac{\frac{C_{mSample}}{C_{ref}}}{\frac{C_{mBackground}}{B_{ref}}}$ $C_{mSample}$—TM content $C_{mBackground}$—geochemical background concentration of TM C_{ref}—concentration of the reference TM in analysed sample B_{ref}—reference TM concentration in the reference environment	EF < 1—no enrichment, 1 $\leq$ EF < 3—minor enrichment, 3 $\leq$ EF < 5—moderate enrichment, 5 $\leq$ EF < 10—moderately severe enrichment, 10 $\leq$ EF < 25—severe enrichment, 25 $\leq$ EF < 50—very severe enrichment, EF $\geq$ 50—ultra-high enrichment	+ dedicated to individual metal + simple in use (easy to calculate) + comparing the sample to the reference contamination status of analysed element and to the normalise element (e.g., Al or Fe) as well as to reference content of analysed TM and normalise element + estimate the anthropogenic impact at all the possibility of contamination assessment of several TMs in reference to normalise element + precise scale + widely used – omits the TM availability and mobility in the environment – omits the natural variability process, grain-size – incorrect selection of *reference* value leads to mistaken results	[60]
Pollution Index (P_i) = Single Pollution Index (*SPI*)	$P_i = SPI = PI = \dfrac{C_{mSample}}{C_{mBackground}}$ $C_{mSample}$—TM content $C_{mBackground}$—geochemical background concentration of TM	P_i < 1—unpolluted, low level of pollution 1 $\leq P_i \leq$ 3—moderate polluted 3 > P_i—strong polluted	+ dedicated to individual metal + simple in use (easy to calculate) + comparing the sample to the background contamination status of analysed element +/– similar to CF – omits the TM availability and mobility in the environment – omits the natural variability process, grain size, and reference elements ratios	[15,54]

Table A2. *Cont.*

Index	Formula	Classification	Description (Pros "+" and Cons "–")	References
Threshold Pollution Index (PI_T)	$PI_T = \dfrac{C_{mSample}}{C_{TL}}$ $C_{mSample}$—TM content C_{TL}—tolerance levels of metal concentration;	$PI_T < 1$—unpolluted $1 \leq PI_T \leq 2$—low polluted $2 \leq PI_T \leq 3$—moderate polluted $3 \leq PI_T \leq 5$—strong polluted $5 \leq PI_T$—very strong polluted	+ dedicated to individual metal + simple in use (easy to calculate) + comparing the sample to the tolerance level of metal concentration (which could be assumed as a comparative value before assessment) +/– freedom in determining the threshold parameter (C_{TL}) (which can be matched with stringent pollution determination regulations) +/– similar to CF – omits the TM availability and mobility in the environment – omits the natural variability process, grain size, and reference elements ratios	[28]

* Preindustrial concentration of TM according to Håkanson [58] [mg/kg]: Hg—0.25, Cd—1.0, As—15, Cu—50, Pb—70, Cr—90, Zn—175.

Appendix C

Table A3. A list of Complex Pollution Indices (CPIs) given in the literature with classification and basic description of strengths and weaknesses.

Index	Formula	Classification	Description (Pros "+" and Cons "–")	References
Pollution Load Index (*PLI*)	$PLI = \sqrt[n]{CF_1 \cdot CF_2 \cdot CF_3 \cdot \ldots \cdot CF_i}$ CF_i—Contamination Factor of i element n—the number of analysed TMs	$PLI < 1$—not polluted $PLI = 1$—baseline levels of pollution $PLI > 1$—polluted	+ allows for identifying the contamination in relation to several trace metals + easy to apply (easy to calculate) + widely used + gives the comprehensive screen of sediment sample + allows comparing samples taken from different locations + used CF (includes C_{mPre})* – omits the TM availability and mobility in the environment – omits the natural variability process, grain size, and reference elements ratios – incorrect selection of C_{mPre} value could lead to mistaken results	[62]

Table A3. *Cont.*

Index	Formula	Classification	Description (Pros "+" and Cons "–")		References
Degree of contamination (C_{deg})	$C_{deg} = \sum_{i=1}^{n} CF_i$ CF_i Contamination Factor of i element n—the number of analysed TMs	$C_{deg} < 8$—low degree of contamination $8 \leq C_{deg} < 16$—moderate degree of contamination $16 \leq C_{deg} < 32$—considerable degree of contamination $C_{deg} \geq 32$—very high degree of contamination	+	allows for identifying the contamination in relation to several trace metals	[58]
			+	easy to apply (easy to calculate)	
			+	precise scale	
			+	assesses a sum of contamination factors	
			+	used CF (includes C_{mPre}) *	
			–	not widely used	
			–	does not include geochemical background	
			–	the preindustrial reference value is necessary	
			–	omits the TMs availability and mobility in the environment	
			–	omits the natural variability process, grain size, and reference elements ratios	
Modified contamination factor (mC_{deg})	$mC_{deg} = \frac{1}{n} \cdot \sum_{i=1}^{n} CF_i$ CF_i Contamination Factor of i element n—the number of analysed TMs	$mC_{deg} < 1.5$—very low $1.5 \leq mC_{deg} < 2$—low $2 \leq mC_{deg} < 4$—moderate $4 \leq mC_{deg} < 8$—high $8 \leq mC_{deg} < 16$—very high $16 \leq mC_{deg} < 32$—extremely high $mC_{deg} \geq 32$—ultra-high	+	allows for identifying the contamination in relation to several trace metals	[64]
			+	easy to apply (easy to calculate)	
			+	precise scale	
			+	assesses a sum of contamination factors	
			+	used CF (includes C_{mPre}) *	
			+	widely used	
			–	does not include geochemical background	
			–	the preindustrial reference value is necessary	
			–	omits the TMs availability and mobility in the environment	
			–	omits the natural variability process, grain size, and reference elements ratios	
Sum of Pollution Index (PI_{sum})	$PI_{sum} = \sum_{i=1}^{n} P_i$ P_i—calculated value for Pollution Index n—the number of analysed TMs	The classification for Pi can be used in PI_{sum}. The values in PI_{sum} should be multiplied by n (count of TMs): $PI_{sum} < 1n$—unpolluted, low level of pollution $1n \leq PI_{sum} \leq 3n$—moderately polluted $3n < PI_{sum}$—heavily polluted	+	defined as the sum of all determined contents of TMs, expressed as Pi	[63]
			+	dedicated to combining all analysed TMs	
			+	based on Pi (includes $C_{mBackground}$)	
			–	does not require the variation of natural processes	
			–	omits the TMs availability and mobility in the environment	
			–	the key is the choice of appropriate $C_{mBackground}$ value	
			–	does not include precise scale	

Table A3. *Cont.*

Index	Formula	Classification	Description (Pros "+" and Cons "–")	References
Average of Pollution Index (PI_{Avg})	$PI_{Avg} = \frac{1}{n} \cdot \sum_{i=1}^{n} P_i$ P_i—calculated value for Pollution Index n—the number of analysed TMs	PI_{Avg} values in excess of 1.0 show a lower quality of the sediments, which is conditioned by high contamination and low quality	+ allows for identifying the contamination in relation to several trace metals + based on Pi (includes $C_{mBackground}$) + easy to apply (easy to calculate) + lack of threshold for maximum values – does not require the variation of natural processes – omits the TMs availability and mobility in the environment – the key is the choice of appropriate $C_{mBackground}$ value – does not include precise scale	[63]
Weighted Average of Pollution Index (PI_{wAvg})	$PI_{wAvg} = \sum_{i=1}^{n} w_i \cdot P_i$ P_i—calculated value for Pollution Index w_i weight of P_i n—the number of analysed TMs	When PI_{wAvg} is used with the $\Sigma w_i = 1$ condition, terminologies can also be used as single indices (the classification for Pi can be applied)	+ dedicated to combining all analysed TMs + based on Pi (includes $C_{mBackground}$) – does not include precise scale – require to establish the weight of P_i values for each TM – the condition $\Sigma w_i = 1$ is not necessary, so the "average" is just for the sake of meaning in terminology	[58,59,63]
Background enrichment factor = New Pollution Index (PIN)	$PIN = \sum_{i=1}^{n} W_i^2 \cdot P_i$ P_i—calculated value for Pollution Index $W_i$class of TM considering degree of contamination (from 1 to 5 basing on P_i) n—the number of analysed TMs	$0 \leq PIN < 7$—clean $7 \leq PIN < 95.1$—trace contaminant $95.1 \leq PIN < 518.1$—lightly contaminant $518.1 \leq PIN < 2548.5$—contaminant $PIN \geq 2548.8$—high contaminant	+ dedicated to integrating contamination into a single value + based on Pi (includes $C_{mBackground}$) + precise scale – not widely used – omits the TMs availability and mobility in the environment – requires the computation of Wi	[65]
Product of Pollution Index (PI_{Prod})	$PI_{Prod} = \prod_{i=1}^{n} P_i$ P_i—calculated value for Pollution Index n—the number of analysed TMs	The classification for Pi can be used in $PI_{Prod.}$ The values in PI_{Prod} should be powered by n (count of TMs): $PI_{Prod} < 1^n$—unpolluted, low level of pollution $1^n \leq PI_{Prod} \leq 3^n$—moderately polluted $3^n < PI_{Prod}$—heavily polluted	+ defined as the product of all determining contents of TMs, expressed as Pi + dedicated to combining all analysed TMs + based on Pi (includes $C_{mBackground}$) – does not require the variation of natural processes – omits the TMs availability and mobility in the environment – the key is choice of appropriate $C_{mBackground}$ value – does not include precise scale – not widely used	[63]

Table A3. *Cont.*

Index	Formula	Classification	Description (Pros "+" and Cons "–")	References
Weighted power product of Pollution Index (PI_{wpProd})	$PI_{wpProd} = \prod\limits_{i=1}^{n} P_i{}^{w_i}$ P_i—calculated value for Pollution Index w_i weight of P_i n—the number of analysed TMs	When PI_{wpProd} is used with the $\Sigma w_i = 1$ condition, terminologies can also be used as single indices (the classification for Pi can be applied)	+ dedicated to combine all analysed TMs + based on Pi (includes $C_{mBackground}$) – does not include precise scale – require to establish the weight of P_i values for each TM – the condition $\Sigma w_i = 1$ is not necessary	[63]
Vector modulus of Pollution Index ($PI_{vectorM}$)	$PI_{vectorM} = \sqrt{\frac{1}{n}\cdot\sum\limits_{i=1}^{n} P_i^2}$ P_i—calculated value for Pollution Index n—the number of analysed TMs	Not specified	+ easy to apply (easy to calculate) + application of $C_{mBackground}$ + based on Pi + dedicated to combine all contaminations in one index – not specified scale – not widely used (not much described in the literature) – does not require the variation of natural processes	[63]
Nemerow Pollution Index ($PI_{Nemerow}$)	$PI_{Nemerow} = \sqrt{\frac{\left(\frac{1}{n}\cdot\Sigma_{i=1}^{n}\,P_i\right)^2 + P_{imax}^2}{2}}$ P_i—calculated value for Pollution Index P_{imax}—the maximum value of the single pollution indices of all TMs n—the number of analysed TMs	$PI_{Nem} < 0.7$—safety domain $0.7 \leq PI_{Nem} < 1$—precaution domain $1 \leq PI_{Nem}$ w< 2—slightly polluted domain $2 \leq PI_{Nem} < 3$—moderately polluted domain $PI_{Nem} > 3$—seriously polluted	+ reflects the sediment environmental pollution + emphasize the maximum value of the single P_i of all TMs + precise scale + dedicated to combine all analysed TMs with reference to the maximum value of the single P_i – does not include weight of P_i values – needs to rank elements	[66]
Integrated Pollution Index ($IntPI$)	$IntPI = mean(P_{ii})$ P_i—calculated value for Pollution Index	$IPI < 1$—low contaminated $1 \leq IPI \leq 2$—moderately contaminated $IPI < 2$—heavily contaminated	+ easy to apply (easy to calculate) + application of $C_{mBackground}$ + based on Pi – not widely used – does not require the variation of natural processes	[67]
Metallic Pollution Index (MPI)	$MPI = \left(Pi_1\cdot Pi_2\cdot\ldots\cdot Pi_n\right)^{1/n}$ P_i—calculated value for Pollution Index n—the number of analysed TMs	$MPI > 1$—indicate pollution $MPI < 1$—indicate no pollution	+ similar to PLI + basing on Pi values + gives the comprehensive screen of sediment sample + allows to compare samples taken from different locations – omits the TM availability and mobility in the environment – omits the natural variability process, grain size, and reference elements ratios	[68,69]

* (includes C_{mPre})—with reference to the literature, the CPIs are related to the IPI, but it should be noted that in the IPI C_{mPre} and $C_{mBackground}$ are two different concepts that are often confused/or taken interchangeably.

Appendix D

Table A4. A list of ecotoxicological indices (EIs) given in the literature with classification and basic description of strengths and weaknesses.

Index	Formula	Classification	Description (Pros "+" and Cons "–")		References
Individual Ecotoxicological Index (IEI)					
Ecological Risk Factor (ER, $E_r{}^i$) Risk Index (RI)	$ER^i = T_r^i \cdot CF^i$ $* T_r^i$ – toxic-response factor for a given element "i" CF^i—contamination factor for a given element "i";	$ER^i < 40$—low potential ecological risk; $40 \leq ER^i < 80$—moderate potential ecological risk; $80 \leq ER^i < 160$—considerable potential ecological risk; $160 \leq ER^i < 320$—high potential ecological risk; $ER^i \geq 320$—very high ecological risk	+ + + – –	dedicated to the toxicity of individual metals simple in use (easy to calculate) uses CF values omits the TM availability and mobility in the environment omits the natural variability process, grain size, and reference elements ratios	[58]
Complex Ecotoxicological Index (CEI)					
Contamination Severity Index (CSI)	$CSI = \sum_{i=1}^{n} w \cdot \left(\left(\frac{C_{mSample}}{ERL} \right)^{\frac{1}{2}} + \left(\frac{C_{mSample}}{ERM} \right)^2 \right)$ w—the computed weight of each TM according to Pejman et al. [86] ** $C_{mSample}$ TM content ERL—Effects Range-Low according to Table 2 ERM—Effects Range-Median according to Table 2 $w = \frac{(loading\ value_i \cdot eigen\ value)}{\sum_{i=1}^{n}(loading\ value_i \cdot eigen\ value)}$ *loading value$_i$* and *eigen value*—determined with the use of PCA/FA	$CSI < 0.5$—uncontaminated $0.5 \leq CSI < 1$—very low severity of contamination $1 \leq CSI < 1.5$—low severity of contamination $1.5 \leq CSI < 2$—low to moderate severity of contamination $2 \leq CSI < 2.5$—moderate severity of contamination $2.5 \leq CSI < 3$—moderate to high severity of contamination $3 \leq CSI < 4$—high severity of contamination $4 \leq CSI < 5$—very high severity of contamination $5 \leq CSI$—ultra-high severity of contamination	+ + + – – –	helpful in determining the limit of toxicity above which adverse impacts on the sediment environment are observed precise scale dedicated to combining all contaminations in one index includes adverse biological effects not widely used needs the ERL and ERM values requires the w values of each TM, which should be calculated using PCA/FA with considered the factors attributed to anthropogenic sources	[86]
Sediment Pollution Index (SPI)	$SPI = \dfrac{\sum_{i=1}^{n} \frac{C_{mSample}{}^i}{C_{mAverage}{}^i} \cdot T_r^i}{\sum_{i=1}^{n} T_r^i}$ $C_{mSample}$ TM content in a sample $C_{mAverage}$—average shale concentration of TM $* T_r^i$—toxic-response factor for a given element "i" n—the number of analysed TMs	$0 \leq SPI < 2$—natural sediments $2 \leq SPI < 5$— low polluted $5 \leq SPI < 10$—moderately polluted $10 \leq SPI < 20$—highly polluted $SPI > 20$—dangerous sediments	+ + + + – – – –	includes the toxicity of individual metal simple in use (easy to calculate) combined index delivers the information of diverse sites contamination omits the TM availability and mobility in the environment omits the natural variability process, grain size, and reference elements ratios neglecting the changes of TM/reference element ratios based on natural processes TM toxicity weights are available for Hg, Cd, As, Cu, Pb, Cr, and Zn	[58,64]

Table A4. *Cont.*

Index	Formula	Classification	Description (Pros "+" and Cons "–")		References
Risk Assessment Code (*RAC*)	The exchangeable and carbonate fractions are determined by a single extraction with a CH_3COOH 0.11 M solution	Percentage extracted by CH_3COOH 0.11 M solution is compared to the following scale: $RAC \leq 1$—no risk; $1 < RAC \leq 10$—low risk; $11 < RAC \leq 30$—medium risk; $31 < RAC \leq 50$: very high risk	+ + –	indicates the potential risk to the ecosystem caused by TMs bounded in fractions weakly associated in the sediments Delivers real risk information the extraction procedure is needed (time-consuming procedure)	[35]
Risk Index (*PERI*)	$PERI = \sum_{i=1}^{n} ER^i$ n—the number of analysed TMs ER^i—calculated value for Ecological Risk Factor	$PERI < 90$—low $9 \leq PERI < 180$—moderate $180 \leq PERI < 360$—strong $360 \leq PERI < 720$—very strong $PERI \geq 720$—highly strong	+ + + –	comprehensive ecological risk and contamination assessment includes the toxicity of analysed metals (basing on ER^i) simple in use (easy to calculate) disadvantages similar to ER^i	[87,88]
The probability of toxicity (*MERMQ*)	$MERMQ = \dfrac{\sum_{i=1}^{n} \frac{C_{mSample}{}^i}{ERM^i}}{n}$ $C_{mSample}$ TM content n—the number of analysed TMs ERM—Effects Range-Median according to Table 2 This index is also found as combined effect of Toxic Metals, means (*mPECQ*): $mPECQ = \sum_{i=1}^{n} \dfrac{\frac{C_{mSample}{}^i}{PEC^i}}{n}$ $C_{mSample}$ TM content n—the number of analysed TMs PEC—probable effect concentration according to Table 2	$MERMQ < 0.1$—low-risk level (probability of toxicity—9%) $0.1 \leq MERMQ < 0.5$—medium risk level (probability of toxicity—21%) $0.5 \leq MERMQ < 1.5$—high risk level (probability of toxicity—49%) $MERMQ > 1.5$—very high-risk level (probability of toxicity—76%)	+ + + + – – +	dedicated to combining all contaminations in one index precise scale helpful in determining the biological effects identification of risk areas not widely used needs the ERM/PEC values does not require the variation of natural processes	[86]

* Toxic response factor for subsequent trace metals according to Håkanson [58]: Hg—40, Cd—30, As—10, Cu—5, Pb—5, Cr—2, Zn—1. ** Weight of each heavy metal according to Pejman et al. [86]: Cu—0.075, Zn—0.075, Cr—0.134, Ni—0.215, Pb—0.251, Cd—0.25.

References

1. Sakson, G.; Brzezinska, A.; Zawilski, M. Emission of heavy metals from an urban catchment into receiving water and possibility of its limitation on the example of Lodz city. *Environ. Monit. Assess.* **2018**, *190*, 281. [CrossRef] [PubMed]
2. Zafra, C.; Temprano, J.; Tejero, I. The physical factors affecting heavy metals accumulated in the sediment deposited on road surfaces in dry weather: A review. *Urban Water J.* **2017**, *9006*, 1–11. [CrossRef]
3. Antoniadis, V.; Shaheen, S.M.; Stärk, H.J.; Wennrich, R.; Levizou, E.; Merbach, I.; Rinklebe, J. Phytoremediation potential of twelve wild plant species for toxic elements in a contaminated soil. *Environ. Int.* **2021**, *146*, 106233. [CrossRef]
4. Liu, L.; Luo, X.B.; Ding, L.; Luo, S.L. *Application of Nanotechnology in the Removal of Heavy Metal from Water*; Elsevier Inc.: New York, NY, USA, 2018; ISBN 9780128148389.
5. Wuana, R.A.; Okieimen, F.E. Heavy metals in contaminated soils: A review of sources, chemistry, risks and best available strategies for remediation. *Int. Sch. Res. Not. Ecol.* **2011**, *2011*, 1–20. [CrossRef]
6. Weissmannová, H.D.; Pavlovský, J. Indices of soil contamination by heavy metals—Methodology of calculation for pollution assessment (minireview). *Environ. Monit. Assess.* **2017**, *189*, 1–25. [CrossRef]
7. Kowalska, J.B.; Mazurek, R.; Gąsiorek, M.; Zaleski, T. Pollution indices as useful tools for the comprehensive evaluation of the degree of soil contamination—A review. *Environ. Geochem. Health* **2018**, *40*, 1–26. [CrossRef]
8. Szmagliński, J.; Nawrot, N.; Pazdro, K.; Walkusz-Miotk, J.; Wojciechowska, E. The fate and contamination of trace metals in soils exposed to a railroad used by Diesel Multiple Units: Assessment of the railroad contribution with multi-tool source tracking. *Sci. Total Environ.* **2021**, *798*, 149300. [CrossRef]
9. Bednarova, Z.; Kuta, J.; Kohut, L.; Machat, J.; Klanova, J.; Holoubek, I.; Jarkovsky, J.; Dusek, L.; Hilscherova, K. Spatial patterns and temporal changes of heavy metal distributions in river sediments in a region with multiple pollution sources. *J. Soils Sediments* **2013**, *13*, 1257–1269. [CrossRef]
10. Bartoli, G.; Papa, S.; Sagnella, E.; Fioretto, A. Heavy metal content in sediments along the Calore river: Relationships with physical-chemical characteristics. *J. Environ. Manag.* **2012**, *95*, S9–S14. [CrossRef] [PubMed]
11. Monrabal-Martinez, C.; Meyn, T.; Muthanna, T.M.; Monrabal-martinez, C.; Meyn, T.; Muthanna, T.M. Characterization and temporal variation of urban runoff in a cold climate—Design implications for SuDS. *Urban Water J.* **2018**, *16*, 1–9. [CrossRef]
12. Sekabira, K.; Origa, H.O.; Basamba, T.A.; Mutumba, G.; Kakudidi, E. Assessment of heavy metal pollution in the urban stream sediments and its tributaries. *Int. J. Environ. Sci. Technol.* **2010**, *7*, 435–446. [CrossRef]
13. Murphy, L.U.; Cochrane, T.A.; O'Sullivan, A. The influence of different pavement surfaces on atmospheric copper, lead, zinc, and suspended solids attenuation and wash-off. *Water Air Soil Pollut.* **2015**, *226*, 1–14. [CrossRef]
14. Charters, F. Characterising and Modelling Urban Runoff Quality for Improved Stormwater Management. Ph.D. Thesis, Department of Civil and Natural Resources Engineering, University of Canterbury, Christchurch, New Zealand, 2016.
15. Nour, H.E.; El-Sorogy, A.S.; Abdel-Wahab, M.; Almadani, S.; Alfaifi, H.; Youssef, M. Assessment of sediment quality using different pollution indicators and statistical analyses, Hurghada area, Red Sea coast, Egypt. *Mar. Pollut. Bull.* **2018**, *133*, 808–813. [CrossRef] [PubMed]
16. Walaszek, M.; Bois, P.; Laurent, J.; Lenormand, E.; Wanko, A. Urban stormwater treatment by a constructed wetland: Seasonality impacts on hydraulic efficiency, physico-chemical behavior and heavy metal occurrence. *Sci. Total Environ.* **2018**, *637–638*, 443–454. [CrossRef] [PubMed]
17. Simpson, S.; Batley, G. *Sediment Quality Assessment a Practical Guide*, 2nd ed.; CSIRO Publishing: Collingwood, Australia, 2016; ISBN 9781486303847.
18. Nawrot, N.; Wojciechowska, E.; Rezania, S.; Walkusz-Miotk, J.; Pazdro, K. The effects of urban vehicle traffic on heavy metal contamination in road sweeping waste and bottom sediments of retention tanks. *Sci. Total Environ.* **2020**, *749*, 141511. [CrossRef]
19. USEPA. *Contaminated Sediment Remediation Guidance for Hazardous Waste Sites*; Official Solid Waste Emergency Response; USEPA: Washington, DC, USA, 2005; p. 236.
20. Dung, T.T.T.; Cappuyns, V.; Swennen, R.; Phung, N.K. From geochemical background determination to pollution assessment of heavy metals in sediments and soils. *Rev. Environ. Sci. Biotechnol.* **2013**, *12*, 335–353. [CrossRef]
21. Fiedler, H.D.; López-Sánchez, J.F.; Rubio, R.; Rauret, G.; Quevauviller, P.; Ure, A.M.; Muntau, H. Study of the stability of extractable trace metal contents in a river sediment using sequential extraction. *Analyst* **1994**, *119*, 1109–1114. [CrossRef]
22. Bonanno, G.; Borg, J.A.; Di Martino, V. Levels of heavy metals in wetland and marine vascular plants and their biomonitoring potential: A comparative assessment. *Sci. Total Environ.* **2017**, *576*, 796–806. [CrossRef]
23. Klink, A.; Macioł, A.; Wisłocka, M.; Krawczyk, J. Metal accumulation and distribution in the organs of Typha latifolia L. (cattail) and their potential use in bioindication. *Limnologica* **2013**, *43*, 164–168. [CrossRef]
24. Kabata-Pendias, A.; Pendias, H. *Biogeochemistry of Trace Elements*; Polish Scientific Publishing Company: Warsaw, Poland, 1999; Volume 2, ISBN 0849315751.
25. Al-Homaidan, A.A.; Al-Otaibi, T.G.; El-Sheikh, M.A.; Al-Ghanayem, A.A.; Ameen, F. Accumulation of heavy metals in a macrophyte Phragmites australis: Implications to phytoremediation in the Arabian Peninsula wadis. *Environ. Monit. Assess.* **2020**, *192*, 1–10. [CrossRef]

26. Relić, D.; Dordević, D.; Popović, A. Assessment of the pseudo total metal content in alluvial sediments from Danube River, Serbia. *Environ. Earth Sci.* **2011**, *63*, 1303–1317. [CrossRef]
27. Ure, A.M.; Davidson, C.M. Chemical speciation in soils and related materials by selective chemical extraction. In *Chemical Speciation in the Environment*, 2nd ed.; Wiley-Blackwell: Malden, MA, USA, 2007; pp. 265–300. [CrossRef]
28. Sun, C.; Zhang, Z.; Cao, H.; Xu, M.; Xu, L. Concentrations, speciation, and ecological risk of heavy metals in the sediment of the Songhua River in an urban area with petrochemical industries. *Chemosphere* **2019**, *219*, 538–545. [CrossRef] [PubMed]
29. Vieira, L.M.; Neto, D.M.; do Couto, E.V.; Lima, G.B.; Peron, A.P.; Halmeman, M.C.R.; Froehner, S. Contamination assessment and prediction of 27 trace elements in sediment core from an urban lake associated with land use. *Environ. Monit. Assess.* **2019**, *191*, 1–20. [CrossRef]
30. Kumar, V.; Sharma, A.; Pandita, S.; Bhardwaj, R.; Thukral, A.K.; Cerda, A. A review of ecological risk assessment and associated health risks with heavy metals in sediment from India. *Int. J. Sediment Res.* **2020**, *35*, 516–526. [CrossRef]
31. Cui, S.; Zhang, F.; Hu, P.; Hough, R.; Fu, Q.; Zhang, Z.; An, L.; Li, Y.F.; Li, K.; Liu, D.; et al. Heavy metals in sediment from the urban and rural rivers in Harbin City, Northeast China. *Int. J. Environ. Res. Public Health* **2019**, *16*, 4313. [CrossRef]
32. Liu, C.; Yin, J.; Hu, L.; Zhang, B. Spatial distribution of heavy metals and associated risks in sediment of the urban river flowing into the Pearl River Estuary, China. *Arch. Environ. Contam. Toxicol.* **2020**, *78*, 622–630. [CrossRef]
33. Chassiot, L.; Francus, P.; De Coninck, A.; Lajeunesse, P.; Cloutier, D.; Labarre, T. Spatial and temporal patterns of metallic pollution in Québec City, Canada: Sources and hazard assessment from reservoir sediment records. *Sci. Total Environ.* **2019**, *673*, 136–147. [CrossRef] [PubMed]
34. Hanfi, M.Y.; Mostafa, M.Y.A.; Zhukovsky, M.V. Heavy metal contamination in urban surface sediments: Sources, distribution, contamination control, and remediation. *Environ. Monit. Assess.* **2020**, *192*, 32. [CrossRef] [PubMed]
35. Xia, P.; Ma, L.; Sun, R.; Yang, Y.; Tang, X.; Yan, D.; Lin, T.; Zhang, Y.; Yi, Y. Evaluation of potential ecological risk, possible sources and controlling factors of heavy metals in surface sediment of Caohai Wetland, China. *Sci. Total Environ.* **2020**, *740*, 140231. [CrossRef]
36. Dhamodharan, A.; Abinandan, S.; Aravind, U.; Ganapathy, G.P.; Shanthakumar, S. Distribution of Metal Contamination and Risk Indices Assessment of Surface Sediments from Cooum River, Chennai, India. *Int. J. Environ. Res.* **2019**, *13*, 853–860. [CrossRef]
37. Siddiqui, E.; Pandey, J. Assessment of heavy metal pollution in water and surface sediment and evaluation of ecological risks associated with sediment contamination in the Ganga River: A basin-scale study. *Environ. Sci. Pollut. Res.* **2019**, *26*, 10926–10940. [CrossRef] [PubMed]
38. Khan, M.H.R.; Liu, J.; Liu, S.; Li, J.; Cao, L.; Rahman, A. Anthropogenic effect on heavy metal contents in surface sediments of the Bengal Basin river system, Bangladesh. *Environ. Sci. Pollut. Res.* **2020**, *27*, 19688–19702. [CrossRef] [PubMed]
39. Dević, G.J.; Ilić, M.V.; Zildzović, S.N.; Avdalović, J.S.; Miletić, S.B.; Bulatović, S.S.; Vrvić, M.M. Investigation of potentially toxic elements in urban sediments in Belgrade, Serbia. *J. Environ. Sci. Health Part A Toxic Hazard. Subst. Environ. Eng.* **2020**, *55*, 765–775. [CrossRef] [PubMed]
40. Wang, Z.; Zhou, J.; Zhang, C.; Qu, L.; Mei, K.; Dahlgren, R.A.; Zhang, M.; Xia, F. A comprehensive risk assessment of metals in riverine surface sediments across the rural-urban interface of a rapidly developing watershed. *Environ. Pollut.* **2019**, *245*, 1022–1030. [CrossRef] [PubMed]
41. Cui, S.; Gao, S.; Zhang, F.; Fu, Q.; Wang, M.; Liu, D.; Li, K.; Song, Z.; Chen, P. Heavy metal contamination and ecological risk in sediment from typical suburban rivers. *River Res. Appl.* **2020**, 1–9. [CrossRef]
42. Barhoumi, B.; Beldean-Galea, M.S.; Al-Rawabdeh, A.M.; Roba, C.; Martonos, I.M.; Bălc, R.; Kahlaoui, M.; Touil, S.; Tedetti, M.; Driss, M.R.; et al. Occurrence, distribution and ecological risk of trace metals and organic pollutants in surface sediments from a Southeastern European river (Someşu Mic River, Romania). *Sci. Total Environ.* **2019**, *660*, 660–676. [CrossRef]
43. Nodefarahani, M.; Aradpour, S.; Noori, R.; Tang, Q.; Partani, S.; Klöve, B. Metal pollution assessment in surface sediments of Namak Lake, Iran. *Environ. Sci. Pollut. Res.* **2020**. [CrossRef]
44. Nargis, A.; Sultana, S.; Raihan, M.J.; Haque, M.E.; Sadique, A.B.M.R.; Sarkar, M.S.I.; Un-Nabie, M.M.; Zhai, W.; Cai, M.; Habib, A. Multielement analysis in sediments of the River Buriganga (Bangladesh): Potential ecological risk assessment. *Int. J. Environ. Sci. Technol.* **2019**, *16*, 1663–1676. [CrossRef]
45. Xia, W.; Wang, R.; Zhu, B.; Rudstam, L.G.; Liu, Y.; Xu, Y.; Xin, W.; Chen, Y. Heavy metal gradients from rural to urban lakes in central China. *Ecol. Process.* **2020**, *9*, 1–11. [CrossRef]
46. Wojciechowska, E.; Nawrot, N.; Walkusz-Miotk, J.; Matej-Łukowicz, K.; Pazdro, K. Heavy metals in sediments of urban streams: Contamination and health risk assessment of influencing factors. *Sustainability* **2019**, *11*, 563. [CrossRef]
47. Nawrot, N.; Wojciechowska, E.; Matej-Łukowicz, K.; Walkusz-Miotk, J.; Pazdro, K. Spatial and vertical distribution analysis of heavy metals in urban retention tanks sediments: A case study of Strzyza Stream. *Environ. Geochem. Health* **2020**, *8*. [CrossRef] [PubMed]
48. Jaskuła, J.; Sojka, M.; Fiedler, M.; Wróżyński, R. Analysis of spatial variability of river bottom sediment pollution with heavy metals and assessment of potential ecological hazard for the Warta river, Poland. *Minerals* **2021**, *11*, 327. [CrossRef]
49. Kostka, A.; Leśniak, A. Natural and anthropogenic origin of metals in lacustrine sediments; assessment and consequences—A case study of Wigry lake (Poland). *Minerals* **2021**, *11*, 1–22. [CrossRef]
50. Ribbe, N.; Arinaitwe, K.; Dadi, T.; Friese, K.; von Tümpling, W. Trace-element behaviour in sediments of Ugandan part of Lake Victoria: Results from sequential extraction and chemometrical evaluation. *Environ. Earth Sci.* **2021**, *80*, 1–14. [CrossRef]

51. Xiao, H.; Shahab, A.; Xi, B.; Chang, Q.; You, S.; Li, J.; Sun, X.; Huang, H.; Li, X. Heavy metal pollution, ecological risk, spatial distribution, and source identification in sediments of the Lijiang River, China. *Environ. Pollut.* **2021**, *269*, 116189. [CrossRef]

52. Castro, M.F.; Almeida, C.A.; Bazán, C.; Vidal, J.; Delfini, C.D.; Villegas, L.B. Impact of anthropogenic activities on an urban river through a comprehensive analysis of water and sediments. *Environ. Sci. Pollut. Res.* **2021**, *28*, 37754–37767. [CrossRef]

53. HELCOM Guideline for the Determination of Heavy Metals in Sediment. 2015. Available online: https://helcom.fi/?s=Guideline+for+the+determination+of+heavy+metals+in+sediment (accessed on 12 May 2021).

54. Mulligan, C.; Fukue, M.; Sato, Y. *Sediments Contamination and Sustainable Remediation*; CRC Press: Boca Raton, FL, USA, 2009; ISBN 9781420062236.

55. Tack, F.M.G.; Verloo, M.G. Chemical speciation and fractionation in soil and sediment heavy metal analysis: A review. *Int. J. Environ. Anal. Chem.* **1995**, *59*, 225–238. [CrossRef]

56. Gałuszka, A. A review of geochemical background concepts and an example using data from Poland. *Environ. Geol.* **2007**, *52*, 861–870. [CrossRef]

57. Turekian, K.K.; Wedepohl, K.H. Distribution of the elements in some major units of the earth's crust. *Geol. Soc. Am. Bull.* **1961**, *72*, 175–192. [CrossRef]

58. Håkanson, L. An ecological risk index for aquatic pollution control.a sedimentological approach. *Water Res.* **1980**, *14*, 975–1001. [CrossRef]

59. Müller, V.G. Pollutants in sediments—Sediments as pollutants [available in German: Schadstoffe in Sedimenten—Sedimente als Schadstoffe]. *Mitt. Osterr. Geol. Ges.* **1986**, *79*, 107–126.

60. Sutherland, R.A. Bed sediment-associated trace metals in an urban stream, Oahu, Hawaii. *Environ. Geol.* **2000**, *39*, 611–627. [CrossRef]

61. Tomlinson, D.L.; Wilson, J.G.; Harris, C.R.; Jeffrey, D.W. Assessment of heavy metal enrichment and degree of contamination around the copper-nickel mine in the Selebi Phikwe Region, Eastern Botswana. *Environ. Ecol. Res.* **2013**, *1*, 32–40. [CrossRef]

62. Tomlinson, D.L.; Wilson, J.G.; Harris, C.R.; Jeffrey, D.W. Problems in the assessment of heavy-metal levels in estuaries and the formation of a pollution index. *Helgol. Meeresunters.* **1980**, *33*, 566–575. [CrossRef]

63. Gong, Q.; Deng, J.; Xiang, Y.; Wang, Q.; Yang, L. Calculating pollution indices by heavy metals in ecological geochemistry assessment and a case study in parks of Beijing. *J. China Univ. Geosci.* **2008**, *19*, 230–241. [CrossRef]

64. Abrahim, G.M.S.; Parker, R.J. Assessment of heavy metal enrichment factors and the degree of contamination in marine sediments from Tamaki Estuary, Auckland, New Zealand. *Environ. Monit. Assess.* **2008**, *136*, 227–238. [CrossRef] [PubMed]

65. Weissmannová, H.D.; Pavlovský, J.; Chovanec, P. Heavy metal contaminations of urban soils in Ostrava, Czech Republic: Assessment of metal pollution and using Principal Component Analysis. *Int. J. Environ. Res.* **2015**, *9*, 683–696.

66. Cheng, J.; Shi, Z.; Zhu, Y. Assessment and mapping of environmental quality in agricultural soils of Zhejiang Province, China. *J. Environ. Sci.* **2007**, *19*, 50–54. [CrossRef]

67. Chen, T.B.; Zheng, Y.M.; Lei, M.; Huang, Z.C.; Wu, H.T.; Chen, H.; Fan, K.K.; Yu, K.; Wu, X.; Tian, Q.Z. Assessment of heavy metal pollution in surface soils of urban parks in Beijing, China. *Chemosphere* **2005**, *60*, 542–551. [CrossRef]

68. Caeiro, S.; Costa, M.H.; Ramos, T.B.; Fernandes, F.; Silveira, N.; Coimbra, A.; Medeiros, G.; Painho, M. Assessing heavy metal contamination in Sado Estuary sediment: An index analysis approach. *Ecol. Indic.* **2005**, *5*, 151–169. [CrossRef]

69. Gong, C.; Ma, L.; Cheng, H.; Liu, Y.; Xu, D.; Li, B.; Liu, F.; Ren, Y.; Liu, Z.; Zhao, C.; et al. Characterization of the particle size fraction associated heavy metals in tropical arable soils from Hainan Island, China. *J. Geochem. Explor.* **2014**, *139*, 109–114. [CrossRef]

70. Lu, X.; Wang, L.; Lei, K.; Huang, J.; Zhai, Y. Contamination assessment of copper, lead, zinc, manganese and nickel in street dust of Baoji, NW China. *J. Hazard. Mater.* **2009**, *161*, 1058–1062. [CrossRef] [PubMed]

71. Chassiot, L.; Francus, P.; De Coninck, A.; Lajeunesse, P.; Cloutier, D.; Labarre, T. Dataset for the assessment of metallic pollution in the Saint-Charles River sediments (Québec City, QC, Canada). *Data Brief* **2019**, *26*, 104256. [CrossRef] [PubMed]

72. Guartán, J.A.; Emery, X. Regionalized classification of geochemical data with filtering of measurement noises for predictive lithological mapping. *Nat. Resour. Res.* **2021**, *30*, 1033–1052. [CrossRef]

73. Bojakowska, I.; Sokołowska, G. Geochemical purity classes of water sediments [available in Polish: Geochemiczne klasy czystości osadów wodnych]. *Przegląd Geol.* **1998**, *46*, 49–54.

74. Ure, A.M.; Berrow, M.L. The elemental constituents of soils. In *Environmental Chemistry*; Bowen, H.J.M., Ed.; Royal Society of Chemistry: London, UK, 1982; Volume 2, pp. 92–204.

75. Lis, J.; Pasieczna, A. *Geochemical Atlas of Gdańsk Region [Available in Polish: Geochemiczny Atlas Region Gdańskiego]*; Wydawnictwo Kartograficzne Polskiej Agencji Ekologicznej: Warsaw, Poland, 1999.

76. Burton, G.A. Sediment quality criteria in use around the world. *Limnology* **2002**, *3*, 65–75. [CrossRef]

77. Wenning, R.J.; Ingersoll, C.G. *Use of Sediment Quality Guidelines and Related Tools for the Assessment of Contaminated Sediments*; SETAC: Pensacola, FL, USA, 2005.

78. Smith, S.; MacDonald, D.; Keenleyside, K.; Ingersoll, C.; Field, J. A preliminary evaluation of sediment quality assessment values for freshwater ecosystems. *J. Great Lakes Res.* **1996**, *22*, 624–638. [CrossRef]

79. Long, E.; Morgan, L. *The Potential for Biological Effects of Sediment-Sorbed Contaminants Tested in the National Status and Trends Program*; NOAA Technical Memorandum NOS OMA 52; National Oceanic and Atmospheric Administration: Seattle, WA, USA, 1991; p. 175.

80. Persaud, D.; Jaagumagi, R.; Hayton, A. *Guidelines for the Protection and Management of Aquatic Sediment Quality in Ontario*; Water Resources Branch, Ontario Ministry of the Environment: Toronto, ON, Canada, 1993.
81. MENVIQ. *Interim Criteria for Quality Assessment of St. Lawrence River Sediment*; St. Lawrence Centre and Ministère de l'Environnement du Québec: Quebec City, QC, Canada, 1992; p. 28.
82. ANZECC; ARMCANZ. Australian and New Zealand guidelines for fresh and marine water quality. In *The Guidelines Australian and New Zealand Environment and Conservation Council Agriculture and Resource Management Council of Australia and New Zealand*; National Water Quality Management Strategy; Australian and New Zealand Environment and Conservation Council, Australian Water Association: Artarmon, Australia, 2000; Volume 1, p. 314.
83. LAWA-Arbeitskreis. *"Qualitative Hydrology of Rivers" Assessment of the Water Quality of Rivers in the Federal Republic of Germany* [Available in German: *"Zielvorgaben" in Zusammenarbeit Mit LAWA-Arbeitskreis "Qualitative Hydrologie der Fliessgewässer" Beurteilung der Wasserbeschaffenheit von Fliessgewaessern in der Bundesrepublik Deutschland]*; Herausgegeben von der Länderarbeitsgemeinschaft Wasser: Berlin, Germany, 1998; ISBN 3889612245.
84. Ahlf, W.; Hollert, H.; Neumann-Hensel, H.; Ricking, M. A guidance for the assessment and evaluation of sediment quality: A German approach based on ecotoxicological and chemical measurements. *J. Soils Sediments* **2002**, *2*, 37–42. [CrossRef]
85. *Directive 2000/60/EC of the European Parliament and of the Council of 23 October 2000 Establishing a Framework for Community Action in the Field of Water Policy*; Publications Office of the European Union: Luxembourg, 23 October 2000; pp. 1–73.
86. Pejman, A.; Bidhendi, G.N.; Ardestani, M.; Saeedi, M.; Baghvand, A. A new index for assessing heavy metals contamination in sediments: A case study. *Ecol. Indic.* **2015**, *58*, 365–373. [CrossRef]
87. Kulbat, E.; Sokołowska, A. Methods of assessment of metal contamination in bottom sediments (case study: Straszyn Lake, Poland). *Arch. Environ. Contam. Toxicol.* **2019**, *77*, 605–618. [CrossRef] [PubMed]
88. Baran, A.; Tarnawski, M.; Koniarz, T. Spatial distribution of trace elements and ecotoxicity of bottom sediments in Rybnik reservoir, Silesian-Poland. *Environ. Sci. Pollut. Res.* **2016**, *23*, 17255–17268. [CrossRef]
89. Von Tümpling, W.; Scheibe, N.; Einax, J.W. Extended sediment quality rating for trace elements in urban waters—Case study Klinke, Germany. *Clean Soil Air Water* **2013**, *41*, 565–573. [CrossRef]
90. Sutherland, R.A. BCR®-701: A review of 10-years of sequential extraction analyses. *Anal. Chim. Acta* **2010**, *680*, 10–20. [CrossRef] [PubMed]
91. Viana, J.L.M.; Menegário, A.A.; Fostier, A.H. Preparation of environmental samples for chemical speciation of metal/metalloids: A review of extraction techniques. *Talanta* **2021**, *226*, 122119. [CrossRef] [PubMed]
92. Kersten, M. Speciation of trace metals in sediments. In *Chemical Speciation in the Environment*, 2nd ed.; Ure, A.M., Davidson, C.M., Eds.; John Wiley & Sons: New York, NY, USA, 2002; pp. 301–321.
93. Tessier, A.; Campbell, P.G.C.; Bisson, M. Sequential extraction procedure for the speciation of particulate trace metals. *Anal. Chem.* **1979**, *51*, 844–851. [CrossRef]
94. Boszke, E. *Specification of Speciation of Selected Heavy Metals (Fe, Mn, Zn, Cd, Ni, Cr, Co) in Bottom Sediments of the Baltic Sea and the Spitsbergen Region*; Institute of Oceanology of the Polish Academy of Science: Sopot, Poland, 1995. (In Polish)
95. Miller, W.P.; Martens, D.C.; Zelany, L.W. Effect of the sequence in extraction of trace metals from soils. Soil. *Soil Sci. Soc. Am. J.* **1986**, *50*, 598–601. [CrossRef]
96. Ure, A.M.; Quevauviller, P.; Muntau, H.; Griepink, B. An account of the improvement and harmonization of extraction techniques undertaken under the auspices of the BCR of the Commission of the European Communities. *Int. J. Environ. Anal. Chem.* **1993**, *51*, 135–151. [CrossRef]
97. Stankovic, S.; Stankovic, A.R. Bioindicators of toxic metals. In *Green Materials for Energy, Products and Depollution*; Springer: Amsterdam, The Netherlands, 2013; pp. 151–228. ISBN 9789400768352.
98. Farias, D.R.; Hurd, C.L.; Eriksen, R.S.; Macleod, C.K. Macrophytes as bioindicators of heavy metal pollution in estuarine and coastal environments. *Mar. Pollut. Bull.* **2018**, *128*, 175–184. [CrossRef] [PubMed]
99. Rayment, G.E.; Barry, G.A. Indicator tissues for heavy metal monitoring—Additional attributes. *Mar. Pollut. Bull.* **2000**, *41*, 353–358. [CrossRef]
100. Salinitro, M.; Tassoni, A.; Casolari, S.; de Laurentiis, F.; Zappi, A.; Melucci, D. Heavy metals bioindication potential of the common weeds *Senecio vulgaris* L. *Polygonum aviculare* L. and *Poa annua* L. *Molecules* **2019**, *24*, 2813. [CrossRef]
101. Singh, S.; Mishra, H.; Suprasanna, P. Evaluation of arsenic remediation, morphological and biochemical response by Vetiveria zizanoides L. plants grown on artificially arsenic contaminated soil: A field study. *Ecol. Eng.* **2021**, *168*, 106267. [CrossRef]
102. Bonanno, G.; Vymazal, J. Compartmentalization of potentially hazardous elements in macrophytes: Insights into capacity and efficiency of accumulation. *J. Geochem. Explor.* **2017**, *181*, 22–30. [CrossRef]
103. Milke, J.; Gałczyńska, M.; Wróbel, J. The Importance of biological and ecological properties of Phragmites Australis (Cav.) Trin. Ex Steud., in phytoremendiation of aquatic ecosystems—The review. *Water* **2020**, *12*, 1770. [CrossRef]
104. Ganjali, S.; Tayebi, L.; Atabati, H.; Mortazavi, S. *Phragmites australis* as a heavy metal bioindicator in the Anzali wetland of Iran. *Toxicol. Environ. Chem.* **2014**, *96*, 1428–1434. [CrossRef]
105. Salam, M.M.A.; Mohsin, M.; Pulkkinen, P.; Pelkonen, P.; Pappinen, A. Effects of soil amendments on the growth response and phytoextraction capability of a willow variety (*S. viminalis* × *S. schwerinii* × *S. dasyclados*) grown in contaminated soils. *Ecotoxicol. Environ. Saf.* **2019**, *171*, 753–770. [CrossRef] [PubMed]

106. Minkina, T.; Fedorenko, G.; Nevidomskaya, D.; Fedorenko, A.; Chaplygin, V.; Mandzhieva, S. Morphological and anatomical changes of Phragmites australis Cav. due to the uptake and accumulation of heavy metals from polluted soils. *Sci. Total Environ.* **2018**, *636*, 392–401. [CrossRef] [PubMed]

107. Nawrot, N.; Wojciechowska, E.; Pazdro, K.; Szmagliński, J.; Pempkowiak, J. Uptake, accumulation, and translocation of Zn, Cu, Pb, Cd, Ni, and Cr by P. australis seedlings in an urban dredged sediment mesocosm: Impact of seedling origin and initial trace metal content. *Sci. Total Environ.* **2021**, *768*, 144983. [CrossRef]

108. Mikołajczak, P.; Borowiak, K.; Niedzielski, P. Phytoextraction of rare earth elements in herbaceous plant species growing close to roads. *Environ. Sci. Pollut. Res.* **2017**, *24*, 14091–14103. [CrossRef]

109. You, Y.; Wang, L.; Ju, C.; Wang, G.; Ma, F.; Wang, Y.; Yang, D. Effects of arbuscular mycorrhizal fungi on the growth and toxic element uptake of Phragmites australis (Cav.) Trin. ex Steud under zinc/cadmium stress. *Ecotoxicol. Environ. Saf.* **2021**, *213*, 112023. [CrossRef]

110. Bayouli, I.T.; Bayouli, H.T.; Dell'Oca, A.; Meers, E.; Sun, J. Ecological indicators and bioindicator plant species for biomonitoring industrial pollution: Eco-based environmental assessment. *Ecol. Indic.* **2021**, *125*, 107508. [CrossRef]

111. Liu, C.; Liu, W.; Huot, H.; Yang, Y.; Guo, M.; Morel, J.L.; Tang, Y.; Qiu, R. Responses of ramie (*Boehmeria nivea* L.) to increasing rare earth element (REE) concentrations in a hydroponic system. *J. Rare Earths* **2021**. [CrossRef]

112. Hussain, Z.; Rasheed, F.; Tanvir, M.A.; Zafar, Z.; Rafay, M.; Mohsin, M.; Pulkkinen, P.; Ruffner, C. Increased antioxidative enzyme activity mediates the phytoaccumulation potential of Pb in four agroforestry tree species: A case study under municipal and industrial wastewater irrigation. *Int. J. Phytoremediat.* **2020**, *23*, 704–714. [CrossRef]

113. Mohsin, M.; Kuittinen, S.; Salam, M.M.A.; Peräniemi, S.; Laine, S.; Pulkkinen, P.; Kaipiainen, E.; Vepsäläinen, J.; Pappinen, A. Chelate-assisted phytoextraction: Growth and ecophysiological responses by Salix schwerinii EL Wolf grown in artificially polluted soils. *J. Geochem. Explor.* **2019**, *205*, 106335. [CrossRef]

114. Nworie, O.E.; Qin, J.; Lin, C. Trace element uptake by herbaceous plants from the soils at a multiple trace element-contaminated site. *Toxics* **2019**, *7*, 3. [CrossRef] [PubMed]

115. Rasheed, F.; Zafar, Z.; Waseem, Z.A.; Rafay, M.; Abdullah, M.; Salam, M.M.A.; Mohsin, M.; Khan, W.R. Phytoaccumulation of Zn, Pb, and Cd in Conocarpus lancifolius irrigated with wastewater: Does physiological response influence heavy metal uptake? *Int. J. Phytoremediat.* **2020**, *22*, 287–294. [CrossRef]

116. Kim, H.T.; Kim, J.G. Seasonal variations of metal (Cd, Pb, Mn, Cu, Zn) accumulation in a voluntary species, Salix subfragilis, in unpolluted wetlands. *Sci. Total Environ.* **2018**, *610*, 1210–1221. [CrossRef]

117. Salam, M.M.A.; Kaipiainen, E.; Mohsin, M.; Villa, A.; Kuittinen, S.; Pulkkinen, P.; Pelkonen, P.; Mehtätalo, L.; Pappinen, A. Effects of contaminated soil on the growth performance of young Salix (Salix schwerinii EL Wolf) and the potential for phytoremediation of heavy metals. *J. Environ. Manag.* **2016**, *183*, 467–477. [CrossRef]

118. Eid, E.M.; Shaltout, K.H. Bioaccumulation and translocation of heavy metals by nine native plant species grown at a sewage sludge dump site. *Int. J. Phytoremediat.* **2016**, *18*, 1075–1085. [CrossRef] [PubMed]

119. Mohsin, M. Potentiality of Four Willow Varieties for Phytoremediation in a Pot Experiment. Master's Thesis, Faculty of Science and Forestry, School of Forest Sciences, University of Eastern Finland, Kuopio, Finland, 2016.

120. Vymazal, J. Concentration is not enough to evaluate accumulation of heavy metals and nutrients in plants. *Sci. Total Environ.* **2016**, *544*, 495–498. [CrossRef]

121. Carpenter, D.; Boutin, C.; Allison, J.E.; Parsons, J.; Ellis, D.M. Uptake and effects of six rare earth elements (REEs) on selected native and crop species growing in contaminated soils. *PLoS ONE* **2015**, *10*, e0129936. [CrossRef]

122. Ladislas, S.; El-Mufleh, A.; Gérente, C.; Chazarenc, F.; Andrès, Y.; Béchet, B. Potential of aquatic macrophytes as bioindicators of heavy metal pollution in urban stormwater runoff. *Water Air Soil Pollut.* **2012**, *223*, 877–888. [CrossRef]

Article

Reducing the Negative Technogenic Impact of the Mining Enterprise on the Environment through Management of the Water Balance

Elena Menshikova , Viacheslav Fetisov, Tatyana Karavaeva, Sergey Blinov, Pavel Belkin * and Sergey Vaganov

Department of Geology, Perm State University, Bukireva St. 15, 614990 Perm, Russia;
menshikova_e@list.ru (E.M.); fetisov.v.v@gmail.com (V.F.); jewel_@list.ru (T.K.); blinov_s@mail.ru (S.B.);
vaganov_ss@mail.ru (S.V.)
* Correspondence: pashabelkin@mail.ru

Received: 1 December 2020; Accepted: 18 December 2020; Published: 21 December 2020

Abstract: In terms of anthropogenic impact of mining and processing enterprises, the adjacent territories are contaminated by upstream tailings dams. The contamination is developed by active seepage of liquid waste through the body of the dams. The authors have analyzed water balance at the Kachkanar Mining and Processing Plant tailings dump (Russia, Ural Region). The company develops vanadium-containing titanium-magnetite iron ores with low ore (15%). This, along with high productivity, has determined the formation of a large number of tailings and significant amount of wastewater. The purpose of the studies is to substantiate the need to manage the seepage discharge process by means of enclosing dams to ensure environmentally safe operation of the tailings dump. The research objectives included field measurements of seepage volumes, their evaluation by computational methods and analysis of anthropogenic geochemical load on natural waters. The obtained results show an increase in seepage discharge volume from 41.91 million m^3 (in 2017) to 81.44 million m^3 (in 2026) as the height of the dams increases. These losses will lead to water shortages in the enterprise's water recycling system. Calculation of pollutants in wastewater with the exception of natural component showed the leading role of technogenic factor in the content of Ti (up to 84%), V (up to 96%), Co (up to 86%) and Mo (up to 93%). Increasing the volume of seepage discharge will lead to an increase in natural water pollution within the area. Ecologically efficient management of the enterprise's water balance is ensured by the use of tailings thickening technology and implementation of closed water supply systems.

Keywords: iron ore deposits; tailing dumps; seepage water; water balance of the enterprise; metals

1. Introduction

The mining industry, characterized by the extraction and processing of minerals, is a powerful factor of environmental transformation and a source of geochemical load on the natural complexes of the adjacent territories. As a result of ore processing at the enrichment plants, after extraction of valuable components, unused waste remains-tailings, which are crushed waste rock. Generally, the volumes of tailings significantly exceed the volumes of extracted useful components [1]. The main way of wet tailings disposal at enrichment plants is their placement in tailing dumps, where solid and liquid components of tailings are separated. Solid particles settle and accumulate, and clarified water is used in the water recycling system of the technological process.

Tailing dumps are created by building enclosing earth dams. Structural features of the tailing dump and enclosing dams depend on the terrain relief, rocks of the base and the necessary volume of the wastes to be placed [2,3]. As the tailing dump fills up, in order to ensure its useful capacity enclosing dams are built up using wastes previously stored in the tailing dump. The main disadvantage of this method is a considerable volume of water discharged through the dam body and discharged at the base [2]. Many researchers point out that seepage through the dams disrupts their stability and is one of the main causes of emergency situations [1,4–11]. Another negative consequence of seepage discharge through the tailing dumps is the significant loss of water used in the recycling water supply system of the ore beneficiation process [12]. In addition, seepage discharge through the dams may lead to contamination of soils and natural waters in the adjacent areas [1,10,11].

Tailings dams are typically designed to seep/leak in order to keep the dam itself from becoming saturated, and ideally to lead to the ultimate dewatering of the tailings themselves. Unsaturated tailings pose less risk to the public if a catastrophic dam failure were to occur. Upstream-type dams pose the greatest risk of dam failure, and saturation of the tailings underlying an upstream dam significantly increases this risk. As a result, reliable seepage collection from a tailings impoundment/dam, either by a downgradient seepage collection dam, or by interception wells, is an important part of tailings impoundment design and management. Some seepage will need to occur in perpetuity, since the impoundment is not designed to stop the infiltration of precipitation. Efficiently collecting this seepage should be the primary goal, while reducing the volume of seepage can provide additional environmental and economic value.

The results of a comprehensive assessment of the technogenic impact of iron ore mining and processing enterprises on the natural and technical systems of the adjacent areas indicate a significant role of upstream tailing dumps as pollution focuses [12–17]. The liquid phase of the pulp entering the tailings dump contains significant amounts of free oxygen and carbon dioxide, creating favorable conditions for intensive leaching of tailings with the transfer to the liquid phase of a significant number of chemical elements and their compounds. Studies conducted at the "Krasnokamensky mine" OJSC (OAO) tailings dump (Krasnoyarsk Region, Russia) found that the clarified waters of the tailings dump by their chemical composition correspond to the liquid phase of the discharged pulp. In the upper layer (6.0–9.0 m) of the technogenic aquifer in the body of the tailings dump, the content of chemical elements was established at the level of their concentration in clarified waters. In the interval of technogenic aquifer from 9.0 to 30.0 m the content of Fe, Cu, Li, Mo increases three times, Mg—seven times. The concentrations of Ba, Co, Mn, Ni, Pb, Sr, V, Zn are substantially increased [14].

The tailings contain a number of trace elements that can migrate to the environment. One of the ways of elements migration is their removal as a part of seepage leaks of technogenic water from the tailings dump system. A group of elements, including Ba, Co, Cr, Cu, Li, Mn, Mo, Ni, Pb, Sb, Ti, V, Zn, has migratory abilities. The share of elements dissolution reaches 1% of the gross content on average. The maximum degree of dissolution—up to 7% is typical for iron. Moving to the technogenic aquifer, the elements retain their active migration abilities and can migrate from the tailing dump as part of technogenic waters. The migration of Fe, Ba, Co, Cr, Cu, Li, Mn, Mo, Ni, Pb, Sb, Ti, V, Zn is confirmed by their high concentrations in the seepage discharge waters of the tailings dump. In terms of chemical composition, seepage discharge waters are comparable with technogenic waters from the base of tailings mass within the beach zone [14].

Being discharged at the base of the tailings dumps, seepage water enters the environment, resulting in geochemical transformation of its components. The negative impact of tailing dumps, typified by the increased content of metals-iron ore accompanying minerals, can be traced at considerable distances. Analysis of the variability pattern of landscape pollution in the area of the tailings dump "Lebedinsky GOK" JSC (AO) (Belgorod Region, Russia) indicates an annual increase in the gross iron content at a distance of up to 10 km from the source of pollution [15].

Studies have shown that, in terms of technogenic loads, the mining and processing plants of the Kursk Magnetic Anomaly affect the environment within a radius of 20 to 30 km, in terms of impact on the quality of agrolandscapes—within a radius of 15 to 20 km [16].

Increased metal concentrations—Ni (coefficient of technogenic concentration (ratio of chemical element content in the studied soil to background content) Kc = 3.7), Mn (Kc = 5), Cr (Kc = 7)—have been identified in soils of transaccumulative landscapes in the lower gleyed horizons, which are influenced by ground waters associated with the tailings area of "Mikhailovsky GOK" PJSC (ПАО) (Kursk Region, Russia) [17].

Environmental safety of tailing dumps operation will be largely determined by the adopted water supply and drainage schemes. The priority task of safe operation of mining enterprises is to manage the water balance in order to reduce the amount of polluted wastewater entering the natural environment. Thus, control of seepage discharge is a necessary requirement for the tailings dump operation.

The purpose of the studies is to substantiate the need to control the seepage discharge process by means of enclosing dams to ensure environmentally safe operation of the tailings dump.

The research objectives included field measurements of seepage water volumes at the tailings dump, assessment of seepage water volumes by calculation methods, forecasting of changes in seepage water volumes until 2026, analysis of technogenic geochemical load on natural complexes.

2. Study Area

EVRAZ Kachkanar Mining and Processing Plant (EVRAZ KGOK) is the largest mining enterprise of the Russian Federation and the only enterprise in the world that develops poor vanadium-containing titanium-magnetite iron ores.

Active deposits belong to the Kachkanar intrusive massif located in the western wing of the Tagil megasynclinorium, which is part of the platinum-bearing belt of the Urals. The massif lies among a powerful complex of volcanogenic, volcanogenic-sedimentary, insignificantly sedimentary, in varying degrees metamorphosed rocks of the Upper Ordovician and Silurian. The ore-bearing Kachkanar gabbro-pyroxenite massif covers an area of about 110 km^2. Geochemical spectra of ore elements containing rocks of titanium-magnetite formation have the following form: Fe, Cr, V, Mn, Ni, Co, Cu. The ores are characterized by an iron content of 16–17% in disseminated and 30–45% in schlieren differences with a content of titanium dioxide of 1.2–2.2% and 3–4%, vanadium pentoxide of 0.13% and 0.25%, respectively [18].

The production capacity of the plant is about 55 million tons of iron ore per year [19,20].

High productivity of the plant and low content of useful component in the developed titanium-magnetite ores (15%) determined the main problems of the plant: the need to use large amounts of water in the technological process, which results in the formation of a large number of tailings and significant amounts of wastewater.

In 2014–2016, the enterprise's water consumption was 32.6–40.8 million m^3/year. Approximately 50% of this volume is discharged into watercourses of adjacent territories with controlled polluted discharge [21].

In the course of processing of mined ore at the stage of wet magnetic separation, wastes of waste (depleted) rock with a dry matter content of about 10% are generated. Waste is disposed in the tailings dump by means of hydraulic transport system. Annually about 45 million tons of tailings are delivered from the processing plant to the tailings dump [22].

The tailings dump is used for placing the tailings of wet magnetic separation, clarification of pulp and wastewater with their subsequent inclusion into the recycling water supply system of the enterprise.

The tailings dump is upstream and hillside and is located in the valleys of the Vyia River and its right-bank tributary—the Rogalevka River, at a distance of 1 km from the processing plant. Operation of the tailings dump began in 1963 [23].

The modern tailings dump consists of three compartments: Rogalevsky, Intermediate and Vyisky (Figure 1), each with its own pond and drainage facilities. The compartments are located in a cascade: Rogalevsky, Intermediate, Vyisky, with a difference in height on water surface in the settling ponds in the direction from south to north. Capacities of Rogalevsky and Intermediate compartments are formed by enclosing upstream dams and, partially, from the north-east, east and south-east, by hillsides. The capacity of the Vyisky compartment is formed by blocking the embankment dam of the Vyia River valley. Above the Vyia River bed, the compartment is limited by the lower water retaining dam of the Lower-Vyia Reservoir, located in a cascade towards the Vyisky compartment of recycled water. The Rogalevsky and Intermediate compartments are intended for tailings storage for beneficiation and clarification of the liquid phase of pulp. The Vyisky compartment is primarily used to receive clarified water from the Rogalevsky and Intermediate compartments of the settling ponds.

Figure 1. Tailings Dump Scheme.

As of the end of 2019, the total area of all the compartments is 19.63 km^2, and the volume of waste placed in the tailings dump reaches 949.75 million m^3.

Currently, there are nine dams along the perimeter of the tailings dump, composed of wet magnetic separation waste. The average height of the dams as of 2017 is 62 m; the average length is 1510 m [24]. The dams are built up by moving the tailings of the tailings dump's beach zone.

The following main components are singled out in the water balance output part of the enterprise's water balance: seepage water discharge to the surface of the slopes and at the base of the tailings dump, seepage through the tailings dump bed, evaporation from the water surface of the tailings dump, water accumulation in the pore space of the tailings of wet magnetic separation, irretrievable water losses in the technological process, as well as water discharge from the Vyisky compartment through the spillway (if the normal level is exceeded).

Seepage discharge at the base of dams № 1, № 2, № 3, № 4, № 5 and East is the main component of the water balance output part, making 66% of its total volume. The seepage waters of the dams bring contaminants into the natural environment changing the natural geochemical situation in the territories adjacent to the tailings dump.

3. Materials and Methods

Quantitative assessment of seepage discharge volumes through enclosing dams of the tailings dump was made based on the results of the annual cycle of hydrometric studies. Field measurements were performed on a monthly basis from August 2016 to July 2017. Hydrometric work included: measuring the width of the filtration water flow, the average depth, instrumental determination of water flow velocities, calculations of the cross-section wet area, average velocity and flow rate. Flow velocities measurements on velocity verticals were performed by the GR-21M hydrometric flowmeter using the main method. Depth, flow velocity and water flow rate measurements were performed in accordance with the requirements of the Russian Federation Hydrometeorological Service. Location of the hydrometric sections is shown in Figure 1.

For better accuracy of measurements in several hydrometric control sections automatic water level and temperature loggers, brand Solinst 3001 LT Levelogger Junior Edge (manufacturer Solinst Canada Ltd., Georgetown, ON, Canada), were installed, which registration data were taken into account when determining the cross-sectional area and flow rate. At the water level loggers installation sites, the depth measurements were made and the corresponding flow cross-sections were constructed. Taking into account the calculated areas of the flow cross-section wet area, diagrams of the dependence of the cross-section area on the water level were drawn, and linear approximation equations were obtained.

There are various calculation methods of quantitative evaluation of seepage through earth dams. Mathematical models based on finite difference methods [25] and finite element methods [26] are widely used for seepage calculations [8,27–30]. Analytical dependencies based on the Darcy's law and the Dupuy's equation are well known and can be used to evaluate seepage discharges through dams composed of sufficiently homogeneous dispersed sediments. Along with field and laboratory methods of research, mathematical methods based on empirical dependencies are used to estimate the hydraulic conductivity of dispersed sediments taking into account their particle size distribution [31,32].

Seepage volumes of water discharge through dams, position of depression curve and hydraulic gradients are determined on the basis of seepage calculations. Seepage calculations through the earth dams have been performed according to accepted assumptions: laminar seepage through the homogeneous and isotropic soils of the dam on a horizontal watertight and pervious base—based on the Darcy's law and Dupuy's formula [33,34].

To estimate seepage through the dams, the hydraulic conductivity of disperse deposits of dam bodies represented by wet magnetic separation tailings was determined taking into account their particle size distribution composition. Samples of the wet magnetic separation tailings were taken from the beach area of the tailings dump along several lines of the dam cross section at distances of 5, 10, 25, 50, 75, 100 and 150 m from the dam crest. The total number of samples taken was 201, including: dam № 1 (42 samples on 6 lines), Eastern dam (47 samples on 4 lines), dam № 2 (42 samples on 6 lines), dam № 3 (35 samples on 5 lines), dam № 4 (35 samples on 5 lines). Methods and results of determination of particle size distribution of wet magnetic separation tailings, calculations of coefficient of grain uniformity of tailings particle size distribution and their porosity as well as calculations of tailings hydraulic conductivity are given in detail [24].

The dependence of the hydraulic conductivity on inhomogeneity of the particle size distribution composition of sediments has been considered in the works of many researchers [35–37].

The results of study of wet magnetic separation tailings indicate that the values of effective grain diameter d_{10} are in the range of 0.06–0.14 mm, with prevailing values of 0.08–0.10 mm. The values of effective grain diameter d_{60} are 0.2–0.6 mm. The values of coefficient of grain uniformity of particle size distribution are in the range from 2 to 8 and mainly do not exceed 5 [24].

Calculation of hydraulic conductivity K was performed according to Hazen [38] and Beyer [39] equations. These values of hydraulic conductivity were used to estimate seepage through dams.

Assessment of geochemical load on the territory's natural complexes was carried out by comparing the concentrations of major pollutants in seepage discharge waters with background values of these components in natural waters. The assessment was carried out for the components, average concentrations of which in seepage waters are two or more times higher than the background values. Background water samples were taken mainly from the territory's main watercourse—the Vyia River (Lower-Vyia Reservoir), upstream of the tailings dump, outside its influence zone. Seepage discharge water samples were taken from dams № 1, East, № 2, № 3, № 4. Sampling points are shown in Figure 1.

Laboratory research of chemical and mineral composition of tailings was carried out in the Nanomineralogy Sector of the "Center for Collective Use of Perm State National Research University". Chemical composition of tailings was determined by X-ray fluorescent method on S8 Tiger spectrometer (Bruker), additionally the estimation of microelement composition was carried out using ICP-MS (Aurora M90, Bruker). Mineral composition of tailings was determined by optical mineralogical analysis (Nikon SMZ 745) with control mineral composition determination using diffractometric method (2D PHASER diffractometer, Bruker). Research on determination of microelements concentration in natural and seepage waters were carried out at the Institute of Geology and Geochemistry of Ural Branch of Russian Academy of Sciences by ICP-MS method (ELAN-9000, Perkin Elmer).

4. Results and Discussion

The results of field hydrometric measurements performed to assess the seepage discharge through the earth dams, as well as the results of determining the seepage discharge by calculation methods are presented in detail [24].

The results of assessment of seepage discharge through the dams of tailings dump by direct hydrometric measurements and based on calculations are presented in Table 1.

Table 1. Seepage discharge volumes through the tailing dumps [24].

Dams	Total Seepage Discharge Per Year, min m^3										
	2017 [a]	2017 [b]	2018	2019	2020	2021	2022	2023	2024	2025	2026
Dam № 1	9.99	10.61	12.52	13.99	15.6	17.38	19.35	20.67	22.03	23.43	24.67
East dam	2.13	2.54	3.12	3.61	3.95	4.29	4.65	5.03	5.42	5.82	6.18
Dam № 2	17.38	17.94	19.5	20.12	20.75	21.38	22.01	23.62	25.28	27	28.52
Dam № 5	0.28	0.20	0.40	0.53	0.66	0.80	0.94	1.16	1.41	1.68	1.94
Dam № 3	5.11	5.15	6.02	6.25	6.49	6.74	7.01	7.85	8.75	9.88	10.06
Dam № 4	6.24	5.47	6.4	6.57	6.77	6.97	7.18	7.99	8.84	9.9	10.07
Total through all dams	41.13	41.91	47.96	51.07	54.22	57.56	61.14	66.32	71.73	77.71	81.44

2017 [a]—for the considered period August 2016–July 2017 made in hydrometric control sections; 2017 [b]—for the considered period August 2016–July 2017 calculated analytically.

The results of the seepage discharge assessment made on the basis of monthly field observations and determined by the calculation method are comparable. This allows using the calculation method to forecast seepage discharge through the dams of the tailings dump for the period up to 2026. To perform the forecast calculations of seepage through the dams, the average annual water levels in the ponds of Rogalevsky and Intermediate compartments were taken into account for each year according to the plan of dams raise. When estimating the average annual water level in the ponds, specifics of its rise during the year were taken into account: the first half of the year is characterized by slower rise of water levels compared to the second half.

Calculations show that as the height of the dam embankment, the height of the tailings dump and, consequently, the water level in the ponds increase, the total volume of seepage discharge increases from 41.91 million m^3 (in 2017) to 81.44 million m^3 (in 2026) [24]. Seepage discharge increases in accordance with Darcy's law: increase of water level in ponds leads to increase of hydraulic gradient and amount of seepage water.

Estimated volumes of seepage discharge through dams were used to forecast the enterprise's water balance up to 2026. Balance calculations showed that irrecoverable losses as a result of seepage discharge through dams will lead to annually increasing water deficit in the enterprise's water recycling system [24].

The increase in water volumes due to other sources of water balance does not fully solve the problem of water shortage. Water withdrawal from the Lower-Vyia Reservoir, which is the main additional source in the recycling water supply system, is limited by the requirements of the Reservoir Use Rules. Thus, construction and commissioning of drainage facilities for collection and return of seepage discharge water through dams of the tailings dump is a necessary condition for the water recycling system operation.

Natural conditions and technogenic factors that change them collectively lead to formation of specific environmental situation (including geochemical) in the territories.

Association of the area with the ore province is the reason of high content of Fe, Ti, V, Cr, Mn, Co, Cu, Zn, Mo in natural waters, and this determines the natural hydrochemical background of the area.

As a technogenic factor in changing the geochemical background, the development of the deposit, accompanied by ore breaking and the formation of tailings of wet magnetic separation, activates transition of some of the elements of the geochemical association into soluble forms upon contact of natural waters and tailings.

Based on the research results, the content of ore elements in the composition of the tailings of wet magnetic separation is naturally lower than in ores; however, their content is significant. The chemical composition of tailings (Tables 2 and 3) is determined by the presence of predominant minerals. In particular, the presence of silica and calcium and magnesium oxides is explained by the abundance of CaMg[Si$_2$O$_6$] diopside among pyroxene grains of the mineral species corresponding in composition. Alumina is mainly present in plagioclases and minerals of the epidote group. Iron is contained in the epidote, amphiboles, goethite and pyrite. The rarest elements are vanadium (in titanomagnetite) and scandium (in pyroxenes). The presence of sulfur is associated with the presence of sulfides, which also contain non-ferrous metals (copper, arsenic, antimony, lead, zinc). Chromium is mainly contained in chromium-spinelids.

Wastewater generated as a result of contact with tailings, including filtration discharge water from dams, enters the environment, leading to contamination of natural waters. Regular observations carried out by EVRAZ KGOK record an increase in V content in the river waters below the tailings dump by 3–6 times compared to the background values. The results of the geoecological studies indicate that the concentrations in the river waters below the tailings dump are higher than the background values: Ti—1.7–9.1; V—4.6–6.0; Co—4.9–6.3; Mo—2.2–2.6 times, which occasionally leads to exceeding the permissible values established for the waters of watercourses of fishery importance. In the waters of the river below the tailings dump, an excess of the permissible level for the content of V was recorded—by 2.3–3.0 times.

Geo-ecological research revealed increased contents of Ti, V, Co, Mo in seepage discharge waters. Content of Ti in seepage discharge waters exceeds background values 1.1–6.4 times; Co—4.0–7.1 times; Mo—5.9–14.8 times; V—7.0–23.1 times.

The total amount of the substance entering the natural environment with seepage discharge waters during the year is given in Table 4. Values of seepage discharge volumes obtained using direct hydrometric measurements are taken for calculation. Calculation was made for each dam taking into account the volume of seepage discharge and concentration of contaminants in its waters with the subsequent summation.

In the total amount of the substance entering the natural environment, only a part of it is a direct contaminant, since in natural conditions these components are also contained in natural waters. Table 5 shows the amounts of pollutants that determine the technogenic component in the total amount of the incoming substance. The number of pollutants is calculated with the exception of the natural component due to the background content of components in natural waters.

Table 2. Chemical composition of tailings according to XRF, wt%.

Component	Sample No. 1	Sample No. 2	Sample No. 3
SiO_2	45.31	44.77	44.41
TiO_2	0.803	0.722	0.834
Al_2O_3	6.29	7.94	6.99
FeO	9.57	10.40	9.49
MnO	0.121	0.117	0.118
CaO	22.51	20.12	20.80
MgO	14.16	13.32	14.07
Na_2O	0.730	1.202	1.030
K_2O	0.045	0.072	0.051
P_2O_5	0.011	0.020	0.015
S	0.021	0.026	0.020
Sc	0.0114	0.0105	0.0111
Ba	0.0008	0.0089	0.0129
Cr	0.0058	0.0104	0.0057
V	0.0288	0.0320	0.0284
Sr	0.0095	0.0202	0.0148
Cu	0.0025	0.0027	0.0017
Zn	0.0040	0.0041	0.0039
Pb	0.0009	0.0010	0.0010
Ni	0.0038	0.0042	0.0037
W	0.0055	0.0050	0.0150
Co	0.0034	0.0034	0.0038
As	0.0010	0.0010	0.0008
Sb	0.0006	0.0009	0.0010
Y	0.0007	0.0007	0.0006
Sn	0.0005	0.0009	0.0009

Table 3. Chemical composition of the tailings according to ICP-MS (26 samples), mg/kg.

Contents	Cr	Co	Ni	Cu	Zn	As	Mo	Sn	Ba	W	Pb
Min	59.704	16.865	53.534	32.267	66.301	7.421	0.143	1.053	23.87	0.508	4.897
Max	146.628	68.029	87.961	78.692	141.059	17.064	1.681	4.416	415.25	5.127	84.424
Average	88.051	56.185	68.165	48.934	98.004	10.902	0.394	1.597	68.96	13.92	19.451

Table 4. Total amount of substances entering the natural environment with seepage discharge waters of dams.

Dams	2017					2026				
	Seepage Discharge Volume, Million m^3	Amount of Substance, kg				Seepage Discharge Volume, Million m^3	Amount of Substance, kg			
		Ti	V	Co	Mo		Ti	V	Co	Mo
№ 1	9.99	124.05	35.42	3.17	12.80	24.67	306.33	87.48	7.82	31.60
East dam	2.13	5.54	23.75	0.88	5.99	6.18	16.06	68.90	2.56	17.37
№ 2	17.38	45.61	203.02	9.28	55.98	28.52	74.84	333.14	15.23	91.86
№ 3	5.11	13.14	35.94	2.86	11.63	10.06	25.87	70.76	5.62	22.89
№ 4	6.24	13.83	60.56	3.52	13.47	10.07	22.32	97.73	5.68	21.73
Total amount of substances, kg	202.16	358.69	19.71	99.86		-	445.42	658.01	36.91	185.45

Table 5. Quantity of contaminants entering the natural environment with seepage discharge waters of dams caused by technogenic load.

Dams	2017						2026				
	Seepage Discharge Volume, Million m³	Contaminants Quantity, kg				Seepage Discharge Volume, Million m³	Contaminants Quantity, kg			
		Ti	V	Co	Mo		Ti	V	Co	Mo
№ 1	9.99	104.67	30.38	2.37	10.63	24.67	258.47	75.02	5.85	26.25
East dam	2.13	1.40	22.67	0.71	5.53	6.18	4.07	65.78	2.06	16.03
№ 2	17.38	11.89	194.24	7.89	52.21	28.52	19.51	318.74	12.95	85.67
№ 3	5.11	3.23	33.36	2.45	10.52	10.06	6.36	65.68	4.82	20.70
№ 4	6.24	1.72	57.41	3.02	12.11	10.07	2.78	92.64	4.87	19.55
Total amount of contaminants, kg		122.91	338.06	16.44	90.99	-	291.19	617.87	30.55	168.20

The ratio of natural and technogenic components in the total amount of incoming substance for each dam and in average in the whole seepage discharge volume is shown in Figures 2 and 3. The analysis of diagrams indicates the determining role of the technogenic factor in the content of Ti, V, Co, Mo in seepage discharge waters of dams. The share of technogenic component of V, Co and Mo in the total amount of incoming substance reaches 86–96%, 75–86% and 83–93%, respectively. The ratio of natural and technogenic component Ti in the total amount of incoming substance change in a wider range. In seepage discharge of the majority of dams, the share of technogenic component is small and makes 12–26%. Only in seepage discharge waters of dam No. 1, the share of technogenic component reaches 84% which, as a result in the total seepage discharge flow of all dams, leads to the dominating role of technogenic factor in Ti input into natural environment. The share of the technogenic component is 61%.

The amount of substance, the content of which is caused by technogenic factors, exceeds the natural content up to 5.4 times for Ti; 6.0–18.2 times for V; 3.0–6.1 times for Co; 4.9–13.8 times for Mo.

The forecasting of increase of seepage discharge volumes allows confirming the growth of pollutants entering the natural environment. Taking into account that the technological process of beneficiation does not change, the concentration of contaminants in seepage waters can be assumed to be constant when calculating. The volume of seepage discharge is accepted according to the results of the forecast for 2026 made using calculation method [33]. The results of calculations given in Tables 4 and 5 indicate 1.6–2.9 times growth of technogenic geochemical load for all analyzed components by 2026.

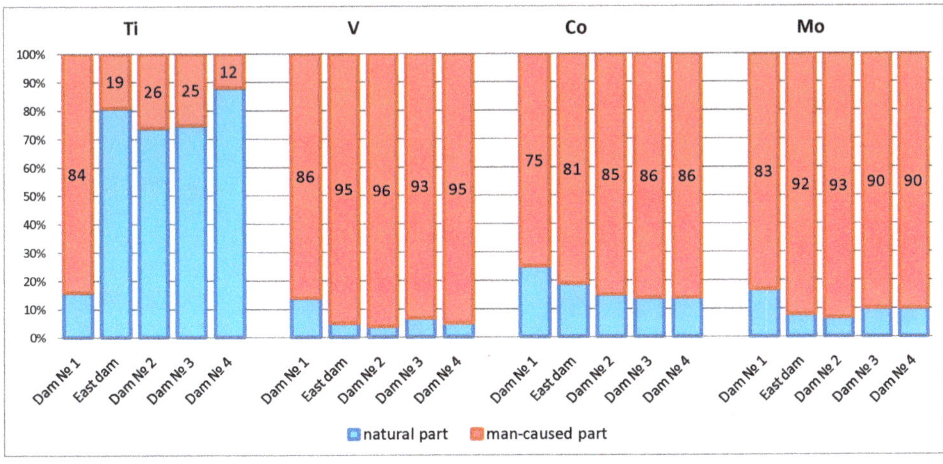

Figure 2. Ratio of natural and technogenic components in the total amount of substance entering the natural environment with seepage discharge waters of dams.

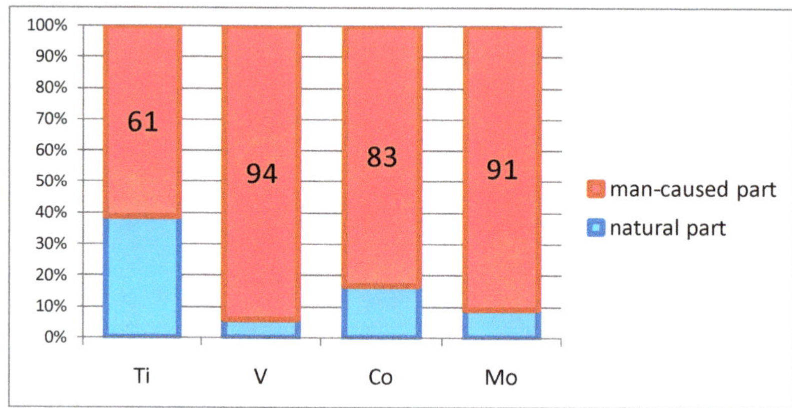

Figure 3. Average ratio of natural and technogenic components in the total amount of substance entering the natural environment with seepage discharge waters of dams.

Despite significant production volumes, the amount of waste and waste water produced by the enterprise has a relatively small negative impact on the environment. This is primarily due to the fact that the mined ores practically do not contain toxic substances that are typical, for example, for nonferrous metal ores, in particular: S, As, Cd, Hg and others. In the process of ore processing and beneficiation, chemical reagents are not used, and imported materials used in the production of agglomerate and pellets—bentonite clay, limestone, coke—are not toxic substances and constitute only 0.7% of the total amount of raw materials used. According to the results of a series of studies conducted by specialists of the Russian Academy of National Economy under the Government of the Russian Federation (Moscow) and employees of the Association "Medicine and Ecology" (Ekaterinburg), the production of EVRAZ KGOK is recognized as non-toxic and safe [40,41]. Nevertheless, one of the main priorities of its management is to improve the environmental efficiency of the enterprise, which can be achieved by managing those aspects of the activity that are associated with the most significant impact on the environment [40]. Ecological efficiency of the water balance management scheme as one of the priority aspects of ecologically directed activity consists of maximum use of recycled water and decrease in the volume of discharged wastewater, including seepage discharge water of dams, and is achieved by thickening the tailing pulp with return of clarified water for technological needs.

Issues related to management of water supply and disposal processes, thickening of stored waste in order to reduce risks of emergency situations at tailing dumps and negative environmental consequences of their operation are discussed in the papers [42–53].

New technological approaches aimed at solving technical, economic and environmental problems that arise during transportation and storage of waste and recycling water supply of beneficiation plants, appeared in the world practice of tailings storage in the middle of the 20th century [51,52]. Storage of tailings in the form of thickened pulp is based on dewatering of tailings material to a condition that does not divide into phases and fractions. Such storage allows using smaller plot of land in comparison with tailings dumps. Currently, almost all projects of tailings facilities in the world consider thickening to a state of paste. The technology of tailings thickening to the state of paste allows successfully storing them in a dump together with overburden rocks. Technological schemes for stacking paste-like tailings together with overburden rocks completely exclude construction of tailing dumps (sludge storage facilities) or allow reducing their vast territories and minimize operating costs [51].

For the first time in the Russian Federation hydrocyclones for thickening and improving the efficiency of tailing dumps were used at Achisai Polymetallic Plant. Subsequently, hydrocyclones were operated at the tailing dumps of Almalyk, Afrikanda, Mirgalimsay, Kantagi, Khaidarkan and other beneficiation plants. Overseas, application of hydrocyclones for tailing dumps reclamation began in the USA in 1950s. Currently, this technology is widely used in Canada, Africa, Chile, Brazil and South Africa.

In order to increase ecological efficiency, EVRAZ KGOK "Mekhanobr Engineering" JSC (AO) (St. Petersburg, Russia) has developed technical solutions for transition to the modern technology of storage of wet magnetic separation tailings and reconstruction of tailing facilities, providing engineering measures for tailings thickening, interception and maximum exclusion of seepage water entering the adjacent territory. When considering and justifying the tailings storage options, special attention was paid to improving the safety level of the tailings dump and minimizing the negative impact on the environment by using the method of preliminary thickening of pulp [23]. The fundamental possibility of obtaining paste for each type of tailings largely depends on the content of particles up to 20 μm in the pulp, which should be at least 20–30% of the total amount [51,54]. In the composition of the tailings of wet magnetic separation, the particle content of less than 0.074 mm reaches 75%, which proves the possibility of their thickening to paste-like state [51].

The technology ensures thickening of the initial tailing pulp of the beneficiation plant from 7–11 to 75% of solids content. The volume of water entering the tailing dump as part of wet magnetic separation decreases 40 times—to 12.19 million m^3/year [55], which is lower than the seepage discharge volumes of the tailing dumps currently observed. The water content of the tailings at 75% thickening is 0.33 m^3 per 1 ton of solid matter, i.e., approximately equal to the volume of pores in the stored tailings. Insignificant volumes of water discharged from the tailings go to the drainage system and further to the water recycling system. Implementation of measures on collection and accumulation of seepage water from the tailings dump with its subsequent use for production needs ensures reduction of geochemical load on natural waters of adjacent territories.

Positive experience of tailings thickening has been obtained, for example, at the Uchalinsky mining and processing plant, which extracts and processes chalcopyrite to produce copper and zinc concentrates. Outotec's paste thickening technology enabled the enterprise to reduce the volume of water in the tailings and increase the solid content from 20% to 70–72%. The solution makes it possible to efficiently extract process water for reuse. It reduces the area occupied by the tailing dump and the risk of environmental pollution. It also allows effective return of process water and reduces power consumption for pulp and water pumping. Thickened tailings mixed with cement can be used for backfill in a used pit. The possibility of placing thickened tailings in the worked-out area reduces the volume of waste stored in the tailings dump, which makes it possible to increase the operating life of the tailings dump from 20 (with the traditional method of tailings storage) to 50 years [56].

Application of the technology of dewatering and thickening of the tailings of wet magnetic separation, creation of closed water supply systems and reduction of wastewater discharge, including water from seepage discharge of dams, provides ecological efficiency of the enterprise's water balance management. As a result, the technogenic load on the environment is considerably reduced, natural water pollution is reduced, and the problem of water shortage for technological needs is resolved. The forecasted ecological efficiency of the implementation of technical solutions is confirmed by the results of the EVRAZ KGOK health risk assessment. The assessment showed that the values of hazard indices calculated for priority pollutants do not exceed 1, which corresponds to the level of acceptable risk. The values of individual carcinogenic risk do not exceed one additional case per 1 million exposed persons, which corresponds to a negligible low level of risk that does not differ from ordinary daily risks [40].

5. Conclusions

The conducted research and obtained results show the necessity to control seepage discharge and to manage the water balance of the enterprise.

Seepage discharge at the base of the tailings dump dams is the main component of the water balance output, accounting for 66% of its total volume.

The forecast of seepage discharge for the period until 2026 shows the increase of seepage water volumes in the course of the tailings dump operation and the increase of the dam height. Balance calculations showed that irrecoverable losses as a result of seepage discharge through the dams will lead to annually increasing water deficit in the recycling water supply system of the enterprise.

The seepage waters of the dams bring pollutants into the natural environment and change the natural geochemical situation in the areas adjacent to the tailings dump. Technogenic factor is the determining factor in the inflow of Ti, V, Co, Mo into the natural environment with seepage water.

Increased seepage discharge will result in technogenic geochemical load increase on the territory's natural complexes.

Ecologically efficient management of the enterprise's water balance can be ensured by applying dewatering and thickening of tailings with wet magnetic separation, creating closed water supply systems and reducing wastewater discharge, including water from seepage discharge of dams. As a result, the technogenic geochemical load on the environment is significantly reduced and the problem of water shortage used in the technological process is solved.

Author Contributions: Conceptualization, T.K., E.M. and V.F.; data curation, V.F.; formal analysis, V.F.; funding acquisition, S.B.; investigation, P.B. and S.V.; methodology, V.F. and T.K.; project administration, E.M. and S.B.; resources, E.M. and S.B.; software, V.F.; supervision, E.M.; validation, V.F. and E.M.; visualization, T.K.; writing—original draft, T.K., E.M. and P.B.; writing—review and editing, E.M., V.F., T.K. and P.B. All authors have read and agreed to the published version of the manuscript.

Funding: The publication has been prepared as part of the implementation of the Program of activities of the world-class scientific and educational center «Rational Subsoil Use» for 2019–2024 with the financial support of the Ministry of Education and Science of the Russian Federation (Order of the Government of the Russian Federation No. 537 dated 30 April 2019).

Conflicts of Interest: The authors declare that there is no conflict of interest.

References

1. Kossoff, D.; Dubbin, W.E.; Alfredsson, M.; Edwards, S.J.; Macklin, M.G.; Hudson-Edwards, K.A. Mine tailings dams: Characteristics, failure, environmental impacts, and remediation. *Appl. Geochem.* **2014**, *51*, 229–245. [CrossRef]

2. Vick, S.G. *Planning, Design and Analysis of Tailings Dams*, 2nd ed.; BiTech Publishers Ltd.: Vancouver, BC, Canada, 1990; pp. 69–91. [CrossRef]

3. *Design and Evaluation of Tailings Dams*; U.S. Environmental Protection Agency, Office of Solid Waste: Washington, DC, USA, 1994; pp. 4–40.

4. Mittal, H.K.; Morgenstern, N.R. Seepage Control in Tailings Dams. *Can. Geotech. J.* **1976**, *13*, 277–293. [CrossRef]

5. Klohn, E.J. Seepage Control for Tailings Dams. In *Mine Drain, Proceedings of the Int Mine Drain Symp, Denver, CO, USA, 1 May 1979*; Miller Freeman Publications: San Francisco, CA, USA, 1979; pp. 671–672.

6. Hu, S.; Chen, Y.; Liu, W.; Zhou, S.; Hu, R. Effect of seepage control on stability of a tailings dam during its staged construction with a stepwise-coupled hydro-mechanical model. *Int. J. Min. Reclam. Environ.* **2015**, *29*, 125–140. [CrossRef]

7. Aboelela, M.M. Control of seepage through earth dams based on pervious foundation using toe drainage systems. *J. Water Resour. Prot.* **2016**, *8*, 1158–1174. [CrossRef]

8. Cheng, P.; Chen, S.; Chen, W. The Finite Element Analysis of Tailings Dam Seepage. In Proceedings of the 2016 2nd International Conference on Architectural, Civil and Hydraulics Engineering (ICACHE 2016), Kunming, China, 29–30 September 2016. [CrossRef]

9. Naeini, M.; Akhtarpour, A. A numerical investigation on hydro-mechanical behaviour of a high centreline tailings dam. *J. S. Afr. Inst. Civ. Eng.* **2018**, *60*, 49–60. [CrossRef]

10. Shen, L.; Luo, S.; Zeng, X.; Wang, H. Review on anti-seepage technology development of tailings pond in China. In Proceedings of the 1st International Symposium on Mine Safety Science and Engineering (ISMSSE), Beijing, China, 26–29 October 2011; pp. 1803–1809. [CrossRef]

11. Lyu, Z.; Chai, J.; Xu, Z.; Qin, Y.; Cao, J. A Comprehensive Review on Reasons for Tailings Dam Failures Based on Case History. *Adv. Civil. Eng.* **2019**, *2019*, 4159306. [CrossRef]

12. Mironenko, V.A.; Rumynin, V.G. *Problems of Hydrogeoecology*, 1st ed.; Publishing House of the Moscow State Mining University: Moscow, Russia, 1999; Volume 3, pp. 205–284. (In Russian)

13. Yanin, Y.P. Environmental impact assessment of Russian iron ore deposits development. *Ecol. Expert.* **2019**, *5*, 2–94. (In Russian) [CrossRef]

14. Tselyuk, D.I.; Tselyuk, I.N. Complex assessment of technogenic influence of alluvial tailing dumps of iron ore objects of Eastern Siberia on environment. *Bull. KSPU Named After V P Astafiev* **2012**, *3*, 328–331. (In Russian)

15. Yarg, L.A.; Zhitinskaya, O.M. Information basis for the optimal exploitation of the natural-technical systems «Ore deposits». *Izv. Vyssh. Uchebn. Zaved. Geol. Razved.* **2017**, *5*, 78–81. (In Russian) [CrossRef]

16. Pashkevich, M.A.; Ponurova, I.K. Geoecological features of anthropogenic pollution of natural ecosystems of the zone of influence of the tailing dumps of Mikhailovsky GOK. *Gorn. Inf. Analyt. Bull.* **2006**, *5*, 349–356. (In Russian)

17. Kaidanova, O.V.; Zamotaev, I.V.; Suslova, S.B.; Shilkrot, G.S. Geochemical transformation of landscapes of Zheleznogorsk industrial area (Kursk oblast). *Izv. Ross. Akad. Nauk Ser. Geogr.* **2018**, *3*, 91–104. [CrossRef]

18. Semyachkov, A.I.; Medvedev, O.A.; Pochechun, V.A.; Arkhipov, M.V. Ecological assessment of the conditions of geological environment based on the engineer ecological and engineer geological tests. *Izv. Vyssh. Uchebn. Zaved. Gornyi Zhurnal* **2012**, *2*, 80–84. (In Russian)

19. EVRAZ KGOK Official Website. Available online: http://rus.evraz.com (accessed on 25 November 2020).

20. Vlokh, Y.V. Development prospects for Kachkanar Mining and Processing Integrated Works. *Gornyi Zhurnal* **2016**, *7*, 46–50. (In Russian) [CrossRef]

21. *State Report on the Status and Protection of the Environment of the Sverdlovsk Region*; U.S. Ministry of Natural Resources and Environment of Sverdlovsk Region: Yekaterinburg, Russia, 2017; pp. 20–43.

22. Deryagina, S.E.; Astaf'eva, O.V.; Medvedev, A.N. State and prospects of changing the environmental impact of JSC "Evraz Kachkanarsky ore-mining and processing plant" during the Sobstvenno-Kachkanarsky deposit development. *Econ. Prir.* **2013**, *1*, 24–30. (In Russian)

23. Kuznetsov, A.G.; Lytin, O.V.; Karelin, A.E. Tailing dump of PJSC "EVRAZ KGOK" and its development prospects. *Gornyi Zhurnal* **2013**, *9*, 17–19. (In Russian)

24. Fetisov, V.V.; Menshikova, E.A. Evaluation and Prediction of Seepage Discharge Through Tailings Dams When Their Height Increases. In *Applied Geology*; De Maio, M., Tiwari, A.K., Eds.; Springer: Cham, Switzerland, 2020; pp. 21–41. [CrossRef]

25. La Touche, G.D.; Garrick, H. Hydraulic performance of liners in tailings management and heap leach facilities. In Proceedings of the International Mine Water Association Symposium, Bunbury, Australia, 30 September–4 October 2012; pp. 371–378.

26. Kealy, C.D.; Busch, R.A. *Determining Seepage Characteristics of Mill-Tailing Dams by the Finite-Element Method*; Report of investigations/United States Department of the Interior, Bureau of Mines: Washington, DC, USA, 1971; 113p.

27. Cowherd, D.C.; Miller, K.C.; Perlea, V.G. Seepage Through Mine Tailings Dams. In Proceedings of the Third International Conference on Case Histories in Geotechnical Engineering, St. Louis, MI, USA, 1–4 June 1993; pp. 463–468.

28. Rykaart, M.; Fredlund, M.; Stianson, J. Solving tailings impoundment water balance problems with 3-D seepage software. *Geotech. News* **2001**, *19*, 50–55.

29. Chong, L.; Zhenzhong, S.; Lei, G.; Liqun, X.; Kailai, Z.; Tian, J. The Seepage and Stability Performance Assessment of a New Drainage System to Increase the Height of a Tailings Dam. *Appl. Sci.* **2018**, *8*, 1840. [CrossRef]

30. Yuan, L.; Lei, J. The Analysis of the Seepage Characteristics of Tailing Dams Based on FLAC^{3D} Numerical Simulation. *Open Civil Eng. J.* **2015**, *9*, 400–407. [CrossRef]

31. Adajar, M.A.Q.; Zarco, M.A.H. An empirical model for predicting hydraulic conductivity of mine tailings. *Int. J. Geomate* **2014**, *7*, 1054–1061. [CrossRef]

32. Aubertin, M.; Bussière, B.; Chapuis, R.P. Hydraulic conductivity of homogenized tailings from hard rock mines. *Can. Geotech. J.* **1996**, *33*, 470–482. [CrossRef]

33. Nedriga, V.P. *Gidrotekhnicheskie Sooruzheniya (Hydrotechnical Constructions)*; Stroyizdat: Moscow, Russia, 1983; pp. 64–152. (In Russian)

34. Rozanov, N.P.; Bochkarev, J.W.; Lapshenkov, V.S. *Gidrotekhnicheskie Sooruzheniya (Hydrotechnical Constructions)*; Agropromizdat: Moscow, Russia, 1985; pp. 17–33. (In Russian)

35. Vukovic, M.; Soro, A. *Determination of Hydraulic Conductivity of Porous Media from Grainsize Composition*; Water Resources Publications: Littleton, CO, USA, 1992.

36. Odong, J. Evaluation of empirical formulae for determination of hydraulic conductivity based on grain-size analysis. *J. Am. Sci.* **2007**, *3*, 54–60.

37. Rosas, J.; Jadoon, K.Z.; Missimer, T.M. New empirical relationship between grain size distribution and hydraulic conductivity for ephemeral streambed sediments. *Environ. Earth Sci.* **2015**, *73*, 1303–1315. [CrossRef]

38. Hazen, A. *Some Physical Properties of Sands and Gravels, with Special Reference to Their Use in Filtration*; Massachusetts State Board of Health: Boston, MA, USA, 1892.

39. Beyer, W. Zur Bestimmung der Wasserdurchlässigkeit von Kiesen und Sanden aus der Kornverteilungskurve. *Wasserwirtschaft-Wassertechnik* **1964**, *14*, 165–168.

40. Medvedev, A.N.; Astafieva, O.V.; Deryagina, S.E. Introduction of modern technologies in mining industry as a way of environmental impact reduction (JSC "Vanadium"). *Reg. Environ. Issues* **2014**, *4*, 233–237. (In Russian)

41. Isakov, O.V.; Kuchin, V.I. Nature protection activitiy of "EVRAZ KGOK" JSC. *Gornyi Zhurnal* **2013**, *9*, 27–30. (In Russian)

42. East, D.; Fernandez, R. Managing Water to Minimize Risk in Tailings Storage Facility Design, Construction, and Operation. *Mine Water Environ.* **2020**. [CrossRef]

43. Edraki, M.; Baumgartl, T.; Manlapig, E.; Bradshaw, D.; Franks, D.M.; Moran, C.J. Designing mine tailings for better environmental, social and economic outcomes: A review of alternative approaches. *J. Clean. Prod.* **2014**, *8*, 411–420. [CrossRef]

44. Yin, S.; Shao, Y.; Wu, A.; Wang, H.; Liu, X.; Wang, Y. A systematic review of paste technology in metal mines for cleaner production in China. *J. Clean. Prod.* **2020**, *247*, 119590. [CrossRef]

45. Wang, C.; Harbottle, D.; Liu, Q.; Xu, Z. Current state of fine mineral tailings treatment: A critical review on theory and practice. *Miner. Eng.* **2014**, *58*, 113–131. [CrossRef]

46. Fourie, A. Preventing catastrophic failures and mitigating environmental impacts of tailings storage facilities. In Proceedings of the 6th International Conference on Mining Science and Technology (ICMST 2009), Xuzhou, China, 18–20 October 2009; Volume 1, pp. 1067–1071. [CrossRef]

47. ScienceDirect. The benefits of dewatering tailings. *World Pumps* **2020**, *2*, 26–27. [CrossRef]

48. Adiansyah, J.S.; Rosano, M.; Vink, S.; Keir, G. A framework for a sustainable approach to mine tailings management: Disposal strategies. *J. Clean. Prod.* **2015**, *108*, 1050–1062. [CrossRef]

49. Silin, V.A. Creation of the Kupol ore mine tailings dum effective management system. *Ratsional'noye Osvoyeniye Nedr* **2018**, *1*, 66–71. (In Russian)

50. Baranov, V.F. Final tailings thickening and disposal systems. (world practice review). *Obogashchenie Rud.* **2013**, *3*, 43–48. (In Russian)

51. Kislyakov, V.; Nikitin, A.; Shershnev, A. On the possibility of paste thickening of the concentrating processing tailings. *Mine Surv. Subsurf. Use* **2012**, *4*, 21–24. (In Russian)

52. Kibirev, V.I. Tailings slurry thickening—A step to "green" technology of tailings disposal. *Obogashchenie Rud* **2010**, *6*, 44–48. (In Russian)

53. Algebraistova, N.K.; Alekseeva, E.A.; Kolesnikova, T.A. On intensification of the tailings thickening process (Ob intensifikatsii protsessa sgushcheniya khvostov). *Gorn. Inf. Analyt. Bull.* **2004**, *5*, 323–327. (In Russian)

54. Vinogorodsky, E.B. Technical and economic assessment of mineral processing tailings slurries thickening. *Obogashchenie Rud.* **2010**, *6*, 39–43. (In Russian)

55. Karavaeva, T.; Ushakova, E.; Shchukova, I. Disposal of Open-Pit and Wastewaters of EVRAZ KGOK. In Proceedings of the International Mine Water Association Symposium, Perm, Russia, 15–19 July 2019; pp. 297–301.

56. OUTOTEC®. Tailings paste thickening complex at Uchalinsky GOK reduced technogenic impact on the environment (Kompleks pastovogo sgushcheniya khvostov OUTOTEC® na Uchalinskom GOKe snizil tekhnogennuyu nagruzku na okruzhayushchuyu sredu). *Gorn. Promyshlennost* **2020**, *1*, 75–76. (In Russian)

Publisher's Note: MDPI stays neutral with regard to jurisdictional claims in published maps and institutional affiliations.

Article

Elemental Concentrations of Major and Trace Elements in the Spring Waters of the Arctic Region of Russia

Andrey I. Novikov [1,*], Anna A. Shirokaya [1] and Marina V. Slukovskaya [1,2,3]

1   I.V. Tananaev Institute of Chemistry and Technology of Rare Elements and Mineral Raw Materials, Kola Science Centre, Russian Academy of Sciences, 184209 Apatity, Russia; a.shirokaia@ksc.ru (A.A.S.); m.slukovskaya@ksc.ru (M.V.S.)
2   Laboratory of Nature-Inspired Technologies and Environmental Safety of the Arctic Region, Kola Science Centre, Russian Academy of Sciences, 184209 Apatity, Russia
3   Agrarian Technological Institute, Peoples' Friendship University of Russia (RUDN University), 117198 Moscow, Russia
*   Correspondence: a.novikov@ksc.ru

Abstract: The Arctic region of Russia is rich with natural water resources. Some residents of this area prefer to use water from spring sources instead of tap water. However, the elemental composition for most of the springs is unknown, making it very important to regularly update water quality data. In this paper, the chemical composition of 24 natural springs near large cities of the Murmansk region were identified via titration, potentiometry, and mass spectrometry analysis with the low detection limits. The concentrations of a considerable number of micro-components have been determined for the first time. Concentrations of some hazardous pollutants have been found in spring water that exceed Russian hygienic limits by 2.5 times. In terms of chemical indicators, it was shown that in accordance with European and national water standards, only half of the tested natural water sources in the Murmansk region can be used for drinking water. The preferential spring was recommended near each major city in the region according to a set of chemical parameters. The results of chemical composition of the spring waters were disseminated publicly by placing the tags with QR codes with the link to the cloud storage near each studied spring.

Keywords: natural water; water quality; Arctic region; elemental composition; springs; QR code

Citation: Novikov, A.I.; Shirokaya, A.A.; Slukovskaya, M.V. Elemental Concentrations of Major and Trace Elements in the Spring Waters of the Arctic Region of Russia. *Minerals* **2022**, *12*, 8. https://doi.org/10.3390/min12010008

Academic Editors: Ana Romero-Freire and Hao Qiu

Received: 30 November 2021
Accepted: 17 December 2021
Published: 22 December 2021

Publisher's Note: MDPI stays neutral with regard to jurisdictional claims in published maps and institutional affiliations.

1. Introduction

According to the World Health Organization (WHO), the main factors that affect human health are living conditions and habits (52–53%), genetic factors (20%), state of the environment (20%), and medical support (7–8%). Human health is closely related to drinking water quality [1–4]. Concentrations of chemical elements contained in drinking water must meet hygienic standards to maintain a water and salt balance for body homeostasis and prevent possible toxic effects [5]. Excessive concentrations of chemical elements in water can cause health problems due to both acute and chronic (months and years) toxic effects. For example, nitrates are constantly present in the human body in small amounts, but problems associated with the formation of methemoglobin, which is unable to carry oxygen, may develop with an increase in the concentration of nitrates, which leads to oxygen starvation [6–9]. This is extremely dangerous when experiencing symptoms from diseases such as COVID-19 [10–12]. A particular danger occurs when high quantities of potentially toxic elements, the effects of which do not appear immediately, are contained in water, because the diseases they cause are chronic [2,13–16].

Most trace elements are potentially toxic to the human body [17] in high concentrations, whereas some of them are essential for humans in small amounts as microelements (Table 1).

Table 1. The physiological functions of certain chemical elements on the human body.

Element	Function	Reference
Ca	Included in the skeleton and teeth in the form of phosphates; participates in the processes of blood clotting and muscle and neuronal reactions	[18,19]
Cr	Essential for the metabolism of sugars and fats	[17,20,21]
Mn	Part of enzymes, an enzyme activator, important in antioxidant action	[17,22]
Fe	Contained in hemoglobin in the enzymes peroxidase, catalase, and cytochrome oxidase; participates in the transformation of cellular energy	[17,23]
Co	A cofactor of several enzymes, such as vitamin B12, which controls the production of red blood cells	[17,24]
Ni	A cofactor in certain enzyme metal functions	[25]
Cu, Zn	Participates in the composition of metalloenzymes	[17,26,27]
Se	Part of the selenocysteine and selenomethionine amino acids	[28]
Mo	A cofactor of enzymes that affect the oxidation of purines and aldehydes, protein synthesis, and the metabolism of certain nutrients	[17,29]
Cd	Probably a metalohormone	[17,30]

The growing pollution of freshwater systems through emissions from industrial enterprises is one of the key environmental problems facing humanity around the world [31,32]. Even though most pollutants are present in low concentrations, they can accumulate in the body and cause serious toxicological problems [33–35]. About 884 million people worldwide do not have basic access to sanitary drinking water sources [36]. Four billion people experience severe water shortages for at least one month per year [37]. Back in 2010, the journal *Nature* showed that almost 80% of the world's population is threatened by water pollution [38]. There is a growing distrust of tap water, so people in cities increasingly choose bottled or spring water for drinking [17,39].

People living in the extreme natural conditions of the Arctic regions experience constant high stress on their health due to factors such as the polar night and polar day, the unstable geomagnetic field, the long winter period, and the small number of sunny days throughout the year [40]. In this regard, issues related to the health of the population in such regions are extremely relevant.

The city of Murmansk, with a population of over 280,000 people, is the largest city in the world above the Arctic Circle [41]. It is located on the Kola Peninsula along the coast of the Barents Sea, which is part of the Arctic Ocean. The intensive development of the region, which has a population of more than 700,000 people, refused from the discovery of numerous deposits of mineral resources, which have been actively developed since the first half of the 20th century. Their extraction, processing, and transportation has a constant negative impact on the environment, including pollution of both soils [42,43] and water bodies [4,44–52]. At the same time, the territory of the peninsula is marked on the map "Global geography of incident threat to human water security and biodiversity" as an area with a low risk of possible pollution. However, the groundwaters of the region are characterized by low protection and can be easily polluted [50]. Therefore, using this groundwater for drinking purposes, without proper controls, can be dangerous to human health. The population is only fragmentarily informed about the quality of spring waters The purpose of this work was to study and make public the chemical composition of the most popular spring waters in the Murmansk region, Russia (Kola peninsula, Russian Arctic zone). The study included the following steps: (i) to sample spring water and to determine the content of both macro-components and previously obscure micro-components, including potentially toxic elements with a low detection limit; (ii) to determine the water quality by comparison the results with international [9,53,54] and Russian national standards [55,56];

(iii) to make the results publicly available by placing the tags with QR codes (with the link to the cloud storage with data set) near each studied spring.

2. Materials and Methods

2.1. Study Area and Sampling

The Murmansk Region is in the northwest of the Russian Federation. It borders Finland and Norway in the west and northwest and is washed by the White and Barents Seas. The Murmansk region's total area is 144,900 km^2. The greatest length from west to east is about 550 km, from north to south, 400 km. Almost the entire territory lies north of the Arctic Circle and is located on the Kola Peninsula. Only the western and southwestern parts of the region connect to the mainland. The Murmansk region is one of the most lacustrine and riverine areas in Russia. The density of the river network is significant; some rivers have lengths of over 200 km. The climate of the Murmansk region is arctic-temperate and maritime, but it is influenced by the warm North Atlantic current, making it relatively mild. In the north of the peninsula the average temperature in winter is -14 °C, in summer, $+14$ °C. In the center and south of the peninsula during the winter months, frosts reach -40 to -50 °C. The average annual rainfall is about 400 mm [57].

The Murmansk region is one of the most urbanized regions of Russia; the population as of 1 January 2021, was 732,000 people. The largest cities in the region are Murmansk (282,851), Apatity (53,847), Monchegorsk (40,675), Kandalaksha (29,750), Kirovsk (25,944), and Olenegorsk (19,887) [39]. In this study, the elemental composition in the water of the most popular springs near or within the territory of large cities in the region was studied (Figure 1). All springs are easily accessible and used daily by the local citizens. The spring map is available by link [58] or by QR code (Figure 1c).

(a) (b) (c)

Figure 1. (a,b) Sampling points within the Murmansk region, Russia; (c) QR code for access to google map with marked selection points.

Overall, 24 springs were examined from 19 September to 4 October 2021 (Table 2).

Table 2. Coordinates, type and location of the investigated springs, air and water temperature, pH and flow rate at the time of sampling.

№	Type	Location	Coordinates	T Air, °C	T Water, °C	Yield of Water, L/s
		Murmansk and Kola district (Figure 2a)				
1	Descending	On the city outskirts, 50 m east of the Chapel	68°58′19.2″ N 33°07′10.7″ E	9.0	5.6	*
2	Descending	200 m west of the roads' junction to Abram-Mys, Murmansk and Nikel	68°58′42.4″ N 32°59′10.9″ E	7.0	6.1	0.04

Table 2. *Cont.*

№	Type	Location	Coordinates	T Air, °C	T Water, °C	Yield of Water, L/s
3	Descending	Between 24 and 25 km of the Murmansk-Verkhnetulomsky highway, on the right, 15 m north of the road	68°47′18.6″ N 32°31′01.3″ E	8.7	2.9	0.27
4	Ascending	On the village outskirts, 150 m north of the Murmansk-Verkhnetulomsky highway	68°49′53.3″ N 32°45′15.7″ E	6.8	3.1	0.65
5	Descending	Between 2 and 3 km of the Kola-Zverosovkhoz highway, 50 m west of the road	68°51′43.4″ N 33°05′01.0″ E	5.0	4.3	0.10
6	Descending	Between 7 and 8 km of the upper airport street, 10 m north of the road	68°48′55.4″ N 32°57′53.2″ E	4.2	3.3	0.11
7	Descending	In the lowland, 40 m east of residential buildings	68°48′43.4″ N 32°48′50.9″ E	3.9	3.0	0.21
8	Descending	Between 7 and 8 km of the ring road around Murmansk, 50 m west of the road bridge	68°53′46.0″ N 33°07′23.5″ E	2.2	4.9	0.05
Monchegorsk district (Figure 2b)						
9	Descending	Southeastern city outskirts, 150 m southeast of the sports school building	67°55′20.9″ N 32°57′28.6″ E	6.9	5.0	0.33
10	Self-flowing well	8 km of the Monchegorsk-Rizh-Guba highway, 200 m north-east of the bus stop	67°52′59.8″ N 33°02′53.8″ E	7.6	5.1	0.40
11	Self-flowing well	8 km of the Monchegorsk-Rizh-Guba highway, 250 m north-east of the bus stop	67°53′01.3″ N 33°02′59.5″ E	8.2	2.9	0.35
12	Ascending	Between 1271 and 1272 km of the E105 highway, 50 m southeast of the highway	68°01′02.7″ N 32°57′19.1″ E	7.4	2.6	**
13	Self-flowing exploration well	250 m east of the roads' junction to the Laplandia station along the E105 highway	68°16′09.4″ N 33°20′28.8″ E	9.2	3.6	0.20
14	Ascending	Near 1253 km of the E105 highway, 120 m south-west of the highway	67°52′22.7″ N 32°47′13.4″ E	6.3	3.8	**
Kandalaksha district (Figure 2c)						
15	Descending	Between 1179 and 1180 km of the E105 highway, 5 m north-west of the highway	67°23′42.8″ N 32°28′54.8″ E	12.2	4.1	0.018
16	Descending	The near-channel part of the left bank of the Niva River, 100 m downstream of the bridge	67°11′15.2″ N 32°28′11.8″ E	11.1	4.4	0.020
17	Descending	Between 6 and 7 km of the Kandalaksha-Umba highway, 5 m north of the highway	67°07′28.6″ N 32°32′36.6″ E	13.2	4.0	0.14
18	Ascending	Between 71 and 72 km of the Kandalaksha-Umba highway, 15 m east of the highway	66°55′58.0″ N 33°49′02.4″ E	12.5	5.4	**
19	Descending	The territory of the farm fellowship	66°42′38.4″ N 34°21′26.8″ E	11.7	3.8	0.26
Apatity-Kirovsk district (Figure 2c)						
20	Self-flowing exploration well	150 m south-west of the nunnery in the foothills of the Khibiny mountains	67°38′11.1″ N 33°43′20.8″ E	12.3	3.4	**
21	Self-flowing exploration well	25 m west of the road in the foothills of the Khibiny mountains	67°38′10.2″ N 33°42′45.3″ E	11.5	3.4	**
22	Mixed	Between 8 and 9 km of the Apatity-Kirovsk highway, 30 m south of the highway	67°34′56.4″ N 33°33′26.9″ E	11.8	3.3	**

Table 2. *Cont.*

№	Type	Location	Coordinates	T Air, °C	T Water, °C	Yield of Water, L/s
23	Self-flowing well	The eastern outskirts of Apatity, in the area of a garage cooperative	67°33′12.5″ N 33°26′09.9″ E	11.1	3.7	0.27
24	Ascending	Southeast of the city of Apatity, between 7 and 8 km of the Apatity-Khibiny airport highway, 1.5 km to the left of the road	67°29′55.6″ N 33°29′17.3″ E	10.0	8.4	**

* it was impossible to determine the flow rate, water accumulates in the well; ** it was impossible to determine the flow rate, too large a head or several flows.

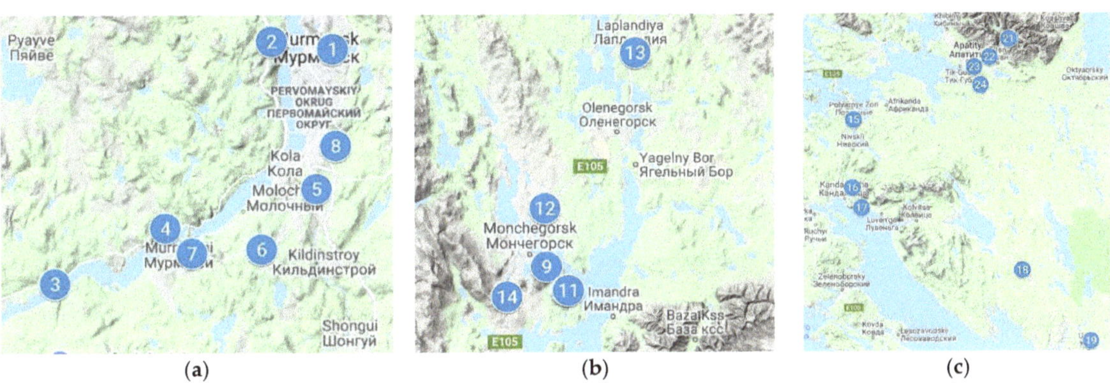

| (a) | (b) | (c) |

Figure 2. Sampling points. (**a**) Murmansk and Kola district; (**b**) Monchegorsk district; (**c**) Kandalaksha district and Apatity-Kirovsk district.

The appearance of various types of springs is shown in Figure 3.

| (a) | (b) | (c) |

Figure 3. Types of springs: descending (**a**), ascending (**b**) and self-flowing well (**c**).

2.2. Sample Collection and Analysis

Water samples for the element analysis were collected into disposable plastic test-tubes with a volume of 50 mL ($n = 2$) and preserved with 0.5 mL nitric acid (puriss. spec. 27-5, Russian government standard GOST 11125-84). For the determination of the ionic compositions, the probes were collected and loaded into pre-prepared disposable plastic bottles with a volume of 1 L ($n = 2$). All collected samples were stored in a portable refrigerator and transported in an automobile at a constant temperature of 5 °C.

The waters were analyzed by a mass spectrometer, with inductively coupled plasma (Perkin Elmer ELAN 9000 DRC-e, Waltham, MA, USA). Multielement calibration solutions from Inorganic Ventures (Christiansburg, VA, USA) (IV-STOCK-29, IV-STOCK-21, IV-STOCK-28) were used for instrument calibration. Standard reference materials were used for quality control: V-STOCK-1643 Trace Elements in Fresh Water (High Purity Standards, Charleston, SC, USA). The pH values of the solutions and the concentration of chloride- and nitrate-ions were determined by the potentiometric method using the ionomer EXPERT-001 (Econix-Expert, Moscow, Russia), which was equipped with two ion-selective electrodes—a nitrate-selective electrode (ELIT-021, Nico Analit, Moscow, Russia) and a chloride-selective electrode (ECOM-Cl, Econix, Moscow, Russia)—in combination with the reference electrode ESr-10101 (Izmeritelnaya tekhnika, Moscow, Russia). As a buffer solution for the determination of nitrate-ions, a potassium alum of 1% and a phosphate buffered saline was used; for the determination of chloride-ions, a 1 M solution of KNO_3 was used. To control the quality of the measurements, a standard sample of nitrate ions GSO 6696-93, chloride ions GSO 6687-93, and GSO composition of water 7886-2001 were used. The hydrocarbonate content in the water was determined by the titrimetric method, and a standard sample GSO 8403-2003 of the hydrocarbonate solution was used for checking the measurement quality. Water and air temperature was measured using a HI 145 thermometer (Hanna Instruments, Woonsocket, RA, USA). The pH was determined directly during sample collecting using a TDS-3 temperature sensor to consider the thermal compensation effect.

3. Results and Discussion

Tables 3 and 4 summarize the results of the analysis of 24 samples collected for the determination of macro- and micro-components, respectively. The standards of the World Health Organization (WHO) [9], United States Environmental Protection Agency (USEPA) [54], European Commission (EC) [53], and Russian Sanitary and Epidemiological Rules and Norms (SanPiN) 1.2.3685-21 (Hygienic standards and requirements for ensuring the safety and (or) harmlessness to humans of environmental factors) [56] and 2.1.4.1116-02 (Drinking water. Hygienic requirements for the quality of bottled water. Quality control) [55] are provided for comparison. It should be noted that SanPiN 2.1.4.1116-02 establishes a higher standard for maximum permissible concentration (MPC) for drinking bottled water of the first and highest categories.

Table 3. Content of macro components, anions, pH, and total hardness in spring water.

	pH	Hardness, mmol/L	HCO^{3-}, mg/L	NO^{3-}, mg/L	Cl^-, mg/L	Na, mg/L	Mg, mg/L	K, mg/L	Ca, mg/L
USEPA	6.5–8.5	-	-	44	250	-	-	-	-
EC	6.5–8.5	-	-	50	250	-	-	12	-
WOH	6.5–8.5	1,2	30–400	50	250	-	-	-	-
MPC *	6–9	5	-	45	350	200	50	-	-
First category **	6.5–8.5	<3.5	<400	<20	<250	<200	<65	<20	<130
Highest category **	6.5–8.5	0.75–3.5	30–400	<5	<150	<20	5–50	2–20	25–80
1	5.28	0.16	<10	0.97	<12	4.2	1.4	0.6	3.8
2	5.99	0.16	18.61	0.47	<12	4.0	1.4	0.7	4.1
3	6.28	0.16	23.92	0.58	<12	3.0	1.3	0.9	4.3

Table 3. *Cont.*

	pH	Hardness, mmol/L	HCO_3^-, mg/L	NO_3^-, mg/L	Cl^-, mg/L	Na, mg/L	Mg, mg/L	K, mg/L	Ca, mg/L
4	6.91	0.47	68.40	1.09	<12	5.3	4.1	2.6	12.0
5	6.02	0.29	20.87	0.52	<12	4.4	2.3	0.9	7.8
6	5.88	0.15	15.29	0.65	<12	2.5	1.4	0.8	3.6
7	7.40	0.43	47.87	9.17	<12	5.1	4.2	2.0	10.4
8	6.28	0.25	20.87	0.88	35.5	21.8	2.0	1.6	6.6
9	7.07	0.44	44.27	0.54	<12	3.0	4.3	1.1	10.6
10	7.59	1.51	69.15	2.14	<12	3.3	5.0	2.3	52.1
11	7.42	1.22	71.33	0.81	<12	3.4	4.5	1.9	41.3
12	7.03	0.27	22.94	0.48	<12	2.7	2.7	1.1	6.6
13	6.30	0.16	10.88	0.49	<12	3.0	1.0	0.6	4.7
14	6.67	0.19	<10	1.30	<12	1.5	1.0	0.3	5.9
15	6.06	1.15	42.56	3.32	156	61.6	6.9	2.8	34.8
16	7.03	0.31	51.64	0.51	<12	4.8	2.6	1.6	8.3
17	6.43	0.20	19.89	1.10	<12	2.4	0.9	0.5	6.5
18	6.97	0.37	52.39	2.10	<12	4.0	3.4	1.6	9.1
19	8.96	0.89	136.2	0.91	36.1	44.0	7.4	4.6	23.4
20	9.26	0.16	59.16	7.19	<12	21.5	0.3	7.1	5.8
21	8.26	0.27	60.66	8.39	<12	20.4	0.8	7.3	9.4
22	7.52	0.39	54.64	1.41	14.4	8.8	1.8	4.6	12.8
23	7.71	1.39	173.3	6.20	16.2	13.5	2.2	3.2	52.2
24	7.79	0.36	47.09	0.23	<12	2.3	1.5	1.8	12.1

* SanPiN 1.2.3685-21; ** SanPiN 2.1.4.1116-02.

Table 4. Content of microelements in spring water.

	Li, µg/L	B, µg/L	Al, µg/L	V, µg/L	Mn, µg/L	Fe, µg/L	Ni, µg/L	Cu, µg/L	Zn, µg/L	Rb, µg/L	Sr, µg/L	Mo, µg/L	Ba, µg/L	Pb, µg/L	Bi, µg/L
USEPA	-	-	200	-	50	300	-	1000	-	-	-	-	2000	15	-
EC	-	-	200	-	50	200	-	-	-	-	-	-	100	-	-
WOH	-	2400	200	-	400	300	70	2000	-	-	-	70	1300	10	100
MPC *	30	500	200	100	100	300	20	1000	5000	100	7000	70	700	10	100
First category **	<30	<500	<200	-	<50	<300	<20	<1000	<5000	-	<7000	<70	<700	<10	-
Highest category **	<30	<300	<100	-	<50	<300	<20	<1000	<3000	-	<7000	<70	<100	<5	-
1	3.5	<10	459.0	<1	14.3	74.3	13.2	5.0	10.4	2.3	65.2	<0.3	18.9	0.16	<0.5
2	2.6	<10	34.0	<1	<3	<20	1.6	<1	1.5	1.3	38.5	0.5	13.9	<0.1	2.4
3	3.1	<10	24.2	<1	<3	<20	1.9	1.3	5.1	3.0	34.9	0.5	10.7	0.70	2.0
4	4.2	<10	5.0	<1	<3	31.3	1.8	<1	2.3	0.2	121.2	1.9	17.3	0.10	1.7
5	1.6	13.3	46.6	<1	<3	45.3	2.7	1.3	2.2	2.2	53.0	0.4	37.8	0.12	1.9
6	1.7	<10	26.4	<1	<3	<20	<1	<1	2.9	2.6	33.2	0.3	16.2	<0.1	<0.5
7	2.4	<10	112.6	2.3	<3	213.2	<1	<1	2.8	0.8	139.1	0.8	25.5	<0.1	2.5
8	1.2	<10	12.5	<1	<3	25.8	8.0	1.9	16.9	4.8	63.0	0.5	26.5	<0.1	2.1
9	1.7	<10	30.8	1.2	<3	51.8	1.1	1.0	3.3	0.3	40.7	0.5	5.9	0.20	1.8
10	2.5	<10	6.8	<1	<3	153.2	2.0	<1	2.9	0.2	169.1	0.8	53.2	<0.1	1.8
11	3.2	<10	6.8	<1	<3	110.4	1.8	<1	5.7	0.2	140.1	0.5	40.2	<0.1	<0.5
12	2.2	<10	5.8	<1	<3	21.8	<1	<1	11.0	0.1	33.0	1.7	17.9	0.13	2.3
13	1.4	<10	9.7	<1	<3	<20	<1	<1	2.4	0.3	38.4	0.4	4.9	<0.1	2.2
14	<1	<10	10.7	<1	<3	27.0	4.4	<1	4.0	0.6	17.1	0.5	2.2	0.26	3.0
15	11.6	<10	43.7	1.6	295.7	186.2	24.3	2.4	5.6	0.6	195.0	0.3	103.1	0.41	1.9
16	5.5	<10	8.2	1.5	<3	<20	<1	<1	1.3	0.2	64.1	0.4	9.6	<0.1	<0.5

Table 4. *Cont.*

	Li, µg/L	B, µg/L	Al, µg/L	V, µg/L	Mn, µg/L	Fe, µg/L	Ni, µg/L	Cu, µg/L	Zn, µg/L	Rb, µg/L	Sr, µg/L	Mo, µg/L	Ba, µg/L	Pb, µg/L	Bi, µg/L
17	1.0	<10	95.6	1.0	<3	49.0	<1	1.6	5.5	0.6	21.5	0.5	4.2	0.17	2.6
18	2.9	<10	6.1	1.1	<3	81.7	1.1	<1	6.7	0.2	46.2	0.6	8.1	<0.1	2.2
19	2.6	38.6	6.4	<1	10.7	59.7	1.1	<1	2.9	0.9	196.3	1.5	21.5	<0.1	2.0
20	1.0	<10	298.6	5.0	<3	73.1	<1	1.3	8.1	3.0	24.7	14.3	1.7	0.15	2.0
21	1.2	<10	24.3	<1	<3	20.7	<1	<1	3.8	8.7	62.9	19.5	0.9	0.15	<0.5
22	<1	<10	5.4	<1	<3	35.1	<1	<1	6.7	2.2	10.1	0.5	0.5	<0.1	2.6
23	<1	18.1	7.2	2.2	<3	124.7	1.7	1.6	5.0	1.0	202.6	0.8	68.5	<0.1	2.2
24	<1	<10	10.2	<1	<3	29.9	<1	<1	3.5	0.3	71.2	1.8	8.7	<0.1	1.9

* SanPiN 1.2.3685-21; ** SanPiN 2.1.4.1116-02.

The pH values for most of the studied samples were in the 6–8 range; however, both more acidic and more alkaline waters were also observed in the 5.28–9.26 pH range (Table 3). Some of the samples with low pH values (samples 1, 2, 5, and 15) had a yellowish tint, which is typical for water with a high concentration of humic acids [17]. Water from samples 1–3, 5, 6, 8, 13, 15, 17, 19, and 20 did not follow the standards of the WHO, USEPA, EC, and the requirements of SanPiN 2.1.4.1116-02 for water pH values. The value of total water hardness varied from 0.15 to 1.5 mmol/L (i.e., waters were mainly of the soft type). All samples met the requirements for drinking water, but only water from samples 10, 11, 15, 19, and 23 corresponded to the highest category. Hydrocarbonates were found in all samples except samples 1 and 14, but only samples 4, 7, 9–12, 15, 16, and 18–24 met the requirements of the WHO and SanPiN 2.1.4.1074-01 for water of the highest category. In terms of nitrate content, all samples corresponded to MPC standards, but samples 7, 20, 21, and 23 did not correspond to the highest category. In terms of chloride ion content, only sample 15 did not correspond to the highest category.

In principle, the cationic composition of the studied samples met the requirements of the WHO and SanPiN 2.1.4.1116-02. However, for samples 1 and 20, concentrations of Al were detected that were 2.5 and 1.5 times higher, respectively, than the MPC. In sample 15, the concentration of Mn was 3 times higher than the MPC and slightly exceeded the MPC of Ba (Table 4). Moreover, samples 8, 15, and 19–20 did not meet the requirements for drinking water of the highest category because they exceeded the MPC of Na, whereas samples 7 and 15 exceeded the MPC of Al and Ni, respectively. Water of the highest category was characterized by a minimum content of Mg, K, and Ca cations [55]. Samples 10, 15, and 19 matched this with their Mg content; samples 4, 7, 10, 15, and 19–23 matched with their K content; and samples 10, 11, 15, and 23 matched with their Ca content.

The concentration of Ag, As, Be, Cd, Co, Cr, Re, Sb, and Tl did not exceed the detection limit (LOD) in all analyzed samples. The LOD of these elements and maximal permissible concentrations according to various standards are shown in Table 5. The exceptions were samples 1 and 8, where the Co concentrations were 4.0 and 0.30 µg/L, respectively. All water samples met the most stringent requirements of SanPiN 2.1.4.1116-02 for these parameters.

Table 5. The detection limits of chemical elements.

	Be	Cr	Co	As	Se	Ag	Cd	Sb	Te	Re	Tl
LOD, µg/L	0.2	1	0.2	1	10	1	0.1	0.2	2	0.1	0.1
WHO	12	50	-	10	40	-	3	20	-	-	-
MPC *	0.2	50	100	10	10	50	1	5	10	-	0.1
Highest category **	<0.2	<30	<100	<6	<10	<25	<1	<5	-	-	-

* SanPiN 1.2.3685-21; ** SanPiN 2.1.4.1116-02.

The results of chemical composition of spring waters are in good agreement with respect to the content of macro-components both with the previously obtained data [59] and with the chemical composition of springs located at the same latitude and similar relief [60–62].

In general, most of the studied springs of the Kola Peninsula are sources of ultra-fresh waters with a salinity of less than 0.2 g/L. Table 6 shows the number of springs that correspond to the MPC standards for the analyzed parameters and meet the requirements for drinking water of the first and highest categories according to this classification [55].

Table 6. Springs' compliance with the standards.

Area	MPC	Drinking Water of the First Category	Drinking Water of the Highest Category
Murmansk and Kola district	4, 7	4, 7	-
Monchegork district	9, 10, 11, 12, 14	9, 10, 11, 12, 14	10 (11 *)
Kandalaksha district	16, 18	16, 18	-
Apatity-Kirovsk district	21, 22, 23, 24	21, 22, 23, 24	-
Totally (24 springs)	13	13	1 (2 *)

* slightly lower in the required indicators for Mg and K.

The chemical composition of the spring waters showed that some sources did not meet the requirements for the MPCs in SanPiN 1.2.3685-21 but are in demand among the population, which is extremely dangerous and can negatively affect their health. Springs 1, 15 and 20 should be noted as especially hazardous, since a significant excess of MPC for Al, Mn and Al, respectively, was found in their waters. However, at least one or several springs can be recommended near each large settlement. Based on the data set obtained, and given the low waters salinity, the most preferred for drinking are spring 4 for the Murmansk and Kola district, spring 11 for the Monchegorsk district, spring 16 for the Kandalaksha district, and spring 24 for the Apatity-Kirovsk district.

In general, springs are better protected from the urban and industry negative effect in comparison with surface waters. Nevertheless, seasonal pollution by surface waters in the Arctic region is possible, which is most pronounced during the period of snow melting [63]. Therefore, regular monitoring of all springs used as the source of drinking water is required.

The results and conclusions of this study will be posted on Yandex.Disk, which will be available via the link and QR code placed near each spring (Figure 4). By selecting the folder with the corresponding spring number, people will be able to find information on the chemical composition of the water of each studied spring.

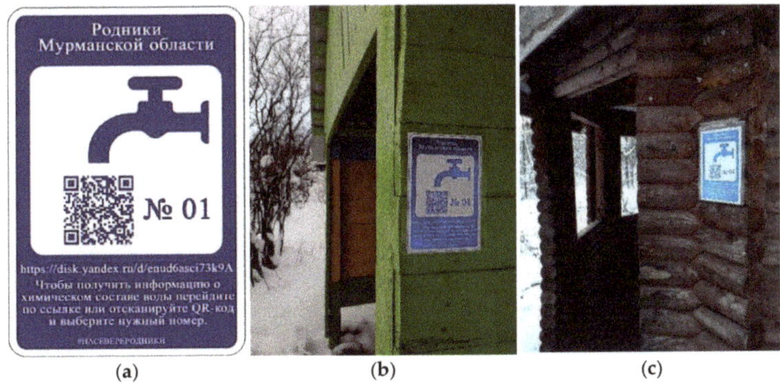

(a) (b) (c)

Figure 4. Plates with the link to information on the chemical composition of the spring water: (**a**) an example of a plate (in Russian) with a QR code; (**b**,**c**) examples of plate placement.

4. Conclusions

The water chemical composition of 24 of the most popular springs near the largest cities of the Murmansk region in Russia were analyzed. Almost half of them did not meet the requirements of drinking water for the determined parameters. Basically, the waters of the region did not correspond to the Russian, European, and US standards in terms of pH, but also, in some springs, significant excess in the concentrations of aluminum and manganese were recorded. The results of the chemical analysis showed that 13 out of 24 investigated springs were potable and corresponded to water of the first category, according to Russian hygienic standards. Finally, the water from one of the sources even met the requirements for water of the highest category. Nevertheless, the regular monitoring of springs used by the population as the source of drinking water, especially for the content of potentially toxic and hazardous microelements, is needed due to the change in groundwater chemical composition after snow melting and the anthropogenic pressure in the industrialized Arctic region.

Author Contributions: Conceptualization, A.I.N., A.A.S. and M.V.S.; methodology, A.I.N. and A.A.S.; software, A.I.N.; validation, A.I.N., A.A.S. and M.V.S.; formal analysis, A.I.N. and A.A.S.; investigation, A.I.N. and A.A.S.; resources, A.I.N. and A.A.S.; data curation, A.I.N.; writing—original draft preparation, A.I.N. and A.A.S.; writing—review and editing, A.I.N., A.A.S. and M.V.S.; visualization, A.I.N. and A.A.S.; supervision, A.I.N. and A.A.S.; project administration, A.I.N.; funding acquisition, A.I.N. All authors have read and agreed to the published version of the manuscript.

Funding: This research was funded by Department of Education and Science of the Murmansk Region, agreement number 224 and the Russian Academy of Sciences theme № AAAA-A18-118030590098-0.

Data Availability Statement: Data set available on request to corresponding authors.

Acknowledgments: The authors thank Dauvalter Vladimir for counseling.

Conflicts of Interest: The authors declare have no conflict of interest. The funders had no role in the design of the study; in the collection, analyses, or interpretation of data; in the writing of the manuscript, or in the decision to publish the results.

References

1. Elpiner, L.I.; Bear, S.A.; Zektser, I.S.; Klige, R.K.; Shapovalov, A.E. The effect of water abundance in a territory on the population health. *Water Resour.* **2007**, *34*, 340–349. [CrossRef]
2. Schwarzenbach, R.P.; Egli, T.; Hofstetter, T.B.; Von Gunten, U.; Wehrli, B. Global Water Pollution and Human Health. *Annu. Rev. Environ. Resour.* **2010**, *35*, 109–136. [CrossRef]
3. Halder, J.; Islam, N. Water Pollution and its Impact on the Human Health. *J. Environ. Hum.* **2015**, *2*, 36–46. [CrossRef]
4. Moiseenko, T.I.; Megorskii, V.V.; Gashkina, N.A.; Kudryavtseva, L.P. Water pollution effect on population health in an industrial northern region. *Water Resour.* **2010**, *37*, 194–203. [CrossRef]
5. Trofimovich, E.M.; Aizman, R.I. System of metabolism of drinking water as a methodological basis for the estimation of its mineral composition. *Hyg. Sanit.* **2019**, *98*, 555–562. [CrossRef]
6. Zhang, Y.; Wu, J.; Xu, B. Human health risk assessment of groundwater nitrogen pollution in Jinghui canal irrigation area of the loess region, northwest China. *Environ. Earth Sci.* **2018**, *77*, 273. [CrossRef]
7. Khan, A.F.; Srinivasamoorthy, K.; Prakash, R.; Gopinath, S.; Saravanan, K.; Vinnarasi, F.; Babu, C.; Rabina, C. Human health risk assessment for fluoride and nitrate contamination in the groundwater: A case study from the east coast of Tamil Nadu and Puducherry, India. *Environ. Earth Sci.* **2021**, *80*, 724. [CrossRef]
8. Chen, J.; Wu, H.; Qian, H.; Gao, Y. Assessing Nitrate and Fluoride Contaminants in Drinking Water and Their Health Risk of Rural Residents Living in a Semiarid Region of Northwest China. *Expo. Health* **2017**, *9*, 183–195. [CrossRef]
9. World Health Organization. *Guidelines for Drinking—Water Quality*, 4th ed.; WHO: Geneva, Switzerland, 2017.
10. Mejía, F.; Medina, C.; Cornejo, E.; Morello, E.; Vásquez, S.; Alave, J.; Schwalb, A.; Málaga, G. Oxygen saturation as a predictor of mortality in hospitalized adult patients with COVID-19 in a public hospital in Lima, Peru. *PLoS ONE* **2020**, *15*, e0244171. [CrossRef]
11. Akhavan, A.R.; Habboushe, J.P.; Gulati, R.; Iheagwara, O.; Watterson, J.; Thomas, S.; Swartz, J.L.; Koziatek, C.A.; Lee, D.C. Risk Stratification of COVID-19 Patients Using Ambulatory Oxygen Saturation in the Emergency Department. *West. J. Emerg. Med.* **2020**, *21*, 5–14. [CrossRef]
12. Rubin, S.J.S.; Falkson, S.R.; Degner, N.R.; Blish, C. Clinical characteristics associated with COVID-19 severity in California. *J. Clin. Transl. Sci.* **2021**, *5*, 1–4. [CrossRef] [PubMed]

13. Rappaport, S.M.; Smith, M.T. Environment and Disease Risks. *Science* **2010**, *330*, 460–461. [CrossRef]
14. Yokel, R.A. The toxicology of aluminum in the brain: A review. *Neurotoxicology* **2000**, *21*, 813–828.
15. Seiler, H.; Sigel, A.; Sigel, H. (Eds.) *Handbook on Metals in Clinical and Analytical Chemistry*; Marcel Dekker: New York, NY, USA, 1994.
16. Kolubaeva, Y.V. Chemical Composition of Underground Water Sources of Non-Centralized Water Supply in the Tomsk Region of the Tomsk Oblast and Its Possible Impact on Human Health. In *Water-Rock Interaction: Geological Evolution*; Buryat Scientific Center SB RAS: Ulan-Ude, Russian, 2020; pp. 159–162.
17. Boyd, C.E. *Water Quality*; Springer International Publishing: Cham, Swizerlands, 2020; ISBN 978-3-030-23334-1.
18. Vannucci, L.; Fossi, C.; Quattrini, S.; Guasti, L.; Pampaloni, B.; Gronchi, G.; Giusti, F.; Romagnoli, C.; Cianferotti, L.; Marcucci, G.; et al. Calcium Intake in Bone Health: A Focus on Calcium-Rich Mineral Waters. *Nutrients* **2018**, *10*, 1930. [CrossRef]
19. Cormick, G.; Belizán, J.M. Calcium Intake and Health. *Nutrients* **2019**, *11*, 1606. [CrossRef]
20. Anderson, R.A. Chromium as an Essential Nutrient for Humans. *Regul. Toxicol. Pharmacol.* **1997**, *26*, S35–S41. [CrossRef] [PubMed]
21. Iskra, R.; Antonyak, H. *Chromium in Health and Longevity*; Springer: Cham, Swizerlands, 2018; pp. 133–162.
22. Aschner, J.L.; Aschner, M. Nutritional aspects of manganese homeostasis. *Mol. Asp. Med.* **2005**, *26*, 353–362. [CrossRef] [PubMed]
23. Abbaspour, N.; Hurrell, R.; Kelishadi, R. Review on iron and its importance for human health. *J. Res. Med. Sci.* **2014**, *19*, 164–174.
24. Yamada, K. Cobalt: Its Role in Health and Disease. *Met. Ions Life Sci.* **2013**, *13*, 295–320. [CrossRef]
25. Spears, J.W. Nickel as a "Newer Trace Element" in the Nutrition of Domestic Animals. *J. Anim. Sci.* **1984**, *59*, 823–835. [CrossRef] [PubMed]
26. Roohani, N.; Hurrell, R.; Kelishadi, R.; Schulin, R. Zinc and its importance for human health: An integrative review. *J. Res. Med. Sci.* **2013**, *18*, 144–157. [PubMed]
27. Araya, M.; Pizarro, F.; Olivares, M.; Arredondo, M.; González, M.; Méndez, M. Understanding copper homeostasis in humans and copper effects on health. *Biol. Res.* **2006**, *39*, 183–187. [CrossRef] [PubMed]
28. Hamilton, S.J. Review of selenium toxicity in the aquatic food chain. *Sci. Total Environ.* **2004**, *326*, 1–31. [CrossRef] [PubMed]
29. Schwarz, G.; Belaidi, A.A. Molybdenum in Human Health and Disease. *Met. Ions Life Sci.* **2013**, *13*, 415–450. [CrossRef]
30. Byrne, C.; Divekar, S.D.; Storchan, G.B.; Parodi, D.A.; Martin, M.B. Cadmium—A metallohormone? *Toxicol. Appl. Pharmacol.* **2009**, *238*, 266–271. [CrossRef]
31. Adewoyin, O.O.; Kayode, O.T.; Omeje, O.; Odetunmibi, O.A. Risk assessment of heavy metal and trace elements contamination in groundwater in some parts of Ogun state. *Cogent Eng.* **2019**, *6*, 1632555. [CrossRef]
32. Kurt, M.A.; Yıldırım, Ü.; Güler, C.; Güven, O. Antimony and arsenic contamination in water from antimonite mineralization: A case study from Turhal (Tokat, Northern Turkey). *Environ. Forensics* **2021**, 1–13. [CrossRef]
33. Schwarzenbach, R.P.; Escher, B.I.; Fenner, K.; Hofstetter, T.B.; Johnson, C.A.; von Gunten, U.; Wehrli, B. The Challenge of Micropollutants in Aquatic Systems. *Science* **2006**, *313*, 1072–1077. [CrossRef]
34. Verma, R.; Dwivedi, P. Heavy Metal Water Pollution-A Case Study. *Sci. Technol.* **2013**, *5*, 98–99.
35. Belkhiri, L.; Tiri, A.; Mouni, L. Assessment of Heavy Metals Contamination in Groundwater: A Case Study of the South of Setif Area, East Algeria. In *Achievements and Challenges of Integrated River Basin Management*; IntechOpen: London, UK, 2018. [CrossRef]
36. United Nations Children's Fund (UNICEF); World Health Organization (WHO). *Progress on Drinking Water, Sanitation and Hygiene: 2017 Update and SDG Baselines*; United Nations Children's Fund (UNICEF); World Health Organization (WHO): Geneva, Switzerland, 2017; ISBN 9789241512893.
37. Mekonnen, M.M.; Hoekstra, A.Y. Sustainability: Four billion people facing severe water scarcity. *Sci. Adv.* **2016**, *2*, e1500323. [CrossRef]
38. Vörösmarty, C.J.; McIntyre, P.B.; Gessner, M.O.; Dudgeon, D.; Prusevich, A.; Green, P.; Glidden, S.; Bunn, S.E.; Sullivan, C.A.; Liermann, C.R.; et al. Global threats to human water security and river biodiversity. *Nature* **2010**, *467*, 555–561. [CrossRef]
39. Rosinger, A.Y.; Patel, A.I.; Weaks, F. Examining recent trends in the racial disparity gap in tap water consumption: NHANES 2011–2018. *Public Health Nutr.* **2021**, 1–7. [CrossRef]
40. Rozhkov, V.P.; Belisheva, N.K.; Martynova, A.A.; Soroko, S.I. Psychophysiological and cardiohemodynamic effects of solar, geomagnetic, and meteorological factors in humans under the conditions of the arctic region. *Hum. Physiol.* **2014**, *40*, 397–409. [CrossRef]
41. Statdata. Ru—Site about Countries and Cities. Available online: http://www.statdata.ru/naselenie/murmanskoj-oblasti (accessed on 30 November 2021).
42. Slukovskaya, M.V.; Vasenev, V.I.; Ivashchenko, K.V.; Dolgikh, A.V.; Novikov, A.I.; Kremenetskaya, I.P.; Ivanova, L.A.; Gubin, S.V. Organic matter accumulation by alkaline-constructed soils in heavily metal-polluted area of Subarctic zone. *J. Soils Sediments* **2021**, *21*, 2071–2088. [CrossRef]
43. Saltan, N.; Slukovskaya, M.; Mikhaylova, I.; Zarov, E.; Skripnikov, P.; Gorbov, S.; Khvostova, A.; Drogobuzhskaya, S.; Shirokaya, A.; Kremenetskaya, I. *Assessment of Soil Heavy Metal Pollution by Land Use Zones in Small Towns of the Industrialized Arctic Region, Russia*; Springer: Singapore, 2021; pp. 100–110. [CrossRef]
44. Mosendz, I.A.; Kremenetskaya, I.P.; Novikov, A.I.; Tereshchenko, S.V. Removing copper and nickel from contaminated water with vermiculite-sungulite materials. *Tsvetnye Met.* **2021**, *2021*, 36–41. [CrossRef]

45. Postevaya, M.A.; Slukovskii, Z.I. Assessment of the Concentration of Heavy Metals in the Water of the Lakes of the City of Murmansk. In *Geography: Development of Science and Education*; Herzen, A.I., Ed.; Russian State Pedagogical University: Saint Petersburg, Russian, 2021; pp. 111–115. (In Russian)
46. Guzeva, A.; Slukovskii, Z.; Dauvalter, V.; Denisov, D. Trace element fractions in sediments of urbanised lakes of the arctic zone of Russia. *Environ. Monit. Assess.* **2021**, *193*, 378. [CrossRef]
47. Dauvalter, V.A.; Kashulin, N.A. Assessment of the Ecological State of the Arctic Freshwater System Based on Concentrations of Heavy Metals in the Bottom Sediments. *Geochem. Int.* **2018**, *56*, 842–856. [CrossRef]
48. Slukovskii, Z.; Medvedev, M.; Mitsukov, A.; Dauvalter, V.; Grigoriev, V.; Kudryavtzeva, L.; Elizarova, I. Recent sediments of Arctic small lakes (Russia): Geochemistry features and age. *Environ. Earth Sci.* **2021**, *80*, 302. [CrossRef]
49. Slukovskii, Z.; Dauvalter, V.; Guzeva, A.; Denisov, D.; Cherepanov, A.; Siroezhko, E. The Hydrochemistry and Recent Sediment Geochemistry of Small Lakes of Murmansk, Arctic Zone of Russia. *Water* **2020**, *12*, 1130. [CrossRef]
50. Dauvalter, V.A.; Dauvalter, M.V. *Groundwater Condition in Monchegorsk Region*; Bulletin of the Kola Science Center RAS, KSC RAS: Apatity, Russian, 2010; pp. 26–33. (In Russian)
51. Ayras, M.; de Caritat, P.; Chekushin, V.A.; Niskavaara, H.; Reimann, C. Ecogeochemical investigation, kola peninsula: Sulphur and trace element content in snow. *Water Air Soil Pollut.* **1995**, *85*, 749–754. [CrossRef]
52. Gregurek, D.; Melcher, E.; Pavlov, V.A.; Reimann, C.; Stumpfl, E.F. Mineralogy and Petrology Mineralogy and Mineral Chemistry of Snow Filter Residues in the Vicinity of the Nickel-Copper Processing Industry, Kola Peninsula, NW Russia. *Mineral. Petrol.* **1999**, *65*, 87–111. [CrossRef]
53. Council of the European Union. [Internet]. Council Directive 98/83/EC of 3 November 1998 on the Quality of Water Intended for Human Consumption. *Off. J. Eur. Communities* **1998**. Available online: https://agris.fao.org/agris-search/search.do?recordID=CS2005000008 (accessed on 23 November 2021).
54. National Primary Drinking Water Regulations I US EPA. Available online: https://www.epa.gov/ground-water-and-drinking-water/national-primary-drinking-water-regulations#Inorganic (accessed on 23 November 2021).
55. SanPiN 2.1.4.1116-02 Drinking Water. Hygienic Requirements for the Quality of Water Packaged in Containers. Quality Control. Available online: https://docs.cntd.ru/document/902225826 (accessed on 28 November 2021). (In Russian).
56. SanPiN 1.2.3685-21 Hygienic Standards and Requirements to Ensure Safety and (or) Harmful for Human Environmental Factors. Available online: https://docs.cntd.ru/document/573500115 (accessed on 28 November 2021). (In Russian).
57. Department of Natural Resources, Ecology and Fisheries of the Murmansk Region Report on the State and Protection of the Environment of the Murmansk Region in 2020. Murmansk 2021. Available online: https://Gov-Murman.Ru/Bitrix/Components/B1team/Govmurman.Element.File/Download.Php?ID=409632&FID=573553 (accessed on 23 November 2021). (In Rassian).
58. Selection Map. Available online: https://www.google.com/mymaps/viewer?mid=10ubQEJJWf2SghqpQMtgcljh6YI8t5hso&hl=en (accessed on 21 December 2021). (In Russian).
59. Springs of the Kola Land. Available online: https://helion-ltd.ru/springs-kola-earth/ (accessed on 25 November 2021). (In Russian).
60. Gunnarsdóttir, M.J.; Gardarsson, S.M.; Jonsson, G.S.; Ármannsson, H.; Bartram, J. Natural background levels for chemicals in Icelandic aquifers. *Hydrol. Res.* **2014**, *46*, 647–660. [CrossRef]
61. Kousa, A.; Komulainen, H.; Hatakka, T.; Backman, B.; Hartikainen, S. Variation in groundwater manganese in Finland. *Environ. Geochem. Health* **2021**, *43*, 1193–1211. [CrossRef]
62. Gunnarsdottir, M.J.; Gardarsson, S.M.; Jonsson, G.S.; Bartram, J. Chemical quality and regulatory compliance of drinking water in Iceland. *Int. J. Hyg. Environ. Health* **2016**, *219*, 724–733. [CrossRef] [PubMed]
63. Scheuer, C.; Boot, E.; Carse, N.; Clardy, A.; Gallagher, J.; Heck, S.; Marron, S.; Martinez-Alvarez, L.; Masarykova, D.; Mcmillan, P.; et al. Physical/Geographical Characteristics of the Artic. In *AMAP Assessment Report: Arctic Pollution Issues*; Arctic Monitoring and Assessment Programme (AMAP), AMAP: Oslo, Norway, 1998; pp. 9–24.

Article

Concentrations of Major and Trace Elements within the Snowpack of Tyumen, Russia

Dmitriy Moskovchenko *, Roman Pozhitkov, Aleksandr Zakharchenko and Aleksandr Tigeev

Tyumen Scientific Centre, Siberian Branch of Russian Academy of Sciences, Malygina st., 86,
625026 Tyumen, Russia; pozhitkov-roma@yandex.ru (R.P.); fic@tmnsc.ru (A.Z.); ttrruubbaa@mail.ru (A.T.)
* Correspondence: moskovchenko1965@gmail.com; Tel.: +7-9224883211

Abstract: A study on the composition of snow allowed for a quantitative determination of pollutants deposited from the atmosphere. Concentrations of dissolved (<0.45 μm) and particulate fractions of 62 chemical elements were determined by ICP–MS and ICP–AES in 41 samples of snow from Tyumen (Russia). The background sites were characterized by a predominance of the dissolved phase of elements, except for Al, Sn, Cr, Co and Zr. The increased concentrations of dissolved Cd, Cu, Zn, Pb, Ni, As and Mo can be explained by a long-range atmospheric transport from the sources located in the Urals. The urban sites showed multiple increases in particulate depositions and a predominance of the particulate phase, with a high degree of enrichment in many heavy metals. Sources of trace elements were determined according to the enrichment factor (EF). Highly enriched elements (Pb, Sb, Cd, Ag, Mo, As, Zn and Cu) with an EF > 100 were emitted from anthropogenic sources. According to the potential ecological risk index (RI), the worst ecological conditions were identified in Tyumen's historical center, industrial zone and along roads with the heaviest traffic. The data obtained in the present study allowed us to identify the most polluted parts of the city, which are located in the center and along the roads with the most intensive traffic. This research could offer a reference for the atmospheric pollution prevention and control in Tyumen.

Keywords: Western Siberia; snow pollution; trace metals and metalloids; atmospheric depositions; solubility

Citation: Moskovchenko, D.; Pozhitkov, R.; Zakharchenko, A.; Tigeev, A. Concentrations of Major and Trace Elements within the Snowpack of Tyumen, Russia. *Minerals* **2021**, *11*, 709. https://doi.org/10.3390/min11070709

Academic Editors: Ana Romero-Freire and Hao Qiu

Received: 1 June 2021
Accepted: 28 June 2021
Published: 30 June 2021

Publisher's Note: MDPI stays neutral with regard to jurisdictional claims in published maps and institutional affiliations.

1. Introduction

Currently industrial production is accompanied by the emission and spread of massive quantities of trace metals and metalloids (TMMs) [1]. An accurate and complete emission inventory for atmospheric trace metals is needed for both the modeler community and policymakers to assess the current levels of environmental contamination by these pollutants, major emission sources and source regions, and the contribution of the atmospheric pathway to the contamination of terrestrial and aquatic environments [2]. Some of Russia's large industrial regions are the sources of hazardous contamination of the environment by TMMs on a global scale. For example, TMMs emissions from the metallurgical plants of the Kola Peninsula and the Norilsk Industrial District, which have been described in several studies [3–6], have been shown to cause pollution of Arctic ecosystems [7,8]. However, there have been very few studies on the contamination processes in other regions of Russia, in particular, Western Siberia and the Urals.

Snowpack is often used in studies of atmospheric pollution. Insoluble aerosol particles, as well as soluble compounds, including various pollutants, are washed out of the atmosphere by snow, which acts as a reservoir of TMMs accumulation during the winter period [9–11]. When surfaces are covered with snow, the influence of soils on the formation of atmospheric dust aerosols is minimized, which can allow for accurate determinations of anthropogenic contributions and assessments of TMMs deposition rates in winter. Thus, the analysis of the chemical composition of snow provides useful information on aerosol chemistry and long-range distribution patterns of anthropogenic substances emitted into the atmosphere [12,13].

Pollution by TMMs creates the greatest ecological hazard in highly populated cities. At the present time, 75% of the Russian population lives in cities [14]. Urbanization leads to the concentration of people, enterprises producing goods and services, and transport flows within cities, which consequently become the centers of environmental pollution. Most Russian cities, where ecological conditions have been assessed, are regarded as geochemical anomalies with highly polluted air, soils, water bodies and bottom sediments [15]. The majority of pollutants within such urban environments is deposited from the atmosphere [16]. Particles suspended in the air have high contents of TMMs which are deposited on the soil surface and cause deterioration of its properties. Atmospheric deposition is considered to be a major source of toxic TMMs, such as Cd, Cu Zn, Hg and Pb, for ecosystems [17,18]. Thus, an understanding of the impurities in natural snow is important in order to determine the air quality and to inform the pollution of the environment [19–21]. Since urban environments are heavily affected by anthropogenic pollutants, the quantitative analysis of TMMs in the snow cover can be of crucial importance [22].

In Siberia, snowpack is present for a period of 5–9 months. Such a long period of TMMs accumulation in snow allows for an accurate evaluation of the levels and sources of pollution. Snowpack is well known to be an informative object for evaluating aerogenic contaminants [23,24]. Unfortunately, the data on trace elements in snow meltwater collected from large and geographically homogeneous territories of Western Siberia are limited [11]. However, those limited data do demonstrate a significant impact of anthropogenic activities on the snow composition. In particular, it has been shown that there are latitudinal gradients of concentrations of certain TMMs within snowpack; for example, concentrations of Pb, Zn, Cd, V, Co and Sr decrease in a northward direction due to the location of the main industrial plants in the south [25]. It has also been suggested that the development of oil and gas fields in Western Siberia has caused TMMs accumulation in snow [26,27]. Anthropogenic impacts result in an enrichment of the particulate fraction of snow by many TMMs as compared to their levels in soils [11,28,29].

To date, the accumulation of major elements and TMMs present within the snowpack of cities of Western Siberia, as well as the interregional transmission of pollutants, has been poorly studied. Several studies of TMMs in the particulate fraction of snow have been conducted in Tomsk and Novosibirsk. It has been shown that Tomsk thermal power plants induce an enrichment of dust aerosols in Hg, Ba, Sb, La, As and Sr [30]. Moreover, in Novosibirsk, high TMMs concentrations in the particulate fraction of snow have been shown to have a technogenic origin and indicate a potentially higher toxicity of wintertime aerosols as compared to summertime aerosols [31].

Tyumen is one of the largest and rapidly developing cities of Western Siberia. The rapid development of Tyumen has been primarily due to the discoveries of numerous deposits of petroleum and natural gas in the north of Western Siberia during the second half of the 20th century. The population of Tyumen City was 150 thousand people in the early 1960s, which increased to around 500 thousand in the early 2000s and to over 800 thousand at the present time. In the late 1990s, it was revealed that the operation of the Tyumen thermal power station had caused the accumulation of Mn, Cr, Ni, Pb and V in the snowpack [32]. Since that time, the city's population has grown significantly and the distribution and specialization of industries have developed with the addition of new enterprises, including ferrous metallurgy and petroleum refining. An assessment of soil composition in Tyumen has shown increased concentrations of V, Cr, Co, Ni, Cu and Zn near to main roads and metalworking plants and anomalies of As and Pb near the operating facilities of electric storage battery plants [33]. The worsening of the environmental conditions in Tyumen necessitates research on the sources of pollutants and an ecological assessment, which can be carried out by analyzing the snowpack.

The present study had the following aims:

- To identify the basic distribution of the major and trace elements of snow,

- To evaluate the ratios of particulate and dissolved forms of TMMs,
- To assess the ecological risks associated with TMMs pollution of the atmosphere.

The originality of the present study is due to the combination of (i) the analysis of the snowpack composition within such a large Siberian industrial center as Tyumen and (ii) the assessment of both dissolved + particulate forms of major and trace elements in snow samples.

The proportions of dissolved and particulate forms of elements in snow meltwater allow for the determination of emission sources [15]. Elements associated with anthropogenic and marine sources are in general soluble upon thawing, elements from crustal sources are insoluble, elements with mixed sources tend to show intermediate behavior with average solubilities ranging from 27% to 96% for different elements [34]. In addition, the proportions of dissolved and particulate forms of elements in snow meltwater predetermine either their accumulation in soil or migration with meltwater into streams and rivers [35], which allows for the prognosis of contamination consequences.

2. Materials and Methods

2.1. Study Area and Sampling

Tyumen is located in the southwestern part of the West Siberian Plain, at the southern margin of the taiga zone (Figure 1).

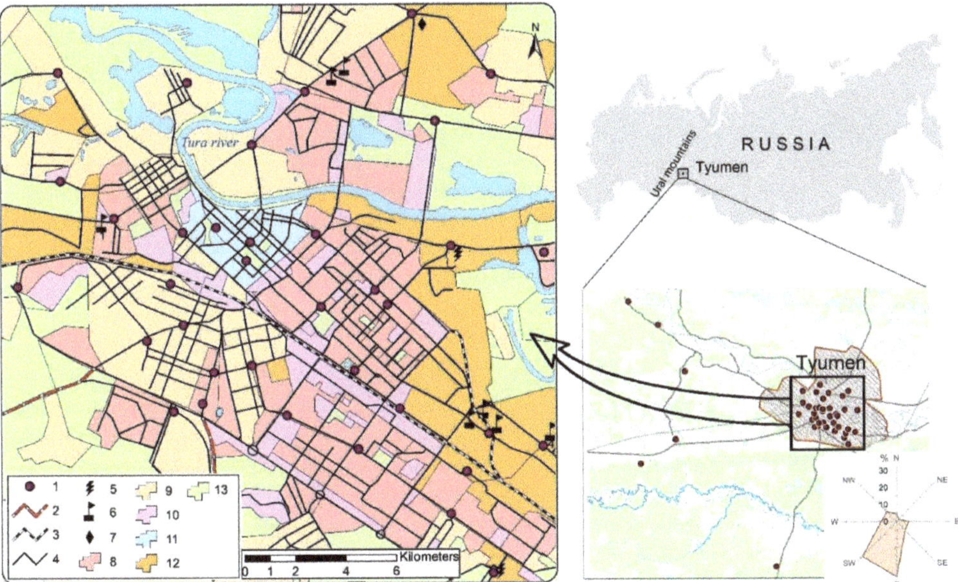

Figure 1. Sampling sites and land-use areas within the city of Tyumen, Russia: 1—sampling sites; 2—main federal roads; 3—Trans-Siberian railway; 4—main city roads; 5—power plants; 6—mechanical engineering and metalworking facilities; 7—oil refining and plastic production facilities; 8—high-rise residential area; 9—low-rise residential area; 10—modern business zones; 11—historical center; 12—industrial zones; 13—recreational and unbuilt zones. The diagram in the lower right corner shows the wind directions within the period from November 2019 to February 2020.

Tyumen has a cold humid continental climate, with mean monthly temperatures ranging from −22 °C in January to +17 °C in July and a mean annual precipitation of 480 mm. On average, stable snow cover is present for five months of the year, from early November to early April. The cold period is characterized by a predominance of southern and western winds, which carry pollutants to the north and northeast [36,37]. The winter season of 2019–2020 was not exceptional in this respect, as it was dominated by southern, southwestern and south–southwestern winds. A flat surface topography within the city minimizes the influence of the orographic factor on the translocation of pollutants. Winds of different directions provide for the redistribution of pollutants within the city and beyond [38].

Tyumen is a large center of transportation and trade with a well-developed building industry. Tyumen now has a number of industrial plants producing machinery, equipment and building materials, petroleum refineries and power stations. Moreover, the city currently has 385 thousand motorized vehicles in use and a network of roads totaling 1241 km in length [39]. Major transport routes pass through Tyumen to northern regions of Western Siberia, where the largest oil and gas fields are located. Exhaust gases from motorized vehicles make up more than 80% of the total emissions of pollutants into the atmosphere within the city [40]. Traffic on the busiest roads reaches 8000 vehicles per hour [41]. Most (89.4%) of the motorized vehicles have gasoline engines. The geochemical condition of the city is estimated as hazardous and ecologically unfavorable residential areas have been identified [39–41].

2.2. Sample Collection

Snow samples were taken within a period from the 17th to 20th of February 2020, when the air temperature in Tyumen was close to $0°$ C, which was significantly above its mean range of -15 to $-10\ °$C. On average, the snow cover of the city disappears on the 9th of April [37]. However, the winter season of 2019–2020 was abnormally warm, with short-term warmings throughout January and February. Therefore, sampling was moved to the earlier dates in order to take samples of snowpack prior to the beginning of its thawing. A total precipitation of 106 mm was recorded over the period from November 2019 to February 2020 [36], which is higher than the mean of 91 mm over the period 1960–2019 [37].

Snow was sampled from within the Tyumen City and from background sites at distances of 20–35 km to the west and southwest from the city (see Figure 1). Taking into account that winds from the south prevailed during the study period, the locations of background sampling sites excluded any contamination of snow by pollutants from Tyumen.

During sampling, the land-use type of each surveyed area was taken into account. The functional zones (different land-use areas) within Tyumen are not clearly separated, but form a mosaic pattern of industrial, residential, commercial and recreational areas within the city. However, different types of land use were recorded at the sampling sites as follows:

(1) The historical center with buildings that have existed since the 17th century and are now used by social and administrative organizations;
(2) Low-rise residential area that have existed since the 19th century;
(3) High-rise residential area that have been constructed within a period from the 1950s to the present time and that now house the major part of the Tyumen population;
(4) Modern business zones;
(5) Industrial zones;
(6) Transport zones affected by road traffic (located between main roads and buildings of various usage). Sampling within transport zones was carried out at distances of more than 15 m from roads in order to exclude direct contamination by road dust. The sampled snowpack was visibly undisturbed by any human activity.

Samples of snow were taken by using a VS-43 snow gauge, which is a widely used instrument for meteorological observations in Russia. This snow gauge consists of an aluminum pipe (sampling cylinder) with the cross-section area of 50 cm^2 and a measurement scale for determining the depth of snow. In order to exclude contamination of samples by soil particles, we removed the lowest 3 cm of snow that was directly adjacent to the ground. Samples were placed in 24 dm^3 plastic containers with lids, which were washed with distilled water prior to sampling. Closed containers with samples were transported to the environmental chemistry laboratory. In total, 35 samples were taken from the city of Tyumen, and 6 samples were from background sites. Locations of sampling sites are shown in Figure 1. A detailed description of the sampling sites is presented in Supplementary Materials Table S1.

2.3. Analytical Procedures

The snowpack samples were melted at room temperature. In the melted samples, pH values were measured by using HI83141 and HM-500 HydroMaster ionometers, and salinity was determined by using a COM-100 conductometer. Afterwards, the partitioning of elements between the dissolved phase and the solid phase was performed in order to separately analyze TMMs in water-soluble mobile forms and forms bound with mineral and organomineral compounds [42]. The partitioning was achieved by passing the melted samples through pre-weighed Millipore ash-less nitrocellulose filters (0.45 μm pore size). Around 1.5–2 liters of meltwater was filtered to obtain a sufficient amount of solid residue from each sample. The filters were dried at 95 °C and weighed again for determination of the mass of trapped solids.

Contents of trace elements (Li, Be, Sc, V, Cr, Ti, Mn, Co, Ni, Cu, Zn, Ga, As, Se, Rb, Sr, Y, Zr, Nb, Mo, Rh, Pd, Ag, Cd, Sn, Sb, Te, Cs, Ba, La, Ce, Pr, Nd, Sm, Eu, Gd, Tb, Dy, Ho, Er, Tm, Yb, Lu, Hf, Ta, W, Re, Ir, Hg, Tl, Pb, Bi, Th and U) and major elements Na, Mg, Al, P, S, K, Ca and Fe (in weight percent oxide for the particulate fraction) were determined by inductively coupled plasma–mass spectrometry ICP–MS (Thermo Elemental-X7 spectrometer, USA) and inductively coupled plasma–atomic emission spectrometry ICP–AES (Thermo Scientific iCAP-6500 spectrometer, USA).

The solids trapped on filters were decomposed by an acid digestion in open system. The samples were placed in Teflon beakers (volume 50 mL), and 0.1 mL of a solution containing 8 μg L^{-1} 145Nd, 61Dy and 174Yb was added and moistened with several drops of deionized water. Then 0.5 mL of $HClO_4$ (perchloric acid fuming 70% Supratur, Merck), 3 mL HF (Hydrofluoric acid 40% GR, ISO, Merck), 0.5 mL of HNO_3 (nitric acid 65%, maximum 0.0000005%% GR, ISO, Merck) and evaporated until intense white vapors appeared. The beakers were cooled, their walls were washed with water, and the solution was again evaporated to wet salts. Then 2 mL of HCl (hydrochloric acid fuming 37% OR, ISO, Merck) and 0.2 mL of 0.1M H_3BO_3 solution (analytical grade) were added and evaporated to a volume of 0.5–0.7 mL. The resulting solutions were transferred into polyethylene bottles, 0.1 mL of a solution containing 10 mg L^{-1} In (internal standard) was added, diluted with deionized water to 20 mL, and analysis was performed. As control samples in Teflon beakers, the above procedures were performed without samples, and the resulting solutions were used as controls.

The accuracy of the analytical procedure was confirmed by analysis of Standard reference material Gabbro Essexit STD-2A (GSO 8670-2005). The methods, recoveries and analytical results of certified reference material for the solid phase are given in Supplementary Materials Table S2.

Element concentrations in filtrates were determined by a quantitative method, using reference solutions with 0.5, 1 and 10 μg L^{-1} concentrations of studied elements. In the studied samples, element concentrations were calculated by the spectrometer software. The accuracy of determinations was verified by using the standard sample of "Trace Metals in Drinking Water" produced by High-Purity Standards (Charleston, VA, USA). In addition, the accuracy of the filtrate analyses was verified by comparing the results of ICP–MS and ICP–AES determinations of Li, Al, Mn, Cu, Zn, Sr and Ba. In all cases, differences between the values determined by the two methods were within the standard errors of those methods. The detection limit (DL) was calculated by using the following equation:

$$DL = Ci + 3SD$$

where Ci is the mean value of element concentration in deionized water s and SD is the standard deviation for repeated measurements. The methods, recoveries and analytical results of certified reference material for the dissolved phase were given in Supplementary Materials Table S3.

2.4. Calculations and Data Analysis

Calculations of statistical parameters (arithmetic and geometric mean, standard deviation) were performed by using Microsoft Excel. Total concentrations of elements were calculated with the use of geometric mean values, because the compilation of Normal Probability Plots showed an absence of normal distribution for most of the analyzed elements. Geometric means have been previously applied in the assessment of regional variations in the snowpack composition within Western Siberia [11]; therefore, geometric mean values in the present study were the most appropriate for the comparability of the data obtained. In order to evaluate the intensity of pollution, we calculated the contamination factor (CF) and the enrichment factor (EF) [43] with the help of the following formulas:

$$CF = Cn/Cb$$

where Cn is the measured concentration of the element in the sample and Cb is the background value of the element.

$$EF = (Cn/CAl)sample/(Cn/CAl)crust$$

where (Cn)sample is the measured concentration of the element of interest, (Cn) crust is the concentration of the same element in the Earth's crust, and CAl is the concentration of the reference element in the same sample and the Earth's crust. Unfortunately, available data on soil composition within the Tyumen area include only very few trace elements, which makes it impossible for us to compare the enrichment of particulate matter of snow with that of local soils. Determination of regional baseline concentrations of TMMs in soils still remains an unresolved problem due to the vastness of the territory of Western Siberia, highly variable TMMs concentrations in soils and only small datasets analyzed to date [44].

For the EF calculations, we used data on average element concentrations in the upper part of the continental Earth's crust according to Reference [45].

The CF and EF indices have different applications. The CF values were calculated with the use of the local background values and allow for the evaluation of the influence of urban pollution sources. The EF calculations were based on comparisons with mean values in the Earth's crust which allowed us to evaluate the pollution level at a global scale. Specifically, following the common practice in this field, the normalization to the upper Earth's crust allowed us to assess the enrichment of the atmospheric aerosol and estimate the long-range atmospheric transport of soluble and particulate forms of elements. According to References [46,47], element sources are classified into three groups depending on EF values as follows: EF < 10 is considered to be of a crustal origin, without enrichment; 10 < EF < 100 has a mixed (crustal and anthropogenic) origin; EF > 100 indicates man-made air pollution.

Concentrations in the solid phase were recalculated in relation to the volume of filtered meltwater to allow for a direct comparison with the solution data. Then the percentage of soluble elements (PSE) was calculated as follows:

$$PSE = (Cn\ dissolved)/(Cn\ total)\cdot100\%$$

Geometric means were used in PSE calculations. The potential ecological risk index E_r^i, which characterizes a degree of the ecological risk of a single element [48], was calculated by using the following equation:

$$E_r^i = CF\cdot T_r^i$$

where CF is the contamination factor and T_r^i is the toxicity response coefficient.

In this study, we used the response T_r^i values according to References [48,49], as follows: Zn, Mn, Fe, W, Sr = 1; Cr, Mo, Sn, Sb = 2 ; Pb, Cu, Co, Ni = 5; As = 10, Cd = 30. For risk assessments we adopted the following gradation: E_r^i < 40 describes low risk; 40 < E_r^i < 80 indicates moderate risk; 80 < E_r^i < 160 indicates considerable risk; 160 < E_r^i < 320 indicates high risk; E_r^i > 320 indicates extreme risk [49,50].

The total potential ecological risk index (*RI*) characterizes the overall degree of the ecological risk of all metals under investigation [48].

$$RI = \sum E_r^i$$

where E_r^ii is a potential ecological risk index of a single element. Risk levels were graded as follows: RI < 150, low; 150 < RI < 300, moderate; 300 < RI < 600, considerable; RI > 600, high ecological risk.

In order to assess the risks associated with different forms of elements, we calculated E_r^i and *RI* values separately for dissolved and particulate forms and their total concentrations.

3. Results and Discussion

3.1. Salinity, pH, and Dust Content

Meltwaters from the background sites had an acid reaction, which is typical for snowpack within the taiga zone. Their mean pH value (4.7) was lower than the average meltwater pH (5.3) within the Khanty-Mansi Autonomous Okrug [26]. The acidification of snow around Tyumen can be explained by long-range atmospheric transport from the Urals' metallurgical plants. It has been shown that sulfur compounds emitted from smelters and also from oil-producing enterprises, where they are synthesized as a by-product of gas combustion, are the major factors responsible for water acidification in Russia [51]. Considering that Tyumen is affected by predominantly southern and southwestern winds during the winter (Figure 1), the acidifying agents were likely to be transported from the Southern Urals, where numerous metallurgical districts (Chelyabinsk, Magnitogorsk, Sibay, etc.) are located at distances of 300–500 km to the southeast of Tyumen. Massive emissions of sulfur dioxide from metallurgical plants of the Urals have been shown to have the potential to cause acidification of atmospheric precipitation [52]. For example, the air within the Karabash copper smelter area is characterized by an SO_2 concentration of 20,000 $\mu g\ m^{-3}$, which is far higher than the standard SO_2 concentration of 500 $\mu g\ m^{-3}$ according to the WHO air quality guidelines [53]. The development of a weakly acid reaction of snow has been reported in the eastern part of the Sverdlovsk (Yekaterinburg) Oblast, which borders the Tyumen Oblast from the west [52]. The long-range atmospheric transport of TMMs-enriched aerosols from the Urals' metallurgical plants has also been shown to affect the area of Arctic seas [54].

The mean meltwater salinity of 9.5 mg L^{-1} was typical for unpolluted areas of Western Siberia. A dissolved salt content of less than 15 mg L^{-1} of atmospheric precipitation can be regarded as a regional background in Russia [55].

The content of particulate matter in meltwaters from the background sites varied from 4 to 10 mg L^{-1}, which is slightly higher than the values reported for the Arctic snow cover [29,56]. It is probable that the dust in the studied background samples is sourced mainly from soils of the steppe areas that are located south of Tyumen, which have a relatively thin snow cover [11]. Deflation of soil particles under such conditions can be quite intensive even during the winter [57]. It is also not excluded that some aerosols can be carried from the industrial regions of the Middle and Southern Urals.

The composition of snow meltwaters within the city of Tyumen was significantly different from that within the background area. There was a significant increase in the mean pH value (6.3) of the city snow as compared to the mean background value (4.7), although the reaction of most of the samples from the city remained weakly acid with only a few samples with neutral and weakly alkaline reactions. The observed alkalinization of the city snow was due to the deposition of dust from building materials, which predominantly consist of carbonates [57].

Snow meltwaters in the city had a mean salinity of 68.1 mg L^{-1}, which was more than 7 times higher than that in the background. The maximal content of dissolved salts (202–564 mg L^{-1}) was observed near the roads treated with de-icing agents, which are mainly composed of technical salt NaCl. For comparison, the salinity of snow meltwater within Moscow is 4 times as high as the mean background value (23 mg L^{-1}) [58]. The content of particulate matter in meltwaters from the city varied from 9 to 121 mg L^{-1}, with a mean value of 37 mg L^{-1}, which was 5 times as high as the background value.

3.2. Elemental Composition

The elemental composition of snow meltwater from the background sites is presented in Table 1. In the dissolved phase, Ca was predominant and other major elements were arranged in the following decreasing order of concentrations: Na > S > Mg > K > Fe > Al. Interestingly, major elements in the global river discharge form a nearly identical order based on the values of the dissolved transport index (DTI) as follows: Na ≈ Ca > Mg > K > P > Fe > Al [59]. The DTI is a ratio of dissolved forms of elements to their total content, which in the global river discharge depends on the solubility of those elements. The similarity of the concentration orders led us to conclude that the element contents in the snowpack from the study area were predetermined by their solubility, with only insignificant influences of anthropogenic and other factors (e.g., a predominance of either continental or oceanic aerosols).

Table 1. The elemental composition (µg L^{-1}) of meltwater from the background sites.

Elements	Dissolved				Particulate				Total	PSE, %
	DL	Mean	SD	Geometric Mean	DL	Mean	SD	Geometric Mean		
Li	0.007	0.14	0.095	0.11	0.0014	0.005	0.0029	0.005	0.12	95.7
Na	5	402	374	282	1.3	2.8	0.6	2.7	284.7	98.9
Mg	4	131	64	119	1.5	12.4	12.5	9.5	128.5	92.4
Al	1	10.2	4.9	8.9	0.9	11.7	5.4	10.9	19.8	45.2
P	16		nd		0.25	0.76	0.64	0.63	0.63	nd
S	15	355	68	350	1.5	7.2	0.5	7.1	357.1	98.0
K	5	72.3	18.7	70.3	0.9	4.1	1.1	3.9	74.2	94.7
Ca	6	595	135	581	2.5	5.6	0.3	5.6	586.6	99.0
Ti	0.8		nd		0.036	0.081	0.075	0.065	0.065	nd
V	0.07	0.152	0.088	0.11	0.025	0.086	0.055	0.072	0.18	59.4
Cr	0.6		nd		0.05	0.27	0.130	0.25	0.25	nd
Mn	0.07	7.3	6.17	5.89	0.05	0.15	0.073	0.14	6.04	97.6
Fe	4	20.2	7.53	18.9	3.0	14.0	9.48	12.2	31.1	60.8
Co	0.1		nd		0.003	0.009	0.006	0.008	0.008	nd
Ni	0.3	0.95	0.46	0.80	0.032	0.161	0.099	0.15	0.95	84.2
Cu	0.3	4.68	2.53	4.09	0.032	0.091	0.046	0.082	4.17	98.1
Zn	0.7	13.2	5.51	12.2	0.14	0.535	0.083	0.530	12.7	96.1
As	0.06	0.505	0.13	0.49	0.031	0.046	0.015	0.045	0.54	90.7
Rb	0.006	0.100	0.023	0.098	0.0023	0.009	0.005	0.008	0.11	89.1
Sr	0.08	1.505	0.37	1.46	0.014	0.044	0.010	0.043	1.51	96.7
Y	0.002	0.004	0.002	0.003	0.0005	0.004	0.002	0.004	0.007	42.9
Zr	0.004		nd		0.0005	0.032	0.012	0.030	0.030	nd
Nb	0.004		nd		0.0005	0.0044	0.0013	0.0042	0.0042	nd
Mo	0.008	0.051	0.016	0.049	0.0002	0.0094	0.0025	0.0092	0.058	84.5
Ag	0.004	0.022	0.016	0.017	0.0002	0.010	0.008	0.008	0.025	68.0
Cd	0.004	0.096	0.030	0.092	0.0004	0.001	0.002	0.001	0.093	99.0
Sn	0.008	0.076	0.043	0.066	0.0034	0.137	0.149	0.091	0.157	42.0
Sb	0.004	0.104	0.029	0.100	0.0015	0.003	0.001	0.003	0.104	96.2
Cs	7·10^{-4}	0.0058	0.0015	0.0057	0.0004	0.0009	0.0004	0.0009	0.0065	87.7

Table 1. *Cont.*

Elements	Dissolved				Particulate				Total	PSE, %
	DL	Mean	SD	Geometric Mean	DL	Mean	SD	Geometric Mean		
Ba	0.05	3.20	1.15	4.69	0.005	0.08	0.04	0.08	4.77	98.3
La	0.003	0.0100	0.0050	0.0090	0.0003	0.0069	0.0027	0.0065	0.016	58.1
Ce	0.002	0.014	0.009	0.012	0.0003	0.013	0.005	0.012	0.024	50.0
Pr	$4 \cdot 10^{-4}$	0.0017	0.0010	0.0015	0.0003	0.0010	0.0005	0.0009	0.0024	62.5
Nd	0.002	0.0072	0.0026	0.0068	0.0003	0.0058	0.0037	0.0049	0.0118	57.6
Sm	$9 \cdot 10^{-4}$	0.0044	0.0042	0.0032	0.0005	0.0008	0.0005	0.0008	0.0039	82.1
W	0.003	0.012	0.007	0.011	0.0005	0.003	0.002	0.002	0.012	91.7
Pb	0.02	1.34	0.76	1.12	0.002	0.13	0.10	0.10	1.23	91.1
Bi	0.001	0.011	0.008	0.008	0.0003	0.0010	0.0004	0.0010	0.0090	88.9
Th	$9 \cdot 10^{-4}$	0.0019	0.0006	0.0019	0.0002	0.0011	0.0006	0.0010	0.0028	67.9
U	$5 \cdot 10^{-4}$	0.0015	0.0005	0.0014	0.0002	0.0005	0.0003	0.0004	0.0019	73.7

Note: concentrations of Be, Sc, Ga, Se, Rh, Pd, Te, Eu, Gd, Hg, Tb, Dy, Ho, Er, Yb, Hf, Ta and Lu were below their detection limits; nd—not detected in more than 80% of samples; total is the sum of the geometric mean.

The majority of elements within the background area occurred mainly in their dissolved forms, the proportions of which varied from 43% (Sn) to 98.9% (Na) (see Table 1). Al and Sn were represented mostly by particulate forms. Aluminum belongs to very weakly soluble elements and Sn is similar to Fe and Al in terms of solubility [60]. Concentrations of Ti, Cr, Co and Zr in the dissolved phase were below their detection limits in more than 80% of samples, which is also indicative of the predominance of their particulate forms. Approximately equal proportions of dissolved and particulate forms were found in Fe, V and some of the rare-earth elements (Ce, Nd and La). A majority of TMMs (Cu, Zn, Ni, As, Mo, Cd, Pb, Sr, Ba and Bi) were found predominantly in the dissolved phase (83–98% of their total concentrations).

According to Reference [11], Na, Ca, Sr and Cd in snow meltwaters from West-Siberian areas to the north of Tyumen are found mostly in dissolved forms; K, As, Zn and Ba are equally divided between the dissolved and particulate phases; and other elements occur predominantly in the particulate phase. The background sites near Tyumen were distinguished by much higher proportions of dissolved forms of elements. We attribute such a discrepancy to the fact that the study [11] included not only background, but also areas affected by local pollution sources. Even low-intensity human impact can result in an increased deposition of dust aerosols and shifts the balance of dissolved and particulate phases.

Studies conducted in different regions have demonstrated that the proportions of dissolved and particulate phases are predetermined by their emission sources. According to Reference [61], Al, Ca, Fe, K, Mg and Na originate from natural sources, whereas trace metals, including Cd, Cu, Ni, Pb and Zn, are primarily associated with anthropogenic activities. An analysis of the literature showed that Al, Fe, Cr and Co generally prevail within conventional background areas located at significant distances from any sources of industrial emissions. For example, atmospheric aerosols from the Northern Atlantic region have been shown to be dominated by the particulate fractions of Al, Fe and Co [62]. A study on snow composition in China showed that the least soluble elements were Fe, Al and Cr with a solubility of less than 30% [46]. Likewise, in Quebec (Canada), a study has shown that Fe, Ni, Cr, Co, V, La, Be, Ce, Tl, Y and Rb are found predominantly in the particulate fraction [10]. In that study the assemblage of elements found mainly in the particulate fraction is very similar to that in the present study, except for Ni, which near Tyumen mostly occurred in the dissolved form, which substantiates the conclusion about its anthropogenic origin.

Assemblages of elements that occur predominantly in dissolved forms in different background areas of the world can vary depending on their anthropogenic emission sources, the composition of local soils and the direction of atmospheric transport. In China, the most soluble elements (with solubilities reaching 60–70 %) are As, Mn, Cu, Zn Cd and Se emitted from regional coal-fired power plants and municipal solid waste incineration [46]. In Canada, the influence of a metal smelter resulted in the predominance of soluble Cu, Zn, Cd, Pb, S, Sr and Sb [10].

It can be stated that background areas, where dust deposition is insignificant, are characterized by the predominance of dissolved fractions of ecologically hazardous TMMs (Cd, Pb, Ni, Cu, Zn, As and Mo), except for Fe, Co and Cr that occur mainly in particulate forms. The low solubility of Fe, Co, Cr and Al is explained by their strong bonds with the crystalline lattice of aluminosilicate minerals. The elements such as Zn, Cu, Mn, Ni and V bound with carbonate minerals are highly soluble [63]. Therefore, the quantities and ratios of dissolved and particulate fractions of trace elements are predetermined by the mineralogical composition of dust aerosols.

In order to assess the local geochemical background of Tyumen, we compared the data obtained from the background sites with average values for Western Siberia [11]. Most of major and trace elements around Tyumen have higher concentrations in the dissolved phase as compared to the regional average (Figure 2). The most significant differences were observed in the elements connected with carbonate dust aerosols of natural origin (Mn and V) and the elements emitted into the atmosphere from industrial plants and transportation (Pb, Cd, Cu, Sb, Ni and As). It should be noted that the metallurgical districts of the Southern Urals are characterized by a similar assemblage of pollutants. The particulate deposition rates and concentrations of Cu, Zn, Pb, Cd, As, Se, Bi, Sb and Sn within the industrial mining districts of the Southern Urals exceed those within conventional background areas by three orders of magnitude [64]. In Chelyabinsk (Southern Urals), Cd, Cu, Mn, Ni, Pb, Sr and Zn enrichment of atmospheric dust is caused by the metallurgical industry [65]. Therefore, high concentrations of dissolved Pb, Cd, As, Cu, Sb and Ni, as the snow acidification within the study area, can be associated with the influence of the industrial regions located in the middle and Southern Urals, taking into account the predominance of southern and southwestern winds within the study area (see Figure 1). Previously, it has been shown that some of the pollutants present in the south of the Tyumen Oblast have been carried from the Urals' large industrial plants [66]. Thus, the concentration of dissolved forms of elements is an effective indicator of the long-range transport of aerosols.

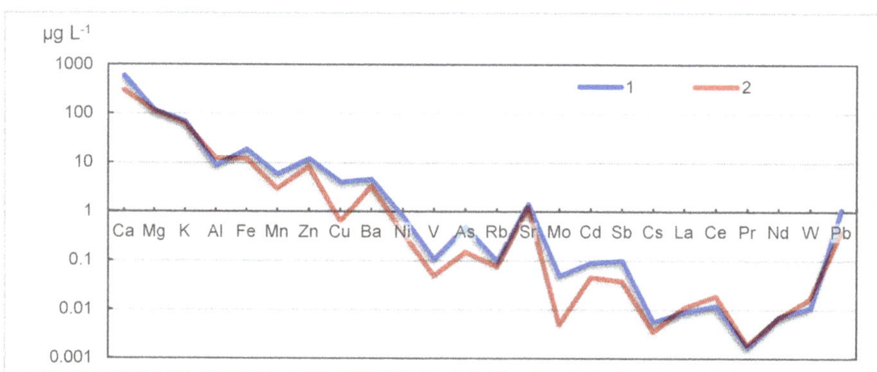

Figure 2. Concentrations of elements in the dissolved phase: 1—background sites near Tyumen (our data); 2—mean regional concentrations within Western Siberia [8].

The elemental composition of snow meltwater within Tyumen, with the partitioning between dissolved and particulate fractions, is presented in Table 2.

Table 2. The elemental composition (μg L^{-1}) of meltwater within the city of Tyumen.

Elements	Dissolved			Particulate			Total	PSE, %
	Mean	SD	Geometric Mean	Mean	SD	Geometric Mean		
Li	0.88	2.73	0.31	0.78	0.95	0.27	0.58	52.5
Be		nd		0.047	0.054	0.018	0.018	nd
Na	15612	24849	9115	375	458	136	9251	98.5
Mg	1473	867	1226	11608	15962	3219	4445	27.6
Al	14	9	11	2472	2937	836	847	1.3
P	19.7	19.2	14.5	82.4	156.6	25.7	40.2	36.1
S	1123	396	1070	187	231	80	1150	93.0
K	393	350	293	345	403	138	431	67.9
Ca	6674	4105	5753	2953	4348	777	6530	88.1
Sc		nd		1.11	1.44	0.33	0.33	nd
Ti		nd		184.7	260.6	54.0	54.0	nd
V	0.16	0.07	0.14	6.71	8.09	2.16	2.30	6.1
Cr		nd		81.2	109.8	24.7	24.7	nd
Mn	1.8	1.4	1.4	98.9	123.4	31.5	32.9	4.3
Fe	10.9	9.3	6.2	5728	7882	1638	1644	0.4
Co		nd		5.71	8.03	1.58	1.58	nd
Ni	4.3	2.5	3.7	93.4	129.9	27.8	31.5	11.7
Cu	4.9	3.6	3.6	24.2	37.1	8.0	11.6	31.3
Zn	10.0	4.0	9.1	59.1	70.6	25.0	34.1	26.7
Ga		nd		0.46	0.56	0.15	0.15	nd
As	0.44	0.18	0.36	1.04	1.15	0.47	0.83	43.4
Rb	0.37	0.30	0.30	1.13	1.34	0.42	0.72	41.1
Sr	30.0	74.5	13.8	6.95	9.08	2.5	16.3	85.2
Y		nd		0.62	0.79	0.21	0.21	nd
Zr		nd		4.36	5.66	1.52	1.52	nd
Nb		nd		0.39	0.52	0.14	0.14	nd
Mo	0.24	0.09	0.22	0.47	0.75	0.16	0.38	57.9
Ag	0.026	0.014	0.021	0.035	0.036	0.018	0.04	52.5
Cd	0.050	0.019	0.045	0.097	0.11	0.038	0.083	56.3
Sn	0.079	0.063	0.054	1.46	2.21	0.55	0.61	8.9
Sb	0.35	0.20	0.28	1.28	1.69	0.46	0.74	38.4
Cs	0.022	0.032	0.015	0.080	0.095	0.030	0.044	34.1
Ba	59.3	96.6	35.9	24.3	31.4	8.2	44.1	81.4
La	0.010	0.006	0.009	0.73	0.89	0.27	0.28	3.2
Ce	0.014	0.010	0.011	1.61	2.0	0.57	0.585	1.9
Pr	0.0015	0.0008	0.0013	0.15	0.18	0.052	0.053	2.5
Nd	0.0059	0.0033	0.0051	0.543	0.65	0.200	0.205	2.5
Sm	0.0026	0.0015	0.0022	0.11	0.135	0.040	0.043	5.1
Gd	0.0015	0.0004	0.0015	0.096	0.12	0.034	0.036	4.2
Er	0.0015	0.00024	0.00075	0.063	0.077	0.022	0.022	3.4
W	0.11	0.11	0.09	1.34	1.81	0.36	0.44	20.5
Pb	0.34	0.30	0.21	13.1	20.8	4.6	4.80	4.4
Bi	0.0081	0.0071	0.0059	0.11	0.16	0.040	0.046	12.8
Th	0.0014	0.0004	0.0014	0.13	0.15	0.050	0.052	2.7
U	0.0043	0.0091	0.0024	0.084	0.105	0.028	0.030	8.0

Note: concentrations of Se, Rh, Pd, Te, Eu, Hg, Tb, Dy, Ho, Tm, Yb, Hf, Re, Ir, Ta Tl and Lu were below their detection limits; nd—not detected in more than 80% of samples; total is the sum of the geometric mean.

Concentrations of major elements decreased in the following order: Na > Ca > Mg > S > K > P > Al > Fe. More than 50% of Na, S, Ca and K were found in dissolved forms, as those elements are known to be the most active water migrants [60]. Hence, major element concentrations in the snowpack samples from both the city and background areas depend primarily on the element solubility. However, it should be noted that some samples had very high concentrations (up to 144 mg L^{-1}) of Na, which were indicative of the use of de-icing agents, i.e., solid technical salt NaCl being the main agent applied to Tyumen's roads and pavements in order to remove snow and ice.

A prevalence of dissolved forms was observed in highly soluble trace elements (Mo, Cd, Li and Sr), as well as in Ba, which has a high solubility when the SO_4^{2-} concentration is low [60]. Other TMMs were characterized by the predominance of particulate fractions; for example, Al, Fe, Ni, Cr, Co, Ti, Pb, Sn and Bi had less than 13% of dissolved forms (Figure 3). Similarly, urban snow of Moscow contains Sn, Ti, Bi, Al, W, Fe, Pb, V, Cr, Rb, Mo, Mn, As, Co, Cu, Sb and Mg mainly in particulate form, and Ca and Na in dissolved form [67]. Due to the increased rate of dust aerosol deposition within Tyumen as compared to that in the background area, the balance between dissolved and particulate forms of elements is shifted towards the predominance of particulates (Figure 3).

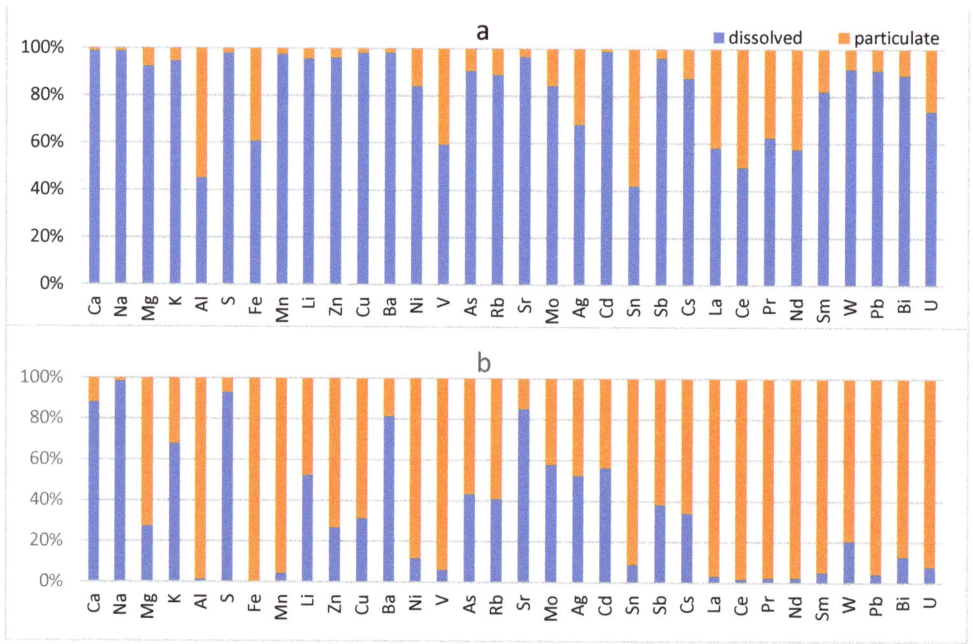

Figure 3. Proportions of dissolved (1) and particulate (2) forms of elements in the background area (**a**) and Tyumen City (**b**).

The values of CF, characterizing the surplus of element concentration within the city as compared to the background, were significantly higher in the particulate phase as compared to the dissolved phase (Figure 4). This is connected with a higher rate of dust deposition within the city, as well as a higher total concentration of TMMs in the dust. It should be mentioned that concentrations of dissolved forms of some elements (Mn, Fe, Co, Cu, Zn, As, Cd, Sn, Pb and Bi) within the city were lower than those in the background, with CF < 1 (Figure 4). This finding is paradoxical, taking into account that the deposition rate of metals in particulate forms had multiplied by tens and hundreds of times within the city. Nevertheless, similar situations have been reported in previous studies. For example, Viklander [68] detected a decrease in the dissolved Pb concentration in the sites affected by heavy traffic in comparison with that in background sites. Possible explanations can include either the coagulation of small (d = 0.8–1.5 µm) particles of dissolved fractions into larger (d = 50–600 µm) insoluble particles [69], or the adhesion of smaller to larger particles that would also decrease the content of dissolved forms [68].

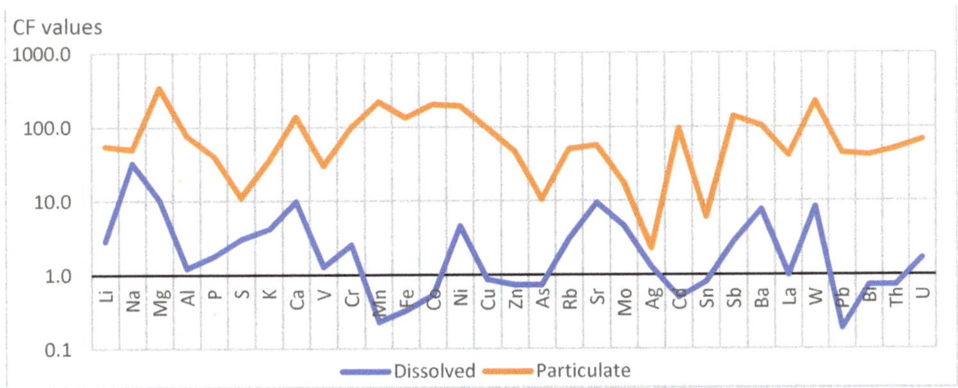

Figure 4. Logarithmic contamination factor values computed by using geometric means.

The EF values for trace and major elements are shown in Figure 5. The background area was characterized by the most significant enrichment in Cd, Ag, Sb, Zn and Cu (EF > 1000) and the minimal enrichment in Fe and V (Figure 5a).

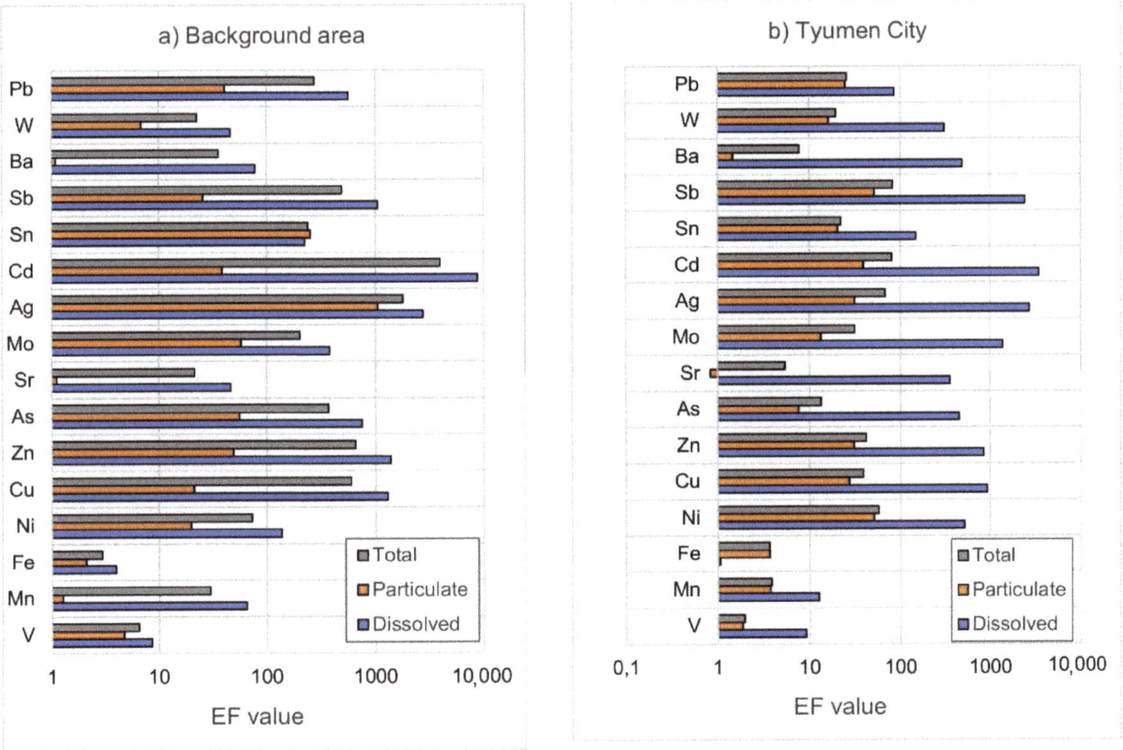

Figure 5. The enrichment-factor values of TMMs in the background area (**a**) and Tyumen City (**b**).

Similar results were obtained in the study by Shevchenko et al. [11] showing that snowpack within Western Siberia has a range of EF values, 1–5 (Fe) to more than 1000 (Sb, Zn and Cd). The enrichment of aerosols in Cd and Zn due to manmade pollution has been detected in many studies in different parts of the globe. Kim et al. [70] noted that Asian and non-Asian dust was strongly or extremely polluted by Cd, Zn, Pb, Cu and As. Dong et al. [71] reported that Zn and Cd were clearly enriched (EF values >100) in surface snow and a snow pit on the Northern Tibetan Plateau of China. In the snow of Teheran (Iran), maximal EF values are reported for Zn, Pb and Cd [72]. In Southern China, atmospheric aerosols are enriched in Se, Cd, Zn, Pb, As, Mo and Cu (EF > 100) [46]. The nonferrous metal industry makes a significant contribution to the emissions of Cd, Zn Pb and Cu into the environment [73]. As discussed above, high concentrations of Pb and Cd in the snowpack around Tyumen City are connected with the wind transportation of polluted air from the Urals, which is the largest metallurgical and metal processing center [66].

Based on the EF values of the TMMs deposited with the snowpack of Tyumen City (Figure 5b), we can distinguish three groups of elements as follows: (1) low enriched (EF 1–10) elements (Rb, Fe and V) sourced from soils; (2) moderately enriched (EF 10–100) elements (W, Ba, Sr, Ni, Co, Cr and Mn) that originate from mixed soil-anthropogenic sources; and (3) highly enriched elements (Pb, Sb, Cd, Ag, Mo, As, Zn and Cu) with EF values > 100, reflecting an important contribution from anthropogenic sources.

Pollution within Tyumen City was primarily associated with local sources. The main source of Pb and Sb emissions in Tyumen was the storage battery plant, around which these elements were found in highest concentrations. Until very recently this plant produced storage batteries from lead-antimony alloys containing Pb, Sb, As, Bi and Sn [74]. It is well known that vehicle emissions and road dust resuspension can be an essential source of Pb, Sb, Cu, Ni and other metals [75–77]. However, leaded gasoline has been banned in Russia since 2006. Concentrations of Pb in the street dust of Tyumen have been shown to be lower than those in the street dust of many large cities and lower than those in soils [39]. Therefore, transport has an insignificant influence on the Pb concentration in the snowpack of Tyumen, but it probably makes significant contributions to the increase in Ni, Sb and Cu concentrations.

The abrasion of tires, metal parts of cars and road markings is known as a source of Sb [76]. As noted by Hjortenkrans et al. [78], on motorway sections where traffic regularly brakes (traffic lights, intersections, etc.) and in other decelerating environments such as roads with traffic lights, roundabouts and intersections, the Sb concentration is more than 8 times the background level.

Principal sources of Cu in the atmosphere include fossil fuel burning, traffic emissions, fuel combustion and industrial combustion [79]. According to Reference [39], concentrations of Cu, Sb, As and Mo in the street dust of Tyumen City are most likely related to traffic pollution, although they can come from other sources.

Emissions of Zn and Cd originate mainly from solid waste incinerators [80]. The abrasion of automobile tires and the manufacturing and dumping of Cd batteries may be other sources of cadmium emissions [81]. Cu and Zn have a close relationship with fossil fuel combustion [82].

According to Reference [83], Ni, Fe and Cr in city dust are mainly traffic related. Diesel engines use fuels with Ni additives [84]. The most important source of Ni is the combustion of petroleum and petroleum products [80]. The global emission of Ni from the combustion of oil products is estimated to range from 10 to more than 40 kt/yr [85].

3.3. Contamination Levels and Risk Assessment

The values of E_r^i calculated for dissolved (1), particulate (2) fractions and the total concentrations of TMM (3) calculated for the city territory are shown in Figure 6. The E_r^i values of particulate forms of elements were many times higher than those of dissolved forms, which was due to the high dust deposition rates within the city that are five times higher than those in the background area. It should be noted that the CF values were also significantly higher in the particulate phase as compared to the dissolved phase (see Figure 4). Dissolved forms of all elements presented low ecological risks ($E_r^i < 40$). Particulate forms of Cd, Cu, Ni, Co, Mn, Pb, W, Sb and Cr were associated with extremely high ecological risks ($E_r^i > 320$). The mean values of E_r^i indices calculated from the total content of particulate and dissolved fractions were shown to increase in the following order: Zn(5.0) < Mn(13.6) < Sn(14.4) < Mo(23.4) < Sr(23.9) < As(26.3) < V(28.8) < Sb(30.3) < Cu(30.4) < Cd(45.3) < Pb(45.4) < W(99.9) < Ni(440) < Cr(604) < Co(637). Therefore, moderate risk is associated with the atmospheric deposition of Cd and Pb; considerable risk is associated with W deposition; high risk is associated with Fe; and very high risk is associated with Ni, Co and Cr.

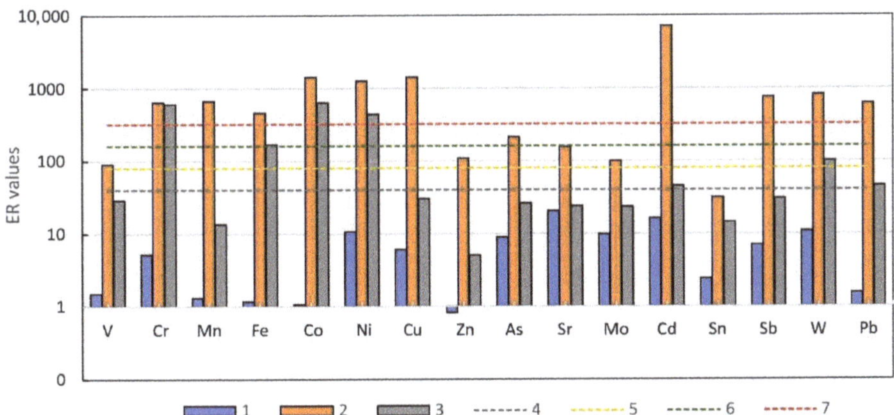

Figure 6. Potential ecological risk index values: 1—dissolved forms; 2—particulate forms; 3—total; 4–7—ecological risk levels: 4—low (<40); 5—moderate (40–80); 6—considerable (80–160); 7—high (320). The values above 320 characterize an extreme ecological risk.

The ranges of RI values within the Tyumen City were very wide due to the diversity of emission sources. Our calculations showed that 31% of sampling sites belong to the low risk category (RI < 80), 20%—moderate risk (80 < RI < 160), 9%—considerable risk (160 < RI < 320), and 40%—high risk (RI > 320) (Figure 7a). The distribution of RI values over different land-use areas within Tyumen showed that transport zones, industrial zones and the historical center had the most unfavorable ecological conditions with RI > 600 (high ecological risk) (Figure 7b). Near the main motorways, there were increased dust deposition rates and increased contents of particulate fractions of elements. The historical center has been affected by long-term soil pollution. Much better ecological conditions were observed in the areas of modern high-rise residential areas and the business zone that appeared between 1970 and 2020.

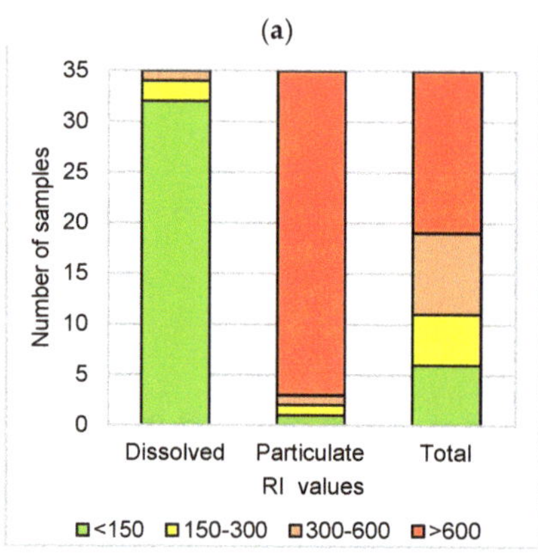
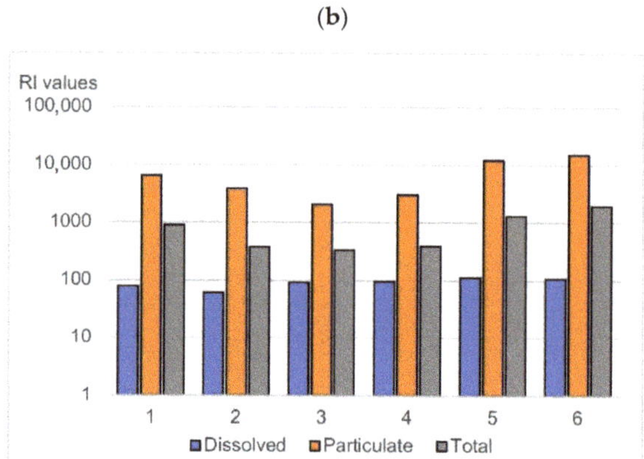

Figure 7. RI distribution in Tyumen City: (**a**) the frequency distribution of RI $_{total}$ values in snow samples and (**b**) mean values of RI within different land-use areas, based on the total content of particulate and dissolved TMMs in the snowpack. Land-use areas: 1—the historical center; 2—low-rise residential area; 3—high-rise residential area; 4—modern business zones; 5—industrial zones; 6—transport zones.

3.4. Comparison of TMMs Concentration with Other Data

In order to assess the level of pollution in Tyumen in relation to other locations, we compared the data obtained from the literature sources [22,24,57–61,67,68]. The rates of TMMs deposition within snow in different cities significantly varies depending on the local and regional anthropogenic emission sources. The range of variations in TMMs concentrations is very wide, reaching 1 to 2 orders of magnitude. Selected data from the literature are shown in Table 3.

Table 3. A comparison of trace element concentrations in snow from different locations ($\mu g \cdot L^{-1}$).

City, Country, Location	Co	Ni	Cu	Zn	As	Cd	Pb	Mn	V	Reference
Tyumen, Russia dissolved	nd	3.7	3.6	9.1	0.36	0.045	0.21	1.4	0.14	
particulate	1.58	27.8	8.0	25.0	0.47	0.038	4.6	31.5	2.16	This study
total	1.58	31.5	11.6	34.1	0.83	0.08	4.8	32.9	2.29	
Moscow, traffic zones, dissolved	0.28	4.6	5.2	16	0.05	0.068	0.35	11	0.92	
particulate	1.6	6.1	1.7	33	0.27	0.053	5	55	12	[67]
total	1.88	10.7	6.9	49	0.32	0.121	5.35	66	12.9	
Vladivostok, Russia (total), city	-	0.64	2.81	32	-	0.11	0.91	36.4	0.69	[86]
suburb	-	0.36	2.16	22.7	-	0.05	1.04	17.3	0.69	
Lake Baikal and the adjacent territory(total)	1.0	1.0	2.0	5.0	-	1.0	1.5	8.0	2.0	[24]
Chernogolovka, Russia(total)	4.1	7.8	5.8	240	-	-	7.2	141	16	[87]
Svirsk, Russia (total)	0.41	2.3	2.3	18	3.7	0.07	0.48	-	3.3	[88]
Poznan, Poland (total)	-	3.77	2.03	13.2	0.71	0.08	4.93	-	-	[22]
Tianjin, China. 2015 year (total)	0.17	1.25	1.96	22.1	1.37	0.66	0.17	13.6	0.34	[89]
Luleo, Sweden, City Center, no-traffic area (total)	-	-	14	50	-	-	16	-	-	[68]

In relation to other cities, Tyumen is characterized by a high concentration of Ni in the snow, which is only comparable with the values reported from Istanbul [11]. Concentrations of other elements are at intermediate levels and fit into the range of variations found in the other cities. For example, the Pb concentration in Tyumen is higher than those in Vladivostok [86] and Svirsk [88] (Russia), but lower than those in Chernogolovka (Russia) [87] and Luleo (Sweden) [68]. However, it should be mentioned that the latter two cities were studied in the late 1990s, when Pb additives to car fuel were widely used. The concentration of Cd, which is mainly emitted into the atmosphere through coal combustion, is quite low in Tyumen, where the thermal power stations and a significant proportion of private households use natural gas as a fuel.

4. Conclusions

The background sites located at distances of 25–35 km from Tyumen City were characterized by increased concentrations of soluble Cd, Cu, Zn, Pb, Ni, As and Mo as compared to their average levels in the snowpack of Western Siberia. A similar assemblage of pollutants is reported in published papers on metallurgical districts of the Southern Urals located at distances of 250–350 to the southwest of Tyumen. Due to the predominance of southern and southwestern winds in the study area during the winter, the increased concentrations of those elements can be explained by a long-range atmospheric transport from the sources located in the Urals. This conclusion is additionally substantiated by the acidification of snow due to the SO_2 emissions from metallurgical plants. Therefore, soluble TMM concentrations in the snowpack provide information on a long-range transport of pollutants, which can be used for modeling of the distribution of those pollutants.

The analysis of the composition of snow meltwater within Tyumen revealed significant human-induced differences as compared to the snowpack of the background areas. In particular, the contents of dissolved salts and dust in Tyumen snow meltwater were multiplied by 7 and 5 times, respectively, and pH increased from 4.7 (in the background) to 6.3 (in the city).

There was a manifold increase in TMM deposition from the atmosphere in Tyumen. The balance of dissolved and particulate fractions of elements also changed: the background area was characterized by the predominance of dissolved forms of the majority of TMMs, while the city was completely dominated by particulate forms. Therefore, a combined assessment of both particulate and dissolved fractions is necessary for a precise determination of pollution level. The TMM deposition from snow within the city was increased by 1 to 2 orders of magnitude in comparison with the background due to the increased rate of atmospheric particulate deposition and higher concentrations of the elements in dust particles. Highly enriched elements (Pb, Sb, Cd, Ag, Mo, As, Zn and Cu) with EF values >100 reflected an important input from anthropogenic sources. Calculations of RI showed that 40% of sampling sites belonged to the category of very high risk. The highest RI values were in the historical center, industrial zone and along the roads with a most intensive automobile traffic. Relatively low pollution levels were observed in the modern residential areas and business zones of Tyumen. The comparison with other cities showed that Tyumen's snowpack had a high content of Ni, while other element concentrations were at intermediate levels.

This research could offer a reference for the atmospheric pollution prevention and control in Tyumen. The data obtained in the present study allowed us to identify the most polluted parts of the city, which are located in the center and along the roads with the most intensive traffic. Ecological monitoring of Tyumen City should prioritize the assessment of particulate forms of Ni, Co Sb, Zn, Cd, Pb, Cu, As, Sn, W and Bi, which can adversely affect human health.

Supplementary Materials: The following are available online at https://www.mdpi.com/article/10.3390/min11070709/s1. Table S1: Description of sampling sites. Table S2: Methods of analysis, analytical results and recovery of certified reference material "Gabbro Essexit STD-2A (GSO 8670-

2005). Table S3: Methods of analysis, analytical results and recovery of certified reference material "Trace Metals in Drinking Water".

Author Contributions: Conceptualization and methodology, D.M.; field sample acquisition, R.P. and A.T.; writing—review and editing, D.M. and A.Z.; visualization, R.P. and A.T. All authors have read and agreed to the published version of the manuscript.

Funding: This research was funded by the Russian Foundation for Basic Research (project no. 19-05-50062) and project no. 121041600045-8 of RAS Siberian Branch.

Data Availability Statement: No applicable.

Acknowledgments: We would like to thank the financial support from the Russian Foundation for Basic Research (19-05-50062). The authors are especially grateful to Vasiliy Karandashev for the element determination.

Conflicts of Interest: The authors declare no conflict of interest.

References

1. Kasimov, N.S.; Vlasov, D. Global and Regional Geochemical Indexes of Production of Chemical Elements. *Geogr. Environ. Sustain.* **2014**, *7*, 52–65. [CrossRef]
2. Pacyna, J.M.; Pacyna, E.G. An assessment of global and regional emissions of trace metals to the atmosphere from anthropogenic sources worldwide. *Environ. Rev.* **2001**, *9*, 269–298. [CrossRef]
3. Zhulidov, A.V.; Robarts, R.D.; Pavlov, D.F.; Kämäri, J.; Gurtovaya, T.Y.; Meriläinen, J.J.; Pospelov, I.N. Long-term changes of heavy metal and sulphur concentrations in ecosystems of the Taymyr Peninsula (Russian Federation) North of the Norilsk Industrial Complex. *Environ. Monit. Assess.* **2011**, *181*, 539–553. [CrossRef]
4. Äyräs, M.; De Caritat, P.; Chekushin, V.A.; Niskavaara, H.; Reimann, C. Ecogeochemical investigation, Kola peninsula: Sulphur and trace element content in snow. *Water Air Soil Pollut.* **1995**, *85*, 749–754. [CrossRef]
5. Gregurek, D.; Melcher, F.; Pavlov, V.A.; Reimann, C.; Stumpfl, E.F. Mineralogy and mineral chemistry of snow filter residues in the vicinity of the nickel-copper processing industry, Kola Peninsula, NW Russia. *Miner. Pet.* **1999**, *65*, 87–111. [CrossRef]
6. Onuchin, A.A.; Burenina, T.A.; Zubareva, O.N.; Trefilova, O.V.; Danilova, I.V. Pollution of snow cover in the impact zone of enterprises in Norilsk Industrial Area. *Contemp. Probl. Ecol.* **2014**, *7*, 714–722. [CrossRef]
7. Shevchenko, V.; Lisitzin, A.; Vinogradova, A.; Stein, R. Heavy metals in aerosols over the seas of the Russian Arctic. *Sci. Total. Environ.* **2003**, *306*, 11–25. [CrossRef]
8. Callaghan, T.V.; Johansson, M.; Brown, R.D.; Groisman, P.Y.; Labba, N.; Radionov, V.; Bradley, R.S.; Blangy, S.; Bulygina, O.N.; Christensen, T.R.; et al. Multiple Effects of Changes in Arctic Snow Cover. *Ambio* **2011**, *40*, 32–45. [CrossRef]
9. Barrie, L.A. Arctic air pollution: An overview of current knowledge. *Atmos. Environ.* **1986**, *20*, 643–663. [CrossRef]
10. Telmer, K.; Bonham-Carter, G.F.; Kliza, D.A.; Hall, G.E.M. The atmospheric transport and deposition of smelter emissions: Evidence from the multi-element geochemistry of snow, Quebec, Canada. *Geochim. Et Cosmochim. Acta* **2004**, *68*, 2961–2980. [CrossRef]
11. Shevchenko, V.P.; Pokrovsky, O.S.; Vorobyev, S.; Krickov, I.V.; Manasypov, R.M.; Politova, N.V.; Kopysov, S.G.; Dara, O.M.; Auda, Y.; Shirokova, L.S.; et al. Impact of snow deposition on major and trace element concentrations and elementary fluxes in surface waters of the Western Siberian Lowland across a 1700 km latitudinal gradient. *Hydrol. Earth Syst. Sci.* **2017**, *21*, 5725–5746. [CrossRef]
12. Tranter, M.; Brimblecombe, P.; Davies, T.D.; Vincent, C.E.; Abrahams, P.W.; Blackwood, I. The composition of snowfall, snowpack and meltwater in the Scottish highlands-evidence for preferential elution. *Atmos. Environ. (1967)* **1986**, *20*, 517–525. [CrossRef]
13. Telloli, C. Metal Concentrations in Snow Samples in an Urban Area in the Po Valley. *Int. J. Geosci.* **2014**, *5*, 1116–1136. [CrossRef]
14. Demographics. Federal State Statistics Service. Available online: https://rosstat.gov.ru/folder/12781 (accessed on 21 May 2021).
15. Kasimov, N.S.; Bityukova, V.R.; Malkhazova, S.M.; Kosheleva, N.E.; Nikiforova, E.M.; Shartova, N.V.; Vlasov, D.V.; Timonin, S.A.; Krainov, V.N. *Regions and Cities of Russia: Integrated Environmental Assessment*; IP Filimonov MV Publ.: Moscow, Russia, 2014; 560p. (In Russian)
16. Saet, Y.E.; Revich, B.A.; Yanin, E.P. *Geochemistry of the Environment*; Nedra Publisher: Moscow, Russia, 1990; p. 335. (In Russian)
17. Lim, J.-H.; Sabin, L.D.; Schiff, K.C.; Stolzenbach, K.D. Concentration, size distribution, and dry deposition rate of particle-associated metals in the Los Angeles region. *Atmos. Environ.* **2006**, *40*, 7810–7823. [CrossRef]
18. Barrie, L.A.; Lindberg, S.E.; Chan, W.C.; Ross, H.B.; Arimoto, R.; Church, T.M. On the concentration of trace metals in precipitation. *Atmos. Environ.* **1987**, *21*, 1133–1135. [CrossRef]
19. Wania, F.; Hoff, J.; Jia, C.; Mackay, D. The effects of snow and ice on the environmental behaviour of hydrophobic organic chemicals. *Environ. Pollut.* **1998**, *102*, 25–41. [CrossRef]
20. Zhang, C.; Wu, G.; Gao, S.; Zhao, Z.; Zhang, X.; Tian, L.; Mu, Y.; Joswiak, D. Distribution of major elements between the dissolved and insoluble fractions in surface snow at Urumqi Glacier No. 1, Eastern Tien Shan. *Atmos. Res.* **2013**, *132*, 299–308. [CrossRef]

21. Baysal, A.; Baltaci, H.; Ozbek, N.; Destanoglu, O.; Ustabaşı, G. Şirin; Gumus, G. Chemical characterization of surface snow in Istanbul (NW Turkey) and their association with atmospheric circulations. *Environ. Monit. Assess.* **2017**, *189*, 275. [CrossRef]

22. Siudek, P.; Frankowski, M.; Siepak, J. Trace element distribution in the snow cover from an urban area in central Poland. *Environ. Monit. Assess.* **2015**, *187*, 225. [CrossRef] [PubMed]

23. Baltrėnaitė, E.; Baltrėnas, P.; Lietuvninkas, A.; Šerevičienė, V.; Zuokaitė, E. Integrated evaluation of aerogenic pollution by air-transported heavy metals (Pb, Cd, Ni, Zn, Mn and Cu) in the analysis of the main deposit media. *Environ. Sci. Pollut. Res.* **2014**, *21*, 299–313. [CrossRef] [PubMed]

24. Belozertseva, I.A.; Vorobyeva, I.B.; Vlasova, N.V.; Lopatina, D.N.; Yanchuk, M.S. Snow pollution in Lake Baikal water area in nearby land areas. *Water Resour.* **2017**, *44*, 471–484. [CrossRef]

25. Ermolov, Y.V.; Makhatkov, I.D.; Khudyaev, S.A. Background concentrations of chemical elements in snow cover of the typical regions of the Western Siberia. *Atmos. Okean Opt.* **2014**, *27*, 790–800. (In Russian)

26. Moskovchenko, D.V.; Babushkin, A.G. Features of formation of the chemical composition of snow cover on the territory of the Khanty-Mansi Autonomous Okrug. *Kriosf. Zemli* **2012**, *14*, 71–81. (In Russian)

27. Pozhitkov, R.; Moskovchenko, D.; Soromotin, A.; Kudryavtsev, A.; Tomilova, E. Trace elements composition of surface snow in the polar zone of northwestern Siberia: The impact of urban and industrial emissions. *Environ. Monit. Assess.* **2020**, *192*, 1–14. [CrossRef] [PubMed]

28. Shevchenko, V.P.; Vorob'ev, S.N.; Kirpotin, S.N.; Kritskov, I.V.; Manasypov, R.M.; Pokrovsky, O.S.; Politova, N.V. Investiga-tions of insoluble particles in the snow cover of Western Siberia from Tomsk to the Ob estuary. *Atmos. Ocean. Opt.* **2015**, *28*, 499–504. (In Russian)

29. Shevchenko, V.; Vorobyev, S.; Krickov, I.; Boev, A.; Lim, A.; Novigatsky, A.; Starodymova, D.; Pokrovsky, O. Insoluble Particles in the Snowpack of the Ob River Basin (Western Siberia) a 2800 km Submeridional Profile. *Atmosphere* **2020**, *11*, 1184. [CrossRef]

30. Talovskaya, A.; Yazikov, E.; Filimonenko, E.; Lata, J.-C.; Kim, J.; Shakhova, T. Characterization of solid airborne particles deposited in snow in the vicinity of urban fossil fuel thermal power plant (Western Siberia). *Environ. Technol.* **2018**, *39*, 2288–2303. [CrossRef]

31. Syso, A.I.; Artamonova, V.S.; Sidorova, M.Y.; Ermolov, Y.V.; Cherevko, A.S. Pollution of the atmosphere, snow and soil cover of Novosibirsk. *Atmos. Ocean. Opt.* **2005**, *18*, 593–599. (In Russian)

32. Guseinov, A.N.; Karabatov, P.A.; Lysova, G.V.; Panfilova, L.A. Ecological and geochemical monitoring in the territory around the thermal power plants of Tyumen City. *Teploenergetika* **1997**, *12*, 31–36. (In Russian)

33. Konstantinova, E.; Minkina, T.; Sushkova, S.; Antonenko, E.; Konstantinov, A. Levels, sources, and toxicity assessment of polycyclic aromatic hydrocarbons in urban topsoils of an intensively developing Western Siberian city. *Environ. Geochem. Health* **2019**, *42*, 325–341. [CrossRef] [PubMed]

34. Jickells, T.; Davies, T.; Tranter, M.; Landsberger, S.; Jarvis, K.; Abrahams, P. Trace elements in snow samples from the Scottish Highlands: Sources and dissolved/particulate distributions. *Atmos. Environ. Part. A. Gen. Top.* **1992**, *26*, 393–401. [CrossRef]

35. Reinosdotter, K.; Viklander, M. A Comparison of Snow Quality in Two Swedish Municipalities—Luleå and Sundsvall. *Water Air Soil Pollut.* **2005**, *167*, 3–16. [CrossRef]

36. Weather Archive in Tyumen. Weather Schedule. Available online: https://rp5.ru (accessed on 22 May 2021).

37. Scientific and applied reference book on the climate of the USSR. In *Series Issue Tyumen and Omsk Regions*; Gidrometeoizdat Publisher: Saint Petersburg, Russia, 1998; p. 702. (In Russian)

38. Guseinov, A.N. *Ecology of Tyumen City: State, Problems*; Slovo Publisher: Tyumen, Russia, 2001; p. 176. (In Russian)

39. Konstantinova, E.; Minkina, T.; Konstantinov, A.; Sushkova, S.; Antonenko, E.; Kurasova, A.; Loiko, S. Pollution status and human health risk assessment of potentially toxic elements and polycyclic aromatic hydrocarbons in urban street dust of Tyumen city, Russia. *Environ. Geochem. Health* **2020**, 1–24. [CrossRef] [PubMed]

40. Krest'yannikova, E.V.; Kozlova, V.V.; Larina, N.S.; Larin, S.I. Chemical and environmental assessment of lead pollution in the atmosphere of the city of Tyumen. *Izv. RAS SamSC* **2015**, *17*, 679–684. (In Russian)

41. Germanova, T.V.; Kernozhitskaya, A.F. On the issue of environmental assessment of the transport system in urbanized territories (on the example of Tyumen City). *Izv. RAS SamSC* **2014**, *16*, 1713–1716. (In Russian)

42. Sorokina, O.I.; Kosheleva, N.E.; Kasimov, N.S.; Golovanov, D.L.; Bazha, S.N.; Dorzhgotov, D.; Enkh-Amgalan, S. Heavy metals in the air and snow cover of Ulan Bator. *Geogr. Nat. Resour.* **2013**, *34*, 291–301. [CrossRef]

43. Buat-Menard, P.; Chesselet, R. Variable influence of the atmospheric flux on the trace metal chemistry of oceanic suspended matter. *Earth Planet. Sci. Lett.* **1979**, *42*, 399–411. [CrossRef]

44. Il'in, V.B.; Syso, A.I.; Baidina, N.L.; Konarbaeva, G.A.; Cherevko, A.S. Background concentrations of heavy metals in soils of southern Western Siberia. *Eurasian Soil Sci.* **2003**, *36*, 494–500.

45. Rudnick, R.L.; Gao, S. Composition of the continental crust. In *Treatise on Geochemistry*, 3rd ed.; The Crust; Elsevier Science: New York, NY, USA, 2003; pp. 1–64.

46. Li, T.; Wang, Y.; Li, W.J.; Chen, J.M.; Wang, T.; Wang, W.X. Concentrations and solubility of trace elements in fine particles at a mountain site, southern China: Regional sources and cloud processing. *Atmos. Chem. Phys. Discuss.* **2015**, *15*, 8987–9002. [CrossRef]

47. Huang, J.; Li, Y.; Li, Z.; Xiong, L. Spatial variations and sources of trace elements in recent snow from glaciers at the Tibetan Plateau. *Environ. Sci. Pollut. Res.* **2018**, *25*, 7875–7883. [CrossRef]

48. Hakanson, L. An ecological risk index for aquatic pollution control.a sedimentological approach. *Water Res.* **1980**, *14*, 975–1001. [CrossRef]

49. Zhang, X.; Wang, Y.; Guo, S.; Li, H.; Liu, J.; Zhang, Z.; Yan, L.; Tan, C.; Yang, Z.; Guo, X. Concentration and speciation of trace metals and metalloids from road-deposited sediments in urban and rural areas of Beijing, China. *J. Soils Sediments* **2020**, *20*, 3487–3501. [CrossRef]

50. Zhao, R.; Li, N. Assessment of Ecological Risk of Soil in Rapid Urbanization Area by Risk Regionalization. In *IOP Conference Series: Earth and Environmental Science*; IOP Publishing: Bristol, UK, 2021; Volume 687, p. 012169.

51. Moiseenko, T.; Gashkina, N.; Dinu, M.I.; Khoroshavin, V.Y.; Kremleva, T.A. Influence of natural and anthropogenic factors on water acidification in humid regions. *Geochem. Int.* **2017**, *55*, 84–97. [CrossRef]

52. Artemov, I.E. The Influence of Urbanization on the Acid-Base Characteristics of Natural Waters. Extended Abstract of Ph.D. Thesis, Institute of Global Climate and Ecology of Roshydromet and RAS, Moscow, Russian, 2010.

53. Williamson, B.J.; Udachin, V.; Purvis, O.W.; Spiro, B.; Cressey, G.; Jones, G.C. Characterisation of airborne particulate pollution in the Cu smelter and former mining town of Karabash, South Ural Mountains of Russia. *Environ. Monit. Assess.* **2004**, *98*, 235–259. [CrossRef]

54. Vinogradova, A.A.; Maksimenkov, L.O.; Pogarskii, F.A. Changes in the atmospheric circulation and environmental pollution in Siberia from the industrial regions of Norilsk and the Urals in the early 21st century. *Atmos. Ocean. Opt.* **2009**, *22*, 396–404. [CrossRef]

55. Svistov, P.F.; Polishchuk, A.I. Atmospheric precipitations over Russian cities and regions. *Priroda* **2014**, *3*, 28–36. (In Russian)

56. Shevchenko, V.P.; Lisitsyn, A.P.; Shtain, R.; Goryunova, N.V.; Klyuvitkin, A.A.; Kravchishina, M.D.; Krivs, M.; Novigatskii, A.N.; Sokolov, V.T.; Filippov, A.S.; et al. Distribution and composition of insoluble particles in Arctic snow. *Probl. Arktiki I Antarkt.* **2007**, *1*, 106–118. (In Russian)

57. Evseeva, N.C.; Kvasnikova, Z.N.; Romashova, T.N.; Osintzeva, N.V. Aeolian erosion of soil during cold period of the year on Tom-Yaiskoe watershed divide (Western Siberia). *Geogr. Nat. Resour.* **2003**, *3*, 101–105.

58. Kasimov, N.S.; Kosheleva, N.E.; Vlasov, D.V.; Terskaya, E.V. Geochemistry of snow cover within the Eastern district of Moscow. *Vestn. Mosk. Un-Ta Geogr.* **2012**, *4*, 14–24. (In Russian)

59. Martin, J.-M.; Maybeck, M. Elemental mass balance of material carried by major world rivers. *Mar Chem.* **1979**, *7*, 173–206. [CrossRef]

60. Perel'man, A.I.; Kasimov, N.S. *Landscape Geochemistry*; Astreja 2000 Publisher: Moscow, Russia, 1999; p. 768. (In Russian)

61. Galfi, H.; Österlund, H.; Marsalek, J.; Viklander, M. Mineral and Anthropogenic Indicator Inorganics in Urban Stormwater and Snowmelt Runoff: Sources and Mobility Patterns. *Water Air Soil Pollut.* **2017**, *228*, 1–18. [CrossRef]

62. López-García, P.; Gelado-Caballero, M.D.; Collado-Sánchez, C.; Hernández-Brito, J.J. Solubility of aerosol trace elements: Sources and deposition fluxes in the Canary Region. *Atmos. Environ.* **2017**, *148*, 167–174. [CrossRef]

63. Desboeufs, K.; Sofikitis, A.; Losno, R.; Colin, J.; Ausset, P. Dissolution and solubility of trace metals from natural and anthropogenic aerosol particulate matter. *Chemosphere* **2005**, *58*, 195–203. [CrossRef]

64. Udachin, V.N. Ecogeochemistry of the Southern Urals mining industry. Ph.D. Thesis, Tomsk Polytechnic University, Tomsk, Russian, 2012.

65. Krupnova, T.G.; Rakova, O.V.; Gavrilkina, S.V.; Antoshkina, E.G.; Baranov, E.O.; Yakimova, O.N. Road dust trace elements contamination, sources, dispersed composition, and human health risk in Chelyabinsk, Russia. *Chemosphere* **2020**, *261*, 127799. [CrossRef] [PubMed]

66. Boev, V.A.; Lezhnina, A.A. Metals in the snow cover of the Tyumen district of the Tyumen region. *Vestn. TyumGU* **2012**, *7*, 41–48. (In Russian)

67. Vlasov, D.; Vasil'Chuk, J.; Kosheleva, N.; Kasimov, N. Dissolved and Suspended Forms of Metals and Metalloids in Snow Cover of Megacity: Partitioning and Deposition Rates in Western Moscow. *Atmosphere* **2020**, *11*, 907. [CrossRef]

68. Viklander, M. Substances in Urban Snow. A comparison of the contamination of snow in different parts of the city of Luleå, Sweden. *Water Air Soil Pollut.* **1999**, *114*, 377–394. [CrossRef]

69. Nakamura, A.; Okada, R. The coagulation of particles in suspension by freezing-thawing. *Colloid Polym. Sci.* **1976**, *254*, 718–725. [CrossRef]

70. Kim, J.-E.; Han, Y.-J.; Kim, P.-R.; Holsen, T.M. Factors influencing atmospheric wet deposition of trace elements in rural Korea. *Atmos. Res.* **2012**, *116*, 185–194. [CrossRef]

71. Dong, Z.; Kang, S.; Qin, X.; Li, X.; Qin, D.; Ren, J. New insights into trace elements deposition in the snow packs at remote alpine glaciers in the northern Tibetan Plateau, China. *Sci. Total. Environ.* **2015**, *529*, 101–113. [CrossRef] [PubMed]

72. Kamani, H.; Hoseini, M.; Safari, G.H.; Jaafari, J.; Mahvi, A.H. Study of trace elements in wet atmospheric precipitation in Tehran, Iran. *Environ. Monit. Assess.* **2014**, *186*, 5059–5067. [CrossRef]

73. Özsoy, T.; Örnektekin, S. Trace elements in urban and suburban rainfall, Mersin, Northeastern Mediterranean. *Atmos. Res.* **2009**, *94*, 203–219. [CrossRef]

74. Pavlov, D.; Dakhouche, A.; Rogachev, T. Influence of arsenic, antimony and bismuth on the properties of lead/acid battery positive plates. *J. Power Sources* **1990**, *30*, 117–129. [CrossRef]

75. Zhang, Y.; Cao, S.; Xu, X.; Qiu, J.; Chen, M.; Wang, D.; Guan, D.; Wang, C.; Wang, X.; Dong, B.; et al. Metals compositions of indoor PM2.5, health risk assessment, and birth outcomes in Lanzhou, China. *Environ. Monit. Assess.* **2016**, *188*, 1–13. [CrossRef] [PubMed]
76. Ozaki, H.; Yoshimura, K.; Asaoka, Y.; Hayashi, S. Antimony from brake dust to the combined sewer collection system via road effluent under rainy conditions. *Environ. Monit. Assess.* **2021**, *193*, 369. [PubMed]
77. Ramírez, O.; de la Campa, A.M.S.; Amato, F.; Moreno, T.; Silva, L.; de la Rosa, J.D. Physicochemical characterization and sources of the thoracic fraction of road dust in a Latin American megacity. *Sci. Total. Environ.* **2019**, *652*, 434–446. [CrossRef] [PubMed]
78. Hjortenkrans, D.; Bergbäck, B.; Häggerud, A. New Metal Emission Patterns in Road Traffic Environments. *Environ. Monit. Assess.* **2006**, *117*, 85–98. [CrossRef] [PubMed]
79. Nriagu, J.O. A History of Global Metal Pollution. *Science* **1996**, *272*, 223. [CrossRef]
80. Nriagu, J.O.; Pacyna, J.M. Quantitative assessment of worldwide contamination of air, water and soils by trace metals. *Nat. Cell Biol.* **1988**, *333*, 134–139. [CrossRef]
81. Mugica-Alvarez, V.; Maubert, M.; Torres-Rodríguez, M.; Muñoz, J.; Rico, E. Temporal and spatial variations of metal content in TSP and PM10 in Mexico City during 1996. *J. Aerosol Sci.* **2002**, *33*, 91–102. [CrossRef]
82. Cong, Z.; Kang, S.; Zhang, Y.; Li, X. Atmospheric wet deposition of trace elements to central Tibetan Plateau. *Appl. Geochem.* **2010**, *25*, 1415–1421. [CrossRef]
83. Yang, T.; Liu, Q.; Li, H.; Zeng, Q.; Chan, L. Anthropogenic magnetic particles and heavy metals in the road dust: Magnetic identification and its implications. *Atmos. Environ.* **2010**, *44*, 1175–1185. [CrossRef]
84. Al-Khashman, O.A. Determination of metal accumulation in deposited street dusts in Amman, Jordan. *Environ. Geochem. Health* **2007**, *29*, 1–10. [CrossRef] [PubMed]
85. Kabata-Pendias, A. *Trace Elements in Soils and Plants: Fourth Edition*; CRC Press: Boca Raton, UK; London, UK; New York, DC, USA, 2011; p. 584.
86. Kondrat'Ev, I.I.; Mukha, D.E.; Boldeskul, A.G.; Yurchenko, S.G.; Lutsenko, T.N. Chemical composition of precipitation and snow cover in the Primorsky krai. *Russ. Meteorol. Hydrol.* **2017**, *42*, 64–70. [CrossRef]
87. Hoffmann, P.; Karandashev, V.K.; Sinner, T.; Ortner, H.M. Chemical analysis of rain and snow samples from Chernogolovka, Russia by IC, TXRF and ICP-MS. Fresenius. *J. Anal. Chem.* **1997**, *357*, 1142–1148.
88. Grebenshchikova, V.I.; Efimova, N.; Doroshkov, A.A. Chemical composition of snow and soil in Svirsk city (Irkutsk Region, Pribaikal'e). *Environ. Earth Sci.* **2017**, *76*, 712. [CrossRef]
89. Wu, G.; Wei, Q.; Sun, C.; Gao, J.; Pan, L.; Guo, L. Determination of major and trace elements in snow in Tianjin, China: A three-heating-season survey and assessment. *Air Qual. Atmos. Heal.* **2016**, *9*, 687–696. [CrossRef]

Article

Authigenic Minerals of the Derbent and South Caspian Basins (Caspian Sea): Features of Forms, Distribution and Genesis under Conditions of Hydrogen Sulfide Contamination

Nina Kozina *, **Liudmila Reykhard and Olga Dara**

Shirshov Institute of Oceanology, Russian Academy of Sciences, 36, Nakhimovskiy Prospect, 117997 Moscow, Russia; mollusc@mail.ru (L.R.); olgadara@mail.ru (O.D.)
* Correspondence: kozina_nina@bk.ru

Abstract: This paper presents the results of complex lithological, mineralogical, and geochemical studies of bottom sediments of deep-water basins of the Caspian Sea (Derbent and South Caspian Basins) in areas contaminated by hydrogen sulfide. In the course of complex studies, numerous manifestations of authigenic mineral formation associated with the stage of early diagenesis have been established. Authigenic minerals belonging to the groups of sulfates (gypsum, barite), chlorides (halite), carbonates (calcite, low Mg-calcite; kutnohorite), and sulfides (framboidal pyrite), as well as their forms and composition, have been identified by a complex of analytical methods (X-ray diffractometry (XRD), scanning electron microscopy (SEM) with energy-dispersive X-ray spectroscopy (EDS); atomic absorption spectroscopy (AAS); coulometric titration (CT)); the nature of their distribution in bottom sediments has been assessed. Carbonates and sulfates are predominant authigenic minerals in the deep-water basins of the Caspian Sea. As a part of the study, differences have been established in the composition and distribution of associations of authigenic minerals in the bottom sediments in the deep-water basins. These are mineral associations characteristic of the uppermost part of the sediments (interval 0–3 cm) and underlying sediments. In the Derbent Basin, in sediments of the interval 3–46 cm, an authigenic association is formed from gypsum, calcite, magnesian calcite, siderite, and framboidal pyrite. An association of such authigenic minerals as gypsum and calcite is formed in sediments of the 0–3 cm interval. In the South Caspian Basin, in sediments of the interval 3–35 cm, an association of such authigenic minerals as gypsum, halite, calcite, magnesian calcite, and framboidal pyrite is formed. The association of such authigenic minerals as gypsum, halite, calcite, magnesian calcite, kutnohorite, and framboidal pyrite is characteristic of sediments of the 0–3 cm interval. We consider the aridity of the climate in the South Caspian region to be the main factor that determines the appearance of such differences in the uppermost layer of sediments of the basins. Judging by the change in the composition of authigenic associations, the aridity of the South Caspian increased sharply by the time of the accumulation of the upper layer of sediments (interval 0–3 cm). Taking into account lithological, mineralogical and geochemical data, the features of the processes of authigenic mineral formation in the deep-water basins of the Caspian Sea under conditions of hydrogen sulfide contamination have been determined. Analysis of the results obtained and published data on the conditions of sedimentation in the Caspian Sea showed that hydrogen sulfide contamination recorded in the bottom layer of the water column of the deep-water basins of the Caspian Sea may affect the formation of authigenic sulfides (framboidal pyrite), sulfates (gypsum), and carbonates (calcite and kutnohorite) associated with the activity of sulfate-reducing bacteria in reducing conditions.

Keywords: bottom sediments; modern sedimentation processes; mineralogy; geochemistry; authigenic minerals; hydrogen sulfide contamination; arid climate; Caspian Sea

Citation: Kozina, N.; Reykhard, L.; Dara, O. Authigenic Minerals of the Derbent and South Caspian Basins (Caspian Sea): Features of Forms, Distribution and Genesis under Conditions of Hydrogen Sulfide Contamination. *Minerals* **2022**, *12*, 87. https://doi.org/10.3390/min12010087

Academic Editors: Ana Romero-Freire and Hao Qiu

Received: 30 November 2021
Accepted: 7 January 2022
Published: 13 January 2022

Publisher's Note: MDPI stays neutral with regard to jurisdictional claims in published maps and institutional affiliations.

1. Introduction

Authigenic mineral formation attracts much attention from lithologists and marine geologists. Parageneses of authigenic minerals are considered the basis for diagnostics of sediment and rock formation stages. The essence of this process has been succinctly formulated by the American lithologist R. Fairbridge in 1985. Authigenic mineral formation (authigenesis) is a process in which minerals are formed in situ [1,2]. This term has been introduced by Kalkowsky [3]; nowadays, it is usually applied to sedimentary rocks in low-temperature conditions [2].

The formation of authigenic minerals in sedimentary basins occurs during sedimentation and subsequent diagenetic transformations of bottom sediments. Authigenesis occurs in the place of burial of sediments as a result of chemical, physicochemical, and biological (vital activity of organisms) processes [4–6]. Therefore, authigenic minerals are the main indicators of the environment, in which the formation of sedimentary deposits takes place, as well as of the processes of sediment transformation [7–12]. Authigenic minerals in the mineral composition of bottom sediments are an important genetic sign. The presence of some authigenic minerals indicates certain physicochemical characteristics of the water column in the sedimentation basin and subsequent diagenetic changes in sedimentary deposits. These changes include the behavior of various forms of chemical elements and the processes of microbial activity [12].

Authigenic minerals can be divided into organogenic minerals and chemogenic minerals [2]. Organogenic minerals are formed during the life of marine organisms (opal, aragonite, magnesian calcite, calcite, apatite, etc.) [2]. Chemogenic minerals are formed in a variety of ways. Minerals such as aragonite, dolomite, gypsum, halite, glauconite, and barite can be formed as a result of precipitation from true or colloidal solutions in sea or pore water [2]. Minerals of Fe-Mn nodules can form as a result of the precipitation of compounds from water and their redistribution in sediments during early diagenesis [2]. Hydrotroilite and pyrite are formed by the synthesis of compounds during diagenesis [2].

Thus, the formation of authigenic carbonates in the bottom sediments of the seas and oceans are considered in the works of Lein [13,14], Astakhova [15,16], Krylov [17,18], Derkachev [19], Kravchishina [10,20], et al. The manifestations of authigenic barite mineralization are considered in the works of Astakhova [21], Derkachev [22], et al.

The study of the complex process of authigenic mineral formation makes it possible to understand the history of the formation of sedimentary deposits from the stage of the mobilization of sedimentary material in the seas and oceans to its burial and transformation in diagenesis. Mineralogical research is a key group of methods used to study this process. In the Caspian Sea (Figure 1a), authigenic minerals have been studied rather poorly. Therefore, further research will significantly supplement the understanding of the history of sedimentation and general issues of the geological development of the Caspian Sea basin. Earlier, the processes of sedimentogenesis in the Caspian Sea, including the presence of authigenic minerals in the bottom sediments of the Caspian Sea, were presented with different generalizations in the works of L.S. Kulakova, L.I. Lebedev, Y.P. Khrustalev, and some others [23–26].

In the last few years, the researchers of the P.P. Shirshov Institute of Oceanology, Russian Academy of Sciences, have performed multidisciplinary research of oceanological, geological, and biochemical processes in the Caspian Sea [27–32]. As a result, new data on the distribution of sedimentary matter flows in the water column have been described [33–36]. In the works of Lukashin [33–36], the first data on the chemical composition of dispersed matter from sediment traps is presented. For the first time, seasonal changes in the fluxes of the biogenic triad were studied, along with the contribution of three main components of sedimentary matter ($SiO_{2amorph}$, $CaCO_3$, OM). It has been established that the annual variability of fluxes demonstrates its seasonal growth in spring and summer [33,34]. As a result, new knowledge has been gained about the biogeochemical processes occurring in the water column and in the near-bottom water layer [37–39]. Lein [32,39] presents the results of an investigation into microbial and biogeochemical processes at the water–sediment interface.

The results of the microbiological and biogeochemical investigation demonstrated that, in spite of the absence of a connection with the ocean and other specific features, the Caspian Sea has the characteristics of a typical marine basin. The maximal sulfate reduction rate, the highest methane concentration, and the highest total microbial abundance (TMA) were observed in the sediment layer at 0.5–1.5 cm (up to 3.5 cm), rather than in the warp. The fact that the most intensive biogeochemical activity was associated not with the warp, but with the underlying sediments made the processes of suspended matter transformation at the bottom water–sediment interface in the Caspian Sea different from those described in the process for the Russian Arctic seas. The microbial population of the sea bottom has the most important effect on sediment formation in the Caspian Sea. Microbial communities actively participate in all biogeochemical processes that take place in the near-bottom water and sediments [32,39]. As a result, new data have shown the distribution of micro- and macroelements, as well as rare-earth elements in surface bottom sediments [40–43]. Maslov and Gordeev [40–43] present an analysis of the distribution of rare and trace elements (Zr, Hf, Th, V, Cr, Co, Ni, Cu, Sr, and Ba) in modern bottom sediments of different subsystems of the Caspian Sea which, as well as their REE systematics, shows a geochemical set of features. The role of clusters, including fine-grained ones, coming from the mountain structures of the Caucasus and the Elburs, is already very small near the coasts, which is largely determined by both the influence of the marginal filter and the large-scale current cycles existing in the Caspian Sea [42,43]. As a result, the mineral composition of surface bottom sediments has already been described [8,9,44–46]. In the works of Kozina [8,9,44–46], the mineral composition of heavy and light subfractions of modern bottom sediments of the Caspian Sea have been studied. Terrigenous-mineralogical provinces have been identified. The main sources of detrital material have been established. The structural features and distribution pattern of framboidal pyrite in the bottom sediments of the South Caspian Basin have been studied. Authigenic minerals in the surface bottom sediments of the South Caspian Basin have been studied. Our study continues this research series and complements the current understanding of modern sedimentation processes and diagenesis in the Caspian Sea.

The Caspian Sea is the largest inland sea with no connection to the ocean. However, specific features of sedimentation processes here are similar to those observed in the sea [16]. The Caspian Sea occupies a deep depression, 27–28 m below the level of the World Ocean. The Basin area is 378,400 km^2 [24,26]. The Basin is a catchment basin of the river systems of the Russian Plain, Caucasus and Elbrus. The catchment basin is 3.6 million km^2 [24]. The main rivers of the catchment basin are the Volga, Ural, Terek, Sulak, Samur, Kura, and Sefidrud. According to the nature of the natural situation and structure, the sea is divided into 3 parts: the Northern, Middle, and Southern Caspian, delimited by the Mangyshlak and Absheron thresholds. There is no river flow in the eastern part of the Middle and Southern Caspian. A characteristic feature of the catchment basin of the sea is its location in the humid, semiarid, and arid zones [24,26]. This feature preconditions the nature of the basin's supply by sedimentary material. Hydrological and sedimentation processes in the Caspian Sea are largely determined by the presence of deep-water basins, Derbent Basin (central part of the Sea), and South Caspian Basin (southern Caspian Sea) (Figure 1). The Derbent Basin is located in the center of the Middle Caspian Sea and is surrounded totally by a narrow shelf. Its maximum depth reaches 788 m [24,26]. The extension of this basin is from northwest to southeast. A powerful cyclonic current dumps a huge amount of thin clastic material into this basin. The clastic material is washed away from the northern regions of the sea or comes with the waters of the Terek, Sulak, and Samur rivers. On the slopes of the Derbent Basin, traces of landslides and suspension flows are often found [24]. The formation of sedimentation in the Derbent Basin is determined by a number of factors: the influence of the powerful flow of the Volga River and the influence of Western Caucasian rivers. The South Caspian Basin occupies a major part of the Southern Caspian. The maximum depth of the basin is 1025 m. The western side is cragged, and the eastern side rises gently and passes into a wide shelf area [24]. The bottom of the South

Caspian Basin is complicated by mud volcanoes, as well as tectonic uplifts, which are expressed in the relief. Geologically, the South Caspian Basin has a number of features: its maximum depth is 1025 m and the bottom relief is uneven and complicated by numerous mud volcanoes; outcrops of active hydrothermal vents are located on the eastern coast of the South Caspian Basin [23–25]. In the South Caspian Basin, the sedimentation processes are preconditioned by their location in the arid zone [34,35].

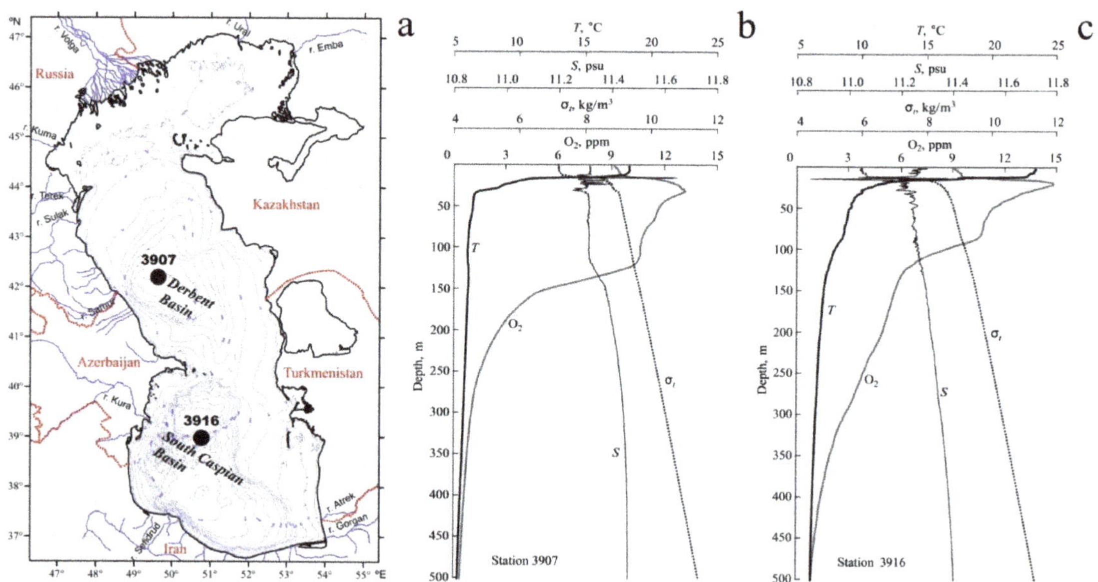

Figure 1. (**a**) Map-scheme of the research area and the location of the sampling of the Derbent (station 3907) and South Caspian (station 3916) Basins; (**b,c**) CTD-profiles of distribution of temperature (*T*), salinity (*S*), conventional density (*δt*), and oxygen (O_2) in the water column of station 3907 and station 3916, modified according to [37].

Regard must be paid to the presence of hydrogen sulfide recorded in the near-bottom waters of the Derbent and South Caspian Basins since the 1930s [28,37,47]. This phenomenon was observed alongside significant fluctuations in sea level, recorded since the 1950s [48] and is associated with both climatic changes and anthropogenic factors [37].

In 2010–2012, after detailed hydrological and microbiological studies had been performed, the presence of dissolved hydrogen sulfide, a process of modern sulfate reduction, and, as a consequence, the anaerobic regime was described under conditions of stable temperature and hydrochemical stratification of the water column, at depths exceeding 650 m and 880 m in the Derbent and South Caspian Basins, respectively [37]. At the same time, the concentration of hydrogen sulfide increases from the upper limit of its detection to the near-bottom water layer, from 20 to 120 μL/L in the Derbent Basin and from 60 to 240 μL/L in the South Caspian Basin (Figure 1). Active processes of bacterial sulfate reduction have also been recorded in the surface layer of bottom sediments (0.0–1.5 cm) [37].

This research continues a series of studies devoted to the issue of authigenic mineral formation in the bottom sediments in areas of hydrogen sulfide contamination of the enclosed and marginal seas of Russia [8,9,49–52]. Here, we aim to study the features of the processes of authigenic mineral formation in the bottom sediments of the deep-water basins of the Caspian Sea under conditions of hydrogen sulfide contamination.

2. Materials and Methods

2.1. Materials

The material was collected during the 39th cruise of the R/V *Rift* in 2012 [28,32,33].

Bottom sediments were sampled in the Derbent Basin at a 720 m depth (station no. 3907; 42°11.680′ N, 49°37.978′ E) and in the South Caspian Basin at a 1000 m depth (station no. 3916; 38°58.528′ N, 50°45.738′ E) (Figure 1a). Sampling was carried out using a KUM multicorer MiniMUC (Germany). This sampling method allowed one to preserve the boundary layer between the water column and bottom sediments, i.e., the upper flocculated layer (fluffy layer), with no disturbance.

Bottom sediments were sampled by a core in the Derbent Basin (46 cm long, 10 cm diameter) and in the South Caspian Basin (35 cm long, 10 cm diameter). The sediment core was subdivided with different discreteness, starting at 0.5 cm sections (upper part of the core) and increasing up to 5.0 cm sections (the deepest sediment layers). The lithological types of bottom sediments were distinguished according to [53].

2.2. Methods

Mineralogical and geochemical studies of bottom sediments were carried out in the P.P. Shirshov Institute of Oceanology, Russian Academy of Sciences (Moscow, Russia). The mineral composition was analyzed on a D8 ADVANCE X-ray diffractometer (Bruker AXC, Karlsruhe, Germany). The identification of clay minerals and the determination of their quantitative ratios was carried out according to widespread methods [54–57]. In accordance with these methods, oriented air-dry preparations were prepared from a suspension of the clay fraction separated from the sample in distilled water. Further, the preparations were saturated with ethylene glycol (for diagnostics of minerals of the smectite group, as well as mixed-layered formations with swelling layers). Then, the preparations were warmed up at 550 °C (for the diagnosis of kaolinite and chlorite). The technique was applied in full or reduced, as required. The visualization of microscopic components and analysis of their chemical composition were performed by a Scanning Electron Microscope Vega 3 Tescan (TESCAN, Brno, Czech Republic) with an X-ray spectral microanalyzer INCA Energy (OXFORD Instruments, High Wycombe, UK), and Scanning Electron Microscope Phenom ProX G6 with an integrated EDS system (Thermo Scientific Phenom, Eindhoven, Netherlands). The total content of micro- and macroelements in bottom sediments was determined by the method of flame Atomic Absorption Spectroscopy in an acetylene-air flame in a Kvant-2A Spectrophotometer (Cortec, Moscow, Russia). The content of total, organic, and carbonate carbon was determined by coulometric titration in an AN-7529 carbon analyzer (Akvilon, Podolsk, Russia).

3. Results

3.1. Lithological, Mineralogical, and Geochemical Characteristics of Bottom Sediments of the Derbent Basin

The bottom sediments in the central part of the Derbent Basin (station no. 3907; sediment core length of 46 cm) are represented by alternating pelitic silts, mainly weakly calcareous (with massive and microlayer texture, expressed in the alternation of thin layers of unoxidized and oxidized silts) and hydrotroilite layers (up to 1 mm thick). The top layer of sediments (0.0–1.0 cm) is represented by a reduced black flocculent silty deposit. A strong smell of hydrogen sulfide is present in bottom sediments down to a depth of 10 cm.

According to microscopic studies, supported by the results of XRD, the mineral composition of the bottom sediments in the Derbent Basin is represented mainly by quartz (18–29%), feldspars (albite, up to 15%; potassium feldspar, up to 11%), and clay minerals (illite, up to 17%; kaolinite, up to 10%; chlorite, up to 16%; and smectite, up to 1%) (Table 1). The sediments have a high content of authigenic minerals of the sulfate groups (gypsum, up to 28%), carbonates (calcite, up to 8%; magnesian calcite, up to 3%; aragonite, up to 2%; siderite and dolomite, up to 1% each), and sulfides (pyrite, up to 3%) (Table 1).

Table 1. Mineral composition in bottom sediments of the Derbent and South Caspian Basins.

Depth, cm	Qz, %	Ab, %	K-fsp, %	Chl, %	Kln, %	Ilt, %	Sme, %	Gp, %	Hl, %	Cal, %	Mg-Cal, %	Dol, %	Sd, %	Arg, %	Kut, %	Py, %
Derbent Basin (CT. 3907)																
0–0.5	18	13	10	15	10	16	1	11	n.d.	5	n.d.	1	n.d.	n.d.	n.d.	n.d.
1.5–2	24	14	8	16	9	17	1	7	n.d.	4	n.d.	n.d.	n.d.	n.d.	n.d.	n.d.
3.5–4	25	15	11	12	7	16	1	5	n.d.	5	n.d.	1	1	n.d.	n.d.	1
6–7	20	12	8	10	6	14	1	20	n.d.	6	1	n.d.	n.d.	n.d.	n.d.	2
9–10	29	14	8	9	6	13	1	4	n.d.	8	3	1	1	2	n.d.	1
12–13	18	15	9	9	7	16	1	13	n.d.	7	2	n.d.	1	1	n.d.	1
15–17	18	12	9	10	6	14	1	19	n.d.	5	3	n.d.	n.d.	1	n.d.	2
21–23	20	13	8	11	6	15	1	13	n.d.	7	3	n.d.	n.d.	2	n.d.	1
29–31	17	12	7	9	5	12	1	28	n.d.	6	2	1	n.d.	n.d.	n.d.	1
37–39	28	15	7	11	7	14	1	4	n.d.	8	2	n.d.	n.d.	2	n.d.	1
44–46	25	15	8	10	7	13	1	8	n.d.	8	2	n.d.	n.d.	n.d.	n.d.	3
South Caspian Basin (CT. 3916)																
0–0.5	17	11	10	10	6	13	tr.	7	16	6	1	n.d.	n.d.	n.d.	2	1
0.5–1	14	13	n.d.	11	7	15	tr.	22	6	6	n.d.	n.d.	n.d.	n.d.	6	n.d.
1–1.5	15	10	6	8	6	20	tr.	25	3	4	6	n.d.	n.d.	n.d.	2	n.d.
2.5–3	25	14	7	10	6	18	tr.	3	1	7	4	n.d.	n.d.	n.d.	2	n.d.
4–4.5	26	13	7	10	6	21	0.3	3	1	7	3	n.d.	n.d.	2	n.d.	1
6–7	22	14	7	14	7	17	0.4	2	1	7	4	n.d.	n.d.	1	n.d.	4
9–10	26	13	7	10	6	16	0.4	3	1	7	9	n.d.	n.d.	n.d.	n.d.	2
12–13	23	14	8	10	6	17	0.4	2	1	8	10	n.d.	n.d.	n.d.	n.d.	1
16–17	28	14	8	8	8	12	0.4	2	1	8	9	n.d.	n.d.	n.d.	n.d.	2
20–22	25	13	6	13	7	17	0.5	1	2	7	8	n.d.	n.d.	n.d.	n.d.	n.d.
26–28	22	12	15	11	6	16	0.3	n.d.	1	7	9	n.d.	n.d.	n.d.	n.d.	1
33–35	**29**	**11**	**8**	**6**	**4**	**10**	**tr.**	**8**	**1**	**8**	**11**	**n.d.**	**n.d.**	**1**	**n.d.**	**3**

Notation: Quartz—Qz, Albite—Ab, K-feldspar—K-fsp, Chlorite—Chl, Kaolinite—Kln, Illite—Ilt, Smectite—Sme, Gypsum—Gp, Halite—Hl, Calcite—Cal, Mg-calcite—Mg-Cal, Dolomite—Dol, Siderite–Sd, Aragonite—Arg, Kutnohorite—Kut, Pyrite—Py, tr.—traces, n.d.—not detected.

Gypsum was found throughout the entire core in the pelitic sediments of the Derbent Basin (station no. 3907). The gypsum content varies at 4–28% without a definite pattern (Table 1). Gypsum is present as crystals of hexagonal and prismatic shapes, as well as their aggregates (Figure 2a,b).

In the deposits of the Derbent Basin, carbonate minerals are found along with the entire core, but mineral associations differ in different layers. In particular, calcite is present in all studied layers. An uneven increase in the content of calcite (from 5 to 8%) and the appearance of Mg-calcite (1–3%) are recorded in the sediments from the 6.0 cm depth and below (Table 1). Calcite occurs as aggregate accumulations of microcrystals with a corroded surface (Figure 2c). Mg-calcite is presented as prismatic crystals with perfect cleavage and crystalline aggregates with traces of severe corrosion on the cleavage (Figure 2d) and crystalline aggregates with traces of severe corrosion on the cleavage planes (Figure 2e). Dolomite, siderite, and aragonite are distributed unevenly along with the entire core; their content does not exceed 2% (Table 1). Siderite is found only in the upper part of the column (Table 1). Opposite to it, aragonite is absent in the upper part of the column and is found only at a depth of 9.0–10.0 cm. Dolomite is presented by rare prismatic crystals, siderite, by micro globular aggregates (Figure 2f). Aragonite was not detected by SEM.

The content of pyrite varies from 1 to 3% throughout the entire core. The pyrite presents as framboids and crystallites, which develop both in zones enriched with detritus of diatom frustules and in the pore space of pelitic sediments (Figure 2g–j). The framboids have a spherical shape (up to 12–15 μm in diameter). They form various crystallites—irregular, polyhedral, and globular.

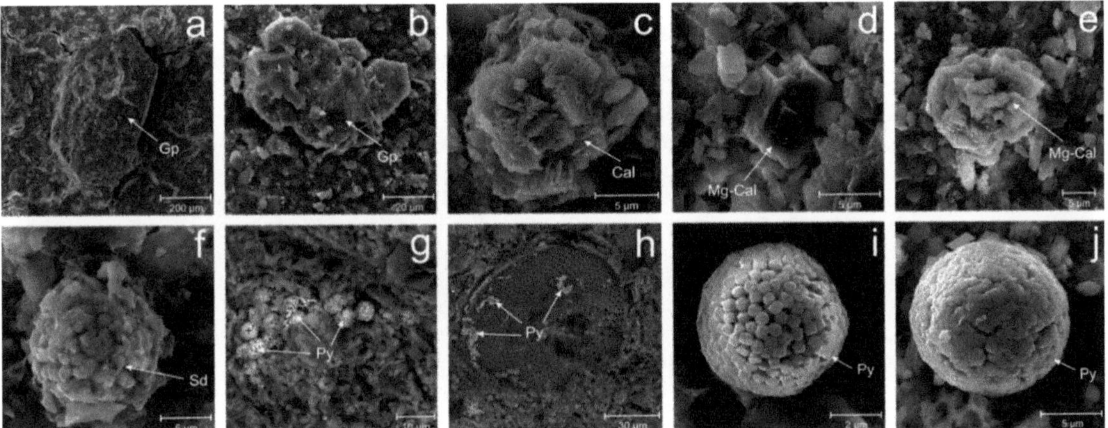

Figure 2. Authigenic minerals in bottom sediments of the Derbent Basin (SEM Images): (**a**) hexagonal gypsum crystal; (**b**) an aggregate of gypsum crystals; (**c**) aggregate of calcite crystals with traces of corrosion; (**d**) a crystal of Mg-calcite; (**e**) highly corroded aggregate of Mg-calcite crystals; (**f**) siderite microglobular aggregate; (**g**) cluster of pyrite framboids and crystallites; (**h**) pyrite crystallites on the surface of the centric diatom frustula; (**i**) pyrite framboid of polygonal crystallites; (**j**) pyrite framboid of irregular and globular crystallites.

Geochemical analyses evidence that the studied bottom sediments in the Derbent Basin are strongly but unevenly enriched in organic carbon and $Fe_{(total)}$ (Figure 3).

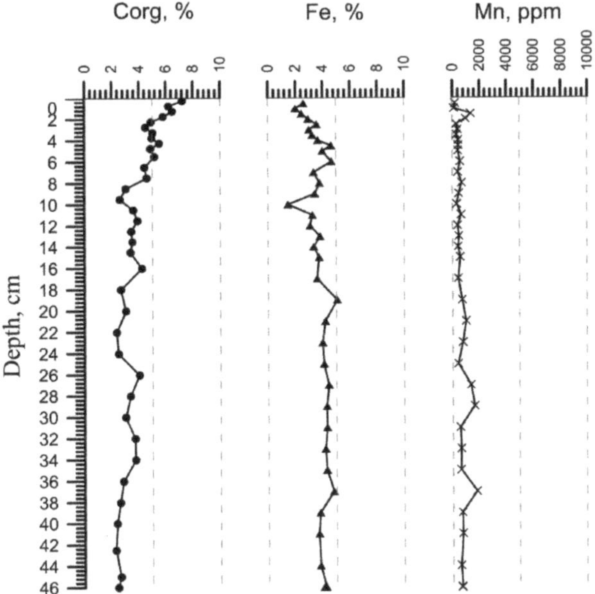

Figure 3. Distribution of Corg, Fe, Mn in the bottom sediments of the Derbent Basin (st. 3907).

$Fe_{(total)}$ content varies from 1.45% to 5.07% through the entire core, tending to increase unevenly from the upper layers down to the lowest, except the 9.0–10.0 cm layer, where iron content decreases and reaches a minimum value of 1.45% (Figure 3).

The upper layers of bottom sediments are highly enriched in organic carbon, C_{org} (Figure 3). In the silty deposit, the organic carbon content reaches its maximum value of 7.2%; down the column, it decreases down to a minimum value of 2.2% (41–44 cm depth).

3.2. Lithological, Mineralogical, and Geochemical Characteristics of the South Caspian Basin

Bottom sediments of the central part of the South Caspian Basin (station no. 3916; core length of 35 cm) are represented by alternating pelitic sediments with an insignificant content of sandy-silty admixtures, mostly weakly calcareous, with a micro layered and porous-cavernous texture, saturated with hydrotroilite to varying degrees by smears, concretions, and thin layers. The upper layer of bottom sediment (0.0–0.5 cm) is represented by reduced silty deposits, presented as black and green flakes. A strong smell of hydrogen sulfide is present in bottom sediments down to a depth of 8 cm.

According to microscopic studies, supported by the XRD method, the mineral composition of bottom sediments is represented by quartz (14–29%), feldspars (albite, up to 14%; feldspar, up to 15%), and clay minerals (illite, up to 21%; kaolinite, up to 8%; chlorite, up to 14%; and smectite, up to 0.5%) in various ratios [8] (Table 1).

A high content of authigenic minerals belonging to the groups of sulfates, carbonates, and sulfides, such as gypsum (up to 25%), halite (up to 16%) calcite (up to 8%), magnesian calcite (up to 11%), kutnohorite (up to 6%), and pyrite (up to 4%) is observed in the core (Table 1).

Gypsum is found in the pelitic sediments of the South Caspian Basin along the entire length of the core, with a maximum content of 7–25% in the uppermost layer at 0.0–1.5 cm. Gypsum is present in the sediments as radial-radiant aggregates (rosettes), hexagonal gypsum crystals, intergrowths of prismatic crystals, prismatic single crystals with perfect cleavage, fibrous aggregates, and crystalline masses [9] (Figure 4a–d). The size of gypsum crystals ranges from 1–3 µm to 50 µm.

Barite, another authigenic mineral of the sulfate group, is presented by single inclusions of micro aggregates of lamellar crystals in the deposits of the South Caspian Basin. Barite was detected only by SEM+EDS (Figure 4e).

Halite was found in the form of aggregates of irregularly shaped crystals (Figure 4f).

Carbonate minerals occur in sediments along with the entire core; they are represented by calcite, Mg-calcite, aragonite, and kutnohorite (Table 1). The contents of calcite and Mg-calcite increase unevenly from a 2.5 cm depth and deeper down to the end of the core.

Calcite is one of the predominant minerals of the carbonate group in the sediments of the South Caspian Basin. The calcite content in the sediments of the uppermost part of the core (0.0–2.5 cm) ranges from 4 to 6% and constitutes 7–8% from a 2.5 cm depth and below (Table 1). Calcite occurs in bottom sediments by parallel-columnar aggregates, continuous fibrous masses, crystals with perfect cleavage along the rhombohedron, microglobular calcite aggregate, and aggregate accumulations of microcrystals (Figure 4g,h) [9]. The content of Mg-calcite varies from 1 to 11% (Table 1). Mg-calcite is presented as prismatic crystals and lamellar aggregates. Dolomite is found as single rhombohedral crystals (Figure 4i).

Aragonite was found only by the XRD method in a very insignificant amount (1–2%) in the upper and lower parts of the core (Table 1).

Kutnohorite was detected in the uppermost layer of the sediment (0.0–3.0 cm), where its content varies from 2 to 6%, reaching a maximum in the layer of 0.5–1.0 cm (Table 1). Kutnohorite is represented by numerous clusters of irregularly shaped microcrystal aggregates (Figure 4j) [8,9,44].

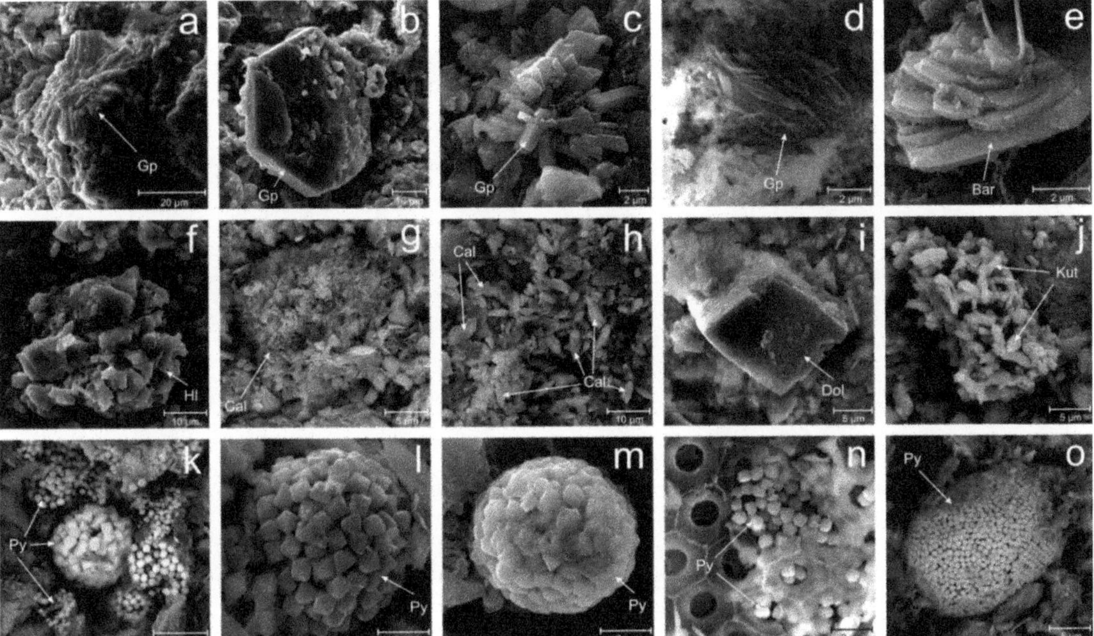

Figure 4. Authigenic minerals in bottom sediments of the South Caspian Basin (SEM Images): (**a**) gypsum radial-radiant aggregate of the "gypsum rose" type; (**b**) a hexagonal gypsum crystal; (**c**) an aggregate of gypsum long-prismatic crystals; (**d**) fibrous gypsum aggregate; (**e**) an aggregate of barite crystals; (**f**) an aggregate of halite crystals; (**g**) microglobular calcite aggregate; (**h**) cluster of calcite microcrystals; (**i**) rhombohedral dolomite crystal; (**j**) an aggregate of kutnohorite microcrystals; (**k**) pyrite framboid and crystallite cluster; (**l**) pyrite framboid from octahedral crystallites; (**m**) pyrite framboid of irregular crystallites; (**n**) pyrite crystallites within the diatom frustula; (**o**) irregular pyrite framboid.

The content of sulfides (represented by pyrite) varies from 1to 4% without a definite pattern. It is important to note that pyrite has already been found in the Caspian Sea and its maximum content is observed at a depth of 6.0–7.0 cm. It was established that pyrite is present in the form of framboids—spherical aggregates of a complex structure, consisting of crystallites, as well as in the form of individual crystallites or their disordered accumulations, which may be the result of the mechanical destruction of framboids (Figure 4k–o). As a rule, framboids have a spherical shape, up to 20 μm in diameter. Framboids of irregular shape (possibly deformed), up to 30 μm × 40 μm in size, were also found. Crystallites have different shapes (octahedral, globular, and irregular) and are randomly located in the structure of framboids. It is important to note that crystallites, even within the same framboid, may have different shapes. Framboidal pyrite forms both on the surface of plant attrite and frustules of diatoms and in the intergranular pore space of pelitic deposits [8].

A very high content of halite (16%) has been found in the uppermost part of the studied bottom sediments. Down the core, the halite content sharply decreases, constituting only 1–2% below the layer of 2.5–3.0 cm (Table 1).

According to geochemical analyses of bottom sediments of the South Caspian Basin, the studied sediments are strongly, but unevenly, enriched in organic matter, $Fe_{(total)}$, and manganese [8,9] (Figure 5).

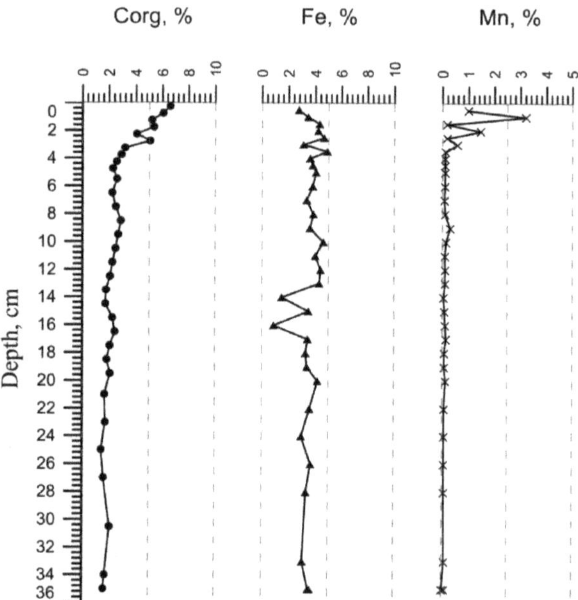

Figure 5. Distribution of Corg, Fe, and Mn in the bottom sediments of the South Caspian Basin (st. 3916).

$Fe_{(total)}$ content varies greatly over the entire core without any regularity. The minimum iron content (0.9%) is observed in the 15.0–16.0 cm layer, and the maximum (4.9%) in the upper part of the core (3.0–3.5 cm) (Figure 5).

The upper layers of bottom sediments are highly enriched in organic carbon. In the silty deposit layer, the C_{org} content reaches a maximum of 6.5%. Down the core, it unevenly decreases to a minimum of 1.4% at a 24.0–26.0 cm depth.

The maximum manganese content is typical for the fluffy layer and the upper layers of sediments, amounting to 1.4% (1.5–2.0 cm) and 3.2% (0.5–1.0 cm). The manganese content drops sharply down to 1830 ppm from a 2.0 cm depth [8,9].

4. Discussion

4.1. Authigenic Minerals and Features of Their Formation in the Bottom Sediments of the Derbent Basin and the South Caspian Basin

In the bottom sediments of the Derbent and South Caspian basins authigenic mineral formation is manifested by authigenic minerals of the sulfate group (gypsum, barite), carbonate group (calcite, Mg-calcite, kutnohorite, dolomite, siderite, and aragonite), chloride group (halite), and sulfide group (framboidal pyrite). Gypsum, calcite, Mg-calcite, and pyrite are the most common authigenic minerals in the Holocene deposits of the studied areas. The presence of these minerals both in the upper and in deeper layers of the sedimentary strata is associated with a combination of special geological, hydrological, and biogeochemical conditions prevailing in these areas [32,33,37,39,47].

In the study areas, authigenic minerals are presented in various forms. Radial-radiant crystalline aggregates, crystals of various habitus and their aggregates are characteristic forms for minerals of the sulfate group (gypsum, barite). Carbonates (calcite, Mg-calcite, kutnohorite, and siderite) are presented as crystals, microcrystalline aggregates, fibrous masses, and globular forms. Sulfides (pyrite) are presented as crystallites and their aggregates (framboids).

Gypsum is an aqueous calcium sulfate. In all studied layers of bottom sediments of the Derbent and South Caspian basins, authigenic precipitation of gypsum is observed.

The formation of gypsum in the bottom sediments of the deep-water basins of the Caspian Sea in areas of hydrogen sulfide contamination presumably occurs in two ways, similar to the formation of sulfates in the sediments of the seas and oceans [2,5]. In particular, the first way is chemogenic, i.e., gypsum is formed simultaneously with the host rocks as a result of chemical precipitation from sulfate-rich aqueous solutions. The second way is biochemical, whereby the content of organic matter and carbonates in sediments is high and accompanied by the activity of sulfate-reducing bacteria. The sulfur cycle of marine sediments is primarily driven by the dissimilatory sulfate reduction to sulfide by anaerobic microorganisms [58,59]. Sulfate-reducing bacteria are anaerobic microorganisms that use sulfate as the final electron acceptor in the oxidation of organic compounds or molecular hydrogen. Under anaerobic conditions, sulfate reducers carry out terminal destruction of organic compounds and produce hydrogen sulfide. Sulfate-reducing bacteria include representatives of Desulfovibrio desulfurcans and Desulfotomaculum [60]. At present, there are very few data on sulfate reducers in the bottom sediments of the deep-water zones of the Caspian Sea. However, there is evidence that sulfate-reducing bacteria from the Desulfobacterales family have been identified in the upper part of the bottom sediments of the southern part of the Caspian Sea [61]. Sulfate-reducing bacteria can contribute to the formation of various minerals, mainly sulfides (pyrite, hydrotroilite). It has been established that the biogenic mechanisms of mineral formation in global biogeochemical cycles by bacteria-induced mineral precipitation, including sulfate-reducing bacteria in the sulfur cycle, are very diverse [62].

In the South Caspian Basin, the maximum gypsum content is observed in the upper layers of the sediment. High concentrations of gypsum in the upper layers are most likely associated with the peculiarities of sedimentation processes in an arid climate [5,9] and the chemogenic formation of gypsum in sediments from pore solutions saturated with sulfur and calcium, as well as with biogeochemical processes occurring in bottom sediments of the South Caspian Basin under conditions of hydrogen sulfide contamination [9,39]. In the Derbent Basin, high gypsum concentrations are observed in all studied layers of bottom sediments.

Barite is found only in the deposits of the South Caspian Basin by single microcrystalline aggregates. Based on the appearance of barite crystals, which differs from the appearance of biogenic marine barite [63], the composition of the authigenic mineral association [49,50,64], and the conditions of sedimentation in the South Caspian Basin, it may be assumed that the origin of barite is early diagenetic, taking place during crystallization of pore solutions supersaturated with S and Ba [9,50].

High halite contents, found only in the upper layer of bottom sediments of the South Caspian Basin, are possibly related to sedimentation under arid conditions typical for this zone [24,26].

Carbonate minerals are the most widespread authigenic minerals in the sediments of the deep-water basins of the Caspian Sea. Calcite and Mg-calcite predominate in the sediments; dolomite, siderite, kutnohorite, and aragonite are found in insignificant quantities. A greater variety of forms of authigenic calcite is observed in the deposits of the South Caspian Basin compared with the Derbent Basin.

In bottom sediments, the formation of authigenic calcite from solutions, which are supersaturated with calcium carbonate, occurs during the decomposition of organic matter by bacteria. The latter is accompanied by the formation of CO_2, hydrogen sulfide, etc. [2,5]. The formation of authigenic carbonate minerals in the sediments of the Derbent and South Caspian basins is greatly influenced by several factors, such as the chemical composition of pore waters, reducing conditions, abundant content of organic matter in sediments, and arid climate. As a result of the interaction of the constituent components of the sediment with pore waters and dissolved organic matter, processes of authigenic formation of carbonate minerals occur with the active influence of sulfate-reducing bacteria.

The Mg-calcite content in the deposits of the South Caspian Basin is several times higher than that of the Derbent Basin (Table 1). The amount of Mg-calcite in the sediments of the South Caspian Basin increases with depth. No Mg-calcite is found in the sediments of the Derbent Basin in the upper layers of the sediment (0.0–6.0 cm).

Authigenic siderite in the Derbent Basin is formed under reduced conditions with an abundant content of organic matter and the active participation of sulfate-reducing bacteria. In a reducing environment, when carbon dioxide reacts with silicate minerals, iron bicarbonates are formed; they transform into siderite during further stages of diagenesis, and further into pyrite in the presence of hydrogen sulfide and hydrogen sulfide bacteria. It is important to note that siderite is found in the sediments of the Derbent Basin. Siderite is found in the layers where the formation of framboidal pyrite is recorded. Siderite has not been recorded at all in the sediments of the South Caspian Basin.

Aragonite was found only by the XRD method as an insignificant rare impurity (1–2%) in the mineral composition of the studied bottom sediments. No crystalline forms were found to identify this mineral by microscopic methods. Therefore, it may be assumed that aragonite has a biogenic origin and is most likely present in single biogenic detrital components (for example, shell detritus) in the bottom sediments of the South Caspian Basin, as noted in [8,9].

Kutnohorite is found only in the surface layer of bottom sediments of the South Caspian Basin (0.0–3.0 cm) and should be attributed to authigenic minerals in the early stage of diagenesis. Pore water supersaturated with Ca^{2+} and Mn^{2+} may be the source of these cations; carbon dioxide, which is a part of kutnohorite, is formed during microbial decomposition of organic matter in sediments [44]. Manganese ores, which are widespread within the Greater Caucasus, and the waters of the Cheleken-Boyadag hydrothermal system enriched in manganese, may also supply Mn to the South Caspian Basin [33,34,44].

Dolomite is found in the deposits of the Derbent Basin at an insignificant admixture (1%), and it is absent in the sediments of the South Caspian Basin. According to some researchers [33,35], dolomite is transported here with aeolian material.

Framboidal pyrite is a widespread authigenic mineral in modern bottom sediments of the Derbent and South Caspian basins. It is present both in deep layers of the sedimentary strata and in the very surface layer of bottom sediments, which is associated with a combination of special geological, hydrological, and geochemical conditions prevailing in this area [8]. Firstly, these are the specific conditions of deep-water stagnant depressions of the closed-sea basin, which contributed to the accumulation of a thick stratum of finely dispersed pelitic sediments (with a pelitic fraction content of more than 90%) enriched in organic matter, iron, sulfates, and carbonates. Secondly, this is the presence of high concentrations of hydrogen sulfide and methane in the bottom layers of the water column and the bottom sediments of the Derbent and South Caspian basins [39]. The latter results in reducing conditions forming already in the upper layers of bottom sediments. It is assumed that the total amount of pyrite that may form in sediments at the stage of early diagenesis is limited by the rate of input of decomposable organic matter, dissolved sulfate, and iron minerals. As it is known, the abundance of organic matter in sediments allows the development of sulfate-reducing bacteria, which reduce seawater sulfates and release hydrogen sulfide. The hydrogen sulfide then reacts with mobile iron, which ultimately leads to the formation of pyrite [65].

In the studied bottom sediments, framboidal pyrite is found in small quantities, although this mineral is an important indicator of the reduction zone of diagenesis. Its presence in the uppermost layer of sediments (in the South Caspian Basin, in the fluffy layer) testifies to the specificity of diagenetic processes under conditions of hydrogen sulfide contamination of the bottom layer of deep-water basins. According to geochemical data, no significant dependence of pyrite amount is found on the maximum contents of iron and organic matter. It can be assumed that the formation of pyrite is regulated by the presence or absence of a certain amount of sulfur in the pore waters of sediments. A significant amount of gypsum containing sulfur indicates a fairly high sulfur content in the sediments. The relationship between the content of gypsum and pyrite in the sediments was not traced. Therefore, it can be assumed that the formation of framboidal pyrite is regulated not by a certain content of sulfur or iron, but by bacterial activity.

The content of C_{org} in the bottom sediments of the Derbent and South Caspian basins is very high, especially in the upper layer, which provides favorable conditions for microorganisms, reaching high numbers here: in the Derbent Basin, $165/520 \times 10^6$ cells cm^{-3} in the silty deposit layer; in the South Caspian Basin, $150/180 \times 10^6$ cells cm^{-3} in the surface layer and $1500/8400 \times 10^6$ cells cm^{-3} in 0.5–1.5 cm layer [39]. Iron comes with terrigenous runoff from the surrounding land, partly with aeolian material, and possibly with the hydrothermal waters of Cheleken Peninsula [33–35]. The high sulfate content in pore waters was noted earlier.

The aridity of the climate and the absence of runoff from the Caspian Sea leads to the concentration in bottom sediments of all components entering the South Caspian Basin. The subsequent transformation of bottom sediments in the process of diagenetic phase differentiation of matter under the influence of hydrogen sulfide contamination leads to the formation of specific associations of authigenic minerals. However, these mineral associations differ from the mineral associations of the Derbent Basin, in which sedimentation and diagenesis also take place under the influence of hydrogen sulfide contamination, but under different geographic, geological, and hydrological conditions.

4.2. Associations of Authigenic Minerals in Bottom Sediments of Deep-Water Basins of the Caspian Sea

Analysis of the distribution of authigenic minerals in the studied bottom sediments of the Caspian Sea makes it possible to distinguish different associations of authigenic minerals for each of the deep-water basins. These are mineral associations characteristic of the uppermost part of the sediments (interval 0–3 cm) and underlying sediments. Primary methodological limitations, such as the asynchrony of data over intervals in different basins and different column powers, do not allow for a more detailed selection. Additional arguments for identifying the lower boundary of the upper interval were obtained as a result of the analysis of data from geochemical studies.

Thus, in the Derbent Basin, in sediments of the 3–46 cm interval, an authigenic association is formed from gypsum, calcite, magnesian calcite, siderite, and framboidal pyrite. The upper boundary of the interval is distinguished by the disappearance of such authigenic minerals as siderite and framboidal pyrite in sediments. An association of such authigenic minerals as gypsum and calcite is formed in sediments of the 0–3 cm interval. According to geochemical data, the sediments of this interval are characterized by maximum C_{org} contents.

In the South Caspian Basin, in sediments of the 3–35 cm interval, an association of such authigenic minerals as gypsum, halite, calcite, magnesian calcite, and framboidal pyrite is formed. The upper boundary of the interval is distinguished by the appearance of kutnohorite in the sediments. The association of such authigenic minerals as gypsum, halite, calcite, magnesian calcite, kutnohorite, and framboidal pyrite is characteristic of sediments of the 0–3 cm interval. According to geochemical data, the bottom sediments of this interval contain the maximum concentrations of C_{org} and Mn.

A comparative analysis of the qualitative and quantitative composition of associations of authigenic minerals in different intervals of bottom sediments of each basin shows the change in the processes of authigenic mineral formation over time.

The most contrasting are the compositions of authigenic associations in the uppermost layer of bottom sediments of the Derbent and South Caspian basins. Judging by the higher total content of authigenic minerals and a more diverse mineral composition, the intensity of authigenic mineral formation during the period of accumulation of this layer of sediments in the South Caspian Basin is much higher than in the Derbent Basin. The very high contents of halite and gypsum, in our opinion, clearly indicate more arid sedimentation conditions in the southern part of the Caspian Sea.

The associations of authigenic minerals in the sediments of different basins are various, indicating differences in the conditions of sedimentation and diagenesis, despite the fact that the water column and bottom sediments of both basins contain high concentrations of hydrogen sulfide and a high content of C_{org}.

We consider the aridity of the climate in the South Caspian region to be the main factor that determines the appearance of such differences in the uppermost layer of sediments of the basins. Judging by the changes in the composition of authigenic associations over time, the aridity in the South Caspian region increased sharply by the time of the accumulation of the upper layer (interval 0–3 cm).

Another feature of sedimentation and diagenetic processes in the South Caspian Basin is the effect of the discharge of hydrothermal waters near the Cheleken Peninsula and emissions of mud volcanoes. It is these sources that could cause the accumulation of high contents of barium and manganese in the sediments. Furthermore, subsequent diagenetic transformations led to the formation of authigenic minerals such as barite and kutnohorite.

5. Conclusions

As a result of the comprehensive lithological, mineralogical, and geochemical studies in the bottom sediments of the Derbent and South Caspian basins, it was possible to establish numerous manifestations of authigenic mineral formation, confined to the stage of early diagenesis. Authigenic minerals are represented by sulfates (gypsum, barite), chlorides (halite), carbonates (calcite, Mg-calcite, siderite, dolomite, kutnohorite), and sulfides (framboidal pyrite). Carbonates and sulfates are the predominant authigenic minerals in the deep basins of the Caspian Sea. The formation of authigenic gypsum is observed in all studied layers of sediments in the Derbent Basin and the South Caspian Basin. Calcite and Mg-calcite are the most common minerals of the carbonate group; dolomite, siderite, and kutnohorite are found in smaller quantities. Gypsum and calcite, represented by a large number of different forms, accumulate in the bottom sediments of the South Caspian depression, which is characterized by an arid climate. Framboidal pyrite is formed both in the uppermost layers of the sediment (fluffy layer), due to the presence of hydrogen sulfide pollution, and in deeper layers, where the processes of reductive sub-water diagenesis take place.

It has been established that hydrogen sulfide contamination recorded in the bottom layer of the water column of the deep-water basins of the Caspian Sea can affect the formation of authigenic sulfides (framboidal pyrite), sulfates (gypsum), and carbonates (calcite, siderite, and kutnohorite), which are related to the activity of sulfate-reducing bacteria under reducing conditions.

It was revealed that specific associations of authigenic minerals in the bottom sediments of the Derbent and South Caspian basins are the product of the interaction of hydrological, biogeochemical, and sedimentation-diagenetic processes. The anaerobic oxidation of methane in combination with bacterial sulfate reduction are key biogeochemical processes occurring in the bottom sediments of these areas. Both processes control the formation of specific authigenic mineralization.

Author Contributions: Conceptualization, N.K. and L.R.; Methodology, N.K. and L.R.; Software, N.K. and L.R.; Validation, N.K. and L.R.; Analytical Studies, N.K., L.R. and O.D.; Investigation, N.K. and L.R.; Resources, N.K.; Data Curation, N.K. and L.R.; Writing—Original Draft Preparation, N.K.; Writing—Review and Editing, N.K. and L.R.; Visualization, N.K. and L.R.; Funding Acquisition, N.K. All authors have read and agreed to the published version of the manuscript.

Funding: Preparation and processing of samples, lithological-mineralogical and geochemical studies, as well as interpretation of the obtained data were supported by the Russian Science Foundation, grant no. 19-77-00015. X-ray diffractometry was performed within the framework of the State Assignment for Institute of Oceanology RAS (theme no. 0128-2021-0006).

Acknowledgments: The authors are grateful to the colleagues from the P.P. Shirshov Institute of Oceanology for their support. Special thanks go to A.K. Ambrosimov, A.A. Klyuvitkin, M.D. Kravchishina, N.V. Politova, A.N. Novigatsky and V.V. Gordeev for invaluable help, to L.V. Demina for C_{org} and C_{total} measurements, and to V.A. Karlov and M.A. Kovalev for assistance in performing scanning electron microscopy.

Conflicts of Interest: The authors declare no conflict of interest.

References

1. Larsen, G.; Chilingar, D. (Eds.) *Diagenesis and Catagenesis of Sedimentary Formations*; Mir: Moscow, Russia, 1971; p. 463. (In Russian)
2. Yapaskurt, O.V. *Genetic Mineralogy and Stadial Analysis of Sedimentary Rock and Ore Formation Processes*; Textbook; ESLAN: Moscow, Russia, 2008; p. 356. (In Russian)
3. Kalkowsky, E. Über die Erforschung der archaeischen Formationen, Neues Jahrb. *Min. Geol. Palaeont. Monatsh* **1880**, *1*, 1–29.
4. Strakhov, N.M. *Selected Works. General Problems of Geology, Lithology and Geochemistry*; Nauka: Moscow, Russia, 1983; p. 640. (In Russian)
5. Strakhov, N.M. *Fundamentals of the Lithogenesis Theory*; Academy of Sciences of the USSR: Moscow, Russia, 1960.
6. Glenn, C.R.; Filippelli, G.M. Authigenic mineral formation in the marine environment: Pathways, processes and products. *Deep. Sea Res. Part II Top. Stud. Oceanogr.* **2007**, *54*, 1141–1146. [CrossRef]
7. Lein, A.Y.; Ivanov, M.V. *Biogeochemical Cycle of Methane in the Ocean*; Nauka: Moscow, Russia, 2009; p. 576. (In Russian)
8. Kozina, N.; Reykhard, L.; Dara, O.; Gordeev, V. Framboidal pyrite formation in the bottom sediments of the South Caspian Basin under conditions of hydrogen sulfide contamination. *Russ. J. Earth Sci.* **2018**, *18*, 1–10. [CrossRef]
9. Kozina, N.V.; Reykhard, L.Y.; Dara, O.M.; Gordeev, V.V. Characteristic property of the formation of authogenic minerals in bottom sediments of the South Caspian Basin under conditions of hydrogen sulfide contamination. *Oceanology* **2021**. *in print*.
10. Kravchishina, M.D.; Lein, A.Y.; Reykhard, L.E.; Dara, O.M.; Flint, M.V.; Savvichev, A.S. Authigenic Mg-calcite at a cold methane seep site in the Laptev Sea. *Oceanology* **2017**, *57*, 174–191. [CrossRef]
11. Glenn, C.R.; Prévôt-Lucas, L.; Lucas, J. (Eds.) *Marine Authigenesis: From Global to Microbial*; Special Publication No. 66; SEPM (Society for Sedimentary Geology): Tulsa, OK, USA, 2000; 525p.
12. Reykhard, L.Y. Pyrite framboid in bottom sediments of the White Sea. In *Materials of the Annual Meeting of the Russian Mineralogical Society*; Russian Mineralogical Society: Saint Petersburg, Russia, 2014; pp. 62–64. (In Russian)
13. Lein, A.Y. Authigenic Carbonate Formation in the Ocean. *Lithol. Miner. Resour.* **2004**, *39*, 1–30. [CrossRef]
14. Lein, A.Y.; Galchenko, V.V.; Pokrovsky, B.G. Marine carbonate nodules as a result of microbial oxidation of methane gas hydrate in the Sea of Okhotsk. *Geochemistry* **1989**, *10*, 1396–1406.
15. Astakhova, N.V. Autigenic carbonates in Upper Pleistocene-Holocene sediments of marginal seas of the northwestern Pacific Ocean. *Pac. Geol.* **1999**, *18*, 41–49. (In Russian)
16. Astakhova, N.V.; Nervov, G.A.; Yakusheva, I.N. Carbonate-barite mineralization in the Deryugin Depression (Sea of Okhotsk). *Pac. Geol.* **1990**, *3*, 37–42. (In Russian)
17. Krylov, A.A.; Khlystov, O.M.; Hachikubo, A.; Minami, H.; Zemskaya, T.; Logvina, E.A.; Lomakina, A.V.; Semenov, P.B. The reconstruction of the mechanisms of problematic authigenic carbonates formation in diagenetic and catagenetic environments associated with the generation/oxidation of hydrocarbons. *Limnol. Freshw. Biol.* **2020**, *4*, 928–930. [CrossRef]
18. Krylov, A.; Khlystov, O.; Zemskaya, T.; Minami, H.; Hachikubo, A.; Kida, M.; Shoji, H.; Naudts, L.; Poort, J.; Pogodaeva, T.P. First discovery and formation process of authigenic siderite from gas hydrate–bearing mud volcanoes in fresh water: Lake Baikal, eastern Siberia. *Geophys. Res. Lett.* **2008**, *35*, 05405. [CrossRef]
19. Derkachev, A.N.; Nikolaeva, N.A. Features of authigenic mineralogy in the sediments of the Sea of Okhotsk. In *Far Eastern Seas of Russia*; Akulichev, V.A., Ed.; Book 3 Geological and Geophysical Research; Nauka: Moscow, Russia, 2007; pp. 223–239. (In Russian)
20. Kravchishina, M.D.; Lein, A.Y.; Flint, M.V.; Baranov, B.V.; Miroshnikov, A.Y.; Dubinina, E.O.; Dara, O.M.; Boev, A.G.; Savvichev, A.S. Methane-Derived Authigenic Carbonates on the Seafloor of the Laptev Sea Shelf. *Front. Mar. Sci.* **2021**, *8*, 690304. [CrossRef]
21. Astakhova, N.V.; Lipkina, M.I.; Melnichenko, Y.I. Hydrothermal barite mineralization in the Deryugin depression of the Sea of Okhotsk. *Dokl. Earth Sci.* **1987**, *295*, 212–215.
22. Derkachev, A.N.; Nikolaeva, N.A.; Mozherovskiy, A.V.; Baranov, B.V.; Barinov, N.N.; Minami, H.; Hachikubo, A.; Shoji, H. Manifestation of carbonate-barite mineralization around methane seeps in the Sea of Okhotsk (the western slope of the Kuril Basin). *Oceanology* **2015**, *55*, 390–399. [CrossRef]
23. *The Caspian Sea. Moscow*; USSR Publishers MSU: Moscow, Russia, 1969; p. 264. (In Russian)
24. *The Caspian Sea: Problems of Sedimentogenesis*; Nauka: Moscow, Russia, 1989; p. 184. (In Russian)
25. Lebedev, L.I.; Mayev, E.G.; Bordovsky, O.K.; Kulakova, L.S. *Bottom Sediments of the Caspian Sea*; Nauka: Moscow, Russia, 1973; p. 188. (In Russian)
26. Khrustalev, Y.P. *Regularities of Sedimentation in the Intracontinental Seas of the Arid Zone*; Nauka: Saint Petersburg, Russia, 1989; p. 261. (In Russian)

27. Ambrosimov, A.K.; Klyuvitkin, A.A.; Kozina, N.V.; Kravchishina, M.D.; Libina, N.V.; Filippov, A.S.; Artamonova, K.V.; Torgunova, N.I.; Baranov, V.I.; Pol'kin, V.V. Complex studies of the Caspian Sea during the 41-st cruise of the R/V Rift. *Oceanology* **2014**, *54*, 671–676. [CrossRef]

28. Ambrosimov, A.K.; Klyuvitkin, A.A.; Goldin, Y.A.; Kravchishina, M.D.; Mutovkin, A.D.; Novigatsky, A.N.; Politova, N.V.; Zakharova, E.E.; Savvichev, A.S.; Korzh, A.O.; et al. Multidisciplinary studies of the Caspian Sea system during cruise 39 of the R/V Rift. *Oceanology* **2014**, *54*, 395–399. [CrossRef]

29. Ambrosimov, A.K.; Klyuvitkina, A.A.; Filippov, A.S.; Kozina, N.V.; Nemirovskaya, I.A.; Zolotykh, E.O.; Bud'ko, D.F.; Serebrennikova, E.A.; Bondar', A.V. Study of the north and central Caspian on the October 2015 expedition of the R/F Nikifor Shurekov. *Oceanology* **2017**, *57*, 756–758. [CrossRef]

30. Ambrosimov, A.K.; Klyuvitkin, A.A.; Filippov, A.S.; Korch, A.O. Multidisciplinary studies of the Caspian Sea System during the cruise of the R/V Tantal in May 2015. *Oceanology* **2017**, *56*, 757–759. [CrossRef]

31. Klyuvitkin, A.A.; Ambrosimov, A.K.; Kravchishina, M.D.; Kozina, N.V.; Budko, D.F.; Dukhova, L.A.; Serebrennikova, E.A.; Korzh, A.O. Comprehensive study of the Caspian Sea system during the second cruise of the Research Vessel Nikifor Shurekov. *Oceanology* **2015**, *55*, 311–314. [CrossRef]

32. Lein, A.Y.; Lisitzin, A.P.; Kravchishina, M.D.; Kozina, N.V.; Dara, O.M. Modern sedimentation: Sedimentogenesis and early diagenesis. In *The System of the Caspian Sea*; Nauchnyi Mir: Moscow, Russia, 2016; pp. 303–399. (In Russian)

33. Lukashin, V.N.; Kravchishina, M.D.; Klyuvitkin, A.A.; Kozina, N.V.; Novigatsky, A.N.; Politova, N.V. Suspended matter. In *The System of the Caspian Sea*; Nauchnyi Mir: Moscow, Russia, 2016; pp. 213–302. (In Russian)

34. Lisitzin, A.P.; Lukashin, V.N. Composition of dispersed sedimentary matter and fluxes in the water column of the Caspian Sea. *Dokl. Earth Sci.* **2015**, *464*, 956–962. [CrossRef]

35. Lukashin, V.N.; Lisitzin, A.P.; Dara, O.M.; Kozina, N.V.; Klyuvitkin, A.A.; Novigatsky, A.N. Mineral composition of sedimentary matter in the Caspian Sea. *Oceanology* **2016**, *6*, 852–862. [CrossRef]

36. Lukashin, V.N.; Lisitzin, A.P. Geochemistry of dispersed sedimentary matter and its fluxes in the water column Caspian Sea. *Oceanology* **2016**, *56*, 675–689. [CrossRef]

37. Ivanov, M.V.; Savvichev, A.S.; Klyuvitkin, A.A.; Zakharova, E.E.; Rusanov, I.I.; Lein, A.Y.; Lisitsyn, A.P.; Chul'tsova, A.L. Resumption of hydrogen sulfide contamination of the water column of deep basin in the Caspian Sea. *Dokl. Earth Sci.* **2013**, *453*, 1094–1099. [CrossRef]

38. Kozina, N.V.; Lein, A.Y.; Dara, O.M. Unknown biomorphic structures—Environmental formations or laboratory artifact? *Priroda* **2016**, *4*, 70–73. (In Russian)

39. Lein, A.Y.; Savvichev, A.S.; Kravchishina, M.D.; Kozina, N.V.; Peresypkin, V.I.; Zakharova, E.E.; Veslopolova, E.F.; Mitskevich, I.N.; Shul'ga, N.A.; Lobus, N.V.; et al. Microbiological and biogeochemical properties of the Caspian Sea sediments and water column. *Microbiology* **2014**, *83*, 648–660. [CrossRef]

40. Gordeev, V.V.; Kozina, N.V.; Baturin, G.N. Distribution of metals (Cu, Zn, Ni, Co, Cr, Pb, Fe and Mn) in surface layer of bottom sediments of the Central and Southern Caspian region. *Okeanol. Issled.* **2019**, *47*, 26–51.

41. Maslov, A.V.; Ronkin, Y.L.; Kozina, N.V.; Klyuvitkin, A.A.; Novigatskii, A.N.; Filippov, A.S.; Shevchenko, V.P.; Kovach, V.P. Rare-earth element distribution and 87Sr/86Sr systematics in modern bottom sediments of the Caspian Sea. *Dokl. Earth Sci.* **2014**, *459*, 1418–1422. [CrossRef]

42. Maslov, A.V.; Kozina, N.V.; Klyuvitkin, A.A.; Novigatsky, A.N.; Filippov, A.S.; Shevchenko, V.P. Disrtibution of some rare and trace elements in modern bottom sediments of the Caspian Sea. *Oceanology* **2016**, *56*, 552–563. [CrossRef]

43. Maslov, A.V.; Kozina, N.V.; Shevchenko, V.P.; Klyuvitkin, A.A.; Sapozhnikov, P.V.; Zavialov, P.O. REE systematics in modern bottom sediments of the Caspian Sea and river Deltas worldwide: Experience of comparison. *Dokl. Earth Sci.* **2017**, *475*, 797–802. [CrossRef]

44. Dara, O.M.; Lein, A.Y.; Kozina, N.V.; Ivanov, M.V. First find of kutnohorite in modern sediments of the *Caspian basin. Dokl. Earth Sci.* **2015**, *465*, 1257–1261. [CrossRef]

45. Kozina, N.V.; Novigatsky, A.N. Composition of fragmental minerals of the surface layer of bottom sediments in the Caspian Sea. *Oceanology* **2014**, *54*, 348–364. [CrossRef]

46. Kozina, N.V. Carbonate accumulation in the Southern Caspian Sea. *J. Comput. Theor. Nanosci.* **2019**, *16*, 134–139. [CrossRef]

47. Sapozhnikov, V.V.; Grashchenkova, O.K.; Kivva, K.K.; Azarenko, A.V. Hydrological and hydrochemical studies of the middle and south Caspian Seafrom the R/V Issledovatel' Kaspia (September 2–17, 2006). *Oceanology* **2007**, *47*, 290–293. [CrossRef]

48. Lebedev, S.A.; Kostianoy, A.G. *Satellite Altimetry of the Caspian Sea*; Publishing Center OCEAN of the International Ocean Institute: Moscow, Russia, 2005; p. 366.

49. Reykhard, L.; Kozina, N.; Dara, O.; Shulga, N. Authigenic barite in the bottom sediments of the seas and oceans: Crystal morphology, composition, indicator properties. Abstract Collection Book. In Proceedings of the 6th World Multidisciplinary Earth Sciences Symposium WMESS 2020, Prague, Czech Republic, 7–11 September 2020; p. 161.

50. Reykhard, L.Y.; Kozina, N.V. Typomorphic features and indicator properties of the authigenic barite from bottom sediments of deep-water basins of the Caspian and Black seas. In Proceedings of the XXIV International Scientific conference (School) on Marine Geology, IO RAS, Moscow, Russia; 2021; pp. 278–281. (In Russian)

51. Kozina, N.V.; Reykhard, L.Y.; Dara, O.M.; Starodymova, D.P.; Kochenkova, A.I. Features of authigenic mineral formations in Holocene bottom sediments of deep-sea depressions of the Black and Caspian Seas in areas of hydrogen sulfide contamination. In *The Collection: Proceedings of the IX International Scientific and Practical Conference "Marine Research and Education (MARESEDU-2020)"*; OOO «PoliPRESS» Collection: Moscow, Russia, 2020; pp. 115–117. (In Russian)
52. Reykhard, L.Y.; Shulga, N.A.; Kozina, N.V.; Novichkova, E.A.; Dara, O.M.; Boev, A.G. Biominerals in aquatic geosystems under extreme conditions. In Proceedings of the BIOMIN XV: 15th International Symposium on Biomineralization, Munich, Germany, 9–13 September 2019; p. 104.
53. Bezrukov, P.L.; Lisitzin, A.P. Classification of bottom sediments in modern basins. *Proceed. Inst. Oceanol. USSR Acad. Sci.* **1960**, *32*, 3–14. (In Russian)
54. Brown, G. (Ed.) *X-ray Methods of Studying and Structure of Clay Minerals*; Mir.: Moscow, Russia, 1965; 600p.
55. Moore, D.M.; Reynolds, R.C. *X-ray Diffraction and the Identification and Analysis of Clay Minerals*, 2nd ed.; Oxford University Press: New York, NY, USA, 1997; 380p.
56. Biscaye, P.E. Mineralogy and sedimentation of recent deep sea clay in the Atlantic Ocean and adjacent seas and oceans. *Geol. Soc. Am. Bull.* **1965**, *76*, 803–832. [CrossRef]
57. Shlykov, V.G. *X-ray Analysis of Dispersed Soils*; M. GEOS.: Moscow, Russia, 2006; 176p. (In Russian)
58. Jørgensen, B.B.; Kasten, S. Sulfur cycling and methane oxidation. In *Marine Geochemistry*; Schulz, H.D., Zabel, M., Eds.; Springer: Berlin/Heidelberg, Germany, 2006; pp. 271–309.
59. Jørgensen, B.B.; Findlay, A.J.; Pellerin, A. The Biogeochemical Sulfur Cycle of Marine Sediments. *Front. Microbiol.* **2019**, *10*, 849. [CrossRef]
60. Volkov, I.I. *Geochemistry of Sulfur in Ocean Sediments*; Nauka: Moscow, Russia, 1984; 272p. (In Russian)
61. Mahmoudi, N.; Robeson, M.S.; Castro, H.F.; Fortney, J.L.; Techtmann, S.M.; Joyner, D.C. Microbial community composition and diversity in Caspian Sea sediments. *FEMS Microbiol. Ecol.* **2015**, *91*, 1–11. [CrossRef]
62. Hoffmann, T.D.; Reeksting, B.J.; Gebhard, S. Bacteria-induced mineral precipitation: A mechanistic review. *Microbiology* **2021**, *167*, 4. [CrossRef]
63. Paytan, A.; Griffith, E.M. Marine barite: Recorder of variations in ocean export productivity. *Deep-Sea Res. II* **2007**, *54*, 687–705. [CrossRef]
64. Reykhard, L.Y.; Kozina, N.V.; Dara, O.M.; Boev, A.G.; Novichkova, E.A.; Shulga, N.A.; Reykhard, A.G. Application of complex microscopic studies for solution of urgent problems of oceanology and marine geology. In Proceedings of the IX International Scientific-Practical Conf. Marine Studies and Education (MARESEDU-2020), Moscow, Russia, 26–30 October 2020; Volume 3, pp. 91–94. (In Russian)
65. Ocean Chemistry. Geochemistry of bottom sediments. In *Oceanology Series*; Nauka: Moscow, Russia, 1979; Volume 2, p. 536. (In Russian)

Article

Reactive, Sparingly Soluble Calcined Magnesia, Tailor-Made as the Reactive Material for Heavy Metal Removal from Contaminated Groundwater Using Permeable Reactive Barrier

Alena Fedoročková *, Pavel Raschman, Gabriel Sučik, Mária Švandová and Agnesa Doráková

Faculty of Materials, Metallurgy and Recycling, Technical University of Košice, Letná 9, 040 01 Košice, Slovakia; pavel.raschman@tuke.sk (P.R.); gabriel.sucik@tuke.sk (G.S.); ferencovamaria@gmail.com (M.Š.); agnesdorakova@gmail.com (A.D.)
* Correspondence: alena.fedorockova@tuke.sk; Tel.: +421-55-602-3109

Citation: Fedoročková, A.; Raschman, P.; Sučik, G.; Švandová, M.; Doráková, A. Reactive, Sparingly Soluble Calcined Magnesia, Tailor-Made as the Reactive Material for Heavy Metal Removal from Contaminated Groundwater Using Permeable Reactive Barrier. *Minerals* **2021**, *11*, 1153. https://doi.org/10.3390/min11111153

Academic Editors: Ana Romero-Freire and Hao Qiu

Received: 7 October 2021
Accepted: 14 October 2021
Published: 20 October 2021

Publisher's Note: MDPI stays neutral with regard to jurisdictional claims in published maps and institutional affiliations.

Abstract: A laboratory method was designed and verified that allows for the testing of alkaline, magnesite-based reactive materials for permeable reactive barriers (PRBs) to remove heavy metals from contaminated groundwater. It was found that caustic calcined magnesia (CCM) with high reactivity and low solubility to remove Cu^{2+}, Zn^{2+}, Ni^{2+}, and Mn^{2+} cations from mixed aqueous solutions can be prepared by calcination at a suitable temperature and residence time. Regarding the solubility of both the reactive material itself and the precipitates formed, the CCM should contain just a limited content of lime. One way is the calcination of a ferroan magnesite at temperatures above 1000 °C. However, the decrease in pH is accompanied by lower efficiency, attributed to the solid-phase reactions of free lime. A different way is the calcination of magnesite under the conditions when $CaCO_3$ is not thermally decomposed. The virtually complete removal of the heavy metals from the model solution was achieved using the CCM characterised by the fraction of carbonates decomposed of approximately 80% and with the highest specific surface area. CCM calcined at higher temperatures could also be used, but this would be associated with higher consumption of crude magnesite. Under the conditions considered in the present work, the product obtained by the calcination at 750 °C for 3 h appeared to be optimal. The full heavy metal removal was observed in this case using less magnesite, and, moreover, at a lower temperature (resulting, therefore, in a lower consumption of energy for the calcination and material handling).

Keywords: groundwater; heavy metal; precipitation; caustic calcined magnesia (CCM); permeable reactive barrier (PRB)

1. Introduction

Groundwater is the main source of drinking water worldwide [1,2]. However, there are numerous sites all over the world characterised by groundwater contaminated with organic and/or inorganic compounds, including toxic metals. Groundwater contamination with heavy metals as a consequence of mining, refining, and industrial operations, such as galvanic processing [3], scrubbing of the flue gases in power plants [4], leather tanning [5], improper waste disposal, or accidents in the production of chemicals, represents an important challenge for the authorities responsible for the environmental protection [6,7].

Although some heavy metals, such as copper (Cu), manganese (Mn), and zinc (Zn), are essential for living organisms at specific concentrations, significant amounts can result in shortness of breath, central nervous system disorders, reproductive failure, gastrointestinal problems, and several types of cancers [8–12]. The assimilation of high concentrations of Mn is reported to cause brain malfunction and lead to neurotic disorders [13]. The presence and accumulation of nickel has a toxic or carcinogenic effect on living organisms [14]. Significant amounts of Zn can affect infertility, kidney, and CNS disorders, while overdoses of Cu cause depression and lung cancer [15,16].

In order to reduce the contents of heavy metals in contaminated groundwater, extensive research has been carried out during the last decades to develop efficient and low-cost remediation methods. The conventional solution to this problem considers treatment plants where groundwater is pumped to the land surface and treated using the controlled addition of chemicals, and then redisposed [17]. These "active" techniques, referred to as "pump-and-treat", require an equipment, maintenance, and supply of energy, carry the risk of exposure to contamination, are unable to remove contaminants adsorbed to the soil [1], and may be impracticable at abandoned sites [18]. An alternative method that is being used increasingly—especially in those areas where the heavy metals are slowly leached from a source [19]—is a permeable reactive barrier (PRB) [4]. In principle, a PRB is a depth filter employing an appropriate granular sparingly soluble reactive material capable of removing the pollutants from the contaminated groundwater as the stream of groundwater passes the PRB [17]. This "passive" treatment system is able to improve water quality by means of precipitation, sorption, or microbial degradation [2,20] using mostly gravitational gradients, practically without any requirements for energy [18].

The most widespread method used to treat acidic metalliferous groundwater is neutralisation (or pH control) [6]. It is well known that the solubility of many cations of metals is reduced in a range of neutral to slightly basic pH, while the solubility can increase in either very acidic or very basic solutions [6,21,22]. Since the minimum solubility of various metals generally occurs at different pH values, the optimum pH for a given contaminated groundwater treatment, which depends on its chemical composition [6], requires a careful control [5,6,23–25]. The minimal solubility of the hydroxides of the majority of metals is reached at a pH between 9.5 and 10.0 [7]. To neutralise an acidic contaminated groundwater, raise its pH, and reach the threshold, the water is brought into contact with an appropriate sparingly soluble alkaline material. It is necessary to guarantee that the optimal pH will be maintained in the barrier during the whole period of its performance [23].

Removing metals via hydroxide precipitation has several advantages: (i) the technology is well-established, simple, and relatively inexpensive; (ii) the ability to achieve the regulatory effluent limits for several metals has been proven; and (iii) non-metal pollutants, such as soaps and fluorides, can be removed [24]. The formation of hydroxides is mostly accompanied by coprecipitation and/or adsorption and gives a mixed precipitate [7,26].

Due to their availability and low cost, limestone [27,28] and lime [2,3,6,24,27] are, in general, the most commonly employed alkaline reactive materials. However, there are certain limitations associated with their use in the PRBs. If, for example, the precipitation of certain metals requires a pH of 8 to 10, limestone is not effective as it only raises the pH to around 8 [6,29]. Moreover, carbon dioxide formed during the limestone dissolution buffers the reactions and makes it difficult to raise the pH even above 6 or 7 [18,30]; although this pH allows for the precipitation of the hydroxides of trivalent metals (Fe^{3+}, Al^{3+}, Cr^{3+}), it is not high enough for divalent metal hydroxides to be formed (Fe^{2+}, Zn^{2+}, Mn^{2+}, Cu^{2+}, Pb^{2+}, Ni^{2+}, Co^{2+}, and Cd^{2+}) [18,31–34]. In contrast, both lime and portlandite are unsuitable for passive treatment [31]. They have high solubility as compared to limestone and can raise the pH up to 12–12.5 [2,31,32], which makes some precipitates more soluble [5,24]. Moreover, the use of these substrates results in excessive sludge production and poor settling [27], and the treated, highly alkaline water might thus be harmful to the environment.

An alternative to lime or portlandite is magnesium oxide (MgO), especially in the form of the caustic calcined magnesia (CCM), MgO-rich material obtained by the calcination of natural magnesite ($MgCO_3$) [33]. In relation to the PRBs, CCM is reported to have the following advantages as compared to lime:

- $Mg(OH)_2$ is much less soluble than $Ca(OH)_2$ in aqueous solutions ($K_{S,Mg(OH)2} = 1 \times 10^{-11}$; $K_{S,Ca(OH)2} = 4 \times 10^{-6}$); it is, therefore, an attractive alternative reagent for passive remediation systems [18].
- Upon contact with water, MgO hydrates to form $Mg(OH)_2$, which buffers aqueous solutions' pH between 8.0–10.5 [18,32,33,35]. In this pH range, the solubility of most

metal hydroxides is very low, so divalent metals cations, such as Zn^{2+}, Cd^{2+}, Ni^{2+}, or Co^{2+}, can be removed by precipitation [18,33,35,36].

- MgO has been used for the immobilization of not only the cations of heavy metals but also anionic species, such as borate [26] and fluoride [37], and a strong chemical afinity to arsenic has also been observed [38].
- The residual sludge volume is much lower, mainly due to the elimination of the gypsum formation from sulphate-rich effluents [7,36]; it is easier to be dewatered [24,36] and is less hazardous and, therefore, cheaper to handle [31,32].

The removal of heavy metals from (acidic) aqueous effluents by MgO has been intensively studied by previous authors using synthetic, high-purity (99%) MgO [5,26,37,38], bulk CCM (coarse or fine-grained) [31,32,39,40], or a mixture of a fine-grained CCM and coarse inert matrix (e.g., wood shavings or gravel), so-called dispersed alkaline substrate (DAS) [18,31,33,35,41,42]. The substrates were examined using both laboratory (batch or column) tests [5,18,26,31–33,37–41] and field experiments [35,41,42].

The aim of the present work was to design a laboratory method that allows for:

(a) testing the reactivity of natural magnesite-based reactive materials for the PRBs;

(b) defining the conditions for magnesite calcination when the reactive material (CCM) with the optimal reactivity for the removal of the selected heavy metals (Cu^{2+}, Zn^{2+}, Ni^{2+}, and Mn^{2+}) from contaminated water is obtained.

Though the calcination temperature of pure synthetic $MgCO_3$ has been found to affect the extent of the MgO hydration [26,37] as well as the immobilization mechanism of borate [26] and fluoride [37], no publications have been published that comprehensively report on how to prepare the MgO from magnesite with the optimal reactivity for the removal of the monitored pollutants.

2. Materials and Methods

2.1. Materials

Bulk breunneritic magnesite concentrate with particle sizes up to 2 mm, supplied by the company SMZ (Jelšava, Slovakia), was used in the present study. For the laboratory tests, raw magnesite with grain size ranging from 80 to 250 μm was obtained by grinding and dry-screening. The content of the main elements in the input (uncalcined) magnesite sample was determined in an external certified chemical laboratory EKOLAB (https://www.ekolab.sk/certificates (accessed on 22 May 2003)) using atomic absorption spectrometry (AAS, contrAA® 700 Analytik Jena, Jena, Germany). The results of chemical analysis are shown in Table 1.

Table 1. Chemical composition of the raw (uncalcined) magnesite.

Element	Mg	Fe	Ca	Si	Al	Loss on Ignition (L.O.I.)
Weight %	25.1	3.0	2.3	0.14	0.10	50.2

2.2. Calcination of Magnesite

First, to find the calcination temperature, thermal decomposition of uncalcined raw magnesite was studied by simultaneous differential thermal analysis and thermogravimetric analysis in air atmosphere (DTA/TGA, NETZSCH STA 449 F3 Jupiter, NETZSCH-Gerätebau GmbH, Selb, Germany), and by high-temperature X-ray powder diffraction analysis in high-temperature chamber using CoK_α radiation (X'Pert PRO XL XRD System from PANalytical, Malvern, UK) as well.

Then, the samples of caustic calcined magnesia (CCM) for the batch reactivity tests were prepared as follows: uncalcined raw magnesite was placed in a quartz crucible with a lid and inserted in a laboratory electric high-temperature chamber oven (NETZSCH Model 417, NETZSCH-Gerätebau GmbH, Selb, Germany) where the sample was heated at a predefined temperature (between 550 °C and 1400 °C) for a selected time (10–240 min) under static air atmosphere and then cooled down naturally to room temperature in closed oven.

After calcination, the weight of each sample was measured to calculate the weight loss. The calcined products were stored in a vacuum desiccator prior to their use in experiments.

The samples of uncalcined and calcined magnesite were characterised by X-ray powder diffraction by Bragg-Bretano Θ—2Θ geometry (XRD, Rigaku, MiniFlex 600 diffractometer, Akishima, Japan) using CuK_α radiation at 40 kV and 15 mA at a step of 0.01° and speed of 2° min^{-1} specified by PDXL2, ver. 2.6.1.2 software with the Powder Diffraction File (PDF-2, ver. 2.1502, Newtown Square, PA, USA) databases from International Centre for Diffraction Data (ICDD). Measurements of the specific surface area were performed using the N_2-adsorption B.E.T. method (NOVA 1000 Quantachrome Instrument, Boynton Beach, FL, USA).

2.3. Batch Precipitation Tests

The reactivity of prepared CCM was tested using its reaction with a model mixed aqueous solution of Cu^{2+}, Zn^{2+}, Ni^{2+}, and Mn^{2+} at 25 °C. The initial concentration of each cation in the used model mixed solution was approx. 50 mg/L, and the initial pH was 5.1. The model mixed solution was prepared by dissolving the sulphates of Cu^{2+}, Zn^{2+}, Ni^{2+}, and Mn^{2+} (special grade, Sigma-Aldrich, St. Louis, MO, USA) in distilled water. In each test, a defined amount (0.2 g) of prepared CCM was added to 40 mL of the model solution (liquid:solid ratio = 200:1) in a plastic bottle and horizontally shaken on a rotary shaker at 50 min^{-1} and 25 °C for 2 h. (In preliminary experiments, it was observed that the equilibrium was achieved within 10 to 90 min under the conditions used). Then, the slurry was filtered (Munktell Filtrak (Quant.) Grade: 389; 84 g/m^2; 8–12 μm pore size), and the final pH was measured in the filtrate by a pH meter (HI4222, Hanna Instruments, Nusfalau, Romania) by ORP electrode (HI3131B, Hanna Instruments, Nusfalau, Romania). The initial concentrations of metal ions in the model mixed solution and the residual concentrations of dissolved metal ions in the filtrate were verified and determined by inductively coupled plasma-atomic emission spectrometry (ICP-OES, Thermo Scientific iCAP 6000 DUO, Thermofisher Scientific, Cambridge, UK). The single element solution standards (1000 ppm in HNO_3, Fisher Scientific UK, Leics, UK) for calibration were used. The efficiency of individual metals removal (E_i) was calculated using Equation (1):

$$E_i = \frac{c_{i,0} - c_{i,r}}{c_{i,0}} \tag{1}$$

where $c_{i,0}$ is the initial concentration (mg/L) and $c_{i,r}$ is the final concentration (mg/L) of each cation $i = Cu^{2+}$, Ni^{2+}, Mn^{2+}, and Zn^{2+} in the model mixed solution. The results are presented as the average of three repeated measurements.

3. Results and Discussion

The major mineral phases identified in the raw magnesite sample by the X-ray powder diffraction method (Figure 1) were: magnesite (00-008-0479 PDF-2 card), ferroan variety of magnesite (00-036-0383 PDF-2 card) representing the solid solution of magnesite ($MgCO_3$) and siderite ($FeCO_3$) and characterised by the formula $(Mg,Fe)CO_3$, where the Mg:Fe atomic ratio ranged from 90:10 to 70:30, and dolomite (00-036-0426 PDF-2 card). The minor phases were calcite (00-004-0636 PDF-2 card) and enstatite (00-002-0520 PDF-2 card).

3.1. Thermal Decomposition of Magnesite

The results of the thermal analysis (Figure 2) showed that the thermal decomposition of the magnesite started at 440 °C, and the maximum endothermic effect (corresponding to the maximum rate of the thermal decomposition of the magnesite mineral) was observed at 660 °C. The second endothermic effect contains two hidden peaks with an onset at 697 °C, indicated by the subsequent thermal decomposition of the dolomitic part of the crude magnesite. The total observed weight loss due to the release of the carbon dioxide was 50.2 wt.%.

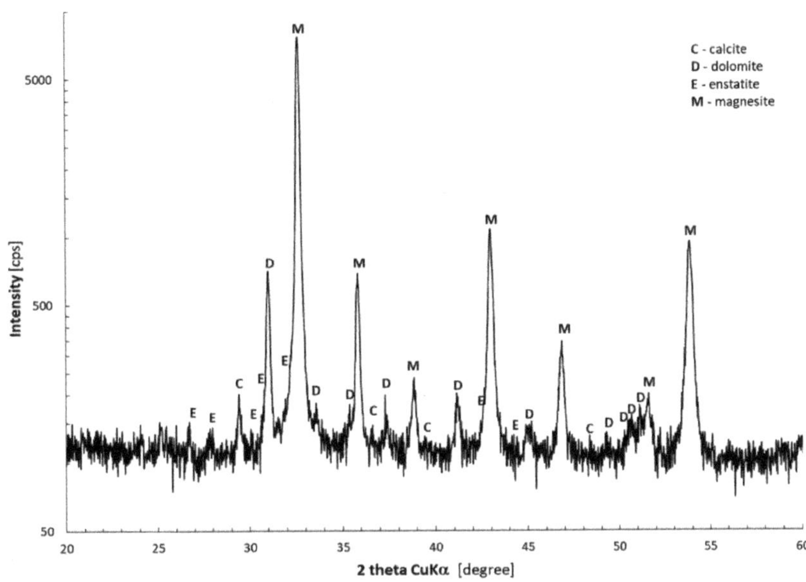

Figure 1. XRD analysis of the raw magnesite used for the experiments.

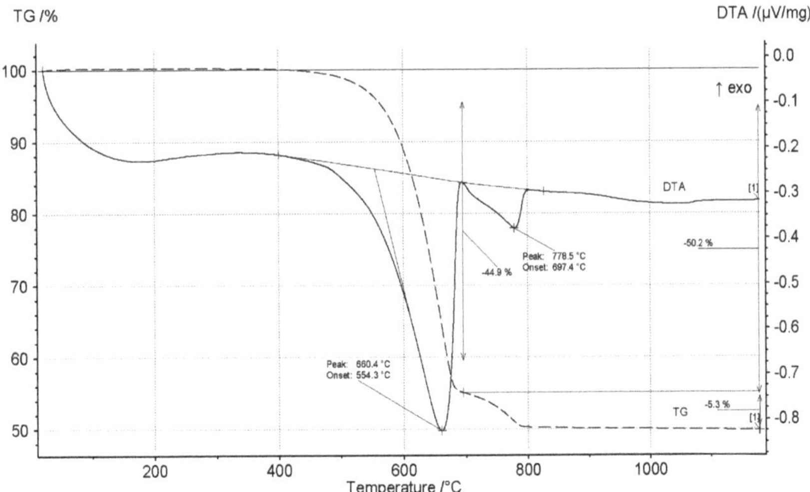

Figure 2. Results of thermal analysis illustrating the thermal decomposition of magnesite and dolomite during heating the raw magnesite.

The influence of the calcination temperature between 400 and 1200 °C on the phase composition and crystallinity is illustrated in Figure 3. Between 600 and 800 °C, the phases of magnesite and dolomite originally present gradually disappeared, and the structure of the calcined sample appeared to be highly disordered. Starting from 600 °C, the XRD count from the magnesite got progressively lower and practically disappeared at 800 °C. The dolomitic part of the raw magnesite decomposed at temperatures above 700 °C, and no dolomite could actually be identified at 800 °C. At temperatures above 700 °C, new phases (mainly periclase, MgO) occurred.

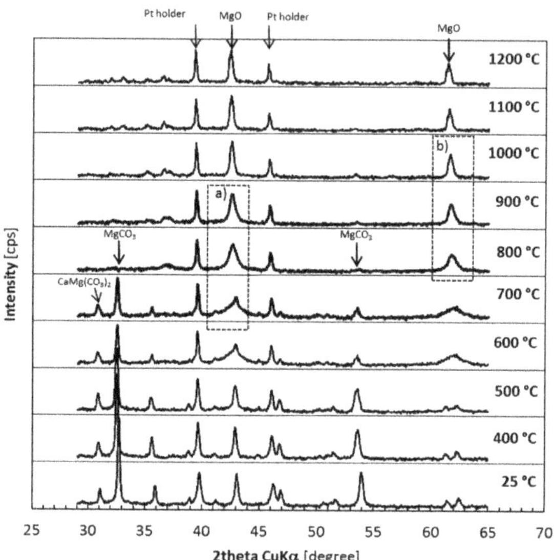

Figure 3. In situ high-temperature X-ray powder diffraction sequence illustrating the changes in phase composition and crystallinity during heating the raw magnesite. (Magnesite and dolomite reflections gradually disappear above 700 °C (frame "a"), and the new phase of periclase is created (frame "b").).

The DTA, TGA, and XRD analyses suggest that the samples should be calcined at temperatures between 600 and 900 °C to produce a reactive form of magnesia.

The effect of the calcination temperature (from 550 °C to 1000 °C) and time (from 10 min to 4 h) on the thermal decomposition of the carbonates present in the crude (uncalcined) magnesite are illustrated in Figure 4.

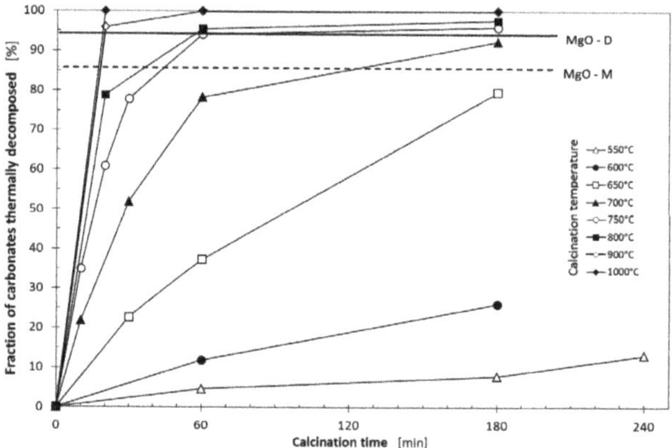

Figure 4. Influence of temperature and time on fraction of carbonates decomposed during heating the raw magnesite.

The fraction of the carbonates thermally decomposed, X_t, was calculated for each calcined sample using the following relationship (Equation (2)):

$$X_t = \frac{100}{w_0} \cdot \frac{m_0 - m_t}{m_0}$$

(2)

where m_0 is the weight of the sample before calcination (equal to 40 g), m_t is the weight of the sample after calcination, and w_0 is the total content of the carbon dioxide chemically bonded in the form of carbonates in the original (uncalcined) magnesite rock (represented by the L.O.I. and equal to 50.2 wt.%).

Two horizontal lines in Figure 4 represent the two limits, which are as follows: the dashed line (MgO-M) = the thermal decomposition of the magnesite was completed and the decomposition of the $MgCO_3$ from the dolomitic part just started; the full line (MgO-D) = the thermal decomposition of the $MgCO_3$ from the dolomitic part was completed and the decomposition of the $CaCO_3$ just started.

Figure 4 shows that, at a temperature of 550 °C, the fraction of the carbonates thermally decomposed, X_t, reached 13% after 4 h. At temperatures above 900 °C, the carbonates were almost completely decomposed within 1 h.

3.2. Removal of Heavy Metals by Precipitation Using the CCM

The influence of the calcination temperature and time on the efficiency (E_i) of the Cu^{2+}, Zn^{2+}, Ni^{2+}, and Mn^{2+} removal from the model mixed solution by the addition of a sample of the prepared CCM as a precipitating agent is illustrated in Figure 5a–d. The accuracy of the measurements expressed as %RSD ranged from 0.2% to 0.8% depending on the wavelengths of the selected emission lines.

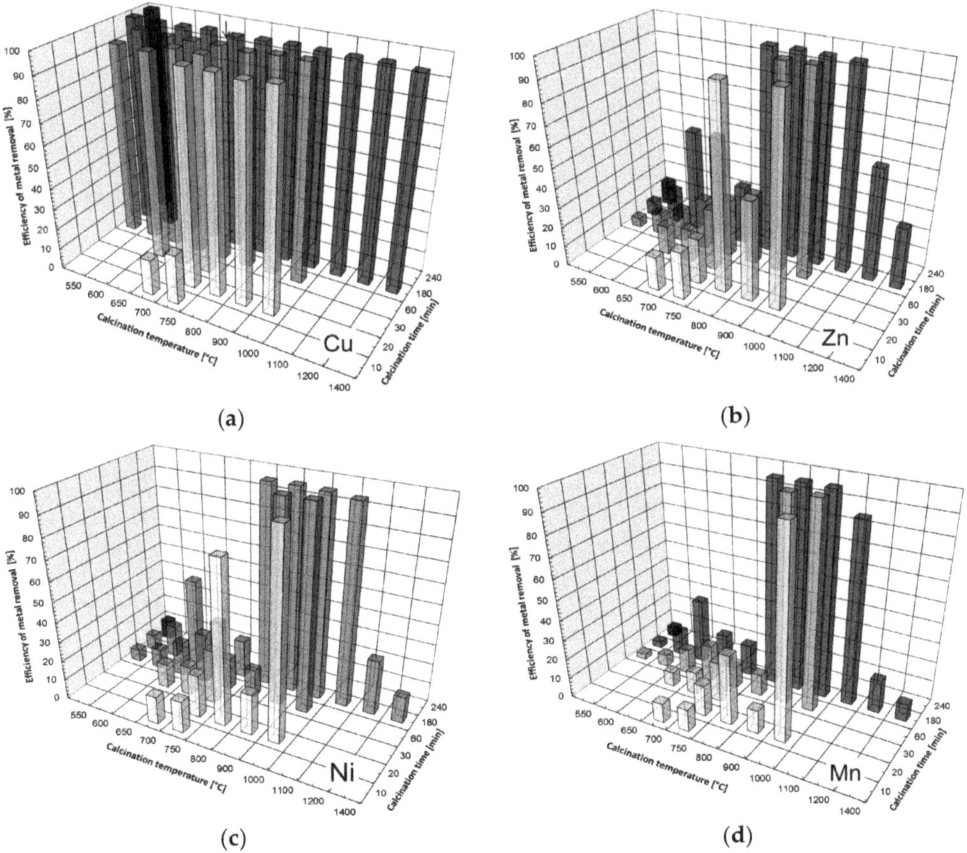

Figure 5. Influence of calcination temperature and time on the efficiency of monitored metal removal from the model solution by addition of the CCM tested: (**a**) Cu^{2+}, (**b**) Zn^{2+}, (**c**) Ni^{2+}, and (**d**) Mn^{2+} using a liquid-to-solid ratio l:s = 200:1.

A comparison of the graphs in Figure 5a–d suggests that the behaviours of Zn, Ni, and Mn were very similar, but quite different from that of Cu. After an addition of the CCM, the copper was completely precipitated (Figure 5a) with the exception of the three cases where the magnesite had not been thermally activated enough due to the low calcination temperature and/or short calcination time (samples 550 °C for 1 h, 700 °C for 10 min, and 750 °C for 10 min). In contrast, the efficiency of the removal of the Zn, Ni, and Mn first grew with the calcination temperature and time and reached approximately 100% at a calcination temperature of 800 °C to 1000 °C and sufficiently long calcination time (from 20 min at 1000 °C up to 3 h at 800 °C), and then decreased at temperatures above 1100 °C (Figure 5b–d).

The comparison of the results in Figure 5 shows that the most reactive material, which would achieve a comprehensive removal of all the monitored metals, can be prepared by calcination at temperatures above 800 °C. These results are in good agreement with the final pH values (measured in the filtrate at the end of each test) that are shown in Figure 6.

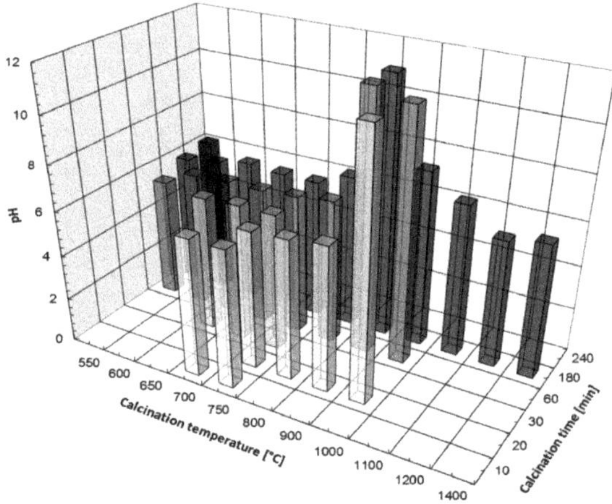

Figure 6. Final pH measured at the end of individual precipitation tests.

The highest final values of the pH in Figure 6 correlate very well with the maximum values of the efficiency of the individual metal removal, E_i, in Figure 5b–d. However, the highest observed final values of the pH (up to 11.5) cannot be attributed to $Mg(OH)_2$, which buffers aqueous solutions to a pH between 8.0–10.5 [18,32,33,35]. It can, therefore, be assumed that these high pH values are due to the presence of free lime, which is formed in the samples calcined above 800 °C as a product of the thermal decomposition of the dolomite and/or calcite originally present in the raw magnesite used.

To be able to explain the changes in the reactivity of the CCM prepared, a comprehensive analysis of the changes taking place in the raw magnesite during the calcination is needed.

The results of the phase analysis (Figure 7a) showed that, at a temperature of 900 °C, the free lime is formed as a product of the thermal decomposition of the dolomite and/or calcite originally present in the used raw magnesite. In addition, from the diffractograms shown in Figure 7a, it can be seen that the calcination at very high temperatures (above 1000 °C) results in the solid-phase reaction of the free lime (CaO) and the formation of dicalcium ferrite ($2CaO \cdot Fe_2O_3$), as illustrated by the graph in Figure 7b, while the XRD counts from the CaO (200) were getting progressively lower with an increase in the calcination temperature; those from the $2CaO \cdot Fe_2O_3$ (141) were simultaneously increased. The conversion of the CaO to $2CaO \cdot Fe_2O_3$ explains the gradual decrease in the reactivity of the products calcined at temperatures above 1000 °C (Figure 5b–d).

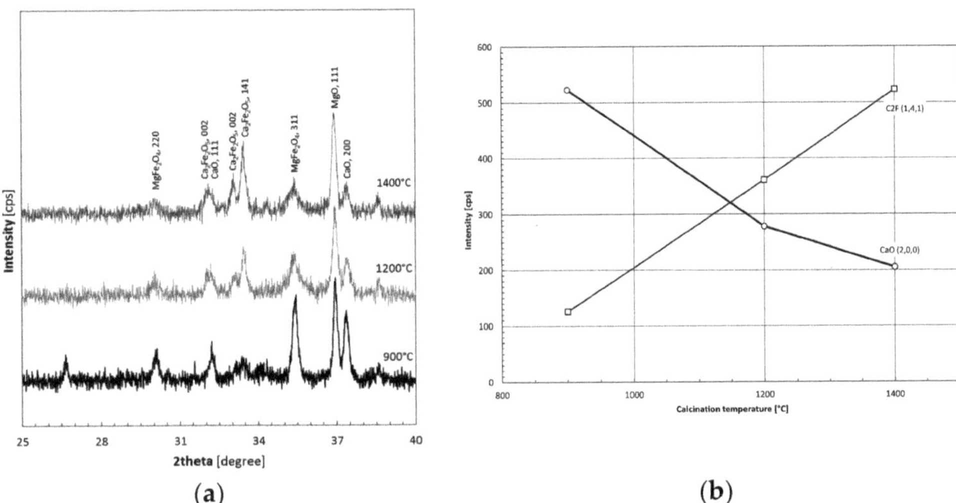

Figure 7. Results of phase analysis illustrating the solid-phase chemical reactions occurring during heating the raw magnesite: (**a**) identified products at various temperatures, (**b**) conversion of free lime to dicalcium ferrite ($2CaO \cdot Fe_2O_3$).

The CCM intended for use as a reactive material in the PRB should contain just a limited content of CaO in the form of lime. This limit is determined by the demands on a proper performance of the PRB. The presence of the lime in the CCM can become significant, especially when raw magnesite with a higher CaO content is used. Although calcination at temperatures up to 1000 °C leads to a seemingly higher efficiency of heavy metal removal (Figure 5), this effect is at the expense of the benefits resulting from the low solubility of MgO. The presence of the lime results in high pH values (pH up to 12–12.5) that make some precipitates more soluble [5,24]. Moreover, highly alkaline water might be harmful to the environment.

One way to reduce the lime content in the CCM is the calcination of the magnesite at very high temperatures; at temperatures above 1000 °C, a decrease in the pH values was observed (Figure 6), accompanied by lower efficiency (Figure 5), which may be attributed to the solid-phase reaction of the free lime (CaO) and the formation of dicalcium ferrite ($2CaO \cdot Fe_2O_3$) (Figure 7a,b).

A different way to avoid overly high lime content in the CCM is the calcination of the magnesite under the conditions when $CaCO_3$ is not thermally decomposed. The graph in Figure 8 shows the removal efficiencies of the individual metals (shown in Figure 5) depending on the amount of crude magnesite that has to be calcined to treat 1 m^3 of contaminated water with the same composition as the model solution used. Maintaining the same conditions (the composition of the liquid and solid phase, l:s ratio = 200:1) as used in the laboratory tests, 5 kg of calcined magnesite would be necessary to treat 1 m^3 of the model solution used. The mass of crude magnesite to produce 5 kg of the CCM under the given calcination temperature and time was calculated using Equation (3):

$$m_{crude\ magnesite} = \frac{500\ \text{kg}}{100 - 0.502 X_t} \tag{3}$$

From Figure 8, it can be seen that the efficiency of the removal of the individual metals by the addition of the calcined magnesite decreased in the order Cu >> Zn > Ni > Mn. This finding is in accordance with the fact that a hydroxide with a lower value of the solubility product is preferably precipitated. The values of the solubility products of the hydroxides of the monitored metals decrease in the following order: $Cu(OH)_2 < Zn(OH)_2 < Ni(OH)_2 < Mn(OH)_2$.

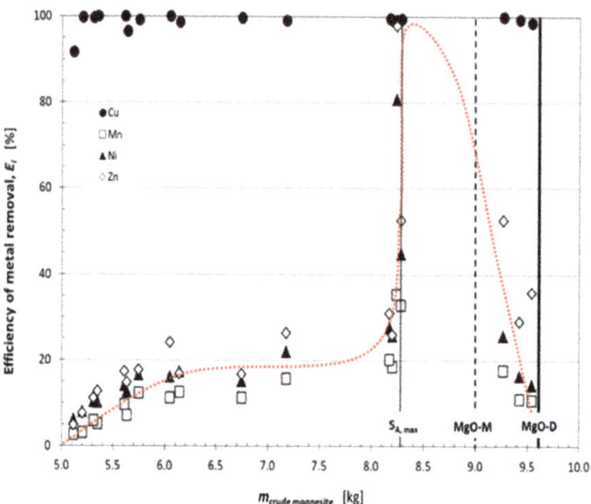

Figure 8. Efficiency of individual metals removal, E_i, in dependence on the amount of crude magnesite, $m_{crude\ magnesite}$, consumed to prepare the CCM under different calcination temperature and time. (Experimental conditions: Total heavy metal concentration at the beginning of each precipitation test = 3.33 mol m^{-3}; l:s ratio = 200:1. Dotted curve = regression curve).

From the course of the dependences $E_i = f_i(m_{magnesite})$ in Figure 8 (and Equation (3): the higher X_t, the higher $m_{crude\ magnesite}$), it follows that E_i was initially increased with an increase in X_t of the added calcined magnesite: the practically complete precipitation of the copper occurred already after the addition of the CCM characterised by X_t of approximately 8%, while the efficiency E_i of Zn, Ni, and Mn was first growing rapidly with increasing X_t, then the rise of E_i was slowed down. At $m_{crude\ magnesite}$ = 8.3 kg (X_t = 79%), E_i reached the maximum (characterised by the step-wise increase in E_i). When the conversion X_t was assigned to the corresponding value of the specific surface area shown in Figure 9, it was found that the maximum efficiency E_i was achieved by the addition of the CCM with the largest specific surface area. This limit is marked in Figure 8 with a full vertical line ($S_{A,MAX}$). (The specific surface area reached the maximum of 112 m^2·g^{-1} at the calcination temperature of 750 °C and time 30 min, where X_t = 79%). Then, until the MgCO$_3$ in the raw magnesite was completely decomposed (this limit is marked with the dashed vertical line (MgO-M) in both Figures 4 and 8), the efficiency E_i remained almost constant and close to its maximum value. At last, despite the increasing content of the MgO, E_{Zn} was very slowly getting higher, while the efficiencies for Ni and Mn were decreasing at the same time. The drop in the efficiency is likely to be caused by: (i) the recrystallisation of the periclase and (ii) the loss of the reactive MgO due to its solid-phase reaction with the Fe$_2$O$_3$ to produce magnesium ferrite (MgO·Fe$_2$O$_3$) (Figure 7a).

The above results show that the virtually complete removal of the heavy metals from the model solution can be achieved with the addition of CCM characterised by the fraction of carbonates decomposed X_t = 79% and the highest specific surface area. Under the conditions of this work, the magnesite was calcined at 750 °C for 3 h. Magnesite calcined at higher temperatures could also be used, but this would be associated with the higher consumption of the crude magnesite to prepare the same amount of the CCM. Approximately 8.3 kg of crude magnesite calcined at 750 °C for 3 h, but up to 10.0 kg of magnesite calcined at 1000 °C for 20 min would be consumed to prepare 5 kg of CCM to remove the heavy metals from 1 m^3 of the model solution. The higher specific consumption of crude magnesite brings increased handling and heating costs, and the higher degree of carbonate decomposition (X_t) increases the calcination costs (thermal decomposition). The product obtained by calcination at 750 °C for 3 h, therefore, appears to be optimal: it is enough to decompose only 80% of the MgCO$_3$ present in the magnesite at a lower temperature (therefore, less energy is needed for calcination), and the same amount of heavy metals can be precipitated by adding 1 t of this material as by adding 1 t of the CCM obtained at higher temperatures.

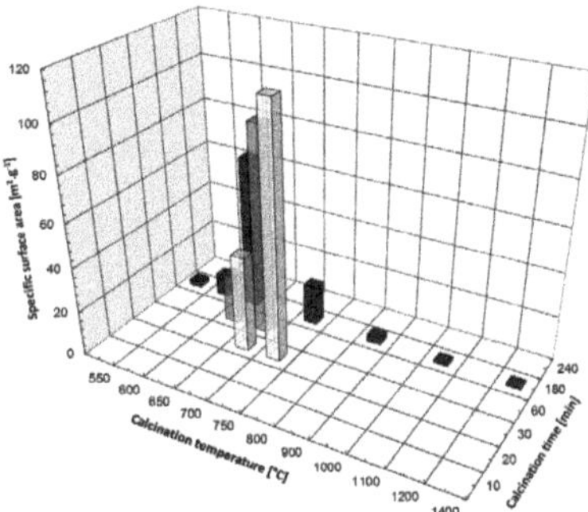

Figure 9. Influence of calcination temperature and time on specific surface area during heating the raw magnesite.

It is assumed that the removal of heavy metals from aqueous solutions by the addition of CCM is based on precipitation, according to the overall chemical equation:

$$MgO(s) + H_2O(l) + Me^{2+}(aq.) = Mg^{2+}(aq.) + Me(OH)_2(s), \qquad (4)$$

eventually,

$$CaO(s) + H_2O(l) + Me^{2+}(aq.) = Ca^{2+}(aq.) + Me(OH)_2(s). \qquad (5)$$

This assumption is based on the fact that, at the end of each test, the Mg^{2+} and Ca^{2+} contents in the final solution were also determined. The results showed that the $Mg^{2+} + Ca^{2+}$ contents corresponded within the experimental error to the total amount of heavy metals removed from the solution during the test. This conclusion differs from the findings of Sasaki et al. [26,37], who used the reactive MgO obtained by the calcination of pure (synthetic) $MgCO_3$ as a reactive material for the removal of fluoride and borate. Sasaki et al. [26,37] found that the efficiency did not depend on a specific surface but the critical factor was the basicity per unit surface area, which was gradually getting higher in the range of the calcination temperature between 600 °C and 1100 °C. The reason(s) for this difference may be:

(a) the differences in the nature of the starting materials used: natural magnesite was used in this work;

(b) the differences between the products of calcination: the calcined magnesite prepared and used in this work was characterised by a high internal surface to external surface ratio;

(c) the different mechanism of the heavy metals removal: while adsorption was identified in the work of Sasaki et al. [17,28], precipitation prevailed in the present work, which was proved by high contents of $Mg^{2+} + Ca^{2+}$ cations in the final solutions when the reaction (phase contact) was stopped.

4. Conclusions

The caustic calcined magnesia (CCM) can be applied as an alkaline reactive material in permeable reactive barriers (PRBs) with dispersed alkaline substrate, designed for the removal of heavy metals from contaminated groundwater. There are two crucial properties that characterise an appropriate CCM for this application: high reactivity and low solubility. The present work reports on how the tailor-made CCM for use in the PRBs can be prepared by the thermal activation of the natural magnesite. It was found out that the CCM with the required reactivity for the removal of the Cu^{2+}, Zn^{2+}, Ni^{2+}, and Mn^{2+} cations from a model mixed aqueous solution can be obtained by a suitable adjustment of the calcination temperature and time.

Regarding the solubility, the CCM should contain just a limited content of CaO in the form of lime, which is produced by the thermal decomposition of $CaCO_3$. This limit is determined by the demands on a proper performance of the PRB. Although the calcination at temperatures up to 1000 °C practically results in full heavy metal removal, this effect is at the expense of the benefits resulting from the low solubility of MgO. The presence of lime results in high pH values (pH up to 12–12.5) that make some precipitates more soluble. Moreover, highly alkaline purified water might be harmful to the environment.

One way to reduce the lime content in the CCM is the calcination of magnesite at very high temperatures. At the temperatures above 1000 °C, a decrease in the pH values was observed. However, the decrease in pH was accompanied by lower efficiency, attributed to the solid-phase reaction of free lime (CaO) to form the dicalcium ferrite ($2CaO \cdot Fe_2O_3$).

A different way to avoid high lime content in the CCM is the calcination of magnesite under the conditions when $CaCO_3$ is not thermally decomposed. The results have shown that the virtually complete removal of the heavy metals from the model solution was achieved using the CCM characterised by the fraction of the carbonates decomposed $X_t = 79\%$ and the highest specific surface area. Under the conditions of this work, the magnesite was calcined at 750 °C for 3 h. Magnesite calcined at higher temperatures could also be used, but this would be associated with a higher consumption of crude magnesite to prepare the same amount of the CCM. The higher specific consumption of crude magnesite brings increased handling and heating costs, and the higher degree of carbonate decomposition increases the calcination costs (thermal decomposition).

Whether to choose the operation conditions for the calcination when the calcium carbonate is also decomposed or not depends on the composition of the magnesite used and the overall cost of the CCM production. However, to verify the designed laboratory method for testing the reactivity of the CCM-based reactive materials for the PRBs, the product obtained by calcination at 750 °C for 3 h appeared to be optimal. Adding 1 t of this material allowed for the same amount of heavy metals precipitated as adding 1 t of the CCM obtained at higher temperatures. At the same time, it was enough to decompose only 80% of the $MgCO_3$ present in the magnesite at a lower temperature (resulting in a lower consumption of energy for the calcination and material handling).

Author Contributions: Conceptualization, P.R., G.S. and A.F.; methodology, P.R.; software, G.S.; validation, A.F., A.D. and M.Š.; formal analysis, A.D.; investigation, A.D., M.Š.; resources, A.F.; data curation, P.R.; writing—original draft preparation, A.F., P.R.; writing—review and editing, P.R., A.F.; visualization, G.S.; supervision, P.R.; project administration, A.F.; funding acquisition, P.R. All authors have read and agreed to the published version of the manuscript.

Funding: This work was financially supported by the Scientific Grant Agency of the Ministry of Education of the Slovak Republic and the Slovak Academy of Sciences (VEGA 1/0176/19).

Acknowledgments: The authors thank Beatrice Plešingerová for specific surface analysis.

Conflicts of Interest: There are no conflict to declare.

Abbreviations

CCM	caustic calcined magnesia
PRB	permeable reactive barrier
HM	heavy metals
DAS	dispersed alkaline substrate, i.e., mixture of a fine-grained caustic calcined magnesia and coarse inert matrix (e.g., wood shavings or gravel)
E_i	efficiency of individual heavy metals removal from solution (%)
X_t	fraction of carbonates thermally decomposed during the raw magnesite heating (%)

References

1. Permeable Reactive Barriers for Inorganic and Radionuclide Contamination. Kate Bronstein National Network of Environmental Management Studies Fellow for U.S. Environmental Protection Agency Office of Solid Waste and Emergency Response Office of Superfund Remediation and Technology Innovation, Washington, DC, August 2005. Available online: https://clu-in.org/download/studentpapers/bronsteinprbpaper.pdf/ (accessed on 30 November 2016).
2. Thiruvenkatachari, R.; Vigneswaran, S.; Naidu, R. Permeable reactive barrier for groundwater remediation. *J. Ind. Eng. Chem.* **2008**, *14*, 145–156. [CrossRef]
3. Fu, F.; Wang, Q. Removal of heavy metal ions from wastewaters: A review. *J. Environ. Manag.* **2011**, *92*, 407–418. [CrossRef]
4. Johnson, D.B.; Hallberg, K.B. Acid mine drainage remediation options: A review. *Sci. Total Environ.* **2005**, *338*, 3–14. [CrossRef]
5. Barbooti, M.M.; Abid, B.A.; Al-Shuwaiki, N.M. Removal of heavy metals using chemicals precipitation. *Eng. Tech. J.* **2011**, *29*, 595–612.
6. Taylor, J.; Pape, S.; Murphy, N. A Summary of Passive and Active Treatment Technologies for Acid and Metalliferous Drainage (AMD), Fifth Australian Workshop on Acid Drainage, 29–31 August 2005 Fremantle, Western Australia. The Australian Centre for Minerals Extension and Research (ACMER), Earth Systems PTY Ltd., Australian Business Number 29 006 227 532. Available online: http://www.earthsystems.com.au/wp-content/uploads/2012/02/AMD_Treatment_Technologies_06.pdf/ (accessed on 30 November 2016).
7. Blais, J.F.; Djedidi, Z.; Ben Cheikh, R.; Tyagi, R.D.; Mercier, G. Metals precipitation from effluents: Review. *Pract. Period. Hazard. Toxic Radioact. Waste Manag.* **2008**, *12*, 135–149. [CrossRef]
8. Cantor, K.P. Drinking water and cancer. *Cancer Causes Control.* **1997**, *8*, 292–308. [CrossRef] [PubMed]
9. Calderon, R.L. The epidemiology of chemical contaminants of drinking water. *Food Chem. Toxicol.* **2000**, *38*, 13–20. [CrossRef]
10. Kavcar, P.; Sofuoglu, A.; Sofuoglu, S.C. A health risk assessment for exposure to trace metals via drinking water ingestion pathway. *Int. J. Hyg. Environ. Health* **2009**, *212*, 216–227. [CrossRef]
11. Kelly, B.C.; Myo, A.N.; Pi, N.; Bayen, S.; Leakhena, P.C.; Chou, M.; Tan, B.H. Human exposure to trace elements in central Cambodia: Influence of seasonal hydrology and food-chain bioaccumulation behaviour. *Ecotox. Environ. Saf.* **2018**, *162*, 112–120. [CrossRef] [PubMed]
12. Wu, G.; Kang, H.; Zhang, X.; Shao, H.; Chu, L.; Ruan, C. A critical review on the bio-removal of hazardous heavy metals from contaminated soils: Issues, progress, eco-environmental concerns and opportunities. *J. Hazard. Mater.* **2010**, *174*, 1–8. [CrossRef] [PubMed]
13. Nthunya, L.N.; Masheane, M.L.; George, M.; Kime, M.B.; Mhlanga, S.D. Removal of Fe and Mn from polluted water sources in Lesotho using modified clays. *J. Water Chem. Technol.* **2019**, *41*, 81–86. [CrossRef]
14. Yang, S.; Li, J.; Shao, D.; Hu, J.; Wang, X. Adsorption of Ni (II) on oxidized multi-walled carbon nanotubes: Effect of contact time, pH, foreign ions and PAA. *J. Hazard. Mater.* **2009**, *166*, 109–116. [CrossRef] [PubMed]
15. Rehman, I.; Ishaq, M.; Ali, L.; Khan, S.; Ahmad, I.; Ud Din, I.; Ullah, H. Enrichment, spatial distribution of potential ecological and human healthrisk assessment via toxic metals in soil and surface water ingestion in the vicinity of Sewakht mines, district Chitral, Northern Pakistan. *Ecotox. Environ. Saf.* **2018**, *154*, 127–136. [CrossRef]
16. Sani, H.A.; Ahmad, M.B.; Hussein, M.Z.; Ibrahim, N.A.; Musa, A.; Saleh, T.A. Nanocomposite of ZnO with montmorillonite for removal of lead and copper ions from aqueous solutions. *Process Saf. Environ. Protect.* **2017**, *109*, 97–105. [CrossRef]
17. Gibert, O.; De Pablo, J.; Cortina, J.L.; Ayora, C. In situ removal of arsenic from groundwater by using permeable reactive barriers of organic mater/limestone/zero-valent iron mixtures. *Environ. Geochem. Health* **2010**, *32*, 373–378. [CrossRef] [PubMed]
18. Rötting, T.S.; Ayora, C.; Carrera, J. Improved passive treatment of high Zn and Mn concentrations using caustic magnesia (MgO): Particle size effects. *Environ. Sci. Technol.* **2008**, *42*, 9370–9377. [CrossRef] [PubMed]
19. Simon, F.; Meggyes, T. Removal of organic and inorganic pollutants from groundwater using permeable reactive barriers. *Land Contam. Reclamat.* **2000**, *8*, 103–116.
20. Hashim, M.A.; Mukhopadhyay, S.; Sahu, J.N.; Sengupta, B. Remediation technologies for heavy metal contaminated ground water. *J. Environ. Manag.* **2011**, *92*, 2355–2388. [CrossRef] [PubMed]
21. Permeable Reactive Barriers: Lessons Learned/New Directions. Interstate Technology & Regulatory Council: Washington, DC, USA, 2005; p. 9. Available online: http://www.itrcweb.org/GuidanceDocuments/PRB-4.pdf/ (accessed on 30 November 2016).
22. Lee, M.K.; Saunders, J.A. Effects of pH on metals precipitation and sorption: Field bioremediation and geochemical modeling approaches. *Vadose Zone J.* **2003**, *2*, 177–185.
23. Obiri-Nyarko, F.; Grajales-Mesa, S.J.; Malina, G. An overview of permeable reactive barriers for in situ sustainable groundwater remediation. *Chemosphere* **2014**, *111*, 243–259. [CrossRef]
24. Army Corps. Engineering Manuals, Manual No. 1110-1-4012: Engineering and Design Precipitation/Coagulation/Flocculation, Department of the Army U.S. Army Corps of Engineers, 2001, Washington, DC 20314-1000. Available online: http://www.publications.usace.army.mil/Portals/76/Publications/EngineerManuals/EM_1110-1-4012.pdf/ (accessed on 30 November 2016).
25. Peters, R.W.; Shem, L. Separation of Heavy Metals: Removal from Industrial Wastewaters and Contaminated Soil. In Proceedings of the Symposium on Emerging Separation Technologies for Metals and Fuels, Palm Coast, FL, USA, 13–18 March 1993. Available online: http://www.osti.gov/scitech/servlets/purl/6504209-mKij3S/ (accessed on 30 November 2016).
26. Sasaki, K.; Moriyama, S. Effect of calcination temperature for magnesite on interaction of MgO-rich phases with boric acid. *Ceram. Int.* **2014**, *40*, 1651–1660.

27. Barakat, M.A. New trends in removing heavy metals from industrial wastewater. *Arab. J. Chem.* **2011**, *4*, 361–377. [CrossRef]
28. Skousen, J. Anoxic limestone drains for acid mine drainage treatment. *Green Lands* **1991**, *21*, 30–35.
29. Golab, A.N.; Peterson, M.A.; Indraratna, B. Selection of potential reactive materials for a permeable reactive barrier for remediating acidic groundwater in acid sulphate soil terrains. *Q. J. Eng. Geol. Hydrogeol.* **2006**, *39*, 209–223. [CrossRef]
30. Akcil, A.; Koldas, S. Acid mine drainage (AMD): Causes, treatment and case studies. *J. Clean. Prod.* **2006**, *14*, 1139–1145. [CrossRef]
31. Cortina, J.L.; Lagreca, I.; De Pablo, J. Passive in situ remediation of metal-polluted water with caustic magnesia: Evidence from column experiments. *Environ. Sci. Technol.* **2003**, *37*, 1971–1977. [CrossRef] [PubMed]
32. Rötting, T.; Cama, J.; Ayora, C.; Cortina, J.L.; De Pablo, J. The Use of Caustic Magnesia to Remove Cadmium from Water. In Proceedings of the 9th International Mine Water Association Congress, Oviedo, Spain, 5–7 September 2005; pp. 635–639. Available online: http://www.imwa.info/docs/imwa_2005/IMWA2005_090_Roetting.pdf/ (accessed on 30 November 2016).
33. Caraballo, M.A.; Rötting, T.S.; Silva, V. Implementation of an MgO-based metal removal step in the passive treatment system of Shilbottle, UK: Column experiments. *J. Hazard. Mater.* **2010**, *181*, 923–930. [CrossRef]
34. Macías, F.; Caraballo, M.A.; Nieto, J.M.; Rötting, T.S.; Ayora, C. Natural pretreatment and passive remediation of highly polluted acid mine drainage. *J. Environ. Manag.* **2012**, *104*, 93–100. [CrossRef] [PubMed]
35. Caraballo, M.A.; Rötting, T.S.; Macías, F.; Nieto, J.M.; Ayora, C. Field multi-step limestone and MgO passive system to treat acid mine drainage with high metal concentrations. *Appl. Geochem.* **2009**, *24*, 2301–2311. [CrossRef]
36. Lin, X.; Burns, R.C.; Lawrance, G.A. Heavy metals in wastewater: The effect of electrolyte composition on the precipitation of cadmium (II) using lime and magnesia. *Water Air Soil Poll.* **2005**, *165*, 131–152. [CrossRef]
37. Sasaki, K.; Fukumoto, N.; Moriyama, S.; Hirajima, T. Sorption characteristics of fluoride on to magnesium oxide-rich phases calcined at different temperatures. *J. Hazard. Mater.* **2011**, *191*, 240–248. [CrossRef]
38. Moore, R.C.; Anderson, D.R. System for Removal of Arsenic from Water. U.S. Patent 6,821,434B1, 23 November 2004.
39. Navarro, A.; Chimenos, J.; Muntaner, D.; Fernández, I. Permeable reactive barriers for the removal of heavy metals: Lab-scale experiments with low-grade magnesium oxide. *Ground Water Monit. Remediat.* **2006**, *26*, 142–152. [CrossRef]
40. Rotting, T.S.; Cama, J.; Ayora, C.; Cortina, J.L.; De Pablo, J. Use of caustic magnesia to remove cadmium, nickel, and cobalt from water in passive treatment systems: Column experiments. *Environ. Sci. Technol.* **2006**, *40*, 6438–6443. [CrossRef] [PubMed]
41. Ayora, C.; Caraballo, M.A.; Macías, F.; Rötting, T.S.; Carrera, J.; Nieta, J.M. Acid mine drainage in the Iberian Pyrite Belt: 2. Lessons learned from recent passive remediation experiences. *Environ. Sci. Pollut. Res.* **2013**, *20*, 7837–7853. [CrossRef]
42. Macías, F.; Caraballo, M.A.; Rötting, T.S.; Pérez-López, R.; Nieto, J.M.; Ayora, C. From highly polluted Zn-rich acid mine drainage to non-metallic waters: Implementation of a multi-step alkaline passive treatment system to remediate metal pollution. *Sci. Total Environ.* **2012**, *433*, 323–330. [CrossRef] [PubMed]

MDPI
St. Alban-Anlage 66
4052 Basel
Switzerland
Tel. +41 61 683 77 34
Fax +41 61 302 89 18
www.mdpi.com

Minerals Editorial Office
E-mail: minerals@mdpi.com
www.mdpi.com/journal/minerals